20 089 426 00

AF616055

2-Oxoglutarate-Dependent Oxygenases

RSC Metallobiology Series

Titles in the Series:
1: Mechanisms and Metal Involvement in Neurodegenerative Diseases
2: Binding, Transport and Storage of Metal Ions in Biological Cells
3: 2-Oxoglutarate-Dependent Oxygenases

How to obtain future titles on publication:
A standing order plan is available for this series. A standing order will bring delivery of each new volume immediately on publication.

For further information please contact:
Book Sales Department, Royal Society of Chemistry, Thomas Graham House, Science Park, Milton Road, Cambridge, CB4 0WF, UK
Telephone: +44 (0)1223 420066, Fax: +44 (0)1223 420247,
Email: booksales@rsc.org
Visit our website at www.rsc.org/books

2-Oxoglutarate-Dependent Oxygenases

Edited by

Robert P. Hausinger
Department of Microbiology, Michigan State University, USA
Email: hausinge@msu.edu

and

Christopher J. Schofield
The Dyson Perrins Laboratory, University of Oxford, UK
Email: christopher.schofield@chem.ox.ac.uk

RSC Metallobiology Series No. 3

Print ISBN: 978-1-84973-950-4
PDF eISBN: 978-1-78262-195-9
ISSN: 2045-547X

A catalogue record for this book is available from the British Library

Published by The Royal Society of Chemistry,
Thomas Graham House, Science Park, Milton Road,
Cambridge CB4 0WF, UK

Registered Charity Number 207890

For further information see our web site at www.rsc.org

Preface

In the 1960s scientists discovered three examples of 2-oxoglutarate (2OG)-dependent oxygenases catalysing the hydroxylation of (i) prolyl residues in collagen precursors, (ii) the free base thymine and (iii) γ-butyrobetaine. During the intervening 50 years or so, a remarkably broad diversity of alternate reactions and substrates has been revealed for the 2OG-dependent oxygenases and extensive advances have been achieved in our understanding of their structures and catalytic mechanisms. These mononuclear Fe(II) enzymes most often catalyse hydroxylation reactions, but other types of chemistries are known including dealkylation, desaturation, cyclization and halogenation reactions. These reactions are employed in a variety of biological roles including the biosynthesis of cellular components, the generation of important small molecules including widely used antibiotics, the disposal of undesired molecules, and in regulatory processes such as hypoxic signalling. 2OG-dependent oxygenases are also important agrochemical targets and are being pursued as therapeutic targets for a wide range of diseases including cancer and anaemia. Given the numerous recent advances and biomedicinal interest in this field, we recognized the utility of compiling a series of in-depth reviews of these enzymes in a single text. Our hope is that such a timely compendium will be a valuable resource and serve to unite and stimulate the field.

This book offers an overview of the rapidly evolving field of research involving 2OG-dependent oxygenases. We begin with four broad summary chapters that highlight critical aspects of this field. Chapter 1 (Hausinger) explores the vast enzymatic landscape associated with the amazingly diverse set of reactions catalysed by 2OG-dependent oxygenases. An overview of the conserved structural platform and metallocentres employed by 2OG-dependent oxygenases is presented in Chapter 2 (Schofield and colleagues). Advances

RSC Metallobiology Series No. 3
2-Oxoglutarate-Dependent Oxygenases
Edited by Robert P. Hausinger and Christopher J. Schofield

Published by the Royal Society of Chemistry, www.rsc.org

in our mechanistic understanding of these enzymes, including information on the key intermediates, are highlighted in Chapter 3 (Bollinger, Krebs and coworkers). Biomimetic analogues that have been helpful in dissecting the mechanism and metallocentre properties, and which may ultimately enable the discovery of novel catalysts, are described in Chapter 4 (Que and colleagues). These overview discussions set the stage for an additional 17 chapters focused on carefully targeted topics.

The substrates for many 2OG-dependent oxygenases are macromolecules, including proteins, lipids and oligonucleotides. Hydroxylases acting on prolyl and lysyl residues in pro-collagen are presented in Chapter 5 (Myllyharju). Chapter 6 (Schofield and coworkers) describes work on other protein hydroxylases that play a central role in hypoxic signalling in animals. The very rapidly developing field, from both biochemical and physiological perspectives, of 2OG-dependent oxygenases (the JmjC enzymes) that remove methyl groups from lysyl-derived side chains *via* oxidative dealkylations is described in Chapter 7 (Cheng and Trievel). Chapter 8 (Müller and Hausinger) summarizes knowledge of the DNA/RNA repair enzyme AlkB that catalyses the dealkylation of modified bases and its ALKBH mammalian homologues. Related oxygenases acting on nucleic acids are then described, *i.e.* two RNA-specific demethylases, FTO and ALKBH5, in Chapter 9 (Zheng and He) and ALKBH8 and its role in synthesis of a modified tRNA in Chapter 10 (Falnes and Ho). The series of ten-eleven translocase (TET) oxygenases that oxidize 5-methylcytosine in DNA, and which have been attracting substantial biomedical interest, is described in Chapter 11 (Aravind and associates). Finally, the conversion of the thymine base within DNA into the novel base J (β-D-glucopyranosyloxymethyluracil) by one of these enzymes in selected protozoa is detailed in Chapter 12 (Sabatini and colleagues).

A multitude of other 2OG-dependent enzymes utilize small molecule substrates and function in diverse roles that include lipid metabolism or synthesis of plant signalling molecules and antibiotics. Chapter 13 (Vaz and van Vlies) details two enzymes involved in the biosynthesis of carnitine, which plays a central role in fatty acid metabolism in animals. The metabolism of the chlorophyll metabolite phytanic acid by a 2OG-dependent hydroxylase is described in Chapter 14 (Wanders and colleagues). The roles of selected enzymes during the biosynthesis of flavonoids and gibberellins are summarized in Chapter 15 (Martens and Matern) and Chapter 16 (Hedden and Phillips), respectively. Chapter 17 (Andersson and Valegård) and Chapter 18 (Smith and Khare) cover roles of 2OG-dependent oxygenases in the biosynthesis of the medicinally important cephalosporin antibiotics and various halogenated molecules, respectively.

Several enzymes are structurally and mechanistically related to the 2OG-dependent oxygenases, but do not utilize 2OG as a cosubstrate. Chapter 19 (Rutledge) summarizes the amazing chemistry of isopenicillin *N* synthase, which catalyses a unique bicyclization reaction. Formation of the agriculturally important plant signalling molecule ethylene is catalysed by 1-aminocyclopropane-1-carboxylate oxidase as described in Chapter 20 (Simaan and

Réglier). Finally, two enzymes that produce distinct products from an alternative 2-oxo acid, 4-hydroxyphenylpyruvate, and their commercial importance, are detailed in Chapter 21 (Shah and Moran).

We sincerely thank all of the authors for their outstanding contributions, which reveal the current states of understanding in this ever-expanding field of research. We apologize for the incomplete coverage due to space constraints. We hope the book will stimulate further research on the functions of 2OG-dependent oxygenases at levels ranging from the biochemical to the physiological and hope it may in a small way help to enable translation of the basic science into medicinal and agricultural benefits. Finally, we feel that there surely must be more roles and amazing reactions catalysed by 2OG-dependent oxygenases waiting to be discovered.

Christopher J. Schofield and
Robert P. Hausinger

Contents

RSC Metallobiology Series No. 3
2-Oxoglutarate-Dependent Oxygenases
Edited by Robert P. Hausinger and Christopher J. Schofield

Published by the Royal Society of Chemistry, www.rsc.org

CHAPTER 1

Biochemical Diversity of 2-Oxoglutarate-Dependent Oxygenases

ROBERT P. HAUSINGER*[a,b]

[a]Department of Microbiology and Molecular Genetics, [b]Department of Biochemistry and Molecular Biology, Michigan State University, East Lansing, MI 48824, USA
*E-mail: hausinge@msu.edu

1.1 Introduction

This chapter summarizes the diverse array of biochemical transformations that are catalysed by Fe(II)- and 2-oxoglutarate (2OG, also known as α-ketoglutarate)-dependent oxygenases.[1–5] As described more comprehensively in Chapter 2, most of these enzymes coordinate their active site metal at one end of a double-stranded β-helix fold (Figure 1.1A)[6–8] using a 2-His-1-carboxylate motif, bind 2OG within the core, and interact with their primary substrates using less conserved regions of the core and additional loops. Most representatives of this enzyme family catalyse hydroxylation reactions (Figure 1.1B), but desaturation, ring formation, ring expansion, halogenation and other types of chemistry are known, as described for several examples in later sections of this chapter. The mechanisms of these enzymes are detailed in Chapter 3, supported by chemical model investigations summarized in Chapter 4. A highly simplified mechanistic scheme for hydroxylases

RSC Metallobiology Series No. 3
2-Oxoglutarate-Dependent Oxygenases
Edited by Robert P. Hausinger and Christopher J. Schofield

Published by the Royal Society of Chemistry, www.rsc.org

based on investigations of TauD (taurine hydroxylase), a well-characterized taurine-degrading enzyme (see Section 1.6.4),[9] is depicted in Figure 1.1C. As illustrated, the binding of 2OG displaces two waters from an octahedral Fe(II) site (converting state a to state b); a third water is lost as the primary substrate binds (generating state c); the newly vacant coordination site is used to react with O_2 to form an Fe(III)–superoxo species (state d); oxidative decarboxylation of 2OG yields an Fe(IV)–oxo species (state e, also known as the ferryl intermediate); this powerful oxidant abstracts a hydrogen atom from the substrate to generate a radical and Fe(III)–hydroxide (state f); and hydroxyl radical rebound (or more complex chemistry)[10] completes

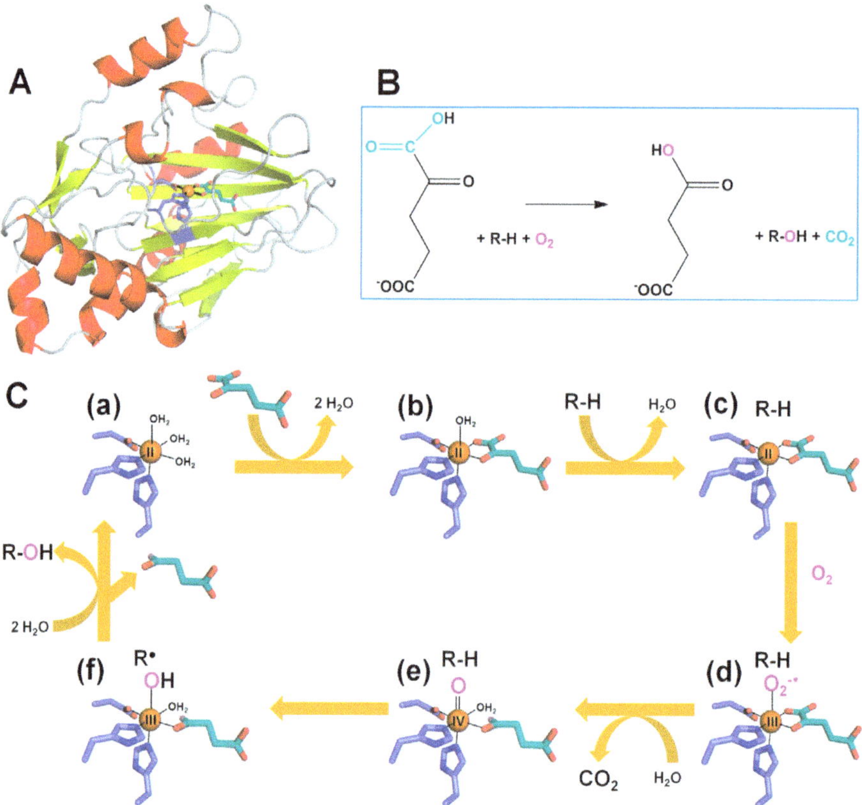

Figure 1.1 Structure and mechanism of a representative 2OG-dependent oxygenase. (A) Cartoon structure of TauD (PDB ID: 1GY9) with red α helices, yellow β strands, grey unstructured regions, Fe (orange sphere), metal ligands (sticks with blue carbons) and 2OG (sticks with cyan carbons). The major grouping of β strands forms a double-stranded β helix core. (B) Overall hydroxylation reaction in which one atom of O_2 is incorporated into substrate (R–H) in a step that is driven by the oxidative decarboxylation of 2OG; CO_2 is derived from the C-1 position of 2OG and succinate incorporates the second atom of O_2. (C) Simplified hydroxylation mechanism.

substrate hydroxylation and recycles the enzyme to the initial Fe(II) state. Such enzymes are dioxygenases because both atoms of oxygen are incorporated into substrates, with one atom ending up in succinate and the second in the hydroxylated product (which decomposes spontaneously in some reactions).[11] These enzymes are correctly referred to as primary substrate:2OG dioxygenases or primary substrate hydroxylases; it is incorrect to cite these enzymes as primary substrate dioxygenases or 2OG dioxygenases because these names imply both atoms of oxygen become incorporated into a single substrate. The other types of chemistries associated with this family of enzymes are also likely to utilize the Fe(IV)-oxo intermediate.

The total number of 2OG-dependent oxygenases found in biology is immense. For example, humans and other animals are thought to possess about 80 such enzymes.[7] Even more are predicted to be present in plants such as the model organism *Arabidopsis thaliana*.[12] Some 2OG-dependent oxygenases are widely distributed throughout aerobic life forms, whereas highly specialized enzymatic catalysts have evolved in various microorganisms or plants for synthesis or degradation of compounds found in their unique niches. In the following sections, the vast array of reactions catalysed by 2OG-dependent oxygenases are divided into seven categories. Section 1.2 describes the 2OG-dependent enzymes acting on proteins. These reactions may lead to structural consequences, have oxygen-sensing functions, alter histone properties, or possess other roles. Section 1.3 summarizes the metabolism of 2OG-dependent oxygenases acting on polynucleotides. The functions of these enzymes include DNA or RNA repair of alkylation damage, roles in transcriptional regulation, biosynthesis of base J in DNA or of specific tRNAs, and demethylation of 5-methylcytosine. Enzymes involved in lipid-related metabolism are discussed in Section 1.4. Examples include reactions related to carnitine biosynthesis, the degradation of phytanic acids, and the decoration of ornithine lipids or lipid A. Plant-specific representatives and close relatives are highlighted in Section 1.5. Of interest are reactions related to synthesis of flavonoids and anthocyanins, gibberellins, alkaloids, and other metabolites found predominantly in plants. Section 1.6 covers enzymes that act on a variety of small molecules including free amino acids, nucleobases or nucleosides, herbicides, sulfonates/sulfates and phosphonates. In some cases, the products resulting from these transformations are incorporated into antibiotics whereas other reactions are involved in metabolite recycling or alternative biochemical pathways. Additional 2OG-dependent oxygenases utilized for antibiotic biosynthesis are provided in Section 1.7. Examples include several halogenating enzymes and other representatives derived from bacterial and fungal sources. Finally, Section 1.8 covers enzymes that are related in structure or mechanism to the 2OG-dependent oxygenases. These include isopenicillin *N* synthase and the plant-specific ethylene-forming enzyme, which contain the 2OG oxygenase fold yet fail to utilize 2OG, and two enzymes that share a distinct fold and use the alternative oxo-acid 4-hydroxyphenylpyruvate.

1.2 2OG-Dependent Dioxygenases that Act on Proteins

This section describes 2OG-dependent dioxygenases that act directly on protein side chains or at modified sites in proteins and catalyse the reactions summarized in Figure 1.2.

1.2.1 Protein Substrates with Structural Roles

The earliest investigations of the 2OG-dependent dioxygenases were focused on enzymes acting on protein substrates with structural roles. For example, seminal studies from the 1960s demonstrated that prolyl 4*R*-hydroxylase (Figure 1.2A) utilizes 2OG as a cosubstrate for its function in collagen synthesis.[13,14] Similar findings were observed for two other enzymes needed to functionalize collagen, prolyl 3*S*-hydroxylase (Figure 1.2B)[15] and procollagen lysyl 5*R*-hydroxylase (PLOD, Figure 1.2C).[16] By 1982, a remarkably prescient mechanism was proposed for prolyl 4*R*-hydroxylase and contained many of the features shown in Figure 1.2C, including coordination of Fe(II) by a facial triad, bidentate coordination of 2OG, and the generation of a reactive metal–oxo intermediate.[17] Collagen, the most abundant protein in mammals, is first synthesized as procollagen, which undergoes extensive modifications including the formation of 4*R*-hydroxyproline (4Hyp, accounting for 10% of its residues), 3*S*-hydroxyproline (3Hyp, ~1% of its residues), and 5*R*-hydroxylysine (Hyl, 0.5–7% of its residues) prior to assembly into its triple helical structure and export to the extracellular matrix.[18] 4Hyp is also found in elastin and other human structural proteins, plant cell wall components, and various other proteins or peptides of algae, selected bacteria, and even a virus.[19] Similarly, 3Hyp and Hyl have been detected in non-collagen proteins. The collagen-specific prolyl 4*R*-hydroxylases form heterodimers with protein disulfide isomerase, but only the homomeric enzymes from the bacterium *Bacillus anthracis* and the alga *Chlamydomonas reinhardtii* have been crystallized.[20–22] These structures reveal the basis for stereospecificity and substrate specificity. The procollagen-specific 2OG-dependent dioxygenases are described in greater detail in Chapter 5.

1.2.2 Protein Substrates with Oxygen-Sensing Roles

An O_2-sensing role for 2OG-dependent dioxygenases was uncovered in 2001.[23,24] A key component of this signalling pathway is the hypoxia-inducible factor (HIF-1) which directs the transcription of several genes under low O_2 conditions. The two subunits of this heterodimer (HIF-α and HIF-β in humans) are constitutively synthesized, but human HIF-α is modified in cells under normoxic conditions by three HIF-α-specific prolyl 4*R*-hydroxylases (PHDs, Figure 1.2A). These enzymes act in a tissue-specific manner to oxidize Pro-402 and Pro-564 of HIF-1α, which results in enhanced affinity of the

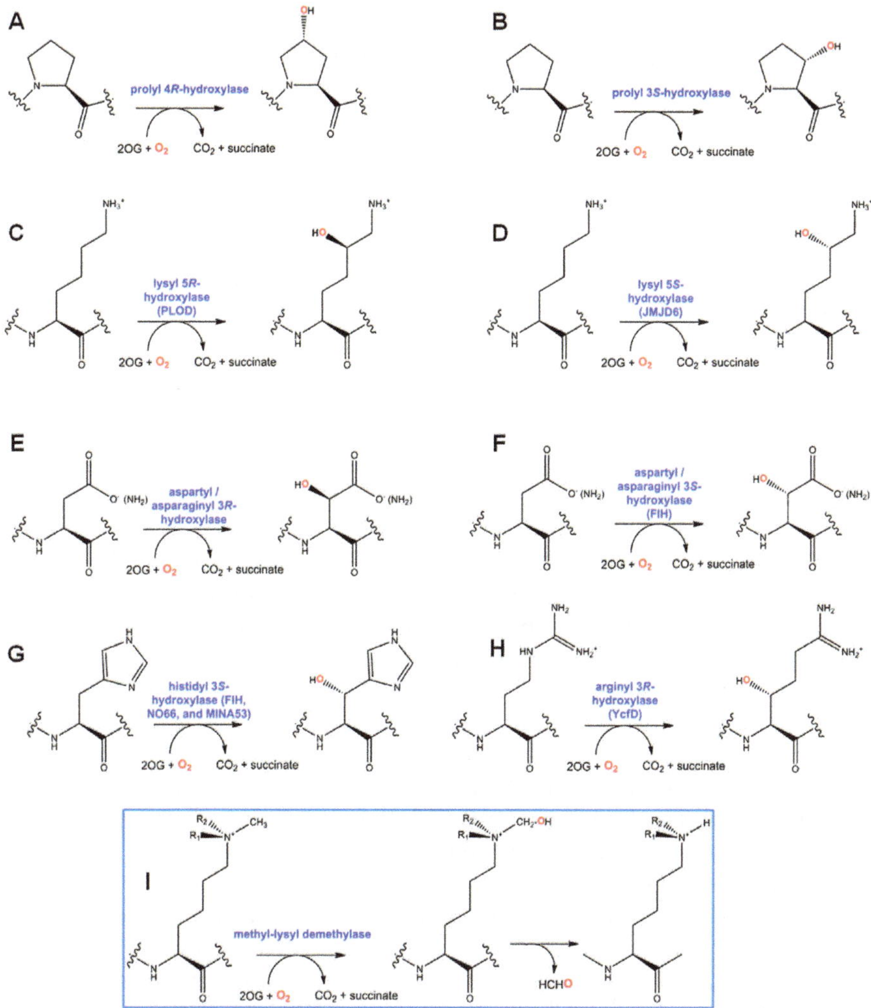

Figure 1.2 2OG-dependent dioxygenase reactions involving protein substrates. (A and B) Prolyl hydroxylases acting at the 4*R* and 3*S* positions. (C and D) Lysyl hydroxylases specific to the 5*R* and 5*S* positions. (E and F) Aspartyl or asparaginyl hydroxylases that insert oxygen at the 3*R* and 3*S* positions. (G) Histidyl 3*S*-hydroxylase. (H) Arginyl 3*R*-hydroxylase. (I) Methyl lysyl demethylases, where R_1 and R_2 are H or CH_3. A lysyl 4-hydroxylase is also known, but this reaction is not depicted because the enantiospecificity has not been reported.

protein towards the von Hipple–Lindau (VHL) tumour suppressor protein, elongin B and elongin C,[25,26] leading to polyubiquitinylation and proteosomal destruction of the transcription factor. The structure has been described for PHD2[27] and for PHD2 in complex with HIF-1α,[28] providing great insight into the geometric aspects of catalysis.

Independent of the PHD-specific process, a separate hydroxylation event involving Asn-803 of HIF-1α occurs in the presence of oxygen.[29] This modification is carried out by another 2OG oxygenase: factor-inhibiting HIF (FIH), an asparaginyl 3*S*-hydroxylase (Figure 1.2F).[30] The addition of a single atom of oxygen to HIF-1α results in loss of interaction with the p300/CBP transcription coactivators and abrogation of hypoxic signalling.[31,32] Several structures have been reported for FIH, including its interaction with a peptide derived from the coactivator.[33–35]

The O_2-sensing functions of the PHD and FIH 2OG-dependent dioxygenases have been summarized in several reviews,[36–38] and their structures and biological activities are described further in Chapters 2 and 6.

1.2.3 Ribosomal Protein Hydroxylases

Hydroxylation of ribosomal proteins has been demonstrated for 2OG-dependent enzymes, as first identified for cells from humans and *Escherichia coli*.[39] The human proteins NO66 and MINA53 catalyse 3*S*-hydroxylation of histidyl residues (Figure 1.2G) in Rpl8 and Prl27a, respectively, while the homologous enzyme YcfD of *E. coli* catalyses arginyl 3*R*-hydroxylation of Rpl16 (Figure 1.2H). The precise biological functions of these modifications are unclear, but when the *ycfD* strain was provided with low nutrient concentrations its growth rate was reduced compared to a wild-type strain. The crystal structure was first determined for the *E. coli* version of YcfD,[40] but additional structures quickly became available for truncated versions of NO66 and MINA53 along with the YcfD homologue of the thermophile *Rhodothermus marinus* and the same YcfD with bound substrate.[41] The structure of the yeast enzyme, known as Tpa1, had also been reported prior to identifying its role in ribosome hydroxylation.[42] A 2OG and Fe(II)-dependent oxygenase domain protein (OGFOD1) is widely found in eukaryotes, where it hydroxylates a particular prolyl residue in small ribosomal protein S23 (RPS23).[43] In mammals, loss of the enzyme leads to the formation of stress granules, stoppage of translation and growth inhibition. Whereas the human enzyme catalyses prolyl 3*S*-hydroxylation (Figure 1.2B), the corresponding enzymes from *Schizosaccharomyces pombe*, *Saccharomyces cerevisiae* (Tpa1, see above), and *Ostreococcus tauri* catalyse dihydroxylation of the cognate residues.[44] Furthermore, a *Drosophila* version of 2OGFOD1, known as Sudestada1, was studied; RNAi-mediated reduction in levels in tissue culture cells leads to smaller cell size, fewer cells and decreased translation efficiency.[45]

1.2.4 Other Protein Substrates

Some 2OG-dependent dioxygenases utilize protein substrates beyond those described above, often without a clear role being identified for the modification. Two Jumonji domain (JMJD)-containing proteins and several other examples are illustrated here.

The nuclear protein JMJD6 hydroxylates the C5 position of a lysyl residue in a protein associated with RNA splicing.[46] Of special interest, the stereospecificity of the enzyme is opposite to that of PLOD, yielding lysyl 5*S*-hydroxylysine (Figure 1.2D).[47] Alternatively, JMJD6 was proposed to catalyse demethylation of methylated argininyl residues,[48] and recent work extends this theme to support such a role in regulation of transcriptional pause release.[49] The crystal structure of the catalytic domain of JMJD6 is known.[50,51] Another Jumonji domain protein, JMJD4, was shown to catalyse C4 hydroxylation of a lysyl residue (not depicted in Figure 1.2 because the enantiospecificity is not reported) in the NIKS motif of eukaryotic release factor 1.[52] Hydroxylation at this site is needed for efficient translation termination.

The hydroxylation of side chains of aspartic acid and asparagine residues (Figure 1.2E) in epidermal growth factor-like domains of various vitamin K-dependent proteins, coagulation factors and complement proteins is also carried out by 2OG-dependent dioxygenases.[53,54] The resulting 3*R*-hydroxyaspartyl (Hya) and 3*R*-hydroxyasparaginyl (Hyn) groups are thought to be important because mice lacking aspartyl (asparaginyl) β-hydroxylase activity exhibit developmental defects and greater susceptibility to intestinal neoplasia.[55] In addition, overexpression of the gene encoding the enzyme is linked to cellular transformation, at least in the case of biliary epithelial cells.[56] Of related interest, a bacterial enzyme, CinX from *Streptomyces cinnamoneus*, catalyses the same type of reaction to modify an aspartyl group in a 15-residue peptide during the synthesis of cinnamycin, an antimicrobial peptide known as a lantibiotic.[57]

Aside from its O_2-sensing function related to hydroxylation of HIF-α Asn-803, FIH hydroxylates ankyrin repeat domains (ARDs) of endogenous Notch receptors.[58] The ARD modifications appear to stabilize the domains and probably affect protein–protein interactions.[59] In these roles, FIH catalyses aspartyl 3*S*-hydroxylase (Figure 1.2F)[60] and histidyl 3*S*-hydroxylase (Figure 1.2G)[61] activities using Asp and His residues in ARD domains. FIH also hydroxylates members of the apoptosis-stimulating p53-binding protein (ASPP) family, such as modification of ASPP2 at Asn-986.[62] An alternative function of type I collagen prolyl 4-hydroxylase involves oxygen addition to Pro-700 of Argonaute 2, an essential component of the RNA-induced silencing complex for RNA interference.[63] This modification is also thought to promote stabilization of the protein.

In addition to catalysing the hydroxylation of target proteins as described above, many 2OG-dependent oxygenases catalyse self-hydroxylation reactions. This process is exemplified by TfdA and TauD, described in Sections 1.6.3 and 1.6.4, which generate hydroxytryptophan and dihydroxyphenylalanine from Trp and Tyr, respectively.[64–68] It is unclear whether such modifications have any beneficial/regulatory role, but the abundance of aromatic residues near the active sites of many 2OG-dependent enzymes led to the proposal of a sacrificial function for these side chains, thus sparing the enzymes from more deleterious chemistry.[69] Other examples of auto-catalysed oxidative modifications are found in procollagen 4*R*-hydroxylase that adds oxygen atoms to unknown sites of four of its own peptides;[70] JMJD6 that catalyses lysyl 5*S*-hydroxylation of one of its own Lys residues in the absence of its RNA splicing factor, perhaps as a regulatory mechanism;[71] FIH that catalyses the formation of hydroxytryptophan when HIF-1α is absent;[72] AlkB and ALKBH3 (both described in Section 1.3) that hydroxylate Trp and Leu residues, respectively, at their active sites.[73,74] Two reviews of autocatalysed oxidative modifications in this enzyme family have been published.[75,76]

1.2.5 Histone Demethylases

Another group of 2OG-dependent dioxygenases using protein substrates are those acting on N^{ε}-methylated lysine residues of histones. Eukaryotic DNA wraps around the histone core (composed of two each of histone H2A, H2B, H3 and H4) leaving the histone amino terminal regions accessible; these regions undergo several types of post-translational modification including the addition of methyl groups on specific lysyl or arginyl residues. The positions and extents of these modification have long been known to play a critical role in gene regulation.[77] In 2006, several research groups demonstrated that 2OG-dependent enzymes containing JmjC (jumonji domain C) domains catalyse demethylation reactions by hydroxylating the target methyl groups, with the hemiaminal intermediates subsequently decomposing with release of formaldehyde (Figure 1.2I).[78–82] The enzymes act on particular residues of specific histones with explicit levels of methylation. Unlike the flavin-dependent N^{ε}-methyl lysyl demethylase, the 2OG oxygenases can act on all three N^{ε}-methylation states of the modified residue. The first crystal structure for one of these enzymes was published that same year,[83] and follow-up studies have clarified the basis of the exquisite substrate specificities.[84–88] A molecular threading mechanism for the peptide substrate was proposed in the case of methylated H3K36 histone-specific KDM2A protein,[89] and the degree of methylation was suggested to influence whether the ferryl group was aligned according to the previously described 'in line' or 'off line' modes.[1] Additional discussion of this important family of enzymes is provided in Chapter 7 and selected reviews.[90–92]

1.2.6 Other Protein Demethylases

Several proteins other than histones are methylated in cells, so corresponding demethylases are likely to be identified. As one example, ALKBH4 was shown to mediate the removal of the methyl group from Lys84 in cytoplasmic actin, a reaction that regulates actin dynamics.[93] Of additional interest, ALKBH4 was also found to associate with several proteins associated with chromatin or involved in transcription, although the functions of these interactions remain unknown.[94]

1.3 2OG-Dependent Dioxygenases that Act on DNA or RNA

1.3.1 Demethylation of Alkylated DNA or RNA Substrates

In 2002, the *E. coli* enzyme AlkB was shown to directly repair alkylation damage to DNA by 2OG-dependent oxidation of 1-methyladenine (1meA) and 3-methylcytosine (3meC) lesions, resulting in the spontaneous loss of formaldehyde and restoration of the native base (Figure 1.3A and B).[95,96] Soon thereafter, AlkB-specific chemistry was expanded to include oxidative repair of alkylated RNA[97] and the range of lesions corrected was extended to 1-methylguanine (1meG, Figure 1.3C),[98] 3-methylthymine (3meT, Figure 1.3D),[98,99] N^6-methyladenine (N^6meA, Figure 1.3E),[100] N^2-methylguanine and N^4-methylcytosine (not shown),[101] as well as bases with slightly larger alkyl groups and exocyclic adducts (not shown). Several structures of AlkB have provided keen insights into the mechanism of substrate binding and substrate specificity.[102–105] Furthermore, the hemiaminal intermediate has been trapped in the protein by binding 3meT and exposing to oxygen.[106]

AlkB-like proteins are widely distributed in other bacteria,[107] viruses[108] and many types of eukaryotes. Of special interest, mammals have nine homologues of AlkB that are referred to as ALKBH (or ABH) followed by numbers 1–8 along with FTO (sometimes referred to as ALKBH9).[109–111] No direct evidence of polynucleotide demethylation has been identified for several of these proteins, including ALKBH4 mentioned earlier in Section 1.2.6. In contrast, ALKBH2 and ALKBH3 were shown to repair 1meA and 3meC in DNA,[112] with ALKBH3 also active with alkylated RNA.[97] Consistent with this differential specificity, knockout mouse studies demonstrated that ALKBH2 is the primary demethylase for repairing alkylated DNA.[113] The structures of ALKBH2 and ALKBH3 reveal the basis of their distinct specificities.[74,103,114–116] ALKBH1 has a more restricted specificity, acting only on 3meC in DNA and RNA,[117] but it also was proposed to be a methylated-histone demethylase[118] and is associated with lyase activity at abasic sites.[119,120] ALKBH5 exhibits N^6meA demethylase activity (Figure 1.3E) for alkylated mRNA.[121] This modification on RNA can affect its processing and has regulatory implications, such as controlling the length of the circadian clock.[122] Human and zebrafish ALKBH5 structures reveal the basis for nucleic acid recognition and catalysis.[123–125] The reaction catalysed by ALKBH8 is described in the next section. Meanwhile, the chemistries catalysed by ALKBH6 and ALKBH7 remain unknown, but the structure of ALKBH7 has been determined and shown to lack a nucleotide recognition lid consistent with a potential protein hydroxylation role.[126]

FTO is named for the fat mass and obesity-associated gene, whose inactivation protects mice from obesity.[127] Early studies reported the enzyme acts on 3meT in DNA,[128] or 3meT and 3-methyluracil (3meU, Figure 1.3F) in DNA and RNA;[129] however, more recent studies have shown that N^6meA in RNA is a major substrate.[130] The RNA-associated N^6-hydroxymethyladenine resulting

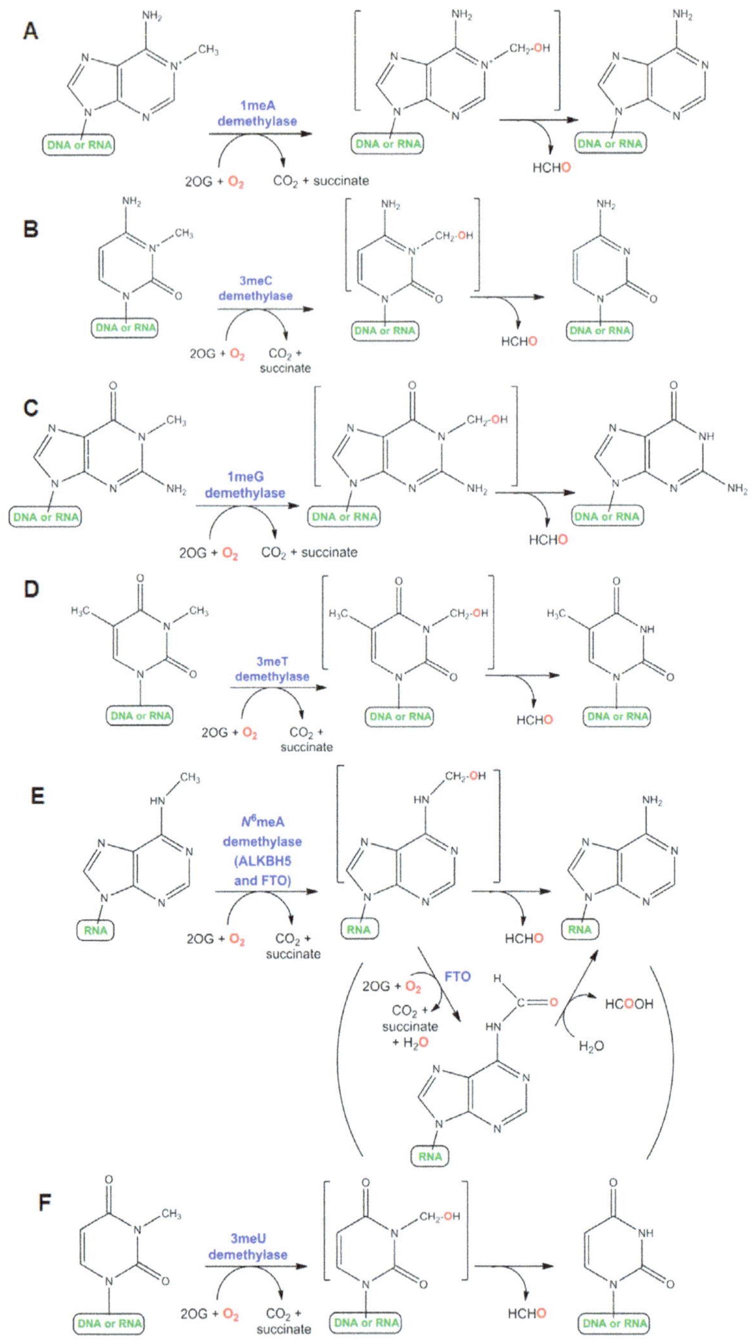

Figure 1.3 2OG-dependent dioxygenase reactions involving demethylations of alkylated DNA and RNA substrates. The reactions reverse lesions related to (A) 1meA, (B) 3meC, (C) 1meG, (D) 3meT, (E) N^6meA and (F) 3meU.

from the latter activity is sufficiently stable that it can be further hydroxylated by FTO to form N^6-formyladenine (the reaction shown in parentheses in Figure 1.3E).[131] The structure of FTO has been determined in the presence of 3meT mononucleotide and with several inhibitors.[132,133]

Further details on polynucleotide repair by AlkB and ALKBH species are found in Chapter 8, while ALKBH5 and FTO are described more fully in Chapter 9.

1.3.2 Other Oxidative Modifications of DNA or RNA

In addition to the oxidative demethylations of alkylated bases in DNA and RNA discussed above, 2OG-dependent dioxygenases catalyse several additional types of reaction using these substrates (Figure 1.4).

Two types of tRNA modifications have been shown to require 2OG-dependent hydroxylases. Working with an accessory protein, ALKBH8 first utilizes its RNA recognition and methyltransferase motifs to convert 5-carboxymethyluridine (cm^5U) to 5-methoxycarbonylmethyluridine (mcm^5U) at position 34 in the anticodon loop of human tRNA.[134] This modified tRNA then serves as the substrate for the hydroxylase domain of ALKBH8 which forms 5*S*-methoxycarbonylhydroxymethyluridine ($mchm^5U$) (Figure 1.4A).[135,136] The structure of the RNA recognition and hydroxylase domains of ALKBH8 has been reported.[137] Curiously, protozoan ALKBH8 catalyses both tRNA modification and DNA repair reactions.[138] Additional discussion of the biosynthesis of this unique base modification can be found in Chapter 10. A second tRNA-related hydroxylase acts on 7-(α-amino-α-carboxypropyl)wyosine known as wybutosine and abbreviated yW (Figure 1.4B), a tricyclic base that abuts the anticodon loop of $tRNA^{Phe}$ in eukaryotes and archaea.[139] Among the numerous derivatives formed from this base are hydroxylated species generated by TYW5 (tRNA-yW-synthesizing enzyme 5), a member of the JmjC domain family (see Section 1.2.4), whose structure has been reported.[140]

A family of ten-eleven translocation (TET) proteins contributes to the epigenetic regulation of a wide range of genes in many eukaryotes by controlling the methylation status of cytosine in CpG islands of DNA. For example, TET1 was shown to convert 5-methylcytosine (5meC) to 5-hydroxymethylcytosine (5hmC) in 2009 (Figure 1.4C).[141] Soon thereafter this result was confirmed and extended to TET2 and TET3.[142] More recently, TET enzymes were shown to catalyse subsequent reactions to create 5-formylcytosine (5fC)[143] and 5-carboxylcytosine (5caC),[144,145] including in a fungus.[146] In addition, the TET-catalysed oxidation of thymine to 5-hydroxymethyluracil in DNA of embryonic stem cells has been reported.[147] The various oxidized derivatives may have regulatory functions,[148] such as the hypoxic gene induction noted in neuroblastoma.[149] Moreover, the series of reactions from 5mC is suggested to play a role in active demethylation in DNA, either *via* (i) excision of 5fC or 5caC by thymine-DNA glycosylase, (ii) deamination of 5hmC followed by glycosylase removal, (iii) retro-aldol chemistry of 5hmC or 5fC with release of formaldehyde or formate, or (iv) decarboxylation of 5caC.[150] The structures of human TET2 in complex

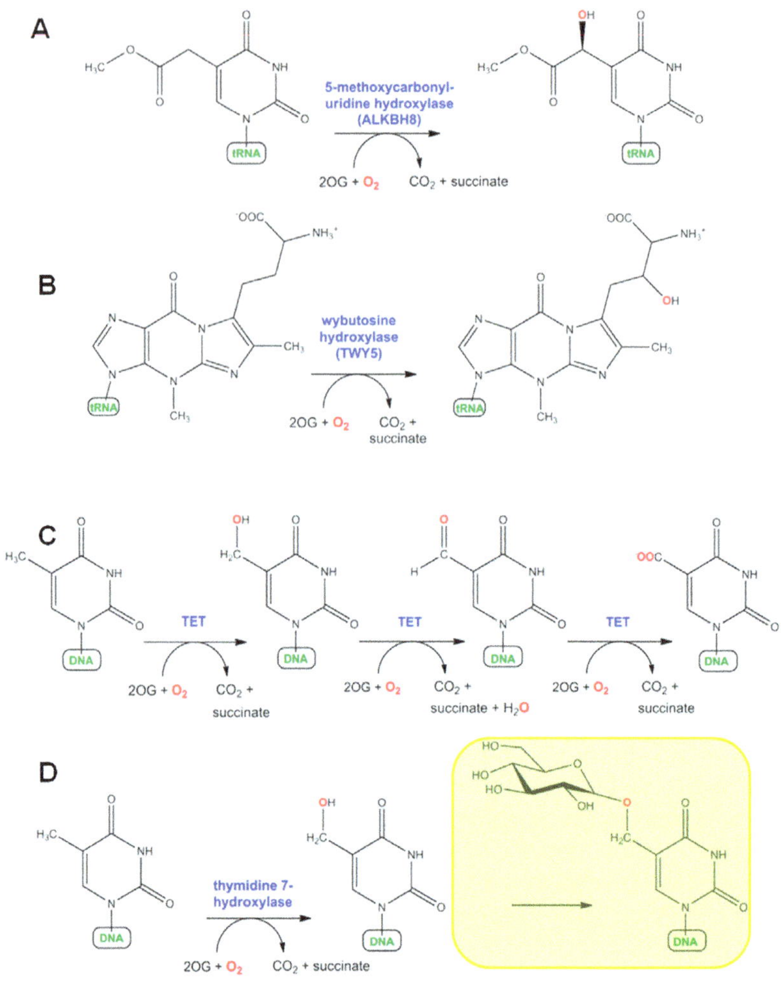

Figure 1.4 Non-demethylase 2OG-dependent dioxygenase reactions utilizing RNA or DNA substrates. (A) Action of ALKBH8 in the synthesis of a modified tRNA. (B) Wybutosine hydroxylase catalysis by TWY5. (C) The sequential reactions of TET proteins acting on 5meC to form 5hmC, 5fC and 5caC. (D) Thymidine 7-hydroxylase activity catalyses the first step in formation of base J (a subsequent glucosyl transfer reaction is highlighted in yellow).

with DNA and a TET family protein from *Naegleria gruberi* in complex with 5meC-containing DNA are reported.[151,152] Additional information on the biology of the TET proteins is provided in Chapter 11 and in reviews.[153,154]

Nuclear genomes of kinetoplastid flagellates and some unicellular flagellates contain β-D-glucopyranosyloxymethyluracil in their telomeric repeats.[155] This modified base, first identified in 1993, is also called β-D-glucosyl-hydroxymethyluracil or base J.[156] Synthesis of base J involves two steps, hydroxylation of thymidine to form hydroxymethyluracil followed by glucosylation (Figure 1.4D).[157,158] JBP1, a protein that binds to and enhances the levels of base J,[159,160] was shown to belong to the 2OG-dependent dioxygenase family and is related to the TET proteins.[161,162] JBP1 along with the related JBP2, which does not bind to base J in DNA,[163] were directly demonstrated to catalyse thymidine hydroxylation.[164] Further information about the role of 2OG-dependent dioxygenases in base J biosynthesis is provided in Chapter 12.

1.4 Lipid-Related Metabolism Involving 2OG-Dependent Oxygenases

Several 2OG-dependent hydroxylase reactions are used in lipid biosynthesis (Figure 1.5).

Carnitine or γ-trimethyl-hydroxybutyrobetaine is a small molecule that becomes linked to fatty acids, allowing for their transport across the inner mitochondrial membrane and degradation in the matrix by the β-oxidation pathway.[165] The synthesis of carnitine initiates with protein-derived ε-*N*-trimethyllysine that undergoes β-hydroxylation, aldolase cleavage to glycine plus trimethylaminobutyraldehyde, dehydrogenation to γ-butyrobetaine, and another hydroxylation (Figure 1.5A). The ε-*N*-trimethyllysine hydroxylase activity was shown to require 2OG in 1978,[166] and the enzyme was purified and characterized in 2001.[167] The hydroxylation of γ-butyrobetaine was found to require 2OG in 1968,[168] and the enzyme was purified from *Pseudomonas* spp. AK1 in 1977 and from calf liver in 1981.[169,170] A detailed comparison of substrate specificities for the human and bacterial enzymes revealed surprisingly large differences in reactivity for several γ-butyrobetaine analogues.[171] The structure of human γ-butyrobetaine hydroxylase has been reported.[172,173] Further discussion of the enzymes involved in carnitine metabolism is provided in Chapter 13.

Phytanic acid (2,6,10,14-tetramethylhexadecanoic acid), a polyisoprenoid derived from the phytol moiety of chlorophyll, cannot be directly degraded *via* the β-oxidation pathway due to its 3-methyl group. To overcome this hurdle, the C-1 carbon is removed by action of a series of four enzymes: a ligase converts the molecule to phytanoyl-coenzyme A (CoA),[174] hydroxylation occurs at the C-2 position,[175] a lyase releases formyl-CoA and methylpentadecanal (pristanal),[176,177] and a dehydrogenase forms pristanic acid (2,6,10,14-tetramethylpentadecanoic acid)[178] which is a substrate for β-oxidation. Interruption of this pathway results in Refsum disease, which is associated with dysfunctions including retinitis pigmentosa, polyneuropathy and ataxia.[179] Phytanoyl-CoA hydroxylase (Figure 1.5B), shown to be a 2OG-dependent enzyme,[175] was purified from rat liver and as the recombinant human protein.[180,181] The human enzyme was structurally characterized, providing an understanding of mutations associated with Refsum disease.[182] Further description of phytanic acid metabolism is found in Chapter 14.

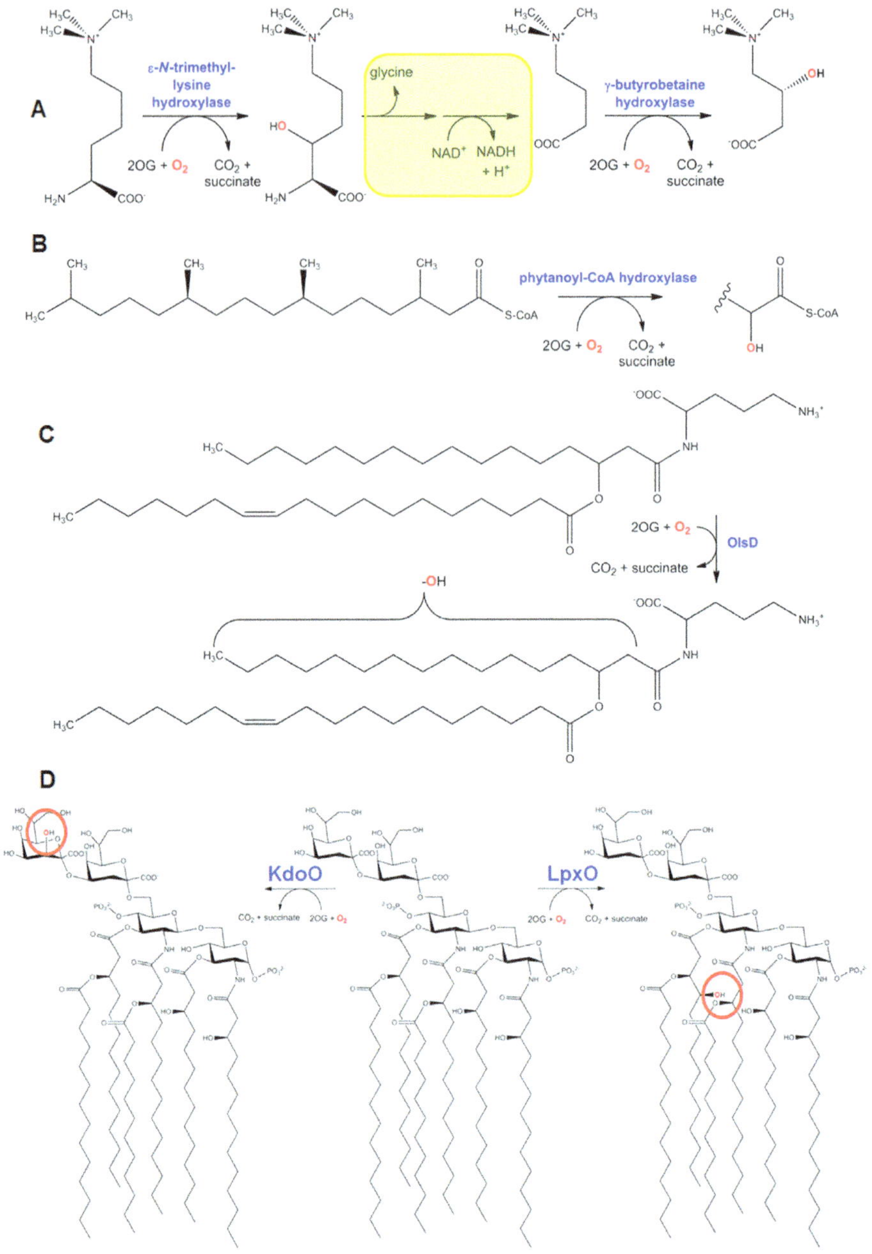

Figure 1.5 2OG-dependent dioxygenases involved in lipid metabolism. (A) Biosynthesis of the acyl group carrier molecule carnitine includes two 2OG-dependent hydroxylases that act on ε-*N*-trimethyllysine and γ-butyrobetaine (with intervening aldolase and dehydrogenase reactions highlighted in yellow). (B) Phytanic acid metabolism includes the activity of phytanoyl-CoA hydroxylase. (C) Hydroxylation of the acyl chain of ornithine lipids is catalysed by OlsD. (D) KdoO and LpxO hydroxylase reactions involving lipid A.

Much like phytanoyl-CoA hydroxylase, OlsD from *Burkholderia cenocepacia* (and probably from other *Burkholderia* and *Serratia* species) catalyses the hydroxylation of an acyl chain (Figure 1.5C) as part of a lipid metabolism pathway.[183] In this case, however, the pathway is biosynthetic rather than degradative. The first step of the pathway involves OlsB, an *N*-acyl transferase that transfers a 3-hydroxy fatty acyl group from the acyl carrier protein to the α-amino group of ornithine. The resulting lyso-ornithine lipid is the substrate of OlsA, an *O*-acyltransferase that transfers a fatty acyl group from the acyl carrier protein to the C-3 hydroxyl group. This ornithine-containing lipid, or its C-2 hydroxylated derivative (not shown), is the substrate of OlsD that hydroxylates somewhere on the amide-linked fatty acyl chain.

Two 2OG-dependent enzymes have been shown to modify lipid A (Figure 1.5D), with one targeting a specific acyl side chain and the other acting on the 3-deoxy-D-*manno*-oct-2-ulosonic (Kdo) moiety. *Salmonella typhimurium*, previously known to synthesize lipid A containing secondary *S*-2-hydroxyacyl chains, was demonstrated to use the O_2- and 2OG-dependent enzyme LpxO to achieve this modification.[184] LpxO was the first integral membrane protein representative of this enzyme family.[185] A Kdo 3-hydroxylase (KdoO) that adds an oxygen atom to the deoxysugar was identified in *Burkholderia ambifaria* and *Yersinia pestis*.[186] Homologues to the genes encoding these proteins are found in other Gram-negative bacteria.

1.5 Plant Metabolite Biosynthesis Using 2OG-Dependent Oxygenases

Plants possess many of the 2OG-dependent oxygenases already mentioned, but in addition they utilize members of this enzyme family for some of their unique biosynthetic needs.[187] Here we illustrate how plants (and, in a few cases, other organisms) use these enzymes for generating flavonoids, gibberellins, alkaloids and other predominantly plant-specific products.

1.5.1 2OG-Dependent Oxygenases in Flavonoid Biosynthesis

The flavonoids are polyphenolic compounds that include flavanones, flavones, isoflavones, flavonols and anthocyanins with more than 9000 such compounds known (see examples in Figure 1.6). These substances are used by plants for defence against pathogens, as signalling molecules in plant–microbe interactions, to minimize photodamage, in flower colour, and other roles.[188] When consumed by humans they function as antioxidants, antimalarials and potential anticancer agents.[189] The entrance to the main pathway shown in Figure 1.6 (indicated by the green arrow) is the flavanone naringenin, the substrate for two distinct 2OG-dependent oxygenases. Flavone synthase I (FNS) catalyses a desaturation reaction at the C-2/C-3 position forming *trans*-dihydroflavonol,[190] while flavanone 3β-hydroxylase (FHT) adds a hydroxyl group to C-3.[191] The cytochrome P450 enzyme flavanone

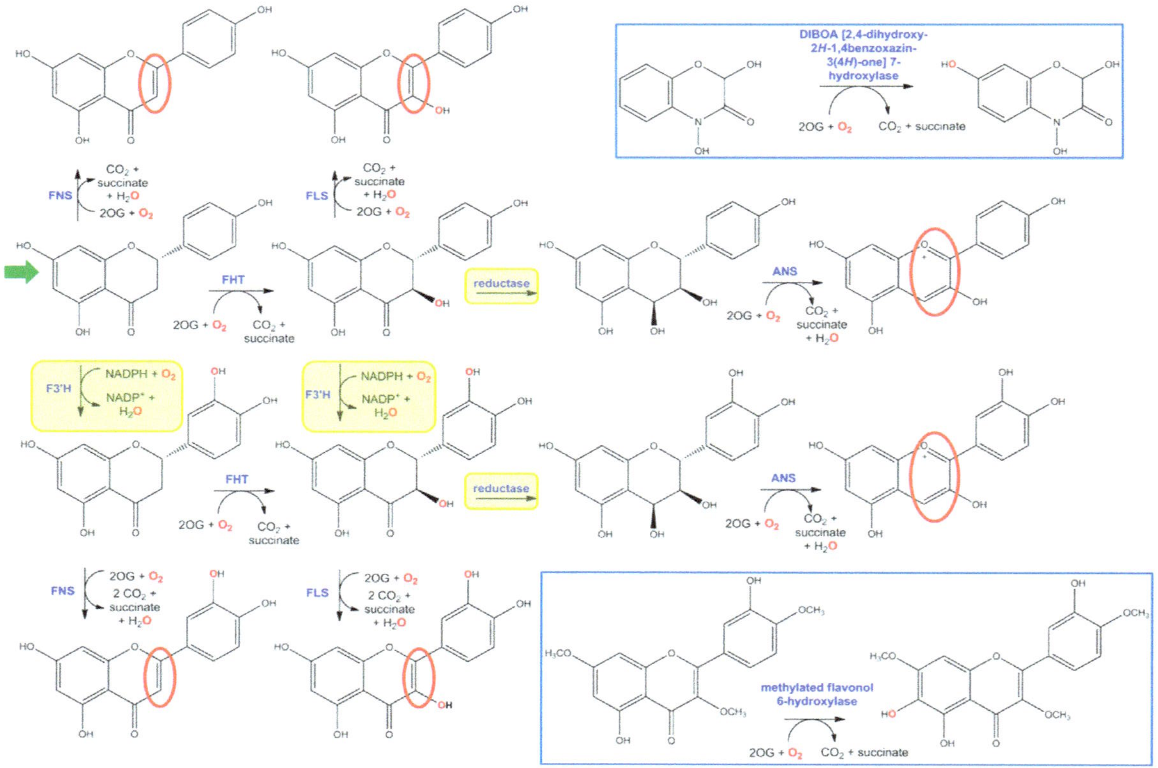

Figure 1.6 Representative 2OG-dependent oxygenases of flavonoid metabolism. Flavanone (indicated by the green arrow) can undergo desaturation by flavone synthase I (FNS), C-3 hydroxylation by flavanone 3β-hydroxylase (FHT), or C-3′ hydroxylation by flavanone 3′-hydroxylase (F3′H, not a 2OG-dependent enzyme as indicated by yellow highlighting), with the product of the latter reaction also serving as substrate for both FNS and FHT. The FHT-derived products can undergo desaturation by flavonol synthase (FLS) or can be reduced by a non-2OG enzyme (also highlighted in yellow). The reduced products are converted to anthocyanins by anthocyanidin synthase (ANS). The boxes to the upper and lower right depict the reactions of related enzymes that hydroxylate at the 7 and 6 positions of flavonoid compounds.

3′-hydroxylase (F3′H) uses naringenin (or its 3β-hydroxy derivative) and converts the appended phenol to a catechol, with this species also being a substrate for FNS and FHT. The FHT-derived *trans*-dihydroflavonol products are substrates for another 2OG-dependent desaturase called flavonol synthase (FLS).[192,193] Alternatively, the *trans*-dihydroflavonols can be reduced *via* non-2OG-dependent enzymes to provide leucoanthocyanidin substrates for anthocyanidin synthase (ANS).[194] The structure of ANS from *Arabidopsis thaliana* has been elucidated with the bound substrate analogue dihydroquercetin and the unnatural substrate naringenin.[192,195] Two other flavonoid-related reactions catalysed by 2OG-dependent hydroxylases are 2,4-dihydroxy-2*H*-1,4-benzoxazin-3(4*H*)-one (DIBOA) 7-hydroxylase and methylated flavonol 6-hydroxylase (see boxes in Figure 1.6).[196–200] Flavonoid biosynthesis is discussed in much greater detail in Chapter 15.[201]

1.5.2 2OG-Dependent Oxygenases of Gibberellin Biosynthesis

Tetracyclic diterpenoids known as gibberellins (GAs) are widely distributed in higher plants and are also found in some lower plants, bacteria and fungi.[202] At least 136 distinct GA structures are reported (commonly referred to as GA_1-GA_{136}; see http://www.plant-hormones.info/gibberellins.htm). A small sampling of such structures is shown in Figure 1.7, which depicts selected 2OG-dependent transformations of these molecules. Many GAs possess a carboxylate at C-7, introduced by oxidation of the GA_{12}-aldehyde to form GA_{12}, with this product undergoing partial oxidization at C-7 by a hydroxylase to form GA_{53}. Depending on their biological sources these reactions can be catalysed by either cytochrome P450 or 2OG-dependent oxygenases (Figure 1.7A).[203] The GA C20 oxidase can catalyse sequential reactions that convert the C-20 methyl group (*e.g.* GA_{12}/GA_{53}) to the alcohol (GA_{15}/GA_{44}), aldehyde (GA_{24}/GA_{19}), and in some cases the carboxylate (GA_{25}/GA_{17}) (Figure 1.7B).[204,205] Alternatively, the same enzyme can catalyse an oxidative transformation that eliminates the carboxylate as CO_2 while forming a γ-lactone (GA_9/GA_{20}).[206] GA 3β oxidase converts these products to the corresponding hydroxylated species (GA_4/GA_1).[207] GA 2β oxidase acts on the same substrates (producing GA_{51}/GA_{29}) or on the products of the prior reaction (producing GA_{34}/GA_8).[208] In addition, other 2OG-dependent enzymes can catalyse several types of desaturation reactions (not shown),[202,209] such as a recently characterized fungal GA_4 desaturase that introduces a double bond between C-1 and C-2.[210] Further information on this remarkable family of enzymes is available in Chapter 16.

1.5.3 2OG-Dependent Oxygenases in Alkaloid Synthesis

Four examples are depicted to illustrate how plants use 2OG-dependent oxygenases for alkaloid biosynthesis. Scopolamine is a hallucinogenic tropane alkaloid produced by *Hyoscyamus niger* (henbane). The last two steps in its synthesis are catalysed by hyoscyamine 6β-hydroxylase that carries out both

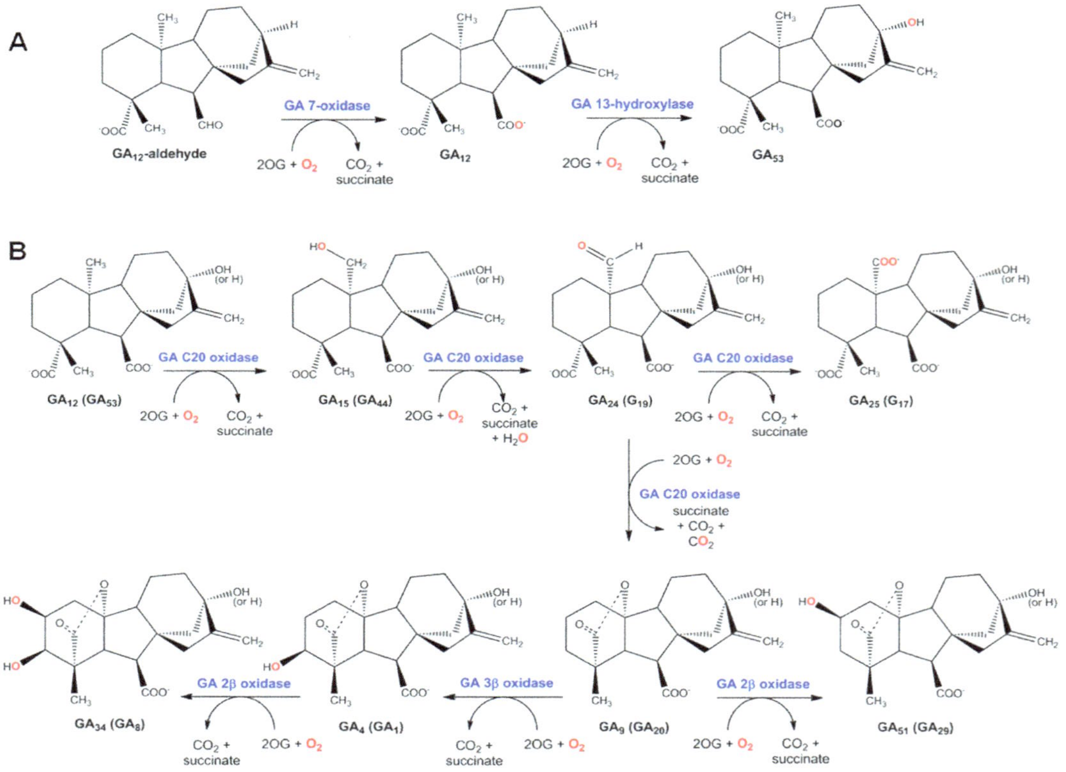

Figure 1.7 Selected 2OG-dependent oxygenases of gibberellin (GA) biosynthesis. (A) Reactions of GA 7-oxidase and GA 13-hydroxylase. (B) Three consecutive reactions of GA C20 oxidase that convert a methyl group to an alcohol, aldehyde and carboxylic acid; a distinct reaction catalysed by GA C20 oxidase to eliminate CO_2 and form a γ-lactone; GA 2β and GA 3β oxidase activities.

hydroxylation and epoxidation steps (Figure 1.8A).[211,212] Vinblastine and vincristine are alkaloids produced by *Caranthus roseus* (periwinkle), which have been used for treatment of Hodgkin's lymphoma and acute leukaemia. These compounds are synthesized by a complex biosynthetic pathway that utilizes a 2OG-dependent hydroxylase to generate deacetylvindoline (Figure 1.8B), which is then further modified.[213,214] Codeine and morphine are important pharmaceuticals obtained from *Papaver somniferum* (opium poppy). Their complex biosynthetic pathway includes an intermediate named thebaine, which can be demethylated at one site by thebaine 6-*O*-demethylase to produce codeinone (Figure 1.8C) or demethylated at a second site by codeine *O*-demethylase to form oripavine (Figure 1.8D).[215] These two reactions plus that catalysed by codeinone reductase yields morphine. A series of additional reactions (not shown) involving alkaloid metabolism in opium poppy are catalysed by these enzymes and the 2OG-dependent dioxygenase protopine *O*-dealkylase, PODA.[216] In the fungus *Claviceps purpurea*, a cyclization reaction (not shown) in the synthesis of D-lysergic acid alkaloid peptides is catalysed by a member of this enzyme family, and the 2OG-bound holoprotein, EasH, was structurally characterized.[217] Another fungal example is found in the synthesis of loline alkaloids by *Epichloë* species, where the 2OG-dependent oxygenase LolO introduces an ether bridge into a pyrrolizidine ring system (Figure 1.8E).[378]

1.5.4 Other Plant-Specific 2OG-Dependent Oxygenases

Additional representatives of the 2OG-dependent oxygenases that are primarily restricted to plants include enzymes involved in the biosynthesis of phytosiderophores, coumarins and glucosinolates, or the degradation of hormonal compounds.

Under iron-deficient conditions, some grasses, cereals and rice produce and secrete iron-binding compounds such as the mugineic acid-related species made by *Hordeum vulgare* (barley). This plant contains two 2OG-dependent dioxygenases, IDS2 and IDS3, that act on 2′-deoxymugineic acid to form 3-epihydroxy-2′-deoxymugineic acid and mugineic acid, respectively (Figure 1.8F).[218] 2′-Deoxymugineic acid is converted to 3-epihydroxymugineic by combined actions of the two enzymes.

Coumarins (1,2-benzopyrones) such as scopoletin and umbelliferone are synthesized by many higher plants where they are used for defence against phytopathogens. A key enzyme in the pathway used by *Arabidopsis thaliana* for generating scopoletin is a 2OG-dependent enzyme that hydroxylates the *ortho* position of feruloyl-CoA,[219] with subsequent isomerization, hydrolysis and lactonization steps providing the product (Figure 1.8G). The same activity was detected using two recombinant proteins from *Ipomoea batatas* (sweet potato), one of which also used *p*-coumaroyl-CoA to form umbelliferone (where H is present at the 3′ position in Figure 1.8G).[220] An enzyme with the latter dual activity has also been characterized from *Ruta graveolens* (the common rue).[221]

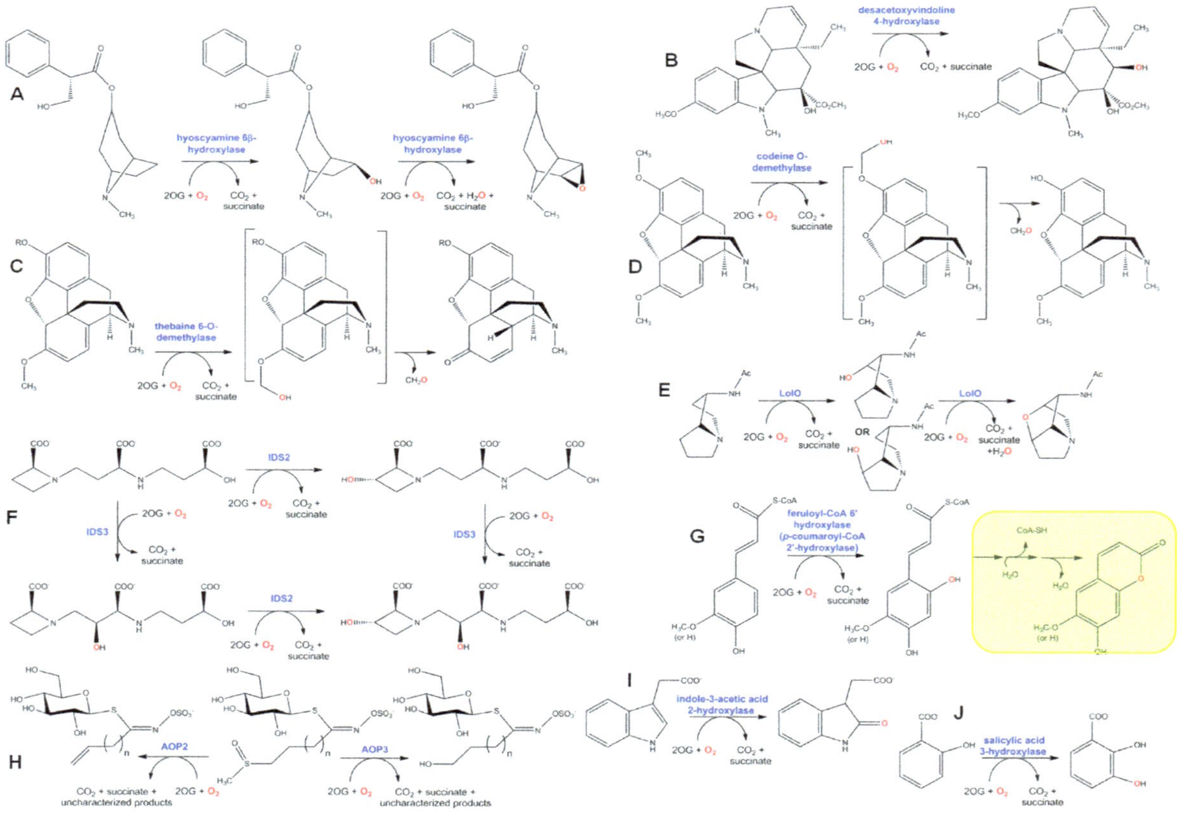

Figure 1.8 Steps in the synthesis of selected alkaloids and other primarily plant-specific products catalysed by 2OG-dependent oxygenases. (A) Hydroxylase and epoxidase activities of hyoscyamine hydroxylase. (B) Desacetoxyvindoline 4-hydroxylase. (C) Thebaine 6-*O*-demethylase. (D) Codeine *O*-demethylase. (E) LolO-catalysed ether bridge formation in loline alkaloid biosynthesis. (F) Mugineic acid phytosiderophore synthesis. (G) Feruloyl-CoA 6′-hydroxylase and *p*-coumaroyl-CoA 2′-hydroxylase in the pathways for synthesis of scopoletin and umbelliferone (yellow highlight). (H) Action of AOP2 and AOP3 on methylsulfinylalkyl glucosinolates. (I) Indole-3-acetic acid 2-hydroxylase. (J) Salicylic acid 3-hydroxylase.

Glucosinolates (over 130 are known) are predominantly associated with the Brassicaceae family of plants where, following tissue disruption, they are decomposed to form compounds with diverse protective roles against herbivores and pathogens.[222] Two sequence-related 2OG-dependent enzymes in *A. thaliana*, named AOP2 and AOP3, have been shown to participate in glucosinolate biosynthesis by converting methylsulfinylalkyl glucosinolate to the alkenyl or hydroxyalkyl species (Figure 1.8H).[223] The chemistry of these reactions has not been well defined.

Indole-3-acetic acid and salicylic acid are plant hormones that play important roles in growth, development, disease resistance or other functions, but until recently their degradation pathways within plants were unclear. Using rice, a dioxygenase for auxin oxidation (DAO) has now been identified and shown to transform this substrate into 2-oxoindole-3-acetic acid (Figure 1.8I).[224] Mutants affecting the corresponding gene display male sterility and produce infertile seeds. Similarly, a salicylic acid 3-hydroxylase (Figure 1.8J) of *Arabidopsis* was identified and mutants in the corresponding *s3h* gene were found to accumulate salicylate species and exhibit early senescence.[225] When assayed *in vitro*, both 2,3- and 2,5-dihydroxybenzoate are formed; however, only the former appears to be made *in vivo*.

1.6 2OG-Dependent Oxygenases Catalysing Reactions with Free Amino Acids, Nucleobases, Herbicides and Sulfur- or Phosphorous-Containing Compounds

The reactions described in this section are diverse, but generally involve rather small-sized molecules. Several products from these reactions are precursors that become incorporated into antibiotics, whereas 2OG-dependent tailoring enzymes that act directly during antibiotic synthesis are described in Section 1.7.

1.6.1 Amino Acid Hydroxylases

2OG-Dependent hydroxylases acting on free amino acids (*i.e.* not as a side chain of a protein) are known. In the case of L-Pro, different enzymes exhibit each of four distinct specificities. Pro 4*R*-hydroxylase (Figure 1.9A, left) from *Streptomyces griseoviridus* P8648 produces the *trans*-isomer of 4-hydroxyproline that is subsequently utilized for etamycin synthesis.[226,227] This enzyme activity is also found in other bacterial strains of *Streptomyces*, *Dactylosporangium* and *Amycolatopsis*,[228] and in the pneumocandin-producing fungus *Glarea lozoyensis*.[229] L-Pro 4*S*-hydroxylase (Figure 1.9A, right), producing the *cis* isomer, has been identified in *Mesorhizobium loti* and *Sinorhizobium meliloti* where it was shown to require 2OG.[230] L-Pro 3*S*-hydroxylase (Figure 1.9B, left) has been reported in the *G. lozoyensis* fungus mentioned

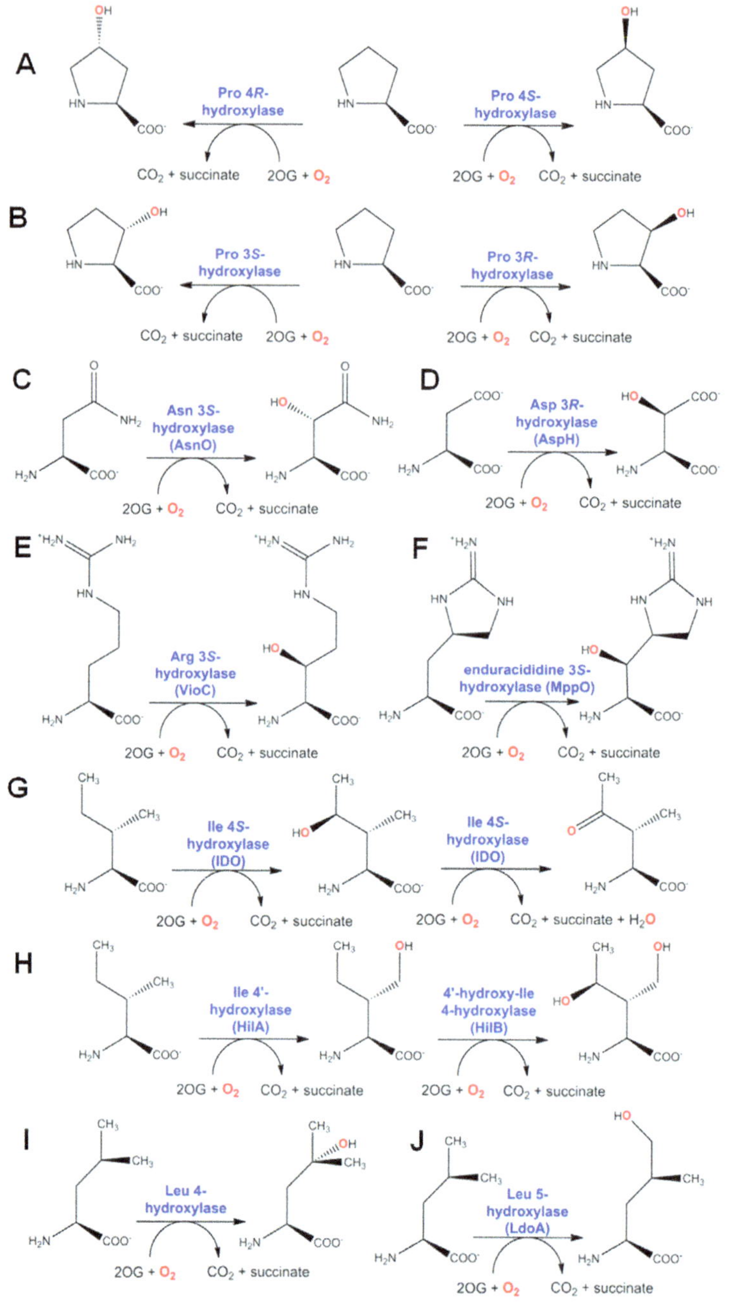

Figure 1.9 2OG-dependent hydroxylases acting on free L-amino acids. (A) Pro 4*R*- and 4*S*-hydroxylases. (B) Pro 3*S*- and 3*R*-hydroxylases. (C) Asn 3*S*-hydroxylase. (D) Asp 3*R*-hydroxylase. (E) Arg 3*S*-hydroxylase. (F) Enduracididine 3-hydroxylase. (G) Ile 4*S*-hydroxylase. (H) Ile 4′- and 4-hydroxylase. (I) Leu 4-hydroxylase. (J) Leu 5-hydroxylase.

above.[229] Enzymes from two actinomycetes were shown to exhibit both 3*S*- and 4*S*-hydroxylase activities using L-Pro.[231] The best studied enzyme using this substrate is L-Pro 3*R*-hydroxylase (Figure 1.9B, right), an activity identified in strains of *Streptomyces* and *Bacillus*.[232] The enzyme was purified from *Streptomyces* sp. strain TH1[233] and the structure of the apoprotein was elucidated.[234]

Several 2OG-dependent hydroxylases act on free L-amino acids with polar side chains. A 3*S*-hydroxylase of L-Asn (creating the *threo* isomer, Figure 1.9C) was characterized from recombinant cells containing *asnO* from *Streptomyces coelicolor*.[235] High-resolution crystal structures of AsnO reveal the basis of this substrate specificity. The product of the reaction is subsequently incorporated (at position nine) into a daptomycin-type lipopeptide called the calcium-dependent antibiotic. A single amino acid substitution (D241N) of AsnO led to use of L-Asp as substrate, forming the *threo* isomer of hydroxyaspartic acid (not shown).[236] By contrast, AspH from *Pseudomonas syringae* catalyses 3*R*-hydroxylation of L-Asp (and L-Asp thioesters, not shown) to form the *erythro* isomers (Figure 1.9D).[237] VioC, a 2OG-dependent Arg 3*S*-hydroxylase (Figure 1.9E), was shown to be used by *Streptomyces vinaceus* to provide a precursor for viomycin biosynthesis.[238] The VioC structure was elucidated and explains the hydroxylation specificity.[239] Enduracididine is an Arg-derived amino acid that undergoes 3*S*-hydroxylation by MppO (Figure 1.9F) to provide β-hydroxyenduracididine, which is incorporated into mannopeptimycins, glycopeptide antibiotics produced by *Streptomyces hydrogroscopicus* NRRL 30439.[240]

Hydroxylases of L-Ile and L-Leu have been widely studied since the first report of 2OG-dependent 4-hydroxyisoleucine synthesis in *Trigonella foenum-graecum* (fenugreek).[241] The recombinant enzyme (IDO) from *Bacillus thuringiensis* was shown to exhibit L-Ile 4*S*-hydroxylase activity,[242] and the enzyme was later shown to also form 2-amino-3-methyl-4-ketopentanoate, probably arising from dihydroxylation of the same carbon position to produce a gem diol that loses water (Figure 1.9G).[243] The products of two adjacent genes in *Pantoea ananatis*, HilA and HilB, catalyse sequential hydroxylation reactions at the 4′ and 4 positions, respectively, of L-Ile (Figure 1.9H).[244] Related family members were identified in a range of other bacteria, in several cases using L-Leu as the preferred substrate and yielding 4-hydroxyleucine (Figure 1.9I).[245] Alternatively, L-Leu 5-hydroxylase (LdoA, Figure 1.9J) was characterized from *Nostoc punctiforme*.[246] All of these enzymes also act on other substrates with reduced catalytic efficiencies (including Leu/Ile substitution), in various cases forming Met sulfoxide, 4-hydroxyvaline and 4-hydroxythreonine (not shown).[243–245]

Four other enzymes are mentioned in this section because they modify amino acid-like substrates or combine another reaction with amino acid hydroxylation. SadA from *Burkholderia ambifaria* acts on several *N*-substituted amino acids with hydrophobic side chains, notably catalysing 3*R* hydroxylation of *N*-succinyl-L-Leu (Figure 1.10A).[246] The structure of SadA has been reported,[247] and variants with altered specificity have been studied.[248]

PvcB of *Pseudomonas aeruginosa* functions in the biosynthesis of the siderophore pyoverdine by participating in the formation of 2-isocyano-6,7-dihydroxycoumarin (reminiscent of the coumarin product shown in Figure 1.8G). The *P. aeruginosa* enzyme catalyses a cyclization reaction (Figure 1.10B),[249] although the mechanistic details remain obscure, and the structure of the protein has been characterized in the absence of ligands.[249] Ectoine is an acid with a secondary amine, so the associated enzyme is described here although this tetrahydropyrimidine could alternatively be included in the next section with nucleobases and nucleosides. Bacteria synthesize ectoine and a variety of other compatible solutes to prevent excessive loss of water when grown in high salinity conditions, and some species convert ectoine

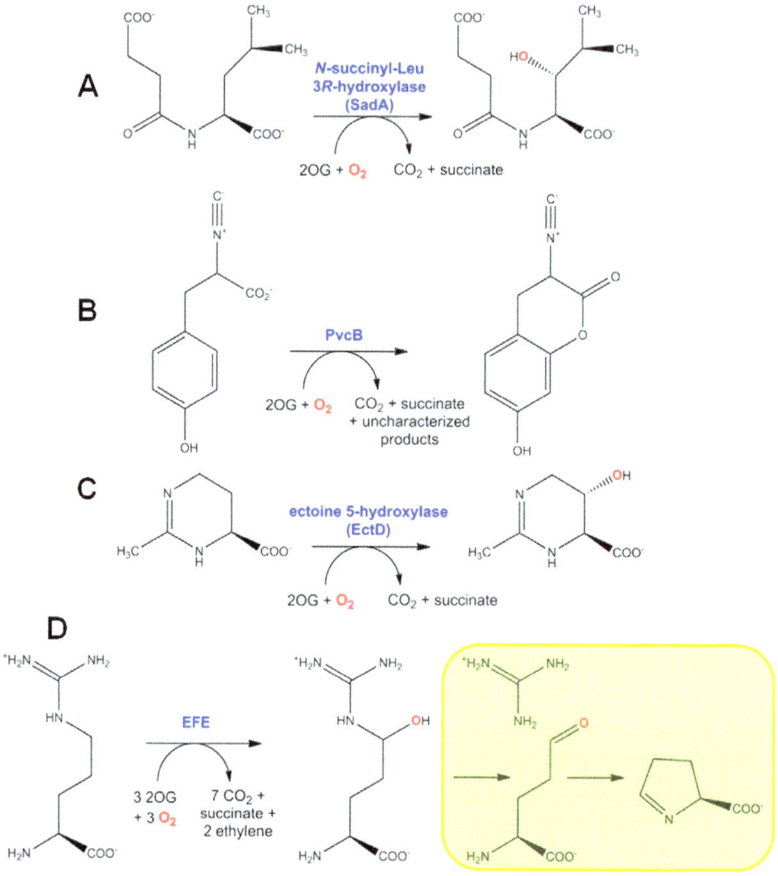

Figure 1.10 2OG oxygenases using substrates resembling amino acids. (A) *N*-succinyl-L-Leu 3*R*-hydroxylase. (B) PvcB reaction in pyoverdine biosynthesis. (C) Ectoine 5-hydroxylase. (D) A poorly characterized bacterial enzyme reported to convert 2OG to ethylene while simultaneously hydroxylating L-Arg, with subsequent spontaneous reactions shown in yellow highlight.

to 5-hydroxyectoine during stationary growth. EctD is a 2OG-dependent enzyme responsible for this ectoine 5-hydroxylase activity (Figure 1.10C).[250] The properties of this protein have been examined from a wide distribution of microorganisms, including extremophiles.[251] Of particular interest, the structure of EctD from *Virgibacillus* (formerly *Salibacillus*) *salexigens* in the absence of substrate has been characterized,[252] and a model of the holoprotein with ectoine and 2OG has been simulated.[253] Finally, *Pseudomonas syringae* pv. *phaseolicola* contains the ethylene-forming enzyme (EFE) which is reported to convert 2OG to ethylene and three molecules of CO_2 while simultaneously converting Arg into guanidine and Δ^1-pyrroline-5-carboxylate.[254] The presumed Arg 5-hydroxylase activity (Figure 1.10D) and formation of the resulting degradation products have not been clearly documented. 2OG-dependent ethylene formation has been observed in other microorganisms, including a mushroom,[255] and the gene encoding EFE is functional when inserted into other hosts,[256,257] including cyanobacteria,[257,258] with the latter microorganisms offering the potential for generating a biofuel from CO_2. The mechanism of 2OG conversion to ethylene by EFE is not understood.

1.6.2 Hydroxylases of Nucleobases and Nucleosides

Two distinct 2OG-dependent oxygenases are known to act on nucleobases. Thymine 7-hydroxylase from the fungus *Neurospora crassa* was among the first 2OG-dependent oxygenases to be characterized.[259,260] The enzyme catalyses sequential oxygen additions to the methyl group of the free base, forming 5-hydroxymethyluracil, 5-formyluracil and 5-carboxyuracil (Figure 1.11A).[261,262] Additional studies have focused on the purified fungal enzyme from *Rhodotorula glutinis*,[263] including analysis of its broad substrate specificity.[264,265] Xanthine hydroxylase, XanA, is a fungal enzyme that catalyses the reaction of Figure 1.11B.[266] This enzyme was discovered by studies involving a mutant strain of *Aspergillus nidulans* that was defective in xanthine dehydrogenase, a widely distributed molybdopterin-containing enzyme, yet was able to grow on xanthine as a nitrogen source. XanA was purified both from the fungal host and as a recombinant protein from *E. coli*; although differing in various types of post-translational modifications, both forms exhibited the activity shown.[267] Although no crystal structure is available for XanA, the likely active site residues were identified from mutagenesis studies and shown to support results of a homology model.[268]

Three types of 2OG-dependent oxygenases have been reported to catalyse reactions with the sugar components of nucleosides. Several fungi contain pyrimidine deoxyribonucleoside 2′-hydroxylases that form the corresponding ribonucleosides (Figure 1.11C).[269,270] Furthermore, *R. glutinis* contains a deoxyuridine or uridine 1′-hydroxylase that forms an unstable intermediate which decomposes with release of the nucleobase and formation of a lactone (Figure 1.11D).[271] Finally, a uridine-5′-monophosphate 5′-hydroxylase (Figure 1.11E) named LipL has been purified from a *Streptomyces* species where it provides a precursor for incorporation into antibiotic A-90289.[272]

Although the enzyme has not been characterized, it appears that a thymine 7-hydroxylase-like activity might be used for nucleoside modification during polyoxin biosynthesis in *Streptomyces avermitilis*.[273] The cells produce 14 distinct forms of polyoxins that possess a common nucleoside core containing 1-(5′-amino-5′-deoxy-β-D-allofurauronosyl)pyrimidine. The SAV_4805 open reading frame of this microorganism was shown to be associated with enhancement of structural diversity at the C-5 position of the pyrimidine ring, consistent with hydroxylation of this site by the encoded protein (Figure 1.11F).

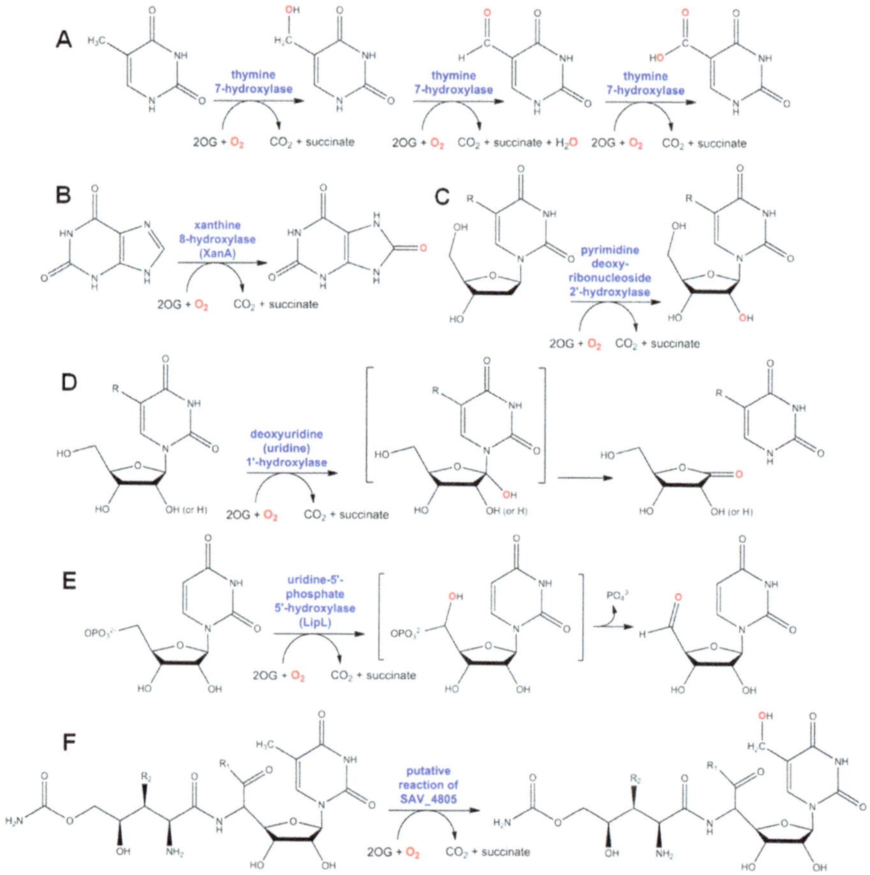

Figure 1.11 2OG-dependent oxygenase reactions involving nucleobases and nucleosides. (A) Thymine 7-hydroxylase. (B) Xanthine hydroxylase. (C) Pyrimidine deoxyribonucleoside 2′-hydroxylase. (D) Deoxyuridine (uridine) 1′-hydroxylase. (E) Uridine-5′-monophosphate 5′-hydroxylase. (F) Possible reaction catalysed by the enzyme associated with the SAV_4805 open reading frame.

1.6.3 Herbicide Degradation by 2OG-Dependent Oxygenases

Phenoxyalkanoic acids are widely used as herbicides for selective control of broad-leaf weeds. An important example is 2,4-dichlorophenoxyacetic acid (2,4-D), for which the biodegradative process has been extensively studied.[274] Bacteria that are capable of growth on 2,4-D as sole carbon source often possess a series of enzymes: TfdA (initially referred to incorrectly as 2,4-D monooxygenase), 2,4-chlorophenol hydroxylase (TfdB), 3,5-dichloro–catechol dioxygenase (TfdC), dichloromuconate cycloisomerase (TfdD), dienelactone hydrolase (TfdE) and maleylacetate reductase (TfdF) yielding 3-oxoadipate which enters intermediary metabolism. Using the enzyme from *Cupriavidus necator* (formerly *Ralstonia eutropha*), the reaction of TfdA was revised to be that of a 2OG-dependent hydroxylase (Figure 1.12A).[275] The properties of TfdA have been extensively characterized, including the specificity towards a diversity of phenoxyalkanoic acids, several spectroscopic features of the protein, and its ability to catalyse self-hydroxylation.[64,276,277] This was the first example of a 2OG-dependent dioxygenase being used for biodegradation of a xenobiotic compound. By incorporating *tfdA*-like genes into transgenic plants, enhanced herbicide tolerance has been obtained.[278]

2-Phenoxypropionic acids differ from phenoxyacetic acids by a single methyl group, resulting in two enantiomeric forms of these compounds. Only the (*R*) enantiomer of 2-(2,4-dichlorophenoxy)propionic acid (dichlorprop) or 2-(4-chloro-2-methyl-phenoxy)propionic acid (mecoprop) are active as herbicides. *Sphingomonas herbicidovorans* MG was shown to possess two proteins, RdpA and SdpA, specific for hydroxylating the (*R*) and (*S*) enantiomers, respectively (Figure 1.12B).[279] The structural basis of the distinct enantiospecificities for these proteins has been assessed by homology modelling, substrate docking and mutagenesis.[280]

The mechanism of herbicidal action of the above aryloxyalkanoic acids depends on the interaction of the compounds with the auxin receptor of plants. It is thus of some interest that rice possesses DAO (see Section 1.5.4), which converts indole-3-acetic acid into 2-oxoindole-3-acetic acid (Figure 1.8I).[224] The distribution of this activity in other plants remains to be defined.

1.6.4 Sulfonate and Sulfate Metabolism by 2OG-Dependent Dioxygenases

TauD is the best characterized 2OG-dependent oxygenase in terms of understanding its catalytic mechanism.[9,281] This *E. coli* enzyme hydroxylates 2-aminoethanesulfonate (taurine), with the hydroxylated sulfonate intermediate spontaneously decomposing to aminoacetaldehyde and sulfite (Figure 1.12C), which is used as a sulfur source by the cells.[282] Crystal structures are available for the *E. coli* protein as well as that from *Pseudomonas putida* KT2440.[283–285] Several of the intermediate states of catalysis

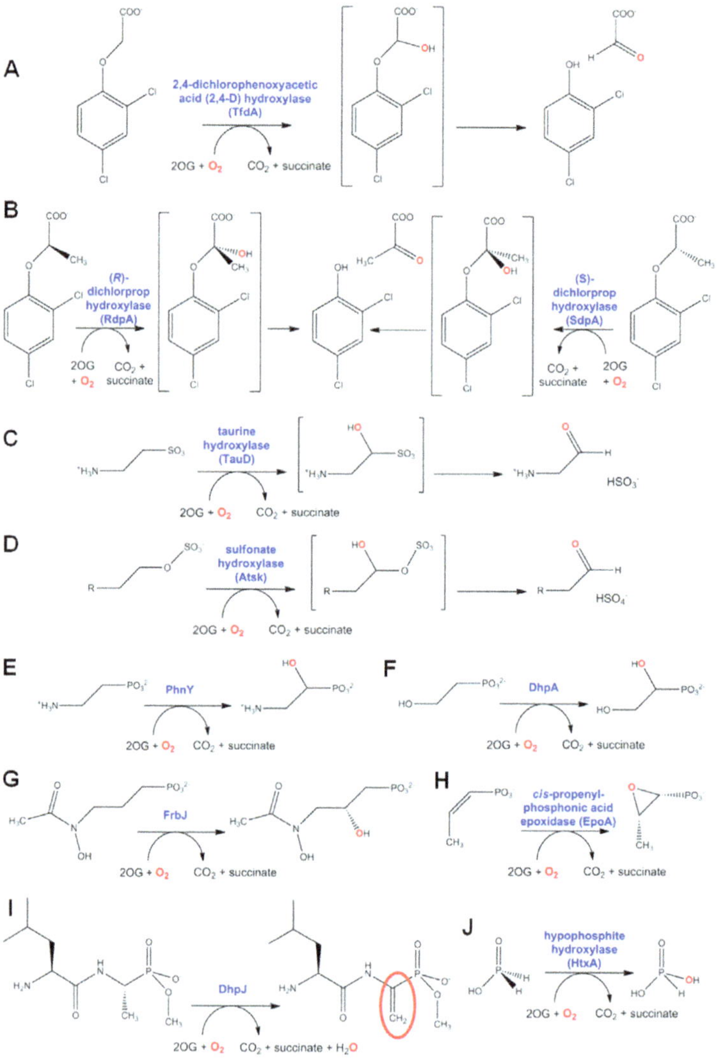

Figure 1.12 2OG-dependent oxygenases involved in metabolism of herbicides, sulfur-containing species and phosphonate compounds. (A) 2,4-D metabolism by TfdA. (B) RdpA and SdpA reactions with enantiomers of dichlorprop. (C) Taurine metabolism by TauD. (D) Sulfate metabolism by AtsK. (E) PhnY catalysed hydroxylation of 2-aminoethylphosphonate. (F) DhpA reaction with hydroxyethylphosphonate. (G) FrbJ catalysed hydroxylation. (H) Epoxidation reaction of EpoA. (I) Desaturation reaction of DhpJ. (J) HtxA conversion of hypophosphite to phosphite.

have been examined spectroscopically,[286,287] including the Fe(IV)-oxo[288–291] and later species.[10] In addition, at least two distinct types of self-hydroxylation chemistries have been studied with TauD, with one case involving a transient tyrosyl radical.[66–68] The enzyme decomposes a variety of other sulfonates (not depicted), including the widely used buffer 3-(*N*-morpholino) propanesulfonic acid, MOPS.[282] A yeast homologue is most active with MOPS among the potential substrates tested, but physiologically it functions to degrade taurocholate – the amide formed between taurine and cholic acid.[292]

Similar to the sulfonate-degrading enzymes discussed above, 2OG-dependent hydroxylases are also used to decompose alkylsulfates (Figure 1.12D). The first such enzyme discovered was AtsK from *P. putida* S-313; this protein is 38% identical to TauD, but it exhibits no activity with taurine.[293] A related enzyme is associated with the Rv3406 locus of *Mycobacterium tuberculosis*.[294] Crystal structures have been obtained with AtsK proteins from both sources.[294–296]

1.6.5 2OG-Dependent Oxygenases in Phosphonate Metabolism

Several members of the 2OG oxygenase enzyme family have been shown to act on various types of phosphonate compounds. The gene encoding PhnY was identified from a screen of ocean-derived metagenomic DNA that allowed *E. coli* cells deficient in C–P lyase (*Δphn*) to grow on 2-aminoethylphosphonic acid as the sole source of phosphorus. The substrate, a close analogue of taurine, is hydroxylated by PhnY (Figure 1.12E)[297] much like the reaction of TauD; however, the hydroxylated phosphonate is stable – unlike the hydroxylated sulfonate. DhpA catalyses a very similar reaction, oxygen addition to hydroxyethylphosphonate (Figure 1.12F) or its O-phosphomonomethyl ester (not shown), in the pathway for biosynthesis of dehydrophos, a vinyl phosphonate tripeptide antibiotic of *Streptomyces luridus*.[298] A hydroxylation reaction is also catalysed by FrbJ of *Streptomyces rubellomurinus*, in this case using the antibiotic FR-900098 as substrate (Figure 1.12G).[299] Rather than hydroxylation, EpoA of *Penicillium decumbens* catalyses epoxidation using *cis*-propenylphosphonic acid (Figure 1.12H) during synthesis of the antibiotic fosfomycin.[300] Another enzyme participating in the synthesis of dehydrophos by *S. luridus* is DhpJ, which catalyses a desaturation reaction of the monomethyl diester of L-Leu-L-1-aminoethylphosphonic acid (Figure 1.12I).[301] While the substrate is not a phosphonate, it is also appropriate to mention here the phosphite-producing hypophosphite hydroxylase activity (Figure 1.12J) of HtxA from *Pseudomonas stutzeri* WM88.[302]

1.7 2OG-Dependent Oxygenases Involved in Antibiotic Biosynthesis

Several 2OG-dependent oxygenases described earlier function in the synthesis of antibiotics, but their actions typically involve provision of small precursor molecules that become incorporated into the final compounds. This section focuses on other family members that function in antibiotic biosynthesis by forming bicyclic β-lactams, tailoring terpenoids, modifying protein-bound *S*-pantetheinyl thioesters compounds, and other roles.

1.7.1 Bicyclic β-Lactam Antibiotic Biosynthesis

This section covers several 2OG-dependent oxygenases that participate in the formation of clinically important bicyclic β-lactam antibiotics.[303] Also of interest for this topic, Section 1.8.1 describes isopenicillin *N* synthase, a structurally-related enzyme that does not use 2OG.

Clavaminate synthase (CAS) is a remarkable trifunctional 2OG-dependent enzyme in the pathway for synthesis of clavulanic acid in *Streptomyces clavuligerus*.[304] Starting with L-Arg and glyceraldehyde-3-phosphate, the cells utilize a thiamine diphosphate-dependent enzyme to form N^2-(2-carboxyethyl)arginine, which cyclizes *via* an ATP-dependent ligase reaction to generate deoxyguanidinoproclavaminate. This compound is the substrate for CAS, which catalyses a 2OG-dependent hydroxylation reaction (Figure 1.13A, left reaction). The guanidine group is removed from the resulting guanidinoproclavaminic acid by a separate hydrolase to yield proclavaminate. This substrate is used in a 2OG-dependent cyclization reaction catalysed by CAS (Figure 1.13A, middle reaction) to form dihydroclavaminic acid, which undergoes desaturation to clavaminic acid in the third 2OG-dependent reaction of CAS (Figure 1.13A, right reaction). Subsequent amino transferase and reductase reactions provide the final clavulanic acid (with an alcohol replacing the amine in the last structure shown). Purified CAS in its various states was studied by several spectroscopic methods, which provided evidence that the binding of both substrates (2OG and the antibiotic precursors) led to the loss of all metal-bound water, thus creating an oxygen binding site on the metal.[305,306] Structures of CAS have been solved in the presence of Fe(II), 2OG and either *N*-α-acetyl-L-Arg, proclavaminic acid, or deoxyguanidinoproclavaminate (with the oxygen analogue NO bound to the metal).[307,308]

Cephalosporin biosynthesis starts with the formation of isopenicillin (see Section 1.8.1), which is epimerized to penicillin, modified by two 2OG-dependent enzymes in prokaryotes or a single dual-function enzyme in eukaryotes, and additionally tailored by other reactions, sometimes including another 2OG family member. The first two 2OG-dependent oxygenases in *S. clavuligerus* are deacetoxycephalosporin C synthase (DAOCS), which catalyses a ring expansion reaction (Figure 1.13B, left), and deacetylcephalosporin C synthase (DACS; Figure 1.13B, right), a hydroxylase.[309–311] In contrast, both reactions are catalysed by the same DAOCS/DACS enzyme from *Cephalosporium acremonium*.[312,313] Several structural studies of DAOCS included the first structure for any 2OG-dependent oxygenase[314] and reveal the modes of substrate and product binding – with the surprising finding of overlapping binding sites of 2OG and penicillin substrates.[315–319] Recent pre-steady state kinetics and binding studies have called into question the interpretations from earlier results and conclude that a ternary complex does form in the protein.[320] Modelling combined with mutagenesis studies suggest that a single residue controls whether the enzyme catalyses ring expansion or hydroxylation, and offers insight into how the *C. acremonium* DAOCS/DACS enzyme catalyses both reactions.[321] The deacetylcephalosporin C resulting

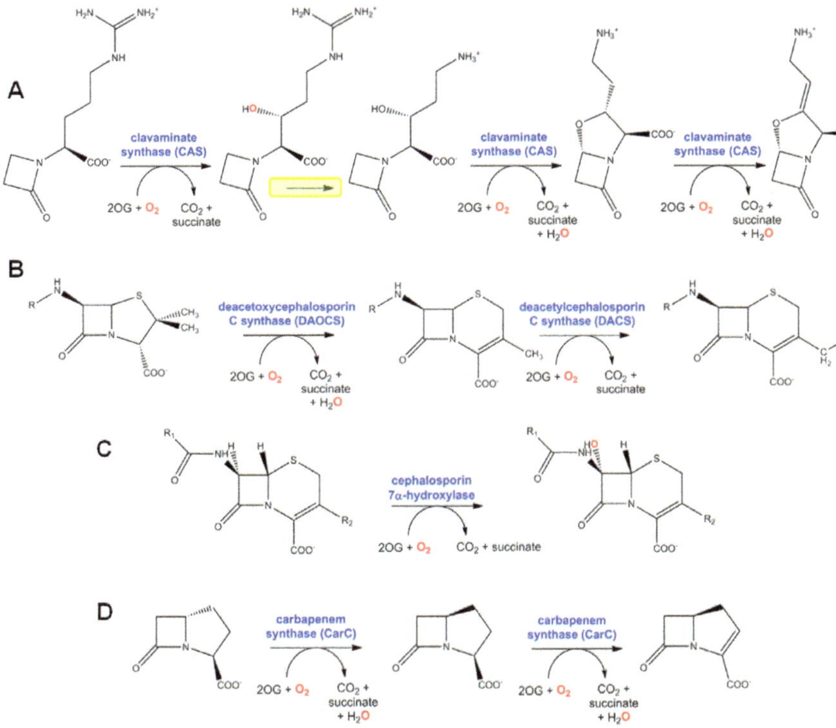

Figure 1.13 2OG-dependent oxygenases for synthesis of bicyclic β-lactam antibiotics. (A) Three reactions of clavaminic acid synthase and a separate non-CAS reaction (yellow highlight). (B) Ring expansion and hydroxylation reactions of cephalosporin biosynthesis. (C) Cephalosporin 7α-hydroxylase. (D) Carbapenem synthase epimerase (an external source of electrons is also needed for this reaction) and desaturation reactions.

from these reactions undergoes further transformations including acylation of the newly introduced hydroxyl group, 2OG-dependent hydroxylation at the 7α position (Figure 1.13C),[322] and methylation of the latter site. These enzymes are described in greater detail in Chapter 17.

Carbapenems are a third grouping of the bicyclic β-lactams that utilize a 2OG-dependent oxygenase in their synthetic pathway. *Pectobacterium carotovorum* (formerly *Erwinia carotovora*) contains the enzyme (2*S*,5*S*)-carboxymethylproline synthase (CarB) that uses glutamate semi-aldehyde and malonyl-CoA to produce the carboxymethylproline derivative. This substrate is subjected to a ligase reaction by (3*S*,5*S*)-carbapenam synthetase (CarA), forming carbapenam-3-carboxylate with its β-lactam ring. Carbapenem synthase (CarC) is a 2OG-dependent oxygenase proposed to catalyse two sequential non-hydroxylase reactions: epimerization to (3*S*,5*R*)-carbapenam-3-carboxylate and desaturation to yield (5*R*)-carbapenem-3-carboxylate (Figure 1.13D).[323] The crystal structure of CarC with bound Fe(II) and 2OG has been elucidated.[324] The stereoinversion reaction appears to involve ferryl abstraction of a substrate hydrogen atom followed by hydrogen atom donation from a tyrosyl side chain, limiting the enzyme to a single turnover unless external electrons are provided.[325]

1.7.2 Synthesis of Terpenoid Antibiotics

The rich biochemistry of terpenoid metabolism includes several examples of reactions catalysed by 2OG-dependent oxygenases. The synthesis of pentalenolactone, a sesquiterpenoid produced by dozens of strains of *Streptomyces*, nicely illustrates this point. A hydroxylase reaction, the conversion of 1-deoxypentalenic acid to 11β-hydroxy-1-deoxypentalenic acid (Figure 1.14A), is catalysed by *S. avermitilis* PltH.[326] The structure of this protein in complex with Fe(II), 2OG and *ent*-1-deoxypentalenic acid (a non-reactive enantiomer), has been defined.[327] Analogous proteins appear to be present in *S. exfolatus* UC5319 and *S. arenae* TU469, where they are named PenH and PntH, respectively.[328] These three strains each possess a second 2OG-dependent oxygenase (PtlD, PenD and PntD, respectively) capable of desaturating pentalenolactone D to pentalenolactone E, with PenD and PntD subsequently forming the epoxide pentalenolactone F (Figure 1.14B). In addition, these same three enzymes can transform neopentalenolactone D to an unstable enollactone desaturation product which is hydrolysed (Figure 1.14C).[328]

Two 2OG-dependent oxygenases utilize the same substrate, fusicocca-2,10(14)-diene-8β,16-diol, to produce distinct diterpene phytohormone-like compounds in fungi.[329] An enzyme from *Alternaria brassicicola* first abstracts a hydrogen atom from the eight-membered ring, the carbon-centered radical migrates, and hydroxyl radical rebound yields the distal hydroxylated product (Figure 1.14D, left) during a step in the synthesis of cotylenin A or brassicicene C. In contrast, an enzyme from *Phomopsis amygdali* catalyses oxidation at C-16 to yield the aldehyde 8β-hydroxyfusicocca-1,10(14)-diene-16-al (Figure 1.14D, right) during fusicoccin A biosynthesis.[329]

A final example of a 2OG-dependent oxygenase involved in synthesis of a terpene glycoside antibiotic is PlaO1 from *Streptomyces* sp. Tü6071.[330] PlaO1 catalyses the critical formation of a γ-butyrolactone during synthesis of phenalinolactone (Figure 1.14E). The chemical mechanism associated with this remarkable reaction has not been detailed, but a cogent hypothesis has been proposed.[331]

1.7.3 2OG-Dependent Oxygenases Acting on Tethered Substrates in Non-Ribosomal Peptide Synthesis

Several 2OG-dependent oxygenases act on protein-tethered *S*-pantetheinyl thioesters of amino acids which undergo non-ribosomal incorporation into peptide-related antibiotics. For example, *Pseudomonas syringae* pv. *syringae* B301D produces the non-ribosomal peptide phytotoxin called syringomycin

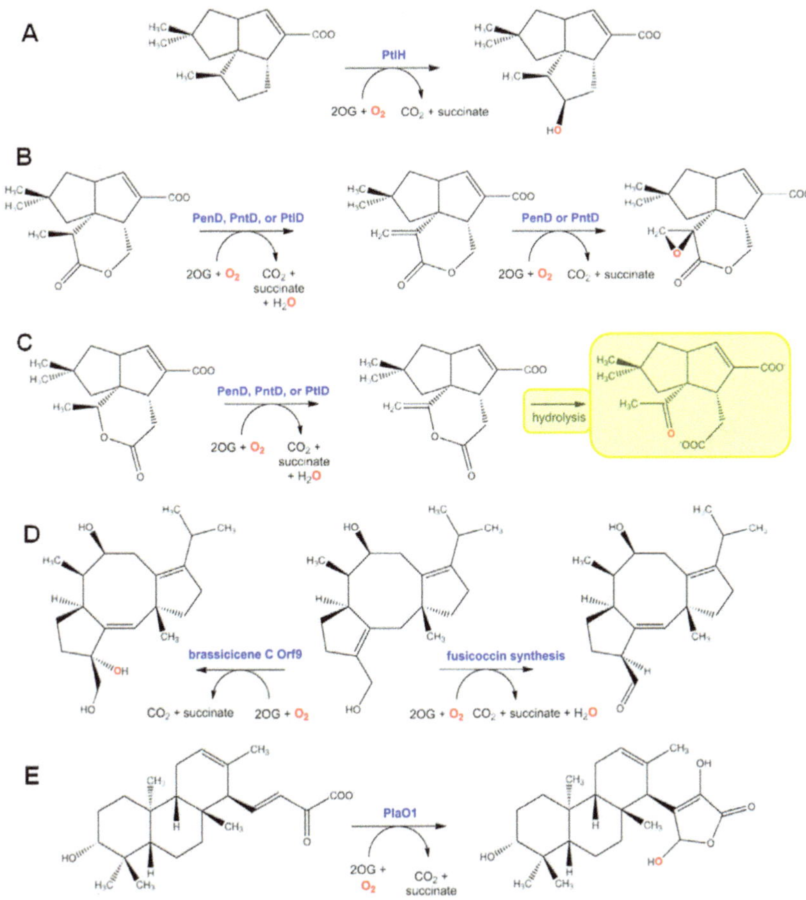

Figure 1.14 Representative 2OG-dependent oxygenases in terpenoid biosynthesis. (A) 1-deoxypentalenic acid 11β-hydroxylase, PtlH. (B) Desaturase and epoxidase reactions of PenD, PntD and PtlD using pentalenolactone D. (C) Neopentalenolactone desaturase reaction by the same enzymes, followed by spontaneous hydrolysis (yellow highlight). (D) Two reactions using fusicocca-2,10(14)-diene-8β,16-diol. (E) PlaO1-catalysed γ-butyrolactone formation in phenalinolactone synthesis.

E by assembling components on a multimodular megasynthetase, SyrE. SyrP is a 2OG-dependent hydroxylase that acts on L-Asp bound to the eighth module of SyrE to yield the *threo* isomer of 3-hydroxyaspartic acid which is inserted into the eighth position of the final phytotoxin (Figure 1.15A).[237]

Two other proteins involved in syringomycin E biosynthesis, SyrB1 and SyrB2, are required for synthesis of the 4-chloro-L-Thr located at position nine of the phytotoxin. Remarkably, SyrB2 was shown to be a 2OG-dependent oxygenase that catalyses a halogenation reaction using L-Thr tethered to SyrB1 (Figure 1.15B).[332] The enzyme also catalyses bromination, dichlorination, nitration and azidation reactions (not depicted).[333,334] The structure

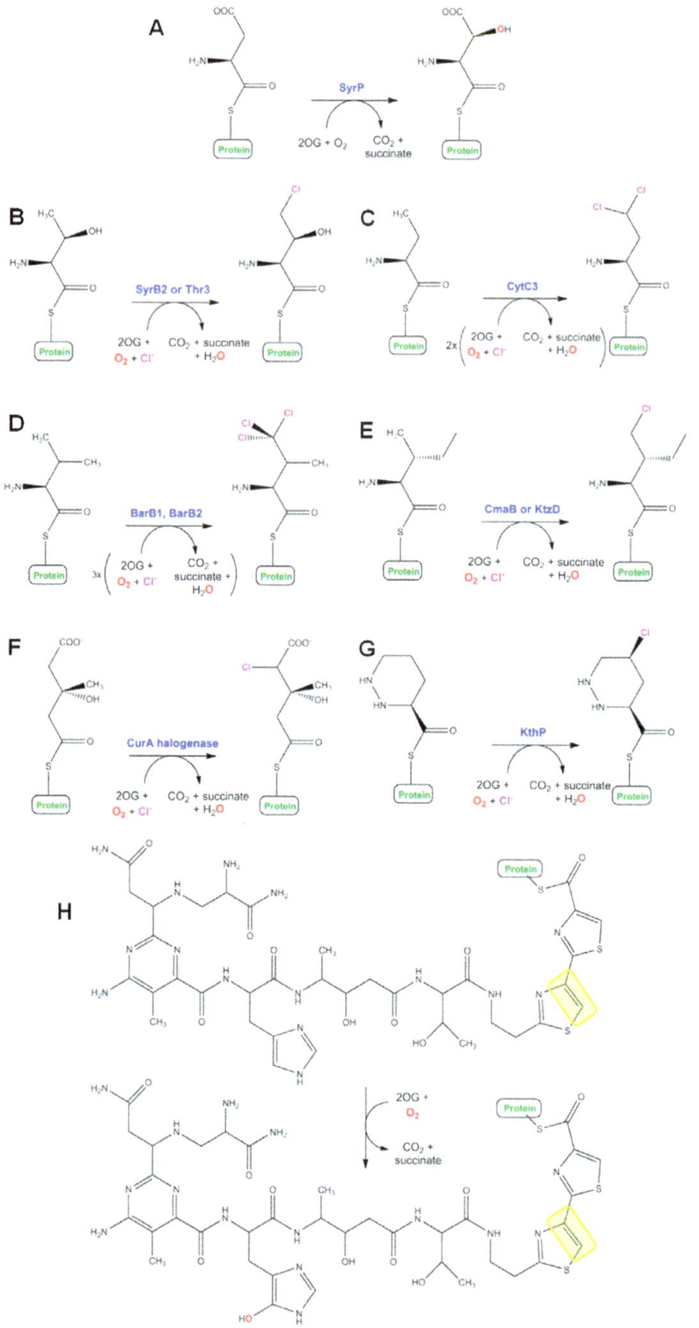

Figure 1.15 Use of protein-tethered substrates by 2OG-dependent oxygenases in antibiotic biosynthesis. (A) SyrP-catalysed formation of a 3-hydroxy-aspartyl group and (B) SyrB2-dependent synthesis of a 3-chlorothreonyl group in syringomycin E synthesis. (C) CytC3 generation of the

of SyrB2 reveals a metallocentre in which Fe(II) is bound only by two histidyl residues (with the typical carboxylate ligand replaced by an alanyl side chain), and coordinated to 2OG and chloride ion.[335] Spectroscopic studies have revealed evidence for a chloroferryl intermediate in catalysis,[336] and studies with substrate analogues suggest that substrate positioning determines whether halogenation or hydroxylation take place.[337] *Streptomyces* sp. OH-5093 produces free 4-chlorothreonine by a similar pathway, in which Thr3 catalyses the halogenation of the tethered amino acid, followed by thioester hydrolysis.[338]

Much like the SyrB1/SyrB2 system just discussed, a cytotrienin-producing *Streptomyces* sp. uses CytC3 to catalyse halogenation of L-2-aminobutyric acid (or L-Val) tethered to CytC2 (Figure 1.15C).[339] Tandem chlorinations followed by thioesterase activity lead to the γ,γ-dichloraminobutyrate antibiotic that is released into the soil. Two interconverting Fe(IV) intermediates were detected in that study, and later work provided evidence for the formation of a bromoferryl intermediate when bromide was used to replace chloride.[340] The structure of CytC3 exhibits the same general architecture as seen for SyrB2, including an Ala residue replacing the typical aspartyl metal ligand.[341]

Five other examples illustrate additional interesting features of 2OG-dependent halogenation enzymes. The marine cyanobacterium *Lyngbya majuscula* produces barbamide, an antibiotic containing trichlorinated leucine that is active against molluscs. BarB2 converts BarA-tethered L-Leu or 4-chloro-L-Leu to the dichlorinated species and BarB1 halogenates the mono- or dichlorinated species to the trichlorinated species (Figure 1.15D).[342] *Pseudomonas syringae* pv. *tomato* DC3000 synthesizes the phytotoxin coronatine which contains 1-amino-1-carboxy-2-ethylcyclopropane (coronamic acid). Synthesis of coronamic acid derives from L-*allo*-isoleucine which is tethered to CmaD, chlorinated by CmaB (Figure 1.15E), and converted to the cyclopropane species by CmaC with elimination of chloride in this cryptic chlorination pathway.[343] An analogous system is present in the ascomycete *Kutzneria*, where KtzD chlorinates L-Ile bound to KtzC, with the product cyclized to the bound (1*S*,2*R*)-allocoronamic acid by KtzA.[344] Another example of cryptic chlorination occurs during curacin A synthesis by *L. majuscula*, the marine bacterium mentioned above. In this case, acyl carrier protein-bound (*S*)-3-hydroxy-3-methylglutarate is chlorinated by CurA (Figure 1.15F), followed by dehydration using CurC, decarboxylation *via* a CurF domain, and

mono- or dichloro L-2-aminobutyryl group in cytotrienin synthesis. (D) Trichlorination of leucine by BarB1 and BarB2 to make barbamide. (E) CmaB- or KtzD-dependent hydroxylation of isoleucyl groups for subsequent cyclization to coronamic acids. (F) CurA chlorination of a 3-hydroxy-3-methylglutaryl group as part of curacin A synthesis. (G) KthP halogenation of a piperazyl group during formation of kutznerides. (H) Hydroxylation of a histidyl group during synthesis of bleomycin, tallysomycin and zorbamycin (yellow highlight is a single bond in the case of zorbamycin).

reductive dechlorination by another CurF domain to yield the protein-bound (1*R*,2*S*)-2-methylcyclopropane-1-carboxylate which is incorporated into curacin A.[345] The structures of CurA in five ligand states have been elucidated.[346] *Kutzneria* sp. 744 chlorinates a protein-bound piperazate residue by using the KthP halogenase (Figure 1.15G) during biosynthesis of a family of antifungal kutznerides.[347] Finally, *Lyngbya majuscula* contains the three-domain protein HctB that catalyses dichlorination (along with hydroxylation and introduction of a vinyl chloride group) on a fatty acyl group attached to its acyl carrier domain region (not depicted).[348] Interestingly, the oxidative reactions of this enzyme are stimulated more than 200-fold by the presence of saturating concentrations of chloride, thus providing an explanation for why hydroxylation does not dominate in the absence of halide salts.[349] Further discussion of 2OG-dependent halogenases can be found in Chapter 18.

Final examples for this section relate to hydroxylation reactions during synthesis of four antitumour natural products made by non-ribosomal peptide synthesis pathways: bleomycin, tallysomycin, zorbamycin and spliceostatin made by *Streptomyces verticillus* ATCC15003, *Streptoalloteichus hindustanus* E465-94, *Streptomyces flavoviridis* ATCC21892 and *Burkholderia* spp., respectively. A 2OG-dependent hydroxylase, potentially associated with open reading frames 1, 10 and 30 of the gene clusters in the first three microbes, is thought to catalyse histidyl group hydroxylation of a large precursor of the final antibiotic species bound to carrier proteins (Figure 1.15H), rather than hydroxylating just the tethered amino acid.[350] The 3-hydroxyhistidyl group is the site of glycation and additional tailoring reactions are used to generate the final antibiotics. The *Burkholderia* enzyme catalyses the synthesis of a hemiketal group (not shown) using the 2OG-dependent mechanism.[351]

1.7.4 Other Roles for 2OG-Dependent Oxygenases in Antibiotic Synthesis

Other 2OG-dependent oxygenases serve important roles in antibiotic biosynthesis, but don't fit into the above categories. These examples are not meant to be comprehensive, but provide useful illustrations of the diversity of reactions catalysed by these enzymes. The just mentioned antibiotic tallysomycin requires TmlH for its synthesis. This enzyme catalyses two 2OG-dependent hydroxylations, at positions C-41 and C-42 (Figure 1.16A).[352] The carbanolamide produced by hydroxylation of C-41 is unstable, but appears to be immediately modified further by TlmK to produce a stable species. It remains unclear whether TlmH acts on the protein-free species as shown or if the modification takes place on the protein-tethered substrate.[350] The fumonisins are mycotoxins produced by *Fusarium verticillioides* and several other filamentous fungi *via* polyketide synthetic routes. Hydroxylation at the C-5 position (Figure 1.16B) is carried out by the 2OG-dependent enzyme Fum3p (formerly associated with the *FUM9* locus).[353,354] Another mycotoxin, verruculogen, is produced by *Aspergillus fumigatus*. The novel creation of an endoperoxide within fumitremorgin B is catalysed by the 2OG-dependent FtmOx1 protein

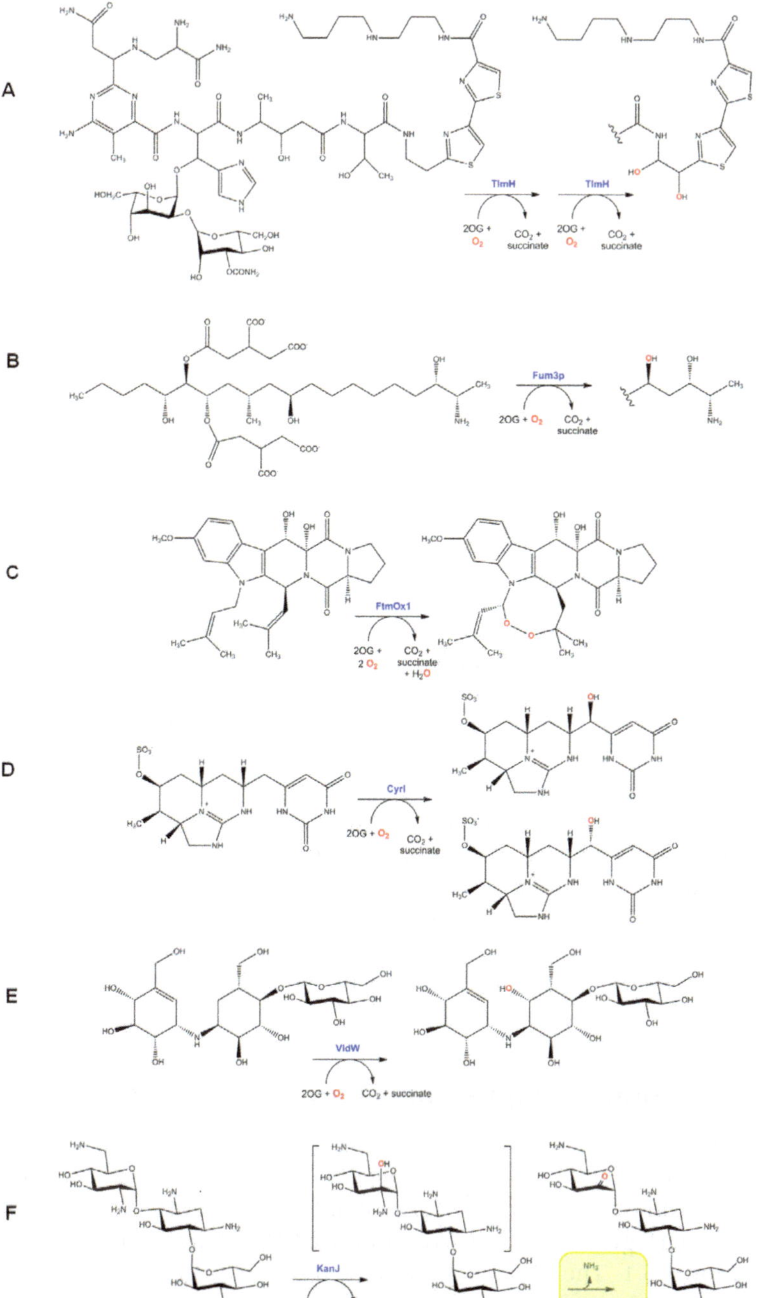

Figure 1.16 More 2OG-dependent oxygenases involved in the biosynthesis of antibiotics. (A) TlmH functioning in tallysomycin biosynthesis. (B) Fum3p modification of a fumiosin. (C) FtmOx1 creation of an endoperoxide in the verruculogen pathway. (D) CyrI production of two cylindrospermopsin products. (E) VldW role in decreasing the toxicity of a validamycin. (F) KanJ hydroxylation and a spontaneous follow-up reaction (yellow highlight) in the penultimate step of kanamycin synthesis.

(Figure 1.16C). Notably, this reaction requires two molecules of oxygen.[355] Several *Oscillatoria* and other cyanobacterial species produce the cyanotoxins cylindrospermopsin and 7-*epi*-cylindrospermopsin. The final step in the synthesis of these compounds is catalysed by the 2OG-dependent hydroxylase Cyrl (Figure 1.16D).[356] *Streptomyces hydrogscopicus* subsp. *Limoneus* produces a family of validamycin compounds with antifungal activity. The main species, validamycin A, is converted to the less effective validamycin B by VldW, a 2OG-dependent hydroxylase (Figure 1.16E).[357] Finally, the penultimate step in kanamycin biosynthesis by *Streptomyces kanamyceticus* is carried out by KanJ. This protein catalyses the hydroxylation of an amino sugar (Figure 1.16F) to yield an unstable hemiaminal which decomposes with release of ammonia to yield 2′-oxokanamycin; the latter compound is the substrate for an NADPH-dependent reductase, KanK, to produce the final antibiotic.

1.8 Related Enzymes

1.8.1 Isopenicillin *N* Synthase

The enzyme isopenicillin *N* synthase (IPNS) catalyses the fascinating transformation of a linear tripeptide, L-δ-(α-aminoadipoyl)-L-cysteinyl-D-valine, into a bicyclic structure, as shown in Figure 1.17A. The resulting product is subsequently metabolized into penicillins and cephalosporins as described earlier in Section 1.7.1.[303] Although IPNS does not utilize 2OG as a cosubstrate, it is related in sequence and structure to the 2OG-dependent oxygenases; indeed, it was the first family member to be structurally characterized.[358] Like the 2OG-dependent enzymes, IPNS coordinates Fe(II) *via* a 2-His-1-carboxylate motif and binds substrate within a double-stranded β helix fold.[359] The thiolate sulfur atom of the substrate forms a ligand to the metallocentre. Snapshots of the reaction were structurally visualized after brief exposure of anaerobic crystals to high pressures of oxygen.[360] This remarkable enzyme is described further in Chapter 19.

1.8.2 1-Aminocyclopropane-1-Carboxylate Oxidase

Another enzyme that does not utilize 2OG, yet is related by sequence and structure to the 2OG-dependent oxygenases, is 1-aminocyclopropane-1-carboxylate oxidase (ACCO).[361,362] This plant enzyme catalyses the synthesis of ethylene (Figure 1.17B) in a reaction that is distinct from that described in Section 1.6.1. The gaseous product is a phytohormone that functions in germination, fruit ripening and senescence. The cyclopropane-containing substrate of ACCO binds to the Fe(II) site *via* its α-amino and α-carboxylate groups according to spectroscopic analyses with isotopically-labelled substrates and substrate analogues;[363,364] such coordination is reminiscent of the bidentate binding of 2OG to the metallocentre of related family members. Ascorbate is required for multiple turnovers of ACCO, but the reductant is not needed for a single turnover.[365] Surprisingly, carbon dioxide is essential for catalysis in addition to being a product, and this molecule is proposed to stabilize the enzyme from undergoing inactivation reactions.[366] Oxidative inaction of ACCO occurs rapidly and can result in cleavage of the peptide backbone.[367,368] Crystal structures have been resolved for the apoprotein and holoprotein forms of ACCO,[369] and modelling studies have led to proposals for the binding sites of substrate, bicarbonate and ascorbate.[370] This enzyme is described in further detail in Chapter 20.

1.8.3 4-Hydroxyphenylpyruvate Dioxygenase and Hydroxymandelate Synthase

Hydroxymandelate synthase (HMS) and 4-hydroxyphenylpyruvate dioxygenase (HPPD) are of special interest because they transform the same substrate into distinct products (Figure 1.17C) by reactions that are mechanistically related to those carried out by the 2OG-dependent oxygenases; however, HMS and HPPD are unrelated to the latter enzymes while being closely related to each other.[371] The substrate for both enzymes is 4-hydroxyphenylpyruvate, a 2-oxo acid, and like 2OG it undergoes oxidative decarboxylation with formation of carbon dioxide and 4-hydroxyphenylacetate. The enzyme intermediate resulting from the C–C cleavage reaction is used either to hydroxylate the

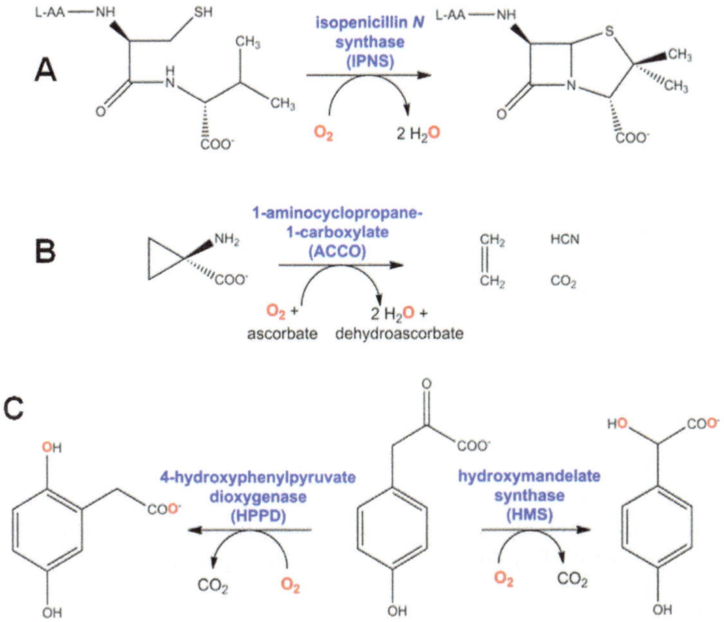

Figure 1.17 Enzymes that are structurally or functionally related to 2OG-dependent oxygenases, but do not utilize 2OG. (A) Isopenicillin *N* synthase. (B) 1-Aminocyclopropane-1-carboxylate oxidase. (C) 4-Hydroxyphenylpyruvate dioxygenase and hydroxymandelate synthase.

methylene group of 4-hydroxyphenylacetate to form hydroxymandelate (in HMS) or it catalyses the 'NIH shift' in which substituent migration and ring hydroxylation produce homogentisate (in HPPD). Isotope effect studies have provided keen insights into how the intermediate partitions to form the two products.[372] Four key residues in the active sites of these enzymes are critical for defining product specificity, and an HPPD has been engineered to exhibit HMS activity.[373] Several catalytic intermediates have been detected by using spectroscopic approaches with these enzymes.[374,375] The structure of HMS from *Amycolatopsis orientalis*[376] exhibits the same fold as found in HPPD, first reported for the enzyme from *Pseudomonas fluorescens*;[377] however, their folds are unrelated to the typical 2OG oxygenase fold. Further information on this pair of enzymes is provided in Chapter 21.

Conspectus: The 2OG-dependent oxygenases catalyze a diverse array of reactions that have profound implications in biology. The structures and mechanisms of these fascinating enzymes are discussed in Chapters 2–4 and representative topics are detailed more fully in Chapters 5–21.

Note added in proof

A large number of publications related to this topic have appeared since submission of this chapter, only three of which are cited here. A prolyl residue in prokaryotic elongation factor Tu is hydroxylated by a 2OG-dependent oxygenase, perhaps serving as an evolutionary precursor to prolyl hydroxylases used in oxygen sensing.[379] A review describing these enzymes in coumarin synthesis has appeared.[380] Finally, Wel05 was shown to be a 2OG-dependent halogenase acting on a free substrate (*i.e.* not tethered to a peptidyl carrier protein).[381]

Acknowledgements

2OG-dependent oxygenase work in the author's laboratory was supported by the National Institutes of Health (GM063584).

References

1. R. P. Hausinger, *Crit. Rev. Biochem. Mol. Biol.*, 2004, **39**, 21–68.
2. V. M. Purpero and G. R. Moran, *J. Biol. Inorg. Chem.*, 2007, **12**, 587–601.
3. C. Loenarz and C. J. Schofield, *Nat. Chem. Biol.*, 2008, **4**, 152–156.
4. J. M. Simmons, T. A. Müller and R. P. Hausinger, *Dalton Trans.*, 2008, **38**, 5132–5142.
5. C. Loenarz and C. J. Schofield, *TIBS*, 2010, **36**, 7–18.
6. I. J. Clifton, M. A. McDonough, D. Ehrismann, N. J. Kershaw, N. Granatino and C. J. Schofield, *J. Inorg. Biochem.*, 2006, **100**, 644–669.
7. M. A. McDonough, C. Loenarz, R. Chowdhury, I. J. Clifton and C. J. Schofield, *Curr. Opin. Struct. Biol.*, 2010, **20**, 1–14.

8. W. Aik, M. A. McDonough, A. Thalhammer, R. Chowdhury and C. J. Schofield, *Curr. Opin. Struct. Biol.*, 2012, **22**, 1–10.
9. D. A. Proshlyakov and R. P. Hausinger, in *Iron-containing Enzymes: Versatile Catalysts of Hydroxylation Reactions in Nature*, ed. D. Kumar and S. P. de Visser, Royal Society of Chemistry, Cambridge, U.K., 2011, pp. 67–87.
10. P. K. Grzyska, E. H. Appelman, R. P. Hausinger and D. A. Proshlyakov, *Proc. Natl. Acad. Sci. U.S.A.*, 2010, **107**, 3982–3987.
11. R. W. Welford, J. M. Kirkpatrick, L. A. McNeill, M. Puri, N. J. Oldham and C. J. Schofield, *FEBS Lett.*, 2005, **579**, 5170–5174.
12. A. G. Prescott and M. D. Lloyd, *Nat. Prod. Rep.*, 2000, **17**, 367–383.
13. J. J. Hutton, Jr., A. L. Trappel and S. Udenfriend, *Biochem. Biophys. Res. Commun.*, 1966, **24**, 179–184.
14. R. E. Rhoads and S. Udenfriend, *Proc. Natl. Acad. Sci. U.S.A.*, 1968, **60**, 1473–1478.
15. J. Risteli, K. Tryggvason and K. I. Kivirikko, *Eur. J. Biochem.*, 1977, **73**, 485–492.
16. U. Puistola, T. M. Turpeenniemi-Hujanen, R. Myllylä and K. Kivirikko, *Biochim. Biophys. Acta*, 1980, **611**, 40–50.
17. H. M. Hanauske-Abel and V. Günzler, *J. Theor. Biol.*, 1982, **94**, 421–455.
18. K. I. Kivirikko and T. Pihlajaniemi, *Adv. Enzymol. Relat. Areas Mol. Biol.*, 1998, **72**, 325–398.
19. K. L. Gorres and R. T. Raines, *Crit. Rev. Biochem. Mol. Biol.*, 2010, **45**, 106–124.
20. M. A. Culpepper, E. E. Scott and J. Limburgn, *Biochemistry*, 2010, **49**, 124–133.
21. M. K. Koski, R. Hieta, C. Böllner, K. I. Kivirikko, J. Myllyharju and R. K. Wierenga, *J. Biol. Chem.*, 2007, **282**, 37112–37123.
22. M. K. Koski, R. Hieta, M. Hirsilä, A. Rönkä, J. Myllyharju and R. K. Wierenga, *J. Biol. Chem.*, 2009, **284**, 25290–25301.
23. P. Jaakkola, D. R. Mole, Y.-M. Tian, M. I. Wilson, J. Gielbert, S. J. Gaskell, A. von Kriegsheim, H. F. Hebestreit, M. Mukherji, C. J. Schofield, P. H. Maxwell, C. W. Pugh and P. J. Ratcliffe, *Science*, 2001, **292**, 468–472.
24. M. Ivan, K. Kondo, H. Yang, W. Kim, J. Valiando, M. Ohh, A. Salic, J. M. Asara, W. S. Lane and W. G. Kaelin, Jr., *Science*, 2001, **292**, 464–467.
25. W.-C. Hon, M. I. Wilson, K. Harlos, T. D. W. Claridge, C. J. Schofield, C. W. Pugh, P. H. Maxwell, P. J. Ratcliffe, D. I. Stuart and E. Y. Jones, *Nature*, 2002, **417**, 975–978.
26. J.-H. Min, H. Yang, M. Ivan, F. Gertler, W. G. Kaelin, Jr. and N. P. Pavletich, *Science*, 2002, **296**, 1886–1889.
27. M. A. McDonough, V. Li, E. Flashman, R. Chowdhury, C. Mohr, B. M. R. Liénard, J. Zondlo, N. J. Oldham, I. J. Clifton, J. Lewis, L. A. McNeill, R. J. M. Kurzeja, K. S. Hewitson, E. Yang, S. Jordan, R. S. Syed and C. J. Schofield, *Proc. Natl. Acad. Sci. U.S.A.*, 2006, **103**, 9814–9819.
28. R. Chowdhury, M. A. McDonough, J. Mecinovíc, C. Loenarz, E. Flashman, K. S. Hewitson, C. Domene and C. J. Schofield, *Structure*, 2009, **17**, 981–989.

29. D. Lando, D. J. Peet, D. A. Whelan, J. J. Gorman and M. L. Whitelaw, *Science*, 2002, **295**, 858–861.
30. L. A. McNeill, K. S. Hewitson, T. D. Claridge, J. F. Seibel, L. E. Horsfall and C. J. Schofield, *Biochem. J.*, 2002, **367**, 571–575.
31. S. A. Dames, M. Martinez-Yamout, R. N. De Guzman, H. J. Dyson and P. E. Wright, *Proc. Natl. Acad. Sci. U.S.A.*, 2002, **99**, 5271–5276.
32. S. J. Freedman, Z.-Y. J. Sun, F. Poy, A. L. Kung, D. M. Livingston, G. Wagner and M. J. Eck, *Proc. Natl. Acad. Sci. U.S.A.*, 2002, **99**, 5367–5372.
33. C. E. Dann, III, R. K. Bruick and J. Deisenhofer, *Proc. Natl. Acad. Sci. U.S.A.*, 2002, **99**, 15351–15356.
34. C. Lee, S. J. Kim, D. G. Jeong, S. M. Lee and S. E. Ryu, *J. Biol. Chem.*, 2003, **278**, 7558–7563.
35. J. M. Elkins, K. S. Hewitson, L. A. McNeill, J. F. Seibel, I. Schlemminger, C. W. Pugh, P. J. Ratcliffe and C. J. Schofield, *J. Biol. Chem.*, 2003, **278**, 1802–1806.
36. C. J. Schofield and P. J. Ratcliffe, *Nat. Rev. Mol. Cell Biol.*, 2004, **5**, 343–354.
37. C. E. Dann, III and R. K. Bruick, *Biochem. Biophys. Res. Commun.*, 2005, **338**, 639–647.
38. N. S. Kenneth and S. Rocha, *Biochem. J.*, 2008, **414**, 19–29.
39. W. Ge, A. Wolf, T. Feng, C.-h. Ho, R. Sekirnik, A. Zayer, N. Granatino, M. E. Cockman, C. Loenarz, N. D. Loik, A. P. Hardy, T. D. W. Claridge, R. B. Hamed, R. Chowdhury, L. Gong, C. V. Robinson, D. C. Trudgian, M. Jiang, M. M. Mackeen, J. S. Mccullagh, Y. Gordiyenko, A. Thalhammer, A. Yamamoto, M. Yang, P. Liu-Yi, Z. Zhang, M. Schmidt-Zachmann, B. M. Kessler, P. J. Ratcliffe, G. M. Preston, M. L. Coleman and C. J. Schofield, *Nat. Chem. Biol.*, 2013, **8**, 960–962.
40. L. van Staalduinen, S. K. Novakowski and Z. Jia, *J. Mol. Biol.*, 2014, **426**, 1898–1910.
41. R. Chowdhury, R. Sekirnik, N. C. Brisset, T. Krojer, C.-h. Ho, S. S. Ng, I. J. Clifton, W. Ge, N. J. Kershaw, G. C. Fox, J. R. C. Muniz, M. Vollmar, C. Phillips, E. S. Pilka, K. L. Kavanagh, F. von Delft, U. Oppermann, M. A. McDonough, A. J. Doherty and C. J. Schofield, *Nature*, 2014, **510**, 422–426.
42. H. S. Kim, H. L. Kim, K. H. Kim, D. J. Kim, S. J. Lee, J. Y. Yoon, H. J. Yoon, H. Y. Lee, S. B. Park, S.-J. Kim, J. Y. Lee and S. W. Suh, *Nucleic Acids Res.*, 2010, **38**, 2099–2110.
43. R. S. Singleton, P. Liu-Yi, F. Formenti, W. Ge, R. Sekirnik, R. Fischer, J. Adam, P. J. Pollard, A. Wolf, A. Thalhammer, C. Loenarz, E. Flashman, A. Yamamoto, M. L. Coleman, B. M. Kessler, P. Wappner, C. J. Schofield, P. J. Ratcliffe and M. E. Cockman, *Proc. Natl. Acad. Sci. U.S.A.*, 2014, **111**, 4031–4036.
44. C. Loenarz, R. Sekirnik, A. Thalhammer, W. Ge, E. Spivakovsky, M. M. Mackeen, M. A. McDonough, M. E. Cockman, B. M. Kessler, P. J. Ratcliffe, A. Wolf and C. J. Schofield, *Proc. Natl. Acad. Sci. U.S.A.*, 2014, **111**, 4019–4024.

45. M. J. Katz, J. M. Acevedo, C. Loenarz, D. Galagovsky, P. Liu-Yi, M. Pérez-Pepe, A. Thalhammer, R. Sekirnik, W. Ge, M. Melani, M. G. Thomas, S. Simonetta, G. L. Boccaccio, C. J. Schofield, M. E. Cockman, P. J. Ratcliffe and P. Wappner, *Proc. Natl. Acad. Sci. U.S.A.*, 2014, **111**, 4025–4030.
46. C. J. Webby, A. Wolf, N. Gromak, M. Dreger, H. Kramer, B. Kessler, M. L. Nielsen, C. Schmitz, D. S. Butler, J. R. Yates, III, C. M. Delahunty, P. Hahn, A. Lengeling, M. Mann, N. J. Proudfoot, C. J. Schofield and A. Böttger, *Science*, 2009, **325**, 90–93.
47. M. Mantri, N. D. Loik, R. B. Hamed, T. D. W. Claridge, J. S. O. McCullagh and C. J. Schofield, *ChemBioChem*, 2011, **12**, 531–534.
48. B. Chang, Y. Chen, Y. Zhao and R. K. Bruick, *Science*, 2007, **318**, 444–447.
49. W. Liu, Q. Ma, K.-B. Wong, W. Li, K. Ohgi, J. Zhang, A. K. Aggarwal and M. G. Rosenfeld, *Cell*, 2013, **155**, 1581–1595.
50. M. Mantri, T. Krojer, E. A. Bagg, C. J. Webby, D. S. Butler, G. Kochan, K. L. Kavanagh, U. Oppermann, M. A. McDonough and C. J. Schofield, *J. Mol. Biol.*, 2010, **401**, 211–222.
51. X. Hong, J. Zang, J. White, C. Wang, C. H. Pan, R. Zhao, R. C. Murphy, S. Dai, P. Henson, J. W. Kappler, J. Hagman and G. Zhang, *Proc. Natl. Acad. Sci. U.S.A.*, 2010, **107**, 14568–14572.
52. T. Feng, A. Yamamoto, S. E. Wilkins, E. Sokolova, L. A. Yates, M. Münzel, P. Singh, R. J. Hopkinson, R. Fischer, M. E. Cockman, J. Shelley, D. C. Trudgian, J. Schödel, J. S. O. McCullagh, W. Ge, B. M. Kessler, R. J. Gilbert, L. Y. Frolova, E. Alkalaeva, P. J. Ratcliffe, C. J. Schofield and J. E. Coleman, *Mol. Cell*, 2014, **53**, 1–10.
53. J. Stenflo, E. Holme, S. Lindstedt, N. Chandramouli, L. H. T. Huang, J. P. Tam and R. B. Merrifield, *Proc. Natl. Acad. Sci. U.S.A.*, 1989, **86**, 444–447.
54. R. S. Gronke, W. J. VanDusen, V. M. Garsky, J. W. Jacobs, M. K. Sardana, A. M. Stern and P. A. Friedman, *Proc. Natl. Acad. Sci. U.S.A.*, 1989, **86**, 3609–3613.
55. J. E. Dinchuk, R. J. Focht, J. A. Kelley, N. L. Henderson, N. I. Zolotarjova, R. Wynn, N. T. Neff, J. Link, R. M. Huber, T. C. Burn, M. J. Rupar, M. R. Cunningham, B. H. Selling, J. Ma, A. A. Stern, G. F. Hollis, R. B. Stein and P. A. Friedman, *Proc. Natl. Acad. Sci. U.S.A.*, 2002, **277**, 12970–12977.
56. N. Ince, S. M. de la Monte and J. R. Wands, *Cancer Res.*, 2000, **60**, 1261–1266.
57. A. Ökesli, L. E. Cooper, E. J. Fogle and W. van der Donk, *J. Am. Chem. Soc.*, 2011, **133**, 13753–13760.
58. M. L. Coleman, M. A. McDonough, K. S. Hewitson, C. Coles, J. Mecinovic, M. Edelmann, K. M. Cook, M. E. Cockman, D. E. Lancaster, B. M. Kessler, R. J. Oldham, N. J. Ratcliffe and C. J. Schofield, *J. Biol. Chem.*, 2007, **282**, 24027–24038.
59. A. P. Hardy, I. Prokes, L. Kelly, I. D. Campbell and C. J. Schofield, *J. Mol. Biol.*, 2009, **392**, 994–1006.
60. M. Yang, W. Ge, R. Chowdhury, T. D. W. Claridge, H. B. Kramer, B. Schmierer, M. A. McDonough, L. Gong, B. M. Kessler, P. J. Ratcliffe, M. L. Coleman and C. J. Schofield, *J. Biol. Chem.*, 2011, **286**, 7648–7660.

61. M. Yang, R. Chowdhury, W. Ge, R. B. Hamed, M. A. McDonough, T. D. W. Claridge, B. M. Kessler, M. E. Cockman, P. J. Ratcliffe and C. J. Schofield, *FEBS J.*, 2011, **278**, 1086–1097.
62. K. Janke, U. Brockmeier, K. Kuhlmann, M. Eisenacher, J. Nolde, H. E. Meyer, H. Mairbäurl and E. Metzen, *J. Cell Sci.*, 2013, **126**, 2629–2640.
63. H. H. Qi, P. P. Ongusaha, J. Myllyharju, D. Cheng, O. Pakkanen, Y. Shi, S. W. Lee, J. Peng and Y. Shi, *Nature (London)*, 2008, **455**, 421–424.
64. A. Liu, R. Y. N. Ho, L. Que, Jr., M. J. Ryle, B. S. Phinney and R. P. Hausinger, *J. Am. Chem. Soc.*, 2001, **123**, 5126–5127.
65. R. Y. N. Ho, M. P. Mehn, E. L. Hegg, A. Liu, M. A. Ryle, R. P. Hausinger and L. Que, Jr., *J. Am. Chem. Soc.*, 2001, **123**, 5022–5029.
66. M. J. Ryle, A. Liu, R. B. Muthukumaran, R. Y. N. Ho, K. D. Koehntop, J. McCracken, L. Que, Jr. and R. P. Hausinger, *Biochemistry*, 2003, **42**, 1854–1862.
67. M. J. Ryle, K. D. Koehntop, A. Liu, L. Que, Jr. and R. P. Hausinger, *Proc. Natl. Acad. Sci. U.S.A.*, 2003, **100**, 3790–3795.
68. K. D. Koehntop, S. Marimanikkuppam, M. J. Ryle, R. P. Hausinger and L. Que, Jr., *J. Biol. Inorg. Chem.*, 2006, **11**, 63–72.
69. M. J. Ryle and R. P. Hausinger, *Curr. Opin. Chem. Biol.*, 2002, **6**, 193–201.
70. Y. Ge, B. G. Lawhorn, M. Elnaggar, S. K. Sze, T. P. Begley and F. W. McLafferty, *Protein Sci.*, 2003, **12**, 2320–2326.
71. M. Mantri, C. J. Webby, N. D. Loik, R. B. Hamed, M. L. Nielsen, M. A. McDonough, J. S. O. McCullagh, A. Böttger, C. J. Schofield and A. Wolf, *MedChemComm*, 2012, **3**, 80–85.
72. Y.-H. Chen, L. M. Comeaux, S. J. Eyles and M. J. Knapp, *Chem. Commun.*, 2008, 4768–4770.
73. T. F. Henshaw, M. Feig and R. P. Hausinger, *J. Inorg. Biochem.*, 2004, **98**, 856–861.
74. O. Sundheim, C. B. Vågbø, M. Bjørås, M. M. L. Sousa, V. Talstad, P. A. Aas, F. Drabløs, H. E. Krokan, J. A. Tainer and G. Slupphaug, *EMBO J.*, 2006, **25**, 3389–3397.
75. E. R. Farquhar, K. D. Koehntop, J. P. Emerson and L. Que, Jr., *Biochem. Biophys. Res. Commun.*, 2005, **338**, 230–239.
76. M. Mantri, Z. Zhang, M. A. McDonough and C. J. Schofield, *FEBS J.*, 2012, **279**, 1563–1575.
77. T. Jenuwein and C. D. Allis, *Science*, 2001, **293**, 1074–1080.
78. Y.-I. Tsukada, J. Fang, H. Erdjument-Bromage, M. E. Warren, C. H. Borchers, P. Tempst and Y. Zhang, *Nature*, 2006, **439**, 811–816.
79. K. Yamane, C. Toumazou, Y.-I. Tsukada, H. Erdjument-Bromage, P. Tempst, J. Wong and Y. Zhang, *Cell*, 2006, **125**, 1–13.
80. J. R. Whetstine, A. Nottke, F. Lan, M. Huart, S. Smolikov, Z. Chen, E. Spooner, E. Li, G. Zhang, M. Colaiacovo and Y. Shi, *Cell*, 2006, **125**, 467–481.
81. R. J. Klose, K. Yamane, Y. Bae, D. Zhang, H. Erdjument-Bromage, P. Tempst, J. Wong and Y. Zhang, *Nature*, 2006, **442**, 312–316.

82. P. A. C. Cloos, J. Christensen, K. Agger, A. Maiolica, J. Rappsilber, T. Antal, K. H. Hansen and K. Helin, *Nature*, 2006, **442**, 307–311.
83. Z. Chen, J. Zang, J. R. Whetstine, X. Hong, F. Davrazou, T. G. Kutateladze, M. Simpson, Q. Mao, C.-H. Pan, S. Dai, J. Hagman, K. Hansen, Y. Shi and G. Zhang, *Cell*, 2006, **125**, 691–702.
84. J.-F. Couture, E. Collazo, P. A. Ortiz-Tello, J. S. Brunzelle and R. C. Trievel, *Nat. Struct. Mol. Biol.*, 2007, **14**, 689–695.
85. S. S. Ng, K. L. Kavanagh, M. A. McDonough, D. Butler, E. S. Pilka, B. M. R. Lienard, J. E. Bray, P. Savitsky, O. Gileadi, F. von Delft, N. R. Rose, J. Offer, J. C. Scheinost, T. Borowski, M. Sundstrom, C. J. Schofield and U. Oppermann, *Nature*, 2007, **448**, 87–91.
86. Z. Chen, J. Zang, J. Kappler, X. Hong, F. Crawford, Q. Wang, F. Lan, C. Jiang, J. R. Whetstine, S. Dai, K. Hansen, Y. Shi and G. Zhang, *Proc. Natl. Acad. Sci. U.S.A.*, 2007, **104**, 10818–10823.
87. J. R. Horton, A. Upadhyay, H. H. Qi, Y. Shi and C. Xiaodong, *Nat. Struct. Mol. Biol.*, 2010, **17**, 38–43.
88. W. W. Yue, V. Hozjan, W. Ge, C. Loenarz, C. D. O. Cooper, C. J. Schofield, K. L. Kavanagh, U. Oppermann and M. A. McDonough, *FEBS Lett.*, 2010, **584**, 825–830.
89. Z. Cheng, P. Cheung, A. J. Kuo, E. T. Yukl, C. M. Wilmot, O. Gozani and D. J. Patel, *Genes Dev.*, 2014, **28**, 1758–1771.
90. C. Johansson, A. Tumber, K. Che, P. Cain, R. Nowak, C. Gileadi and U. Oppermann, *Epigenomics*, 2014, **6**, 89–120.
91. C. C. Thinnes, K. S. England, A. Kawamura, R. Chowdhury, C. J. Schofield and R. J. Hopkinson, *Biochim. Biophys. Acta,* 2014, **1839**, 1416–1432.
92. N. Mosammaparast and Y. Shi, *Annu. Rev. Biochem,* 2010, **79**, 155–179.
93. M.-M. Li, A. Nilsen, Y. Shi, M. Fusser, Y.-H. Ding, Y. Fu, B. Liu, Y. Niu, Y.-S. Wu, C.-M. Huang, M. Olofsson, K.-X. Jin, Y. Lv, X.-Z. Xu, C. He, M.-Q. Dong, J. M. R. Danielsen, A. Klungland and Y.-G. Yang, *Nat. Commun.*, 2013, **4**, 1832.
94. L. G. Bjørnstad, T. J. Meza, M. Otterlei, S. M. Olafsrud, L. A. Meza-Zepeda and P. Ø. Falnes, *PLoS One*, 2012, **7**, e49045.
95. S. C. Trewick, T. F. Henshaw, R. P. Hausinger, T. Lindahl and B. Sedgwick, *Nature*, 2002, **419**, 174–178.
96. P. Ø. Falnes, R. F. Johansen and E. Seeberg, *Nature*, 2002, **419**, 178–182.
97. P. A. Aas, M. Otterlei, P. Ø. Falnes, C. B. Vågbø, F. Skorpen, M. Akbari, O. Sundheim, M. Bjøras, G. Slupphaug, E. Seeberg and H. E. Krokan, *Nature*, 2003, **421**, 859–863.
98. J. C. Delaney and J. M. Essigmann, *Proc. Natl. Acad. Sci. U.S.A.*, 2004, **101**, 14051–14056.
99. P. Koivisto, P. Robins, T. Lindahl and B. Sedgwick, *J. Biol. Chem.*, 2004, **279**, 40470–40474.
100. D. Li, J. C. Delaney, C. M. Page, X. Yang, A. S. Chen, C. Wong, C. L. Drennan and J. M. Essigmann, *J. Am. Chem. Soc.*, 2012, **134**, 8896–8901.

101. D. Li, B. I. Fedeles, N. Shrivastav, J. C. Delaney, X. Yang, C. Wong, C. L. Drennan and J. M. Essigmann, *Chem. Res. Toxicol.*, 2013, **26**, 1182–1187.
102. B. Yu, W. C. Edstrom, Y. Hamuro, P. C. Weber, B. R. Gibney and J. F. Hunt, *Nature*, 2006, **439**, 879–884.
103. C.-G. Yang, C. Yi, E. M. Duguid, C. T. Sullivan, X. Jian, P. A. Rice and C. He, *Nature*, 2008, **452**, 961–965.
104. B. Yu and J. F. Hunt, *Proc. Natl. Acad. Sci. U.S.A.*, 2009, **106**, 14315–14320.
105. P. J. Holland and T. Hollis, *PLoS One*, 2010, **5**, e8680.
106. C. Yi, G. Jia, G. Hou, Q. Dai, W. Zhang, G. Zheng, X. Jian, C.-G. Yang, Q. Cui and C. He, *Nature*, 2010, **468**, 330–333.
107. E. Van den Born, A. Bekkelund, M. N. Moen, M. V. Omelchenko, A. Klungland and P. Ø. Falnes, *Nucleic Acids Res.*, 2009, **37**, 7124–7136.
108. E. van den Born, M. V. Omelchenko, A. Bekkelund, V. Leihne, E. V. Koonin, V. V. Dolja and P. Ø. Falnes, *Nucleic Acids Res.*, 2008, **36**, 5451–5456.
109. M. A. Kurowski, A. S. Bhagwat, G. Papaj and J. M. Bujnicki, *BMC Genomics*, 2003, **4**, 48.
110. L. Sanchez-Pulido and M. A. Andrade-Navarro, *BMC Biochem.*, 2007, **8**, 23.
111. D. Mielecki, D. L. Zugaj, A. Muszewska, J. Piwowarski, A. Chojnacka, M. Mielecki, J. Nieminuszczy, M. Grynberg and E. Grzesiuk, *PLoS One*, 2012, **7**, e30588.
112. T. Duncan, S. C. Trewick, P. Koivisto, P. A. Bates, T. Lindahl and B. Sedgwick, *Proc. Natl. Acad. Sci. U.S.A.*, 2002, **99**, 16660–16665.
113. J. Ringvoll, L. M. Nordstrand, C. B. Vågbø, V. Talstad, K. Reite, P. A. Aas, K. H. Lauritzen, N. B. Liabakk, A. Bjork, R. W. Doughty, P. Ø. Falnes, H. E. Krokan and A. Klungland, *EMBO J.*, 2006, **25**, 2189–2198.
114. L. Lu, C. Yi, X. Jian, G. Zheng and C. He, *Nucleic Acids Res.*, 2010, **38**, 4415–4425.
115. C. Yi, B. Chen, B. Qi, W. Zhang, G. Jia, C. J. Li, A. R. Diner, C.-G. Yang and C. He, *Nat. Struct. Mol. Biol.*, 2012, **19**, 671–676.
116. V. T. Monsen, O. Sundheim, P. A. Aas, M. P. Westbye, M. M. L. Sousa, G. Slupphaug and H. E. Krokan, *Nucleic Acids Res.*, 2010, **38**, 6447–6455.
117. M. P. Westbye, E. Feyzi, P. A. Aas, C. B. Vågbø, V. A. Talstad, B. Kavli, L. Hagen, O. Sundheim, M. Akbari, N.-B. Liabakk, G. Slupphaug, M. Otterlei and H. E. Krokan, *J. Biol. Chem.*, 2008, **283**, 25046–25056.
118. R. Ougland, D. Lando, I. Jonson, J. A. Dahl, M. N. Moen, L. M. Nordstrand, T. Rognes, J. T. Lee, A. Klungland, T. Kouzarides and E. Larsen, *Stem Cells*, 2012, **30**, 2672–2682.
119. T. A. Müller, K. Meek and R. P. Hausinger, *DNA Repair*, 2010, **9**, 58–65.
120. T. A. Müller, M. M. Andrzejak and R. P. Hausinger, *Biochem. J.*, 2013, **452**, 509–518.
121. G. Zheng, J. A. Dahl, Y. Niu, P. Fedorcsak, C.-H. Huang, C. J. Li, C. B. Vågbø, Y. Shi, W.-L. Wang, S.-H. Song, Z. Lu, R. P. G. Bosmans, Q. Dai, Y.-J. Hao, X. Yang, W.-M. Zhao, W.-M. Tong, X.-J. Wang, F. Bogdan, K. Furu, Y. Fu, G. Jia, X. Zhao, J. Liu, H. E. Krokan, A. Klungland, Y.-G. Yang and C. He, *Mol. Cell*, 2013, **49**, 18–29.

122. J.-M. Fustin, M. Doi, Y. Yamaguchi, H. Hida, S. Nishimura, M. Yoshida, T. Isagawa, M. S. Morioka, H. Kakeya, I. Manabe and H. Okamura, *Cell*, 2013, **155**, 793–806.
123. W. Aik, J. S. Scotti, H. Choi, L. Gong, M. Demetriades, C. J. Schofield and M. A. McDonough, *Nucleic Acids Res.*, 2014, **42**, 4741–4754.
124. C. Feng, Y. Liu, G. Wang, Z. Deng, Q. Zhang, W. Wu, Y. Tong, C. cheng and Z. Chen, *J. Biol. Chem.*, 2014, **289**, 11571–11583.
125. W. Chen, L. Zhang, G. Zheng, Y. Fu, Q. Ji, F. Liu, H. Chen and C. He, *FEBS Lett.*, 2014, **588**, 892–898.
126. G. Wang, Q. He, C. Feng, Y. Liu, Z. Deng, Z. Qi, W. Wu, P. Mei and Z. Chen, *J. Biol. Chem.*, **289**, 27924–27936.
127. J. Fischer, L. Koch, C. Emmerling, J. Vierkotten, T. Peters, J. C. Brüning and U. Rüther, *Nature*, 2009, **458**, 894–898.
128. T. Gerken, C. A. Girard, Y. C. Tung, C. J. Webby, V. Saudek, K. S. Hewitson, G. S. Yeo, M. A. McDonough, S. Cunliffe, L. A. McNeill, J. Galvanovskis, P. Rorsman, P. Robins, X. Prieur, A. P. Coll, M. Ma, Z. Jovanovic, I. S. Farooqi, B. Sedgwick, I. Barroso, T. Lindahl, C. P. Ponting, F. M. Ashcroft, S. O'Rahilly and C. J. Schofield, *Science*, 2007, **318**, 1469–1472.
129. G. Jia, C.-G. Yang, S. Yang, X. Jian, C. Yi, Z. Zhou and C. He, *FEBS Lett.*, 2008, **582**, 3313–3319.
130. G. Jia, Y. Fu, X. Zhao, Q. Dai, G. Zheng, Y. Yang, C. Yi, T. Lindahl, Y.-G. Yang and C. He, *Nat. Chem. Biol.*, 2011, **7**, 885–887.
131. Y. Fu, G. Jia, X. Pang, R. N. Wang, X.-J. Wang, C. J. Li, S. Smemo, Q. Dai, K. A. Bailey, M. A. Nobrega, K.-L. Han, Q. Cui and C. He, *Nat. Commun.*, 2013, **4**, 1798.
132. Z. Han, T. Niu, J. Chang, X. Lei, M. Zhao, Q. Wang, W. Cheng, J. Wang, Y. Feng and J. Chai, *Nature*, 2010, **464**, 1205–1209.
133. W. Aik, M. Demetriades, M. K. K. Hamdan, E. A. L. Bagg, K. K. Yeoh, C. Lejeune, Z. Zhang, M. A. McDonough and C. J. Schofield, *J. Med. Chem.*, 2013, **56**, 3680–3688.
134. D. Fu, J. A. N. Brophy, C. T. Y. Chan, K. A. Atmore, U. Begley, R. S. Paules, P. C. Dedon, T. J. Begley and L. D. Samson, *Mol. Cell. Biol.*, 2010, **30**, 2449–2459.
135. Y. Fu, Q. Dai, W. Zhang, J. Ren, T. Pan and C. He, *Angew. Chem., Int. Ed.*, 2010, **49**, 8885–8888.
136. E. van den Born, C. B. Vågbø, L. Songe-Møller, V. Leihne, G. F. Lien, G. Leszczynska, A. Malkiewicz, H. E. Krokan, F. Kirpekar, A. Klungland and P. Ø. Falnes, *Nat. Commun.*, 2011, **2**, DOI: 10.1038/ncomms1173.
137. C. Pastore, I. Topalidou, F. Forouhar, A. C. Yan, M. Levy and J. F. Hunt, *J. Biol. Chem.*, 2012, **287**, 2130–2143.
138. D. Zdzalik, C. B. Vågbø, F. Kirpekar, E. Davydova, A. Puscian, A. M. Maciejewska, H. E. Krokan, A. Klungland, B. Tudek, E. van den Born and P. Ø. Falnes, *PLoS One*, 2014, **9**, e98729.
139. V. De Crécy-Lagard, C. Brochier-Armanet, J. Ubronavicius, B. Fernandez, G. Phillips, B. Lyons, A. Noma, S. Alvarez, L. Droogmans, J. Armengaud and H. Grosjean, *Mol. Biol. Evol.*, 2010, **27**, 2062–2077.

140. M. Kato, Y. Araiso, A. Noma, A. Nagao, T. Suzuki, R. Ishitani and O. Nureki, *Nucleic Acids Res.*, 2010, **39**, 1576–1585.
141. M. Tahiliani, K. P. Koh, Y. Shen, W. A. Pastor, H. Bandukwala, Y. Brudno, S. Agarwal, L. M. Iyer, D. R. Liu, L. Aravind and A. Rao, *Science*, 2009, **324**, 930–935.
142. S. Ito, A. C. D'Alessio, O. V. Taranova, K. Hong, L. C. Sowers and Y. Zhang, *Nature*, 2010, **466**, 1129–1133.
143. T. Pfaffeneder, B. Hackner, M. Truß, M. Münzel, M. Müller, C. A. Deiml, C. Hagemeier and T. Carell, *Angew. Chem., Int. Ed.*, 2011, **50**, 7008–7012.
144. S. Ito, L. Shen, Q. Dai, S. C. Wu, L. B. Collins, J. A. Swenberg, C. He and Y. Zhang, *Science*, 2011, **333**, 1300–1303.
145. Y.-F. He, B.-Z. Li, Z. Li, P. Liu, Y. Wang, Q. Tang, J. Ding, Y. Jia, Z. Chen, L. Li, Y. Sun, X. Li, Q. Dai, C.-X. Song, K. Zhang, C. He and G.-L. Xu, *Science*, 2011, **333**, 1303–1307.
146. L. Zhang, W. Chen, L. M. Iyer, J. Hu, G. Wang, Y. Fu, M. Yu, Q. Dai, L. Aravind and C. He, *J. Am. Chem. Soc.*, 2014, **136**, 4801–4804.
147. T. Pfaffeneder, F. Spada, M. Wagner, C. Brandmayr, S. K. Laube, D. Eisen, M. Truss, J. Steinbacher, B. Hackner, O. Kotljarova, D. Schuermann, S. Michalakis, O. Kosmatchev, S. Schiesser, B. Steigenberger, N. Raddaoui, G. Kashiwazaki, U. Müller, C. G. Spruijt, M. Vermeulen, H. Leonhardt, P. Schär, M. Müller and T. Carell, *Nat. Chem. Biol.*, 2014, **10**, 574–581.
148. C.-H. Song and C. He, *TIBS*, 2013, **38**, 480–484.
149. C. J. Mariani, A. Vasanthakumar, J. Madzo, A. Yesilkanal, T. Bhagat, Y. Yu, S. Bhattacharyya, R. H. Wenger, S. L. Cohn, J. Nanduri, A. Verma, N. R. Prabhakar and L. A. Godley, *Cell Rep.*, 2014, **7**, 1–10.
150. S. Schiesser, B. Hackner, T. Pfaffendeder, M. Müller, C. Hagemeier, M. Truss and T. Carell, *Angew. Chem., Int. Ed.*, 2012, **51**, 6516–6520.
151. L. Hu, Z. Li, J. Cheng, Q. Rao, W. Gong, M. Liu, Y. G. Shi, J. Zhu, P. Wang and Y. Xu, *Cell*, 2013, **155**, 1545–1555.
152. H. Hashimoto, J. E. Pais, X. Zhang, L. Saleh, Z.-Q. Fu, N. Dai, I. R. Corrèa, Jr., Y. Zheng and X. Cheng, *Nature*, 2014, **506**, 391–395.
153. Y.-C. Huang and A. Rao, *Trends Genet.*, 2014, **30**, 464–474.
154. L. Shen, C.-X. Song, C. He and Y. Zhang, *Annu. Rev. Biochem.*, 2014, **83**, 585–614.
155. P. Borst and R. Sabatini, *Annu. Rev. Microbiol.*, 2008, **62**, 235–251.
156. J. H. Gommers-Ampt, F. Van Leeuwen, A. L. de Beer, J. F. Vliegenthart, M. Dizdaroglu, J. A. Kowalak, P. F. Crain and P. Borst, *Cell*, 1993, **75**, 1129–1136.
157. J. H. Gommers-Ampt, A. J. Teixeira, G. van de Werken, W. J. van Dijk and P. Borst, *Nucleic Acids Res.*, 1993, **21**, 2039–2043.
158. F. van Leeuwen, R. Kieft, M. Cross and P. Borst, *Mol. Cell. Biol.*, 1998, **18**, 5643–5651.
159. M. Cross, R. Kieft, R. Sabatini, M. Wilm, M. de Kort, G. A. van der Marel, J. H. van Boom, F. van Leeuwen and P. Borst, *EMBO J.*, 1999, **18**, 6573–6581.

160. M. Cross, R. Kieft, R. Sabatini, A. Dirks-Mulder, I. Chaves and P. Borst, *Mol. Microbiol.*, 2002, **46**, 37–47.
161. Z. Yu, P. A. Genest, B. Ter Riet, K. Sweeney, C. DiPaolo, R. Kieft, E. Christodoulou, A. Perrakis, J. M. Simmons, R. P. Hausinger, H. G. Van Luenen, D. J. Rigden, R. Sabatini and P. Borst, *Nucleic Acids Res.*, 2007, **35**, 2107–2115.
162. L. M. Iyer, D. Zhang, R. F. de Souza, P. J. Pukkila, A. Rao and L. Aravind, *Proc. Natl. Acad. Sci. U.S.A.*, 2014, **111**, 1676–1683.
163. R. Kieft, V. Brand, D. K. Ekanayake, K. Sweeney, C. DiPaolo, W. S. Reznikoff and R. Sabatini, *Mol. Biochem. Parasitol.*, 2007, **156**, 24–31.
164. L. J. Cliffe, G. Hirsch, J. Wang, D. Ekanayake, W. Bullard, M. Hu, Y. Wang and R. Sabatini, *J. Biol. Chem.*, 2012, **287**, 19886–19895.
165. F. M. Vaz and R. J. A. Wanders, *Biochem. J.*, 2002, **361**, 417–429.
166. J. D. Hulse, S. R. Ellis and L. M. Henderson, *J. Biol. Chem.*, 1978, **253**, 1654–1659.
167. F. M. Vaz, R. Ofman, K. Westinga, J. W. Back and R. J. A. Wanders, *J. Biol. Chem.*, 2001, **276**, 33512–33517.
168. G. Lindstedt, S. Lindstedt, B. Olander and M. Tofft, *Biochim. Biophys. Acta*, 1968, **158**, 503–505.
169. G. Lindstedt, S. Lindstedt and I. Nordin, *Biochemistry*, 1977, **16**, 2181–2188.
170. A. Kondo, J. S. Blanchard and S. Englard, *Arch. Biochem. Biophys.*, 1981, **212**, 338–346.
171. A. M. Rydzik, I. K. H. Leung, G. T. Kochan, N. D. Loik, L. Henry, M. A. McDonough, T. D. W. Claridge and C. J. Schofield, *Organic and Biomolecular Chemistry*, 2014, **12**, 6354–6358.
172. K. Tars, J. Rumnieks, A. Zeltins, A. Kazaks, S. Kotelovica, A. Leonciks, J. Saripo, A. Viksna, J. Kuka, E. Liepinsh and M. Dambrova, *Biochem. Biophys. Res. Commun.*, 2010, **398**, 634–639.
173. I. K. H. Leung, T. J. Krojer, G. T. Kochan, L. Henry, F. von Delft, T. D. W. Claridge, U. Oppermann, M. A. McDonough and C. J. Schofield, *Chem. Biol.*, 2010, **17**, 1316–1324.
174. P. A. Watkins, A. E. Howard and S. J. Mihalik, *Biochim. Biophys. Acta*, 1994, **1214**, 288–294.
175. S. J. Mihalik, A. M. Rainville and P. A. Watkins, *Eur. J. Biochem.*, 1995, **232**, 545–551.
176. K. Croes, P. P. Van Veldhoven, G. P. Mannaerts and M. Casteels, *FEBS Lett.*, 1997, **407**, 197–200.
177. K. Croes, M. Casteels, S. Asselberghs, P. Herdewijn, G. P. Mannaerts and P. P. Van Veldhoven, *FEBS Lett.*, 1997, **412**, 643–645.
178. N. M. Verhoeven, D. S. M. Schor, H. J. ten Brink, R. J. A. Wanders and C. Jakobs, *Biochem. Biophys. Res. Commun.*, 1997, **237**, 33–36.
179. R. J. A. Wanders, J. Komen and S. Ferdinandusse, *Biochim. Biophys. Acta*, 2011, **1811**, 498–507.
180. G. A. Jansen, R. Ofman, S. Denis, S. Ferdinandusse, E. M. Hogenhout, C. Jakobs and R. J. A. Wanders, *J. Lipid Res.*, 1999, **40**, 2244–2254.

181. M. Mukherji, W. Chien, N. J. Kershaw, I. J. Clifton, C. J. Schofield, A. S. Wierzbicki and M. D. Lloyd, *Hum. Mol. Genet.*, 2001, **10**, 1971–1982.
182. M. A. McDonough, K. L. Kavanagh, D. Butler, T. Searls, U. Oppermann and C. J. Schofield, *J. Biol. Chem.*, 2005, **280**, 41101–41110.
183. N. González-Silva, I. López-Lara, R. Reyes-Lamothe, A. M. Taylor, D. Sumpton, J. Thomas-Oates and O. Geiger, *Biochemistry*, 2011, **50**, 6396–6408.
184. H. S. Gibbons, S. Lin, R. J. Cotter and C. R. H. Raetz, *J. Biol. Chem.*, 2000, **275**, 32940–32949.
185. H. S. Gibbons, C. M. Reynolds, Z. Guan and C. R. H. Raetz, *Biochemistry*, 2008, **47**, 2814–2825.
186. H. S. Chung and C. R. H. Raetz, *Proc. Natl. Acad. Sci. U.S.A.*, 2011, **108**, 510–515.
187. Y. Kawai, E. Ono and M. Mizutani, *Plant J.*, 2014, **78**, 328–343.
188. B. A. Bohm, *Introduction to Flavonoids*, Harwood Academic Publishers, Amsterdam, The Netherlands, 1998.
189. L. Le Marchand, *Biomed. Pharmacother.*, 2002, **56**, 296–301.
190. L. Britsch, *Arch. Biochem. Biophys.*, 1990, **282**, 152–160.
191. L. Britsch and H. Grisebach, *Eur. J. Biochem.*, 1986, **156**, 569–577.
192. R. C. Wilmouth, J. J. Turnbull, R. W. D. Welford, I. J. Clifton, A. G. Prescott and C. J. Schofield, *Structure*, 2002, **10**, 93–103.
193. F. Wellman, R. Lukacin, T. Moriguchi, L. Britsch, E. Schiltz and U. Matern, *Eur. J. Biochem.*, 2002, **269**, 4134–4142.
194. K. Saito, M. Kobayashi, Z. Gong, Y. Tanaka and M. Yamazaki, *Plant J.*, 1999, **17**, 181–189.
195. R. W. D. Welford, I. J. Clifton, J. J. Turnbull, S. C. Wilson and C. J. Schofield, *Org. Biomol. Chem.*, 2005, **3**, 3117–3126.
196. M. Frey, K. Huber, W. J. Park, D. Sicker, P. Lindberg, R. B. Meeley, C. R. Simmons, N. Yalpani and A. Gierl, *Phytochemistry*, 2003, **62**, 371–376.
197. M. Frey, K. Schullehner, R. Dick, A. Feiesselmann and A. Gierl, *Phytochemistry*, 2009, **70**, 1645–1651.
198. D. Anzellotti and R. K. Ibrahim, *Arch. Biochem. Biophys.*, 2000, **382**, 161–172.
199. D. Anzellotti and R. K. Ibrahim, *BMC Plant Biol.*, 2004, **4**, 20.
200. A. Berim and D. R. Gang, *J. Biol. Chem.*, 2013, **288**, 1795–1805.
201. A.-X. Cheng, X.-J. Han, Y.-F. Wu and H.-X. Lou, *Int. J. Mol. Sci.*, 2014, **15**, 1080–1095.
202. P. Hedden and S. G. Thomas, *Biochem. J.*, 2012, **444**, 11–25.
203. T. Lange, A. Schweimer, D. A. Ward, P. Hedden and J. E. Graebe, *Planta*, 1994, **195**, 98–107.
204. T. Lange, *Planta*, 1994, **195**, 108–115.
205. T. Lange, P. Hedden and J. E. Graebe, *Proc. Natl. Acad. Sci. U.S.A.*, 1994, **91**, 8552–8556.
206. J. L. Ward, P. Gaskin, R. G. S. Brown, G. S. Jackson, P. Hedden, A. L. Phillips, C. L. Willis and M. H. Beale, *J. Chem. Soc., Perkin Trans.*, 2002, **1**, 232–241.
207. J. Williams, A. L. Phillips, P. Gaskin and P. Hedden, *Plant Physiol.*, 1998, **117**, 559–563.

208. S. G. Thomas, A. L. Phillips and P. Hedden, *Proc. Natl. Acad. Sci. U.S.A.*, 1999, **96**, 4698–4703.
209. P. Hedden and A. L. Phillips, *Trends Plant Sci.*, 2000, **5**, 523–528.
210. A. Bhattacharya, S. Kourmpetli, D. A. Ward, S. G. Thomas, F. Gong, S. J. Powers, E. Carrera, B. Taylor, F. N. de Caceres Gonzalez, B. Tudzynski, A. L. Phillips, M. R. Davey and P. Hedden, *Plant Physiol.*, 2012, **160**, 837–845.
211. T. Hashimoto and Y. Yamada, *Eur. J. Biochem.*, 1987, **164**, 277–285.
212. T. Hashimoto, J. Matsuda and Y. Yamada, *FEBS Lett.*, 1993, **329**, 35–39.
213. E. De Carolis, F. Chan, J. Balsevich and V. De Luca, *Plant Physiol.*, 1990, **94**, 1323–1329.
214. E. De Carolis and V. De Luca, *J. Biol. Chem.*, 1993, **268**, 5504–5511.
215. J. M. Hagel and P. J. Facchini, *Nat. Chem. Biol.*, 2010, **6**, 273–275.
216. S. C. Farrow and P. J. Facchini, *J. Biol. Chem.*, 2013, **288**, 28997–29012.
217. J. Havermann, D. Vogel, B. Loll and U. Keller, *Chem. Biol.*, 2014, **21**, 146–155.
218. H. Nakanishi, H. Yamaguchi, T. Sasakuma, N. K. Nishizawa and S. Mori, *Plant Mol. Biol.*, 2000, **44**, 199–207.
219. K. Kai, M. Mizutani, N. Kawamura, R. Yamamoto, M. Tamai, H. Yamaguchi, K. Sakata and B.-I. Shimizum, *Plant J.*, 2008, **55**, 989–999.
220. S. Matsumoto, M. Mizutani, K. Sakata and B.-I. Shimizu, *Phytochemistry*, 2012, **74**, 49–57.
221. G. Vialart, A. Hehn, A. Olry, K. Ito, C. Krieger, R. Larbat, C. Paris, B. Shimizu, Y. Sugimoto, M. Mizutani and F. Bourgaud, *Plant J.*, 2012, **70**, 460–470.
222. N. Agerbirk and C. E. Olsen, *Phytochemistry*, 2012, **77**, 16–45.
223. D. J. Kliebenstein, V. M. Lambrix, M. Reichelt, J. Gershenzon and T. Mitchell-Olds, *Plant Cell*, 2001, **13**, 681–693.
224. Z. Zhao, Y. Zhang, X. Liu, X. Zhang, S. Liu, X. Yu, Y. Ren, X. Zheng, K. Zhou, L. Jiang, X. Guo, Y. Gai, C. Wu, H. Zhai, H. Wang and J. Wan, *Dev. Cell*, 2013, **27**, 113–122.
225. K. Zhang, R. Halitschke, C. Yin, C.-J. Liu and S.-S. Gan, *Proc. Natl. Acad. Sci. U.S.A.*, 2013, **110**, 14807–14812.
226. J. E. Baldwin, R. A. Field, C. C. Lawrence, K. D. Merritt and C. J. Schofield, *Tetrahedron Lett.*, 1993, **34**, 7489–7492.
227. C. C. Lawrence, W. J. Sobey, R. A. Field, J. E. Baldwin and C. J. Schofield, *Biochem. J.*, 1996, **313**, 185–191.
228. T. Shibasaki, H. Mori, S. Chiba and A. Ozaki, *Appl. Environ. Microbiol.*, 1999, **65**, 4028–4031.
229. L. Petersen, R. Olewinski, P. Salmon and N. Conners, *Appl. Microbiol. Biotechnol.*, 2003, **62**, 263–267.
230. R. Hara and K. Kino, *Biochem. Biophys. Res. Commun.*, 2009, **379**, 882–886.
231. R. Hara, N. Uchiumi and K. Kino, *J. Biotechnol.*, 2014, **172**, 55–58.
232. H. Mori, T. Shibasaki, Y. Uozaki, K. Ochiai and A. Ozaki, *Appl. Environ. Microbiol.*, 1996, **62**, 1903–1907.

233. H. Mori, T. Shibasaki, K. Yano and A. Ozaki, *J. Bacteriol.*, 1997, **179**, 5677–5683.
234. I. J. Clifton, L.-C. Hsueh, J. E. Baldwin, K. Harlos and C. J. Schofield, *Eur. J. Biochem.*, 2001, **268**, 6625–6636.
235. M. Strieker, F. Kopp, C. Mahlert, L.-O. Essen and M. A. Marahiel, *ACS Chem. Biol.*, 2007, **2**, 187–196.
236. M. Strieker, L.-O. Essen, C. T. Walsh and M. A. Marahiel, *ChemBioChem*, 2008, **9**, 374–376.
237. G. M. Singh, P. D. Fortin, A. Koglin and C. T. Walsh, *Biochemistry*, 2008, **47**, 11310–11320.
238. X. Yin and T. M. Zabriskie, *ChemBioChem*, 2004, **5**, 1274–1277.
239. V. Helmetag, S. A. Samel, M. G. Thomas, M. A. Marahiel and L.-O. Essen, *FEBS J.*, 2009, **276**, 3669–3682.
240. B. Haltli, Y. Tan, N. A. Magarvey, M. Wagenaar, X. Yin, M. Greenstein, J. A. Hucul and T. M. Zabriskie, *Chem. Biol.*, 2005, **12**, 1163–1168.
241. C. Haefelè, C. Bonfils and Y. Sauvaire, *Phytochemistry*, 1997, **44**, 563–566.
242. S. V. Smirnov, T. Kodera, N. N. Samonova, V. A. Kotlyarova, N. Y. Rushkevich, A. D. Kivero, P. M. Sokolov, M. Hibi, J. Ogawa and B. Shimizu, *Appl. Microbiol. Biotechnol.*, 2010, **88**, 719–726.
243. M. Hibi, T. Kawashima, T. Kodera, S. V. Smirnov, P. M. Sokolov, M. Sugiyama, S. Shimizu, K. Yokozeki and J. Ogawa, *Appl. Environ. Microbiol.*, 2011, **77**, 6926–6930.
244. S. V. Smirnov, P. M. Sokolov, V. A. Kotlyarova, N. N. Samsonova, T. Kodera, M. Sugiyama, T. Torii, M. Hibi, S. Shimizu, K. Yokozeki and J. Ogawa, *MicrobiologyOpen*, 2013, **2**, 471–478.
245. S. V. Smirnov, P. M. Sokolov, M. Kodera, M. Sugiyama, M. Hibi, S. Shimizu, K. Yokozeki and J. Ogawa, *FEMS Microbiol. Lett.*, 2012, **331**, 97–104.
246. M. Hibi, T. Kawashima, P. M. Sokolov, S. V. Smirnov, M. Kodera, M. Sugiyama, S. Shimizu, K. Yokozeki and J. Ogawa, *Appl. Microbiol. Biotechnol.*, 2013, **97**, 2467–2472.
247. H.-M. Qin, T. Miyakawa, M. Z. Jia, A. Nakamura, J. Ohtsuka, Y.-L. Xue, T. Kawashima, T. Kasahara, M. Hibi, J. Ogawa and M. Tanokura, *PLoS One*, 2013, **8**, e63996.
248. H.-M. Qin, T. Miyakawa, A. Nakamura, M. Hibi, J. Ogawa and M. Tanokura, *Biochem. Biophys. Res. Commun.*, 2014, **450**, 1458–1461.
249. E. J. Drake and A. M. Gulick, *J. Mol. Biol.*, 2008, **384**, 193–205.
250. J. Bursy, A. J. Pierik, N. Pica and E. Bremer, *J. Biol. Chem.*, 2007, **282**, 31147–31155.
251. N. Widderich, A. Höppner, M. Pittlekow, J. Heider, S. H. J. Smits and E. Bremer, *PLoS One*, 2014, **9**, e93809.
252. K. Reuter, M. Pittelkow, J. Bursy, A. Heine, T. Craan and E. Bremer, *PLoS One*, 2010, **5**, e10647.
253. N. Widderich, M. Pittlekow, A. Höppner, D. Mulnaes, W. Buckel, H. Gohlke, S. H. J. Smits and E. Bremer, *J. Mol. Biol.*, 2014, **426**, 586–600.

254. H. Fukuda, T. Ogawa, M. Tazaki, K. Nagahama, T. Fujii, S. Tanase and Y. Morino, *Biochem. Biophys. Res. Commun.*, 1992, **188**, 483–489.
255. D. Meng, L. Shen, R. Yang, X. Zhang and J. Sheng, *Biochim. Biophys. Acta*, 2014, **1840**, 120–128.
256. J.-P. Wang, L.-X. Wu, F. Xu, J. Lv, H.-J. Jin and S.-F. Chen, *Bioresour. Technol.*, 2010, **101**, 6404–6409.
257. F. Guerrero, V. Carbonell, M. Cossu, D. Correddu and P. R. Jones, *PLoS One*, 2012, **7**, e50470.
258. J. Ungerer, L. Tao, M. Davis, M. Ghirardi, P.-C. Maness and J. Yu, *Energy Environ. Sci.*, 2012, **5**, 8998–9006.
259. M. T. Abbott, E. K. Schandl, R. F. Lee, T. S. Parker and R. J. Midgett, *Biochim. Biophys. Acta*, 1967, **132**, 525–528.
260. E. Holme, G. Lindstedt, S. Lindstedt and M. Tofft, *Biochim. Biophys. Acta*, 1970, **212**, 50–57.
261. E. Holme, G. Lindstedt, S. Lindstedt and M. Tofft, *J. Biol. Chem.*, 1971, **246**, 3314–3319.
262. C. K. Liu, C. A. Hsu and M. T. Abbott, *Arch. Biochem. Biophys.*, 1973, **159**, 180–187.
263. B. J. Warn-Cramer, L. A. Macrander and M. T. Abbott, *J. Biol. Chem.*, 1983, **258**, 10551–10557.
264. L. D. Thornburg, M.-T. Lai, J. S. Wishnok and J. Stubbe, *Biochemistry*, 1993, **32**, 14023–14033.
265. J. W. Neidigh, A. Darwanto, A. A. Williams, N. R. Wall and L. C. Sowers, *Chem. Res. Toxicol.*, 2009, **22**, 885–893.
266. A. Cultrone, C. Scazzocchio, M. Rochet, G. Montero-Morán, C. Drevet and R. Fernández-Martín, *Mol. Microbiol.*, 2005, **57**, 276–290.
267. G. M. Montero-Moràn, M. Li, E. Rendóm-Huerta, F. Jourdan, D. J. Lowe, A. W. Stumpff-Kane, M. Feig, C. Scazzocchio and R. P. Hausinger, *Biochemistry*, 2007, **46**, 5293–5304.
268. M. Li, T. A. Müller, B. A. Fraser and R. P. Hausinger, *Arch. Biochem. Biophys.*, 2008, **470**, 44–53.
269. L. Bankel, G. Lindstedt and S. Lindstedt, *J. Biol. Chem.*, 1972, **247**, 6128–6134.
270. L. M. Wondrack, C. A. Hsu and M. T. Abbott, *J. Biol. Chem.*, 1978, **253**, 6511–6515.
271. J. Stubbe, *J. Biol. Chem.*, 1985, **260**, 9972–9975.
272. Z. Yang, X. Chi, M. Funabashi, S. Baba, K. Nonaka, P. Pahari, J. Unrine, J. M. Jacobsen, G. I. Elliott, J. Rohr and S. G. Van Lanen, *J. Biol. Chem.*, 2011, **286**, 7885–7892.
273. C. Zhao, T. Huang, W. Chen and Z. Deng, *Appl. Environ. Microbiol.*, 2010, **76**, 7343–7347.
274. R. P. Hausinger, F. Fukumori, D. A. Hogan, T. M. Sassanella, Y. Kamagata, H. Takami and R. E. Saari, in *Microbial Diversity and Genetics of Biodegradation*, ed. K. Horikoshi, M. Fukuda and T. Kudo, Japan Scientific Press, Tokyo, 1997, pp. 35–51.
275. F. Fukumori and R. P. Hausinger, *J. Bacteriol.*, 1993, **175**, 2083–2086.

276. F. Fukumori and R. P. Hausinger, *J. Biol. Chem.*, 1993, **268**, 24311–24317.
277. E. L. Hegg, A. K. Whiting, R. E. Saari, J. McCracken, R. P. Hausinger and L. Que, Jr., *Biochemistry*, 1999, **38**, 16714–16726.
278. S. L. Griffin, J. A. Godbey, T. J. Oman, S. K. Embrey, A. Karnoup, K. Kuppannan, B. W. Barnett, G. Lin, N. V. J. Harpham, A. N. Juba, B. W. Schafer and R. M. Cicchillo, *J. Agric. Food Chem.*, 2013, **61**, 6589–6596.
279. K. Nickel, M. J.-F. Suter and H.-P. E. Kohler, *J. Bacteriol.*, 1997, **179**, 6674–6679.
280. T. A. Müller, M. I. Zavodszky, M. Feig, L. A. Kuhn and R. P. Hausinger, *Protein Sci.*, 2006, **15**, 1356–1368.
281. J. M. Bollinger, Jr., J. C. Price, L. M. Hoffart, E. W. Barr and C. Krebs, *Eur. J. Inorg. Chem.*, 2005, 4245–4254.
282. E. Eichhorn, J. R. van der Ploeg, M. A. Kertesz and T. Leisinger, *J. Biol. Chem.*, 1997, **272**, 23031–23036.
283. J. M. Elkins, M. J. Ryle, I. J. Clifton, J. C. Dunning Hotopp, J. S. Lloyd, N. I. Burzlaff, J. E. Baldwin, R. P. Hausinger and P. L. Roach, *Biochemistry*, 2002, **41**, 5185–5192.
284. J. R. O'Brien, D. J. Schuller, V. S. Yang, B. D. Dillard and W. N. Lanzilotta, *Biochemistry*, 2003, **42**, 5547–5554.
285. S. H. Knauer, O. Hartl-Spiegelhauer, S. Schwarzinger, P. Hänzelmann and H. Dobbek, *FEBS J.*, 2012, **279**, 816–831.
286. M. J. Ryle, R. Padmakumar and R. P. Hausinger, *Biochemistry*, 1999, **38**, 15278–15286.
287. R. B. Muthakumaran, P. K. Grzyska, R. P. Hausinger and J. McCracken, *Biochemistry*, 2007, **46**, 5951–5959.
288. J. C. Price, E. W. Barr, B. Tirupati, J. M. Bollinger, Jr. and C. Krebs, *Biochemistry*, 2003, **42**, 7497–7508.
289. P. J. Riggs-Gelasco, J. C. Price, R. B. Guyer, J. H. Brehm, E. W. Barr, J. M. Bollinger, Jr. and C. Krebs, *J. Am. Chem. Soc.*, 2004, **126**, 8108–8109.
290. D. A. Proshlyakov, T. F. Henshaw, G. R. Monterosso, M. J. Ryle and R. P. Hausinger, *J. Am. Chem. Soc.*, 2004, **126**, 1022–1023.
291. P. K. Grzyska, M. J. Ryle, G. R. Monterosso, J. Liu, D. P. Ballou and R. P. Hausinger, *Biochemistry*, 2005, **44**, 3845–3855.
292. D. A. Hogan, T. A. Auchtung and R. P. Hausinger, *J. Bacteriol.*, 1999, **181**, 5876–5879.
293. A. Kahnert and M. A. Kertesz, *J. Biol. Chem.*, 2000, **275**, 31661–31667.
294. K. M. Sogi, Z. J. Gartner, M. A. Breidenbach, M. J. Appel, M. W. Schelle and C. R. Bertozzi, *PLoS One*, 2013, **8**, e65080.
295. I. Müller, A. Kahnert, T. Pape, G. M. Sheldrick, W. Meyer-Klaucke, T. Dierks, M. Kertesz and I. Usòn, *Biochemistry*, 2004, **43**, 3075–3088.
296. I. Müller, C. Stückl, J. Wakeley, M. Kertesz and I. Usòn, *J. Biol. Chem.*, 2004, **280**, 5716–5723.
297. F. R. McSorley, P. B. Wyatt, A. Martinez, E. F. DeLong, B. Hove-Jensen and D. L. Zechel, *J. Am. Chem. Soc.*, 2012, **134**, 8364–8367.
298. B. T. Circello, A. C. Eliot, J.-H. Lee, W. A. van der Donk and W. W. Metcalf, *Chem. Biol.*, 2010, **17**, 402–411.

299. M. A. DeSieno, W. van der Donk and H. Zhao, *Chem. Commun.*, 2011, **47**, 10025–10027.
300. M. Watanabe, N. Sumida, S. Murakami, H. Anzai, C. J. Thompson, Y. Tateno and T. Murakami, *Appl. Environ. Microbiol.*, 1999, **65**, 1036–1044.
301. D. J. Bougioukou, S. Mukherjee and W. A. van der Donk, *Proc. Natl. Acad. Sci. U.S.A.*, 2013, **110**, 10952–10957.
302. A. K. White and W. W. Metcalf, *J. Biol. Chem.*, 2002, **277**, 38262–38271.
303. N. J. Kershaw, M. E. C. Caines, M. C. Sleeman and C. J. Schofield, *Chem. Commun.*, 2005, 4251–4263.
304. R. W. Busby, M. D.-T. Chang, R. C. Busby, J. Wimp and C. A. Townsend, *J. Biol. Chem.*, 1995, **270**, 4262–4269.
305. E. G. Pavel, J. Zhou, R. W. Busby, M. Gunsior, C. A. Townsend and E. I. Solomon, *J. Am. Chem. Soc.*, 1998, **120**, 743–753.
306. J. Zhou, W. L. Kelly, B. O. Bachmann, M. Gunsior, C. A. Townsend and E. I. Solomon, *J. Am. Chem. Soc.*, 2001, **123**, 7388–7398.
307. Z. Zhang, J. Ren, D. K. Stammers, J. E. Baldwin, K. Harlos and C. J. Schofield, *Nat. Struct. Biol.*, 2000, **7**, 127–133.
308. Z. Zhang, J.-s. Ren, K. Harlos, C. H. McKinnon, I. J. Clifton and C. J. Schofield, *FEBS Lett.*, 2002, **517**, 7–12.
309. S. Kovacevic, B. J. Weigel, M. B. Tobin, T. D. Ingolia and J. R. Miller, *J. Bacteriol.*, 1989, **171**, 754–760.
310. J. E. Dotzlaf and W.-K. Yeh, *J. Biol. Chem.*, 1989, **264**, 10219–10227.
311. B. J. Baker, J. E. Dotzlaf and W.-K. Yeh, *J. Biol. Chem.*, 1991, **266**, 5087–5093.
312. J. E. Dotzlaf and W.-K. Yeh, *J. Bacteriol.*, 1987, **169**, 1611–1618.
313. J. E. Baldwin, R. M. Adlington, J. B. Coates, M. J. C. Crabbe, N. P. Crouch, J. W. Keeping, G. C. Knight, C. J. Schofield, H.-H. Ting, C. A. Vallejo, M. Thorniley and E. P. Abraham, *Biochem. J.*, 1987, **245**, 831–841.
314. K. Valegård, A. C. Terwisscha van Scheltinga, M. D. Lloyd, T. Hara, S. Ramaswamy, A. Perrakis, A. Thompson, W.-J. Lee, J. E. Baldwin, C. J. Schofield, J. Hajdu and I. Andersson, *Nature (London)*, 1998, **394**, 805–809.
315. M. D. Lloyd, H.-J. Lee, K. Harlos, Z.-H. Zhang, J. E. Baldwin, C. J. Schofield, J. M. Charnock, C. D. Garner, T. Hara, A. C. Terwisscha Van Scheltinga, K. Valegård, J. A. C. Viklund, J. Hajdu, I. Andersson, A. Danielsson and R. Bhikhabhai, *J. Mol. Biol.*, 1999, **287**, 943–960.
316. H. J. Lee, M. D. Lloyde, I. J. Clifton, J. E. Baldwin and C. J. Schofield, *J. Mol. Biol.*, 2001, **308**, 937–948.
317. H.-J. Lee, M. D. Lloyd, I. J. Clifton, K. Harlos, A. Dubus, J. Baldwin, E., J.-M. Frere and C. J. Schofield, *J. Biol. Chem.*, 2001, **276**, 18290–18295.
318. K. Valegård, A. C. Terwisscha van Scheltinga, A. Dubus, G. Ranghino, L. M. Öster, J. Hajdu and I. Andersson, *Nat. Struct. Mol. Biol.*, 2004, **11**, 95–101.
319. L. M. Öster, A. C. Terwisscha van Scheltinga, K. Valegård, A. M. Hose, A. Dubus, J. Hajdu and I. Andersson, *J. Mol. Biol.*, 2004, **343**, 157–171.

320. H. Tarhonskaya, A. Szöllössi, I. K. H. Leung, J. T. Bush, L. Henry, R. Chowdhury, A. Iqbal, T. D. W. Claridge, C. J. Schofield and E. Flashman, *Biochemistry*, 2014, **53**, 2483–2493.
321. M. D. Lloyd, S. J. Lipscomb, K. S. Hewitson, C. M. H. Hensgens, J. E. Baldwin and C. J. Schofield, *J. Biol. Chem.*, 2004, **297**, 1540–1546.
322. X. Xiao, S. Wolfe and A. L. Demain, *Biochem. J.*, 1991, **280**, 471–474.
323. R. M. Phelan and C. A. Townsend, *J. Am. Chem. Soc.*, 2013, **135**, 7496–7502.
324. I. J. Clifton, L. X. Doan, M. C. Sleeman, M. Topf, H. Suzuki, R. C. Wilmouth and C. J. Schofield, *J. Biol. Chem.*, 2003, **278**, 20843–20850.
325. W.-c Chang, Y. Guo, C. Wang, S. E. Butch, A. C. Rosenzweig, A. K. Boal, C. Krebs and J. M. Bollinger, Jr., *Science*, 2014, **343**, 1140–1144.
326. Z. You, S. Omura, H. Ikeda and D. E. Cane, *J. Am. Chem. Soc.*, 2006, **126**, 6566–6567.
327. Z. You, S. Omura, H. Ikeda, D. E. Cane and G. Jogl, *J. Biol. Chem.*, 2007, **282**, 36552–36560.
328. M.-J. Seo, D. Zhu, S. Endo, H. Ikeda and D. E. Cane, *Biochemistry*, 2011, **50**, 1739–1754.
329. Y. Ono, A. Minami, M. Noike, Y. Higuchi, T. Toyoumasu, T. Sassa, N. Kato and T. Diari, *J. Am. Chem. Soc.*, 2011, **133**, 2548–2555.
330. M. Daum, H.-J. Schnell, S. Herrman, A. Günther, R. Murillo, R. Müller, P. Bisel, M. Müller and A. Bechthold, *ChemBioChem*, 2010, **11**, 1383–1391.
331. C. Dürr, H.-J. Schnell, A. Luzhetskyy, R. Murillo, M. Weber, K. Wetzel, A. Vente and A. Bechthold, *Chem. Biol.*, 2006, **13**, 365–377.
332. F. H. Vaillancourt, J. Yin and C. T. Walsh, *Proc. Natl. Acad. Sci. U.S.A.*, 2005, **102**, 10111–10116.
333. F. H. Vaillancourt, D. A. Vosburg and C. T. Walsh, *ChemBioChem*, 2006, **7**, 748–752.
334. M. L. Matthews, W.-c. Chang, A. P. Layne, L. A. Miles, C. Krebs and J. M. Bollinger, Jr., *Nat. Chem. Biol.*, 2014, **10**, 299–315.
335. L. C. Blasiak, F. H. Vaillancourt, C. T. Walsh and C. L. Drennan, *Nature*, 2006, **440**, 368–371.
336. M. M. Matthews, C. Krest, E. W. Barr, F. Vaillancourt, C. T. Walsh, M. Green, C. Krebs and J. M. Bollinger, Jr., *Biochemistry*, 2009, **48**, 4331–4343.
337. M. L. Matthews, C. S. Neumann, L. A. Miles, T. L. Grove, S. J. Booker, C. Krebs, C. T. Walsh and J. M. Bollinger, Jr., *Proc. Natl. Acad. Sci. U.S.A.*, 2009, **106**, 17723–17728.
338. M. R. Fullone, A. Paiardini, R. Miele, S. Marsango, D. C. Gross, S. Omura, E. Ros-Herrera, M. C. Bonaccorsi di Patt, A. Laganà, S. Pacarella and I. Grgurina, *FEBS J.*, 2012, **279**, 4269–4282.
339. D. P. Galonić, E. W. Barr, C. T. Walsh, J. M. Bollinger, Jr. and C. Krebs, *Nat. Chem. Biol.*, 2007, **3**, 113–116.
340. D. G. Fujimori, E. W. Barr, M. L. Matthews, G. M. Koch, J. R. Yonce, C. T. Walsh, J. M. Bollinger, Jr., C. Krebs and P. J. Riggs-Gelasco, *J. Am. Chem. Soc.*, 2007, **129**, 13408–13409.

341. C. Wong, D. G. Fujimori, C. T. Walsh and C. L. Drennan, *J. Am. Chem. Soc.*, 2009, **131**, 4872–4879.
342. D. P. Galonić, F. H. Vaillancourt and C. T. Walsh, *J. Am. Chem. Soc.*, 2006, **128**, 3900–3901.
343. F. H. Vaillancourt, E. Yeh, D. A. Vosburg, S. E. O'Conner and C. T. Walsh, *Nature*, 2005, **436**, 1191–1194.
344. C. S. Neumann and C. T. Walsh, *J. Am. Chem. Soc.*, 2008, **130**, 14022–14023.
345. L. Gu, B. Wang, A. Kulkarni, T. W. Geders, R. V. Grindberg, L. Gerwick, K. Håkansson, P. Wipf, J. L. Smith, W. H. Gerwick and D. H. Sherman, *Nature*, 2009, **459**, 731–735.
346. D. Khare, B. Wang, L. Gu, J. Razelun, D. H. Sherman, W. H. Gerwick, K. Håkansson and J. L. Smith, *Proc. Natl. Acad. Sci. U.S.A.*, 2010, **107**, 14099–14104.
347. W. Jiang, J. R. Heemstra, Jr., R. R. Forseth, C. S. Neumann, S. Manaviazar, F. C. Schroeder, K. J. Hale and C. T. Walsh, *Biochemistry*, 2011, **50**, 6063–6072.
348. S. M. Pratter, J. Ivkovic, R. Birner-Gruenberger, R. Breinbauer, K. Zangger and G. D. Straganz, *ChemBioChem*, 2014, **15**, 567–574.
349. S. M. Pratter, K. M. Light, E. I. Solomon and G. D. Straganz, *Biochemistry*, 2014, **136**, 9385–9395.
350. U. Galm, E. Wendt-Pienkowski, L. Wang, X.-X. Huang, C. Unsin, M. Tao, J. M. Coughlin and B. Shen, *J. Nat. Prod.*, 2011, **74**, 526–536.
351. A. S. Eustáquio, J. E. Janso, A. S. Ratnayake, C. J. O'Donnell and F. E. Koehn, *Proc. Natl. Acad. Sci. U.S.A.*, 2014, **111**, E3376–E3385.
352. M. Tao, L. Wang, E. Wendt-Pienkowski, N. Zhang, D. Yang, U. Galm, J. M. Coughlin, Z. Xu and B. Shen, *Mol. Biosyst.*, 2010, **6**, 349–356.
353. J.-A. Seo, R. H. Proctor and R. D. Plattner, *Fungal Genet. Biol.*, 2001, **34**, 155–165.
354. Y. Ding, R. S. Bojja and L. Du, *Appl. Environ. Microbiol.*, 2004, **70**, 1931–1934.
355. N. Steffan, A. Grundmann, S. Afiyatullov, H. Ruan and S.-M. Li, *Org. Biomol. Chem.*, 2009, **7**, 4082–4087.
356. R. Mazmouz, F. Chapuis-Hugon, V. Pichon, A. Méjean and O. Ploux, *ChemBioChem*, 2011, **12**, 858–862.
357. K. H. Almabruk, S. Asamizu, A. Chang, S. G. Varghese and T. Mahmud, *ChemBioChem*, 2012, **13**, 2209–2211.
358. P. L. Roach, I. J. Clifton, V. Fülöp, K. Harlos, G. J. Barton, J. Hajdu, K. Andersson, C. J. Schofield and J. E. Baldwin, *Nature (London)*, 1995, **375**, 700–704.
359. P. L. Roach, I. J. Clifton, C. M. H. Hensgens, N. Shibata, C. J. Schofield, J. Hajdu and J. E. Baldwin, *Nature (London)*, 1997, **387**, 827–830.
360. N. I. Burzlaff, P. J. Rutledge, I. J. Clifton, C. M. H. Hensgens, M. Pickford, R. M. Adlington, P. L. Roach and J. E. Baldwin, *Nature*, 1999, **401**, 721–724.
361. P. John, *Physiol. Plant.*, 1997, **100**, 583–592.
362. D. R. Dilley, Z. Wang, D. K. Dadirjan-Kalbach, F. Ververidis, R. Beaudry and K. Padmanabhan, *AoB Plants*, 2013, **5**, plt031.

363. A. M. Rocklin, D. L. Tierney, V. Kofman, N. M. W. Brunhuber, B. M. Hoffman, R. E. Christoffersen, N. O. Reich, J. D. Lipscomb and L. Que, Jr., *Proc. Natl. Acad. Sci. U.S.A.*, 1999, **96**, 7905–7090.
364. D. L. Tierney, A. M. Rocklin, J. D. Lipscomb, L. Que, Jr. and B. Hoffman, *J. Am. Chem. Soc.*, 2005, **127**, 7005–7013.
365. A. M. Rocklin, K. Kato, H.-w. Liu, L. Que, Jr. and J. D. Lipscomb, *J. Biol. Inorg. Chem.*, 2004, **9**, 171–182.
366. J. Zhou, A. M. Rocklin, J. D. Lipscomb, L. Que, Jr. and E. I. Solomon, *J. Am. Chem. Soc.*, 2002, **124**, 4602–4609.
367. J. N. Barlow, Z. Zhang, P. John, J. E. Baldwin and C. J. Schofield, *Biochemistry*, 1997, **36**, 3563–3569.
368. Z. Zhang, J. N. Barlow, J. E. Baldwin and C. J. Schofield, *Biochemistry*, 1997, **36**, 15999–16007.
369. Z. Zhang, J.-S. Ren, I. J. Clifton and C. J. Schofield, *Chem. Biol.*, 2004, **11**, 1383–1394.
370. L. Brisson, N. El Bakkali-Taheri, M. Giorgi, A. Fadel, J. Kaizer, M. Réglier, T. Tron, E. H. Ajandouz and A. J. Simaan, *J. Biol. Inorg. Chem.*, 2012, **17**, 939–949.
371. P. He and G. R. Moran, *Curr. Opin. Chem. Biol.*, 2009, **13**, 443–450.
372. D. D. Shah, J. A. Conrad and G. R. Moran, *Biochemistry*, 2013, **52**, 6097–6107.
373. M. Gunsior, J. Ravel, G. L. Challis and C. A. Townsend, *Biochemistry*, 2004, **43**, 663–674.
374. K. Johnson-Winters, V. M. Purpero, M. Kavana and G. R. Moran, *Biochemistry*, 2005, **44**, 7189–7199.
375. P. He, J. A. Conrad and G. R. Moran, *Biochemistry*, 2010, **49**, 1998–2007.
376. J. Brownlee, P. He, G. R. Moran and D. H. T. Harrison, *Biochemistry*, 2008, **47**, 20002–22013.
377. L. Serre, A. Sailland, D. Sy, P. Boudec, A. Rolland, E. Pebay-Peyroula and C. Cohen-Addad, *Structure*, 1999, 7, 977–988.
378. J. Pan, M. Bharwaj, J. R. Faulkner, P. Nagabhyru, N. D. Charlton, R. M. Higashi, A.-F. Miller, C. A. Young, R. B. Grossman and C. L. Schardl, *Phytochemistry*, 2014, **98**, 60–68.
379. J. S. Scotti, I. K. H. Leung, M. A. Bentley, J. Paps, H. B. Kramer, J. Lee, W. Aik, H. Choi, S. M. Paulsen, L. A. H. Bowman, N. D. Loik, S. Horita, C.-h. Ho, N. J. Kershaw, C. M. Tang, T. D. W. Claridge, G. M. Preston, M. A. McDonough and C. J. Schofield, *Proc. Natl. Acad. Sci. U.S.A.*, 2014, **111**, 13331–13336.
380. B.-I. Shimizu, *Front. Plant Sci.*, 2014, **5**, 549.
381. M. L. Hillig and X. Liu, *Nature Chem. Biol.*, 2014, **10**, 921–923.

CHAPTER 2

Introduction to Structural Studies on 2-Oxoglutarate-Dependent Oxygenases and Related Enzymes

WEI SHEN AIK[a], RASHEDUZZAMAN CHOWDHURY[a], IAN J. CLIFTON[a], RICHARD J. HOPKINSON[a], THOMAS LEISSING[a], MICHAEL A. MCDONOUGH[a], RADOSŁAW NOWAK[b], CHRISTOPHER J. SCHOFIELD*[a], AND LOUISE J. WALPORT[a]

[a]The Chemistry Research Laboratory, Mansfield Road, Oxford OX1 3TA, UK; [b]Botnar Research Centre, Oxford Biomedical Research Unit, Oxford OX3 7LD, UK

*E-mail: Christopher.schofield@chem.ox.ac.uk

2.1 Introduction to Structural Studies on 2-Oxoglutarate-Dependent Oxygenases

2.1.1 Background

The first Fe(II)- and 2-oxoglutarate (2OG)-dependent oxygenases to be identified were the animal pro-collagen hydroxylases (see Chapter 5).[1–3] However, to date the oxygenase domains of these enzymes have proven to be rather recalcitrant with respect to high-resolution structural studies. Instead, the first crystal structure of a member of this superfamily (henceforth simplified

RSC Metallobiology Series No. 3
2-Oxoglutarate-Dependent Oxygenases
Edited by Robert P. Hausinger and Christopher J. Schofield

Published by the Royal Society of Chemistry, www.rsc.org

to the 2OG oxygenases) came from work on enzymes involved in the biosynthesis of the β-lactam antibiotics.[4–6] This work was enabled by contemporary advances in molecular biology, which allowed the production of unnaturally large amounts of protein in recombinant hosts. The first structure of an enzyme to be studied from the superfamily was that of isopenicillin N synthase (IPNS), which catalyses the unusual four electron oxidation of a dipeptide unit to give the bicyclic penicillin nucleus in a single step (Figure 2.1; also see Chapter 19).[6–9] The crystallographic studies on IPNS unexpectedly revealed the presence of a double-stranded β-helix (DSBH), cupin, or jelly-roll core fold.[7] The IPNS structure also revealed the presence of a non-haem Fe(II) bound at the active site, consistent with spectroscopic studies. Three protein-bound Fe(II) ligands were identified: an Asp and two His residues, which constitute a His-X-Asp/Glu…His (HXD/E…H) triad (often referred to as a facial triad) of residues, which is conserved in most, but not all, members of the 2OG oxygenase superfamily.[10] The initial work on IPNS identified a fourth residue, a glutamine in the region to the *C*-terminal side of the DSBH that was close to the active site metal ion.[6,11,12] This residue was subsequently shown not to be essential for catalysis, but this aspect of the initial work was important because it implied conformational changes during catalysis. Subsequent structures of IPNS, combined with solution studies, have revealed the details of how it catalyses oxidative bicyclisation, and are described in Chapter 19.[8,9] Although the mechanistic details of the IPNS reaction differ substantially from those of typical 2OG oxygenases, it is one of the best characterised enzymes from the superfamily (at least in crystallographic terms).[6–8,13] Time-resolved studies on IPNS employing the exposure of anaerobically prepared crystals with O_2, enabled by the accessibility of high-resolution crystal diffraction and IPNS's lack of requirement for a 2OG cosubstrate,[8,14] have revealed details of conformational changes in the enzyme on substrate binding and during formation of the penicillin bicycle.

IPNS, like 1-aminocyclopropane-1-carboxylic acid oxidase (ACCO)[15] (Chapter 20) and (*S*)-2-hydroxypropylphosphonic acid epoxidase (HPPE),[16] has a similar fold to the typical 2OG oxygenases (Figure 2.2), but it does not use 2OG as a cosubstrate.[17,18] Sequence analysis in the light of the IPNS structure revealed the probably very widespread nature of the 2OG oxygenase subfamily. Confirmation that *bona fide* 2OG oxygenases share the same fold as IPNS came with the crystal structures of the 2OG-dependent enzymes deacetoxycephalosporin C synthase (DAOCS),[4] taurine/2OG oxygenase (TauD),[19] clavaminic acid synthase (CAS)[5] and a proline-3-hydroxylase (P3H).[20] These enzymes catalyse two electron oxidations, including hydroxylation reactions which are typical of 2OG oxygenases (although CAS and DAOCS also catalyse other reactions – see Chapter 17 for a detailed description of DAOCS).[5,21–24] The crystal structures of these enzymes have revealed clear conservation of the triad of metal binding residues, as observed in IPNS, and subsequently in many 2OG oxygenases (including some histone demethylases);[25–29] in some cases, such as CAS, the carboxylate is supplied from a Glu rather than an Asp residue.[13,30,31] The structures with 2OG also defined conserved modes

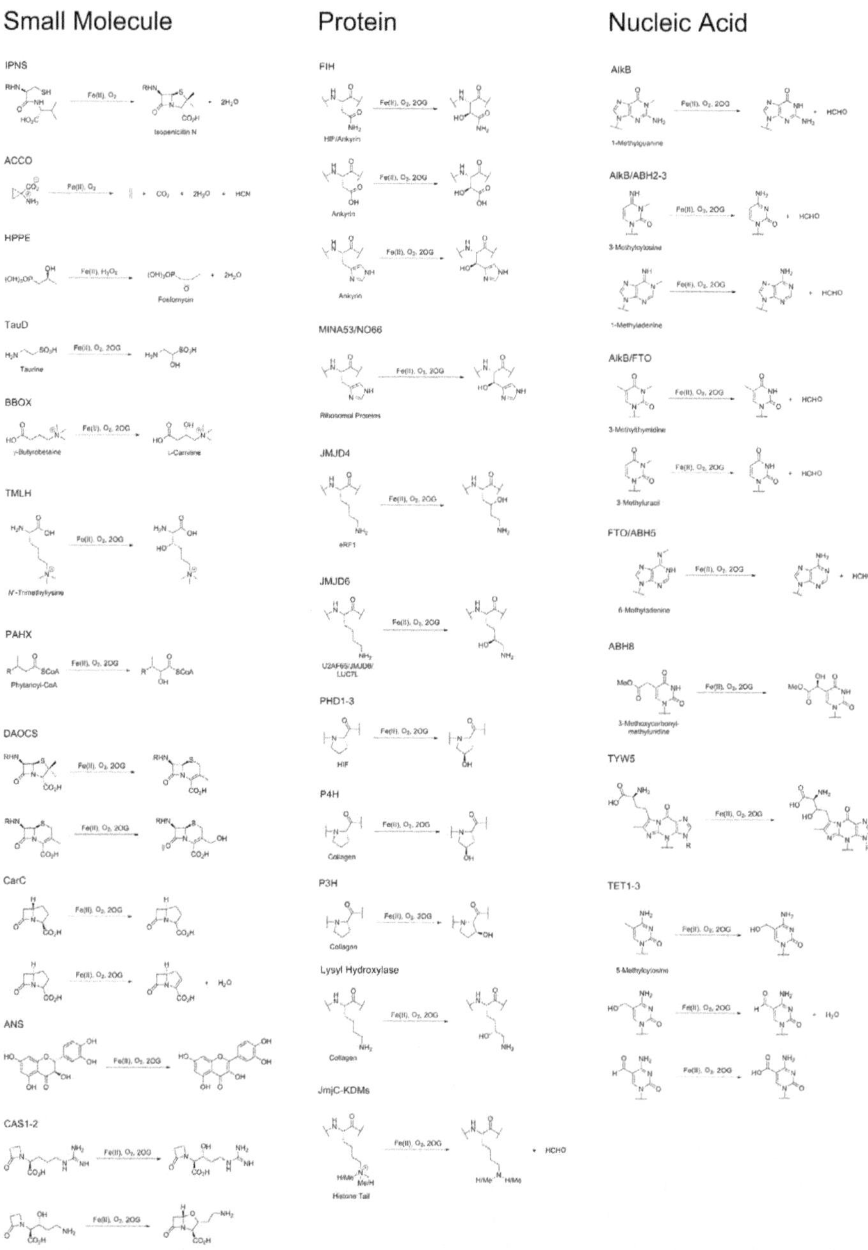

Figure 2.1 Reactions catalysed by 2OG oxygenases and related enzymes of known structure as described in this review. 2OG oxygenases and related enzymes catalyse multiple reaction types (hydroxylations, demethylations, desaturations, epoxidations, epimerizations, ring closure/expansion), on a variety of substrates including small molecules, fatty acids, proteins and nucleic acids. The abbreviations are defined in the main text.

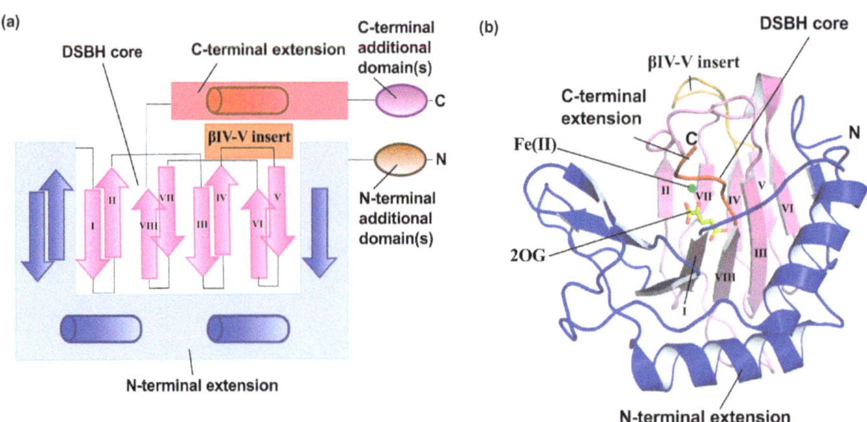

Figure 2.2 Overall structure of the 2OG oxygenases. (a) General two-dimensional topology of a 2OG oxygenase. (b) Three-dimensional representation of the 2OG oxygenase fold adapted from a crystal structure of AlkB (PDB ID: 3BIE).

of cosubstrate binding and provided insights into the spatial relationship between the substrate C–H bond that is oxidized and the active site iron during catalysis (see below).[13,25,30]

Multiple structures have subsequently been reported for 2OG oxygenases including representatives from the subfamilies that catalyse the oxidation of small molecules (amino acids and many other compounds), peptides/proteins, nucleic acids and lipids.[13,25,30,32] Following from pioneering work on AlkB,[33–35] which catalyses DNA repair *via* *N*-demethylation, structural work on 2OG oxygenases catalysing the hydroxylation/demethylation of proteins and nucleic acids has been particularly productive in recent years, reflecting the biomedicinal importance of many of these enzymes (see Chapters 5–12).[13,25,30,32,36,37]

Some 2OG oxygenases are presently being targeted for pharmaceutical development and some are herbicide targets (see Chapters 6, 16, 21).[38–42] Hence, the number of structural investigations on the superfamily is growing, with a particular emphasis on human enzymes. A realistic medium-term ambition is the definition of structures for representatives of all human subfamilies of 2OG oxygenases, if not all human 2OG oxygenases.

The purpose of this chapter is to provide an introduction and overview of structural studies on 2OG oxygenases. Rather than being comprehensive we give illustrated examples of crystallographically characterized enzymes. This chapter should be read in conjunction with related chapters, in particular those describing the reaction mechanisms of 2OG oxygenases (Chapters 1 and 3) and the related enzymes IPNS (Chapter 19) and ACCO (Chapter 20). Other reviews of structural studies on 2OG oxygenases are also available.[13,25,30,32]

2.1.2 2OG Oxygenases Have a Double-Stranded β-Helix Core Fold

All 2OG oxygenases have a characteristic 'distorted' DSBH fold (Figure 2.2) that was first described in virally encoded proteins (tomato bushy stunt virus)[43] and has subsequently been observed in multiple other protein families.[25,44] Aside from the 2OG oxygenases, the DSBH is also characteristic of the plant cupin-fold proteins[45] and the Jumonji C (JmjC) chromatin-associated proteins;[29,46] here we use the terms DSBH and, where appropriate, JmjC. Of the four possible forms of the DSBH fold (two hands of the β-strand helix, two directions to trace the structure), only one (right-handed class I) is observed naturally, suggesting the possibility of divergent evolution from a common ancestor.[13,25]

The DSBH fold is made up of eight β-strands (identified here by Roman numerals for β-strands I–VIII (Figure 2.2).[13,25,32] The eight β-strands occur in an antiparallel manner ('double strands') and fold in a helical manner ('β-helix') to form two β-sheets – in the case of 2OG oxygenases, a major β-sheet (β-strands I, VIII, III and VI) and a minor β-sheet (β-strands II, VII, IV and V). The β-sheets form a squashed barrel type structure, at the more open end of which is located the entrance to the active site leading to the 2OG and metal binding pockets. The 'barrel' nature of the DSBH fold, coupled to the location of the active site at one end, leads to the prediction that DSBH β-I and β-VIII will (normally) be at the more open end of the barrel. Together with β-II and the loops linking β-I/β-II and β-II/β-III, it is predicted that β-I and β-VIII will be in close proximity to the primary substrate binding site. Indeed, these elements are often involved in substrate recognition (Figure 2.3).[32] For all 2OG oxygenases the core DSBH fold is augmented by additional secondary structural elements and, in many cases, additional domains (see below). As shown by the topology comparisons, specific modifications to the DSBH are characteristic of different subfamilies of 2OG oxygenases (Figures 2.3 and 2.4).

The major β-sheet of 2OG oxygenases is usually extended by additional parallel or antiparallel strands from the *N*-terminus, while the minor β-sheet is less frequently extended and normally not by more than one additional β-strand parallel or antiparallel to β-V.[13,32] As in other modifications to the DSBH core fold, the additional β-strands are subfamily characteristic, *e.g.* in the IPNS/ACCO/DAOCS structural subfamily, β-I of the DSBH is extended by two β-strands in an antiparallel fashion, and another two β-strands extend from β-VI. In the CAS/TauD structural subfamily, the extensions to β-VI, but not β-1, are observed (Figures 2.3 and 2.4).[13,32]

In the 2OG oxygenases, DSBH β-II does not always have a well-defined β-strand secondary structure, even though the ϕ/ψ angles (of β-II) are within the β-region of the Ramachandran plot. In a few identified cases, such as in phytanoyl-CoA hydroxylase (PAHX), the loop linking β-I/β-II (in part) forms a helical structure.[47] The opening of the active site is positioned at the end of the DSBH closest to the iron binding residues which are located on β-II (the

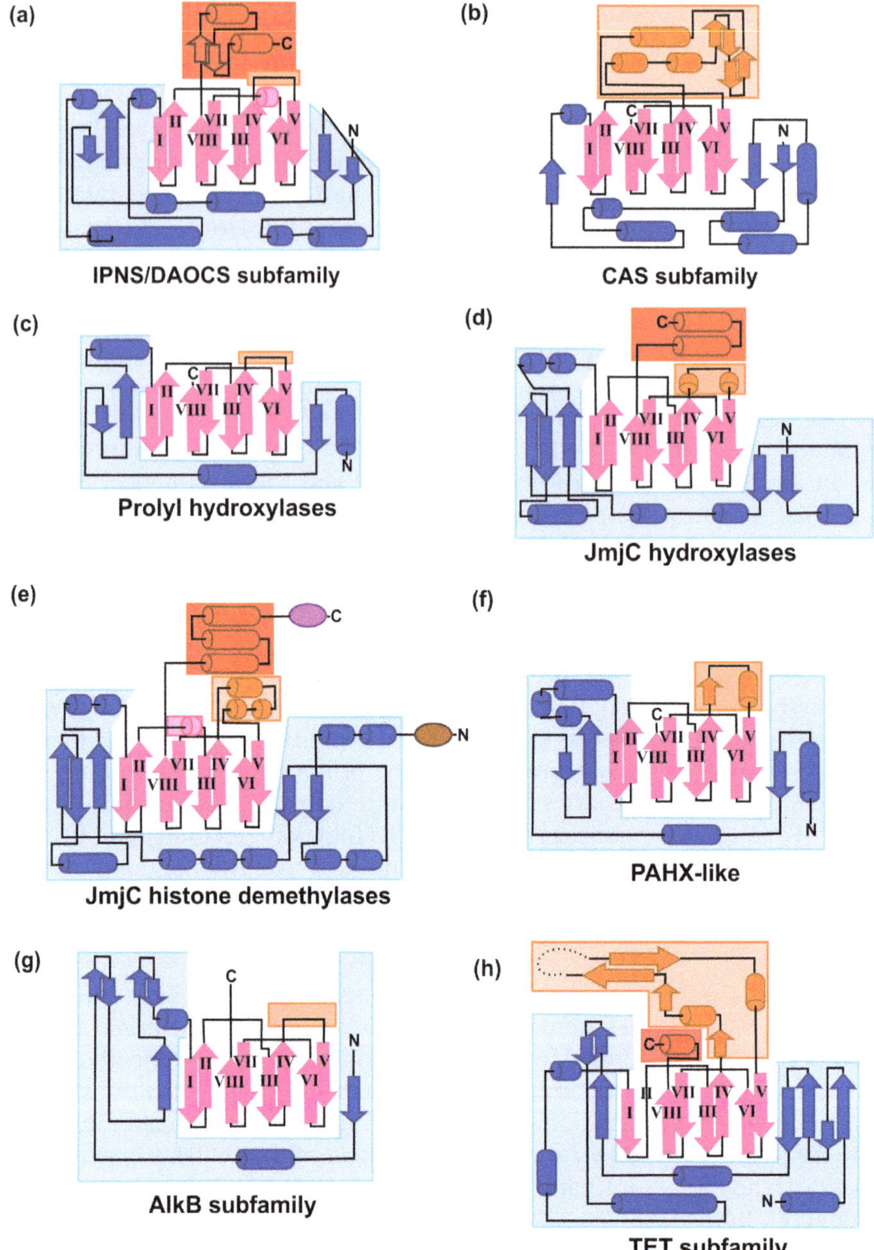

Figure 2.3 Representative topologies from different subfamilies of 2OG oxygenases and related enzymes. All subfamily members contain a distorted double-stranded β-helix core (DSBH) fold (pink strands) comprised of eight strands (β-I–β-VIII) forming two β-sheets within which a well-protected 2OG and iron binding pocket resides. At least two *N*-terminal helices flank the larger β-sheet providing structural support. Varying

HXD of the iron binding triad) and on β-VII (the *C*-terminal/distal His of the HXD/E...H triad).

Insertions in the loops between the β-strands of the DSBH occur most frequently between strands IV and V, probably due to the folding mechanisms of the DSBH fold (Figure 2.3).[32,44] In some cases, *e.g.* DAOCS,[4] IPNS,[6] anthocyanidin synthase (ANS)[48] and the hypoxia inducible factor (HIF) prolyl hydroxylase isoform 2 (PHD2/EGLN1),[49] the β-IV/β-V lining sequences form a tight turn/loop of <10 residues.[25] In others, the insert is much more substantial and can be involved in substrate binding as observed for some small-molecule hydroxylases/oxygenases (*e.g.* CAS,[31] TauD,[19] asparagine hydroxylase (AsnO)[50]) and in JmjC-domain containing hydroxylases (*e.g.* the Asn hydroxylase known as factor inhibiting HIF or FIH)[51–53] (Figure 2.3). In addition to substrate binding, the β-IV/β-V insert can be involved in dimerization as observed for NO66 (Figure 2.5).[19] An apparently extreme example of a β-IV/β-V insert occurs in the ten-eleven translocation oxygenases (TETs) where the insert can be as long as >300 residues, probably forming a separate domain (Figures 2.3 and 2.6).[54–56]

In some cases, *e.g.* in selected JmjC-methyllysyl demethylases (KDMs), the core DSBH/JmjC-fold is one of the multiple domains, although this seems rare outside of the 2OG oxygenases involved in the regulation of protein biosynthesis.[10,57] The additional domains can be involved in dimerization as observed for a domain *C*-terminal to the DSBH in FIH[51–53,58] and the

---

secondary structure elements surrounding the DSBH are often flexible and are involved in primary substrate reconition. (a) The IPNS/DAOCS subfamily of small molecule oxygenases typically contains an extended *C*-terminal region (red) with mixed strand and helix secondary structure. (b) The CAS subfamily has a large β-IV–β-V insert (orange) with mixed strand and helix secondary structure which is involved in substrate recognition. (c) The prolyl hydroxylase subfamily uses both the *N*-terminal region (β-2/β-3 finger loop) and the *C*-terminus in substrate recognition. (d) The JmjC hydroxylases, with the exception of JMJD6, are different from the demethylases in that they have a relatively simpler β-IV/β-V insert, i.e. mostly loop. Many, but not all, members of this subfamily are known to dimerize through a *C*-terminal helical region. (e) The JmjC demethylase subfamily can be multidomain enzymes and vary widely among each other; however, they are unified by a common *N*-terminal region not identified in other oxygenases. (f) The PAHX-like subfamily is relatively compact with flexible *N*-terminal (blue) and β-IV–β-V (orange) regions positioned close to the active site opening. (g) The AlkB subfamily members all share a 'nucleotide recognition lid' (NRL) motif located in the *N*-terminal region (blue) near the active site opening. (h) The TET subfamily of 5mC oxidases is the most complex group of oxygenases, in terms of structure, observed to date. Human TETs are often multidomain enzymes and have an unusually large β-IV–β-V insert. The insert is dramatically shortened in the *Naeglaria* TET. A unique feature observed in TETs is an extended major β-sheet comprised of nine or more strands. The abbreviations are defined in the text.

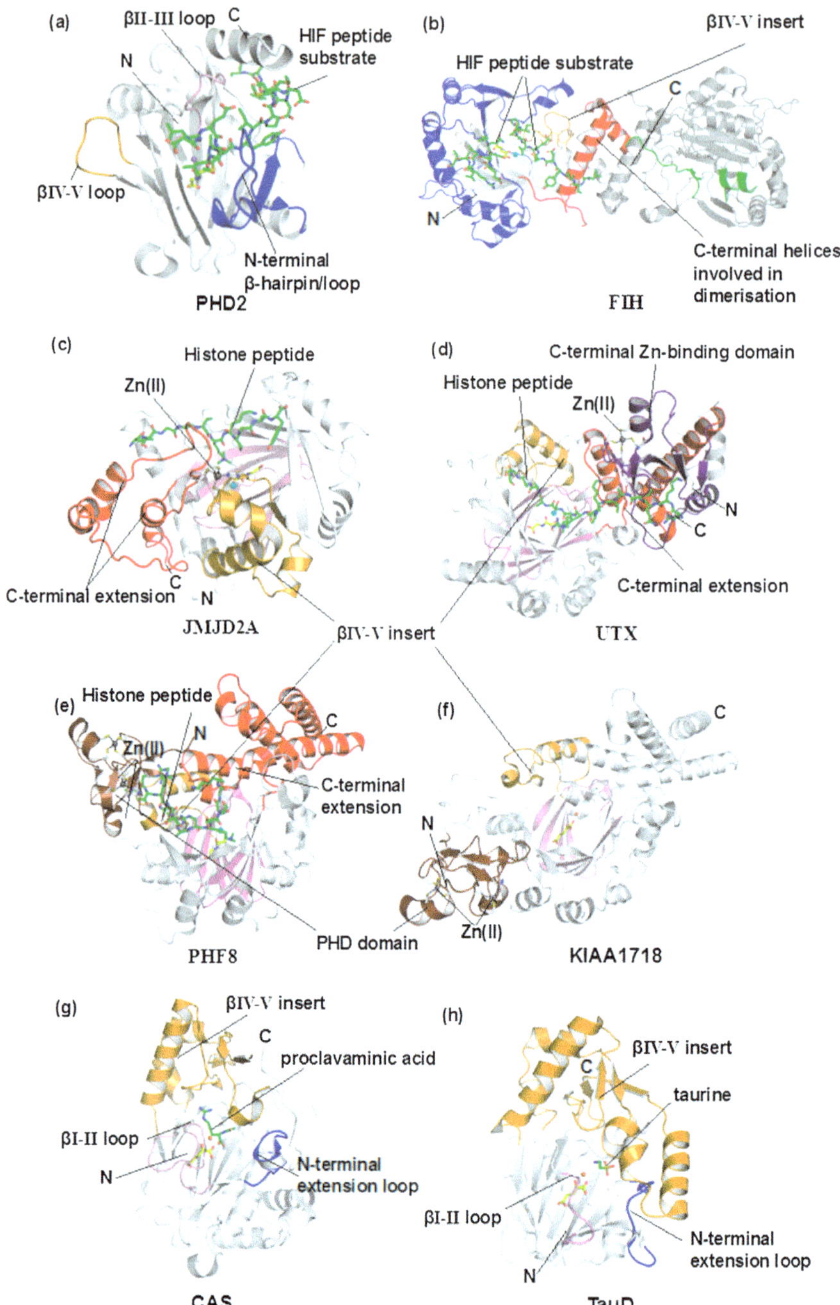

Figure 2.4 Representative crystal structures of 2OG oxygenases. Views of crystal structures for (a) HIF prolyl hydroxylase 2, PHD2 (PDB ID: 3HQR), (b) HIF asparaginyl hydroxylase, FIH (PDB ID: 1H2L), (c) histone demethylase, JMJD2A (KDM4A):H3K36me3 peptide complex (PDB ID: 2OS2), (d) histone demethylase, UTX:H3K27me3 peptide complex (PDB ID: 3AVR), (e) histone demethylase, PHF8:H3K4me3K9me2 peptide complex (PDB ID: 3KV4), (f) histone demethylase, KDM7A (KIAA1718, PDB ID: 3KV6), (g) CAS (PDB ID: 1DRT) and (h) TauD (PDB ID: 1OS7). Specific regions are highlighted using the colour scheme as defined in Figure 2.2. 2OG/*N*-oxalylglycine (NOG), yellow sticks; substrate, green sticks; Fe, orange sphere; Ni, cyan sphere; Mn, purple sphere. The abbreviations are defined in the text.

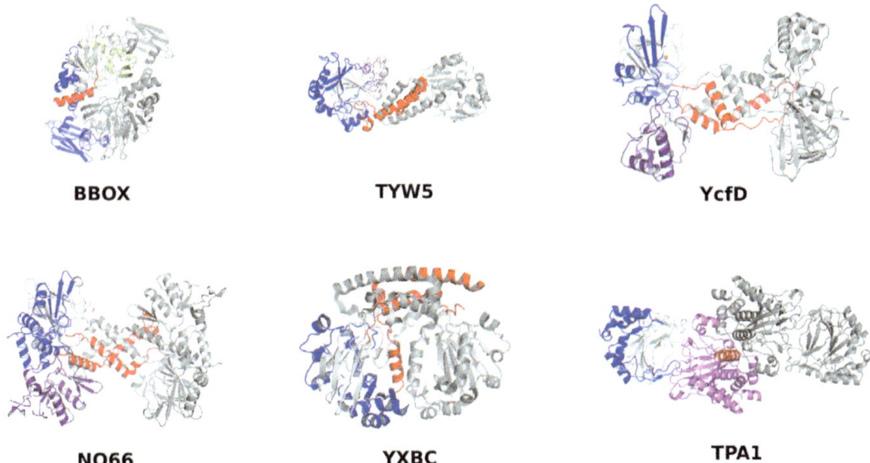

Figure 2.5 Modes of dimerization in representative 2OG oxygenase structures. Within each dimer, one monomer is shown entirely in grey. The other monomer is coloured as follows: DSBH core in white, *N*-terminal region in blue, *C*-terminal dimer interface helices in red, strand 4–5 insert where prominent in yellow, other domains in violet. BBOX (PDB ID: 3MS5); TYW5 (PDB ID: 3AL5); YcfD (PDB ID: 4MUB); NO66 (PDB ID: 4DIQ); YXBC (PDB ID: 1VRB); TPA1 (PDB ID: 3MGU); with abbreviations defined in the text.

related ribosomal oxygenases (ROX) (Figure 2.5),[59,60] or they may enable substrate recognition as observed for plant homeobox domain/PHD/tudor domains present in some KDMs[26,57] and for the 'winged-helix'[61] domain in ROXs.[59] Alternatively, the extra domains may have other catalytic activities, *e.g.* pro-collagen lysyl hydroxylases are linked to a glycosyl transferase[62] and human AlkB homologue 8 (ALKBH8)[63] contains an RNA recognition motif and methyltransferase domains flanking its DSBH fold. The additional domains can contain zinc-binding motifs, *e.g.* as in the dimerization domain of γ-butyrobetaine hydroxylase (BBOX)[64] and the selectivity influencing domain of the KDM plant homeodomain finger protein/PHF8 (located *N*-terminal to the DSBH)[57,65] and the KDM JMJD3 (*C*-terminal to the DSBH) (Figure 2.6).[66]

The regions immediately to the *N*-terminus of the DSBH of 2OG oxygenases apparently play roles in stabilizing the core DSBH fold (this is not necessarily the case for other DSBH fold proteins). Thus, in addition to the β-strands extended from the DSBH present in the major β-sheet, helices α2 and α3 from the *N*-terminal region pack against the outer face of the major β-sheet as observed in multiple crystal structures of 2OG oxygenases including the prolyl hydroxylases and nucleic acid demethylases (NADMs) and as predicted by sequence alignments (Figures 2.3 and 2.4).[25,32,67]

The regions immediately to the *C*-terminus of the DSBH normally have more variation than those at the *N*-terminus, *i.e.* highly conserved secondary structural elements (compared to helices α2 and α3 at the *N*-terminal region)

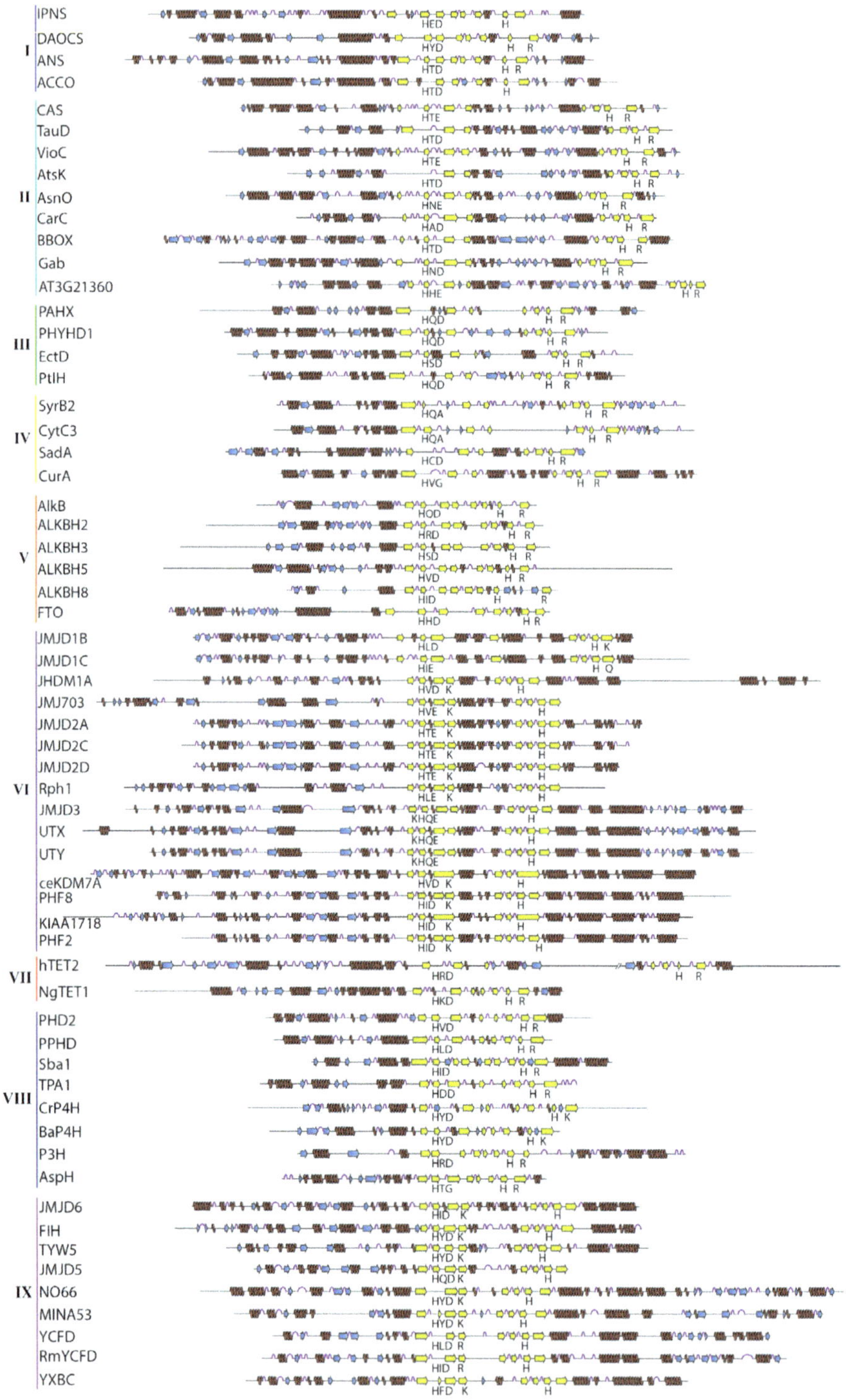

I
IPNS
DAOCS
ANS
ACCO

II
CAS
TauD
VioC
AtsK
AsnO
CarC
BBOX
Gab
AT3G21360

III
PAHX
PHYHD1
EctD
PtlH

IV
SyrB2
CytC3
SadA
CurA

V
AlkB
ALKBH2
ALKBH3
ALKBH5
ALKBH8
FTO

VI
JMJD1B
JMJD1C
JHDM1A
JMJ703
JMJD2A
JMJD2C
JMJD2D
Rph1
JMJD3
UTX
UTY
ceKDM7A
PHF8
KIAA1718
PHF2

VII
hTET2
NgTET1

VIII
PHD2
PPHD
Sba1
TPA1
CrP4H
BaP4H
P3H
AspH

IX
JMJD6
FIH
TYW5
JMJD5
NO66
MINA53
YCFD
RmYCFD
YXBC

(Figure 2.3).[25,32,67] The *C*-terminal regions often contain α-helical elements, which have roles in substrate binding (*e.g.* as in DAOCS,[22] IPNS[7] or PHD2)[68] and/or dimerization (*e.g.* the helical bundles in FIH[51–53,58] and ROXs[59,69,70]). In some cases (*e.g.* DAOCS[22] and PHD2[68]) there is evidence that their *C*-termini are mobile and can undergo conformational changes during substrate binding.[22,68]

2.2 Iron Binding Sites in 2OG Oxygenases and Related Enzymes

Following from the early structures of IPNS[6] and 2OG oxygenases acting on small molecules (DAOCS,[4] TauD[19] and CAS[5]), multiple other 2OG oxygenase structures have revealed a highly conserved stereotypical HXD/E…H iron binding motif (Figure 2.7), which has been described as a 'facial triad' of iron ligating residues.[10,13,30] Very similar binding motifs are observed in other transition metal binding enzymes, including both mono- and di-iron oxygenases from non-DSBH fold structural families and hydrolytic enzymes (*e.g.* metallo β-lactamases).[71,72]

Crystal structures of 2OG oxygenases in the absence of substrate/2OG reveal that the 'open' Fe(II) coordination sites are normally occupied by two to three water molecules, *e.g.* as observed in DAOCS (1RXF),[4] TauD (1OTJ)[73]

Figure 2.6 2OG oxygenases and related enzymes with known structures are shown as a linear representation of their secondary structure. Structurally (and often functionally) related proteins are grouped and marked by coloured lines. I–IV show small molecule oxygenases, V shows the ALKBH family, VI shows the KDMs, VII shows the TET enzymes, VIII shows the prolyl hydroxylases and IX shows the JmjC-only oxygenases. Note that some of the proteins are of unassigned function. Protein name, PDB ID, Uniprot ID: IPNS 1BK0 P05326; DAOCS 1UOF P18548; ANS 1GP6 Q96323; ACCO 1W9Y Q08506; CAS 1DRT Q05581; TauD 1OS7 P37610; VioC 2WBO Q6WZB0; AtsK 1OIK Q9WWU5; AsnO 2OG7 Q9Z4Z5; CarC 1NX8 Q9XB59; BBOX 3O2G O75936; Gab 1JR7 P76621; AT3G21360 1Y0Z Q9LIG0; PAHX 2A1X O14832; PHYHD1 3OBZ Q5SRE7; EctD 3EMR Q2TDY4; PtlH 2RDN Q82IZ1; SyrB2 2FCT Q9RBY6; CytC3 3GJB D0VX22; SadA 3W21 Q0B2N4; CurA 3NNF Q6DNF2; AlkB 3BIE P05050; ALKBH2 3BUC Q6NS38; ALKBH3 2IUW Q96Q83; ALKBH5 4NJ4 Q6P6C2; ALKBH8 3THT Q96BT7; FTO 3LFM Q9C0B1; JMJD1B 4C8D Q7LBC6; JMJD1C 2YPD Q15652; JHDM1A 2YU1 Q9Y2K7; JMJ703 4IGQ Q53WJ1; JMJD2A 2OQ6 O75164; JMJD2C 2XML Q9H3R0; JMJD2D 4HON Q6B0I6; Rph1 3OPT P39956; JMJD3 4EZH Q5NCY0; UTX 3AVR O15550; UTY 3ZLI O14607; ceKDM7A 3N9Q Q9GYI0; PHF8 3KV4 Q9UPP1; KIAA1718 3KV6 Q6ZMT4; PHF2 3PU8 O75151; hTET2 4NM6 Q6N021; NgTET1 4LT5 D2W6T1; PHD2 3HQR Q9GZT9; PPHD 4IW3 Q88CM1; Sba1 3DKQ A3D8P6; TPA1 3MGU P40032; CrP4H 3GZE A8J7D3; BaP4H 3ITQ Q81LZ8; P3H 1E5S O09345; AspH 3RCQ Q12797; JMJD6 3K2O Q6NYC1; FIH 1H2K Q9NWT6; TYW5 3AL5 A2RUC4; JMJD5 4GJZ Q8N371; NO66 4DIQ Q9H6W3; MINA53 2XDV Q8IUF8; YCFD 4CCL P27431; RmYCFD 4CUG D0MK34; YXBC 1VRB P46327.

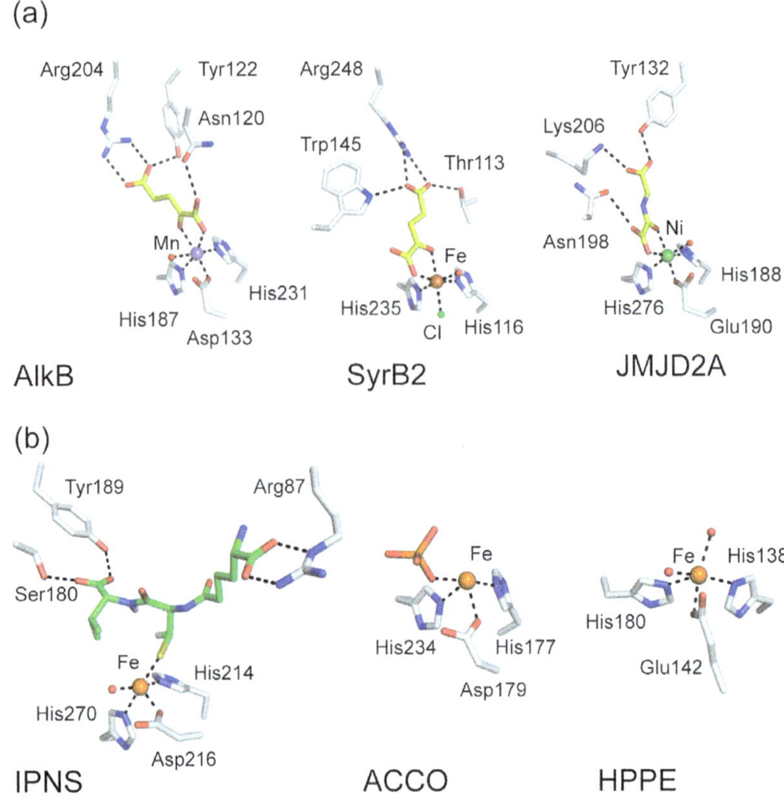

Figure 2.7 Views from crystal structures showing the active sites of 2OG oxygenases and related enzymes. The metal used is specified in the figure. (a) 2OG binding site of AlkB (PDB ID: 3BIE), SyrB2 halogenase (PDB ID: 2FCT), JmjC histone demethylase JMJD2A (PDB ID: 2OQ6). (b) Non-haem and non-2OG, iron-dependent oxidases: *Aspergillus nidulans* IPNS: delta-(L-alpha-aminoadipoyl)-L-cysteinyl-D-valine complex (PDB ID: 1BK0); ACCO: phosphate complex (PDB ID: 1WA6); HPPE (PDB ID: 1ZZ9).

and AlkB homologue 5 (ALKBH5, 4NJ4).[74] There are relatively fewer structures of 2OG oxygenases in the absence of an active site metal,[13,25,32] probably reflecting the stabilizing role of an active site metal on the fold as shown by NMR studies on AlkB.[75,76] Changes in the side chain conformations of the HXD/E…H residues are observed in the absence of metals. Many structures have also been obtained in the presence of 2OG or inhibitors (mostly 2OG analogues, most commonly *N*-oxalylglycine, NOG) which probably stabilise the fold for crystallization.[13,25,32,41,49]

In the 2OG oxygenases, the Fe(II) binding HXD/E…H motif is located at the more 'open end' of the DSBH core.[4–6,32] The Fe(II) binding sites of the DSBH and oxygen-dependent enzymes that have a very close relationship to the 2OG oxygenases, but which do not employ a 2OG cosubstrate (*i.e.* IPNS,[6]

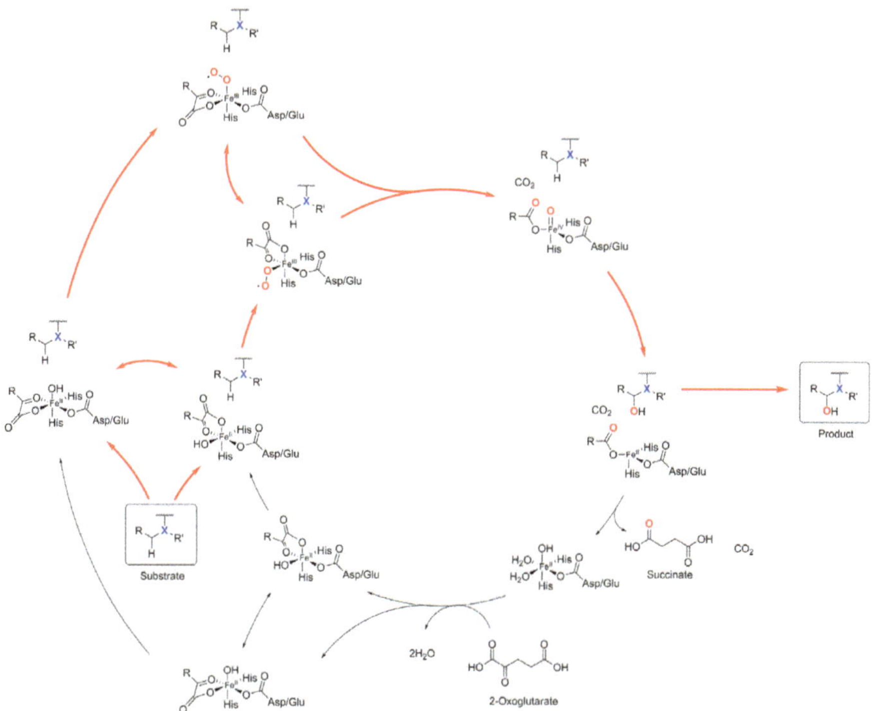

Figure 2.8 Outline mechanism of 2OG oxygenases and related enzymes. The enzyme-bound Fe(II) is coordinated by 2OG prior to binding of the 'primary substrate' in the active site. Molecular oxygen then coordinates the iron (often displacing a water molecule) and induces oxidative decarboxylation of 2OG, forming succinate, carbon dioxide and a highly reactive Fe(IV)–oxo intermediate. This intermediate is able to abstract a hydrogen atom from the primary substrate followed, in most cases, by rapid hydroxylation. When the position of hydroxylation is adjacent to a heteroatom (X = N, O), demethylation occurs. It is likely that the oxidized product dissociates from the active site prior to succinate release (see Chapter 3).

ACCO[15] and HPPE),[77] are similar to those of the 2OG oxygenases and, in the cases of IPNS[6] and ACCO,[31] are indistinguishable from closely related 2OG oxygenases (*e.g.* DAOCS).[4] HPPE is unusual in that it has an HXXXE metal binding motif (instead of HXE) with the first His on the β-II/III loop (instead of β-II), the Glu on β-III (instead of β-II), and the distal His on β-VII.[77] The relatively buried nature of the Fe(II) binding site is consistent with the consensus kinetic mechanism of the 2OG oxygenases (Figure 2.8) in which Fe(II) binding is followed by that of 2OG, primary substrate, and finally oxygen (see Chapters 1 and 3).

The presence of an Asp rather than a Glu residue is much more common in the HXD/E motif (~80% of reported structures contain an Asp residue). Interestingly, there is some evidence that Glu residues may be more common in 2OG oxygenases that catalyse more than one, or unusual, oxidation reactions. Examples include CAS, which catalyses three different types of oxidative reactions,[5,31] and some KDMs such as the KDM4/JMJD2 subfamily which act on all three N^{ε}-methyl lysine methylation states and on at least two histone lysine residues (*i.e.* histone H3 K9 and K36).[26,27] These findings suggest that the probably increased conformational flexibility of Glu compared to Asp coordination could be of catalytic relevance. However, at least some apparently unusual 2OG oxygenases, *e.g.* carbapenem synthase (CarC),[78,79] contain an Fe(II) ligating Asp, and some apparently more normal representatives (*e.g.* AsnO)[50] contain a Glu, hence at present no clear correlation can be made between the presence of Asp or a Glu iron-ligating residue.

One apparent role of the DSBH is to provide a relatively rigid scaffold to support binding of the Fe(II) cofactor (Figure 2.7). However, evidence is emerging from inhibitor studies that perhaps the Fe(II) may not be as 'locked' as most crystallographic analyses would suggest. For example, binding of 5-carboxy-8-hydroxyquinoline (IOX1) to the JmjC oxygenases FIH, KDM4A and KDM6B induces movement of the metal by >1.5 Å.[66,80] Solution studies indicate that most 2OG oxygenases do not bind Fe(II) tightly (in particular compared to metals in porphyrin rings). There is thus potential for 2OG oxygenases to act as Fe(II) sensors, and for other metals to compete with Fe(II) binding *in vivo*; the mechanism by which Co(II) ions cause upregulation of erythropoiesis is proposed to be *via* competition with Fe(II) for binding to the HIF prolyl hydroxylases (see Chapter 6).

The available structures of 2OG oxygenases reveal that the conformations of the two His of the HXD/E…H triad are very similar, with coordination occurring *via* the N^{ε}-nitrogen of the imidazole rings, which are probably deprotonated at this position (Figure 2.7). However, the Fe(II) ligating acid (Asp/Glu) coordinates Fe(II) with one carboxylate oxygen and in most oxygenase structures (with a single exception, *i.e.* PHF2, which has a different active site chemistry),[81] is positioned to hydrogen bond to the metal-bound water using its other carboxylate oxygen. It is likely that the position of the metal-bound water is connected to variations in the observed conformations of the metal-ligating Asp/Glu residue. The conformations adopted by the Fe(II) ligated Asp in many representatives (*e.g.* PAHX,[47] DAOCS,[4] prolyl hydroxylases[68,82] and most 2OG oxygenases acting on nucleic acids[32,83]) are very similar but different from those observed in other examples (*e.g.* the JmjC-hydroxylases including FIH,[51] TYW5,[84] JMJD5,[85] JMJD6[86] and the ROXs[59,69,70]). In the cases of FIH and the ROXs, the iron coordinating Asp is unusual in that it not only coordinates the iron using one of its carboxylate oxygens but also uses the other carboxylate oxygen to form a hydrogen bond to a backbone amide of their protein substrates (in addition to forming hydrogen bonds to the metal-ligated water molecule).

Crystal structures of the 2OG-dependent halogenases SyrB2[87] and CytC3,[88] together with analyses on FIH,[89] have revealed that in some cases only two protein-based residues are required for Fe(II) binding and catalysis (Figure 2.7). SyrB2 and CytC3 catalyse halogenations (and other reactions) of secondary metabolites. In the proposed halogenation mechanism the lack of the Asp/Glu (there is an His-X-Ala motif in both SyrB2 and CytC3) compared to normal 2OG oxygenases enables the halide ion to directly ligate to the iron (in *cis*-geometry, see Chapter 18); oxidative halogenation is mediated *via* ferryl formation in the consensus manner (Chapter 3), followed by hydrogen abstraction from the substrate then formation of the carbon halogen bond with concomitant reduction of the Fe(III) intermediate.[90–92] In the case of the promiscuous protein hydroxylase FIH, which has the typical HXD…H motif, it was found that substitution of the distal His for a Gly, but not an Ala, residue led to a viable enzyme with only two protein-based ligands, as demonstrated by crystallographic analyses.[89] This observation implies caution should be taken in assuming that substitutions to the HXD/E…H motif necessarily produce inactive oxygenases, as is sometimes the case in the cell biology literature. Sequence analyses combined with structural studies indicate that there are exceptions to the HXD/E…H motif in some 2OG oxygenases, especially with respect to the Asp/Glu. The 2OG oxygenase that catalyses hydroxylation of Asp and Asn residues in epidermal growth factor domains (AspH)[93] is predicted to be missing the Asp,[94] as verified by recent crystallographic analyses (PDB: 3RCQ). Linked to the unusual reactions catalysed by 2OG oxygenase-related enzymes (*e.g.* carbon–sulphur bond formation as catalysed by IPNS),[6] the findings that 2OG oxygenases can operate with only two protein ligands and can catalyse halogenations suggest that naturally occurring superfamily members may catalyse unanticipated reactions or that protein engineering of the family may be productive.

Some proteins appear to have evolved from catalytically active 2OG oxygenases, but they probably do not contain 'normal' iron binding sites. There is a 2OG oxygenase conserved from yeast (TPA1) through to humans (OGFOD1) that catalyses prolyl C-3 hydroxylation of a ribosomal protein (RPS23).[95–97] Unusually for a 2OG oxygenase, OGFOD1 contains two DSBH-containing domains, only one of which (the *N*-terminal domain) has catalytic activity; its *C*-terminal DSBH domain is proposed to have arisen by gene duplication from the catalytic domain (Figure 2.5). Whilst the non-catalytic domain contains some of the residues involved in iron/2OG binding in the catalytic domain, crystallographic analyses of TPA1 reveal it does not bind iron or 2OG.[96–98] Several JmjC 2OG oxygenases also contain unusual iron binding sites. For example, PHF2 is of interest within the JmjC KDM subfamily because the distal His residue is replaced by a Tyr, which coordinates the iron *via* its phenolic oxygen. PHF2 : Fe(II) has been shown to bind 2OG and an N^{ε}-trimethylated H3 K4 substrate, but has been reported not to have catalytic activity in isolated recombinant form.[81]

2.3 2OG Binding by 2OG Oxygenases

Multiple structures of 2OG oxygenases complexed with 2OG (or close 2OG analogues such as NOG, succinate, fumarate or 2-hydroxyglutarate), and a metal (Fe or a surrogate) have been reported (Figure 2.7).[13,25,30,40,41] NOG is a near isosteric analogue of 2OG, but it is not reactive with respect to catalysis. Although the use of alternative metal ions or 2OG surrogates has the potential to enable variations in coordination chemistry, there is agreement on the general mode of crystallographically observed 2OG binding, which is supported by spectroscopic analyses (see Chapter 3). The 2OG binding site is directly adjacent to that of Fe(II) and is buried within the major and minor β-sheets of the DSBH.[13,25,30,32] The degree of accessibility of the 2OG binding pocket varies significantly between different 2OG oxygenases. In most cases substrate binding substantially occludes access to the 2OG binding pocket, helping to enable the ordered sequential mechanism in which substrate binding succeeds that of 2OG and in which succinate release occurs after that of product (Figures 2.7 and 2.8). However in some cases (including some KDMs where sequential reactions are catalysed), it may be that 2OG binding and succinate release can occur without (complete) dissociation of the product from the enzyme. NMR studies on AlkB reveal that iron and 2OG binding substantially stabilizes its structure,[75,76] although the extent to which this is a general phenomenon is unclear. There is also evidence that, at least in some cases, 2OG binding preorganises the structure for primary substrate binding.

The oxalyl group of 2OG chelates the Fe(II) in a bidentate manner *via* its ketone oxygen and one of its C-1 carboxylate oxygens to form an octahedral complex. 2OG binding displaces two water molecules from the iron as shown by crystallographic and solution studies (see Chapter 3).[13,30,32] In reported enzyme : 2OG, or enzyme : 2OG : substrate complex structures, the sixth coordination position is in general occupied by a water molecule. Substrate binding weakens the coordination of this water, enabling reaction with O_2 at the same site (Figure 2.8).[13,32] The non-coordinating oxygen of the 2OG C-1 carboxylate is often positioned to interact electrostatically with an alcohol or a basic/polar residue close to the iron binding site, *e.g.* Arg-270 in TauD,[51] Arg-252 in PHD2 (3HQR),[68] Arg-297 in CAS,[5] arginines in most NADMs,[83,99,100] Thr-255 in Myc-induced nuclear antigen mass 53 kDa (MINA53), Thr-419 in nucleolar protein 66 (NO66)[59] or Asn-120 in AlkB.[35] It is possible that this interaction (particularly that involving Arg) promotes the release of carbon dioxide from the active site which, at least in the case of TauD, occurs after reaction of iron-bound 2OG and oxygen, but prior to substrate hydroxylation (see Chapter 3).[19,101]

In reported structures of 2OG-bound complexes, the position of the 2-oxo ketone oxygen of 2OG relative to the protein ligands is (near) invariant, being located *trans*, or approximately *trans*, to the carboxylate of the HXD/E…H motif (Figure 2.7).[13,32] However, the position of the 2OG C-1 carboxylate and metal ligating water has been observed to be either *trans* to the proximal (*i.e.* the more *N*-terminal) or *trans* to the distal (*i.e.* the more *C*-terminal) His of the HXD/E…H motif (Figure 2.8).[13,25,102] Therefore, the position at which O_2 binds after loss of the metal-bound water may vary.

In the case of CAS and KDM7/PHF8, addition of substrate was observed to rearrange the iron-ligated water molecule in the enzyme : 2OG complex,[5,31,57] consistent with spectroscopic analyses on CAS that revealed substrate binding primed the iron site for oxygen binding.[103] The variation in the coordination position of the 2OG C1-carboxylate is probably of mechanistic significance.[102] When the 2OG C1-carboxylate is *trans* to the proximal His, the water-occupied coordination site *trans* to the distal His is adjacent to or projects towards the substrate C–H bond that is oxidized. Coordination of dioxygen to this coordination site should, after reaction with 2OG, generate a ferryl intermediate[92,104] in an appropriate position to affect substrate oxidation. Observation of the 2OG C1-carboxylate in the position *trans* to the distal His suggests that it would have to rearrange to the position *trans* to the proximal His in order for productive dioxygen binding to occur. In some cases it is proposed that this rearrangement, together with removal of a water molecule, may be rate limiting in dioxygen binding (see Chapter 6).

Alternatively, in some cases it is possible that initial dioxygen coordination to Fe(II) occurs at the coordination site projecting away from the substrate with the 2OG C-1 carboxylate in the position *trans* to the distal His. Support for this possibility comes from studies on the binding of nitric oxide, acting as a dioxygen analogue, to the CAS : Fe : 2OG : substrate complex;[31] the nitric oxide was observed to coordinate to the position *trans* to the proximal His and the 2OG C-1 carboxylate was observed to rearrange from *trans* to the proximal His to *trans* to the distal His. It is possible that dioxygen can bind in either or both available coordination sites depending upon the enzyme/substrate combination. However, binding of dioxygen as observed for nitric oxide in the crystallographic work on CAS has the potential advantage that the reactive oxidizing species formed will be buried within the active site and so more isolated from the environment. If dioxygen does coordinate to iron in this manner and the ferryl intermediate is formed at this position it may need to rearrange (flip) subsequent to or concomitant with loss of carbon dioxide from the iron to be in the correct position to affect substrate oxidation.

The methylene carbon atoms of 2OG/NOG adopt extended conformations in most, but not all, reported structures (Figure 2.7). This binding mode is supported by structures with conformationally constrained inhibitor analogues of 2OG, such as pyridine-2,4-dicarboxylate, fumarate and other inhibitors (Section 2.5).[40,41] Structures have also been reported for some 2OG oxygenases with the coproduct succinate, which usually adopts the same general binding mode as 2OG, with one or two of its carboxylate oxygens positioned to coordinate the metal.[35,50,105] In the DAOCS : succinate complex crystal structures, an alternative (although still metal coordinating) binding mode has been reported.[22,106,107] In some cases, these may represent modes of binding relating to inhibition by succinate which is observed for many 2OG oxygenases. Cosubstrate inhibition by high concentrations of 2OG is also commonly observed, but as yet there are no structural insights into its mechanism. Inhibition of 2OG oxygenase activity by the tricarboxylic acid intermediate 2-hydroxyglutarate is due to competition with 2OG, and is proposed to be of pathophysiological relevance in some cancers.[105,108–111]

In all reported structures, the 2OG/NOG side chain projects towards the interior of the DSBH 'barrel' and the 2OG C-5 carboxylate is positioned to interact electrostatically with a basic residue (either Arg or Lys) located at the end of the pocket (Figure 2.7). At least one other polar neutral residue, commonly containing an alcohol side chain (Tyr/Thr/Ser) is also involved in binding the 2OG C-5 carboxylate in a subfamily characteristic manner (Figures 2.6 and 2.7).

In many members of the large structural subfamily of 2OG oxygenases typified by DAOCS, ANS, an algal prolyl-4-hydroxylase, AlkB homologue 8 (ALKBH8)[112,113] and the fat mass and obesity protein (FTO),[114–116] the 2OG C-5 carboxylate binding residues comprise an Arg-X-Ser motif (located on β-VIII) that is also present in related oxygenases (*e.g.* IPNS[7] and ACCO[15]) that do not use 2OG as a cosubstrate. In the case of IPNS the Ser of this motif is involved in binding the tripeptide substrate (Chapter 19).[7] In the case of ACCO,[15] the role of the motif is not yet established, but it is proposed to be involved in binding the bicarbonate cofactor (Chapter 20).

The ubiquitous JmjC subfamily of 2OG oxygenases provides interesting examples of variations in the 2OG C5-carboxylate binding. Many of the JmjC subfamily members, including both JmjC-KDMs (*e.g.* the KDM2s,[117] KDM3s, KDM4s,[26] KDM7s[57]) and the JmjC-hydroxylases (*e.g.* FIH,[51] JMJD5,[85] JMJD6[86,118] and TYW5[84]) employ a Lys residue located at the *N*-terminal end of β-IV to bind the C-5 2OG carboxylate, with an exception, the KDM6 (UTX[119]/UTY[120]/JMJD3[66]) subfamily that uses a Lys located on β-I for this purpose (Figures 2.4 and 2.9). Within the JmjC-ROXs, there appears to have been evolution from the use of an Arg located on β-IV in prokaryotic members (*i.e.* YcfD[70]) to a Lys on β-IV as observed in human homologues of YcfD (MINA53 and NO66[69]) and in most JmjC-hydroxylases.[59] Further, although the general orientation of the 2OG binding pocket in the ROXs is maintained, the coordination position from which hydroxylation occurs varies.

2.4 Substrate Binding by 2OG Oxygenases

Despite sharing a similar DSBH core fold and Fe(II)/2OG binding sites, 2OG oxygenases have evolved to recognize and oxidize a remarkably wide range of substrates including both macromolecules (carbohydrates, lipids, nucleic acids and proteins) and multiple types of small molecules such as amino acids/peptides, flavonoids, and many secondary metabolites (see Chapter 1).[13,30,121] The 2OG oxygenase superfamily members appear to have evolved into distinct subfamilies with characteristic substrate recognition elements. Multiple crystal structures (Figure 2.9), coupled with mutagenesis studies, have informed on how 2OG oxygenases bind their substrates.

In all cases, the entrance to the active site is positioned at the more open end of the DSBH, with the iron and 2OG binding sites sandwiched between the two β-sheets of the DSBH, specific elements of which are directly involved in substrate binding (Figure 2.9). A recent review has identified elements of

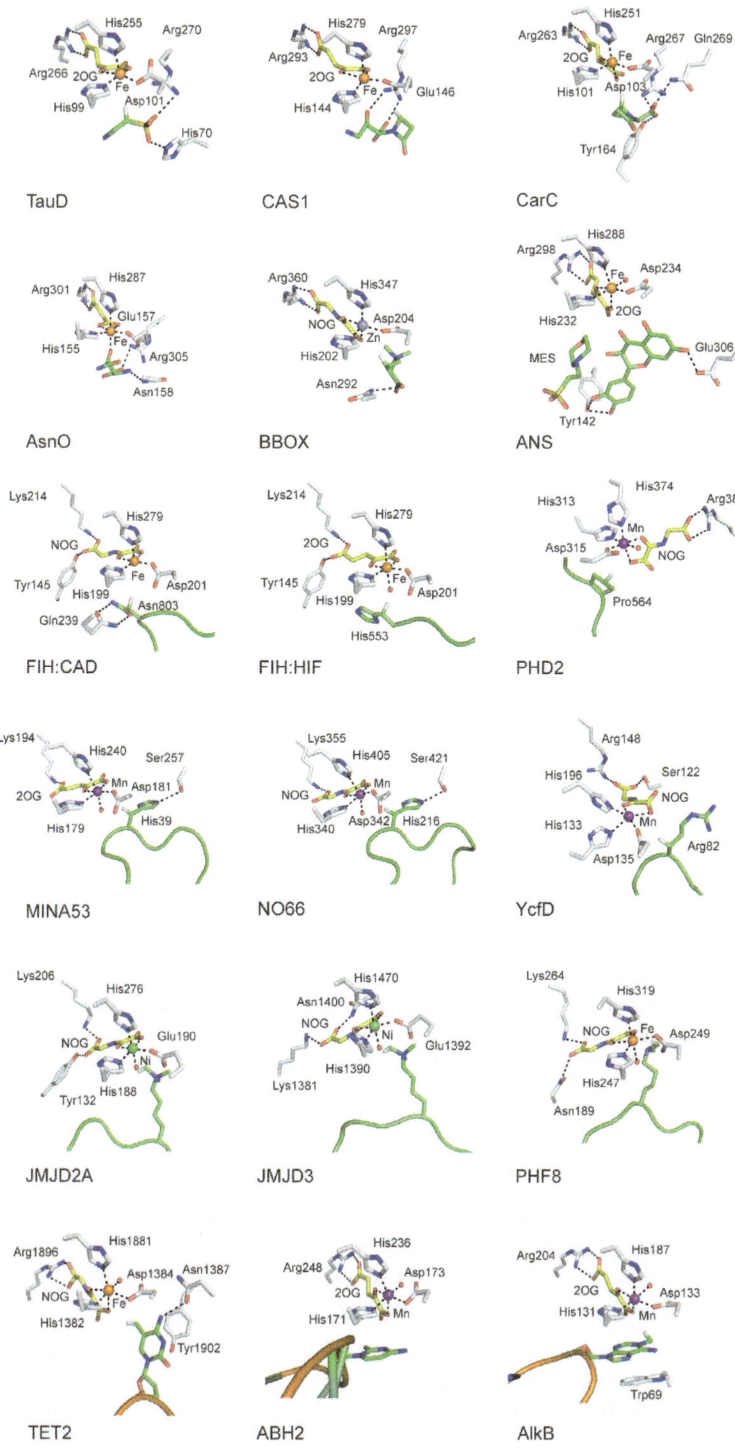

His255
Arg270
Arg266
2OG
Fe
Asp101
His99
His70
TauD
His279
Arg297
Arg293
2OG
Fe
Glu146
His144
CAS1
His251
Arg263
Arg267
Gln269
2OG
Fe
His101
Asp103
Tyr164
CarC
Arg301
His287
Glu157
Fe
His155
Arg305
Asn158
AsnO
Arg360
His347
NOG
Asp204
Zn
His202
Asn292
BBOX
His288
Arg298
Fe
Asp234
2OG
His232
MES
Glu306
Tyr142
ANS
Lys214
His279
NOG
Fe
Tyr145
His199
Asp201
Asn803
Gln239
FIH:CAD
Lys214
His279
2OG
Fe
Tyr145
His199
Asp201
His553
FIH:HIF
His374
His313
Arg383
Mn
Asp315
NOG
Pro564
PHD2
Lys194
His240
Ser257
Mn
Asp181
2OG
His39
His179
MINA53
Lys355
His405
Ser421
Mn
NOG
His340
Asp342
His216
NO66
Arg148
His196
Ser122
NOG
Mn
His133
Arg82
Asp135
YcfD
Lys206
His276
NOG
Glu190
Ni
His188
Tyr132
JMJD2A
His1470
Asn1400
NOG
Ni
Glu1392
His1390
Lys1381
JMJD3
Lys264
His319
NOG
Fe
Asp249
His247
Asn189
PHF8
His1881
Arg1896
Asp1384
Asn1387
NOG
Fe
His1382
Tyr1902
TET2
His236
Arg248
Asp173
2OG
Mn
His171
ABH2
Arg204
His187
2OG
Asp133
His131
Mn
Trp69
AlkB

the DSBH and adjunct regions involved in substrate binding by 2OG oxygenases.[32] The results support general roles for strands β-I, β-II and β-VIII of the DSBH, and the loops linking β-I/II, β-II/III and β-IV/V in substrate binding; the relative importance of these binding elements can be subfamily characteristic. For example in some cases (*e.g.* DAOCS[4] and PHD2[68]), the residues linking β-IV/β-V form a relatively tight turn (see Section 2.1.2), whereas in other cases they comprise a substantial element which can be involved in substrate binding (*e.g.* TauD,[19] CAS[5] and BBOX[64,122]) (Figure 2.9). Some multi-domain 2OG oxygenases use additional domains located *N*- and/or *C*-terminal of the DSBH domain for substrate recognition, as observed with prolyl hydroxylases and some JmjC-KDMs (see below).

The substrates are bound at the active site by a combination of hydrogen bonding, hydrophobic and electrostatic interactions such that the oxidized carbon–hydrogen bond projects towards the metal.[32] In some cases stereoelectronic effects are proposed to weaken the oxidized carbon–hydrogen bond.[13,123] In most cases the substrate atoms which are proposed to undergo reaction with the ferryl intermediate are observed to be adjacent to the active site iron (C-metal distance: normally <4.5 Å for hydroxylases and NADMs,[25,32,83,99,100,124] but can be as high as 5.8 Å for KDMs[27,41]), consistent with substrate oxidation proceeding *via* initial H-atom abstraction followed by rapid hydroxylation (see Chapter 3). ANS (Chapter 15) is an interesting case, not just because it can catalyse both hydroxylations and desaturations (depending on the precise flavonoid substrate)[125,126] but because some structures have revealed, at least in the crystalline state, that its active site can simultaneously accommodate two substrate molecules (Figure 2.9).

Catalytically productive positioning of the substrate with respect to the iron-bound reactive intermediate is important in order to avoid uncoupling

Figure 2.9 Views from crystal structures of 2OG oxygenase:substrate complexes. Note in some cases products are shown. 2OG or NOG and the metal used are specified in the figure. TauD : taurine complex (PDB ID: 1GY9); CAS : proclavaminic acid complex (PDB ID: 1DRT); BBOX1 : γ-butyrobetaine complex (PDB ID: 3O2G); CarC : (3*S*, 5*S*)-carbapenam complex (PDB ID: 4OJ8); AsnO : (2*S*, 3*S*)-3-hydroxyasparagine : succinate complex (PDB ID: 2OG7); ANS : naringenin complex (PDB ID: 2BRT) note only one of two binding naringenin molecules at the active site is shown. FIH : HIF-α peptide complex (PDB ID: 1H2K) and tankyrase-2 peptide complex (PDB ID: 2Y0I); PHD2 : HIF-α CODD peptide complex (PDB ID: 3HQR); *Rhodothermus marinus* YcfD-like hydroxylase : ribosomal protein L16 peptide complex (PDB ID: 4CUG); MINA53 : ribosomal protein L27A peptide complex (PDB ID: 4BXF); NO66 : ribosomal protein L8 peptide complex (PDB ID: 4CCO). JMJD2A : H3K9me3 peptide complex (PDB ID: 2OQ6); PHF8 : H3K9me2 peptide complex (PDB ID: 3KV4); *Mus musculus* JMJD3 : H3K27me3 peptide complex (PDB ID: 4EZH); TET2 : 5′ methyl cytosine DNA fragment complex (PDB ID: 4NM6); ALKBH2 : DNA complex (PDB ID: 3BU0); *E. coli* : *N*′-methyl adenine base substrate complex (PDB ID: 3BIE).

of 2OG and substrate oxidation, achieve selective oxidation of the correct substrate atoms, and avoid self-oxidation of the 2OG oxygenase which can lead to inactivation *via* oxidative modification, often involving fragmentation.[127–129] The control of such reactivity may be anticipated to be particularly challenging, especially in cases where the 2OG oxygenase catalyses multiple reactions. A striking example of the flexibility of 2OG oxygenase catalysis is provided by CAS, which catalyses hydroxylation, oxidative cyclization and desaturation reactions during biosynthesis of the β-lactamase inhibitor clavulanic acid, which at least in a cellular context is a highly efficient process.[9,17,18] Crystallographic studies on CAS have provided insights into how the selectivity is achieved and have served as the basis for further mechanistic studies including by modelling.[5,31] A related enzyme to CAS, CarC, catalyses epimerization and desaturation reactions during biosynthesis of the carbapenem antibiotics.[9,17,18,79] Recent crystallographic studies on CarC have revealed that in the unusual epimerization step the phenolic group of a Tyr residue acts as a hydrogen source following initial abstraction by a ferryl intermediate and stereoinversion.[78]

In an increasing number of cases there is evidence that binding of substrate in the active site region involves an induced fit, as observed for the 2OG oxygenase-related enzyme IPNS[7] (see Chapter 19), the HIF prolyl hydroxylase, PHD2,[68] the *E. coli* alkylation repair protein, AlkB,[35,99] and the TETs.[54,55] The only reported detailed NMR studies for a 2OG oxygenase are on AlkB;[75] these, together with other solution studies (*e.g.* using MS)[130] suggest that the crystallographic studies may underestimate the degree of conformational changes in catalysis by many 2OG oxygenases, in particular those catalysing the oxidation of oligomeric substrates such as proteins and nucleic acids. In some cases evidence for substantial conformational changes in 2OG oxygenases and/or their oligomeric substrates have been accrued crystallographically (*i.e.* in different crystal forms). Constraints of substrate conformation are also important in the oxidation of small molecules to give strained ring products, such as occurs in IPNS-catalysed bicyclic ring-forming reactions during biosynthesis of the penicillin/cephalosporin and clavam type antibiotics.[13]

Many structures of 2OG oxygenases in complex with their substrates reveal significant conformational changes involving non-conserved labile structural elements, both at residue and domain levels. Structures of PHD2,[68] PHF8,[57] prokaryotic YcfD,[59] AsnO[50] and algal prolyl-4-hydroxylase[82] provide evidence for substantial conformational changes during catalysis at domain levels. Notably, PHD2 structures with and without substrate reveal that binding involves movement of a region *N*-terminal to the DSBH, which adopts a partially disordered 'loop-like' conformation as observed in the PHD2 : 2OG/NOG complexes. Concomitant with substrate binding the protein forms a stable loop that, together with other elements including the *C*-terminal region,[68,131] almost entirely encloses the hydroxylated substrate prolyl and adjacent residues. Similar changes have been observed with a collagen-like sequence complexed with a prolyl-4-hydroxylase (3GZE).[82]

For 2OG oxygenases that act on proteins and nucleic acids, the enzyme can alter the conformation of the substrate in order to enable catalytically productive active site binding. Crystallographic analyses on a prolyl hydroxylase from *Pseudomonas putida* (PPHD), which is homologous to the animal PHDs, but which uses elongation factor Tu (EF-Tu) as a substrate, have revealed that the EF-Tu substrate undergoes a major conformational change compared to its uncomplexed state; its Switch 1 loop region (which is crucial for its role in translation) moves by >20 Å, such that the hydroxylated prolyl residue becomes enclosed in the active site; there is analogous use of the β-2/β-3 'loops' of the prolyl hydroxylases in enclosing the substrate.[132] Induced fit is also observed with FIH, both in terms of relatively small movements on substrate binding at the enzyme residue level (involving Trp-296) and in terms of its substrates.[51,133–135] FIH is interesting in that it not only catalyses β-hydroxylation of an Asn residue in the *C*-terminal transcriptional activation domain (CAD) of its HIF-α substrates, but also that it accepts multiple other substrates from the ankyrin repeat domain (ARD) fold family.[133,134,136] Unlike the HIF-α CAD, which is disordered in solution,[137] ARD proteins have very well-characterised structures comprising repeats (typically 4–6, but often more) of a 33-residue motif comprising two α-helices separated by loops.[138] Crystallographic and NMR studies have shown that FIH unwinds the stereotypical ARD fold in order to enable analogous binding to that observed for HIF-α CAD.[139,140] For its ankyrin repeat domain substrates (but not HIF-α), in cells FIH not only catalyses β-hydroxylation of Asn, but also of Asp and His residues.[133–136] With isolated protein the substrate selectivity of FIH extends even further to encompass hydrophobic residues.[135] Overall the work with 2OG oxygenase : protein complexes suggests that, at least in some cases, the interaction should be considered as protein : protein interactions in addition to classical enzyme : substrate interactions.

In catalysis by AlkB and related nucleic acid oxygenses, regions *N*-terminal to the DSBH are involved in binding the nucleic acid phosphate backbone and flipping the methylated base substrate into the active site.[32,35,99] Three residues on an *N*-terminal 'finger' of AlkB force the DNA backbone to flip the *N*-methylated base into the active site.[32,35,99] One human homologue of AlkB, ALKBH2, adopts a different strategy, using a phenylalanine residue to intercalate into the position of the *N*-methylated base (Figures 2.3 and 2.9).[32,99]

In small molecule hydroxylases, substrate binding can also induce conformational changes as a feature of catalysis as observed in BBOX. The multiple BBOX structures with and without substrate imply that a flexible loop (residues 183–199, 'β-I/β-II loop') encloses the enzyme active site upon γ-butyrobetaine (GBB) binding to give an apparently 'closed' conformation that binds the trimethylammonium group of GBB, including by π-cation interactions in an 'aromatic cage' (Figure 2.9).[64,122,141]

In many cases, especially protein hydroxylases and the KDMs, the 2OG oxygenase domains are located within very large multidomain proteins (Figures 2.3 and 2.4).[29,59] There is evidence that at least some of these other domains, which in the case of the KDMs (see Chapter 7) often bind at specific histone marks, are involved in determining substrate selectivity. A nice example is provided by the case of the PHF8/KDM7 subfamily. Like other 2OG-dependent JmjC-KDMs, PHF8 catalyses demethylation of N^{ε}-methyl lysine residues.[57,65,142] A PHD domain located adjacent to the JmjC catalytic domain binds one histone methylation mark (*i.e.* N^{ε}-trimethylated lysine 4 of histone 3, H3K4me3), 'guiding' the JmjC domain to the targeted position for demethylation (H3K9me2) (Figure 2.9).[57,143] Biochemical and structural studies of KIAA1718 (another KDM7),[57] which is related to PHF8 but has a longer linker between its PHD and JmjC domains, have revealed that KIAA1718 is selective for H3K27me2 due to spatial restriction determined by the position of the PHD domain when it is bound to H3K4me3. It is likely that such targeting by non-catalytic modules occurs with other oxygenases. These include most JmjC KDMs,[27,28] ROXs (where winged helix domains are involved in substrate binding),[59] and pro-collagen prolyl hydroxylases[144] and the epidermal growth factor like domain hydroxylase,[93] where tetratricopeptide (TPR) domains are located adjacent to the 2OG oxygenase domains.

The combined structural analyses on 2OG oxygenases reveal specific features that distinguish subfamilies (i) the size of the β-IV/β-V insert, (ii) the positioning of 2OG C-5 carboxylate and the metal-ligated water at the active site, (iii) the extensions to and nature of the major and minor sheets within the DSBH fold, and (iv) the presence of extra-DSBH domains or helical bundles for dimerization and/or substrate binding. Together with the structural conservation of the 2OG binding sites within subfamilies and for closely related isoforms, exploitation of (at least) some of these characteristics may provide useful means for identification of selective inhibitors using structure-based approaches.

2.5 Structural Aspects of Inhibition of 2OG Oxygenases

There is presently considerable interest in the development of selective inhibitors of 2OG oxygenases for therapeutic application, including for the treatment of anaemia, cancer and cardiovascular diseases.[145–148] Most of the inhibitors in clinical and pre-clinical development (Figure 2.10) are active site iron binding or (probably) compete with 2OG; many, but not all, also probably hinder substrate binding. However, some inhibitors binding in the 2OG site, *e.g.* NOG, can promote, or at least not block, substrate binding. This is the case for 2OG oxygenase inhibition by elevated levels of the tricarboxylic acid cycle compounds succinate and fumarate (and the related compound 2-hydroxyglutarate), as occurs in some tumours.[108] Whether or not substrate competition occurs depends on the precise binding mode, as shown by studies on the PHDs, and crystallography may not always serve as a good guide for this.[149] Furthermore, there may be differences in the binding mode, at least with respect to the metal, depending on whether iron or an iron surrogate is used. Interestingly, crystallographic studies suggest that, in some cases, JmjC KDM inhibitors can induce movement of the metal at the active site.[66,80]

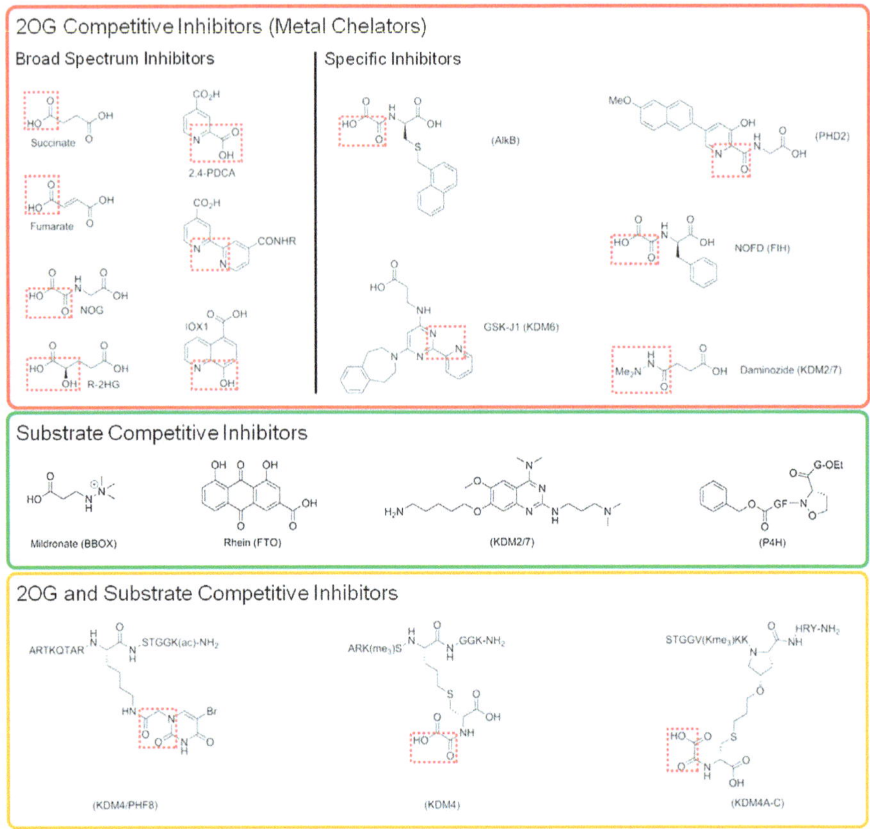

Figure 2.10 Structures of representative inhibitors of 2OG oxygenases. The majority of oxygenase inhibitors identified to date are competitive with 2OG; such examples include TCA cycle intermediates (succinate, fumarate) and bidentate iron chelators (oxalates, pyridinedicarboxylic acids, bipyridines and 8-hydroxyquinolines). Most of these compounds inhibit multiple oxygenases due to the relative similarity of the 2OG binding sites. However, some selective compounds have been reported (top right). More recent studies have identified primary substrate competitive inhibitors of some oxygenases (*e.g.* Mildronate®, rhein), and also both 2OG and substrate competitive inhibitors for KDMs by linking 2OG competitors to substrate peptides.

The conserved general modes of iron and 2OG binding may initially suggest that achieving clinically useful selectivity by such compounds may be difficult. However, studies *in vitro* suggest that differences in the active site pockets can be exploited in the generation of selective inhibitors. Indeed the relatively limited number of human 2OG oxygenases coupled to the diversity of their substrates and consequently their active sites means that achieving selectivity for sets of 2OG oxygenases may be rather easier than, for example, kinases or proteases.

Various template 2OG oxygenase inhibitors of differing breadths of selectivity have been identified that bind in a similar manner to 2OG itself, including *N*-oxalyl amino acids, 2,2′-pyridinedicarboxylates, substituted hydroxamic acids, bipyridines and 8-hydroxyquinolines (Figure 2.10).[80,108,150,151] Most of these inhibitors bind to the iron in a bidentate manner. Some of these compounds (including *N*-oxalylglycine)[152] were developed as inhibitors of the collagen prolyl hydroxylases in pioneering studies aimed at developing antifibrotic agents prior to structural knowledge of the 2OG oxygenases. As shown by structural and kinetic analyses, these compounds act as non-reactive analogues of 2OG. Functionalization of the templates can enable selective inhibition, either by increasing or decreasing binding to particular 2OG oxygenases. By way of example, the HIF prolyl hydroxylases (PHD or EGLN) have a relatively snug 2OG binding pocket, whereas the HIF Asn hydroxylase, FIH, has a larger pocket. Use of C-α functionalized derivatives of NOG with bulky side chains (*e.g.* *N*-oxalyl-(D)-phenylalanine, NOFD) inhibit FIH, but much less potently than the PHDs (see Chapter 6).[153] Analogous approaches have been used to develop selective inhibitors of the JmjC KDMs, where many iron chelating templates are being explored as inhibitors. In the case of the KDM4/JMJD2 subfamily, which has a relatively open 2OG binding pocket, inhibitors that crosslink or occupy the substrate and 2OG binding sites have been found to be highly selective (Figures 2.10 and 2.11).[154,155] Selective inhibitors of other JmjC subfamilies have been developed for use as functional probes.[66] Several PHD inhibitors are now in clinical trials for the treatment of anaemia – almost all probably bind iron in a bidentate manner and compete with 2OG.

The inhibition of plant 2OG oxygenases, especially those involved in gibberellin metabolism, is of major importance in agrochemistry (see Chapter 16). Various inhibitors of gibberellin-related 2OG oxygenases have been developed for widespread use in agriculture. Although as yet there is no detailed structural information on their mode of binding, most are probably iron chelators and 2OG competitors, notably prohexadione, trinexapac-ethyl and related 'triketone' compounds. One plant 2OG oxygenase inhibitor, daminozide (succinyl *N,N*-dimethyl hydrazide) has been used in horticulture as a growth retardant;[156] daminozide is notable because of its small size and because it has also been shown to inhibit members of the human KDM2/7 JmjC subfamily *via* a mechanism involving iron chelation using its hydrazide carbonyl and dimethylamino groups (Figure 2.11).[157]

Various substrate-competitive inhibitors have also been reported, including the tricyclic compound rhein for the nucleic acid demethylase FTO[158] and a number of peptide-based inhibitors for prolyl hydroxylases[159] and JmjC KDMs.[154,155,160] Cyclic peptides in particular show promise as selective inhibitors.[160] However selective, low molecular weight inhibitors of the 2OG oxygenases are rare. One apparently interesting exception is Mildronate® (3-(2,2,2-trimethylhydrazine)propionate (THP, also known as MET88), which in some countries is given to patients after myocardial infarction to inhibit carnitine biosynthesis and hence fatty acid metabolism.[161] Mildronate® inhibits

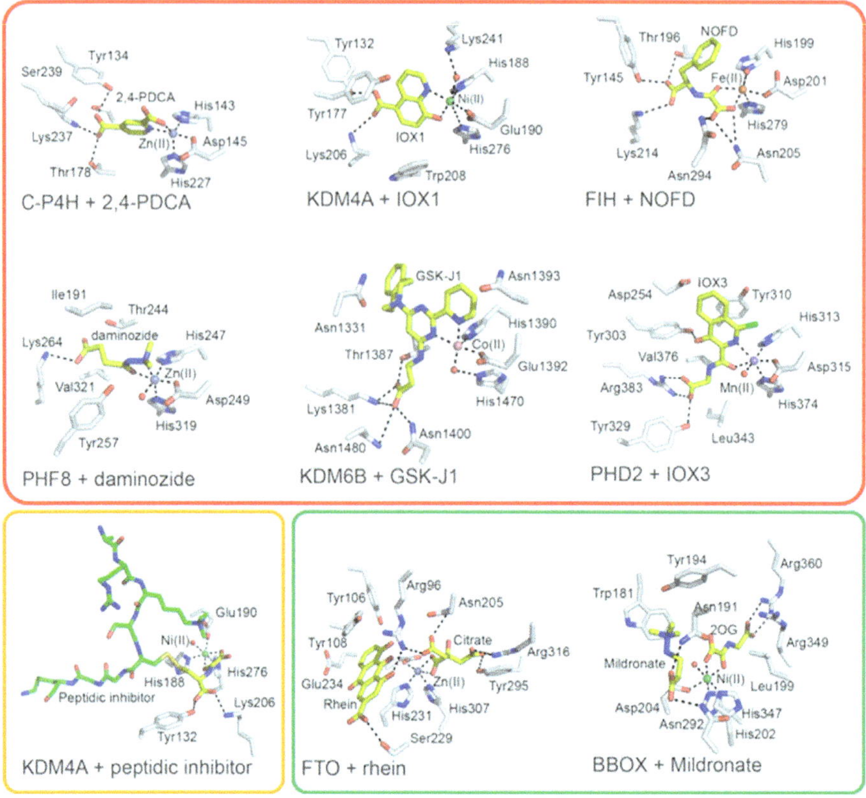

Figure 2.11 Views from crystal structures showing representative 2OG oxygenase : inhibitor complexes. The majority show inhibitors occupying the 2OG cosubstrate binding site: C-PH4 : Zn(II) : 2,4-PDCA complex (PDB ID: 2JIG), KDM4A : Ni(II) : IOX1 complex (PDB ID: 3NJY), FIH : Fe(II) : NOFD complex (PDB ID: 1YCI), PHF8 : Zn(II) : daminozide complex (PDB ID: 4DO0), KDM6B : Co(II) : GSK-J1 complex (PDB ID: 4ASK), PHD2 : Mn(II) : IOX3 complex (PDB ID: 4BQX). Some inhibitors are both primary substrate and 2OG competitors: KDM4A : Ni(II) : peptidic inhibitor complex (PDB ID: 3NJY), and others are primary substrate competitive: FTO : Zn(II) : citrate : rhein complex (PDB ID: 4IE7), BBOX : Ni(II) : 2OG : Mildronate® complex (PDB ID: 3MS5). Inhibitor structures are shown in Figure 2.10.

BBOX, which catalyses the final step in carnitine biosynthesis. Although crystallographic studies reveal that Mildronate® has a very similar binding mode to γ-butyrobetaine, solution studies have revealed it is a competitive substrate, undergoing oxidation to give various products including some arising from an oxidative rearrangement related to the Stevens-type rearrangement in organic synthesis.[162] Recently several reports have described structure-based efforts that have resulted in more potent and selective BBOX inhibitors operating *via* more conventional inhibition mechanisms (*i.e.* iron chelation).[163] Nonetheless, the mechanism of action of Mildronate® suggests that mechanism-based or novel mode of action inhibitors of 2OG oxygenases may be of interest and, although it may have other modes of action, Mildronate® remains the one 2OG oxygenase inhibitor specifically used as a pharmaceutical.

Acknowledgements

We thank the Wellcome Trust, the Biotechnology and Biological Sciences Research Council, Cancer Research UK and the British Heart Foundation for funding our research. RJH acknowledges a William R Miller Junior Research Fellowship, St. Edmund Hall, Oxford. We apologise for incomplete citations due to space constraints – we have focused citations on structural work. We thank our colleagues and collaborators for discussion and support. The author list is alphabetical – contributions should be regarded as being equal.

References

1. K. I. Kivirikko and D. J. Prockop, *Proc. Natl. Acad. Sci. U.S.A.*, 1967, **57**, 782–789.
2. J. Myllyharju, *Matrix Biol.*, 2003, **22**, 15–24.
3. J. Myllyharju, *Ann. Med.*, 2008, **40**, 402–417.
4. K. Valegard, A. C. van Scheltinga, M. D. Lloyd, T. Hara, S. Ramaswamy, A. Perrakis, A. Thompson, H. J. Lee, J. E. Baldwin, C. J. Schofield, J. Hajdu and I. Andersson, *Nature*, 1998, **394**, 805–809.
5. Z. Zhang, J. Ren, D. K. Stammers, J. E. Baldwin, K. Harlos and C. J. Schofield, *Nat. Struct. Biol.*, 2000, **7**, 127–133.
6. P. L. Roach, I. J. Clifton, V. Fulop, K. Harlos, G. J. Barton, J. Hajdu, I. Andersson, C. J. Schofield and J. E. Baldwin, *Nature*, 1995, **375**, 700–704.
7. P. L. Roach, I. J. Clifton, C. M. Hensgens, N. Shibata, C. J. Schofield, J. Hajdu and J. E. Baldwin, *Nature*, 1997, **387**, 827–830.
8. N. I. Burzlaff, P. J. Rutledge, I. J. Clifton, C. M. Hensgens, M. Pickford, R. M. Adlington, P. L. Roach and J. E. Baldwin, *Nature*, 1999, **401**, 721–724.
9. R. B. Hamed, J. R. Gomez-Castellanos, L. Henry, C. Ducho, M. A. McDonough and C. J. Schofield, *Nat. Prod. Rep.*, 2013, **30**, 21–107.
10. E. L. Hegg and L. Que, Jr, *Eur. J. Biochem.*, 1997, **250**, 625–629.
11. O. Landman, I. Borovok, Y. Aharonowitz and G. Cohen, *FEBS Lett.*, 1997, **405**, 172–174.
12. M. Sami, T. J. Brown, P. L. Roach, C. J. Schofield and J. E. Baldwin, *FEBS Lett.*, 1997, **405**, 191–194.
13. I. J. Clifton, M. A. McDonough, D. Ehrismann, N. J. Kershaw, N. Granatino and C. J. Schofield, *J. Inorg. Biochem.*, 2006, **100**, 644–669.
14. P. L. Roach, I. J. Clifton, C. M. Hensgens, N. Shibata, A. J. Long, R. W. Strange, S. S. Hasnain, C. J. Schofield, J. E. Baldwin and J. Hajdu, *Eur. J. Biochem.*, 1996, **242**, 736–740.
15. Z. Zhang, J. S. Ren, I. J. Clifton and C. J. Schofield, *Chem. Biol.*, 2004, **11**, 1383–1394.

16. D. Yun, M. Dey, L. J. Higgins, F. Yan, H. W. Liu and C. L. Drennan, *J. Am. Chem. Soc.*, 2011, **133**, 11262–11269.
17. C. A. Townsend, *Chem. Biol.*, 1997, **4**, 721–730.
18. C. A. Townsend, *Curr. Opin. Chem. Biol.*, 2002, **6**, 583–589.
19. J. M. Elkins, M. J. Ryle, I. J. Clifton, J. C. Dunning Hotopp, J. S. Lloyd, N. I. Burzlaff, J. E. Baldwin, R. P. Hausinger and P. L. Roach, *Biochemistry*, 2002, **41**, 5185–5192.
20. I. J. Clifton, L. C. Hsueh, J. E. Baldwin, K. Harlos and C. J. Schofield, *Eur. J. Biochem.*, 2001, **268**, 6625–6636.
21. C. J. Schofield and Z. Zhang, *Curr. Opin. Struct. Biol.*, 1999, **9**, 722–731.
22. K. Valegard, A. C. Terwisscha van Scheltinga, A. Dubus, G. Ranghino, L. M. Oster, J. Hajdu and I. Andersson, *Nat. Struct. Mol. Biol.*, 2004, **11**, 95–101.
23. N. J. Kershaw, M. E. Caines, M. C. Sleeman and C. J. Schofield, *Chem. Commun.*, 2005, 4251–4263.
24. S. E. Jensen, *J. Ind. Microbiol. Biotechnol.*, 2012, **39**, 1407–1419.
25. M. A. McDonough, C. Loenarz, R. Chowdhury, I. J. Clifton and C. J. Schofield, *Curr. Opin. Struct. Biol.*, 2010, **20**, 659–672.
26. S. S. Ng, K. L. Kavanagh, M. A. McDonough, D. Butler, E. S. Pilka, B. M. Lienard, J. E. Bray, P. Savitsky, O. Gileadi, F. von Delft, N. R. Rose, J. Offer, J. C. Scheinost, T. Borowski, M. Sundstrom, C. J. Schofield and U. Oppermann, *Nature*, 2007, **448**, 87–91.
27. L. J. Walport, R. J. Hopkinson and C. J. Schofield, *Curr. Opin. Chem. Biol.*, 2012, **16**, 525–534.
28. C. Johansson, A. Tumber, K. Che, P. Cain, R. Nowak, C. Gileadi and U. Oppermann, *Epigenomics*, 2014, **6**, 89–120.
29. R. J. Klose, E. M. Kallin and Y. Zhang, *Nat. Rev. Genet.*, 2006, **7**, 715–727.
30. R. P. Hausinger, *Crit. Rev. Biochem. Mol. Biol.*, 2004, **39**, 21–68.
31. Z. Zhang, J. Ren, K. Harlos, C. H. McKinnon, I. J. Clifton and C. J. Schofield, *FEBS Lett.*, 2002, **517**, 7–12.
32. W. Aik, M. A. McDonough, A. Thalhammer, R. Chowdhury and C. J. Schofield, *Curr. Opin. Struct. Biol.*, 2012, **22**, 691–700.
33. T. Duncan, S. C. Trewick, P. Koivisto, P. A. Bates, T. Lindahl and B. Sedgwick, *Proc. Natl. Acad. Sci. U.S.A.*, 2002, **99**, 16660–16665.
34. P. O. Falnes, R. F. Johansen and E. Seeberg, *Nature*, 2002, **419**, 178–182.
35. B. Yu, W. C. Edstrom, J. Benach, Y. Hamuro, P. C. Weber, B. R. Gibney and J. F. Hunt, *Nature*, 2006, **439**, 879–884.
36. A. Thalhammer, Z. Bencokova, R. Poole, C. Loenarz, J. Adam, L. O'Flaherty, J. Schodel, D. Mole, K. Giaslakiotis, C. J. Schofield, E. M. Hammond, P. J. Ratcliffe and P. J. Pollard, *PLoS One*, 2011, **6**, e16210.
37. A. Thalhammer, W. Aik, E. A. L. Bagg and C. J. Schofield, *Drug Discovery Today: Ther. Strategies*, 2012, **9**, e91–e100.
38. B. M. Johnson, B. A. Stier and S. Caltabiano, *Clin. Pharmacol. Drug Dev.*, 2013, **3**, 109–117.
39. W. Rademacher, *Plant Physiol.*, 1992, **100**, 625–629.

40. C. C. Thinnes, K. S. England, A. Kawamura, R. Chowdhury, C. J. Schofield and R. J. Hopkinson, *Biochim. Biophys. Acta*, 2014, **1839**, 1416–1432.
41. N. R. Rose, M. A. McDonough, O. N. F. King, A. Kawamura and C. J. Schofield, *Chem. Soc. Rev.*, 2011, **40**, 4364–4397.
42. L. Yan, V. J. Colandrea and J. J. Hale, *Expert Opin. Therapeutic Patents*, 2010, **20**, 1219–1245.
43. T. J. Gibson and P. Argos, *J. Mol. Biol.*, 1990, **212**, 7–9.
44. K. S. Hewitson, N. Granatino, R. W. Welford, M. A. McDonough and C. J. Schofield, *Philos. Trans. R. Soc., A*, 2005, **363**, 807–828; discussion 1035–1040.
45. J. M. Dunwell, A. Culham, C. E. Carter, C. R. Sosa-Aguirre and P. W. Goodenough, *Trends Biochem. Sci.*, 2001, **26**, 740–746.
46. P. M. Clissold and C. P. Ponting, *Trends Biochem. Sci.*, 2001, **26**, 7–9.
47. M. A. McDonough, K. L. Kavanagh, D. Butler, T. Searls, U. Oppermann and C. J. Schofield, *J. Biol. Chem.*, 2005, **280**, 41101–41110.
48. M. K. Koski, R. Hieta, C. Bollner, K. I. Kivirikko, J. Myllyharju and R. K. Wierenga, *J. Biol. Chem.*, 2007, **282**, 37112–37123.
49. M. A. McDonough, V. Li, E. Flashman, R. Chowdhury, C. Mohr, B. M. Lienard, J. Zondlo, N. J. Oldham, I. J. Clifton, J. Lewis, L. A. McNeill, R. J. Kurzeja, K. S. Hewitson, E. Yang, S. Jordan, R. S. Syed and C. J. Schofield, *Proc. Natl. Acad. Sci. U.S.A.*, 2006, **103**, 9814–9819.
50. M. Strieker, F. Kopp, C. Mahlert, L. O. Essen and M. A. Marahiel, *ACS Chem. Biol.*, 2007, **2**, 187–196.
51. J. M. Elkins, K. S. Hewitson, L. A. McNeill, J. F. Seibel, I. Schlemminger, C. W. Pugh, P. J. Ratcliffe and C. J. Schofield, *J. Biol. Chem.*, 2003, **278**, 1802–1806.
52. C. E. Dann, 3rd, R. K. Bruick and J. Deisenhofer, *Proc. Natl. Acad. Sci. U.S.A.*, 2002, **99**, 15351–15356.
53. C. Lee, S. J. Kim, D. G. Jeong, S. M. Lee and S. E. Ryu, *J. Biol. Chem.*, 2003, **278**, 7558–7563.
54. L. Hu, Z. Li, J. Cheng, Q. Rao, W. Gong, M. Liu, Y. G. Shi, J. Zhu, P. Wang and Y. Xu, *Cell*, 2013, **155**, 1545–1555.
55. H. Hashimoto, J. E. Pais, X. Zhang, L. Saleh, Z. Q. Fu, N. Dai, I. R. Correa, Jr., Y. Zheng and X. Cheng, *Nature*, 2014, **506**, 391–395.
56. M. Tahiliani, K. P. Koh, Y. Shen, W. A. Pastor, H. Bandukwala, Y. Brudno, S. Agarwal, L. M. Iyer, D. R. Liu, L. Aravind and A. Rao, *Science*, 2009, **324**, 930–935.
57. J. R. Horton, A. K. Upadhyay, H. H. Qi, X. Zhang, Y. Shi and X. Cheng, *Nat. Struct. Mol. Biol.*, 2010, **17**, 38–43.
58. D. E. Lancaster, L. A. McNeill, M. A. McDonough, R. T. Aplin, K. S. Hewitson, C. W. Pugh, P. J. Ratcliffe and C. J. Schofield, *Biochem. J.*, 2004, **383**, 429–437.
59. R. Chowdhury, R. Sekirnik, N. C. Brissett, T. Krojer, C. H. Ho, S. S. Ng, I. J. Clifton, W. Ge, N. J. Kershaw, G. C. Fox, J. R. Muniz, M. Vollmar, C. Phillips, E. S. Pilka, K. L. Kavanagh, F. von Delft, U. Oppermann, M. A. McDonough, A. J. Doherty and C. J. Schofield, *Nature*, 2014, **510**, 422–426.

60. W. Ge, A. Wolf, T. Feng, C. H. Ho, R. Sekirnik, A. Zayer, N. Granatino, M. E. Cockman, C. Loenarz, N. D. Loik, A. P. Hardy, T. D. Claridge, R. B. Hamed, R. Chowdhury, L. Gong, C. V. Robinson, D. C. Trudgian, M. Jiang, M. M. Mackeen, J. S. McCullagh, Y. Gordiyenko, A. Thalhammer, A. Yamamoto, M. Yang, P. Liu-Yi, Z. Zhang, M. Schmidt-Zachmann, B. M. Kessler, P. J. Ratcliffe, G. M. Preston, M. L. Coleman and C. J. Schofield, *Nat. Chem. Biol.*, 2012, **8**, 960–962.
61. M. Teichmann, H. Dumay-Odelot and S. Fribourg, *Transcription*, 2012, **3**, 2–7.
62. J. Myllyharju and K. I. Kivirikko, *Ann. Med.*, 2001, **33**, 7–21.
63. D. Fu, J. A. Brophy, C. T. Chan, K. A. Atmore, U. Begley, R. S. Paules, P. C. Dedon, T. J. Begley and L. D. Samson, *Mol. Cell. Biol.*, 2010, **30**, 2449–2459.
64. I. K. Leung, T. J. Krojer, G. T. Kochan, L. Henry, F. von Delft, T. D. Claridge, U. Oppermann, M. A. McDonough and C. J. Schofield, *Chem. Biol.*, 2010, **17**, 1316–1324.
65. C. Loenarz, W. Ge, M. L. Coleman, N. R. Rose, C. D. Cooper, R. J. Klose, P. J. Ratcliffe and C. J. Schofield, *Hum. Mol. Genet.*, 2010, **19**, 217–222.
66. L. Kruidenier, C. W. Chung, Z. Cheng, J. Liddle, K. Che, G. Joberty, M. Bantscheff, C. Bountra, A. Bridges, H. Diallo, D. Eberhard, S. Hutchinson, E. Jones, R. Katso, M. Leveridge, P. K. Mander, J. Mosley, C. Ramirez-Molina, P. Rowland, C. J. Schofield, R. J. Sheppard, J. E. Smith, C. Swales, R. Tanner, P. Thomas, A. Tumber, G. Drewes, U. Oppermann, D. J. Patel, K. Lee and D. M. Wilson, *Nature*, 2012, **488**, 404–408.
67. L. M. Iyer, S. Abhiman, R. F. de Souza and L. Aravind, *Nucleic Acids Res.*, 2010, **38**, 5261–5279.
68. R. Chowdhury, M. A. McDonough, J. Mecinovic, C. Loenarz, E. Flashman, K. S. Hewitson, C. Domene and C. J. Schofield, *Structure*, 2009, **17**, 981–989.
69. Y. Tao, M. Wu, X. Zhou, W. Yin, B. Hu, B. de Crombrugghe, K. M. Sinha and J. Zang, *J. Biol. Chem.*, 2013, **288**, 16430–16437.
70. L. M. van Staalduinen, S. K. Novakowski and Z. Jia, *J. Mol. Biol.*, 2014, **426**, 1898–1910.
71. M. Costas, M. P. Mehn, M. P. Jensen and L. Que, *Chem. Rev.*, 2004, **104**, 939–986.
72. T. Palzkill, *Ann. New York Acad. Sci.*, 2013, **1277**, 91–104.
73. J. R. O'Brien, D. J. Schuller, V. S. Yang, B. D. Dillard and W. N. Lanzilotta, *Biochemistry*, 2003, **42**, 5547–5554.
74. W. Aik, J. S. Scotti, H. Choi, L. Gong, M. Demetriades, C. J. Schofield and M. A. McDonough, *Nucleic Acids Res.*, 2014, **42**, 4741–4754.
75. B. Bleijlevens, T. Shivarattan, E. Flashman, Y. Yang, P. J. Simpson, P. Koivisto, B. Sedgwick, C. J. Schofield and S. J. Matthews, *EMBO Rep.*, 2008, **9**, 872–877.
76. B. Ergel, M. L. Gill, L. Brown, B. Yu, A. G. Palmer, 3rd and J. F. Hunt, *J. Biol. Chem.*, 2014, **289**, 29584–29601.

77. C. Wang, W. C. Chang, Y. Guo, H. Huang, S. C. Peck, M. E. Pandelia, G. M. Lin, H. W. Liu, C. Krebs and J. M. Bollinger, Jr, *Science*, 2013, **342**, 991–995.
78. W. C. Chang, Y. Guo, C. Wang, S. E. Butch, A. C. Rosenzweig, A. K. Boal, C. Krebs and J. M. Bollinger, Jr, *Science*, 2014, **343**, 1140–1144.
79. I. J. Clifton, L. X. Doan, M. C. Sleeman, M. Topf, H. Suzuki, R. C. Wilmouth and C. J. Schofield, *J. Biol. Chem.*, 2003, **278**, 20843–20850.
80. R. J. Hopkinson, A. Tumber, C. Yapp, R. Chowdhury, W. Aik, K. H. Che, X. S. Li, J. B. L. Kristensen, O. N. F. King, M. C. Chan, K. K. Yeoh, H. Choi, L. J. Walport, C. C. Thinnes, J. T. Bush, C. Lejeune, A. M. Rydzik, N. R. Rose, E. A. Bagg, M. A. McDonough, T. J. Krojer, W. W. Yue, S. S. Ng, L. Olsen, P. E. Brennan, U. Oppermann, S. Muller, R. J. Klose, P. J. Ratcliffe, C. J. Schofield and A. Kawamura, *Chem. Sci.*, 2013, **4**, 3110–3117.
81. J. R. Horton, A. K. Upadhyay, H. Hashimoto, X. Zhang and X. Cheng, *J. Mol. Biol.*, 2011, **406**, 1–8.
82. M. K. Koski, R. Hieta, M. Hirsila, A. Ronka, J. Myllyharju and R. K. Wierenga, *J. Biol. Chem.*, 2009, **284**, 25290–25301.
83. P. A. Aas, M. Otterlei, P. O. Falnes, C. B. Vagbo, F. Skorpen, M. Akbari, O. Sundheim, M. Bjoras, G. Slupphaug, E. Seeberg and H. E. Krokan, *Nature*, 2003, **421**, 859–863.
84. M. Kato, Y. Araiso, A. Noma, A. Nagao, T. Suzuki, R. Ishitani and O. Nureki, *Nucleic Acids Res.*, 2011, **39**, 1576–1585.
85. P. A. Del Rizzo, S. Krishnan and R. C. Trievel, *Mol. Cell. Biol.*, 2012, **32**, 4044–4052.
86. M. Mantri, T. Krojer, E. A. Bagg, C. A. Webby, D. S. Butler, G. Kochan, K. L. Kavanagh, U. Oppermann, M. A. McDonough and C. J. Schofield, *J. Mol. Biol.*, 2010, **401**, 211–222.
87. L. C. Blasiak, F. H. Vaillancourt, C. T. Walsh and C. L. Drennan, *Nature*, 2006, **440**, 368–371.
88. C. Wong, D. G. Fujimori, C. T. Walsh and C. L. Drennan, *J. Am. Chem. Soc.*, 2009, **131**, 4872–4879.
89. K. S. Hewitson, S. L. Holmes, D. Ehrismann, A. P. Hardy, R. Chowdhury, C. J. Schofield and M. A. McDonough, *J. Biol. Chem.*, 2008, **283**, 25971–25978.
90. H. J. Kulik, L. C. Blasiak, N. Marzari and C. L. Drennan, *J. Am. Chem. Soc.*, 2009, **131**, 14426–14433.
91. M. L. Matthews, C. S. Neumann, L. A. Miles, T. L. Grove, S. J. Booker, C. Krebs, C. T. Walsh and J. M. Bollinger, Jr, *Proc. Natl. Acad. Sci. U.S.A.*, 2009, **106**, 17723–17728.
92. S. D. Wong, M. Srnec, M. L. Matthews, L. V. Liu, Y. Kwak, K. Park, C. B. Bell, 3rd, E. E. Alp, J. Zhao, Y. Yoda, S. Kitao, M. Seto, C. Krebs, J. M. Bollinger, Jr. and E. I. Solomon, *Nature*, 2013, **499**, 320–323.
93. S. Srikanth, M. Jew, K. D. Kim, M. K. Yee, J. Abramson and Y. Gwack, *Proc. Natl. Acad. Sci. U.S.A.*, 2012, **109**, 8682–8687.
94. D. E. Lancaster, M. A. McDonough and C. J. Schofield, *Biochem. Soc. Trans.*, 2004, **32**, 943–945.

95. M. J. Katz, J. M. Acevedo, C. Loenarz, D. Galagovsky, P. Liu-Yi, M. Perez-Pepe, A. Thalhammer, R. Sekirnik, W. Ge, M. Melani, M. G. Thomas, S. Simonetta, G. L. Boccaccio, C. J. Schofield, M. E. Cockman, P. J. Ratcliffe and P. Wappner, *Proc. Natl. Acad. Sci. U.S.A.*, 2014, **111**, 4025–4030.
96. C. Loenarz, R. Sekirnik, A. Thalhammer, W. Ge, E. Spivakovsky, M. M. Mackeen, M. A. McDonough, M. E. Cockman, B. M. Kessler, P. J. Ratcliffe, A. Wolf and C. J. Schofield, *Proc. Natl. Acad. Sci. U.S.A.*, 2014, **111**, 4019–4024.
97. R. S. Singleton, P. Liu-Yi, F. Formenti, W. Ge, R. Sekirnik, R. Fischer, J. Adam, P. J. Pollard, A. Wolf, A. Thalhammer, C. Loenarz, E. Flashman, A. Yamamoto, M. L. Coleman, B. M. Kessler, P. Wappner, C. J. Schofield, P. J. Ratcliffe and M. E. Cockman, *Proc. Natl. Acad. Sci. U.S.A.*, 2014, **111**, 4031–4036.
98. J. Henri, D. Rispal, E. Bayart, H. van Tilbeurgh, B. Seraphin and M. Graille, *J. Biol. Chem.*, 2010, **285**, 30767–30778.
99. C. G. Yang, C. Yi, E. M. Duguid, C. T. Sullivan, X. Jian, P. A. Rice and C. He, *Nature*, 2008, **452**, 961–965.
100. C. Yi, G. Jia, G. Hou, Q. Dai, W. Zhang, G. Zheng, X. Jian, C. G. Yang, Q. Cui and C. He, *Nature*, 2010, **468**, 330–333.
101. J. C. Price, E. W. Barr, B. Tirupati, J. M. Bollinger, Jr. and C. Krebs, *Biochemistry*, 2003, **42**, 7497–7508.
102. J. A. Hangasky, C. Y. Taabazuing, M. A. Valliere and M. J. Knapp, *Metallomics*, 2013, **5**, 287–301.
103. J. Zhou, W. L. Kelly, B. O. Bachmann, M. Gunsior, C. A. Townsend and E. I. Solomon, *J. Am. Chem. Soc.*, 2001, **123**, 7388–7398.
104. L. M. Hoffart, E. W. Barr, R. B. Guyer, J. M. Bollinger, Jr. and C. Krebs, *Proc. Natl. Acad. Sci. U.S.A.*, 2006, **103**, 14738–14743.
105. K. S. Hewitson, B. M. Lienard, M. A. McDonough, I. J. Clifton, D. Butler, A. S. Soares, N. J. Oldham, L. A. McNeill and C. J. Schofield, *J. Biol. Chem.*, 2007, **282**, 3293–3301.
106. H. J. Lee, M. D. Lloyd, K. Harlos, I. J. Clifton, J. E. Baldwin and C. J. Schofield, *J. Mol. Biol.*, 2001, **308**, 937–948.
107. M. D. Lloyd, H. J. Lee, K. Harlos, Z. H. Zhang, J. E. Baldwin, C. J. Schofield, J. M. Charnock, C. D. Garner, T. Hara, A. C. Terwisscha van Scheltinga, K. Valegard, J. A. Viklund, J. Hajdu, I. Andersson, A. Danielsson and R. Bhikhabhai, *J. Mol. Biol.*, 1999, **287**, 943–960.
108. R. Chowdhury, K. K. Yeoh, Y. M. Tian, L. Hillringhaus, E. A. Bagg, N. R. Rose, I. K. Leung, X. S. Li, E. C. Woon, M. Yang, M. A. McDonough, O. N. King, I. J. Clifton, R. J. Klose, T. D. Claridge, P. J. Ratcliffe, C. J. Schofield and A. Kawamura, *EMBO Rep.*, 2011, **12**, 463–469.
109. L. Dang, D. W. White, S. Gross, B. D. Bennett, M. A. Bittinger, E. M. Driggers, V. R. Fantin, H. G. Jang, S. Jin, M. C. Keenan, K. M. Marks, R. M. Prins, P. S. Ward, K. E. Yen, L. M. Liau, J. D. Rabinowitz, L. C. Cantley, C. B. Thompson, M. G. Vander Heiden and S. M. Su, *Nature*, 2009, **462**, 739–744.

110. P. J. Pollard, N. C. Wortham and I. P. Tomlinson, *Ann. Med.*, 2003, **35**, 632–639.
111. M. A. Selak, S. M. Armour, E. D. MacKenzie, H. Boulahbel, D. G. Watson, K. D. Mansfield, Y. Pan, M. C. Simon, C. B. Thompson and E. Gottlieb, *Cancer Cell*, 2005, 7, 77–85.
112. C. Pastore, I. Topalidou, F. Forouhar, A. C. Yan, M. Levy and J. F. Hunt, *J. Biol. Chem.*, 2012, **287**, 2130–2143.
113. E. van den Born, C. B. Vagbo, L. Songe-Moller, V. Leihne, G. F. Lien, G. Leszczynska, A. Malkiewicz, H. E. Krokan, F. Kirpekar, A. Klungland and P. O. Falnes, *Nat. Commun.*, 2011, **2**, 172.
114. T. Gerken, C. A. Girard, Y. C. Tung, C. J. Webby, V. Saudek, K. S. Hewitson, G. S. Yeo, M. A. McDonough, S. Cunliffe, L. A. McNeill, J. Galvanovskis, P. Rorsman, P. Robins, X. Prieur, A. P. Coll, M. Ma, Z. Jovanovic, I. S. Farooqi, B. Sedgwick, I. Barroso, T. Lindahl, C. P. Ponting, F. M. Ashcroft, S. O'Rahilly and C. J. Schofield, *Science*, 2007, **318**, 1469–1472.
115. Z. Han, T. Niu, J. Chang, X. Lei, M. Zhao, Q. Wang, W. Cheng, J. Wang, Y. Feng and J. Chai, *Nature*, 2010, **464**, 1205–1209.
116. W. Aik, M. Demetriades, M. K. Hamdan, E. A. Bagg, K. K. Yeoh, C. Lejeune, Z. Zhang, M. A. McDonough and C. J. Schofield, *J. Med. Chem.*, 2013, **56**, 3680–3688.
117. Z. Han, P. Liu, L. Gu, Y. Zhang, H. Li, S. Chen and J. Chai, *Front. Sci.*, 2007, 52–61.
118. X. Hong, J. Zang, J. White, C. Wang, C. H. Pan, R. Zhao, R. C. Murphy, S. Dai, P. Henson, J. W. Kappler, J. Hagman and G. Zhang, *Proc. Natl. Acad. Sci. U.S.A.*, 2010, **107**, 14568–14572.
119. T. Sengoku and S. Yokoyama, *Genes Dev.*, 2011, **25**, 2266–2277.
120. L. J. Walport, R. J. Hopkinson, M. Vollmar, S. K. Madden, C. Gileadi, U. Oppermann, C. J. Schofield and C. Johansson, *J. Biol. Chem.*, 2014, **289**, 18302–18313.
121. E. Flashman and C. J. Schofield, *Nat. Chem. Biol.*, 2007, **3**, 86–87.
122. A. M. Rydzik, R. Chowdhury, G. T. Kochan, S. T. Williams, M. A. McDonough, A. Kawamura and C. J. Schofield, *Chem. Sci.*, 2014, **5**, 1765–1771.
123. C. Loenarz, J. Mecinovic, R. Chowdhury, L. A. McNeill, E. Flashman and C. J. Schofield, *Angew. Chem.*, 2009, **48**, 1784–1787.
124. O. Sundheim, C. B. Vagbo, M. Bjoras, M. M. Sousa, V. Talstad, P. A. Aas, F. Drablos, H. E. Krokan, J. A. Tainer and G. Slupphaug, *EMBO J*, 2006, **25**, 3389–3397.
125. R. C. Wilmouth, J. J. Turnbull, R. W. Welford, I. J. Clifton, A. G. Prescott and C. J. Schofield, *Structure*, 2002, **10**, 93–103.
126. R. W. Welford, I. J. Clifton, J. J. Turnbull, S. C. Wilson and C. J. Schofield, *Org. Biomol. Chem.*, 2005, **3**, 3117–3126.
127. M. Mantri, Z. Zhang, M. A. McDonough and C. J. Schofield, *FEBS J.*, 2012, **279**, 1563–1575.
128. A. Liu, R. Y. Ho, L. Que, Jr., M. J. Ryle, B. S. Phinney and R. P. Hausinger, *J. Am. Chem. Soc.*, 2001, **123**, 5126–5127.

129. M. J. Ryle and R. P. Hausinger, *Curr. Opin. Chem. Biol.*, 2002, **6**, 193–201.
130. C. J. Stubbs, C. Loenarz, J. Mecinovic, K. K. Yeoh, N. Hindley, B. M. Lienard, F. Sobott, C. J. Schofield and E. Flashman, *J. Med. Chem.*, 2009, **52**, 2799–2805.
131. E. Flashman, E. A. Bagg, R. Chowdhury, J. Mecinovic, C. Loenarz, M. A. McDonough, K. S. Hewitson and C. J. Schofield, *J. Biol. Chem.*, 2008, **283**, 3808–3815.
132. J. S. Scotti, I. K. Leung, W. Ge, M. A. Bentley, J. Paps, H. Kramer, J. Lee, W. Aik, H. Choi, S. M. Paulsen, L. A. H. Bowman, N. D. Loik, S. Horita, C.-h. Ho, N. J. Kershaw, C. M. Tang, T. D. Claridge, G. M. Preston, M. A. McDonough and C. J. Schofield, *Proc. Natl. Acad. Sci. U.S.A.*, 2014, **111**, 13331–13336.
133. M. Yang, R. Chowdhury, W. Ge, R. B. Hamed, M. A. McDonough, T. D. Claridge, B. M. Kessler, M. E. Cockman, P. J. Ratcliffe and C. J. Schofield, *FEBS J.*, 2011, **278**, 1086–1097.
134. M. Yang, W. Ge, R. Chowdhury, T. D. Claridge, H. B. Kramer, B. Schmierer, M. A. McDonough, L. Gong, B. M. Kessler, P. J. Ratcliffe, M. L. Coleman and C. J. Schofield, *J. Biol. Chem.*, 2011, **286**, 7648–7660.
135. M. Yang, A. P. Hardy, R. Chowdhury, N. D. Loik, J. S. Scotti, J. S. McCullagh, T. D. Claridge, M. A. McDonough, W. Ge and C. J. Schofield, *Angew. Chem., Int. Ed. Engl.*, 2013, **52**, 1700–1704.
136. M. E. Cockman, D. E. Lancaster, I. P. Stolze, K. S. Hewitson, M. A. McDonough, M. L. Coleman, C. H. Coles, X. Yu, R. T. Hay, S. C. Ley, C. W. Pugh, N. J. Oldham, N. Masson, C. J. Schofield and P. J. Ratcliffe, *Proc. Natl. Acad. Sci. U.S.A.*, 2006, **103**, 14767–14772.
137. S. A. Dames, M. Martinez-Yamout, R. N. De Guzman, H. J. Dyson and P. E. Wright, *Proc. Natl. Acad. Sci. U.S.A.*, 2002, **99**, 5271–5276.
138. J. Li, A. Mahajan and M. D. Tsai, *Biochemistry*, 2006, **45**, 15168–15178.
139. M. Yang, R. Chowdhury, W. Ge, R. B. Hamed, M. A. McDonough, T. D. Claridge, B. M. Kessler, M. E. Cockman, P. J. Ratcliffe and C. J. Schofield, *FEBS J.*, 2011, **278**, 1086–1097.
140. M. L. Coleman, M. A. McDonough, K. S. Hewitson, C. Coles, J. Mecinovic, M. Edelmann, K. M. Cook, M. E. Cockman, D. E. Lancaster, B. M. Kessler, N. J. Oldham, P. J. Ratcliffe and C. J. Schofield, *J. Biol. Chem.*, 2007, **282**, 24027–24038.
141. K. Tars, J. Rumnieks, A. Zeltins, A. Kazaks, S. Kotelovica, A. Leonciks, J. Sharipo, A. Viksna, J. Kuka, E. Liepinsh and M. Dambrova, *Biochem. Biophys. Res. Commun.*, 2010, **398**, 634–639.
142. L. Yu, Y. Wang, S. Huang, J. Wang, Z. Deng, Q. Zhang, W. Wu, X. Zhang, Z. Liu, W. Gong and Z. Chen, *Cell Res.*, 2010, **20**, 166–173.
143. W. W. Yue, V. Hozjan, W. Ge, C. Loenarz, C. D. Cooper, C. J. Schofield, K. L. Kavanagh, U. Oppermann and M. A. McDonough, *FEBS Lett.*, 2010, **584**, 825–830.
144. J. Anantharajan, M. K. Koski, P. Kursula, R. Hieta, U. Bergmann, J. Myllyharju and R. K. Wierenga, *Structure*, 2013, **21**, 2107–2118.

145. T. Jokilehto and P. M. Jaakkola, *J. Cell. Mol. Med.*, 2010, **14**, 758–770.
146. P. Vachal, S. Miao, J. M. Pierce, D. Guiadeen, V. J. Colandrea, M. J. Wyvratt, S. P. Salowe, L. M. Sonatore, J. A. Milligan, R. Hajdu, A. Gollapudi, C. A. Keohane, R. B. Lingham, S. M. Mandala, J. A. DeMartino, X. Tong, M. Wolff, D. Steinhuebel, G. R. Kieczykowski, F. J. Fleitz, K. Chapman, J. Athanasopoulos, G. Adam, C. D. Akyuz, D. K. Jena, J. W. Lusen, J. Meng, B. D. Stein, L. Xia, E. C. Sherer and J. J. Hale, *J. Med. Chem.*, 2012, **55**, 2945–2959.
147. R. Natarajan, F. N. Salloum, B. J. Fisher, L. Smithson, J. Almenara and A. A. Fowler Iii, *Vasc. Pharmacol.*, 2009, **51**, 110–118.
148. J. W. Hojfeldt, K. Agger and K. Helin, *Nat. Rev. Drug Discovery*, 2013, **12**, 917–930.
149. L. Poppe, C. M. Tegley, V. Li, J. Lewis, J. Zondlo, E. Yang, R. J. Kurzeja and R. Syed, *J. Am. Chem. Soc.*, 2009, **131**, 16654–16655.
150. O. N. F. King, X. S. Li, M. Sakurai, A. Kawamura, N. R. Rose, S. S. Ng, A. M. Quinn, G. Rai, B. T. Mott, P. Beswick, R. J. Klose, U. Oppermann, A. Jadhav, T. D. Heightman, D. J. Maloney, C. J. Schofield and A. Simeonov, *PLoS One*, 2010, **5**, e15535.
151. K.-H. Chang, O. N. F. King, A. Tumber, E. C. Y. Woon, T. D. Heightman, M. A. McDonough, C. J. Schofield and N. R. Rose, *ChemMedChem*, 2011, **6**, 759–764.
152. C. J. Cunliffe, T. J. Franklin, N. J. Hales and G. B. Hill, *J. Med. Chem.*, 1992, **35**, 2652–2658.
153. M. A. McDonough, L. A. McNeill, M. Tilliet, C. A. Papamicaël, Q.-Y. Chen, B. Banerji, K. S. Hewitson and C. J. Schofield, *J. Am. Chem. Soc.*, 2005, **127**, 7680–7681.
154. E. C. Y. Woon, A. Tumber, A. Kawamura, L. Hillringhaus, W. Ge, N. R. Rose, J. H. Y. Ma, M. C. Chan, L. J. Walport, K. H. Che, S. S. Ng, B. D. Marsden, U. Oppermann, M. A. McDonough and C. J. Schofield, *Angew. Chem., Int. Ed.*, 2012, **51**, 1631–1634.
155. B. Lohse, A. L. Nielsen, J. B. L. Kristensen, C. Helgstrand, P. A. C. Cloos, L. Olsen, M. Gajhede, R. P. Clausen and J. L. Kristensen, *Angew. Chem., Int. Ed.*, 2011, **50**, 9100–9103.
156. J. A. Riddell, H. A. Hageman, C. M. J'Anthony and W. L. Hubbard, *Science*, 1962, **136**, 391.
157. N. R. Rose, E. C. Y. Woon, A. Tumber, L. J. Walport, R. Chowdhury, X. S. Li, O. N. F. King, C. Lejeune, S. S. Ng, T. Krojer, M. C. Chan, A. M. Rydzik, R. J. Hopkinson, K. H. Che, M. Daniel, C. Strain-Damerell, C. Gileadi, G. Kochan, I. K. H. Leung, J. Dunford, K. K. Yeoh, P. J. Ratcliffe, N. Burgess-Brown, F. von Delft, S. Muller, B. Marsden, P. E. Brennan, M. A. McDonough, U. Oppermann, R. J. Klose, C. J. Schofield and A. Kawamura, *J. Med. Chem.*, 2012, **55**, 6639–6643.
158. B. Chen, F. Ye, L. Yu, G. Jia, X. Huang, X. Zhang, S. Peng, K. Chen, M. Wang, S. Gong, R. Zhang, J. Yin, H. Li, Y. Yang, H. Liu, J. Zhang, H. Zhang, A. Zhang, H. Jiang, C. Luo and C.-G. Yang, *J. Am. Chem. Soc.*, 2012, **134**, 17963–17971.

159. H.-s. Moon and T. Begley, *Biotechnol. Bioprocess Eng.*, 2000, **5**, 61–64.
160. U. Leurs, B. Lohse, K. D. Rand, S. Ming, E. S. Riise, P. A. Cole, J. L. Kristensen and R. P. Clausen, *ACS Chem. Biol.*, 2014, **9**, 2131–2138.
161. M. Dambrova, E. Liepinsh and I. Kalvinsh, *Trends Cardiovasc. Med.*, 2002, **12**, 275–279.
162. I. K. H. Leung, T. J. Krojer, G. T. Kochan, L. Henry, F. von Delft, T. D. W. Claridge, U. Oppermann, M. A. McDonough and C. J. Schofield, *Chem. Biol.*, 2010, **17**, 1316–1324.
163. A. M. Rydzik, R. Chowdhury, G. T. Kochan, S. T. Williams, M. A. McDonough, A. Kawamura and C. J. Schofield, *Chem. Sci.*, 2014, **5**, 1765–1771.

CHAPTER 3

Mechanisms of 2-Oxoglutarate-Dependent Oxygenases: The Hydroxylation Paradigm and Beyond

J. MARTIN BOLLINGER JR.*[a,b], WEI-CHEN CHANG[a], MEGAN L. MATTHEWS[a,c], RYAN J. MARTINIE[a], AMIE K. BOAL[a,b], AND CARSTEN KREBS*[a,b]

[a]Department of Chemistry, The Pennsylvania State University, University Park, Pennsylvania 16802, USA; [b]Department of Biochemistry and Molecular Biology, The Pennsylvania State University, University Park, Pennsylvania 16802, USA; [c]Department of Chemical Physiology, The Scripps Research Institute, La Jolla, California 92037, USA
*E-mail: jmb21@psu.edu, ckrebs@psu.edu

3.1 Introduction

One of nature's most potent and versatile strategies for transformation of unactivated carbon centres involves activation of O_2 at an Fe(II) cofactor to form an Fe(IV)-oxo (ferryl) intermediate that can either react as an electrophile or abstract a hydrogen atom (H•) to initiate the installation of a diverse array of functional groups. A large and biologically important subset of enzymes that employ this strategy generate their key ferryl intermediates by coupling

RSC Metallobiology Series No. 3
2-Oxoglutarate-Dependent Oxygenases
Edited by Robert P. Hausinger and Christopher J. Schofield

Published by the Royal Society of Chemistry, www.rsc.org

activation of O_2 at non-haem Fe(II) cofactors to the oxidative decarboxylation of 2-oxoglutarate (2OG) to succinate and CO_2.[1–3] In humans, all such Fe(II)- and 2OG-dependent (Fe/2OG) oxygenases characterized to date are of the *di*oxygenase variety and mediate *hydroxylation* of unactivated aliphatic carbon centres, although in some cases the initial hydroxylated product breaks down non-enzymatically (*e.g.* to formaldehyde in *N*-demethylation reactions)[4–9] and in others it can be further oxidized in reactions that are not formally hydroxylations (*e.g.* the Tet-like 5-methylpyrimidine oxygenases).[10,11] Because they play crucial roles in the biosynthesis of connective tissue (Chapter 5),[12] homeostasis of iron,[13] sensing of oxygen (Chapter 6),[14–16] regulation of body mass (Chapter 9),[9] repair of one class of DNA damage (Chapter 8),[4,5] and control of transcription, differentiation, development and epigenetic inheritance (Chapters 7 and 11),[6–8,10,11] these human Fe/2OG dioxygenases have been intensively studied by biomedical researchers, beginning with the collagen-synthesizing prolyl-4-hydroxylase that becomes insufficiently active under conditions of ascorbic acid (vitamin C) deficiency to cause the connective tissue disease known as scurvy.[12] Plants and microbes also employ Fe/2OG hydroxylases for metabolism and regulation.[2] Through investigations of several of the microbial enzymes, the mechanism of hydroxylation, which involves abstraction of H• by the high-spin (S = 2) ferryl complex and 'rebound' of the Fe(III)-coordinated hydroxyl ligand to the substrate radical, has become reasonably well understood.[17] Bacteria have further adapted the Fe/2OG platform for a remarkable array of divergent reactivities,[2] including olefin epoxidations,[18] aliphatic halogenations,[19,20] olefin-installing 1,2-dehydrogenations,[21,22] oxacycle-installing 1,3- and 1,5-dehydrogenations[23–27] and a redox-neutral stereoinversion.[28] Products of these reactions are, or are further processed to, important drug compounds (*e.g.* antibiotics and anaesthetics).[2,29] An understanding of the mechanisms leading to this manifold of transformations, and the means by which the individual enzymes direct them, could guide the design of new chemical catalysts, the selective inhibition of biomedically relevant members of the enzyme family, and the development of novel bacterially- or chemo-enzymatically-derived natural product drug compounds. In this chapter, we first summarize our understanding of the mechanism of Fe/2OG hydroxylases and then review recent advances in the elucidation of two of the 'alternative' reactivities (halogenation and stereoinversion). Finally, we discuss the remaining, less well understood dehydrogenation reactions, highlighting potentially problematic aspects of published mechanistic hypotheses, presenting alternatives to these published mechanisms, and briefly outlining experiments by which the operant mechanisms might be established.

3.2 Mechanism of the Fe/2OG Hydroxylases

The first comprehensive reaction mechanism for an Fe/2OG hydroxylase was proposed in 1982 by Hanauske-Abel and Günzler and has been termed the HAG mechanism.[30] Despite having been formulated from relatively limited data, it has held up well to experimental scrutiny. An adapted version of the HAG mechanism, which takes into account experimental and computational data obtained since its original proposal, is shown in Scheme 3.1. Important features of this mechanism include: (i) bidentate coordination of 2OG to the Fe(II) centre, which the protein coordinates facially by one carboxylate (Asp or Glu) and two histidine ligands,[31,32] (ii) generation of an open coordination site on the Fe(II) upon binding of the primary substrate,[1,33] (iii) oxidative addition of O_2 to the Fe(II) centre to yield an Fe(III)-superoxo intermediate, (iv) attack of the distal oxygen atom from O_2 on C2 of 2OG to initiate O–O cleavage, decarboxylation and formation of a ferryl intermediate, (v) abstraction of the target H atom from the substrate by the ferryl to yield a Fe(III)–OH and substrate radical (R•), and (vi) formation of the hydroxylated product and a coordinatively unsaturated Fe(II) site by formal recombination of R• and the coordinated hydroxyl radical (oxygen or HO• rebound).[34] Prior to studies on taurine:2OG dioxygenase (TauD) by our group and by Hausinger and coworkers,[35-40] none of the states after addition of O_2 had been directly detected.

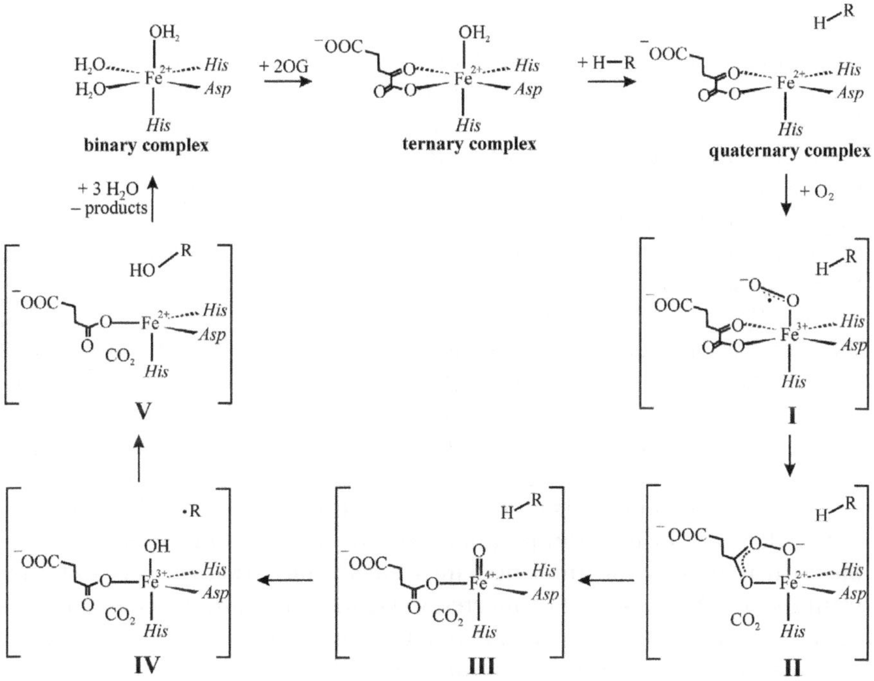

Scheme 3.1 Conserved mechanism of the Fe/2OG dioxygenases.

3.2.1 Evidence for Two Intermediates by Stopped-Flow Absorption Spectroscopy

Stopped-flow absorption (SF-Abs) experiments, in which the deoxygenated reactant TauD:Fe(II):2OG:taurine complex was mixed with a buffer solution containing O_2, provided the first evidence for the accumulation of two intermediate states in the TauD catalytic cycle. The first state (state **III** in Scheme 3.1), which was ultimately shown to contain the ferryl complex, absorbs at 320 nm, whereas the second state (**V**), an Fe(II)-containing product complex, is distinguished by its lack of any absorption features in the UV/visible spectrum. Its accumulation therefore leads to a loss of the metal-to-ligand charge-transfer (MLCT) band at 520 nm from the reactant complex.[41] SF-Abs measurements using these spectroscopic handles provided insight into the kinetics of the reaction. By varying the concentrations of O_2 and the reactant complex, it was found that state **III** (the ferryl complex) forms in a second-order reaction with rate constant 1.5×10^5 $M^{-1}s^{-1}$ at 5 °C.[41] This second-order behaviour implies that states **I** and **II** fail to accumulate to kinetically significant levels.

3.2.2 Evidence that the First Intermediate is a High-Spin Fe(IV) Complex

Further insight into the state of the cofactor was provided by freeze-quench (FQ) Mössbauer spectroscopy. This technique is particularly useful in Fe/2OG enzymes, because all states of the proposed mechanism have an integer electron-spin ground state and therefore give rise to quadrupole doublet features that can readily be detected and quantified.[42] A new quadrupole doublet with isomer shift (δ) of 0.30 mm s^{-1} and quadrupole splitting parameter (ΔE_Q) of 0.88 mm s^{-1} was observed in studies on TauD.[41] Quantification of the associated iron species from the intensity of this feature in spectra of samples quenched at different reaction times allowed the kinetics of the intermediate to be defined. The analysis revealed that this quadrupole doublet is associated with the 320 nm absorbing intermediate detected in the SF-Abs experiments. The Mössbauer parameters, in particular the low isomer shift, suggested that the Fe centre is formally in the +IV oxidation state. This oxidation state was confirmed by cryo-reduction experiments, which revealed that disappearance of the new quadrupole doublet is associated with formation of a high-spin Fe(III) centre detected by electron paramagnetic resonance (EPR) spectroscopy.[41] More insight into the intermediate's electronic structure was obtained from its Mössbauer spectra collected in strong externally applied magnetic fields.[41,42] These experiments revealed that the Fe(IV) centre is in the high-spin ($S = 2$) configuration. This result was unexpected because all enzymatic and inorganic Fe(IV) complexes known before the discovery of the TauD complex had $S = 1$ ground states.[43–47]

3.2.3 Evidence that the Fe(IV) Complex Cleaves the C–H Bond of the Substrate

Single-turnover SF-Abs and FQ-Mössbauer experiments with the substrate isotopologue 1,1-d_2-taurine, in which the target C–H bond is replaced by deuterium, provided evidence that the Fe(IV) complex is the C–H-cleaving intermediate.[36] This isotopic substitution extends the lifetime of the intermediate (*i.e.* $k_{obs,H} = 13\ s^{-1}$ and $k_{obs,D} = 0.35\ s^{-1}$ at 5 °C). The ratio $k_{obs,H}/k_{obs,D} = 37$ represents a lower limit for the intrinsic primary deuterium kinetic isotope effect (D-KIE), which is the ratio of the intrinsic rate constants for C–H/D cleavage (k_{C-H}/k_{C-D}). The magnitude significantly exceeds the classical limit, demonstrating that hydrogen tunnelling contributes to the C–H cleavage step.[48] A more rigorous analysis of the products of the reaction uncovered the existence of an unproductive ferryl-decay pathway, which does not involve abstraction of the target hydrogen and proceeds with a rate constant of $k_{abortive} = 0.1\ s^{-1}$.[49] The experimentally observed rate constant for decay of the ferryl species should be the sum of the rate constants of the individual decay pathways ($k_{obs,H(D)} = k_{C-H(D)} + k_{abortive}$). From this analysis, the true intrinsic D-KIE was estimated to be ~50.[49]

3.2.4 Evidence that the Fe(IV) Complex is a Ferryl [Fe(IV)–Oxo] Species

In the HAG mechanism, the C–H-cleaving complex was suggested to be a ferryl complex (state **III**). Direct evidence for this assignment was provided by a series of spectroscopic experiments. First, Hausinger and coworkers detected the Fe–O stretching mode at 821 cm^{-1} in samples prepared with unlabelled O_2 (>99% $^{16}O_2$) and at 789 cm^{-1} upon use of $^{18}O_2$ in cryogenic, continuous-flow resonance Raman spectroscopic experiments.[37] These features were observed after the quaternary TauD:Fe(II):2OG:taurine complex was reacted with O_2 at −38 °C for 0.22 s.[36] Fe K-edge X-ray absorption spectroscopic experiments carried out on samples enriched in the intermediate provided additional confirmation that it is a ferryl complex.[38] The absorption edge was seen to shift to higher energy relative to that of the reactant Fe(II)-containing complex, as expected for the oxidation of the Fe centre. More importantly, analysis of the X-ray absorption fine structure (EXAFS) region of the spectrum revealed the presence of an Fe ligand at a short distance of 1.62 Å, consistent with the Fe–O distances seen in inorganic ferryl complexes.[45,46,50,51]

More insight into the geometric and electronic structures of high-spin ferryl complexes in general and in TauD in particular was obtained computationally.[52,53] These results revealed that the unusual high-spin ground state is due to the weak ligand field in the x–y plane, which results in lowering of the d_{x2-y2} orbital energy. Importantly, spectroscopic parameters (in particular Mössbauer) were calculated. The results showed that high-spin ferryl complexes have a rather large Mössbauer isomer shift of ~ +0.3 mm s^{-1}, a negative Mössbauer quadrupole splitting of moderate magnitude, and an axial, anisotropic ^{57}Fe hyperfine coupling tensor with a large negative z-component. Whereas the x- and y-components of the ^{57}Fe **A**-tensor had been determined with good precision from field-dependent 4.2 K Mössbauer spectra, the z-component could not be determined from these measurements because the axial positive zero-field splitting of the $S = 2$ electron-spin ground state results in negligible expectation value in the z-direction ($\langle S_z \rangle$) of the $M_S = 0$ ground state of the $S = 2$ manifold.[42] However, the excited $M_S = \pm 1$ and ± 2 states of the $S = 2$ manifold have a large $\langle S_z \rangle$. High-field Mössbauer spectra collected at elevated temperatures, at which these states become populated, allowed for an upper limit of A_z to be determined experimentally.[53] The results confirmed the computational prediction that A_z is significantly more negative than A_{xy}. These studies serve as a good example of the synergistic application of computational and experimental methods to elucidate the structure of reactive intermediates.

3.2.5 Evidence that the Second Intermediate is an Fe(II) Product Complex

The second intermediate state that accumulates in the reaction of the TauD quaternary complex with O_2 was assigned as the Fe(II)-containing product complex (**V** in Scheme 3.1). The high-spin Fe(II) oxidation state of the complex was readily determined from FQ-Mössbauer experiments, which showed that features associated with Fe(II) are formed upon decay of the ferryl intermediate.[39] The features of this second intermediate, although indicative of high-spin Fe(II), are distinct from those of the various stable Fe(II) complexes (binary, ternary and quaternary), and the intermediate was therefore assigned as the product Fe(II) complex. Comparison of the observed rates with steady-state kinetics indicated that decay of the second intermediate by release of the product(s) is the rate-limiting step in the steady state. Consistent with this notion, the steady-state rate was seen to depend on the solvent viscosity.[39] By contrast, when 1,1-d_2-taurine was used in the reaction, decay of the ferryl complex became the rate-limiting step owing to the large D-KIE, and the steady-state rate was then found not to depend on solvent viscosity.[39] Intriguingly, although additional intermediates have not been detected at 5 °C, data from further cryogenic (~ −35 °C) continuous-flow resonance Raman studies were interpreted to imply the accumulation of Fe(III)-oxo and Fe(II)-alkoxo complexes;[40] such an interpretation would suggest that oxygen transfer occurs by a mechanism distinct from radical recombination.[34] This assignment has yet to be corroborated by additional methods, and the steps following H• abstraction by the ferryl complex remain ripe for further inquiry.

3.2.6 Evidence for Generality of the Two Intermediates and Their Relative Kinetics

The reaction of the quaternary complex with O_2 results in accumulation of two intermediates. The first is the C–H-cleaving ferryl intermediate, and the second is the Fe(II)-containing product complex. Importantly, the same two intermediates, exhibiting similar kinetic and spectroscopic features, were observed for the Fe/2OG-dependent dioxygenase prolyl-4-hydroxylase (P4H) from *Paramecium bursaria Chlorella virus 1*.[54] The fact that the same intermediates accumulate in this distantly related Fe/2OG dioxygenase was taken to imply that the hydroxylases employ a single conserved mechanism.[54]

3.3 Fe/2OG Aliphatic Halogenases

Discovered in 2005 by Walsh and coworkers, the Fe/2OG aliphatic halogenases (Chapter 18)[19] are so-called 'tailoring' enzymes of non-ribosomal peptide synthetase (NRPS) and polyketide synthase complexes. The founding member of this subclass of Fe/2OG oxygenases, SyrB2, catalyses chlorination at the C4 position of L-threonine on the pathway to syringomycin, a phytotoxin produced by *Pseudomonas syringae* B301D. Like other tailoring enzymes, SyrB2 does not act on the free amino acid; rather, it requires a carrier protein to deliver the halogenation target to the enzyme's active site. The carrier protein, SyrB1, comprises two domains. The *N*-terminal adenylation domain activates L-threonine (using ATP, which is hydrolysed to AMP and pyrophosphate) and then appends it *via* a thioester linkage to the phosphopantetheine cofactor of the *C*-terminal thiolation domain. (For brevity, such aminoacyl-SyrB1 substrates for SyrB2 will hereafter be denoted by the bolded three-letter code of the amino acid, *e.g.* **Thr**). The crystal structure of SyrB2 revealed that the enzyme-supplied carboxylate ligand present in all other previously characterized Fe/2OG oxygenases is absent in SyrB2, replaced in the sequence by a non-coordinating alanine (Ala-118).[55] This effective ligand deletion vacates a site to which the halide coordinates [PDB IDs: 2FCT (chloride coordination) and 2FCV (bromide coordination)]. This structural insight suggested a mechanism for SyrB2 in which a *cis*-haloferryl intermediate, similar to the ferryl species in the related hydroxylases, abstracts a hydrogen atom from the target position of the substrate to form a substrate radical and a *cis*-halo-hydroxo-ferric species (Scheme 3.2).[55] Subsequently,

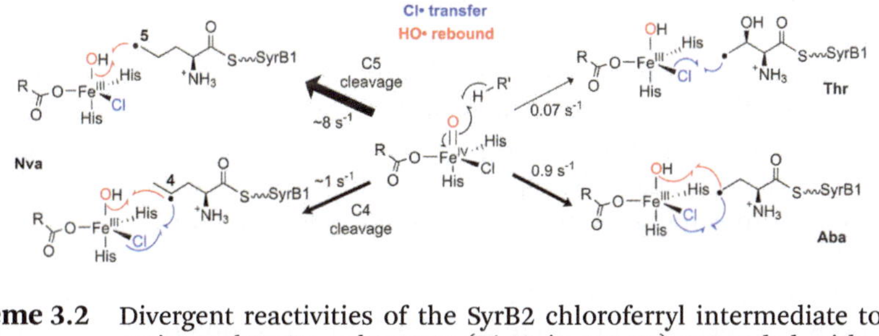

Scheme 3.2 Divergent reactivities of the SyrB2 chloroferryl intermediate toward aminoacyl-SyrB1 substrates (R′–H in centre) appended with: threonine (**Thr**), top right; aminobutyric acid (**Aba**), bottom right; norvaline (**Nva**), left top and bottom. Red arrows represent halogenation, whereas blue arrows represent hydroxylation. R = CH_2CH_2COOH.

the halogenase mechanism diverges from that for hydroxylation by the transfer of the coordinated *halogen* (as Cl• or Br•), rather than the HO•, to the substrate radical. This 'alternative group transfer' mechanism is analogous to that proposed by Baldwin and coworkers[56] and recently experimentally validated by our group[57,58] for closure of the thiazolidine ring in the reaction of isopenicillin *N* synthase (IPNS).[57,58]

3.3.1 Evidence for H•-Abstracting Haloferryl Complexes in the Halogenases

By the methods developed in the previous studies on the hydroxylases, the proposed H•-abstracting haloferryl complex was directly detected first for the halogenase, CytC3, which effects dichlorination of L-2-aminobutyrate tethered to the single-domain carrier (thiolation) protein, CytC2,[59] and subsequently for SyrB2.[17,60–62] Upon rapid mixing of either halogenase:Fe(II) : Cl^- (Br^-) : 2OG:aminoacyl-carrier protein complex with O_2, an intermediate that absorbs at 320 nm (analogous to the ferryl intermediate in TauD) was seen to accumulate rapidly. The marked (20-fold) slowing of the decay of this feature upon use of the appropriately deuterium-substituted amino acid confirmed that the associated complex abstracts hydrogen.[17,62] Importantly, the rate constants for decay of the ferryl complexes in both halogenases were much (~100–1000-fold) less than those for the hydroxylases. The remarkable sluggishness in H• abstraction and adaptive suppression of unproductive decay pathways would become crucial clues to the mechanism by which the enzymes direct their halogenation outcome. FQ-Mössbauer experiments showed that, in all cases, the 320-nm feature develops and decays in association with *two* partially resolved quadrupole doublets with isomer shifts ($\delta = 0.25–0.30$ mm s^{-1}) similar to those of the ferryl complexes in TauD and P4H. High-field spectra again showed that both associated species have $S = 2$ ground states with nearly axial, positive zero-field splittings. The Mössbauer data were thus consistent with assignment of the complexes as Fe(IV). The synchronous development and decay of the two doublets implies that the complexes rapidly interconvert within a single kinetic state.[17,60,62] A computational study by Borowski *et al.* suggested, first, that the two complexes could be coordination isomers of the haloferryl complex, in which the oxide and chloride ligands exchange sites and, second, that this isomerism could be crucial to control of the reaction.[67] However, the significance of the presence of two complexes with respect to the enzymes' direction of their halogenation outcomes remains to be established experimentally.

EXAFS analysis of the intermediate states in CytC3 and SyrB2 indicated the presence of an O/N scatterer at the short distance (1.61 and 1.66 Å for CytC3 and SyrB2, respectively) expected of the ferryl oxo moiety.[61,63] Importantly, both studies also provided evidence for the presence of the halide ligand, Br^- in the CytC study and the native Cl^- in the SyrB2 work, in the form of a scatterer with appropriate intensity at a distance of ~2.3–2.4 Å.

3.3.2 Analysis of Substrate Triggering in the Halogenases

The EXAFS study of the SyrB2 *cis*-chloroferryl complex[62] began with examination of the structural determinants of 'substrate triggering' in the halogenase. All Fe/2OG oxygenases studied to date are subject to this phenomenon, a marked acceleration of the reaction of the Fe(II) cofactor with O_2 upon binding of the primary substrate. Solomon and coworkers showed for several hydroxylases and the halogenase CytC3 that this activation arises, at least in part, from dissociation of the water ligand from the Fe(II) cofactor in response to substrate binding.[1,64,65] A unique aspect of the halogenase substrate is its tripartite nature: in principle, the tethered amino acid, phosphopantetheine arm and carrier protein could all make contacts with the halogenase to impart triggering. An analysis of the effect found, first, that the reaction of SyrB2 with O_2 is accelerated by its native **Thr** substrate by a remarkable ~8000-fold relative to the reaction lacking a carrier–protein substrate.[62] This 'triggering factor' represents the largest seen for an Fe/2OG oxygenase and afforded a large dynamic range to assess its structural determinants. By examination of the SyrB2 reaction with SyrB1 and CytC2 carrier protein substrates appended by either the native L-threonine or other non-native amino acids, determinants of triggering were resolved. The L-threonine-appended aminoacyl-CytC2 carrier–protein substrate triggered formation of the ferryl complex in SyrB2, but with a significantly (~40-fold) diminished efficacy and a concentration dependence indicative of relatively weak binding. By contrast, the native SyrB1 appended by the non-native amino acid, L-valine, exhibited "saturation kinetics" indicative of tighter binding, although it also triggers less effectively than the native **Thr** substrate at saturating concentrations. These observations imply that the majority of the binding determinants are between the two proteins. Nevertheless, SyrB1 having the phosphopantetheine arm but no appended amino acid fails to trigger formation of the ferryl complex to a significant extent, even in the presence of free, untethered amino acid. More strikingly, the triggering efficacy varies by more than 20-fold with the identity of the appended amino acid. Not surprisingly, the native **Thr** has the greatest efficacy, but SyrB1 appended by 2-aminobutyrate, valine, alanine and cyclopropylglycine (in each case the L isomer) all trigger formation of the ferryl complex by at least a factor of 300 compared to the carrier protein lacking an appended amino acid. The conclusions are that the carrier protein directs the amino acid into the halogenase active site by specific protein–protein interactions, and then the halogenase sensitively responds to the structure of the amino acid that caps the phosphopantetheine arm. The resultant experimental flexibility to vary the chemical nature of the target almost at will without losing the ability to interrogate the reactivity of the ferryl complex had not previously been seen for an Fe/2OG oxygenase and ultimately made it possible to understand how the halogenases specify their unique outcome. It is notable that the development of this toolkit ran counter to a previous report that self-charging by SyrB1 is highly specific for L-Thr.[19] This prior study had relied on an indirect assay for exchange of radioactive pyrophosphate into unlabelled ATP as a consequence of reversible adenylation. The development of a simpler, more direct assay (based on the classic Ellman's reagent) for protection of the phosphopantetheine thiol group allowed for preparation of non-native substrates in high yield.

3.3.3 Clues as to How SyrB2 Directs Halogenation in Preference to Hydroxylation

Verification of the proposed H•-abstracting haloferryl complexes in the CytC3 and SyrB2 reactions raised an intriguing question: how do the halogenases prevent the substrate radical from coupling with the hydroxyl ligand, as readily occurs in the related hydroxylases? At that time, several reports had suggested that the halogenases do not hydroxylate their substrates to an appreciable extent;[19,20] they were thought to be highly selective for halogenation. Although our studies on CytC3 and SyrB2 subsequently showed that both enzymes do, in fact, hydroxylate their native substrates to some extent,[63] the question of control of the fate of the substrate radical/*cis*-halo-hydroxo-ferric intermediate state nevertheless became the central mechanistic question and captivated the computational chemistry community. Multiple studies suggested elaborate chemical mechanisms for the radical-transfer selectively.[66–68] These hypotheses included proton shuttling to the hydroxyl ligand to convert it to water and preclude its transfer to the substrate radical, and capture of the hydroxyl ligand (as hydroxide) by the CO_2 produced upon decarboxylation of 2OG. The study probing substrate triggering in SyrB2 provided the first experimental clues to the correct answer to this question.[62] Firstly, decay of the ferryl complex in the presence of SyrB1 presenting L-2-aminobutyrate (**Aba**) was found to be faster than with the native substrate, **Thr**, by ~10-fold. In general, it is quite unusual for an enzyme to process a non-native substrate (one that is not chemically activated) more efficiently than it does the native substrate. Secondly, SyrB1 presenting an amino acid that could not undergo H• abstraction either because of the absence of the target C4 carbon (L-alanine) or because it possesses a very strong (106 kcal mol^{-1} BDE) target C–H bond (L-cyclopropylglycine) was found to stabilize the haloferryl state to an unprecedented extent (half-life of 0.5–2 h at 0 °C).[62] The two observations hinted at (i) a programmed inefficiency in H• abstraction by the haloferryl complex and (ii) protein scaffolds co-adapted to succeed in spite of such sluggish reactions by protecting the reactive intermediate from unproductive decay pathways to a remarkable extent.

3.3.4 Evidence that Substrate Positioning Ensures Halogenation Rather than Hydroxylation

Comparison of the kinetics and outcomes of the SyrB2 reactions with **Thr**, **Aba** and a third substrate, L-norvalinyl-SyrB1 (**Nva**), having an additional methylene unit in its side chain, allowed the physical basis for halogenase chemoselectivity to be revealed. As reported in the literature, **Thr** is almost exclusively chlorinated (although hydroxylation was detected for the first time in this study). By contrast, **Aba** was found to undergo hydroxylation and chlorination to comparable extents (Scheme 3.2).[63] **Nva** extended the telling correlation between rate of H• abstraction and outcome: it was found to support H• abstraction at ~130 times the rate with **Thr** (~10 times that with **Aba**) and to be *almost exclusively hydroxylated*. With two different carbons, the C4 methylene and C5 methyl, potentially able to react with the haloferryl complex, it was important to determine the site(s) of modification. By use of regio-specifically deuterated **Nva** substrates, it was found that abstraction of H• from C5 is favoured over abstraction from C4 by a factor of ~6, and that the C5 radical undergoes exclusively HO• rebound. No C5-chlorinated product could be detected. The minor competing chlorination was found to occur exclusively at C4, which also undergoes hydroxylation to a lesser extent. Remarkably, deuterium substitution of (only) C5 redirects the haloferryl complex to C4 by virtue of the large ^{2}H-KIE, and the result is a *switch in the primary outcome from C5 hydroxylation to C4 chlorination*. The observations imply that Cl• transfer requires a specific disposition of the substrate radical relative to the halo-hydroxo-ferric complex, one that is more favorable for Cl• transfer than for the default HO• rebound. Restraining the target carbon closer to the halogen and farther from the oxygen in this state would, *a priori*, make H• abstraction by the haloferryl complex in the preceding state less favorable. In other words, the halogenases bind their substrates so as to sacrifice proficiency in H• abstraction for selectivity in the subsequent radical group transfer step. Modifications to the appended amino acid that subvert this strict positioning (removal of a tethering hydroxyl group in **Aba** and extension of the side chain in **Nva**) unleash the inherent H•-abstraction potency of the haloferryl complex but consequently also disable the enzyme's control of radical-transfer selectivity.

In silico analyses of the SyrB2 reactions have generally supported the primary role of substrate positioning in directing halogenation.[66–68] The most recent of these studies further suggested that the orientation of the substrate's target C–H bond with respect to the haloferryl complex results in H• abstraction by a π-pathway rather than a σ-pathway and that the engagement of different frontier orbitals in this first step then favours Cl• transfer over HO• rebound in the subsequent step.[69] These studies have necessarily relied on computational models for the halogenase:substrate complex, and their conclusions are, therefore, inherently speculative. Thus, direct structural evidence for the proposed trends in substrate positioning remains an important goal for the future. Such evidence could be obtained by direct measurement of active site distances using hyperfine coupling between specifically deuterium labelled substrates and a SyrB2-$\{Fe–NO\}^7$ complex in pulse EPR spectroscopy, as previously applied to TauD.[70] Ultimately, the solution of high-resolution X-ray crystal structures of the enzyme with its native and non-native substrates remains an important objective.

A more recent study on SyrB2 showed that the mechanistic logic employed by the halogenases and IPNS – abstraction of H• by a ferryl complex and radical-group transfer of a ligand coordinated *cis* to the oxygen – can be extended to installation of other functional groups.[71] It was discovered that, just as binding of the aminoacyl-SyrB1 substrate triggers O_2 activation for ferryl complex formation, so does binding of anion to the Fe(II) cofactor. A number of non-halide ligands were shown to bind, with detectable shifts in the MLCT absorption band at ~500 nm. All anions seen to bind could also support formation of the ferryl complex. Two non-halide anions, nitrite and azide, were found to undergo unprecedented C–N bond coupling reactions to the C4 carbon in **Aba**. The 4-nitro- and 4-azido-**Aba** products were produced in modest yields by the wild-type protein, but yields of both products were enhanced by the A118G substitution, which was anticipated to expand the anion pocket. The ability of SyrB2 to install a variety of functional groups onto completely unactivated aliphatic carbons in a regiospecific fashion could impact the biotechnology and pharmaceutical industries by permitting more direct and economical preparation of starting materials. Although achieving such a goal will require significant additional developments, these studies lay the foundation for such inquiry. With respect to the goal of evolving halogenases for synthetic utility, the requirement for a carrier protein to present the substrate would represent a significant limitation to the scope of possible applications. It is therefore an exciting development that the first small-molecule (carrier protein-independent) aliphatic halogenase has just been reported.[72]

3.4 The Epimerization Catalysed by Carbapenem Synthase

The global emergence of multidrug-resistant pathogens is increasingly recognized as a major threat to human health.[73] Among the hundreds of clinically used antibiotics, carbapenems are often considered a last-line antibiotic to fight these resistant pathogens due to their broad spectrum of activity against both Gram-positive and Gram-negative bacteria including *Staphylococcus epidermidis*, *S. aureus* and *Pseudomonas aeruginosa*.[21,74] Although carbapenems are natural products isolated from bacteria such as *Streptomyces cattleya* and *Pectobacterium carotovorum*, their low concentrations in the host strains and instability during lengthy purification processes renders isolation from the native organism impractical for large-scale drug production.[21,75] As a result, the synthetic preparation of carbapenems has assumed greater importance in the past two decades.[76] Unfortunately, the introduction of multiple chiral centres and installation of an olefin group onto the five-membered ring of the strained bicyclo-[3.2.0] skeleton of carbapenam makes convenient and economical chemical synthesis of carbapenem drugs very challenging. However, developments in bioengineering have the potential not only to furnish more time- and cost-efficient production of known carbapenems but also to provide routes to new unnatural analogues to combat the growing resistance problem. Such biotechnological exploitation will be possible only with a robust understanding of the natural biosynthetic pathways to carbapenems.

3.4.1 The Carbapenem Biosynthesis Gene Cluster

Bioinformatics analysis of the gene cluster associated with (5*R*)-carbapenem production and *in vivo* experiments suggest that only three enzymes are required for biosynthesis of (5*R*)-carbapenems.[28] CarB, carboxymethylproline synthase, uses malonyl–coenzyme A (CoA) as a cosubstrate to attach a carboxymethylene unit to 1-pyrroline-5-carboxylate to form *trans*-5-carboxymethyl-L-proline. CarA, carbapenam synthetase, then uses ATP to drive β-lactam ring closure, generating (3*S*,5*S*)-carbapenam.[21] The stereochemical configuration of C5 set by CarB is opposite to that found in all of the >45 carbapenems isolated to date.[21] Maturation of (3*S*,5*S*)-carbapenam to (5*R*)-carbapenem is carried out by the Fe/2OG oxygenase, carbapenem synthase (CarC), and involves stereoinversion of C5 and desaturation across the C2–C3 bond.[77,78] The transformations catalysed by CarB and CarA are well studied and have been engineered to produce various (3*S*,5*S*)-carbapenam precursors.[21,79] Understanding of the last two transformations had been thwarted by both protein and substrate/product instability.[75,80]

By contrast to the oxidative reactions (*e.g.* hydroxylation, halogenation, cyclization and desaturation) catalysed by other Fe/2OG oxygenases,[1,2,17] the CarC-catalysed stereoinversion reaction is redox neutral with respect to its primary substrate. This outcome has been rationalized by mechanisms invoking H• abstraction from C5 by the canonical ferryl species and subsequent transfer of H• to the opposite face of the resulting substrate radical by, presumably, a donor amino acid.[77,78] *In silico* analysis provided support for such a mechanism and implied the involvement of a dedicated H• donor.[81] It had been suggested that Tyr-67 serves as the dedicated H• donor, on the basis of both an X-ray crystal structure of CarC with the substrate analogue, *N*-acetyl-L-proline, bound to the active site (PDB ID: 1NX8) and amino acid substitution studies coupled to a fluorescence-based *in vivo* assay for carbapenem production.[80–84] Recently, the mechanism of C5 stereoinversion was unambiguously established by a combination of rapid-kinetic and spectroscopic techniques along with the first structure of CarC in complex with its authentic substrate, (3*S*,5*S*)-carbapenam-3-carboxylate (PDB ID: 4OJ8).[85]

3.4.2 The Mechanism of Fe/2OG-Mediated Stereoinversion

The proposed mechanism for the CarC-mediated C5 stereoinversion reaction features abstraction of H• from C5 by a high-spin ferryl complex (Scheme 3.3, state **II**), in analogy to other Fe/2OG oxygenases.[17] The accumulation of this intermediate was demonstrated by FQ-Mössbauer experiments. The spectrum of a sample prepared by mixing the CarC:Fe(II):2OG:substrate complex with excess O_2 at 5 °C and freeze-quenching at a reaction time of

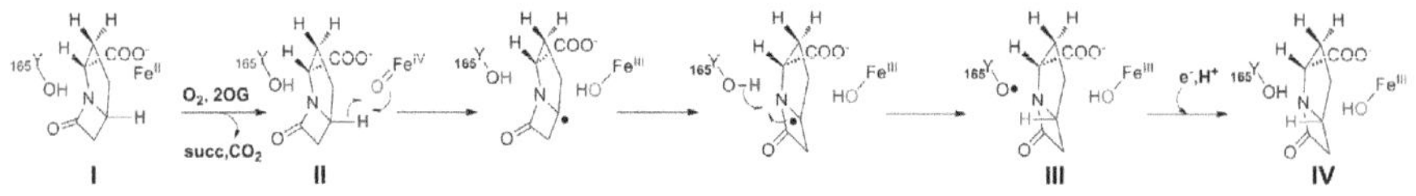

Scheme 3.3 Mechanism of the CarC stereoinversion reaction.

0.15 s exhibited a quadrupole doublet with isomer shift (δ) of 0.28 mm s^{-1} and quadrupole splitting (ΔE_Q) of 0.87 mm s^{-1}, parameters very similar to those of high-spin (S = 2) ferryl complexes in other Fe/2OG oxygenases.[17] The decay of this intermediate was accompanied by formation of a high-spin (S = 5/2) Fe(III) species, consistent with formation of states **III** and **IV** in Scheme 3.3.

SF-Abs experiments initiated by mixing CarC:Fe(II):2OG:substrate with excess O_2 showed that decay of the ferryl complex is associated with formation of a tyrosyl radical (Scheme 3.3, state **III**). A sharp absorption feature at 410 nm and a broader peak at 390 nm, unmistakable hallmarks of a tyrosyl radical (Y•),[86] reached their maximum intensity at a reaction time of ~3 s and then decayed slowly.

X-band EPR spectra of samples freeze-quenched at reaction times of 0.5, 3.0 and 30 s provided further evidence for the Fe(III)- and Y•-containing intermediate (state **III** of Scheme 3.3). Spectra collected at 10 K exhibited two signals with the same kinetics as the 410-nm feature of the tyrosyl radical observed by SF-Abs. The first, a broad absorption-like signal centered at g = 6.95, was attributed to a high-spin Fe(III) centre (S = 5/2) with positive axial zero-field splitting parameter, D, and a rhombicity, E/D, of ~0.05. The second signal at g = 2.0 was assigned to the tyrosyl radical. The fact that the line shape of the g = 2.0 signal is atypical of a magnetically isolated Y• was rationalized by dipolar coupling with the nearby paramagnetic Fe(III) centre, and spectra collected at 100 K corroborated this interpretation. The signal at g = 6.95 was seen almost to vanish at 100 K, owing to more rapid spin relaxation of the Fe(III) centre. The signal at g = 2 became markedly sharper and the hydrogen hyperfine couplings typical of a Y• became discernible, because at this elevated temperature the faster spin relaxation of the Fe(III) species effectively decouples the spins of the two paramagnetic centres on the EPR timescale.

Liquid chromatography–mass spectrometric (LC–MS) analysis of chemically quenched samples confirmed that C5 stereoinversion occurs in the conversion of **II** to **III** (Scheme 3.3). LC-MS analysis of a sample of the CarC : Fe(II) : 2OG : substrate complex not exposed to O_2 gave a single sharp peak corresponding to the substrate. The chromatogram from a sample that was acid-quenched after reacting for 0.15 s (when state **II** predominates) had only a small shoulder for the product. By 3 s, the time at which Y• was seen to accumulate maximally in SF-Abs experiments, the intensity of the product peak had reached its maximum intensity; the trace for the 10 s sample revealed little additional conversion.

3.4.3 Identity of the Hydrogen Atom Donor

The identity of the H•-donating tyrosine was established by the structure of CarC in complex with its authentic (3*S*,5*S*)-carbapenam substrate and site-directed mutagenesis. The structure, solved at a resolution of 2.1 Å by X-ray diffraction experiments, permitted unambiguous assignment of the location and orientation of the substrate (PDB ID: 4OJ8).[85] The bridgehead C5 resides 4.4 Å from the Fe(II), with its hydrogen atom directed toward the cofactor in appropriate geometry for its abstraction by the ferryl intermediate. With substrate bound, clear electron density was discerned for two previously disordered loops, residues 67 to 78 (loop 1) and 162 to 172 (loop 2), that close over the active site, thereby isolating the substrate from bulk solvent. The most mechanistically important change relative to published structures was the ordering of loop 2 to position Tyr-165 directly above C5, opposite the iron centre, with its hydroxyl group located 4.8 Å from the substrate C5 atom. In this position, Tyr-165 is ideally poised to donate H• to the C5 radical to generate the (3*S*,5*R*) epimer, whereas other residues suggested in previous work to serve in this capacity[80,84] are positioned inappropriately.

The implication that Tyr-165 is the hydrogen atom donor was confirmed by SF-Abs experiments using variants of CarC; the Y67F, Y164F and Y191F variants were all shown to form the sharp 410 nm peak, although the amplitudes and kinetics were perturbed to varying extents by the substitutions. By contrast, no 410-nm signal was seen in the reaction of the Y165F variant, and no stereoinversion was detected. Catalytic rather than stoichiometric consumption of the substrate by the Y165F variant was interpreted to imply that removal of the H• donor reverts CarC to a functional, catalytic oxygenase. However, the nature of the altered outcome has not yet been elucidated.

The results suggest that the presence and location of Tyr-165 may be necessary and sufficient to direct the redox-neutral stereoinversion outcome in preference to the default hydroxylation. However, by contrast to the case of the halogenases, in which the enzyme itself is solely responsible for directing the outcome, the CarC reaction may have an additional control element built into the *substrate*. Computational analyses have suggested that the substrate radical formed by abstraction of H• from C5 would have an approximately planar geometry.[81] With other parts of the substrate held in place by interactions with the protein, such a conformational change in the substrate would probably move the C5• away from the Fe(III)–OH complex, disfavouring hydroxylation. Such a change would also move the carbon *toward* Tyr-165, further optimizing for H• donation at the opposite face of the stereocentre.[81] Thus, in CarC as in SyrB2, an 'alternative' reactivity is achieved by presentation of a different radical-coupling partner (H• from Tyr-165 or Cl•) and precise substrate positioning to divert the reaction from the default hydroxylation.

3.5 Oxacyclizations Mediated by the Fe/2OG Oxygenases and Related Enzymes

A number of bioactive natural products have heterocyclic rings that contain oxygen (oxacycles). Three examples – the bacterially produced β-lactamase inhibitor, clavulanic acid, the plant-derived anaesthetic, scopolamine, and the fungal insecticide, norloline – are shown in Scheme 3.4A. In the pathway to each of these compounds, an Fe/2OG oxygenase first hydroxylates a methylene carbon and subsequently couples the nascent hydroxyl group to a nearby unactivated methylene carbon to close the oxacycle.[23–27,29,87] The enzymes catalysing these reactions are known as clavaminic acid synthase (CAS), hyoscyamine 6β-hydroxylase (H6H) and *N*-acetylnorloline synthase (LolO), respectively. For the case of CAS, these hydroxylation and oxacyclization reactions are the first and second of *three* distinct reaction types mediated by the same remarkable enzyme. The third reaction, an olefin-installing 1,2-dehydrogenation reaction, is discussed in Section 3.6. In each of the oxacyclization reactions of Scheme 3.4B, a carbon–hydrogen bond and an oxygen–hydrogen bond are broken and a new carbon–oxygen bond is formed to close a 3- or 5-membered ring. The reactions are thus formally 1,3- or 1,5-dehydrogenations. The mechanisms of the oxacyclization reactions are poorly understood, but all published hypotheses of which we are aware have invoked ferryl complexes as the initiating intermediates.[1–3,17] In light of the direct evidence for ferryl intermediates in the reactions of the hydroxylases, halogenases and CarC, we consider this hypothesis to be sound.

Two other mononuclear non-haem iron enzymes, IPNS (see Chapter 19)[56] and (*S*)-2-hydroxypropyl-1-phosphonate epoxidase (HppE),[88] mediate analogous heterocyclization steps *without requiring a cosubstrate* (Scheme 3.4C). These enzymes produce the core that is further elaborated to produce all penicillin antibiotics (IPNS) and the specific antibiotic fosfomycin (HppE). In these cases, somewhat more is known about the reaction mechanisms. As for the reactions of the Fe/2OG enzymes, ferryl complexes are strongly implicated as the initiating intermediates. For the case of IPNS, the two additional electrons needed to balance the formation of the ferryl complex from the resting Fe(II) cofactor and O_2 are extracted directly from the substrate in the first cyclization step that closes the β-lactam ring of the bicyclic penicillin core. Here, an Fe(III)–superoxo complex abstracts the *pro-R* hydrogen of the β-carbon of the Cys moiety of the L-δ-aminoadipoyl-L-cysteinyl-D-valine (ACV) substrate. An inner-sphere single electron transfer then produces an electrophilic thioaldehyde, which permits β-lactam closure *via* nucleophilic attack of the D-Val amide nitrogen on the L-Cys β-carbon, and leads to formation of the ferryl complex. The ferryl complex removes hydrogen (presumably as H•) from the β-position of the D-Val to initiate closure of the five-membered thiazolidine ring, which completes the bicyclic penicillin nucleus.[56,58] For the case of HppE, we recently showed that the oxidizing cosubstrate is not, as had been thought for more than a decade, O_2 but is rather the reduced form of O_2, H_2O_2.[89] Reaction of the resting Fe(II) cofactor with H_2O_2 is balanced for direct formation of

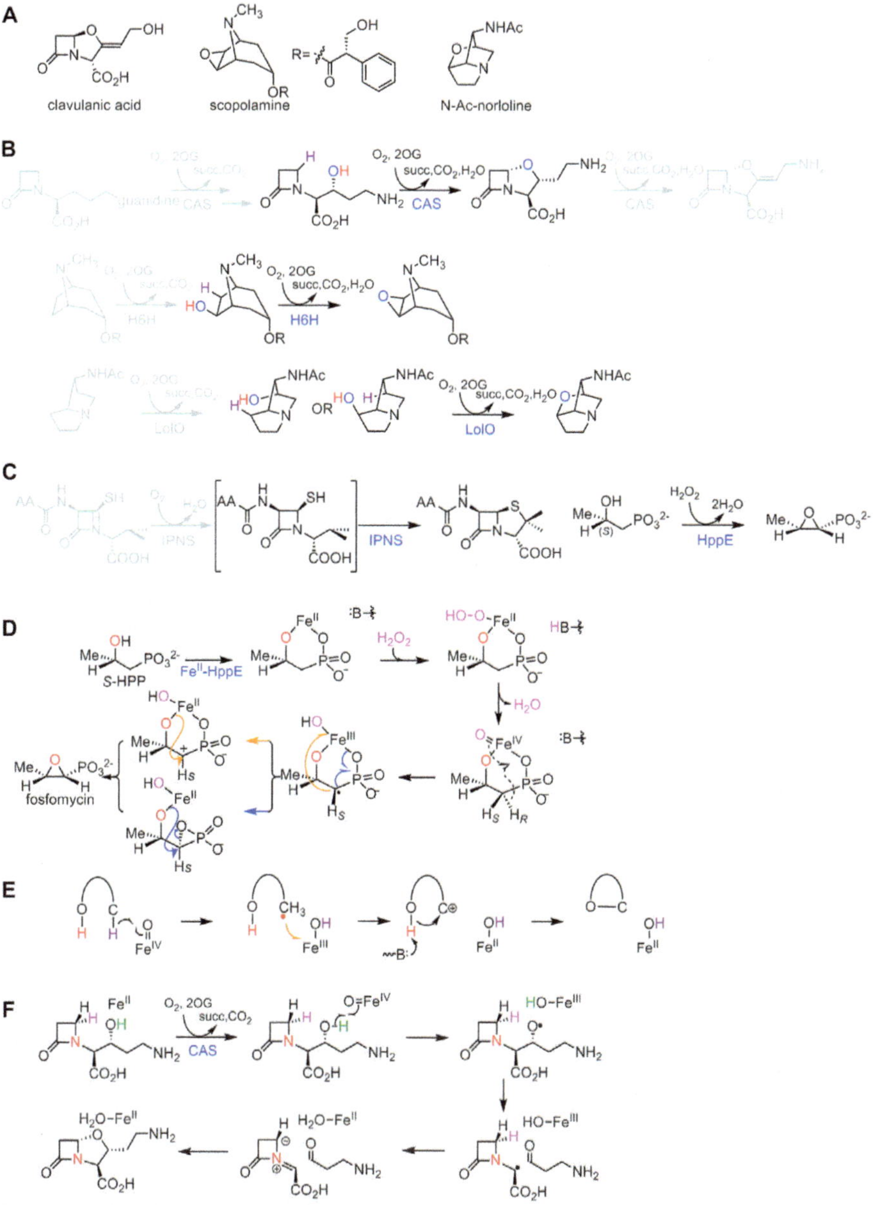

Scheme 3.4 (A) Three examples of oxacyles constructed by the Fe/2OG oxygenases CAS, H6H and LolO, respectively. (B) The CAS (top), H6H (middle) and LolO (bottom) reactions. (C) Ring closure reactions catalysed by the 2OG-*independent* iron(II)-dependent enzymes, IPNS (left) and HppE (right), in isopenicillin *N* and fosfomycin biosynthesis. (D) Current working hypothesis for the epoxidation reaction catalysed by HppE. (E) General carbocation mechanism for Fe/2OG-oxygenase-mediated oxacyclization reactions. (F) Hypothetical mechanism for CAS reaction suggested by Borowski and coworkers.[96]

the presumptive ferryl complex (which to date has not been directly detected). The novel peroxidative nature of the HppE reaction thus obviates a source of the two reducing equivalents that would have been required for the O_2-driven epoxidation and had been sought since the enzyme's discovery. The presumptive ferryl complex abstracts hydrogen (again, we presume as H•) from the *pro-R* position of C1.[90]

In considering possible mechanisms of the ring-closure steps in the IPNS and HppE reactions, it seems crucial that, in the reactant complex of each enzyme, the heteroatom (sulfur for IPNS and oxygen for HppE) that undergoes coupling to the carbon atom coordinates directly to the Fe(II) cofactor at a position that is *cis* to the expected location of the oxo group in the presumptive ferryl complex.[91,92] Both reactions can thus be envisaged to emerge from halogenase-like 'alternative' radical group transfer mechanisms, in which the carbon radicals generated by the ferryl complexes attack the coordinated heteroatoms. However, for HppE, it has been suggested that ring closure following C1 radical formation is a two-step process, involving either (i) formation of a C1 carbocation by electron transfer (ET) from the C1 radical to the Fe(III)–OH complex and a polar $C2–O^- \rightarrow C1^+$ coupling or (ii) an initial radical coupling by a phosphonate O atom followed by polar attack of the C2 alkoxide O atom to open the strained C–O–P three-membered ring and close the epoxide ring of the fosfomycin product (Scheme 3.4D).[89] Either ring-closure mechanism could explain the unusual inversion of the C1 configuration associated with epoxide closure. The first possibility may be particularly relevant to the mechanistic conundrum posed by the 2OG-dependent oxacyclizations (discussed below).

An important distinction between the 2OG-*independent* IPNS and HppE cyclization reactions and those catalysed by the 2OG-*dependent* oxygenases is that, in each of the latter cases, the cosubstrate occupies two coordination sites of the Fe(II) cofactor in its O_2 reactive state.[1–3] The cosubstrate and the 2-histidine/1-carboxylate 'facial triad' of iron ligands supplied by the protein thus occupy a total of five sites. On the basis of the now classic studies of Solomon and coworkers on CAS and other Fe/2OG oxygenases,[28,87] one would anticipate that a water molecule occupies the sixth site and dissociates upon binding of the primary substrate. The resultant square-pyramidal Fe(II) centre would have only a single open site, to which O_2 would then add to initiate ferryl formation. The relevant point is that the oxygen atom of the substrate that undergoes coupling to carbon is unlikely to coordinate directly to the cofactor and, indeed, in the X-ray crystal structure of the CAS:proclavaminate (PCA) complex (the only relevant structure available), the oxygen does not coordinate.[93] Without coordination of the heteroatom, a halogenase-like radical-coupling mechanism for ring closure is implausible: it is unlikely that a bare C–O• could form in order to couple to the C• generated by the ferryl complex. The halogenase solution to the problem of site insufficiency – absence of the carboxylate ligand of the facial triad to open a site for heteroatom coordination – is not employed here: the published structures of CAS and alignments of its amino acid sequence with those of H6H and LolO imply that the three enzymes retain the carboxylate ligand. The halogenase/IPNS paradigm of alternative radical-group transfer seems unlikely to be operant in the oxacyclization reactions.

An alternative mechanism, suggested by the aforementioned studies on HppE[94] and proposed even earlier for CAS,[95] provides one possible resolution to this apparent conundrum: carbocation formation *via* electron transfer (ET) from the carbon radical to the Fe(III)–OH species, followed by polar R–O$^-$ → C$^+$ coupling, could potentially rationalize all three of the 2OG-dependent oxacyclization reactions (Scheme 3.4E). Without coordination of the O atom to the Lewis-acidic cofactor to promote alkoxide formation (as in HppE), general-base assistance might be important in the ring closure steps. In principle, the Fe(II)–OH cofactor produced by the ET step might be basic enough to serve in this capacity.

Although this carbocation mechanism is appealing because it can potentially rationalize all three of the reactions in Scheme 3.4B, the aforementioned structure of the CAS:PCA complex engenders doubt. The expectation that the C4′ carbon atom that undergoes coupling to the oxygen should be closest to the iron cofactor to permit facile hydrogen atom transfer (HAT) to the ferryl complex, initiating C4′ carbocation formation and subsequent polar ring closure, is not borne out. C4′ is positioned a distant 5.4 Å from the iron and out of line with the expected location of the ferryl oxo moiety (Figure 3.1).[93] Surprisingly, the *oxygen* atom to be coupled to C4′ is positioned just 4.2 Å from the cofactor and in line with the expected location of the ferryl oxo group. The stark implication of the structure is that the presumptive ferryl complex might actually abstract H• from the heteroatom! Indeed, Borowski *et al.* proposed just such a mechanism initiated by O–H rather than C–H cleavage. In this mechanism (Scheme 3.4F),[96] formation of the oxygen radical initiates a radical β-scission reaction, producing the carboxylato(2-oxoazetidin-1-yl)methyl radical and 3-aminopropanal. Coupled transfer of a C4′ proton and an electron (PCET) from the radical species to the Fe(III)–OH cofactor then produces a zwitterionic (ylide) intermediate that re-couples with the 3-aminopropanal in a [3 + 2] dipolar cycloaddition that

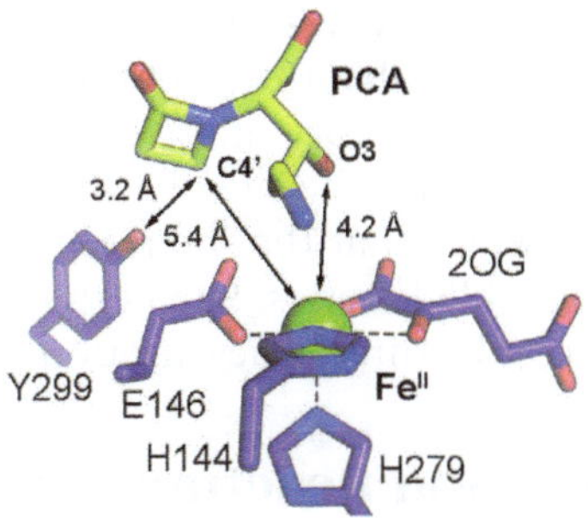

Figure 3.1 Model of the CAS:PCA complex from X-ray crystallographic analysis (PDB ID: 1DRT). The distances of the Fe cofactor to the atoms from which hydrogen may be removed during oxacyclization (O3 and C4′) are indicated.

simultaneously repairs the fragmented C–C bond and forms the new C–O bond. DFT calculations favoured this unusual pathway over the previously proposed mechanisms initiated by HAT from C4′.[87,89]

Attempts to formulate cognate mechanisms initiated by O–H rather than C–H cleavage for the H6H and LolO reactions lead to steps that range from unlikely to seemingly impossible. Thus, the CAS:PCA structure[93] and computational work of Borowski *et al.*[96] challenge the notion, inspired by Occam's Razor, that this apparently common reaction type might proceed by a conserved mechanism. Answering the general question of whether a given reaction is initiated by C–H or O–H cleavage would seem feasible following the precedents of the aforementioned studies on the hydroxylases, halogenases and CarC. Structures of H6H and LolO in complex with 2OG and their oxacyclization substrates should indicate whether the carbon atom to be coupled into the ring is projected toward the iron cofactor or, as in the CAS:PCA structure, the *oxygen* atom is more favorably disposed for HAT. More definitively, transient-state kinetic and spectroscopic analysis coupled with product quantification should permit two key predictions to be evaluated. First, if the reactions proceed *via* HAT from carbon, then substitution of that position of the substrate with deuterium should increase the lifetime of the presumptive ferryl complex. Conversely, for a reaction initiated by HAT from the O–H group, which undergoes facile exchange with solvent water, the lifetime of the ferryl intermediate should be enhanced in ^2H_2O solvent. The initiating step might thus be immediately deduced by transient-state kinetic and spectroscopic experiments akin to those discussed above for the hydroxylases, halogenases and CarC. Second, if any of the enzymes exhibits even modest uncoupling of 2OG decarboxylation from substrate oxidation (*i.e.* an experimental oxacycle:succinate stoichiometry less than unity) when reacting with its all-protium substrate, then challenging it with appropriately deuterium-labelled substrate (for a C–H cleavage mechanism) or ^2H_2O (for an O–H cleavage mechanism) should lead to an even greater fraction of uncoupled ferryl-decay events and a diminished oxacycle:succinate stoichiometry. This phenomenon was illustrated most starkly for the case of the CarC C5 stereoinversion reaction. Here, the ferryl intermediate was seen to have a ~20% 'failure rate' with the all-protium substrate, but substitution of C5 with deuterium pushed the failure rate to ~85% as a consequence of the large intrinsic D-KIE on the initiating HAT step. By this analysis, the site of HAT to the ferryl complex could be deduced, even in the event that the intermediate fails to accumulate as a consequence of unfavorable kinetics. Such experiments should clarify the initiating events in each of the three oxacyclization reactions in Scheme 3.4B.

3.6 C–C Desaturation Reactions Mediated by the Fe/2OG Oxygenases

Conversion of a C–C single bond into a double bond (olefin) by a 1,2-dehydrogenation reaction occurs in the biosynthesis of unsaturated fatty acids and a number of more complex bioactive natural products.[21] In many cases, the olefin is itself essential to the compound's bioactivity, whereas in other cases, the olefin permits subsequent couplings to other atoms within and between biosynthetic intermediates. Fe/2OG oxygenases are known to mediate such reactions in, for example, the biosyntheses of clavulanic acid, carbapenems, and 4,5-dehydro-L-arginine precursor to tetrahydroisoquinoline natural products (Scheme 3.5A).[65,91]

Desaturation across a C–C single bond between completely unactivated carbons requires cleavage of two strong C–H bonds. For example, fatty acyl desaturases introduce olefins into unfunctionalized alkyl chains. These enzymes use non-haem di-iron cofactors to activate oxygen and cleave the two adjacent C–H bonds.[97] Although the mechanisms of these desaturation reactions are incompletely understood, it has been suggested that an intermediate similar to the di-iron(IV) complex, **Q**, from the reaction of soluble methane monooxygenase (sMMO) could initiate desaturation by abstracting the first H•.[97] Given that **Q** in sMMO can cleave the stronger C–H bond of methane, with its homolytic bond dissociation energy (BDE) of 104 kcal mol^{-1}, an analogous complex in the desaturase should be competent to cleave the weaker methylene C–H bond in a fatty acid substrate (BDE ~98 kcal mol^{-1}). This step would leave an Fe_2(III/IV) complex to abstract the second H•, either in a single step or perhaps in sequential ET and proton transfer

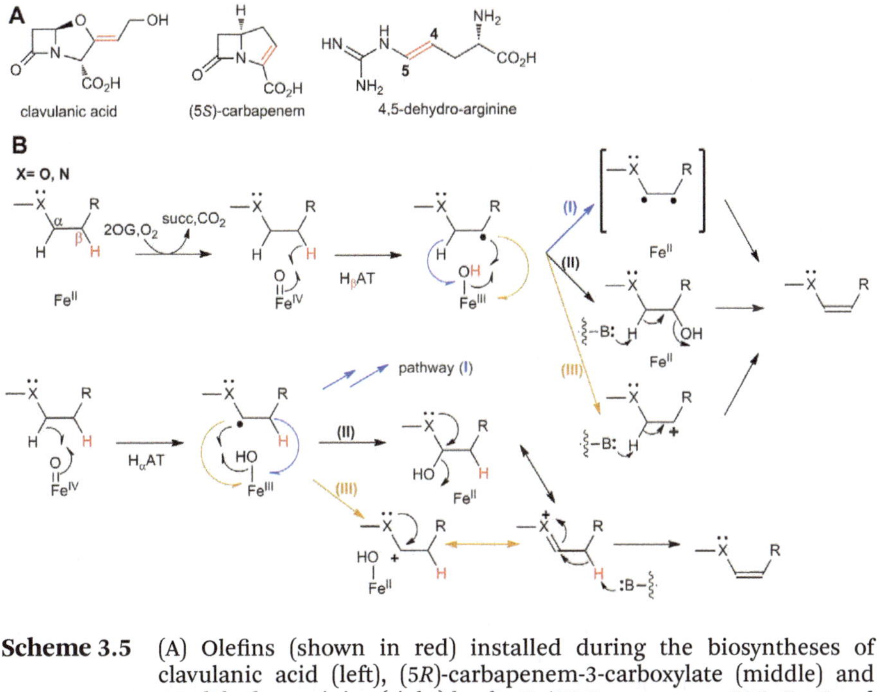

Scheme 3.5 (A) Olefins (shown in red) installed during the biosyntheses of clavulanic acid (left), (5*R*)-carbapenem-3-carboxylate (middle) and 4,5-dehydro-arginine (right) by the Fe/2OG oxygenases CAS, CarC and Cya18 (or NapI), respectively. (B) Possible mechanisms for the desaturation reactions.

(PT) steps (ET/PT). A complex of this oxidation state in the β subunit of class Ia ribonucleotide reductase (cluster **X**) is sufficiently potent to generate the enzyme's catalytically essential tyrosyl radical by cleaving the phenolic O–H bond (BDE of ~87 kcal mol^{-1}).[86] Given the anticipated activating effect of the α radical on the second C–H bond to be cleaved in such reactions, desaturation between carbons devoid of activating functionality is readily rationalized within this manifold for O_2 activation. Similarly, haem-containing cytochrome P450 enzymes, with their defining Fe(IV)-oxo/ligand radical [formally Fe(V)] intermediate known as compound I, also appear poised to mediate even relatively difficult desaturations.[98] HAT to compound I leaves a still potent Fe(IV)-(hydr)oxo complex (compound II) to abstract the second H•, again either in a single HAT or PCET step or by a two-step ET/PT mechanism. Indeed, such enzymes are also known to mediate 1,2-dehydrogenations.

The Fe/2OG oxygenases seem less well equipped for the consecutive H• abstractions (or HAT followed by ET/PT) envisioned in the aforementioned desaturation reactions. The Fe(III)–OH complex resulting from HAT to the ferryl complex would seem to be, at best, a relatively mild oxidant/H• abstractor. Nevertheless, a hypothesis that has been advanced in multiple publications does indeed invoke HAT or PCET from the initial substrate radical produced by the ferryl complex to the Fe(III)–OH form of the cofactor.[64,95,99] As we interpret them, these published proposals appear to envisage the target C–C bond of the substrate as alkyl-like: they do not explicitly depict a role for activating functional groups. However, inspection of the three reactions in Scheme 3.5A reveals that a heteroatom with a lone pair is always bonded to one of the two carbon atoms between which the olefin is installed. The presence of this heteroatom may enable reaction pathways not involving a second HAT or PCET to the modestly potent Fe(III)–OH complex, as outlined in (Scheme 3.5B). We suggest that the heteroatoms may play crucial roles in enabling 1,2-dehydrogenation outcomes *via* the thermodynamically limited Fe(II)/Fe(IV) manifold of the Fe/2OG oxygenases and, conversely, that desaturation of substrates lacking such activating functionality by Fe/2OG oxygenases might not be encountered.

With the two carbon atoms of the nascent olefin differentiated by the presence of the heteroatom, and under the assumption that the reactions are initiated in the usual way by the canonical ferryl intermediate, one can classify the conceivable mechanisms into those initiated by HAT from the β position relative to the heteroatom (Scheme 3.5B, top) and those initiated by HAT from the α position (Scheme 3.5B, bottom). In either case, the carbon radical then has three possible fates: (**I**) as previously proposed, it could undergo a subsequent PCET from the adjacent carbon to form the olefin directly; (**II**) it could undergo a radical coupling with the OH ligand of the Fe(III)–OH complex to produce a hydroxylated intermediate, with a subsequent polar dehydration (1,2-elimination) step giving the desaturated product; (**III**) it could undergo ET from the radical to the Fe(III)–OH cofactor to produce a formal carbocation, with a subsequent proton transfer (PT) step generating the olefin. If the Fe(II)–OH cofactor resulting from the ET step were to act as the base in the PT step, then the third possibility would become just a variant on the PCET mechanism in which the ET and PT steps are uncoupled.

For the H_βAT mechanisms, all conceivable pathways raise questions. In **I**, the potency of the Fe(III)–OH species to abstract the second (α) H• is in doubt, as noted above. In **II**, the acidity of the α hydrogen in the β–hydroxylated intermediate might be insufficient to permit the envisaged 1,2-elimination of (H)OH from the hydroxylated intermediate. In **III**, it is not clear that the stability of the secondary β-carbocation and the reduction potential of the Fe(III)–OH cofactor would make the ET step feasible. For the H_αAT mechanisms, the thermodynamic feasibility questions in pathways **II** and **III** would seem to be mitigated by the proximity of the heteroatom with its lone pair. Thus, elimination of (H)OH from the intermediate could be facilitated by formation of an imminium/oxonium-like intermediate. The ET step in the ET/PT mechanism of pathway **III** could give the same intermediate more economically, in a single step from the initial substrate C_α radical. For these reasons, we favour pathways **II** and **III** of the H_αAT for at least two of the three reactions in Scheme 3.5. Of the three reactions, this proposal is the least comfortable for the desaturation reaction catalysed by CarC. Here, the amide nitrogen of the fused ring system that is bonded to C_α would appear to be less able to assist either dehydration of the α-hydroxylated intermediate (pathway **II**) or ET from the C_α radical (pathway **III**). Thus, as for the oxacyclization reactions of Section 3.5, it is entirely possible that the individual examples of what appears to be a single, unified reaction type might not follow a single conserved mechanism.

As for the oxacyclization reactions, examination of reactions with specifically deuterium-labelled substrates should allow the order of events to be elucidated. Substitution of deuterium for the hydrogen atom that is abstracted by the ferryl species should either extend the lifetime of the ferryl intermediate (observable in FQ-Mössbauer experiments), cause uncoupling of 2OG decarboxylation from desaturation (decreasing the olefin:succinate product stoichiometry), or both. These experiments are in progress in our group.

Acknowledgements

These studies have been supported by the National Institutes of Health (GM-69657 to C.K. and J.M.B.) and the National Science Foundation (MCB-642058 and CHE-724084 to C.K. and J.M.B.). We thank our coworkers and collaborators, whose work is summarized in this review.

References

1. E. I. Solomon, T. C. Brunold, M. I. Davis, J. N. Kemsley, S. K. Lee, N. Lehnert, F. Neese, A. J. Skulan, Y.-S. Yang and J. Zhou, *Chem. Rev.*, 2000, **100**, 235–349.
2. R. P. Hausinger, *Crit. Rev. Biochem. Mol. Biol.*, 2004, **39**, 21–68.

3. M. Costas, M. P. Mehn, M. P. Jensen and L. Que, Jr., *Chem. Rev.*, 2004, **104**, 939–986.
4. S. C. Trewick, T. F. Henshaw, R. P. Hausinger, T. Lindahl and B. Sedgwick, *Nature*, 2002, **419**, 174–178.
5. P. Ø. Falnes, R. F. Johansen and E. Seeberg, *Nature*, 2002, **419**, 178–182.
6. Y.-i. Tsukada, J. Fang, H. Erdjument-Bromage, M. E. Warren, C. H. Borchers, P. Tempst and Y. Zhang, *Nature*, 2006, **439**, 811–816.
7. P. A. C. Cloos, J. Christensen, K. Agger, A. Maiolica, J. Rappsilber, T. Antal, K. H. Hansen and K. Helin, *Nature*, 2006, **442**, 307–311.
8. R. J. Klose, K. Yamane, Y. Bae, D. Zhang, H. Erdjument-Bromage, P. Tempst, J. Wong and Y. Zhang, *Nature*, 2006, **442**, 312–316.
9. T. Gerken, C. A. Girard, Y. C. L. Tung, C. J. Webby, V. Saudek, K. S. Hewitson, G. S. H. Yeo, M. A. McDonough, S. Cunliffe, L. A. McNeill, J. Galvanovskis, P. Rorsman, P. Robins, X. Prieur, A. P. Coll, M. Ma, Z. Jovanovic, I. S. Farooqi, B. Sedgwick, I. Barroso, T. Lindahl, C. P. Ponting, F. M. Ashcroft, S. O'Rahilly and C. J. Schofield, *Science*, 2007, **318**, 1469–1472.
10. S. Ito, L. Shen, Q. Dai, S. C. Wu, L. B. Collins, J. A. Swenberg, C. He and Y. Zhang, *Science*, 2011, **333**, 1300–1303.
11. Y.-F. He, B.-Z. Li, Z. Li, P. Liu, Y. Wang, Q. Tang, J. Ding, Y. Jia, Z. Chen, L. Li, Y. Sun, X. Li, Q. Dai, C.-X. Song, K. Zhang, C. He and G.-L. Xu, *Science*, 2011, **333**, 1303–1307.
12. K. I. Kivirikko and J. Myllyharju, *Matrix Biol.*, 1998, **16**, 357–368.
13. A. Donovan, C. A. Lima, J. L. Pinkus, G. S. Pinkus, L. I. Zon, S. Robine and N. C. Andrews, *Cell Metab.*, 2005, **1**, 191–200.
14. M. Ivan, K. Kondo, H. Yang, W. Kim, J. Valiando, M. Ohh, A. Salic, J. M. Asara, W. S. Lane and W. G. Kaelin, Jr., *Science*, 2001, **292**, 464–468.
15. P. Jaakkola, D. R. Mole, Y.-M. Tian, M. I. Wilson, J. Gielbert, S. J. Gaskell, A. von Kriegsheim, H. F. Hebestreit, M. Mukherji, C. J. Schofield, P. H. Maxwell, C. W. Pugh and P. J. Ratcliffe, *Science*, 2001, **292**, 468–472.
16. A. C. R. Epstein, J. M. Gleadle, L. A. McNeill, K. S. Hewitson, J. O'Rourke, D. R. Mole, M. Mukherji, E. Metzen, M. I. Wilson, A. Dhanda, Y.-M. Tian, N. Masson, D. L. Hamilton, P. Jaakkola, R. Barstead, J. Hodgkin, P. H. Maxwell, C. W. Pugh, C. J. Schofield and P. J. Ratcliffe, *Cell*, 2001, **107**, 43–54.
17. C. Krebs, D. Galonić Fujimori, C. T. Walsh and J. M. Bollinger, Jr., *Acc. Chem. Res.*, 2007, **40**, 484–492.
18. M. A. Hollenhorst, S. B. Bumpus, M. L. Matthews, J. M. Bollinger, Jr., N. L. Kelleher and C. T. Walsh, *J. Am. Chem. Soc.*, 2010, **132**, 15773–15781.
19. F. H. Vaillancourt, J. Yin and C. T. Walsh, *Proc. Natl. Acad. Sci. U. S. A.*, 2005, **102**, 10111–10116.
20. F. H. Vaillancourt, E. Yeh, D. A. Vosburg, S. Garneau-Tsodikova and C. T. Walsh, *Chem. Rev.*, 2006, **106**, 3364–3378.
21. R. B. Hamed, J. R. Gomez-Castellanos, L. Henry, C. Ducho, M. A. McDonough and C. J. Schofield, *Nat. Prod. Rep.*, 2013, **30**, 21–107.
22. T. Hiratsuka, K. Koketsu, A. Minami, S. Kaneko, C. Yamazaki, K. Watanabe, H. Oguri and H. Oikawa, *Chem. Biol.*, 2013, **20**, 1523–1535.

23. S. P. Salowe, E. N. G. Marsh and C. A. Townsend, *Biochemistry*, 1990, **29**, 6499–6508.
24. S. P. Salowe, W. J. Krol, D. Iwata-Reuyl and C. A. Townsend, *Biochemistry*, 1991, **30**, 2281–2292.
25. T. Hashimoto and Y. Yamada, *Plant Physiol.*, 1986, **81**, 619–625.
26. T. Hashimoto, J. Matsuda and Y. Yamada, *FEBS Lett.*, 1993, **329**, 35–39.
27. C. L. Schardl, R. B. Grossman, P. Nagabhyru, J. R. Faulkner and U. P. Mallik, *Phytochemistry*, 2007, **68**, 980–996.
28. R. F. Li, A. Stapon, J. T. Blanchfield and C. A. Townsend, *J. Am. Chem. Soc.*, 2000, **122**, 9296–9297.
29. N. J. Kershaw, M. E. C. Caines, M. C. Sleeman and C. J. Schofield, *Chem. Commun.*, 2005, 4251–4263.
30. H. M. Hanauske-Abel and V. Günzler, *J. Theor. Biol.*, 1982, **94**, 421–455.
31. L. Que, Jr., *Nat. Struct. Biol.*, 2000, **7**, 182–184.
32. K. D. Koehntop, J. P. Emerson and L. Que, Jr., *J. Biol. Inorg. Chem.*, 2005, **10**, 87–93.
33. E. G. Pavel, J. Zhou, R. W. Busby, M. Gunsior, C. A. Townsend and E. I. Solomon, *J. Am. Chem. Soc.*, 1998, **120**, 743–753.
34. J. T. Groves, *J. Chem. Ed.*, 1985, **62**, 928–931.
35. J. M. Bollinger, Jr., J. C. Price, L. M. Hoffart, E. W. Barr and C. Krebs, *Eur. J. Inorg. Chem.*, 2005, **2005**, 4245–4254.
36. J. C. Price, E. W. Barr, T. E. Glass, C. Krebs and J. M. Bollinger, Jr., *J. Am. Chem. Soc.*, 2003, **125**, 13008–13009.
37. D. A. Proshlyakov, T. F. Henshaw, G. R. Monterosso, M. J. Ryle and R. P. Hausinger, *J. Am. Chem. Soc.*, 2004, **126**, 1022–1023.
38. P. J. Riggs-Gelasco, J. C. Price, R. B. Guyer, J. H. Brehm, E. W. Barr, J. M. Bollinger, Jr. and C. Krebs, *J. Am. Chem. Soc.*, 2004, **126**, 8108–8109.
39. J. C. Price, E. W. Barr, L. M. Hoffart, C. Krebs and J. M. Bollinger, Jr., *Biochemistry*, 2005, **44**, 8138–8147.
40. P. K. Grzyska, E. H. Appelman, R. P. Hausinger and D. A. Proshlyakov, *Proc. Natl. Acad. Sci. U. S. A.*, 2010, **107**, 3982–3987.
41. J. C. Price, E. W. Barr, B. Tirupati, J. M. Bollinger, Jr. and C. Krebs, *Biochemistry*, 2003, **42**, 7497–7508.
42. C. Krebs, J. C. Price, J. Baldwin, L. Saleh, M. T. Green and J. M. Bollinger, Jr, *Inorg. Chem.*, 2005, **44**, 742–757.
43. P. G. Debrunner, *Phys. Bioinorg. Chem. Ser.*, 1989, **4**, 137–234.
44. C. A. Grapperhaus, B. Mienert, E. Bill, T. Weyhermüller and K. Wieghardt, *Inorg. Chem.*, 2000, **39**, 5306–5317.
45. J.-U. Rohde, J.-H. In, M. H. Lim, W. W. Brennessel, M. R. Bukowski, A. Stubna, E. Münck, W. Nam and L. Que, Jr., *Science*, 2003, **299**, 1037–1039.
46. M. H. Lim, J.-U. Rohde, A. Stubna, M. R. Bukowski, M. Costas, R. Y. N. Ho, E. Münck, W. Nam and L. Que, Jr., *Proc. Natl. Acad. Sci. U. S. A.*, 2003, **100**, 3665–3670.
47. L. Que, Jr., *Acc. Chem. Res.*, 2007, **40**, 493–500.
48. E. V. Anslyn and D. A. Dougherty, *Modern Physical Organic Chemistry*, University Science Books, Sausalito, California, 2006.

49. J. M. Bollinger, Jr. and C. Krebs, *J. Inorg. Biochem.*, 2006, **100**, 586–605.
50. J. E. Penner-Hahn, K. S. Eble, T. J. McMurry, M. Renner, A. L. Balch, J. T. Groves, J. H. Dawson and K. O. Hodgson, *J. Am. Chem. Soc.*, 1986, **108**, 7819–7825.
51. T. Wolter, W. Meyer-Klaucke, M. Müther, D. Mandon, H. Winkler, A. X. Trautwein and R. Weiss, *J. Inorg. Biochem.*, 2000, **78**, 117–122.
52. F. Neese, *J. Inorg. Biochem.*, 2006, **100**, 716–726.
53. S. Sinnecker, N. Svensen, E. W. Barr, S. Ye, J. M. Bollinger, Jr., F. Neese and C. Krebs, *J. Am. Chem. Soc.*, 2007, **129**, 6168–6179.
54. L. M. Hoffart, E. W. Barr, R. B. Guyer, J. M. Bollinger, Jr. and C. Krebs, *Proc. Natl. Acad. Sci., U. S. A.*, 2006, **103**, 14738–14743.
55. L. C. Blasiak, F. H. Vaillancourt, C. T. Walsh and C. L. Drennan, *Nature*, 2006, **440**, 368–371.
56. J. E. Baldwin and M. Bradley, *Chem. Rev.*, 1990, **90**, 1079–1088.
57. C. Krebs, E. Y. Tamanaha and J. M. Bollinger, Jr., *FASEB J.*, 2011, **25**, 195.1.
58. E. Y. Tamanaha, *PhD Thesis*, The Pennsylvania State University, 2013.
59. M. Ueki, D. P. Galonić, F. H. Vaillancourt, S. Garneau-Tsodikova, E. Yeh, D. A. Vosburg, F. C. Schroeder, H. Osada and C. T. Walsh, *Chem. Biol.*, 2006, **13**, 1183–1191.
60. D. P. Galonić, E. W. Barr, C. T. Walsh, J. M. Bollinger, Jr. and C. Krebs, *Nature Chem. Biol.*, 2007, **3**, 113–116.
61. D. Galonić Fujimori, E. W. Barr, M. L. Matthews, G. M. Koch, J. R. Yonce, C. T. Walsh, J. M. Bollinger, Jr., C. Krebs and P. J. Riggs-Gelasco, *J. Am. Chem. Soc.*, 2007, **129**, 13408–13409.
62. M. L. Matthews, C. M. Krest, E. W. Barr, F. H. Vaillancourt, C. T. Walsh, M. T. Green, C. Krebs and J. M. Bollinger, Jr., *Biochemistry*, 2009, **48**, 4331–4343.
63. M. L. Matthews, C. S. Neumann, L. A. Miles, T. L. Grove, S. J. Booker, C. Krebs, C. T. Walsh and J. M. Bollinger, Jr., *Proc. Natl. Acad. Sci., U. S. A.*, 2009, **106**, 17723–17728.
64. J. Zhou, W. L. Kelly, B. O. Bachmann, M. Gunsior, C. A. Townsend and E. I. Solomon, *J. Am. Chem. Soc.*, 2001, **123**, 7388–7398.
65. M. L. Neidig, C. D. Brown, K. M. Light, D. G. Fujimori, E. M. Nolan, J. C. Price, E. W. Barr, J. M. Bollinger, Jr., C. Krebs, C. T. Walsh and E. I. Solomon, *J. Am. Chem. Soc.*, 2007, **129**, 14224–14231.
66. S. P. de Visser and R. Latifi, *J. Phys. Chem. B*, 2009, **113**, 12–14.
67. T. Borowski, H. Noack, M. Radon, K. Zych and P. E. M. Siegbahn, *J. Am. Chem. Soc.*, 2010, **132**, 12887–12898.
68. H. J. Kulik and C. L. Drennan, *J. Biol. Chem.*, 2013, **288**, 11233–11241.
69. S. D. Wong, M. Srnec, M. L. Matthews, L. V. Liu, Y. Kwak, K. Park, C. B. Bell, E. E. Alp, J. Y. Zhao, Y. Yoda, S. Kitao, M. Seto, C. Krebs, J. M. Bollinger Jr. and E. I. Solomon, *Nature*, 2013, **499**, 320–323.
70. T. M. Casey, P. K. Grzyska, R. P. Hausinger and J. McCracken, *J. Phys. Chem. B*, 2013, **117**, 10384–10394.

71. M. L. Matthews, W.-c. Chang, A. P. Layne, L. A. Miles, C. Krebs and J. M. Bollinger, Jr., *Nat. Chem. Biol.*, 2014, **10**, 209–215.
72. M. L. Hillwig and X. Liu, *Nat. Chem. Biol.*, 2014, **10**, 921–923.
73. Editorial, *Nature*, 2013, 495, 141.
74. S. J. Coulthurst, A. M. Barnard and G. P. Salmond, *Nat. Rev. Microbiol.*, 2005, **3**, 295–306.
75. W. L. Parker, M. L. Rathnum, J. S. Wells, Jr., W. H. Trejo, P. A. Principe and R. B. Sykes, *J. Antibiot.*, 1982, **35**, 653–660.
76. K. M. Papp-Wallace, A. Endimiani, M. A. Taracila and R. A. Bonomo, *Antimicrob. Agents Chemother.*, 2011, **55**, 4943–4960.
77. A. Stapon, R. F. Li and C. A. Townsend, *J. Am. Chem. Soc.*, 2003, **125**, 8486–8493.
78. M. C. Sleeman, P. Smith, B. Kellam, S. R. Chhabra, B. W. Bycroft and C. J. Schofield, *ChemBiochem*, 2004, **5**, 879–882.
79. R. B. Hamed, J. R. Gomez-Castellanos, A. Thalhammer, D. Harding, C. Ducho, T. D. Claridge and C. J. Schofield, *Nat. Chem.*, 2011, **3**, 365–371.
80. R. M. Phelan and C. A. Townsend, *J. Am. Chem. Soc.*, 2013, **135**, 7496–7502.
81. T. Borowski, E. Broclawik, C. J. Schofield and P. E. M. Siegbahn, *J. Comput. Chem.*, 2006, **27**, 740–748.
82. M. Topf, G. M. Sandala, D. M. Smith, C. J. Schofield, C. J. Easton and L. Radom, *J. Am. Chem. Soc.*, 2004, **126**, 9932–9933.
83. I. J. Clifton, L. X. Doan, M. C. Sleeman, M. Topf, H. Suzuki, R. C. Wilmouth and C. J. Schofield, *J. Biol. Chem.*, 2003, **278**, 20843–20850.
84. R. M. Phelan, B. J. DiPardo and C. A. Townsend, *ACS Chem. Biol.*, 2012, **7**, 835–840.
85. W.-c. Chang, Y. S. Guo, C. Wang, S. E. Butch, A. C. Rosenzweig, A. K. Boal, C. Krebs and J. M. Bollinger, Jr., *Science*, 2014, **343**, 1140–1144.
86. J. M. Bollinger, Jr., D. E. Edmondson, B. H. Huynh, J. Filley, J. R. Norton and J. Stubbe, *Science*, 1991, **253**, 292–298.
87. J. E. Baldwin, R. M. Adlington, J. S. Bryans, A. O. Bringhen, J. B. Coates, N. P. Crouch, M. D. Lloyd, C. J. Schofield, S. W. Elson, K. H. Baggaley, R. Cassels and N. Nicholson, *Tetrahedron*, 1991, **47**, 4089–4100.
88. P. H. Liu, K. Murakami, T. Seki, X. M. He, S. M. Yeung, T. Kuzuyama, H. Seto and H.-w. Liu, *J. Am. Chem. Soc.*, 2001, **123**, 4619–4620.
89. C. Wang, W.-c. Chang, Y. S. Guo, H. Huang, S. C. Peck, M. E. Pandelia, G. M. Lin, H.-w. Liu, C. Krebs and J. M. Bollinger Jr., *Science*, 2013, **342**, 991–995.
90. F. Hammerschmidt and H. Kählig, *J. Org. Chem.*, 1991, **56**, 2364–2370.
91. P. L. Roach, I. J. Clifton, C. M. H. Hensgens, N. Shibta, C. J. Schofield, J. Hajdu and J. E. Baldwin, *Nature*, 1997, **387**, 827–830.
92. D. Yun, M. Dey, L. J. Higgins, F. Yan, H.-w. Liu and C. L. Drennan, *J. Am. Chem. Soc.*, 2011, **133**, 11262–11269.
93. Z. H. Zhang, J. S. Ren, D. K. Stammers, J. E. Baldwin, K. Harlos and C. J. Schofield, *Nat. Struct. Biol.*, 2000, **7**, 127–133.

94. W.-c. Chang, M. Dey, P. H. Liu, S. O. Mansoorabadi, S. J. Moon, Z. B. K. Zhao, C. L. Drennan and H.-w. Liu, *Nature*, 2013, **496**, 114–118.
95. D. Iwata-Reuyl, A. Basak and C. A. Townsend, *J. Am. Chem. Soc.*, 1999, **121**, 11356–11368.
96. T. Borowski, S. de Marothy, E. Broclawik, C. J. Schofield and P. E. M. Siegbahn, *Biochemistry*, 2007, **46**, 3682–3691.
97. B. G. Fox, K. S. Lyle and C. E. Rogge, *Acc. Chem. Res.*, 2004, **37**, 421–429.
98. B. Meunier, S. P. de Visser and S. Shaik, *Chem. Rev.*, 2004, **104**, 3947–3980.
99. E. I. Solomon, A. Decker and N. Lehnert, *Proc. Natl. Acad. Sci. U. S. A.*, 2003, **100**, 3589–3594.

CHAPTER 4

Synthetic Models of 2-Oxoglutarate-Dependent Oxygenases

CALEB J. ALLPRESS[a], SCOTT T. KLEESPIES[a], AND LAWRENCE QUE JR*[a]

[a]Department of Chemistry and Center for Metals in Biocatalysis, University of Minnesota, Minneapolis, Minnesota 55455, USA
*E-mail: larryque@umn.edu

4.1 Introduction

The 2-oxoglutarate (2OG)-dependent oxidases and oxygenases represent the largest family of non-haem iron enzymes.[1] They employ 2OG as a cosubstrate to carry out a variety of metabolic transformations, many of which involve the cleavage of a substrate C–H bond. A consensus mechanism has emerged for 2OG-dependent enzymes leading to the formation of an Fe(IV)=O species[2] that is capable of a wide array of substrate oxidations, including hydroxylation, halogenation, desaturation, electrophilic aromatic substitution and epoxidation.[3] Mechanistic studies, particularly by the Bollinger/Krebs group (see Chapter 3),[3] have led to the proposed catalytic cycle shown in Figure 4.1.

The most detailed investigation has been carried out on the enzyme TauD, which catalyses the hydroxylation of the C_α position of taurine as a means

RSC Metallobiology Series No. 3
2-Oxoglutarate-Dependent Oxygenases
Edited by Robert P. Hausinger and Christopher J. Schofield

Published by the Royal Society of Chemistry, www.rsc.org

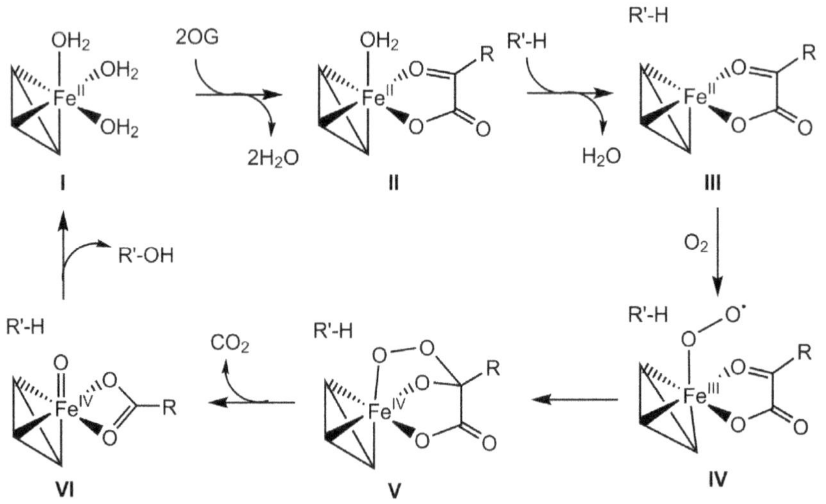

Figure 4.1 Consensus mechanism for 2OG-dependent oxygenases.

of recycling sulfur in the cell. Three of the species in the catalytic cycle have been crystallographically characterized.[4] In the TauD•Fe(II) complex, the Fe(II) centre is bound to two histidine and one aspartate residue in a facial arrangement that is referred to as the 2-His-1-carboxylate facial triad, a common motif found among mononuclear non-haem iron enzymes.[5] The three remaining coordination sites in this octahedral complex are occupied by water molecules. Addition of 2OG results in its binding to iron as a bidentate ligand, displacing two water molecules. Taurine binding at the active site, but not to the iron centre, expels the third water molecule, resulting in a ternary complex with a five-coordinate Fe centre. O_2 binding to the coordinately unsaturated Fe(II) centre then forms an Fe(III)-superoxo adduct, and the nascent superoxide attacks the electrophilic 2-oxo carbon atom of the 2OG ligand to afford an alkylperoxo-Fe(IV) intermediate. Subsequent O–O and C–C bond cleavage results in the formation of CO_2, succinate and an Fe(IV)=O species that is capable of C–H bond cleavage. The hydroxylated product is then produced and the iron centre returns to its original oxidation state to start another catalytic turnover. Of the three proposed O_2-derived intermediates, only the Fe(IV)=O species has been trapped.[6–8] This intermediate, called **J** in the TauD cycle, has been shown to have a high-spin ($S = 2$) Fe(IV)=O centre[6] with an Fe–O distance of 1.62 Å (by EXAFS)[9] and a ν(Fe=O) of 821 cm^{-1} (by resonance Raman spectroscopy).[10] Computational studies suggest that the iron(IV) centre of intermediate **J** may be either five- or six-coordinate.[11] The rate-determining step of the catalytic cycle is the attack of the substrate C–H bond by the Fe(IV)=O moiety, which exhibits an H/D kinetic isotope effect (KIE) of ~50.[7,11] This large KIE allowed TauD-**J** to accumulate in sufficient amounts so as to be characterized by a variety of spectroscopic techniques.

The use of model complexes allows chemists to retain the most pertinent part of the enzyme, the active site, while removing the complexity and synthetic difficulty associated with the rest of the protein. Key features of the active site include the primary coordination sphere, which in 2OG-dependent oxygenases is typically a conserved 2-His-1-carboxylate facial triad,[5] and secondary sphere steric and hydrogen bonding interactions. In model complexes these features may be approximated and tuned by appropriate ligand design choices. Model complexes also allow greater control of the reaction rate through the manipulation of temperature and the use of a wide range of solvents. By controlling the rate of the reaction, intermediates can be trapped and characterized, leading to a greater understanding of the reaction mechanism for the model and subsequent extrapolation to the enzymatic system of interest.

4.2 Functional Models

There has long been an interest in using purely chemical systems to provide an understanding of the reactions catalysed by 2OG-dependent enzymes.[12] Thus, early investigations led to systems that were capable of oxidizing the C–H bonds of substrates by a combination of an iron centre with either hydrogen peroxide or a reducing agent and dioxygen.[13–15] Likewise, the iron-mediated oxidative decarboxylation of 2-oxoacids had also been reported.[16,17] However, it was not until the late 1980s that the first example of a system of relevance to 2OG-dependent enzymes was reported, which catalyses both the decarboxylation of a 2-oxoacid and the two-electron oxidation of a substrate utilizing dioxygen as a terminal electron acceptor. This system coupled the oxidative decarboxylation of pyruvic acid to the oxidation of cyclohexane using an iron catalyst.[18,19] Unfortunately, the iron catalyst was poorly defined and a relationship between the structure and reactivity of the iron-containing species was not developed. Studies over the last two decades have provided structurally well-defined functional models of 2OG-dependent enzymes that utilize combinations of iron, a supporting ligand and model substrates and provide significant mechanistic insights.

4.2.1 Model Complexes with a Tris(2-pyridylmethyl)amine-Based Ligand

The first crystallographically characterized synthetic Fe(II)-2-oxoacid complexes utilized the tetradentate ligands tris(2-pyridylmethyl)amine (TPA) and 6-Me_3-TPA (Figure 4.2) with benzoylformate (BF) as the 2-oxoacid.[20,21] [(TPA)Fe^{II}(BF)]$^+$ is yellow in colour (λ_{max} = 385 nm) and its crystal structure showed the BF to be bound monodentate *via* a carboxylate oxygen, with the remaining coordination site occupied by a solvent molecule. In non-coordinating solvents like CH_2Cl_2, however, the BF becomes bidentate and the complex is green in colour (λ_{max} = 550–610 nm). For the 6-Me_3-TPA complex, the crystal

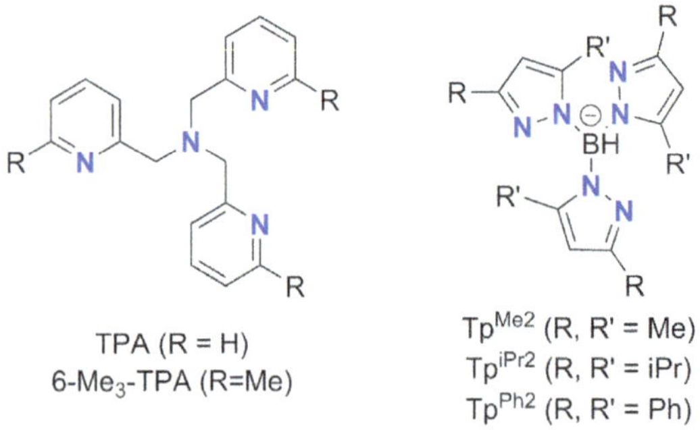

Figure 4.2 Supporting polydentate ligands used to model the active site of 2OG-dependent oxidases and oxygenases.

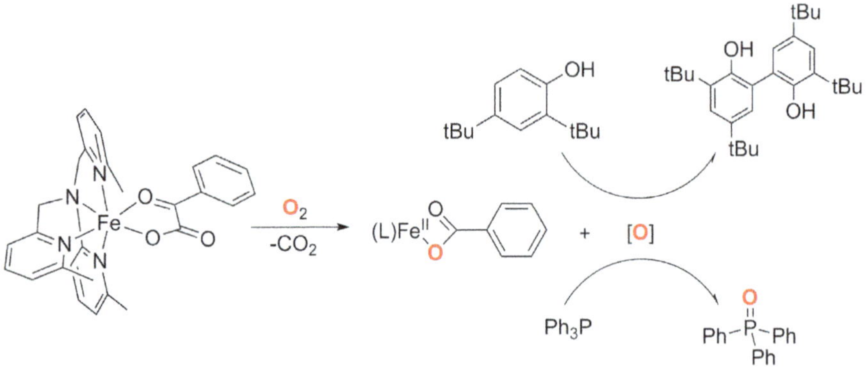

Figure 4.3 Reaction scheme for $[(6\text{-Me}_3\text{TPA})\text{Fe}^{\text{II}}(\text{BF})]^+$.

structure showed that the BF bound in a bidentate fashion (Figure 4.3), giving rise to a characteristic purple colour (λ_{max} = 544 nm) due to a metal-to-ligand charge transfer (MLCT) transition from a filled d orbital of the Fe(II) to a π^* orbital of the keto group.[22] Similar MLCT features have been seen in 2OG-bound enzymes.[23–26]

Both of these model complexes react with dioxygen, albeit over the course of days, to form the corresponding benzoate complexes, which are the products of oxidative decarboxylation of the 2-oxoacid (Figure 4.3). This slow reaction rate is postulated to arise from the fact that the iron centre is coordinatively saturated, so ligand dissociation is required to provide an open site for O_2 to bind and allow the reaction to proceed. This is consistent with the consensus mechanism for 2OG-dependent enzymes (Figure 4.1), in which the sequential binding of 2OG and substrate results in the formation of a five-coordinate ternary complex (**III**) with a vacant binding location for dioxygen that is adjacent to the 2OG and proximal to the substrate. The ligand dissociation step is likely to be rate determining, as no direct evidence for any iron–oxygen intermediates could be found. Interception experiments suggest the formation of an O_2-derived oxidant in the oxidative decarboxylation reaction of these complexes that can perform H-atom abstraction from 2,4-di-*tert*-butylphenol and oxygen-atom transfer to PPh_3.[21]

4.2.2 Model Complexes with a Tris(pyrazolyl)borate-Based Ligand

Switching a tetradentate ligand to a tridentate ligand should in principle leave an open coordination site on the iron centre for O_2 binding and thus result in a faster rate of O_2 activation. The first examples of synthetic Fe(II)-2-oxoacid complexes with a tridentate ligand were obtained through the use of the hydrotris(pyrazolyl)borate ($Tp^{R,R'}$) ligand (Figure 4.2), which more closely mimics the monoanionic facial tridentate ligand set of the '2-His-1-carboxylate facial triad' motif.

[(Tp^{Me2})Fe^{II}(BF)] represents the first example of a synthetic N3 facial Fe(II)-2-oxoacid complex.[27] This complex exhibits the characteristic MLCT feature found in the corresponding TPA complexes, indicating bidentate binding of BF. Unlike with the tetradentate TPA complexes, [(Tp^{Me2})Fe^{II}(BF)] reacted with O_2 in a matter of minutes and afforded benzoate as a product. The inclusion of cyclohexene or *cis*-stilbene during the oxygenation led to their epoxidation. Of note is the retention of configuration found for the *cis*-stilbene reaction, strongly implicating the formation of a metal-based oxidant like that proposed for the 2OG-dependent enzymes.

The replacement of the methyl groups of the Tp^{Me2} ligand with larger substituents led to the formation of [(Tp^{R2})Fe^{II}(2-oxoacid)] complexes that had somewhat slower rates of oxygenation but provided important mechanistic insights. Indirect evidence for the involvement of an iron(III)-superoxo species was found in studies of [(Tp^{iPr2})Fe^{II}(PRV)].[28] Oxygenation of this complex in MeCN at −40 °C resulted in the formation of an O_2 adduct. This species exhibits a visible absorption band at 682 nm, a ν(O–O) of 889 cm^{-1} obtained from a resonance Raman experiment, and a Mössbauer doublet with an isomer shift δ of 0.65 mm s^{-1} and a quadrupole splitting ΔE_Q of 1.35 mm s^{-1}. These spectroscopic features are reminiscent of those for the O_2 adducts of [(Tp^{iPr2})Fe^{II}(O_2CR)] complexes first reported by Kitajima[29] and later characterized crystallographically by Kim and Lippard,[30] which have a (μ-1,2-peroxo) diiron(III) core that is supported by two carboxylate bridges. In the oxygenation of [(Tp^{iPr2})Fe^{II}(PRV)], addition of 2,4,6-tri-*tert*-butylphenol produced the phenoxyl radical in 80% yield and prevented the formation of the peroxo-bridged dimer (Figure 4.4). As the peroxo-bridged dimer reacted 70-fold more slowly with the added phenol, a reactive precursor to this peroxo species was implicated. This reactive precursor is very likely an initially formed Fe(III)-superoxo species, which abstracts the H atom from the O–H bond of the phenol to generate the phenoxyl radical.

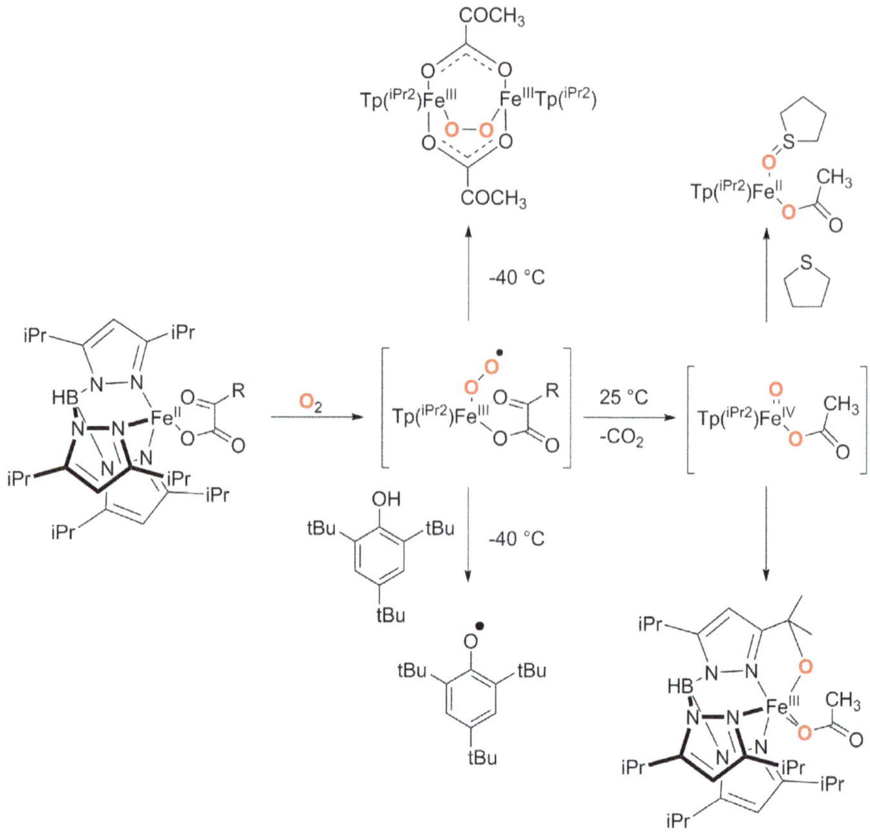

Figure 4.4 Reaction scheme for $[(Tp^{iPr2})Fe^{II}(PRV)]$ (R = Me).

On the other hand, room temperature oxygenation of $[(Tp^{iPr2})Fe^{II}(PRV)]$ gave different results. An oxidized product was observed that exhibited an ESI-MS peak at *m/z* 536, which was formulated to be $\{(Tp^{iPr2})Fe + O\text{–}H\}^+$, suggesting the ligand had been oxygenated. The site of oxygenation was determined to be the tertiary C–H of an isopropyl group by a ^{1}H-NMR analysis of the pyrazoles obtained by treatment of the product with strong acid. These results demonstrated that an oxidant was formed that was capable of cleaving a C–H bond on the Tp ligand. When tetrahydrothiophene was added to the $[(Tp^{iPr2})Fe^{II}(PRV)]$ solution before the start of the reaction, the hydroxylated ligand was no longer detected, and instead tetrahydrothiophene was found to be oxidized to its sulfoxide (Figure 4.4). Even one equivalent of tetrahydrothiophene was sufficient for this interception to occur.[28] This potent oxidant that is capable of both oxidizing a C–H bond and oxygenating tetrahydrothiophene is likely to be an Fe(IV)-oxo species, similar to the intermediates proposed in the catalytic cycle of TauD and other 2OG-dependent enzymes.

Experiments with $[(Tp^{Ph2})Fe^{II}(BF)]$ further enhanced our mechanistic understanding.[31] Oxygenation of $[(Tp^{Ph2})Fe^{II}(BF)]$ resulted in the conversion

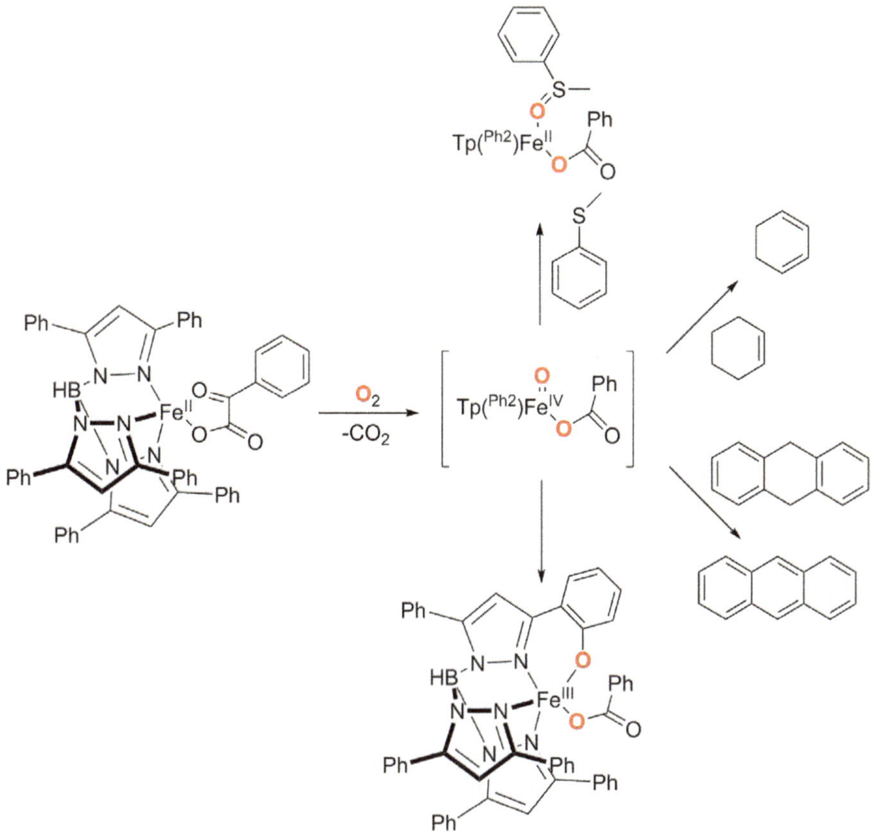

Figure 4.5 Reaction scheme for $[(Tp^{Ph2})Fe^{II}(BF)]$.

of its purple colour (λ_{max} = 531 nm) to green (λ_{max} = 650 nm) over the course of one hour. Characterization of the green product showed that the Tp^{Ph2} ligand had undergone hydroxylation at the ortho position of one of the 3-phenyl rings with the green colour arising from a phenolate-to-Fe(III) charge transfer transition. Benzoate was also present in the final reaction mixture, showing that the BF had undergone decarboxylation (Figure 4.5). A crystal structure of the green product confirmed its formulation. Furthermore, the use of $^{18}O_2$ showed a high percentage of ^{18}O incorporation into both the phenolate and the benzoate, thus making this complex an excellent model for the dioxygenase reactivity seen in non-haem iron oxygenases.

Kinetic studies demonstrated that the rate of formation of the green chromophore was first order with respect to dioxygen and irreversible.[31] Further experiments using substituted BF cofactors showed that electron-withdrawing substituents increased the rate of reaction. A Hammett ρ value of +1.3 was obtained, indicative of a nucleophilic mechanism. The reaction was deduced to have a rate-determining step involving the initially formed superoxo species acting as a nucleophile to attack the electrophilic α-keto group of the BF ligand. Thus the rate-determining step in the activation of O_2 by $[(Tp^{Ph2})Fe^{II}(BF)]$ is early in the reaction mechanism and likely to precede the formation of the Fe(IV)=O oxidant.

The kinetic results for the oxygenation of the $[(Tp^{Ph2})Fe^{II}(BF)]$ complexes make it highly unlikely that the putative Fe(IV)=O oxidant can be isolated and characterized. To learn more about the reactivity of this transient species, Mukherjee *et al.* added potential substrates into the reaction solution to compete with the intramolecular ligand self-hydroxylation, using the deep green chromophore of the self-hydroxylated product as a visual probe to assess the effectiveness of the added substrates at intercepting the proposed oxidant.[32] For example, the addition of 10 equivalents of thioanisole prevented the appearance of the green chromophore and resulted instead in the formation of $[(Tp^{Ph2})Fe^{II}(O_2CPh)(PhS(O)Me)]$ in 70% yield with respect to Fe, showing that intermolecular oxo-transfer can be more facile than intramolecular ligand hydroxylation. Intermolecular H-atom abstraction was also demonstrated. With the addition of 100 equivalents of cyclohexene, only 20% of the green chromophore was observed together with cyclohexadiene as the organic product. The use of cyclohexene-d_{10} afforded the same extent of interception, and the competitive oxidation of cyclohexene and cyclohexene-d_{10} revealed a product kinetic isotope effect of 10. Substrates with C–H bonds as strong as 95 kcal mol^{-1} were also oxidized, but with varying interception efficiencies. These results indicate that quite a powerful oxidant can be formed in the oxygenation of $[(Tp^{Ph2})Fe^{II}(BF)]$. Density functional theory (DFT) calculations suggest that this oxidant is likely to be a high-spin ($S = 2$) Fe(IV)=O complex (Figure 4.6).[32,33]

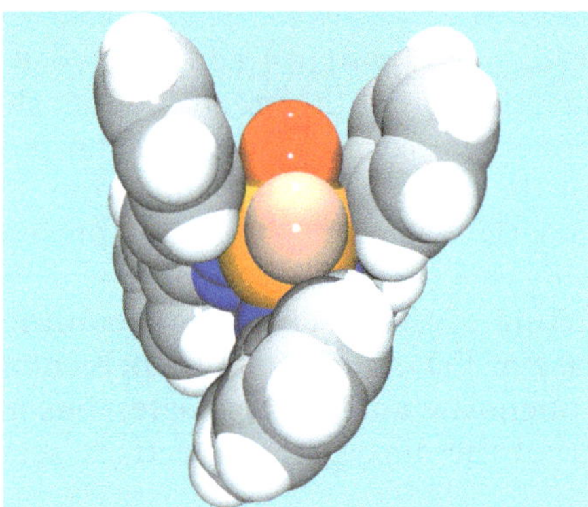

Figure 4.6 DFT-calculated structure for the putative $[Fe^{IV}(O)(Tp^{Ph2})(O_2CPh)]$ oxidant derived from the reaction of $[(Tp^{Ph2})Fe^{II}(BF)]$ with O_2. The red atom represents the oxo, while the peach-coloured atom is the coordinated oxygen atom of the benzoate ligand. The benzoyl moiety has been removed to provide a clearer view of the Fe=O pocket. Reprinted with permission from the publisher of Mukherjee *et al.*[32]

4.3 Models of Reaction Intermediates

Small molecular model systems have the potential to provide significant insight into the structure and reactivity of proposed reaction intermediates such as species **IV–VI** (Figure 4.1). To date, the only one of these intermediates that has been detected in 2OG-dependent enzymes is an Fe(IV)=O species capable of abstracting a hydrogen atom from strong C–H bonds. A number of model complexes of this intermediate have been generated, and together provide significant insight into how such a species may be tuned as a potent oxidant.

4.3.1 Structural and Spectroscopic Properties of the Fe(IV)=O Unit

Since the initial reports of intermediate **J** in the catalytic cycle of TauD,[6–8] several other non-haem Fe(IV)=O intermediates have been identified in the catalytic cycles of 2OG-dependent enzymes, including for prolyl-4-hydroxylase (P4H),[34] CytC3[35,36] and SyrB2.[37] The latter two enzymes are halogenases, and are structurally distinct from other 2OG-dependent enzymes in that the protein-derived carboxylate ligand is replaced with a halide ion.[38] Fe(IV)=O intermediates have been observed in the catalytic cycles of the pterin-dependent oxygenases phenylalanine hydroxylase (PheH) and tyrosine hydroxylase (TyrH), which are functionally related to 2OG-dependent enzymes, albeit utilizing tetrahydropterin instead of 2OG as a two-electron reductant to generate the Fe(IV)=O species.[39,40] Like TauD-**J**, all of these species are high-spin ($S = 2$) and characterized by an absorption feature at ~320 nm, an Fe–O bond of ~1.62 Å, as determined by X-ray absorption spectroscopy (XAS) studies, and an isomer shift of ~0.2–0.3 mm s^{-1} (Table 4.1). However, intermediate **J** remains the only enzymatic species for which a direct probe of the Fe=O bond strength has been identified (ν_{Fe-O} = 821 cm^{-1} by resonance Raman spectroscopy).[10]

4.3.1.1 *High-Spin Five-Coordinate Fe(IV)=O Complexes*

One strategy to generate high-spin Fe(IV)=O complexes has been to use ligands that enforce a trigonal bipyramidal geometry. As can be seen in Figure 4.7, in a trigonal bipyramidal (C_{3v}) ligand field, the d_{x2-y2} and d_{xy} orbitals become degenerate, leading to a default high-spin ground state. Thus, synthetic approaches have focused on the use of tetradentate tripodal ligands to generate five-coordinate Fe(IV)=O complexes, wherein the tripod occupies one axial and all three equatorial positions while the oxo unit occupies the second axial position. The binding of a sixth ligand – either an anion or solvent molecule – is prevented by the use of bulky substituents at the equatorial positions of the tripod. By using these design principles, five-coordinate high-spin Fe(IV)=O complexes have been successfully synthesized, although the use of bulky substituents has proven detrimental to studying trends in their reactivity (Section 4.3.2).

Table 4.1 Spectroscopic and structural parameters of selected Fe(IV)–oxo species.

	λ_{max}, nm (ε, $M^{-1}cm^{-1}$)	δ ($mm\ s^{-1}$)	ΔE_q ($mm\ s^{-1}$)	D (cm^{-1})	r_{Fe} = O (Å)	$\nu_{Fe=O}$ (cm^{-1})	$\Delta^{16/18}O$ (cm^{-1})	S	References
Enzyme Fe(IV)=O's									
TauD 'J'	318 (1550)	0.31	−0.88	10.5	1.62	821	34	2	6–10
P4H	320 (1500)	0.30	−0.82	15.5	—	—	—	2	34
CytC3-Cl	318	0.30, 0.22	−1.09, −0.70	8.1	—	—	—	2	35,36
CytC3-Br	—	0.31, 0.23	−1.06, −0.81	12.5	1.62	—	—	2	35,36
SyrB2	318	0.30, 0.23	1.09, 0.76	—	1.64	—	—	2	37
TyrH	—	0.25	1.27	12.5	—	—	—	2	39
PheH	—	0.28	1.26	—	—	—	—	2	40
High-spin model complexes									
$[Fe^{IV}O(TMG_3tren)]^{2+}$	400 (9800), 825 (260)	0.09	−0.29	5.0	1.661	843	33	2	41,42
$[Fe^{IV}O(H_3buea)]^{-}$	440 (3100), 550 (1900), 808 (280)	0.02	0.43	4.0	1.680	799	31	2	48
$[Fe^{IV}O(tpa^{Ph})]^{-}$	400, 900	0.09	0.51	4.3	1.62	850	36	2	47
$[Fe^{IV}O(TMG_2dien)(CH_3CN)]^{2+}$	380 (8200), 805 (270)	0.08	0.58	4.5	1.65	807	34	2	46
$[Fe^{IV}O(TMG_2dien)(N_3)]^{+}$	412 (9700), 827 (290)	0.12	−0.30	4.6	—	833	38	2	46
$[Fe^{IV}O(TMG_2dien)(Cl)]^{+}$	385 (7800), 803 (290)	0.08	0.41	4.0	1.65	810	35	2	46
$[Fe^{IV}O(H_2O)_5]^{2+}$	~320 (~500)	0.38	0.33	9.7	—	—	—	2	53
Selected intermediate-spin model complexes									
$[Fe^{IV}O(TMC)(CH_3CN)]^{2+}$	820 (400)	0.17	1.24	29	1.646	834	34	1	58–61
$[Fe^{IV}O(TPA)(CH_3CN)]^{2+}$	724 (300)	0.01	0.92	28	1.67	—	—	1	63,64
$Fe^{IV}O(N4Py)]^{2+}$	695 (400)	−0.04	0.93	22	1.639	841	35	1	75,79
$[Fe^{IV}(O)(^{n}Bu\text{-}P2DA)]$	770 (200)	0.04	1.13	27	1.66	—	—	1	56

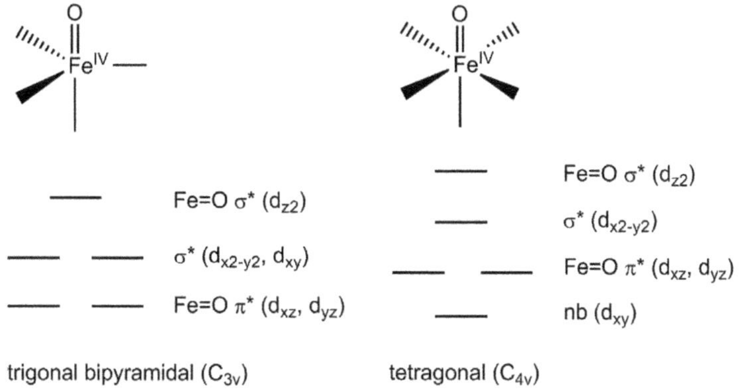

Figure 4.7 d-Orbital splitting diagrams for Fe(IV)=O complexes.

$[Fe^{IV}O(TMG_3tren)]^{2+}$ is the first example of a trigonal bipyramidal S = 2 Fe(IV)=O complex (Figure 4.8) and was synthesized by reaction of its iron(II) precursor with an iodosoarene.[41] The crystal structure of $[Fe^{IV}O(d_{36}$-$TMG_3tren)](OTf)_2$ shows the iron-oxo bond length to be 1.662 Å,[42] which is slightly longer than those determined by EXAFS analysis for iron-oxo bond distances in 2OG-dependent enzymes (1.62 Å) but still within the range observed in synthetic porphyrin-supported Fe(IV)=O units (1.62–1.67 Å).[43,44] Excitation at 514.5 nm identified an isotope-sensitive Raman band at 843 cm^{-1} assigned to an iron-oxo stretching mode, which is somewhat higher energy than that observed for TauD **J** (821 cm^{-1}).[10] The differences in the iron-oxo bond between enzymatic systems and this synthetic example are probably due to the differences in donor ability or negative charge of histidine/carboxylate *versus* tertiary amine and imine ligands, although EXAFS analysis of $[Fe^{IV}O(TMG_3tren)]^{2+}$ also slightly underestimated the iron-oxo bond distances at 1.65 Å. XAS studies also reveal two discernable pre-edge features with a combined area of 27 units, a feature predicted to be characteristic for high-spin Fe(IV)=O species, arising from splitting of the α and β d_{z2} orbitals.[45] This pre-edge feature is a forbidden 1s–3d transition, and so an intense feature is consistent with the expected deviation from centrosymmetry afforded by the strong bonding to the oxo unit relative to the other iron bonding interactions. $[Fe^{IV}O(TMG_3tren)]^{2+}$ has weak absorption features in the near IR region at 825 and 866 nm, which are assigned to d–d transitions that have not yet been observed in enzymatic systems. There is also an intense absorption band at 400 nm that may correspond to the band at ~320 nm observed for the enzymatic Fe(IV)=O species. $[Fe^{IV}O(TMG_3tren)]^{2+}$ exhibits a quadrupole doublet in its Mössbauer spectrum with an isomer shift of 0.09 mm s^{-1}, which is notably shifted relative to the enzymatic systems, probably due to the exclusively nitrogen donors in the TMG_3tren ligand.

As a start towards gaining insight into the role of steric and electronic perturbations in an otherwise isostructural series of complex, the use of the facially-binding TMG_2dien as a ligand afforded a series of Fe(IV)=O species in

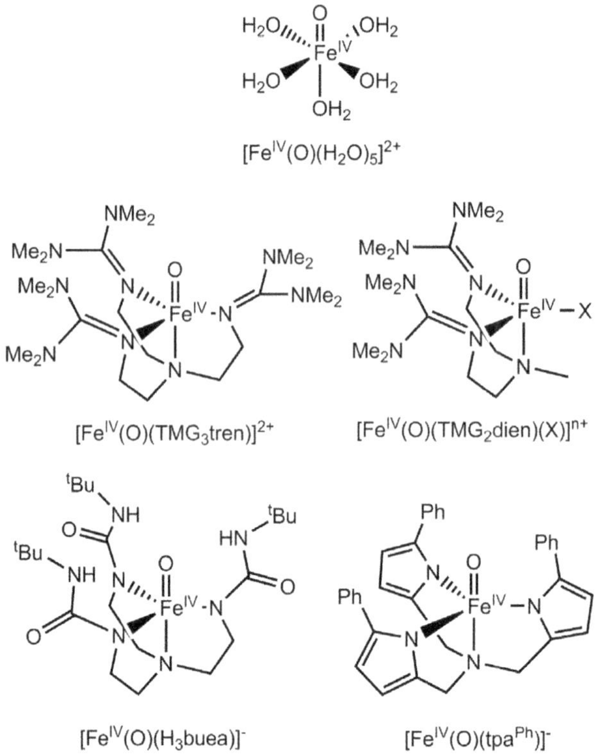

$[Fe^{IV}(O)(H_2O)_5]^{2+}$

$[Fe^{IV}(O)(TMG_3tren)]^{2+}$ $[Fe^{IV}(O)(TMG_2dien)(X)]^{n+}$

$[Fe^{IV}(O)(H_3buea)]^-$ $[Fe^{IV}(O)(tpa^{Ph})]^-$

Figure 4.8 High-spin Fe(IV)=O complexes.

which a solvent molecule (CH_3CN) or anion (Cl^- or N_3^-) was bound in the fifth coordination position *cis* to the oxo unit (Figure 4.8).[46] These complexes have spectroscopic features similar to that of $[Fe^{IV}O(TMG_3tren)]^{2+}$, but the limited number of coordinating groups precludes any strong conclusions being drawn about the influence of their electronic properties on the spectroscopy of the Fe(IV)=O unit.

Chang and coworkers have reported an Fe(IV)=O complex supported by an alternative bulky tripodal ligand (tpa^{Ph}, Figure 4.8) based on pyrrolide donors that was generated by addition of trimethylamine N-oxide to the ferrous precursor.[47] In contrast to TMG_3tren, which has neutral donors, the tpa^{Ph} ligand has three anionic donors, decreasing the overall charge on the complex. This species has similar UV-vis, Mössbauer and Raman parameters to those reported for $[Fe^{IV}O(TMG_3tren)]^{2+}$, but the EXAFS analysis revealed a contracted Fe–O bond distance of 1.62 Å, more in line with the distances observed for the enzymatic systems.

Another high-spin Fe(IV)=O complex was obtained by Borovik and coworkers using a conceptually distinct synthetic route in which ferrocenium ion was utilized to effect the one-electron oxidation of a unique Fe(III)–O precursor, the oxo ligand being stabilized by hydrogen bonding interactions with the trianionic tripodal ligand, H_3buea (Figure 4.8).[48] Perhaps due to the persistence of hydrogen bonding interactions, this species has a weakened Fe=O bond relative to the other high-spin Fe(IV)=O complexes, as evidenced by the ν_{Fe-O} of 799 cm^{-1} and an elongated Fe–O bond length of 1.680 Å determined by X-ray crystallography.

The well-defined structure of the synthetic $[Fe^{IV}(O)(TMG_3tren)]^{2+}$ complex has provided an opportunity for evaluating DFT-predicted spectral features. Nuclear resonance vibrational spectroscopy (NRVS) studies on $[Fe^{IV}(O)(TMG_3tren)]^{2+}$ have allowed the assignment of normal modes associated with a C_{3v} high-spin Fe=O, including the Fe–O stretch, which was observed at 820 cm^{-1} (as compared to 843 cm^{-1} by resonance Raman).[49] The intensity of peaks in the NRVS spectrum is proportional to the contribution in motion of the iron atom, and thus evidence of the 'steric wall' provided by the TMG groups was observed. Fe(IV)=O intermediates in the reaction cycle of the halogenase SyrB2, trapped utilizing the non-native substrate L-cyclopropylglycine, have also been characterized by NRVS.[50] The normal modes observed for this intermediate, in the presence of either Cl^- or Br^-, were more consistent with a five-coordinate iron centre in a trigonal bipyramidal geometry, exhibiting a peak distribution pattern that paralleled that observed for $[Fe^{IV}(O)(TMG_3tren)]^{2+}$. Prior to this work, other computational studies, in the absence of the experimentally-calibrated spectroscopic evidence provided by NRVS, had predicted a six-coordinate Fe(IV)=O intermediate.[11,51,52]

4.3.1.2 High-Spin Six-Coordinate Fe(IV)=O Complexes

The generation of a high-spin six-coordinate Fe(IV)=O complex requires the use of weak-field ligands, particularly in the equatorial plane. In terms of d-orbital occupancy, the difference between attaining an $S = 2$ and an $S = 1$ spin state is the energy separation of the d_{x2-y2} and d_{xy} orbitals (Figure 4.7). This separation increases with stronger field ligands, favouring a $d_{xy}^2 d_{xz}^1 d_{yz}^1 d_{x2-y2}^0$ configuration and an $S = 1$ spin state. As the ligand field weakens, the fourth d electron would prefer to occupy the d_{x2-y2} orbital rather than pair up with the first electron in the d_{xy} orbital, generating the $S = 2$ spin state. However the weak-field nature of the ligand environment makes the complex quite unstable, thereby limiting the number of complexes characterized to have high-spin six-coordinate Fe(IV)=O units. To date in fact, only a single six-coordinate high-spin Fe(IV)=O complex that is spectroscopically accessible has been reported.[53]

Investigations into the potential role of ferryl ions in Fenton chemistry led to the proposal that an Fe(IV)=O unit is formed in the reaction of Fe(II) ions with O_3 at low pH in aqueous solutions.[54,55] This species was found to have a short half-life of ~10 s at room temperature and to exhibit a broad absorption band centered above 300 nm, similar to the ~320 nm feature associated with the Fe(IV)=O intermediates of 2OG-dependent enzymes. In 2005, Bakac and coworkers provided the first spectroscopic evidence for the generation of $[Fe^{IV}(O)(H_2O)_5]^{2+}$ when O_3 was bubbled into an aqueous solution of

$[Fe(H_2O)_6]^{2+}$ at pH 1.[53] Due to its high reactivity (decaying even in the frozen state at 220 K), the aqueous Fe(IV)=O unit could only be characterized by XAS and Mössbauer spectroscopy. Its Mössbauer spectrum shows a quadrupole doublet with $\delta = 0.38$ mm s^{-1} and $\Delta E_Q = 0.33$ mm s^{-1}. The isomer shift of 0.38 mm s^{-1} is the largest for a non-haem Fe(IV)=O unit to date, and even larger than those of enzymatic Fe(IV)=O units, probably due to the weak donating ability of the aqua ligands. Its XANES spectrum shows an Fe K-edge at 7126 eV, which is 1–3 eV higher than other Fe(IV)=O complexes, and a pre-edge feature with a large area of 60–70 units that reflects a highly distorted Fe(IV) centre. DFT calculations bear out the experimentally determined $S = 2$ ground state and high edge energy, and further predict a six-coordinate iron centre with an oxo ligand at 1.62 Å.

There have been other attempts to synthesize $S = 2$ six-coordinate Fe(IV)=O complexes by using an appropriate ligand with a weak-field donor set, although none thus far has yielded such a species that is spectroscopically accessible. One notable attempt was the synthesis of $[Fe^{IV}(O)(^{n}Bu\text{-}P2DA)]$[56] (Figure 4.9), which employed a ligand set containing two carboxylate donors designed to serve as a structural analogue of intermediate **VI** (Figure 4.1). Treatment of $[Fe(^{n}Bu\text{-}P2DA)(Me_2Im)]$ with *m*-CPBA at −95 °C led to the formation of a species with near IR features at 770 and 920 nm and a Mössbauer isomer shift of 0.04 mm s^{-1}, which are indicative of an $S = 1$ Fe(IV)=O species. EXAFS analysis showed the presence of an oxygen scatterer at 1.66 Å, consistent with the Fe(IV)=O assignment. Interestingly, DFT calculations predicted an $S = 1$ ground state, albeit with a low-lying $S = 2$ excited state, thus validating the strategy of utilizing a weak-field carboxylate donor to reduce the splitting of the d_{x2-y2} and d_{xy} orbitals. The generation of additional six-coordinate $S = 2$ complexes remains an important synthetic challenge.

4.3.1.3 S = 1 Fe(IV)=O Complexes

Well-characterized intermediate-spin ($S = 1$) Fe(IV)=O complexes outnumber their $S = 2$ counterparts ten-fold and thus provide a useful point of comparison. These $S = 1$ Fe(IV)=O units are supported by a variety of polydentate ligands that have linear, macrocyclic, tripodal or tetrapodal motifs.[57] The first of these to be structurally and spectroscopically well defined was $[Fe^{IV}(O)(TMC)(CH_3CN)](OTf)_2$ (Figure 4.9),[58–61] which was reported in the same year that intermediate **J** of TauD was identified.[6] The Fe=O bond length was determined by X-ray diffraction to be 1.646 Å and the Fe=O stretching frequency was found to be 834 cm^{-1} by FTIR[58] and 839 cm^{-1} by resonance Raman spectroscopy, which is similar to that found for the high-spin $[Fe^{IV}O(TMG_3tren)]^{2+}$ (843 cm^{-1}).[59] That the Fe(IV)=O bond is insensitive to the spin state of the iron centre is not unexpected, as the d_{xy} and d_{x2-y2} orbitals that differ in electron population depending on spin state are non-bonding with respect to the oxo unit. $[Fe^{IV}(O)(TMC)(CH_3CN)]^{2+}$ has a characteristic near IR band at 824 nm that has been assigned by VT-MCD studies as arising from three d–d transitions of the $S = 1$ Fe(IV)=O unit,[62] but its relatively high molar extinction

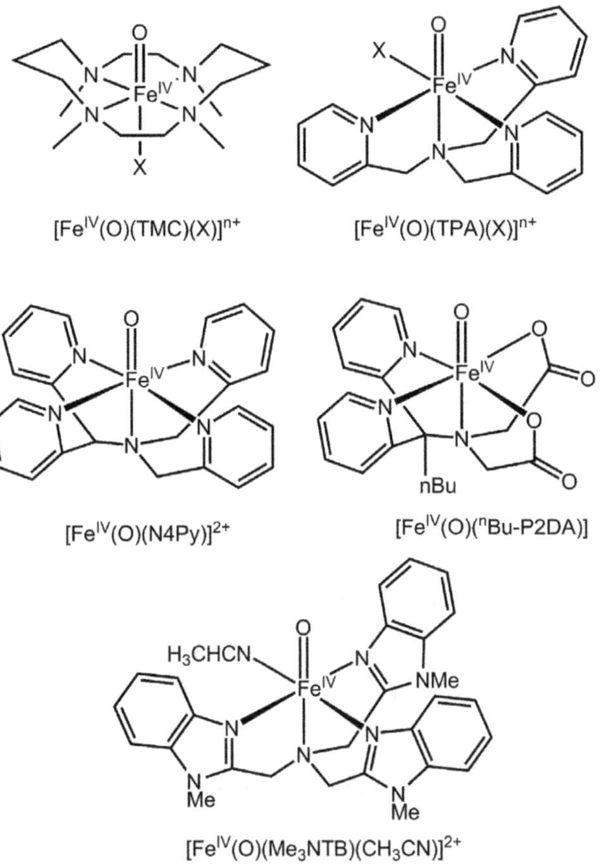

$[Fe^{IV}(O)(TMC)(X)]^{n+}$ $[Fe^{IV}(O)(TPA)(X)]^{n+}$

$[Fe^{IV}(O)(N4Py)]^{2+}$ $[Fe^{IV}(O)(^{n}Bu\text{-}P2DA)]$

$[Fe^{IV}(O)(Me_3NTB)(CH_3CN)]^{2+}$

Figure 4.9 Selected intermediate-spin Fe(IV)=O complexes.

coefficient of 400 suggests the possible admixture of some charge transfer character. This feature has become a convenient spectroscopic signature for identifying this class of complexes.

The plethora of well-characterized intermediate-spin Fe(IV)=O species has allowed extensive studies of the roles that electronic, steric and structural modifications play in modulating their spectroscopic properties. For example, an isostructural series of $[Fe^{IV}(O)(TMC)(X)]^+$ complexes ($X = CF_3CO_2^-$, OH^-, NCO^-, NCS^-, N_3^-, CN^-) has been obtained by metathesis from $[Fe^{IV}(O)(TMC)(CH_3CN)]^{2+}$.[59] Replacement of the neutral CH_3CN ligand led to red-shifting of the near-IR band and an increased separation in energy of the three d–d transitions. The $\nu_{Fe\text{-}O}$ frequencies varied from 814 cm^{-1} for the strongly donating N_3^- ligand up to 854 cm^{-1} for $CF_3CO_2^-$, indicating the clear influence of the *trans* ligand on the Fe=O bond. Conversely, only modest variation of the Fe=O distance between 1.64 and 1.68 Å was observed by EXAFS analysis. In the Mössbauer spectra, the isomer shifts remained within a narrow window of 0.15–0.20 mm s^{-1}, but the quadrupole splitting varied markedly from 0.16

to 1.39 mm s^{-1} as a function of basicity of the *trans* ligand. An analogous study carried out for an isostructural series of $[Fe^{IV}(O)(TPA)(X)]^{n+}$ complexes (TPA = tris(pyridyl-2-methyl)amine (Figure 4.9); $X = CH_3CN$, $CF_3CO_2^-$, Cl^-, Br^-) showed smaller effects for the variable *cis*-ligand on the properties of the Fe(IV)=O unit.[63,64]

4.3.2 Reactivity of Fe(IV)=O Complexes

The ubiquitous Fe(IV)=O unit in the catalytic cycles of 2OG-dependent enzymes is proposed to carry out a diverse array of substrate oxidations, including hydroxylation, dehydrogenation, oxygen and halogen atom transfer, and heterocyclic ring formation.[3] The vast majority of 2OG-dependent oxygenases catalyse reactions that involve an initial hydrogen atom transfer (HAT) step from substrate to the Fe(IV)=O unit to generate a substrate radical and an Fe(III)–OH species. In general, the latter two undergo oxygen rebound to form a C–OH bond, corresponding to substrate hydroxylation (Figure 4.10). However for halogenases and a handful of enzymes that form heterocyclic rings, a C–X bond is formed, where the X atom is ligated to the high-valent iron centre. Heterocyclic ring formation can be construed as an intramolecular variation of the halogenation reaction, where the heteroatom is part of the substrate.

While model systems that are functional models of the entire reaction of 2OG-dependent enzymes have provided important insights into the reactivities of transient Fe(IV)=O intermediates (see above), the reactivities of Fe(IV)=O complexes that are well defined structurally and spectroscopically also have an important role to play in understanding the diverse reactions available to this species in 2OG-dependent enzymes. Synthetic Fe(IV)=O units have been reported to be capable of oxidizing substrates in reactions functionally similar to those carried out in many 2OG-dependent reactions. Interestingly, while all enzymatic Fe(IV)=O units have been reported to be high spin, both intermediate- and high-spin synthetic Fe(IV)=O compounds

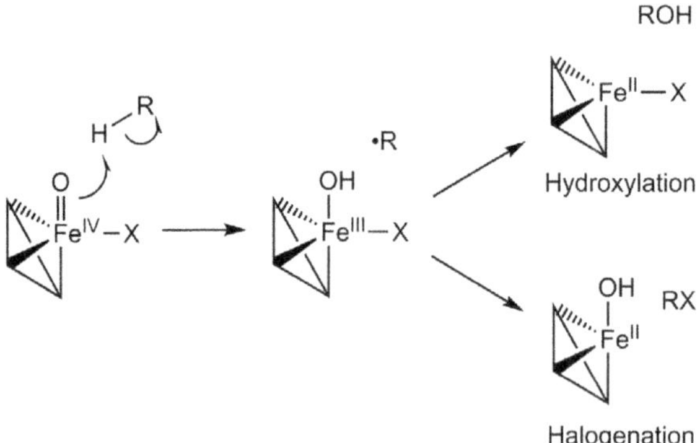

Figure 4.10 Proposed HAT and rebound steps.

have been found to effect aliphatic and aromatic hydroxylation, dehydrogenation of CH_2–CH_2 moieties, and oxygen-atom transfer to phosphines and sulfides. Alkene epoxidation, however, has thus far only been reported for three intermediate-spin Fe(IV)=O compounds.[65–67] With little difference in reaction scope as a function of spin state, the question arises as to whether nature exclusively utilizes high-spin Fe(IV)=O intermediates as a coincidence of the available weak-field ligands, or due to some inherent higher reactivity afforded by the high-spin state.

The majority of studies on the reactivity of synthetic Fe(IV)=O complexes have focused on HAT reactions. Among high-spin Fe(IV)=O complexes, $[Fe^{IV}O(TMG_3tren)]^{2+}$ and $[Fe^{IV}O(tpa^{Ph})]^-$ were shown to oxidize 1,4-cyclohexadiene (BDE = 77 kcal mol^{-1}) at −30 °C with comparable second-order rate constants of 1.2 $M^{-1}s^{-1}$ and 1.4 $M^{-1}s^{-1}$, respectively.[41,47] However, the oxidation of the bulkier substrate DHA (78 kcal mol^{-1}) by $[Fe^{IV}O(TMG_3tren)]^{2+}$ occurred at a 13-fold slower rate, despite having a BDE similar to that of CHD, and $[Fe^{IV}O(tpa^{Ph})]^-$ was found to be unreactive towards DHA. These observations suggest that the bulky ligands used to form these high-spin complexes provide a significant steric hindrance for substrate access to the Fe(IV)=O unit. Consistent with this idea, there was a significant increase in HAT rates for CHD and DHA oxidation (18 and 57 $M^{-1}s^{-1}$, respectively) for the less sterically bulky $[Fe^{IV}O(TMG_2dien)(CH_3CN)]^{2+}$ relative to the TMG_3tren complex, presumably a consequence of the more accessible Fe(IV)=O unit.[46] Thus a 600-fold increase in the DHA oxidation rate was realized by removing one arm of the tripodal TMG_3tren ligand. No documented C–H bond reactivity was reported for $[Fe^{IV}O(H_3buea)]^-$, although it was suggested that its self-decay in DMF at room temperature (half-life of 2.2 h) may occur *via* homolysis of a C–H bond of the DMF solvent (BDE = 82 kcal mol^{-1}).[48]

Additional insight was derived from studying the self-decay of $[Fe^{IV}O(TMG_3tren)]^{2+}$ where intramolecular C–H bond activation was observed (Figure 4.11).[42] Analysis of the product of this reaction revealed the hydroxylation of a ligand methyl group, demonstrating that the $S = 2$ Fe(IV)=O unit in this complex was capable of attacking a C–H bond (~92 kcal mol^{-1}) of comparable strength as the target C–H bonds for some 2OG-dependent enzymes. Isotope labelling experiments showed that the oxygen atom of the self-hydroxylated product was derived from the Fe(IV)=O unit under either an N_2 or O_2 atmosphere. This observation strongly supports oxygen rebound between the nascent N–CH_2• radical and the putative Fe(III)–OH species formed after the initial HAT step, as proposed for 2OG-dependent hydroxylases. This reaction should form a self-hydroxylated Fe(II) product, but it was shown to have an Fe(III) centre even when the reaction was carried out in the absence of O_2. Interestingly, the Fe(III) self-hydroxylated product was found to be present in a 1 : 1 ratio with an $[Fe^{III}(OH)(TMG_3tren)]^{2+}$ by-product. These products were rationalized to form *via* the reaction of the nascent Fe(II) product with residual Fe(IV)=O complex, which was much more facile than the initial HAT step in the self-decay.

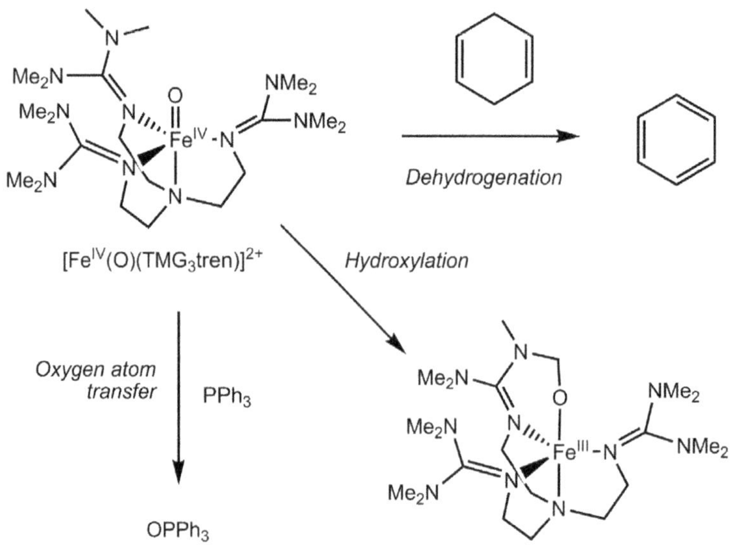

Figure 4.11 Reactivity of $[Fe^{IV}(O)(TMG_3tren)]^{2+}$.

Deuteration of all 12 methyl groups of the TMG_3tren ligand significantly retarded the self-decay rate of the $[Fe^{IV}O(TMG_3tren)]^{2+}$ (KIE = 24 at 25 °C), which turned out to be the key to growing single crystals of the complex for X-ray diffraction studies.[42] A KIE of 18 was also observed for the oxidation of DHA by $[Fe^{IV}O(TMG_3tren)]^{2+}$.[41] The large magnitude of these KIEs indicates that both the inter- and intramolecular C–H bond activation processes probably occur *via* a tunnelling mechanism and are consistent with the KIEs found for enzymatic Fe(IV)=O intermediates.[3,7,11,34]

It has been proposed that nature utilizes high-spin Fe(IV)=O complexes as they are inherently more reactive towards substrates than their intermediate-spin counterparts. This is a notion that has independently been arrived at by several groups through the use of DFT calculations.[62,68–72] In the exchange-enhanced reactivity model proposed by Shaik, a high-spin Fe(IV)=O unit exhibits greater reactivity towards C–H bonds by carrying out H-atom abstraction *via* the lowest unoccupied σ^* molecular orbital, because such a step results in an increased number of exchange interactions that stabilize the transition state. On the other hand, Solomon emphasizes the stereoelectronic features of the frontier molecular orbitals (FMOs). In both approaches, C–H bond attack *via* the σ^* molecular orbital of the Fe(IV)=O unit would entail an Fe–O–H angle of 180°, while a π^* attack would require an Fe–O–H angle closer to 130°. Both σ^* and π^* attack pathways are available for the $S = 2$ Fe(IV)=O unit, while only a π^* attack would be feasible with an $S = 1$ Fe(IV)=O unit.[62,73,74]

Solomon used a spectroscopy-calibrated DFT approach to gain insight into the HAT reactivity of $S = 1$ $[Fe^{IV}(O)(N4Py)]^{2+}$ and $S = 2$ $[Fe^{IV}O(TMG_3tren)]^{2+}$, which have been found to oxidize CHD with similar rates (1.2 $M^{-1}s^{-1}$ and 1.3

$M^{-1}s^{-1}$) at −30 °C.[75–77] It was found that the σ* pathway for $[Fe^{IV}O(TMG_3tren)]^{2+}$ was comparable in energy to the π* pathway for $[Fe^{IV}O(N4Py)]^{2+}$, due to the presence of steric barriers for the approach of substrate in either pathway. However for $[Fe^{IV}(O)(N4Py)]^{2+}$, the reactive β π*-FMO has 42% O $p_{x/y}$ character, which is significantly higher than the 27% O p_z character of the reactive α σ* FMO of $[Fe^{IV}O(TMG_3tren)]^{2+}$ (see Figure 4.7), so there is a larger bonding contribution that lowers the barrier for reaction on the triplet surface. The π* pathway can also be operative in $[Fe^{IV}O(TMG_3tren)]^{2+}$ and leads to the observed intramolecular ligand hydroxylation.

The large number of structurally similar intermediate-spin Fe(IV)=O complexes available has allowed some insights into the effect of ligand basicity on the reactivity of the oxo unit. An interesting study involved a series of $[Fe^{IV}(O)(TMC)(X)]^{+/2+}$ complexes where the X ligand (X = $NCCH_3$, $O_2CCF_3^-$, N_3^- or SR^-) *trans* to the oxo unit was varied in basicity. These complexes showed a decrease in the OAT rate to PPh_3 with an increase in the basicity of the X ligand, in line with a decrease in the electrophilicity of the Fe(IV)=O unit.[78] This trend was supported by observations on a series of related S = 1 Fe(IV)=O complexes supported by pentadentate N5 ligands where the OAT rate to PhSMe was correlated with the Fe(IV/III) redox potentials of the complexes.[79]

In contrast, the HAT rate of the $[Fe^{IV}(O)(TMC)(X)]^{+/2+}$ complexes was found to increase with increasing ligand basicity.[78] This 'anti-electrophilic' trend in intermediate-spin complexes was rationalized using a two-state reactivity (TSR) model, which posits that HAT occurs *via* the S = 2 excited state reaction surface.[80] The more basic axial ligand gives rise to a smaller energy gap between the S = 1 and S = 2 surfaces and consequently a faster HAT rate, as observed. However, exceptions that challenge the ubiquity of the 'anti-electrophilic' trend have been found in subsequent studies of other $[Fe^{IV}(O)(TMC)(X)]^{+/2+}$ complexes,[81] as well as for the series of S = 1 Fe(IV)=O complexes supported by pentadentate N5 ligands.[79] The relevance of the TSR model to 2OG-dependent enzymes, which all employ high-spin Fe(IV)=O units for HAT, remains unclear.

Consideration of the body of information available on non-haem Fe(IV)=O reactivity suggests that a complete understanding of the interplay of factors that control rates of HAT reactivity is still lacking. The most reactive of the intermediate-spin complexes to date is $[Fe^{IV}O(Me_3NTB)(CH_3CN)]^{2+}$ (Figure 4.9) reported by Nam.[82] It reacts with CHD at −40 °C at a rate three orders of magnitude faster than $[Fe^{IV}O(N4Py)]^{2+}$ and can even oxidize cyclohexane at this temperature with a reasonably fast rate. Indeed the Me_3NTB complex is even more reactive than $[Fe(O)(TDCPP)]^+$, an oxo-Fe(IV) porphyrin radical complex. Why the Me_3NTB complex is so reactive is not clear at the present time.

One of the challenges in understanding the HAT reactivity of various Fe(IV)=O complexes has been to sort out the effects of spin state from specific ligand effects because of differences in ligand structure, making this exercise akin to comparing apples with oranges. At present, there is only one pair of complexes with a common Fe(IV)=O unit that is supported by the same polydentate ligand but with different spin states, namely $[(TPA^*)Fe^{IV}(O)(\mu\text{-}O)Fe^{III}(OH)(TPA^*)]^{2+}$ and $[(TPA^*)Fe^{IV}(O)(\mu\text{-}O)Fe^{IV}(OH)(TPA^*)]^{3+}$. These two complexes are related to each other by a difference of one electron at the Fe–OH unit. The former has an $S = 2$ Fe(IV)=O unit and oxidizes DHA three orders of magnitude faster than the latter, which has an $S = 1$ Fe(IV)=O unit.[83] Replacement of the Fe(III)–OH fragment with an Fe(III)–F fragment results in the loss of the hydrogen bond between the Fe(IV)=O unit and the adjacent Fe(III)–OH unit in $[(TPA^*)Fe^{IV}(O)\ (\mu\text{-}O)Fe^{III}(OH)\ (TPA^*)]^{2+}$ and gives rise to the corresponding $[(TPA^*)Fe^{IV}(O)\ (\mu\text{-}O)Fe^{III}(F)\ (TPA^*)]^{2+}$ complex, which is even more reactive by an additional factor of 10.[84] These results provide the strongest argument for the higher HAT reactivity of the $S = 2$ Fe(IV)=O unit.

4.3.3 Models for Other Intermediates in the 2OG-Dependent Enzyme Catalytic Cycle

Intermediates corresponding to the superoxo-Fe(III) species **IV** and alkylperoxo-Fe(IV) species **V** have not yet been detected in the catalytic cycles of 2OG-dependent enzymes. However indirect evidence has been found for the involvement of an Fe(III)-superoxo species analogous to **IV** prior to the decarboxylation reactions in some model systems described in Section 4.2.[32] At the time of writing of this chapter, the first example of a synthetic mononuclear non-haem Fe(III)-superoxo complex had just been reported. This species forms reversibly from the reaction of O_2 with an iron(II) centre supported by a linear pentadentate N_3O_2 ligand at −80 °C (Figure 4.12).[85] It exhibits a ν(O–O) at 1125 cm^{-1}, a value that falls in the middle of the range found for metal-superoxo complexes.[86–89] Unlike its diamagnetic haem counterparts,[90] this adduct is paramagnetic due to the presence of a high-spin iron(III) centre as revealed by its Mössbauer spectrum. The $S = 5/2$ iron centre is coupled to the $S = ½$ superoxo ligand, but it has not yet been established whether the coupling is ferromagnetic or antiferromagnetic. There is precedence for both ferromagnetic and antiferromagnetic coupling in mononuclear non-haem iron(III)-superoxo adducts identified in enzymatic

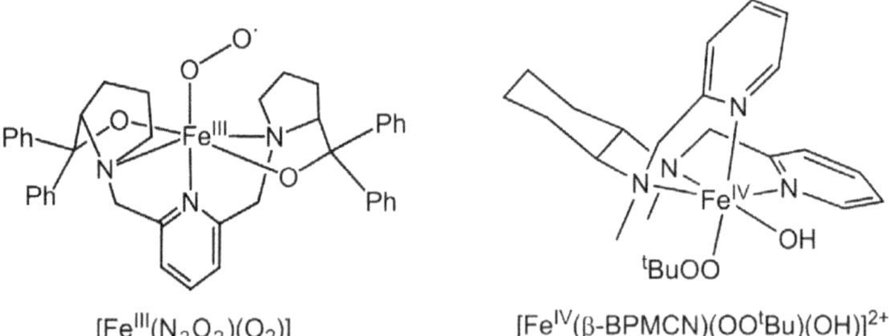

Figure 4.12 Proposed structures of superoxoiron(III) and alkylperoxoiron(IV) complexes that may serve as models for corresponding species in the catalytic cycles of 2OG-dependent enzymes.

systems.[91,92] Given the pentadentate nature of the supporting ligand, it is assumed that the superoxo ligand is bound end-on. As this effort develops, it will be interesting to learn about how the reactivity of the bound superoxide can be affected by the superoxo binding mode, the iron(III) spin state, and the nature of the coupling interaction between the iron(III) and the superoxo ligand.

With respect to modelling alkylperoxo-Fe(IV) species **V**, there is at present only one well-defined example of an alkylperoxo–Fe(IV) complex.[93,94] $[Fe^{IV}(\beta\text{-BPMCN})(OO^tBu)(OH)]^{2+}$ was generated by the reaction of $[Fe^{II}(\beta\text{-BPMCN})]^{2+}$ with ^tBuOOH in CH_2Cl_2 at −80 °C (Figure 4.12) and shown by Mössbauer spectroscopy to have an $S = 1$ Fe(IV) centre. It was found to decay by Fe–O bond homolysis, which is quite distinct from the O–O bond lysis required for the oxidative decarboxylation step in the 2OG enzyme catalytic cycle. Thus, obtaining a suitable model complex that corresponds to the alkylperoxo-Fe(IV) intermediate in the 2OG enzyme catalytic cycle remains an unmet challenge.

4.4 Perspectives and Outlook

The past decade has seen a significant jump in our mechanistic understanding of the chemistry of the 2OG-dependent enzymes, from studies of the enzymes themselves as well as from the synthesis of relevant model complexes. This chapter focuses on insights obtained from the latter efforts.

Several model complexes that have been synthesized and characterized are structurally comparable to either the binary or ternary complexes (**II** and **III**, respectively, in Figure 4.1) of 2OG-dependent enzymes and serve as functional models of these enzymatic reactions. The significantly higher reactivity of the five-coordinate model complexes compared to their six-coordinate counterparts emphasizes the importance of having a vacant coordination site for efficient O_2 binding and subsequent reactivity, as borne out by the observed displacement of water in the enzyme active site that occurs only upon substrate binding in order to prime the iron centre in the 2OG-dependent enzyme for catalysis. The absence of this control would result in the wasteful uncoupling of 2OG oxidative decarboxylation from substrate oxidation.

The oxidations carried out by these functional models encompass a number of enzymatically relevant reaction types including dehydrogenation, hydroxylation, epoxidation and electrophilic aromatic hydroxylation. Mechanistic studies have indicated that all of these reactions are probably carried out by a transient Fe(IV)=O species (**VI** in Figure 4.1), which exhibits remarkable versatility in terms of its reaction scope. Although this transient species has not yet been observed or spectroscopically characterized in any of the functional models, the trapping of such a reaction intermediate would be a phenomenal achievement for the synthetic field. Additionally, interception experiments have provided indirect evidence for the initial formation of an Fe(III)-superoxo moiety (**IV** in Figure 4.1) prior to the formation of the

Fe(IV)=O oxidant, providing the first experimental evidence for such a species in a catalytic cycle of relevance to 2OG-dependent enzymes.

Structurally characterized synthetic Fe(IV)=O complexes began to appear within the same time frame as the corresponding high-valent enzymatic intermediates and have provided significant insight into the physical and electronic structure of these elusive species. Although much has been learned, more remains to be uncovered, particularly about the various factors that modulate the reactivity of the Fe(IV)=O unit. It was recently reported that a halogenase enzyme can convert the target C–H bond into a C–*X* bond with $X = N_3$ or NO_2, in addition to the native Cl and Br groups, by introducing these ions into the first coordination sphere of the iron centre.[95] This development challenges the biomimetic community to develop analogous transformations for potential synthetic applications. Although spectroscopically characterized Fe(IV)=O complexes with *cis*-halide ligands have been reported,[46,63,96] these complexes have not yet been shown to halogenate C–H bonds. On the other hand, two iron complexes of tetradentate ligands have been found to carry out halogenation of cyclohexane either stoichiometrically or catalytically with peroxides or PhIO in the presence of halides.[97,98] High-valent oxoiron species were implicated, but not directly observed. The questions raised in this chapter about the reactivity of oxo-Fe(IV) complexes emphasize the richness of the chemistry that still awaits discovery and comprehension.

Acknowledgments

The authors thank the National Science Foundation (Grant CHE-1361773 to LQ) for financial support of this project.

References

1. R. P. Hausinger, *Crit. Rev. Biochem. Mol. Biol.*, 2004, **39**, 21.
2. H. M. Hanauske-Abel and V. Günzler, *J. Theor. Biol.*, 1982, **94**, 421.
3. C. Krebs, D. Galonić Fujimori, C. T. Walsh and J. M. Bollinger Jr., *Acc. Chem. Res.*, 2007, **40**, 484.
4. J. M. Elkins, M. J. Ryle, I. J. Clifton, J. C. Dunning Hotopp, J. S. Lloyd, N. I. Burzlaff, J. E. Baldwin, R. P. Hausinger and P. L. Roach, *Biochemistry*, 2002, **41**, 5185.
5. K. D. Koehntop, J. P. Emerson and L. Que, Jr., *J. Biol. Inorg. Chem.*, 2005, **10**, 87.
6. J. C. Price, E. W. Barr, B. Tirupati, J. M. Bollinger, Jr. and C. Krebs, *Biochemistry*, 2003, **42**, 7497.
7. J. C. Price, E. W. Barr, T. E. Glass, C. Krebs and J. M. Bollinger, Jr., *J. Am. Chem. Soc.*, 2003, **125**, 13008.
8. J. C. Price, E. W. Barr, L. M. Hoffart, C. Krebs and J. M. Bollinger, Jr., *Biochemistry*, 2005, **44**, 8138.

9. P. J. Riggs-Gelasco, J. C. Price, R. B. Guyer, J. H. Brehm, E. W. Barr, J. M. Bollinger, Jr. and C. Krebs, *J. Am. Chem. Soc.*, 2004, **126**, 8108.
10. D. A. Proshlyakov, T. F. Henshaw, G. R. Monterosso, M. J. Ryle and R. P. Hausinger, *J. Am. Chem. Soc.*, 2004, **126**, 1022.
11. S. Sinnecker, N. Svensen, E. W. Barr, S. Ye, J. M. Bollinger, Jr., F. Neese and C. Krebs, *J. Am. Chem. Soc.*, 2007, **129**, 6168.
12. M. Costas, M. P. Mehn, M. P. Jensen and L. Que, Jr., *Chem. Rev.*, 2004, **104**, 939.
13. J. T. Groves and M. J. Van Der Puy, *Am. Chem. Soc.*, 1976, **98**, 5290.
14. M. Chvapil and J. Hurych, *Nature*, 1959, **184**, 1145.
15. G. A. Hamilton, R. J. Workman and L. Woo, *J. Am. Chem. Soc.*, 1964, **86**, 3390.
16. B. Siegel and J. Lanphear, *J. Am. Chem. Soc.*, 1979, **101**, 2221.
17. B. Siegel and J. Lanphear, *J. Org. Chem.*, 1979, **44**, 942.
18. V. S. Kulikova, A. M. Khenkin and A. E. Shilov, *Kinetika I Kataliz*, 1988, **29**, 1278.
19. A. M. Khenkin and A. E. Shilov, *New J. Chem.*, 1989, **13**, 659.
20. Y.-M. Chiou and L. Que, Jr., *J. Am. Chem. Soc.*, 1992, **114**, 7567.
21. Y.-M. Chiou and L. Que, Jr., *J. Am. Chem. Soc.*, 1995, **117**, 3999.
22. R. Y. N. Ho, M. P. Mehn, E. L. Hegg, A. Liu, M. J. Ryle, R. P. Hausinger and L. Que, Jr., *J. Am. Chem. Soc.*, 2001, **123**, 5022.
23. E. G. Pavel, J. Zhou, R. W. Busby, M. Gunsior, C. A. Townsend and E. I. Solomon, *J. Am. Chem. Soc.*, 1998, **120**, 743.
24. E. L. Hegg, A. K. Whiting, R. E. Saari, J. McCracken, R. P. Hausinger and L. Que, Jr., *Biochemistry*, 1999, **38**, 16714.
25. M. J. Ryle, R. Padmakumar and R. P. Hausinger, *Biochemistry*, 1999, **38**, 15278.
26. S. C. Trewick, T. F. Henshaw, R. P. Hausinger, T. Lindahl and B. Sedgwick, *Nature*, 2002, **419**, 174.
27. E. H. Ha, R. Y. N. Ho, J. F. Kisiel and J. S. Valentine, *Inorg. Chem.*, 1995, **34**, 2265.
28. A. Mukherjee, M. A. Cranswick, M. Chakrabarti, T. K. Paine, K. Fujisawa, E. Münck and L. Que, Jr., *Inorg. Chem.*, 2010, **49**, 3618.
29. N. Kitajima, N. Tamura, H. Amagai, H. Fukui, Y. Moro-oka, Y. Mizutani, T. Kitagawa, R. Mathur, K. Heerwegh, C. A. Reed, C. R. Randall, L. Que, Jr. and K. Tatsumi, *J. Am. Chem. Soc.*, 1994, **116**, 9071.
30. K. Kim and S. J. Lippard, *J. Am. Chem. Soc.*, 1996, **118**, 4914.
31. M. P. Mehn, K. Fujisawa, E. L. Hegg and L. Que, Jr., *J. Am. Chem. Soc.*, 2003, **125**, 7828.
32. A. Mukherjee, M. Martinho, E. L. Bominaar, E. Münck and L. Que, Jr., *Angew. Chem., Int. Ed.*, 2009, **48**, 1780.
33. D. Janardanan, Y. Wang, P. Schyman, L. Que, Jr. and S. Shaik, *Angew. Chem., Int. Ed.*, 2010, **49**, 3342.
34. L. M. Hoffart, E. W. Barr, R. B. Guyer, J. M. Bollinger, Jr. and C. Krebs, *Proc. Natl. Acad. Sci.*, 2006, **103**, 14738.

35. D. P. Galonić, E. W. Barr, C. T. Walsh, J. M. Bollinger, Jr. and C. Krebs, *Nat. Chem. Biol.*, 2007, **3**, 113.
36. D. Galonić Fujimori, E. W. Barr, M. L. Matthews, G. M. Koch, J. R. Yonce, C. T. Walsh, J. M. Bollinger, Jr., C. Krebs and P. J. Riggs-Gelasco, *J. Am. Chem. Soc.*, 2007, **129**, 13408.
37. M. L. Matthews, C. M. Krest, E. W. Barr, F. H. Vaillancourt, C. T. Walsh, M. T. Green, C. Krebs and J. M. Bollinger, Jr., *Biochemistry*, 2009, **48**, 4331.
38. L. C. Blasiak, F. H. Vaillancourt, C. T. Walsh and C. L. Drennan, *Nature*, 2006, **440**, 368–371.
39. B. E. Eser, E. W. Barr, P. A. Frantom, L. Saleh, J. M. Bollinger, Jr., C. Krebs and P. F. Fitzpatrick, *J. Am. Chem. Soc.*, 2007, **129**, 11334.
40. A. J. Panay, M. Lee, C. Krebs, J. M. Bollinger, Jr. and P. F. Fitzpatrick, *Biochemistry*, 2011, **50**, 1928.
41. J. England, M. Martinho, E. R. Farquhar, J. R. Frisch, E. L. Bominaar, E. Münck and L. Que, Jr., *Angew. Chem., Int. Ed.*, 2009, **48**, 3622.
42. J. England, Y. Guo, E. R. Farquhar, V. G. Young, E. Münck and L. Que, Jr., *J. Am. Chem. Soc.*, 2010, **132**, 8635.
43. J. E. Penner-Hahn, K. S. Eble, T. J. McMurry, M. Renner, A. L. Balch, J. T. Groves, J. H. Dawson and K. O. Hodgson, *J. Am. Chem. Soc.*, 1986, **108**, 7819.
44. T. Wolter, W. Meyer-Klaucke, M. Muther, D. Mandon, H. Winkler, A. X. Trautwein and R. Weiss, *J. Inorg. Biochem.*, 2000, **78**, 117.
45. J. F. Berry, S. DeBeer and F. Neese, *Phys. Chem. Chem. Phys.*, 2008, **10**, 4361.
46. J. England, Y. Guo, H. K. M. Van, M. A. Cranswick, G. T. Rohde, E. L. Bominaar, E. Munck and L. Que, Jr., *J. Am. Chem. Soc.*, 2011, **133**, 11880.
47. J. P. Bigi, W. H. Harman, B. Lassalle-Kaiser, D. M. Robles, T. A. Stich, J. Yano, R. D. Britt and C. J. Chang, *J. Am. Chem. Soc.*, 2012, **134**, 1536.
48. D. C. Lacy, R. Gupta, K. L. Stone, J. Greaves, J. W. Ziller, M. P. Hendrich and A. S. Borovik, *J. Am. Chem. Soc.*, 2010, **132**, 12188.
49. S. D. Wong, C. B. Bell III, L. V. Liu, Y. Kwak, J. England, E. Alp, J. Zhao, L. Que, Jr. and E. I. Solomon, *Angew. Chem., Int. Ed.*, 2011, **50**, 3215.
50. S. D. Wong, M. Srnec, M. L. Matthews, L. V. Liu, Y. Kwak, K. Park, C. B. Bell III, E. E. Alp, J. Zhao, Y. Yoda, S. Kitao, M. Seto, C. Krebs, J. M. Bollinger, Jr. and E. I. Solomon, *Nature*, 2013, **499**, 320.
51. H. J. Kulik, L. C. Blasiak, N. Marzari and C. L. Drennan, *J. Am. Chem. Soc.*, 2009, **131**, 14426.
52. T. Borowski, H. Noack, M. Rado, K. Zych and P. E. M. Siegbahn, *J. Am. Chem. Soc.*, 2010, **132**, 12887.
53. O. Pestovsky, S. Stoian, E. L. Bominaar, X. Shan, E. Münck, L. Que, Jr. and A. Bakac, *Angew. Chem., Int. Ed.*, 2005, **44**, 6871.
54. T. Logager, J. Holcman, K. Sehested and T. Pederson, *Inorg. Chem.*, 1992, **31**, 3523.
55. O. Pestovsky and A. Bakac, *J. Am. Chem. Soc.*, 2004, **126**, 13757.
56. A. R. McDonald, Y. Guo, V. V. Vu, E. L. Bominaar, E. Münck and L. Que, Jr., *Chem. Sci.*, 2012, **3**, 1680.

57. A. R. McDonald and L. Que, Jr., *Chem. Rev.*, 2013, **257**, 414.
58. J.-U. Rohde, J. –H. In, M. H. Lim, W. W. Brennessel, M. R. Bukowski, A. Stubna, E. Münck, W. Nam and L. Que, Jr., *Science*, 2003, **299**, 1037.
59. T. A. Jackson, J. –U. Rohde, M. S. Seo, C. V. Sastri, R. DeHont, T. Ohta, T. Kitagawa, E. Münck, W. Nam and L. Que, Jr., *J. Am. Chem. Soc.*, 2008, **130**, 12394.
60. J. –U. Rohde and L. Que, Jr., *Angew. Chem., Int. Ed.*, 2005, **44**, 2255.
61. C. V. Sastri, M. J. Park, T. Ohta, T. A. Jackson, A. Stubna, M. S. Seo, J. Lee, J. Kim, T. Kitagawa, E. Munck, L. Que, Jr. and W. Nam, *J. Am. Chem. Soc.*, 2005, **127**, 12494.
62. A. Decker, J.-U. Rohde, E. J. Klinker, S. D. Wong, L. Que, Jr. and E. I. Solomon, *J. Am. Chem. Soc.*, 2007, **129**, 15983.
63. J.-U. Rohde, A. Stubna, E. L. Bominaar, E. Münck, W. Nam and L. Que, Jr., *Inorg. Chem.*, 2006, **45**, 6435.
64. M. H. Lim, J. U. Rohde, A. Stubna, M. R. Bukowski, M. Costas, R. Y. N. Ho, E. Münck, W. Nam and L. Que, Jr., *Proc. Natl. Acad. Sci.*, 2003, **100**, 3665.
65. V. Balland, M.-F. Charlot, F. Banse, J.-J. Girerd, T. A. Mattioli, E. Bill, J.-F. Bartoli, P. Battioni and D. Mansuy, *Eur. J. Inorg. Chem.*, 2004, 301.
66. J. Annaraj, S. Kim, M. S. Seo, Y.-M. Lee, Y. Kim, S.-J. Kim, Y. S. Choi and W. Nam, *Inorg. Chim. Acta*, 2009, **362**, 1031.
67. W. Ye, D. M. Ho, S. Friedle, T. D. Palluccio and E. V. Rybak-Akimova, *Inorg. Chem.*, 2012, **51**, 5006.
68. H. Hirao, D. Kumar, L. Que, Jr. and S. Shaik, *J. Am. Chem. Soc.*, 2006, **128**, 8590.
69. L. Bernasconi, M. J. Louwerse and E. J. Baerends, *Eur. J. Inorg. Chem.*, 2007, 3023.
70. S. Ye and F. Neese, *Curr. Opin. Chem. Biol.*, 2009, **13**, 89.
71. S. Shaik, H. Chen and D. Janardanan, *Nature Chem.*, 2011, **3**, 19.
72. E. I. Solomon, S. D. Wong, L. V. Liu, A. Decker and M. S. Chow, *Curr. Opin. Chem. Biol.*, 2009, **13**, 99.
73. A. Decker, M. D. Clay and E. I. Solomon, *J. Inorg. Biochem.*, 2006, **100**, 697.
74. M. L. Neidig, A. Decker, O. W. Choroba, F. Huang, M. Kavana, G. R. Moran, J. B. Spencer and E. I. Solomon, *Proc. Natl. Acad. Sci.*, 2006, **103**, 12966.
75. J. Kaizer, E. J. Klinker, N. Y. Oh, J. U. Rohde, W. J. Song, A. Stubna, J. Kim, E. Münck, W. Nam and L. Que, Jr., *J. Am. Chem. Soc.*, 2004, **126**, 472.
76. S. D. Wong, C. B. Bell III, L. V. Liu, Y. Kwak, J. England, E. E. Alp, J. Zhao, L. Que, Jr. and E. I. Solomon, *Angew. Chem., Int. Ed.*, 2011, **50**, 3215.
77. M. Srnec, S. D. Wong, J. England, L. Que, Jr. and E. I. Solomon, *Proc. Natl. Acad. Sci.*, 2012, **109**, 14326.
78. C. V. Sastri, J. Lee, K. Oh, Y. J. Lee, J. Lee, T. A. Jackson, K. Ray, H. Hirao, W. Shin, J. A. Halfen, J. Kim, L. Que, Jr., S. Shaik and W. Nam, *Proc. Natl. Acad. Sci.*, 2007, **104**, 19181.
79. D. Wang, K. Ray, M. J. Collins, E. R. Farquhar, J. R. Frisch, L. Gomez, T. A. Jackson, M. Kerscher, A. Waleska, P. Comba, M. Costas and L. Que, Jr., *Chem. Sci.*, 2013, **4**, 282.
80. H. Hirao, L. Que, Jr., W. Nam and S. Shaik, *Chem. Eur. J.*, 2008, **14**, 1740.

81. J. England, J. O. Bigelow, K. M. Van Heuvelen, E. R. Farquhar, M. Martinho, K. K. Meier, J. R. Frisch, E. Münck and L. Que, Jr., *Chem. Sci.*, 2014, **5**, 1204.
82. M. S. Seo, N. H. Kim, K. B. Cho, J. E. So, S. K. Park, M. Clemancey, R. Garcia-Serres, J. M. Latour, S. Shaik and W. Nam, *Chem. Sci.*, 2011, **2**, 1039.
83. G. Xue, R. De Hont, E. Münck and L. Que, Jr., *Nature Chem.*, 2010, **2**, 400.
84. G. Xue, C. Geng, S. Ye, A. T. Fiedler, F. Neese and L. Que, Jr., *Inorg. Chem.*, 2013, **52**, 3976.
85. C.-W. Chiang, S. T. Kleespies, H. D. Stout, K. K. Meier, P.-Y. Li, E. L. Bominaar, L. Que, Jr., E. Münck and W.-Z. Lee, *J. Am. Chem. Soc.*, 2014, **136**, 10846.
86. M. Schatz, V. Raab, S. P. Foxon, G. Brehm, S. Schneider, M. Reiher, M. C. Holthausen, J. Sundermeyer and S. Schindler, *Angew. Chem., Int. Ed.*, 2004, **43**, 4360.
87. R. L. Peterson, R. A. Himes, H. Kotani, T. Suenobu, L. Tian, M. A. Siegler, E. I. Solomon, S. Fukuzumi and K. D. Karlin, *J. Am. Chem. Soc.*, 2011, **133**, 1702.
88. J. Cho, H. Y. Kang, L. V. Liu, R. Sarangi, E. I. Solomon and W. Nam, *Chem. Sci.*, 2013, **4**, 1502.
89. J. Cho, J. Woo and W. Nam, *J. Am. Chem. Soc.*, 2010, **132**, 5958.
90. M. Momenteau and C. A. Reed, *Chem. Rev.*, 1994, **94**, 659.
91. M. M. Mbughuni, M. Chakrabarti, J. A. Hayden, E. L. Bominaar, M. P. Hendrich, E. Münck and J. D. Lipscomb, *Proc. Natl. Acad. Sci.*, 2010, **107**, 16788.
92. J. A. Crawford, W. Li and B. S. Pierce, *Biochemistry*, 2011, **50**, 10241.
93. M. P. Jensen, M. Costas, R. Y. N. Ho, J. Kaizer, A. Mairata i Payeras, E. Münck, L. Que, Jr., J. U. Rohde and A. Stubna, *J. Am. Chem. Soc.*, 2005, **127**, 10512.
94. A. T. Fiedler and L. Que, Jr., *Inorg. Chem.*, 2009, **48**, 11038.
95. M. L. Matthews, W. Chang, A. P. Layne, L. A. Miles, C. Krebs and J. M. Bollinger, Jr., *Nat. Chem. Biol.*, 2014, **10**, 209.
96. O. Planas, M. Clémancey, J.-M. Latour, A. Company and M. Costas, *Chem. Commun.*, 2014, **50**, 10887.
97. T. Kojima, R. A. Leising, S. Yan and L. Que, Jr., *J. Am. Chem. Soc.*, 1993, **115**, 11328.
98. P. Comba and S. Wunderlich, *Chem. Eur. J.*, 2010, **16**, 7293.

CHAPTER 5

Collagen Hydroxylases

JOHANNA MYLLYHARJU*[a]

[a]Oulu Center for Cell-Matrix Research, Biocenter Oulu and Faculty of Biochemistry and Molecular Medicine, University of Oulu, Finland
*E-mail: johanna.myllyharju@oulu.fi

5.1 Collagen Hydroxylases and Their Roles in Collagen Synthesis

The 2-oxoglutarate (2OG)-dependent dioxygenase superfamily includes three types of hydroxylase that have essential roles in collagen synthesis. These enzymes are collagen prolyl 4-hydroxylases (C–P4Hs), lysyl hydroxylases (LHs) and prolyl 3-hydroxylases (P3Hs), with each subfamily consisting of three isoenzymes in vertebrates. Each oxygenase acts on procollagen polypeptide chains located within the endoplasmic reticulum (ER) lumen in a co- and post-translational manner before the procollagen polypeptides assemble and fold into a triple-helical procollagen molecule (Figure 5.1).[1–5]

C–P4Hs catalyse the hydroxylation of Y position prolyl residues in the repeating (Gly–X–Y)$_n$ sequences of procollagen polypeptide chains (Figure 5.1). For example, the ~1000-amino-acid processed chain of type I collagen contains ~100 4-hydroxyproline (4Hyp) residues in the Y positions.[1–4] The number of 4Hyp residues in collagen chains is quite invariable and essentially all Y position prolines become 4-hydroxylated.[2,3] Collagen molecules consist of three collagen chains that coil around each other to form a triple helix. The 4Hyp residues have a critical role in stabilizing the triple-helical structure at physiological temperatures.[1–4] The hydroxyl group in the 4Hyp

RSC Metallobiology Series No. 3
2-Oxoglutarate-Dependent Oxygenases
Edited by Robert P. Hausinger and Christopher J. Schofield

Published by the Royal Society of Chemistry, www.rsc.org

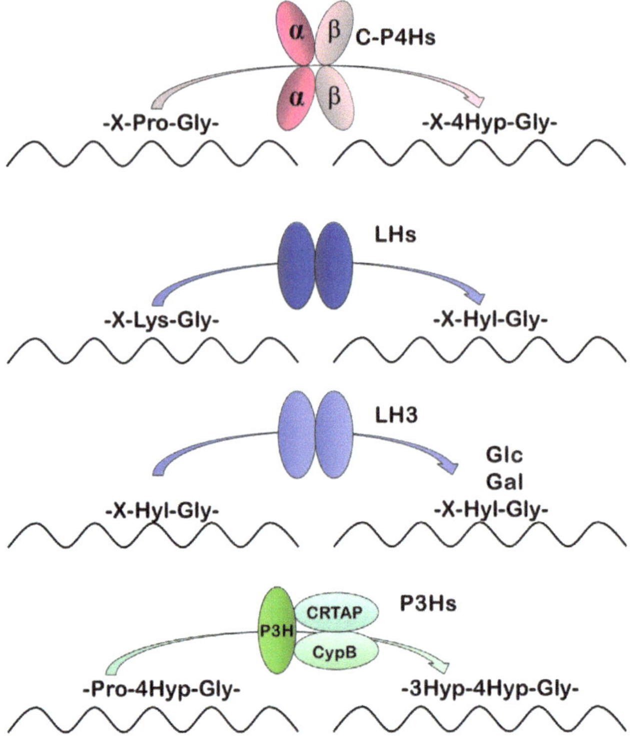

Figure 5.1 Collagen hydroxylases act on repeating $(\text{Gly–X–Y})_n$ sequences of procollagen polypeptide chains within the ER lumen in a co- and post-translational manner before the procollagen polypeptides assemble and fold into the triple-helical procollagen molecule. Collagen prolyl 4-hydroxylases (C–P4Hs) hydroxylate the Y position proline in X–Pro–Gly sequences, lysyl hydroxylases (LHs) hydroxylate the Y position lysine in X–Lys–Gly sequences, and prolyl 3-hydroxylases hydroxylate the X position proline in Pro–4Hyp–Gly sequences. Vertebrate C–P4Hs are tetramers consisting of two catalytic α subunits and two β subunits that are identical to protein disulfide isomerase, LHs are homodimers and P3Hs are multiprotein complexes consisting of the P3H protein, cartilage-associated protein (CRTAP) and cyclophilin B (CypB) in a 1 : 1 : 1 ratio.

is in the *R* configuration and its stabilizing influence is proposed to be mediated, at least in part, *via* stereoelectronic effects.[6]

LHs catalyse the hydroxylation of Y position lysyl residues in collagen polypeptide chains (Figure 5.1).[1,2,7] In contrast to 4Hyp residues, the number of hydroxylysine (Hyl) residues varies in a collagen-type and tissue-specific manner, ranging from 5 to 70 Hyl residues per 1000 amino acids.[2,7] The Hyl residues have at least two functions. First, they determine the chemical nature and strength of crosslinks generated between individual collagen molecules in their spontaneously formed supramolecular structures, such as fibrils.[2,7,8] Initiation of crosslink formation is catalysed by lysyl oxidase (LOX), a copper-dependent amine oxidase, in the extracellular space.[9,10] Second, the hydroxyl groups serve as target sites for glycosylation.[2,7] The function of collagen glycosylation is not yet understood in detail, but it has a role in the assembly and secretion of the triple-helical collagen molecules and it regulates collagen fibril diameter.[11,12]

P3Hs catalyse the formation of 3-hydroxyproline (3Hyp) in the X position prolyl residue of Pro–4Hyp–Gly sequences (Figure 5.1).[2–5] The number of 3Hyp residues in collagen chains is typically low, with 1–2 in the type I collagen chain, and the greatest number being seen in the basement membrane (BM) type IV collagen, with ~10 residues per chain.[2,3] However, several additional 3Hyp sites of low occupancy in specific tissues have recently been identified in fibrillar collagens.[13,14] The biological function of 3Hyp residues in collagen chains is not yet understood in detail, but suggested functions include destabilization or minor added stability of the triple helix and provision of a site of interaction that aligns collagen molecules in an optimal arrangement in collagen fibril assembly for efficient crosslinking.[13,15–17] Nevertheless, the importance of 3Hyp in collagen is indisputable, as manifested by diseases caused by the lack or reduction of 3Hyp in collagens (see Section 5.4.3).

This review concentrates on vertebrate collagen hydroxylases, but C-PHs and LH have also been identified in nematodes and fruit fly and shown to have important roles in their collagen synthesis.[18–36]

5.2 Molecular Composition and Expression of Vertebrate Collagen Hydroxylases

5.2.1 Vertebrate Collagen Prolyl 4-Hydroxylases are $\alpha_2\beta_2$ Heterotetramers

Vertebrate C–P4Hs are $\alpha_2\beta_2$ tetramers with a molecular weight of about 240 kDa.[2,4,37–39] The α subunits contain the catalytic sites and protein disulfide isomerase (PDI) serves as the β subunit. PDI is a multifunctional protein that catalyses disulfide bond formation and acts as a chaperone in protein folding.[2,4,37–39] Interestingly, although the catalytic C–P4H α subunits contain critical intrachain disulfide bonds,[40,41] the disulfide isomerase activity of PDI is not required for their formation.[42] Instead, the importance of PDI as the β subunit of C–P4H is regarded to be related to its chaperone function and ability to keep the highly insoluble α subunit in a soluble and functional conformation.[42] Furthermore, the ER retention signal of PDI is responsible for retaining the C–P4H tetramer within the ER lumen. Besides being required for the generation of the thermally stable triple-helical collagen molecules *via* catalysis of the 4Hyp formation, C–P4H also functions in the quality control of collagen secretion; it retains unassembled procollagen chains within the ER until the triple helical structure is formed.[43] The molecular composition of nematode C–P4Hs is more variable, including $\alpha_2\beta_2$ tetramers with either identical or different α subunits, $\alpha\beta$ dimers and even α_4 tetramers.[18,26,29,31,36]

Three vertebrate C–P4H isoenzymes differing in their α subunits are currently known, *i.e.* C–P4Hs–I, –II and –III with the molecular compositions of $\alpha(I)_2\beta_2$, $\alpha(II)_2\beta_2$ and $\alpha(III)_2\beta_2$, respectively.[44–50] At the amino acid sequence level, the 517-residue processed human α(I) and 514-residue human α(II) subunits are 65% identical, with the greatest identity (~80%) being observed between the catalytic *C*-terminal domains (see below). The amino acid sequence of the human α(III) subunit is more diverse, with the overall identity between the α(I) and α(II) subunits being 35–37% and that between the catalytic domains 56–57%. C–P4H–I is expressed ubiquitously, while expression of C–P4H–II is restricted to selected cell types and tissues, being especially prominent in chondrocytes and cartilage patterns.[51] C–P4H–II expression is also detected in osteoblasts, endothelial cells and in cells of epithelial structures.[52] C–P4H–III seems to be expressed rather ubiquitously, but the level of the α(III) mRNA is markedly lower than that of the α(I) mRNA in all tissues studied.[49]

In addition to hydroxylation of collagens, C–P4H–I has another interesting biological role in the regulation of RNA interference. C–P4H–I interacts with and hydroxylates Argonaute2, an important component of the RNA-induced silencing complex.[53] Prolyl 4-hydroxylation of Argonaute2 increases its stability and affects the RNA interference efficiency.[53,54]

5.2.2 Vertebrate Lysyl Hydroxylases are Homodimers

Three LH isoenzymes have been identified in vertebrates.[55–61] The amino acid sequences of the processed 709-residue LH1 and 712-residue LH2 polypeptides are 75% identical and greatest identity is observed between the *C*-terminal catalytic regions. The processed 714-residue LH3 polypeptide is 57–59% identical to LH1 and LH2, with its catalytic region sharing 69–72% identity. Besides the catalytically critical four residues required for binding iron and 2OG (see Section 5.3), no significant amino acid sequence identities or similarities are shared between the LH, C–P4H α subunit or P3H sequences. All LH isoenzymes are generally expressed in the same tissues and cell types and although certain distinctive patterns have been observed, LH isoenzymes do not seem to have tissue-specific expression patterns.[7] The expression level of LH3 mRNA is rather constant in various human cell lines, whereas the expression levels of LH1 and LH2 mRNAs vary depending on the cell type.[62] The expression of LH isoenzymes during mouse development is rather ubiquitous, although the expression level of an individual isoenzyme varies depending on the developmental time point.[63,64]

LH isoenzymes are homodimers with molecular weights of about 180 kDa.[65] They are present in the ER lumen despite lacking an ER retention signal, and LH1 and LH3 are associated with the ER membrane.[66] In the case of LH1, the region responsible for the ER membrane association has been mapped to a 32-residue *C*-terminal region that is likely to form an extended loop within the catalytic domain that is readily accessible for binding.[66] LH3 is unique among the LH isoenzymes in that it is a multifunctional enzyme that, in addition to LH activity, possesses collagen galactosyltransferase and glucosyltransferase activities (see Section 5.3.1).[65,67–69] LH3 is unique also in that it is the only LH isoenzyme that is secreted into the extracellular space, found on cell surfaces in certain tissues and cell types, and is present in serum *in vivo*.[70] Secretion of LH3 requires the site responsible for LH3 glycosyltransferase activities and utilizes two secretory pathways. LH3 that is found in the extracellular space is secreted *via* the Golgi complex, while LH3 that is localized to the cell surface bypasses the Golgi complex.[71] Interestingly, extracellular LH3 has been shown to be able to modify extracellular proteins in their native, non-denatured state,[70] and the extracellular glycosyltransferase activities of LH3 are important for cell viability and growth.[72]

5.2.3 Vertebrate Prolyl 3-Hydroxylases are Multiprotein Complexes

Of the P3H isoenzymes, P3H1 has been characterized in most detail. P3H1 exists in a multifunctional protein complex that additionally contains the cartilage-associated protein (CRTAP) and cyclophilin B (CypB) in a 1:1:1 ratio.[73,74] The molecular weights of P3H1, CRTAB and CypB are about 90 kDa, 46 kDa and 21 kDa, respectively.[73] CypB has peptidyl-prolyl *cis–trans* isomerase activity and thus controls the rate of collagen triple helix formation when it is present in the P3H1 complex.[74] In addition, the whole complex acts as a chaperone and disulfide isomerase.[74,75]

The amino acid sequences of the processed 736-residue P3H1, 708-residue P3H2 and 736-residue P3H3 polypeptides are 38–46% identical and each of them contains a *C*-terminal ER retention signal.[73] Both unique and common sites of P3H1, P3H2 and P3H3 mRNA expression are detected in mouse tissues.[73,76–78] P3H1 expression typically localizes to tissues rich in fibrillar collagens, such as tendons, cartilage and dermis. P3H2 has been shown to be highly expressed in tissues that are rich in BMs, *e.g.* the kidney, where P3H1 expression is almost absent.

5.3 Structural and Enzymatic Properties of Collagen Hydroxylases

Like other 2OG-dependent dioxygenases the collagen hydroxylases require Fe(II), O_2 and 2OG in their reactions.[37,38] In addition, they require ascorbate that functions, at least in part, as a reductant in so-called uncoupled reaction cycles where decarboxylation of 2OG occurs without subsequent hydroxylation of the collagen polypeptide.[37,38] Collagen hydroxylases catalyse uncoupled reaction cycles even in the presence of saturating substrate concentrations, with the rate of uncoupled reaction cycle being 0.7% of the full reaction cycle in the case of C-P4H.[79] The catalytic sites of collagen hydroxylases contain four conserved critical residues that are common to the entire

2OG-dependent dioxygenase family, *i.e.* a His–X–Asp···His triad that binds the Fe(II) and a Lys or Arg side chain that binds the C5 carboxyl group of 2OG.[41,73,79–81]

5.3.1 Structural Properties

Crystal structures of collagen hydroxylases are not yet available, but as in other 2OG-dependent dioxygenases their catalytic sites are likely to be located in a double-stranded β-helix structure (see Chapter 3), a jelly-roll motif.[82] It has been shown that the C–P4H α subunits and LH polypeptides consist of distinct functional domains and crystal structures of some of the C–P4H α subunit domains as well as the crystal structure of a related algal P4H have been solved (see below). In the following text the numbering of the amino acid residues is according to the processed polypeptides after cleavage of the signal peptides.

The catalytic C–P4H α subunit consists of three domains, an *N*-terminal domain (residues 1–143 in human C–P4H α(I)), a middle peptide-substrate-binding domain (PSB domain, residues 144–244), and a *C*-terminal catalytic domain (residues 245–517) that contains the catalytically critical residues His-412, Asp-414, His-483 and Lys-493 in the human α(I) subunit.[79,83] Binding affinities of various C–P4H peptide substrates and inhibitors to the C–P4H α(I) PSB domain are highly similar to the measured K_m and K_i values of the C–P4H–I tetramer for these peptides.[84] Crystal structure of the PSB domain showed that it consists of two tetratricopeptide repeat motifs formed by five α helices and an additional solvating helix.[85] The PSB domain contains an aromatic groove lined by conserved tyrosyl residues and an Arg–Asp salt bridge that bind peptide substrates and inhibitors in a poly(L-proline) type II helix conformation, with a 3-fold X–Pro–Gly repeat fitting optimally into the binding groove.[83,85,86] The *N*-terminal domain acts as a dimerization domain between the α subunits and thus has an important function in the assembly of the C–P4H tetramer.[86] It consists of six α helices, of which α1–α4 form an extended four-helix bundle dimerization motif where the long α1 helix forms an antiparallel coiled-coil structure with the corresponding helix of the other *N*-terminal domain of the dimer.[86] The α5 and α6 helices and the 10-residue loop following them serve as a linker between the *N*-terminal and PSB domains, with the α5 helix also participating in dimerization contacts by providing a stabilization bridge between the *N*-terminal dimerization domain and the PSB domain.[86]

The *C*-terminal catalytic domain of the C–P4H α subunits is insoluble and thus not amenable to structural studies. However, the crystal structure of a related monomeric P4H from the green alga *Chlamydomonas reinhardtii* (that has 26% amino acid sequence identity) has been solved.[87–89] This enzyme hydroxylates Pro-rich peptides representing 4Hyp-rich proteins required in cell wall synthesis.[87] The structure contains the typical jelly-roll core fold comprised of a double-stranded β-helix structure formed by eight β strands (β6–β18 = βI–βVIII of the jelly-roll motif).[88] The peptide substrate is bound in a poly(L-proline) type II conformation within a shallow groove covered by two flexible loops, β3–β4 and βII–βIII, the catalytic site being located in a pocket in the middle of this groove.[88,89]

LH polypeptides consist of three domains with approximate molecular weights of 30, 37 and 16 kDa.[65] The *N*-terminal 30 kDa domain is not required for LH activity, but it is responsible for the glycosyltransferase activities of LH3; the Cys-120 residue and a five-residue Asp motif in positions 163–167 of this domain being required for these activities.[65,68,69] The 37 kDa middle domain contains a six-residue conserved region (amino acids 517–523 in LH3) that has been shown to be responsible for the homodimerization of LH1 and LH3.[90] Dimerization was found to be necessary for the LH activity, but not glycosyltransferase activities, of LH3.[90] The four catalytically critical residues (His-638, Asp-640, His-690 and Arg-700 in human LH1) involved in Fe(II) and 2OG binding are located in the *C*-terminal 16 kDa domain and a recombinant protein consisting of only the middle and *C*-terminal domains is a fully active LH.[65] Whether the middle domain contributes to the LH activity is currently unknown because, like the *C*-terminal catalytic domain of the C–P4H α subunits, that of LHs is also insoluble and thus not suitable for functional or structural studies.[65]

5.3.2 Catalytic and Inhibitory Properties

C–P4Hs hydroxylate synthetic X–Pro–Gly tripeptides, but not Gly–X–Pro or Pro–Gly–X tripeptides *in vitro* and likewise LHs hydroxylate X–Lys–Gly tripeptides, but not Lys–Gly–X tripeptides.[91] C–P4Hs hydroxylate prolyl residues almost exclusively in the X–Pro–Gly sequence, the only exception being the observation that synthetic peptides Pro–Pro–Ala–Pro and Pro–Pro–Glu–Pro are hydroxylated at a low rate, in agreement with the rare observation of 4Hyp in the X–4Hyp–Ala sequence in some collagen polypeptides.[91] Likewise, although Hyl is predominantly present in the X–Lys–Gly sequences of collagen polypeptides, it is also observed in the X–Lys–Ala and X–Lys–Ser sequences present in the collagen telopeptides (see below).[91] 3Hyp has been detected only in the 3Hyp–4Hyp–Gly sequence. The hydroxylation efficiencies of X–Pro–Gly, X–Lys–Gly and Pro–4Hyp–Gly sequences by C–P4Hs, LHs and P3Hs, respectively, are also affected by the amino acid present in the X position and by other neighbouring amino acids.[91,92] Furthermore, the chain length of the peptide substrate has a marked effect on the K_m values, which decrease with increasing peptide length.[91] For example, the K_m values of C–P4H–I for $(Pro–Pro–Gly)_5$ and $(Pro–Pro–Gly)_{10}$ are 150–250 μM and 18 μM, respectively.[2] Certain differences between the hydroxylation efficiencies of the C–P4H isoenzymes have also been observed. The K_m values of C–P4H–I and C–P4H–III for $(Pro–Pro–Gly)_{10}$ are very similar (18 and 20 μM, respectively), while that of C–P4H–II is higher (95 μM).[47,49] Distinct differences also exist in the binding of the competitive peptide inhibitor, poly(L-proline), to the C–P4H isoenzymes. C–P4H–I is inhibited very effectively by poly(L-proline), with a K_i for poly(L-proline) (M_r 5000–7000) of 0.5 μM, while that for C–P4H–II is much greater, 95 μM.[47] Therefore, certain structural differences are likely to be present in the PSB domains of C–P4H–I and C–P4H–II.

Table 5.1 Selected kinetic parameter values of human C–P4Hs I–III, LH1, LH3 and P3H2 for reaction components. The values for C–P4Hs, LHs and P3H2 were determined using synthetic peptides $(Pro–Pro–Gly)_{10}$, $(Ile–Lys–Gly)_3$ and $(Gly–Pro–4Hyp)_5$ as substrates, respectively. N.D. = not determined.

Cosubstrate	C–P4H–I[a]	C–P4H–II[b]	C–P4H–III[c]	LH1[d]	LH3[d]	P3H2[e]
K_m, μM						
2-Oxoglutarate	20	20–40	20	100	100	80
O_2	40	N.D.	N.D.	45	N.D.	N.D.
	Concentration for half-maximal activity, μM					
Fe(II)	2	2	0.5	2	2	0.5
Ascorbate	300	300–340	370	350	300	110

[a]refs. 79,95.
[b]refs. 47,48.
[c]ref. 49.
[d]refs. 60,91.
[e] ref. 77.

Formation of the most stable pyridinoline collagen crosslinks requires the presence of Hyl in the short non-triple-helical *N*- and *C*-terminal telopeptides of fibril-forming collagens.[8] The telopeptide Hyl is converted to hydroxyallysine by LOX and reacts with a lysyl or Hyl residue within the triple helix of an adjacent collagen molecule, resulting in the formation of a lysylpyridinoline or hydroxylysylpyridinoline crosslink, respectively.[8–10] Biochemical data from patients with mutations in the genes encoding LH1 and LH2 have indicated that LH1 is responsible for hydroxylation of lysyl residues in the triple-helical region of collagens, while LH2 hydroxylates the telopeptide lysines in X–Lys–Ala and X–Lys–Ser sequences (see Section 5.4.2). Using recombinant LHs, it was verified that LH2 is the only isoenzyme capable of hydroxylating the telopeptide lysyl residues, while all three isoenzymes can act on the collagenous triple-helical sequences.[93] Studies using an extensive peptide array have indicated that the amino acid sequence surrounding the target lysyl residue has an effect on the binding efficiency of LHs.[92] No strict isoenzyme specificity was observed, however, although certain preferences were found to exist.

Data from human patients (see Section 5.4.3) clearly indicate that P3H1 is required for 3-hydroxylation of the main target prolyl residue (Pro-986 in the α1 chain of type I collagen) in fibrillar collagens. Recombinant P3H2 hydroxylates more efficiently the synthetic peptides representing the collagen IV hydroxylation sites than a peptide representing the primary hydroxylation site in collagen I.[77] However, suppression of P3H2 by RNA interference also decreases prolyl 3-hydroxylation of fibrillar collagens with the exception of the main target site, which thus seems to be solely hydroxylated by P3H1.[94]

The K_m values of 2OG and O_2 and the concentrations of Fe(II) and ascorbate required for half-maximal activity are fairly similar for C–P4Hs, LHs and P3Hs (Table 5.1). The K_m of O_2 for C–P4H–I is much lower than that of the hypoxia-inducible factor (HIF) P4Hs (see Chapter 6), indicating that C–P4Hs are not effective oxygen sensors.[95] Because of the central role of prolyl 4-hydroxylation in the synthesis of all collagen types, C–P4Hs have been

Table 5.2 Inhibition of recombinant human C–P4H–I by certain metals and 2OG analogues[a].

Inhibitor	IC_{50}, μM
Zn^{2+}	0.6
Co^{2+}	14
Ni^{2+}	37
Inhibitor	K_i, μM
Pyridine 2,4-dicarboxylate	2
Pyridine 2,5-dicarboxylate	0.8
3-Hydroxypyridine-2-carbonyl-glycine	0.4
N-Oxalylglycine	1.9
3,4-Dihydroxybenzoic acid	5
3-Carboxy-4-oxo-3,4-dihydro-1,10-phenanthroline	2
N-((3-Hydroxy-6-chloroquinolin-2-yl)carbonyl)glycine	0.06
Fumarate	190
Succinate	400

[a]refs. 95–97.

regarded as potential targets for antifibrotic therapy. Many compounds, especially metal ions and 2OG mimetics, which inhibit collagen hydroxylases in a competitive manner with respect to the Fe(II) and 2OG, respectively, are known (Table 5.2),[2,91,95–97] but no collagen hydroxylase inhibitor is yet in clinical use. The inhibitory potency of 2OG analogues is generally somewhat weaker towards LH1 than C–P4H–I or P3H1, reflecting the higher K_m of LH1 for 2OG (see Table 5.1).[98] Distinct differences have been observed in the inhibitory potency of several metals and 2OG analogues towards the two P4H classes, the collagen and HIF P4Hs, which indicates that selective inhibitors can be developed.[95–97]

5.4 Human Diseases and Gene-Modified Mouse Models

Mutations in P3Hs and LHs are known to cause heritable connective tissue diseases in humans, while no diseases associated with C–P4H mutations are yet known.[1–5] Upregulation of collagen hydroxylase expression is often associated with fibrotic conditions, thus C–P4Hs have been regarded as potential anti-fibrotic drug targets.[1–5] Of LHs, especially elevated LH2 expression is associated with fibrosis, which is in accordance with the observations of increased lysyl hydroxylation of collagen telopeptides and hydroxylysylpyridinoline crosslinks in fibrotic tissues.[99–102] Interestingly, the genes encoding C–P4H α(I) and α(II) subunits and LH2 are induced during hypoxia by HIF and have been shown to be essential for breast cancer metastasis.[103–108] On the other hand, C–P4Hs and P3Hs have been shown to be downregulated in lymphoma and P3Hs in breast cancer.[109,110] Furthermore, p53 has been shown to increase C–P4H–II as well as type IV and XVIII collagen expression, which leads to increased release of their anti-angiogenic arresten and endostatin fragments, respectively, *via* a matrix metalloproteinase-dependent mechanism.[111,112]

5.4.1 Collagen Prolyl 4-Hydroxylases

As stated above, no human diseases caused by mutations in the genes encoding C–P4H α subunits have been identified to date. Genetic inactivation of the C–P4H α(I) subunit in mouse leads to early embryonic lethality at E10.5.[113] The embryos suffered from an overall developmental delay and frequent rupture of capillary walls.[113] Collagen IV was essentially absent from the BMs that were disrupted.[113] C–P4H–II null mice are viable and apparently normal, but they display very mild abnormalities during skeletogenesis that are more severe in C–P4H–I$^{+/-}$;C–P4H–II$^{-/-}$ mice (Aro, Salo, Khatri, Finnilä, Miinalainen, Sormunen, Pakkanen, Holster, Soininen, Prein, Clausen-Schaumann, Aszódi, Tuukkanen, Kivirikko, Schipani and Myllyharju, unpublished data).

5.4.2 Lysyl Hydroxylases

Mutations have been identified in all three human genes encoding the LH isoenzymes and they lead to severe connective tissue disorders. Human LH1 mutations can cause a specific autosomal-recessive subtype of the Ehlers–Danlos syndrome (EDS), EDS VIA (also known as the kyphoscoliotic subtype), which is characterized by joint hypermobility, kyphoscoliosis, hyperextensible skin, and risk of arterial rupture.[114–117] Biochemical findings in EDS VIA patients include marked reductions in the LH activity and hydroxylysylpyridinoline crosslinks resulting from a decreased amount of Hyl in the collagen triple helix, but the extent of these deficiences varies markedly depending on the tissue and collagen type.[115,117–119] Over 20 mutations have been identified in the gene encoding LH1, the most common one being a duplication of exons 10–16 resulting from recombination of intronic Alu sequences.[114,117,120]

LH1 knockout mice are viable and fertile.[121] Unlike EDS VIA patients, LH1 null mice do not suffer from kyphoscoliosis, but they have muscular hypotonia and sudden deaths caused by aortic rupture are also detected.[121] LH1 null mice display tissue-specific changes in collagen fibril morphology, Hyl content, and collagen crosslinking pattern.[121] The large tissue-specific differences in the Hyl deficiency (14–78%) in LH1 null mice indicate that the ability of LH2 and LH3 to hydroxylate lysines in the triple-helical region of collagens varies in different tissues, most probably reflecting the expression levels of the LH2 and LH3 isoenzymes.[121]

Mutations in the human gene coding for LH2 cause an extremely rare autosomal-recessive connective tissue disorder, Bruck syndrome type 2 (BS2).[99,116,122,123] BS2 patients suffer from osteoporosis, fragile bones, congenital joint contractures, short stature and progressive kyphoscoliosis.[116,124] Surprisingly, LH2 null mice are embryonic lethal (Hyry, Soininen, Bank, Sormunen, Miinalainen, Talsta, Lantto and Myllyharju, unpublished data). Therefore, it is possible that the residual activity observed in recombinant LH2 polypeptides harbouring human BS2 mutations may be critical for the survival of BS2 patients.[125]

The first human connective tissue disorder shown to be caused by LH3 mutations was identified in 2008.[126] The patient had a unique phenotype with features overlapping with many other connective tissue disorders, including a general growth retardation and craniofacial, diaphragmatic and skeletal problems.[126] The patient was found to be a compound heterozygote with recessive inheritance. One of the mutations abolished the LH activity of LH3, while the other severely reduced the glycosyltransferase activities of LH3, and an overall reduction in the LH3 protein level was detected in lymphoblastoid cells of the patient.[126] In addition, a transcriptional defect in one LH3 allele has been shown to lead to deficient glucosyltransferase activity in an epidermolysis bullosa simplex family.[127]

Complete inactivation of LH3 in mouse leads to early embryonic lethality at E9.5.[63,128] Collagen IV is not correctly deposited into BMs but instead accumulates within the ER or is found in extracellular aggregates, which lead to rupture of BMs in the LH3 null embryos.[63] Data from a knock-in mouse line, where only the LH activity of LH3 was specifically disrupted by mutating the codon for the iron-binding Asp residue, showed that the LH activity of LH3 is not required for survival of the mice, although certain defects in BM and collagen fibril assembly are detected in the skin and lungs of newborn mice.[128] Similar defects are also detected in LH3$^{+/-}$ mice.[127] The early embryonic lethality in LH3 null mice was found to be specifically caused by the absence of LH3 glycosyltransferase activities, and evidence from hypomorphic LH3 mice shows that the severity of the BM abnormalities correlates with the degree of reduction in LH3 glycosyltransferase activities.[128] Besides collagen IV, underglycosylation of collagen VI in the LH3 mutant mice has been shown to cause abnormal distribution and aggregation of this collagen in the skeletal muscle, leading to similar ultrastructural muscular alterations that are observed in collagen VI null mice and in Ullrich congenital muscular dystrophy caused by collagen VI mutations.[129]

5.4.3 Prolyl 3-Hydroxylases

The importance of the P3H1/CRTAP/CypB complex and 3Hyp residues in collagen has been highlighted in recent years by the discovery that lack or decrease of 3Hyp in only one position of the α1(I) (Pro-986) or α2(I) (Pro-707) chains of type I collagen causes autosomal-recessive lethal/severe osteogenesis imperfecta (OI). The first human mutations in the P3H1/CRTAP/CypB complex were identified in CRTAP and they lead to a recessive OI ranging from a neonatal lethal phenotype to a severe OI with rhizomelia, the severity of the phenotype depending on whether the mutation leads to complete or partial inactivation of the P3H1/CRTAP/CypB complex.[130,131] Soon after this, the first P3H1 mutations were identified in patients suffering from autosomal-recessive lethal/severe OI with rhizomelia.[132] Common biochemical findings in the patients with CRTAP or P3H1 mutations include reduced or absent 3-hydroxylation of Pro-986 in the α1(I) chain, excess lysyl hydroxylation and glycosylation of type I collagen, and slow secretion and ER retention of type

I collagen.[130–132] Mutations in the gene coding for the third component of the complex, CypB, have been shown to cause recessive OI without rhizomelia.[133–136] Interestingly, 3-hydroxylation of Pro-986 in these patients is either normal or reduced and both normal or delayed folding and secretion of type I collagen has been reported depending on the mutation.[133–136] Since these pioneering discoveries several families with recessive lethal/severe OI caused by different CRTAP, P3H1 and CypB mutations have been identified.[116,137,138]

CRTAP null mice suffer from osteochondrodysplasia characterized by severe osteoporosis, rhizomelia, kyphosis and osteopenia.[130] Biochemical findings in the CRTAP null mice include the complete lack of 3Hyp from the main hydroxylation site in type I and II collagens, overmodification and altered production rate of collagens I and II, increased fibril diameter and production, and increased matrix mineralization rate in osteoblasts.[130] In addition to skeletal defects, CRTAP null mice have a generalized connective tissue disease affecting many tissues.[139] P3H1 null mice also suffer from an OI-like phenotype as they are small and have several bone defects including kyphoscoliosis, decreased bone density, shortened long bones and rhizomelia.[140] In addition, these mice suffer from a hearing impairment, which is a common problem in OI.[141] 3-Hydroxylation of Pro-986 in type I collagen is completely absent in all P3H1 null mouse tissues studied, whereas the 3-hydroxylation level of other target prolyl residues of type I collagen vary depending on the tissue.[140,142] Type I collagen is over-modified and has a slower secretion rate in P3H1 null tissues, and the overall shape and diameter of collagen fibrils are abnormal.[140,142] Loss of one component of the CRTAP/P3H1/CypB complex leads to the absence of the whole complex.[139,143] The resulting loss of functions include prolyl 3-hydroxylation, peptidyl-prolyl *cis–trans* isomerase, chaperone and protein disulfide isomerase activities that assist in the correct folding of the collagen molecules. Therefore, to study specifically the outcome of the absence of P3H1 activity without affecting the other functions of the complex, a knock-in mouse model with a targeted mutation affecting the P3H1 catalytic site has been generated.[144] 3-Hydroxylation of Pro-986 in type I and II collagens is absent in these mice, but over-modification of collagen molecules is not observed.[144] The mice have no overall skeletal deformities, exhibit normal cartilage growth plate histology, and possess normal length of the long bones.[144] Certain differences in the biomechanical properties of the long bones are observed, although they are partly different from those observed in P3H1 null mice.[144] CypB null mice also have severe OI with kyphosis and osteoporosis,[145] and they have abnormal loose skin.[145] 3-Hydroxylation of Pro-986 of collagen I is absent in the CypB null mice, the collagen fibrils have a wider diameter, and abnormalities in the localization of procollagen I to the secretory pathway are observed.[145]

Inactivating mutations in human P3H2 have been shown to cause autosomal-recessive non-syndromic high myopia.[146,147] In contrast, P3H2 null mice are embryonic lethal already at E6.5.[148] Collagen IV is deficient in 3Hyp in the P3H2 null embryos, and the molecular mechanism behind the early death involves the interaction of the 3Hyp deficient collagen IV with the maternal platelet-specific glycoprotein VI (GPVI), which results in aggregation of maternal platelets and death of the P3H2 null embryos because of thrombosis of the maternal blood.[148] These findings agree with the enriched expression of P3H2 in BM-rich tissues and preferential hydroxylation of collagen IV sequences by recombinant P3H2.[73,76–78] The discrepancy between the effects of human and mouse P3H2 mutations may be explained by GPVI polymorphism as one of the two common human GPVI alleles causes lower GPVI density on the platelet surface and reduced thrombogenicity in response to collagen.[148]

References

1. J. Myllyharju and K. I. Kivirikko, *Trends Genet.*, 2004, **20**, 1.
2. J. Myllyharju, *Topics in Current Chemistry: Collagen*, ed. J. Brinckmann, H. Notbohm and P. K. Müller, Springer, Berlin Heidelberg New York, 2005, 247, 115–147.
3. D. M. Hudson and D. R. Eyre, *Connect. Tissue Res.*, 2013, **54**, 245.
4. Y. Ishikawa and H. P. Bächinger, *Biochim. Biophys. Acta*, 2013, **1833**, 2479.
5. J. C. Marini, W. A. Cabral, A. M. Barnes and W. Chang, *Cell Cycle*, 2007, **6**, 1675.
6. M. D. Shoulders and R. T. Raines, *Annu. Rev. Biochem.*, 2009, **78**, 929.
7. R. Myllylä, C. Wang, J. Heikkinen, A. Juffer, O. Lampela, M. Risteli, H. Ruotsalainen, A. Salo and L. Sipilä, *J. Cell. Physiol.*, 2007, **212**, 323.
8. D. R. Eyre, J.-J. Wu, *Topics in Current Chemistry: Collagen*, ed. J. Brinckmann, H. Notbohm and P. K. Müller, Springer, Berlin Heidelberg New York, 2005, 247, 207–229.
9. J. M. Mäki, *Histol. Histopathol.*, 2009, **24**, 651.
10. H. A. Lucero and H. M. Kagan, *Cell. Mol. Life Sci.*, 2006, **63**, 2304.
11. H. Notbohm, M. Nokelainen, J. Myllyharju, P. P. Fietzek, P. K. Müller and K. I. Kivirikko, *J. Biol. Chem.*, 1999, **274**, 8988.
12. M. Sricholpech, I. Perdivara, M. Yokoyama, H. Nagaoka, M. Terajima, K. B. Tomer and M. Yamauchi, *J. Biol. Chem.*, 2012, **287**, 22998.
13. M. A. Weis, D. M. Hudson, L. Kim, M. Scott, J. J. Wu and D. R. Eyre, *J. Biol. Chem.*, 2010, **285**, 2580.
14. C. Yang, A. C. Park, N. A. Davis, J. D. Russell, B. Kim, D. D. Brand, M. J. Lawrence, Y. Ge, M. S. Westphall, J. J. Coon and D. S. Greenspan, *J. Biol. Chem.*, 2012, **287**, 40598.
15. C. L. Jenkins, L. E. Bretscher, I. A. Guzei and R. T. Raines, *J. Am. Chem. Soc.*, 2003, **125**, 6422.
16. K. Mizuno, D. H. Peyton, T. Hayashi, J. Engel and H. P. Bächinger, *FEBS J.*, 2008, **275**, 5830.
17. D. M. Hudson, L. S. Kim, M. Weis, D. H. Cohn and D. R. Eyre, *Biochemistry*, 2012, **51**, 2417.
18. J. Veijola, P. Koivunen, P. Annunen, T. Pihlajaniemi and K. I. Kivirikko, *J. Biol. Chem.*, 1994, **269**, 26746.

19. J. Veijola, P. Annunen, P. Koivunen, A. P. Page, T. Pihlajaniemi and K. I. Kivirikko, *Biochem. J.*, 1996, **317**, 721.
20. P. Annunen, P. Koivunen and K. I. Kivirikko, *J. Biol. Chem.*, 1999, **274**, 6790.
21. L. Friedman, J. J. Higgin, G. Moulder, R. Barstead, R. T. Raines and J. Kimble, *Proc. Natl. Acad. Sci. U. S. A.*, 2000, **97**, 4736.
22. A. D. Winter and A. P. Page, *Mol. Cell. Biol.*, 2000, **20**, 4084.
23. K. L. Hill, B. D. Harfe, C. A. Dobbins and S. W. L'Hernault, *Genetics*, 2000, **155**, 1139.
24. K. R. Norman and D. G. Moerman, *Dev. Biol.*, 2000, **227**, 690.
25. A. Merriweather, V. Guenzler, M. Brenner and T. R. Unnasch, *Mol. Biochem. Parasitol.*, 2001, **116**, 185.
26. J. Myllyharju, L. Kukkola, A. D. Winter and A. P. Page, *J. Biol. Chem.*, 2002, **277**, 29187.
27. P. Riihimaa, R. Nissi, A. P. Page, A. D. Winter, K. Keskiaho, K. I. Kivirikko and J. Myllyharju, *J. Biol. Chem.*, 2002, **277**, 18238.
28. E. W. Abrams and D. J. Andrew, *Mech. Dev.*, 2002, **112**, 165.
29. A. D. Winter, J. Myllyharju and A. P. Page, *J. Biol. Chem.*, 2003, **278**, 2554.
30. E. W. Abrams, W. K. Mihoulides and D. J. Andrew, *Development*, 2006, **133**, 3517.
31. A. D. Winter, K. Keskiaho, L. Kukkola, G. McCormack, M. A. Felix, J. Myllyharju and A. P. Page, *Matrix Biol.*, 2007, **26**, 382.
32. A. D. Winter, G. McCormack and A. P. Page, *Dev. Biol.*, 2007, **308**, 449.
33. K. Keskiaho, L. Kukkola, A. P. Page, A. D. Winter, J. Vuoristo, R. Sormunen, R. Nissi, P. Riihimaa and J. Myllyharju, *J. Biol. Chem.*, 2008, **283**, 10679.
34. S. Bunt, B. Denholm and H. Skaer, *Gene Expression Patterns*, 2011, **11**, 72.
35. U. Topf and R. Chiquet-Ehrismann, *Mol. Biol. Cell*, 2011, **22**, 3331.
36. A. D. Winter, G. McCormack, J. Myllyharju and A. P. Page, *J. Biol. Chem.*, 2013, **288**, 1750.
37. J. Myllyharju, *Matrix Biol.*, 2003, **22**, 15.
38. J. Myllyharju, *Ann. Med.*, 2008, **40**, 402.
39. K. L. Gorres and R. T. Raines, *Crit. Rev. Biochem. Mol. Biol.*, 2010, **45**, 106.
40. D. C. John and N. J. Bulleid, *Biochemistry*, 1994, **33**, 14018.
41. A. Lamberg, T. Pihlajaniemi and K. I. Kivirikko, *J. Biol. Chem.*, 1995, **270**, 9926.
42. K. Vuori, T. Pihlajaniemi, R. Myllylä and K. I. Kivirikko, *EMBO J.*, 1992, **11**, 4213.
43. A. R. Walmsley, M. R. Batten, U. Lad and N. J. Bulleid, *J. Biol. Chem.*, 1999, **274**, 14884.
44. T. Helaakoski, K. Vuori, R. Myllylä, K. I. Kivirikko and T. Pihlajaniemi, *Proc. Natl. Acad. Sci. U. S. A.*, 1989, **86**, 4392.
45. J. A. Bassuk, W. W. Kao, P. Herzer, N. L. Kedersha, J. Seyer, J. A. DeMartino, B. L. Daugherty, G. E. Mark and R. A. Berg, *3rd Proc. Natl. Acad. Sci. U. S. A.*, 1989, **86**, 7382.

46. T. Helaakoski, P. Annunen, K. Vuori, I. A. MacNeil, T. Pihlajaniemi and K. I. Kivirikko, *Proc. Natl. Acad. Sci. U. S. A.*, 1995, **92**, 4427.
47. P. Annunen, T. Helaakoski, J. Myllyharju, J. Veijola, T. Pihlajaniemi and K. I. Kivirikko, *J. Biol. Chem.*, 1997, **272**, 17342.
48. M. Nokelainen, R. Nissi, L. Kukkola, T. Helaakoski and J. Myllyharju, *Eur. J. Biochem.*, 2001, **268**, 5300.
49. L. Kukkola, R. Hieta, K. I. Kivirikko and J. Myllyharju, *J. Biol. Chem.*, 2003, **278**, 47685.
50. C. Van Den Diepstraten, K. Papay, Z. Bolender, A. Brown and J. G. Pickering, *Circulation*, 2003, **108**, 508.
51. P. Annunen, H. Autio-Harmainen and K. I. Kivirikko, *J. Biol. Chem.*, 1998, **273**, 5989.
52. R. Nissi, H. Autio-Harmainen, P. Marttila, R. Sormunen and K. I. Kivirikko, *J. Histochem. Cytochem.*, 2001, **49**, 1143.
53. H. H. Qi, P. P. Ongusaha, J. Myllyharju, D. Cheng, O. Pakkanen, Y. Shi, S. W. Lee, J. Peng and Y. Shi, *Nature*, 2008, **455**, 421.
54. C. Wu, J. So, B. N. Davis-Dusenbery, H. H. Qi, D. B. Bloch, Y. Shi, G. Lagna and A. Hata, *Mol. Cell. Biol.*, 2011, **31**, 4760.
55. R. Myllylä, T. Pihlajaniemi, L. Pajunen, T. Turpeenniemi-Hujanen and K. I. Kivirikko, *J. Biol. Chem.*, 1991, **266**, 2805.
56. T. Hautala, M. G. Byers, R. L. Eddy, T. B. Shows, K. I. Kivirikko and R. Myllylä, *Genomics*, 1992, **13**, 62.
57. H. N. Yeowell, V. Ha, L. C. Walker, S. Murad and S. R. Pinnell, *J. Invest. Dermatol.*, 1992, **99**, 864.
58. M. Valtavaara, H. Papponen, A. M. Pirttilä, K. Hiltunen, H. Helander and R. Myllylä, *J. Biol. Chem.*, 1997, **272**, 6831.
59. M. Valtavaara, C. Szpirer, J. Szpirer and R. Myllylä, *J. Biol. Chem.*, 1998, **273**, 12881.
60. K. Passoja, K. Rautavuoma, L. Ala-Kokko, T. Kosonen and K. I. Kivirikko, *Proc. Natl. Acad. Sci. U. S. A.*, 1998, **95**, 10482.
61. H. Ruotsalainen, L. Sipilä, E. Kerkelä, H. Pospiech and R. Myllylä, *Matrix Biol.*, 1999, **18**, 325.
62. C. Wang, M. Valtavaara and R. Myllylä, *DNA Cell Biol.*, 2000, **19**, 71.
63. K. Rautavuoma, K. Takaluoma, R. Sormunen, J. Myllyharju, K. I. Kivirikko and R. Soininen, *Proc. Natl. Acad. Sci. U. S. A.*, 2004, **101**, 14120.
64. A. M. Salo, C. Wang, L. Sipilä, R. Sormunen, M. Vapola, P. Kervinen, H. Ruotsalainen, J. Heikkinen and R. Myllylä, *J. Cell. Physiol.*, 2006, **207**, 644.
65. K. Rautavuoma, K. Takaluoma, K. Passoja, A. Pirskanen, A. P. Kvist, K. I. Kivirikko and J. Myllyharju, *J. Biol. Chem.*, 2002, **277**, 23084.
66. M. Suokas, O. Lampela, A. H. Juffer, R. Myllylä and S. Kellokumpu, *Biochem. J.*, 2003, **370**, 913.
67. J. Heikkinen, M. Risteli, C. Wang, J. Latvala, M. Rossi, M. Valtavaara and R. Myllylä, *J. Biol. Chem.*, 2000, **275**, 36158.
68. C. Wang, M. Risteli, J. Heikkinen, A. K. Hussa, L. Uitto and R. Myllylä, *J. Biol. Chem.*, 2002, **277**, 18568.

69. C. Wang, H. Luosujärvi, J. Heikkinen, M. Risteli, L. Uitto and R. Myllylä, *Matrix Biol.*, 2002, **21**, 559.
70. A. M. Salo, C. Wang, L. Sipilä, R. Sormunen, M. Vapola, P. Kervinen, H. Ruotsalainen, J. Heikkinen and R. Myllylä, *J. Cell. Physiol.*, 2006, **207**, 644.
71. C. Wang, M. M. Ristiluoma, A. M. Salo, S. Eskelinen and R. Myllylä, *J. Cell. Physiol.*, 2012, **227**, 668.
72. C. Wang, V. Kovanen, P. Raudasoja, S. Eskelinen, H. Pospiech and R. Myllylä, *J. Cell. Mol. Med.*, 2009, **13**, 508.
73. J. A. Vranka, L. Y. Sakai and H. P. Bächinger, *J. Biol. Chem.*, 2004, **279**, 23615.
74. Y. Ishikawa, J. Wirz, J. A. Vranka, K. Nagata and H. P. Bächinger, *J. Biol. Chem.*, 2009, **284**, 17641.
75. Y. Ishikawa and H. P. Bächinger, *J. Biol. Chem.*, 2013, **288**, 31437.
76. S. Järnum, C. Kjellman, A. Darabi, I. Nilsson, K. Edvardsen and P. Aman, *Biochem. Biophys. Res. Commun.*, 2004, **337**, 209.
77. P. Tiainen, A. Pasanen, R. Sormunen and J. Myllyharju, *J. Biol. Chem.*, 2008, **283**, 19432.
78. J. Vranka, H. S. Stadler and H. P. Bächinger, *Cell Struct. Funct.*, 2009, **34**, 97.
79. J. Myllyharju and K. I. Kivirikko, *EMBO J.*, 1997, **16**, 1173.
80. A. Pirskanen, A. M. Kaimio, R. Myllylä and K. I. Kivirikko, *J. Biol. Chem.*, 1996, **271**, 9398.
81. K. Passoja, J. Myllyharju, A. Pirskanen and K. I. Kivirikko, *FEBS Lett.*, 1998, **434**, 145.
82. M. A. McDonough, C. Loenarz, R. Chowdhury, I. J. Clifton and C. J. Schofield, *Curr. Opin. Struct. Biol.*, 2010, **20**, 659.
83. J. Myllyharju and K. I. Kivirikko, *EMBO J.*, 1999, **18**, 306.
84. R. Hieta, L. Kukkola, P. Permi, P. Pirilä, K. I. Kivirikko, I. Kilpeläinen and J. Myllyharju, *J. Biol. Chem.*, 2003, **278**, 34966.
85. M. Pekkala, R. Hieta, U. Bergmann, K. I. Kivirikko, R. K. Wierenga and J. Myllyharju, *J. Biol. Chem.*, 2004, **279**, 52255.
86. J. Anantharajan, M. K. Koski, P. Kursula, R. Hieta, U. Bergmann, J. Myllyharju and R. K. Wierenga, *Structure*, 2013, **21**, 2107.
87. K. Keskiaho, R. Hieta, R. Sormunen and J. Myllyharju, *Plant Cell*, 2007, **19**, 256.
88. M. K. Koski, R. Hieta, C. Böllner, K. I. Kivirikko, J. Myllyharju and R. K. Wierenga, *J. Biol. Chem.*, 2007, **282**, 37112.
89. M. K. Koski, R. Hieta, M. Hirsilä, A. Rönkä, J. Myllyharju and R. K. Wierenga, *J. Biol. Chem.*, 2009, **284**, 25290.
90. J. Heikkinen, M. Risteli, O. Lampela, P. Alavesa, M. Karppinen, A. H. Juffer and R. Myllylä, *Matrix Biol.*, 2011, **30**, 27.
91. R. Myllylä, K. I. Kivirikko, T. Pihlajaniemi, *Post-translational Modification of Proteins*, ed. J. J. Harding, M. J. C. Crabbe, CRC Press, Boca Raton, 1992, 1–51.
92. M. Risteli, O. Niemitalo, H. Lankinen, A. H. Juffer and R. Myllylä, *J. Biol. Chem.*, 2004, **279**, 37535.

93. K. Takaluoma, J. Lantto and J. Myllyharju, *Matrix Biol.*, 2007, **26**, 396.
94. R. J. Fernandes, A. W. Farnand, G. R. Traeger, M. A. Weis and D. R. Eyre, *J. Biol. Chem.*, 2011, **286**, 30662.
95. M. Hirsilä, P. Koivunen, V. Günzler, K. I. Kivirikko and J. Myllyharju, *J. Biol. Chem.*, 2003, **278**, 30772.
96. M. Hirsilä, P. Koivunen, L. Xu, T. Seeley, K. I. Kivirikko and J. Myllyharju, *FASEB J.*, 2005, **19**, 1308.
97. P. Koivunen, M. Hirsilä, A. M. Remes, I. E. Hassinen, K. I. Kivirikko and J. Myllyharju, *J. Biol. Chem.*, 2007, **282**, 4524.
98. K. Majamaa, T. M. Turpeenniemi-Hujanen, P. Latipää, V. Günzler, H. M. Hanauske-Abel, I. E. Hassinen and K. I. Kivirikko, *Biochem. J.*, 1985, **229**, 127.
99. A. J. van der Slot, A.-M. Zuurmond, A. F. J. Bardoel, C. Wigmenga, H. E. H. Pruijs, D. O. Sillence, J. Brinckmann, D. J. Abraham, C. M. Black, N. Verzijl, J. DeGroot, R. Hanemaaijer, J. M. TeKoppele, T. W. J. Huizinga and R. A. Bank, *J. Biol. Chem.*, 2003, **278**, 40967.
100. A. J. van der Slot, A.-M. Zuurmond, A. J. van den Bogaerdt, M. M. W. Ulrich, E. Middelkoop, W. Boers, H. K. Ronday, J. DeGroot, T. W. J. Huizing and R. A. Bank, *Matrix Biol.*, 2004, **23**, 251.
101. J. Brinckmann, S. Kim, J. Wu, D. P. Reinhardt, C. Batmunkh, E. Metzen, H. Notbohm, R. A. Bank, T. Krieg and N. Hunzelmann, *Matrix Biol.*, 2005, **24**, 459.
102. D. F. G. Remst, E. N. Blaney Davidson, E. L. Vitters, A. B. Blom, R. Stoop, J. M. Snabel, R. A. Bank, W. B. van den Berg and P. M. van der Kraan, *Osteoarthritis Cartilage*, 2013, **21**, 157.
103. E. Aro, R. Khatri, R. Gerard-O'Riley, L. Mangiavini, J. Myllyharju and E. Schipani, *J. Biol. Chem.*, 2012, **287**, 37134.
104. T. S. Eisinger-Mathason, M. Zhang, Q. Qiu, N. Skuli, M. S. Nakazawa, T. Karakasheva, V. Mucaj, J. E. Shay, L. Stangenberg, N. Sadri, E. Puré, S. S. Yoon, D. G. Kirsch and M. C. Simon, *Cancer Discovery*, 2013, **3**, 1190.
105. D. M. Gilkes, S. Bajpai, P. Chaturvedi, D. Wirtz and G. L. Semenza, *J. Biol. Chem.*, 2013, **288**, 10819.
106. D. M. Gilkes, S. Bajpai, C. C. Wong, P. Chaturvedi, M. E. Hubbi, D. Wirtz and G. L. Semenza, *Mol. Cancer Res.*, 2013, **11**, 456.
107. D. M. Gilkes, P. Chaturvedi, S. Bajpai, C. C. Wong, H. Wei, S. Pitcairn, M. E. Hubbi, D. Wirtz and G. L. Semenza, *Cancer Res.*, 2013, **73**, 3285.
108. G. Xiong, L. Deng, J. Zhu, P. G. Rychahou and R. Xu, *BMC Cancer*, 2014, **14**, 1.
109. R. Shah, P. Smith, C. Purdie, P. Quinlan, L. Baker, P. Aman, A. M. Thompson and T. Crook, *Br. J. Cancer*, 2009, **100**, 1687.
110. E. Hatzimichael, C. Lo Nigro, L. Lattanzio, N. Syed, R. Shah, A. Dasoula, K. Janczar, D. Vivenza, M. Monteverde, M. Merlano, A. Papoudou-Bai, M. Bai, P. Schmid, J. Stebbing, M. Bower, M. J. Dyer, L. E. Karran, C. ElguetaKarstegl, P. J. Farrell, A. Thompson, E. Briasoulis and T. Crook, *Br. J. Cancer*, 2012, **107**, 1423.
111. J. G. Teodoro, A. E. Parker, X. Zhu and M. R. Green, *Science*, 2006, **313**, 968.

112. S. Assadian, W. El-Assaad, X. Q. Wang, P. O. Gannon, V. Barrès, M. Latour, A. M. Mes-Masson, F. Saad, Y. Sado, J. Dostie and J. G. Teodoro, *Cancer Res.*, 2012, **72**, 1270.
113. T. Holster, O. Pakkanen, R. Soininen, R. Sormunen, M. Nokelainen, K. I. Kivirikko and J. Myllyharju, *J. Biol. Chem.*, 2007, **282**, 2512.
114. H. N. Yeowell and L. C. Walker, *Mol. Genet. Metab.*, 2000, **71**, 212.
115. B. Steinmann, P. M. Royce, A. Superti-Furga, *Connective Tissue and its Heritable Disorders*, 2nd edn, ed. PM Royce, B Steinmann, Wiley-Liss Inc., New York, 2002, 431–523.
116. P. H. Byers and S. M. Pyott, *Annu. Rev. Genet.*, 2012, **46**, 475.
117. P. H. Byers and M. L. Murray, *J. Invest. Dermatol.*, 2012, **132**, E6.
118. A. Ihme, T. Krieg, A. Nerlich, U. Feldmann, J. Rauterberg, R. W. Glanville, G. Edel and P. K. Müller, *J. Invest. Dermatol.*, 1984, **83**, 161.
119. D. Eyre, P. Shao, M. A. Weis and B. Steinmann, *Mol. Genet. Metab.*, 2002, **76**, 211.
120. J. Heikkinen, T. Toppinen, H. Yeowell, T. Krieg, B. Steinmann, K. I. Kivirikko and R. Myllylä, *Am. J. Hum. Genet.*, 1997, **60**, 48.
121. K. Takaluoma, M. Hyry, J. Lantto, R. Sormunen, R. A. Bank, K. I. Kivirikko, J. Myllyharju and R. Soininen, *J. Biol. Chem.*, 2007, **282**, 6588.
122. R. Ha-Vinh, Y. Alanay, R. A. Bank, A. B. Campos-Xavier, A. Zankl, A. Superti-Furga and L. Bonafé, *Am. J. Med. Genet., Part A*, 2004, **131**, 115.
123. M. T. Puig-Hervás, S. Temtamy, M. Aglan, M. Valencia, V. Martínez-Glez, M. J. Ballesta-Martínez, V. López-González, A. M. Ashour, K. Amr, V. Pulido, E. Guillén-Navarro, P. Lapunzina, J. A. Caparrós-Martín and V. L. Ruiz-Perez, *Hum. Mutat.*, 2012, **33**, 1444.
124. D. Viljoen, G. Versfeld and P. Beighton, *Clin. Genet.*, 1989, **36**, 122.
125. M. Hyry, J. Lantto and J. Myllyharju, *J. Biol. Chem.*, 2009, **284**, 30917.
126. A. M. Salo, H. Cox, P. Farndon, C. Moss, H. Grindulis, M. Risteli, S. P. Robins and R. Myllylä, *Am. J. Hum. Genet.*, 2008, **83**, 495.
127. M. Risteli, H. Ruotsalainen, A. M. Salo, R. Sormunen, L. Sipilä, N. L. Baker, S. R. Lamandé, L. Vimpari-Kauppinen and R. Myllylä, *J. Biol. Chem.*, 2009, **284**, 28204.
128. H. Ruotsalainen, L. Sipilä, M. Vapola, R. Sormunen, A. M. Salo, L. Uitto, D. K. Mercer, S. P. Robins, M. Risteli, A. Aszodi, R. Fässler and R. Myllylä, *J. Cell Sci.*, 2006, **119**, 625.
129. L. Sipilä, H. Ruotsalainen, R. Sormunen, N. L. Baker, S. R. Lamandé, M. Vapola, C. Wang, Y. Sado, A. Aszodi and R. Myllylä, *J. Biol. Chem.*, 2007, **282**, 33381.
130. R. Morello, T. K. Bertin, Y. Chen, J. Hicks, L. Tonachini, M. Monticone, P. Castagnola, F. Rauch, F. H. Glorieux, J. Vranka, H. P. Bächinger, J. M. Pace, U. Schwarze, P. H. Byers, M. Weis, R. J. Fernandes, D. R. Eyre, Z. Yao, B. F. Boyce and B. Lee, *Cell*, 2006, **127**, 291.

131. A. M. Barnes, W. Chang, R. Morello, W. A. Cabral, M. Weis, D. R. Eyre, S. Leikin, E. Makareeva, N. Kuznetsova, T. E. Uveges, A. Ashok, A. W. Flor, J. J. Mulvihill, P. L. Wilson, U. T. Sundaram, B. Lee and J. C. Marini, *N. Engl. J. Med.*, 2006, **355**, 2757.
132. W. A. Cabral, W. Chang, A. M. Barnes, M. Weis, M. A. Scott, S. Leikin, E. Makareeva, N. V. Kuznetsova, K. N. Rosenbaum, C. J. Tifft, D. I. Bulas, C. Kozma, P. A. Smith, D. R. Eyre and J. C. Marini, *Nat. Genet.*, 2007, **39**, 359.
133. F. S. van Dijk, I. M. Nesbitt, E. H. Zwikstra, P. G. Nikkels, S. R. Piersma, S. A. Fratantoni, C. R. Jimenez, M. Huizer, A. C. Morsman, J. M. Cobben, M. H. van Roij, M. W. Elting, J. I. Verbeke, L. C. Wijnaendts, N. J. Shaw, W. Högler, C. McKeown, E. A. Sistermans, A. Dalton, H. Meijers-Heijboer and G. Pals, *Am. J. Hum. Genet.*, 2009, **85**, 521.
134. A. M. Barnes, E. M. Carter, W. A. Cabral, M. Weis, W. Chang, E. Makareeva, S. Leikin, C. N. Rotimi, D. R. Eyre, C. L. Raggio and J. C. Marini, *N. Engl. J. Med.*, 2010, **362**, 521.
135. S. M. Pyott, U. Schwarze, H. E. Christiansen, M. G. Pepin, D. F. Leistritz, R. Dineen, C. Harris, B. K. Burton, B. Angle, K. Kim, M. D. Sussman, M. Weis, D. R. Eyre, D. W. Russell, K. J. McCarthy, R. D. Steiner and P. H. Byers, *Hum. Mol. Genet.*, 2011, **20**, 1595.
136. A. M. Barnes, G. Duncan, M. Weis, W. Paton, W. A. Cabral, E. L. Mertz, E. Makareeva, M. J. Gambello, F. L. Lacbawan, S. Leikin, A. Fertala, D. R. Eyre, S. J. Bale and J. C. Marini, *Hum. Mutat.*, 2013, **34**, 1279.
137. J. C. Marini and A. R. Blissett, *J. Clin. Endocrinol. Metab.*, 2013, **98**, 3095.
138. A. Forlino, W. A. Cabral, A. M. Barnes and J. C. Marini, *Nat. Rev. Endocrinol.*, 2011, **7**, 540.
139. D. Baldridge, J. Lennington, M. Weis, E. P. Homan, M. M. Jiang, E. Munivez, D. R. Keene, W. R. Hogue, S. Pyott, P. H. Byers, D. Krakow, D. H. Cohn, D. R. Eyre, B. Lee and R. Morello, *PLoS One*, 2010, **5**, e10560.
140. J. A. Vranka, E. Pokidysheva, L. Hayashi, K. Zientek, K. Mizuno, Y. Ishikawa, K. Maddox, S. Tufa, D. R. Keene, R. Klein and H. P. Bächinger, *J. Biol. Chem.*, 2010, **285**, 17253.
141. E. Pokidysheva, S. Tufa, C. Bresee, J. V. Brigande and H. P. Bächinger, *Matrix Biol.*, 2013, **32**, 39.
142. E. Pokidysheva, K. D. Zientek, Y. Ishikawa, K. Mizuno, J. A. Vranka, N. T. Montgomery, D. R. Keene, T. Kawaguchi, K. Okuyama and H. P. Bächinger, *J. Biol. Chem.*, 2013, **288**, 24742.
143. W. Chang, A. M. Barnes, W. A. Cabral, J. N. Bodurtha and J. C. Marini, *Hum. Mol. Genet.*, 2010, **19**, 223.
144. E. P. Homan, C. Lietman, I. Grafe, J. Lennington, R. Morello, D. Napierala, M. M. Jiang, E. M. Munivez, B. Dawson, T. K. Bertin, Y. Chen, R. Lua, O. Lichtarge, J. Hicks, M. A. Weis, D. Eyre and B. H. Lee, *PLoS Genet.*, 2014, **10**, e1004121.

145. J. W. Choi, S. L. Sutor, L. Lindquist, G. L. Evans, B. J. Madden, H. R. Bergen, T. E. Hefferan, M. J. Yaszemski and R. J. Bram, *3rd.PLoS Genet.*, 2009, **5**, e1000750.
146. S. Mordechai, L. Gradstein, A. Pasanen, R. Ofir, K. El Amour, J. Levy, N. Belfair, T. Lifshitz, S. Joshua, G. Narkis, K. Elbedour, J. Myllyharju and O. S. Birk, *Am. J. Hum. Genet.*, 2011, **89**, 438.
147. H. Guo, P. Tong, Y. Peng, T. Wang, Y. Liu, J. Chen, Y. Li, Q. Tian, Y. Hu, Y. Zheng, L. Xiao, W. Xiong, Q. Pan, Z. Hu, K. Xia, *Clin. Genet.*, 2014, **86**, 575.
148. E. Pokidysheva, S. Boudko, J. Vranka, K. Zientek, K. Maddox, M. Moser, R. Fässler, J. Ware and H. P. Bächinger, *Proc Natl Acad Sci U. S. A.*, 2014, **111**, 161.

CHAPTER 6

The Role of 2-Oxoglutarate-Dependent Oxygenases in Hypoxia Sensing

SARAH E. WILKINS[a], EMILY FLASHMAN[a], JOHN S. SCOTTI[a], RICHARD J. HOPKINSON[a], RASHEDUZZAMAN CHOWDHURY[a], AND CHRISTOPHER J. SCHOFIELD*[a]

[a]Chemistry Research Laboratory, University of Oxford, Mansfield Road, Oxford OX1 3TA, UK
*E-mail: christopher.schofield@chem.ox.ac.uk

6.1 Introduction to the Hypoxic Response in Animals

6.1.1 Background to Biochemistry of the Hypoxia Inducible Factor System

All animals require O_2 for survival in order to maintain respiration and consequently all vital functions. Higher animals have evolved complex respiratory and circulatory systems to ensure the sufficient delivery of O_2 to their cells and tissues. Chronic limitations in the supply of adequate O_2 to tissues results in the state of hypoxia.[1] It has been known for over a century that decreased concentrations of atmospheric O_2 induce physiologically profound hypoxic responses, perhaps most famously involving the increased production of red blood cells.[2,3] Subsequent studies have defined multiple links between hypoxia and diseases, in particular cancer and cardiovascular

RSC Metallobiology Series No. 3
2-Oxoglutarate-Dependent Oxygenases
Edited by Robert P. Hausinger and Christopher J. Schofield

Published by the Royal Society of Chemistry, www.rsc.org

disorders including heart disease and stroke. However, it is only relatively recently that the underlying molecular processes involved in the chronic hypoxic response in animals have started to be unravelled. Unexpectedly, 2-oxoglutarate (2OG)-dependent oxygenases (henceforth simplified to 2OG oxygenases) have been found to play central roles in the processes by which cells sense and respond to hypoxia. Here we review recent advances in our understanding of the chronic hypoxic response, focusing on the roles of 2OG oxygenases (for other recent reviews on the biology of the hypoxic response and its roles in disease see those by Cassavaugh and Lounsbury,[4] Semenza,[5] Mole and Ratcliffe[6] and Palmer and Clegg[7]).

A breakthrough was reported by Semenza and Wang in 1992 in their work on the regulation of erthyropoietin (EPO), a protein hormone that regulates erythropoiesis, *i.e.* red blood cell production.[8] They identified a novel heterodimeric α,β-transcription factor, hypoxia inducible factor (HIF), that binds to promoter regions of the *EPO* gene and induces the transcription of EPO. Crucially, they observed that active HIF is apparently induced under hypoxic conditions.[9] Moreover, they observed that levels of the HIF-α, but not the HIF-β subunit, increase in response to hypoxia.[10,11] Subsequent work has shown that both the activity and levels of HIF-α are regulated by its direct reaction with O_2 (see below).[12–15]

HIF binds to gene response elements characterized by the presence of a 5-nucleotide sequence 5′-G/ACGTG-3′ (the hypoxia response element, HRE), which is thought to be required for binding of HIF to all its target genes.[16] Under hypoxic conditions, HIF-α dimerizes with HIF-β and the heterodimeric complex recruits p300/CBP (CBP: cAMP response element binding protein) transcriptional coactivator proteins, which also promote transcription of numerous other transcription factors including p53.[17–20] HIF-β is identical in sequence to the aryl hydrocarbon receptor nuclear translocator (ARNT), a transcription factor containing basic helix-loop-helix (bHLH) and PAS (Period circadian protein, ARNT, Single-minded protein) domains.[21] ARNT has other roles, but when it heterodimerises with HIF-α, which also has bHLH and PAS domains, it forms the HIF transcription factor and binds to HREs *via* its bHLH regions.[9,22] HIF induces transcription of not only EPO, but of a large set of genes which work to ameliorate the effects of hypoxia at levels ranging from the cellular to the physiological.[23]

HIF has often been characterized as a 'master regulator' of the hypoxic response in animals that in effect 'reprogrammes' cellular biochemistry to accommodate different O_2 availabilities.[24–27] It is important to note that the acute (*i.e.* on a timescale shorter than the rate of new protein biosynthesis) hypoxic response in animals probably does not directly involve HIF. Furthermore, multiple non-HIF processes are involved in the chronic hypoxic response, but the extent of the HIF-mediated response is such that it is often difficult to dissect these away from those involving HIF. The HIF system is probably only present in animals; alternative hypoxia sensing/response mechanisms exist in lower eukaryotes, plants and microorganisms.[28–32] However, as outlined below, 2OG oxygenases related to those involved in the HIF system are conserved in lower organisms and play roles in protein biosynthesis.[33–35]

Many HIF target genes work to increase anaerobic respiration *via* upregulation of glycolysis, whereas others work to increase O_2 supply to tissues by promoting erythropoiesis, vasodilation and angiogenesis.[36–38] Work on understanding factors that regulate the sets of HIF target genes expressed in specific contexts is at a relatively early stage (as for any pleiotropic eukaryotic transcription factor), and is beyond the scope of this enzyme-focused review. There are three human forms of HIF-α (HIF-1α through HIF-3α), of which HIF-1α and HIF-2α are the best characterized. Roles for HIF-1α and HIF-2α in the expression of specific (sometimes overlapping) sets of HIF target genes have been identified.[39–41] For example, genes encoding glycolysis-related proteins (*e.g.* pyruvate dehydrogenase kinase) appear to be predominantly regulated by HIF-1α, whereas HIF-2α is more important in the regulation of EPO production in the kidneys.[42,43] There is a great deal of complexity involved in regulation of HIF target gene expression, such as the emerging evidence for 'cross-talk' between HIF and other transcription factors including nuclear factor-κB (NF-κB).[44,45] Furthermore, 2OG oxygenases (in addition to those that directly modify HIF) play roles in the regulation of HIF target genes, including N^{ε}-methyllysyl residue histone demethylases (KDMs, see Chapter 7).[46–48] Some KDMs are themselves HIF-regulated and there is preliminary evidence that some have the potential to play roles in hypoxic sensing.[49–55]

During hypoxia the levels of HIF-α rise and it forms a dimer with HIF-β in the nucleus, binds to HREs and promotes HIF target gene expression. In contrast, in sufficiently well-oxygenated conditions (normoxia), HIF-1α levels are reduced (HIF-1β levels are not regulated by O_2) resulting in a lack of the active α,β-form of HIF and suppression of the hypoxic response (Figure 6.1). The precise extent of hypoxia required to induce HIF-α in intact animals is difficult to measure; in cell-culture studies, levels of atmospheric O_2 lower than 4% are used to induce HIF-α. In most normal healthy cells the levels of HIF-α are very low or not detected.[10] Levels of HIF-α are elevated in many diseases associated with hypoxia and ischaemia, including heart disease, stroke and cancer.[56,57] There is thus significant medicinal interest in the HIF system, both for understanding the role of the hypoxic response in disease and for its manipulation for therapeutic benefit. From both of these perspectives, identification of the O_2-dependent mechanisms that regulate the levels of HIF-α has been of considerable interest since the discovery of this protein.

6.1.2 The HIF Hydroxylases

An important step in defining a hypoxia sensing component of the HIF system was the finding that HIF-1α is targeted for O_2-dependent proteasomal degradation in normoxia by the von Hippel–Lindau tumour suppressor protein (pVHL) E3 ubiquitin ligase complex.[58–60] Hereditary mutations in the gene encoding for pVHL correlate with VHL disease, which is associated with increased risk of cancer characterized by HIF upregulation and highly

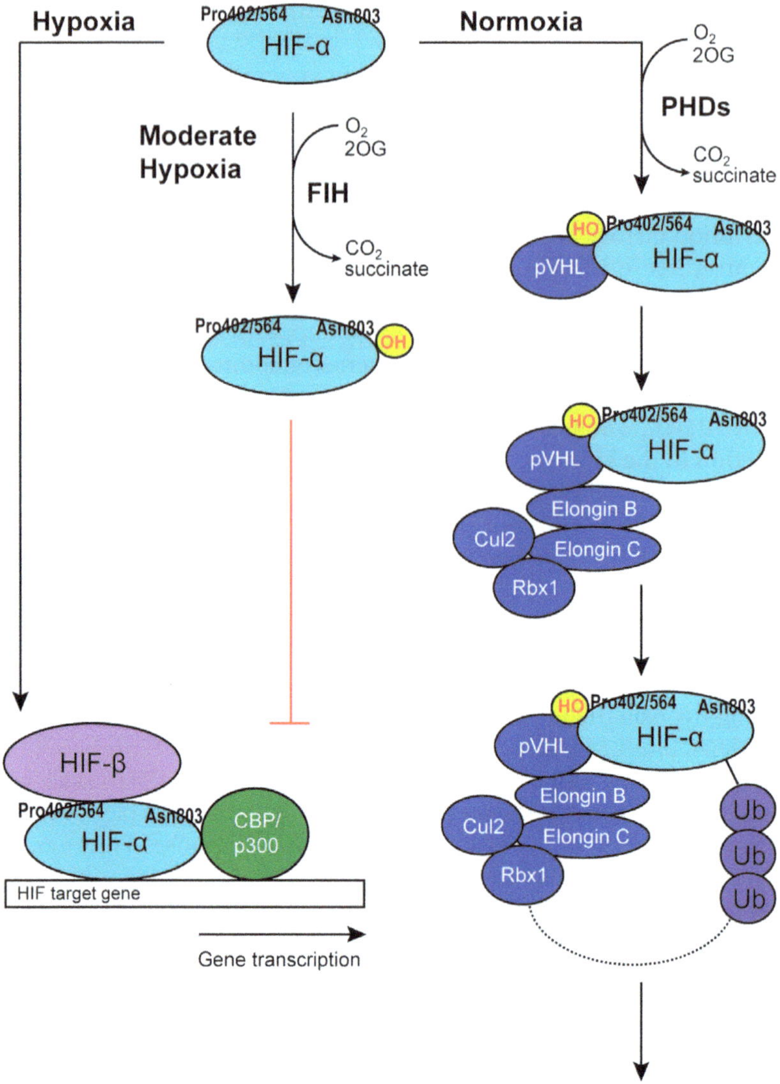

Figure 6.1 Roles of 2OG-dependent oxygenases in the human hypoxic response. In normoxia, HIF-α prolyl residues (Pro-402/Pro-564 in human HIF-1α) undergo hydroxylation catalysed by the PHDs, thus signalling HIF-α for proteasomal degradation by the pVHL E3 ubiquitin ligase complex. In a second method of oxygenase-mediated regulation, FIH catalyses hydroxylation of the HIF-α at a conserved asparaginyl residue (Asn-803 in human HIF-1α), a modification that hinders HIF-α from associating with p300/CBP transcriptional coactivator proteins. In hypoxia, HIF-α is less subject to these O_2-dependent post-translational modifications and translocates to the nucleus where it dimerizes with HIF-β and together with p300/CBP promotes transcription of genes involved in the human hypoxic response. Note that other 2OG oxygenases are probably involved in regulating the sets of HIF target genes expressed in different contexts.

vascularized tumours.[61] pVHL is the targeting component of an E3 ubiquitin ligase complex, also including elongins B and C, Cul2 (Cullin box 2) and Rbx1 (RING box protein 1); the VCB complex (pVHL, elongin C and elongin B) was shown to catalyse ubiquitylation of HIF-1α and HIF-2α, causing efficient degradation *via* proteasome-catalysed hydrolysis.[58–60] However, the mechanism by which increased O_2 levels result in the recruitment of pVHL and subsequent destruction of HIF-1α had yet to be determined. Clues to the mechanism were provided by knowledge that iron chelators or $CoCl_2$ can upregulate HIF and/or induce the hypoxic response.[62,63] Mutation studies led to the identification of two regions of HIF-1α that are responsible for promoting its O_2-dependent degradation: the *N*- and *C*-terminal O_2-dependent degradation domains (NODD and CODD, respectively).[64] Subsequently, two prolyl residues, one in each of NODD and CODD (Pro-402 and Pro-564, respectively, in human HIF-1α), were shown to undergo *trans*-4-hydroxylation catalysed by the 2OG-dependent HIF prolyl hydroxylases (PHDs) (Figure 6.2).[65–68] Either of the NODD or CODD prolyl hydroxylations is sufficient, at least under many conditions, to target HIF-1α (or HIF-2α) for proteasomal degradation *via* the pVHL E3 ligase complex (Figure 6.1).[64,67] Prolyl hydroxylation causes a very substantial (>800-fold) increase in the binding affinity of HIF-α for the VCB complex,[69] a process that is reasonably well understood from a chemical perspective (see below).[70] The catalytic domains of the PHDs are related to (but different from) those of the procollagen prolyl 4-hydroxylases. Importantly, the selectivities of the three human PHDs (PHD 1–3) and procollagen-specific enzymes do not overlap (at least at detectable levels).[66]

Crystal structures of the VCB complex bound to hydroxylated HIF-1α CODD peptide revealed the structural basis for the O_2-dependent mechanism of HIF-1α recognition by pVHL.[69,71] The structures show that Hyp-564 (Hyp: hydroxyproline) is buried in a cleft in the β-domain of pVHL. The Hyp pyrrolidine ring is positioned in a hydrophobic region in pVHL and interacts with Trp-88, Tyr-98 and Trp-117 (Figure 6.3). The hydroxyl group of Hyp-564 displaces a water molecule from pVHL, and is positioned to form hydrogen bonds to the hydroxyl of Ser-111 and the Nδ of His-115.[70] These hydrogen bonds are, at least in part, responsible for the increased binding affinity of the hydroxylated relative to the unhydroxylated form of HIF-α to pVHL.[69]

A second type of 2OG oxygenase mediates a 'switch-like' effect on HIF regulation, but this one 'breaks' rather than 'makes' a protein : protein interaction as in HIF prolyl hydroxylation.[72] HIF-mediated transcription is promoted by interaction of its *C*-terminal transcriptional activation domain (CTAD) with the CBP/p300 transcriptional coactivators. Lando *et al.* revealed that hydroxylation of Asn-803 in the CTAD substantially reduces the interaction of the HIF-α CTAD with CBP/p300.[72] HIF-α Asn hydroxylation was shown to be catalysed by a single enzyme, factor inhibiting HIF (FIH, a somewhat confusing acronym),[73,74] which belongs to a different subfamily of 2OG oxygenases compared to the PHDs.[75–77] Although less well understood than the effect of prolyl hydroxylation, FIH-catalysed hydroxylation at the β-carbon of Asn-803 introduces a steric clash in an otherwise

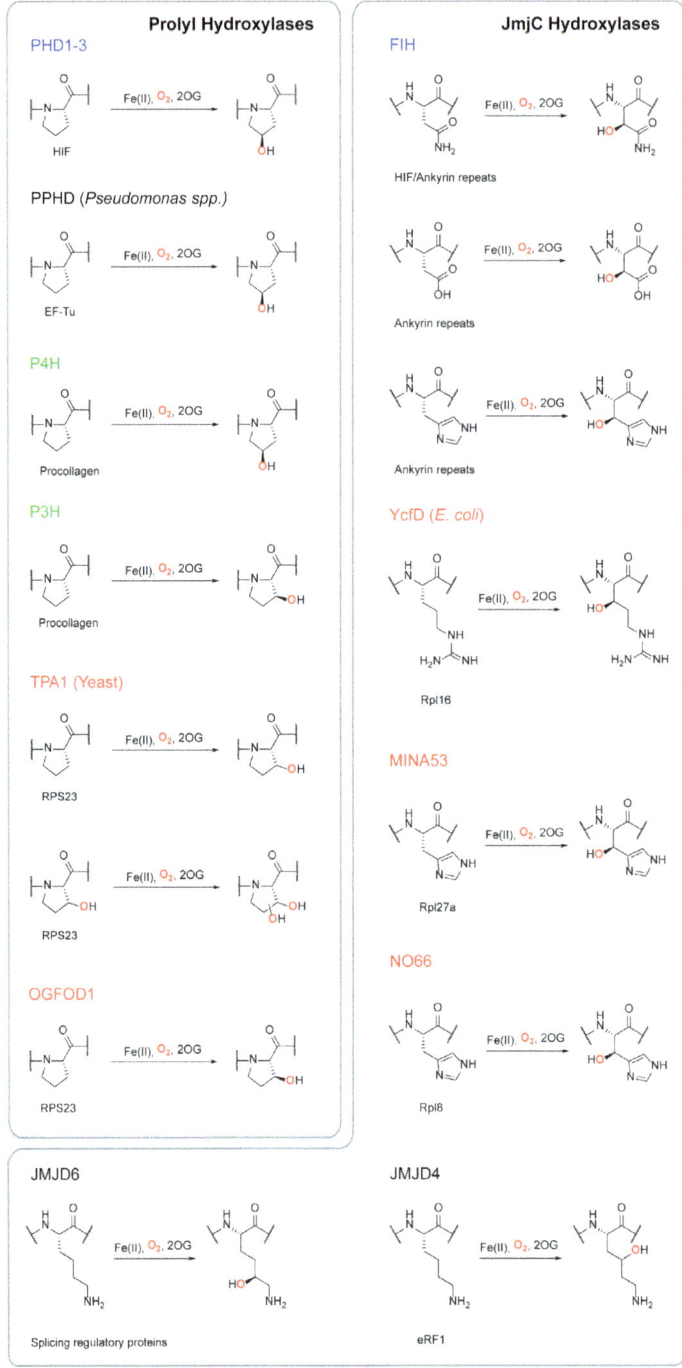

Figure 6.2 Reactions catalysed by the HIF hydroxylases and related enzymes (human unless otherwise specified). Enzyme acronyms are coloured based on their substrates: blue = HIF hydroxylases, green = pro-collagen hydroxylases, red = ribosomal oxygenases, black = miscellaneous.

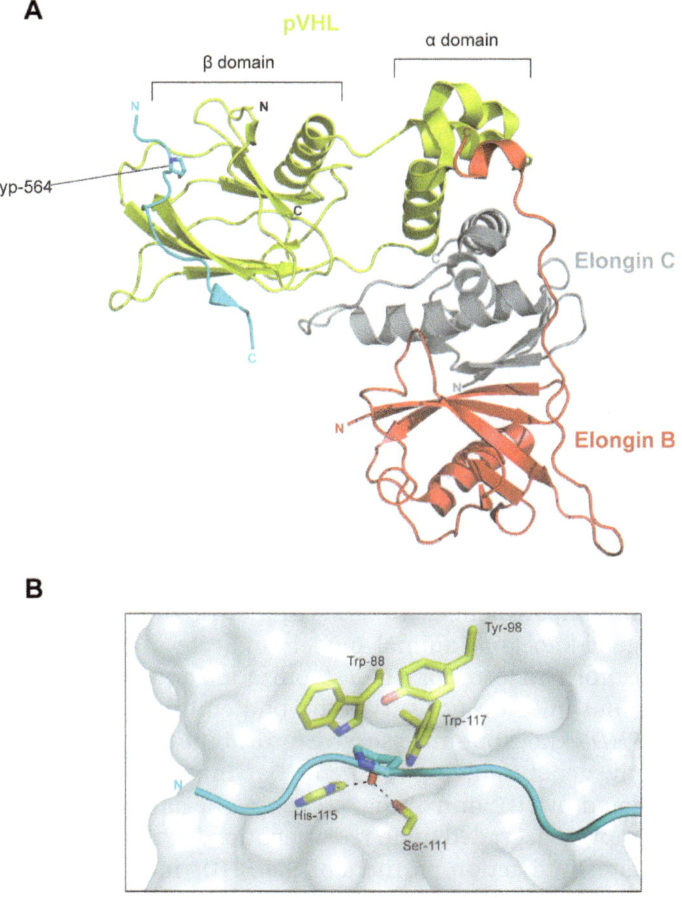

Figure 6.3 Structural overview of the HIF-α:VHL interaction. View from a crystal structure of the VCB complex bound to the hydroxylated (Hyp-564) HIF-1α CODD. (**A**) HIF-1α CODD (residues 561–575) binds to the β-domain of pVHL in the VCB complex, consisting of pVHL, elongin B and elongin C. (**B**) Hyp-564 is positioned in a hydrophobic cleft isolated from solvent, where it makes van der Waals contacts with Trp-88, Trp-117 and Tyr-98 and is positioned to form hydrogen bonds to Ser-111 and His-115 (PDB ID: 1LM8).

hydrophobic interaction that reduces HIF-α binding to p300/CBP coactivator proteins.[78,79]

Thus, two different types of 2OG oxygenase (*i.e.* the PHDs and FIH) play central roles in the HIF-mediated hypoxic response by catalysing hydroxylations that have profound effects on gene expression. Such 'switch-like' effects of hydroxylation/oxygenase catalysis on signalling were unprecedented and raised the possibility of a wider role for 2OG oxygenases in signalling and protein biosynthesis. In subsequent sections we discuss knowledge of the HIF hydroxylases and related enzymes in more detail.

6.2 The HIF Prolyl Hydroxylases

There are three HIF prolyl hydroxylases known to be involved in hypoxic sensing in humans, PHDs 1, 2 and 3 (EGLNs 2, 1 and 3, respectively).[66–68,80] All three PHDs are 2OG oxygenases and are proposed to act as a direct link between cellular O_2 availability and HIF levels by promoting O_2-dependent HIF-α degradation (Figure 6.1).[81] A fourth putative 2OG-dependent HIF prolyl hydroxylase has been identified,[82] although its cellular location (bound to the ER membrane) suggests it is unlikely to be directly involved in the HIF-mediated hypoxic response,[83] and further work is required to validate its biological role. Here we focus on the biochemical and biophysical properties of the HIF PHDs in relation to their proposed role as hypoxia sensors.

Many insights into the biological functions of the PHDs have come from cellular and animal studies with specific, but often overlapping, roles beginning to emerge for the different PHD isoforms.[13,84–86] Their mRNA expression patterns indicate that *PHD2* is ubiquitously expressed, whereas *PHD1* is predominant in the testes and *PHD3* in the heart.[87–89] The intracellular distribution of the PHDs also differs, at least in some cells, with PHD2 being located predominantly in the cytoplasm, PHD1 in the nucleus, and PHD3 found in both cellular compartments.[90] In terms of the HIF-driven hypoxic response, RNAi and gene-knockout animal models indicate that PHD2 is the most important isoform in terms of HIF-α stabilization and hypoxia sensing;[81,88,91] *PHD2* knockout mice do not survive beyond embryonic day 14.5, whereas *PHD1* and *PHD3* knockout mice survive into adulthood (see Myllyharju[92] for reviews on the genetics/biology of the PHDs). Importantly, PHD2 and PHD3 are upregulated by HIF in hypoxia, creating a negative feedback loop such that HIF is rapidly degraded upon reoxygenation.[67,88,90,93–95] Alternative (non-HIF) substrates/interaction partners of the PHDs include IκB kinase-β,[96–98] RNA polymerase II,[99] β2-adrenergic receptor,[100] PKM2,[101] Cep-192[102] and non-muscle actin,[103] amongst many others (see reviews).[104,105] Some effects of the HIF hydroxylases on these substrates are O_2-dependent, although prolyl hydroxylation of endogenous proteins has not been observed in all cases and the physiological roles of these non-HIF PHD partners have yet to be completely resolved. It is possible that in some cases they impact the hypoxic response by competing with HIF-α for binding to the PHDs, as proposed for FIH (see Section 6.3.2).

6.2.1 Biochemistry of the HIF Prolyl Hydroxylases

Sequence comparisons of the catalytic domains of human PHDs1–3 suggest they contain the typical elements of the 2OG oxygenase family.[67] Their *C*-terminally positioned catalytic domains show a high degree of conservation and homology; crystallographic data for PHD2 (see below) reveal the stereotypical double-stranded β-helix (DSBH or jelly-roll) structure.[106,107] The *N*-terminal regions of the PHDs, however, differ significantly: this region of PHD1 is predicted to be disordered, PHD3 lacks an extended *N*-terminus (it consists of the catalytic domain alone), while the *N*-terminus of PHD2 contains a MYND-type zinc finger domain. This domain is apparently conserved in at least one PHD in all animals,[32] consistent with studies indicating that PHD2 is the most important of the three PHD isoforms in higher animals.[81,88,91]

PHD2 forms an unusually stable complex with Fe(II) and 2OG.[108] This stability was first observed during 2OG decarboxylation assays where, in some preparations of PHD2, addition of exogenous Fe(II)/2OG was not required for activity (indicating they were present endogenously in the active site) and by the larger than expected size measured by non-denaturing mass spectrometry.[108] Subsequent studies using NMR and UV-vis spectroscopy have confirmed this result quantitatively: K_D values for metals (Fe(II) in anaerobic conditions or Mn(II) in aerobic conditions) were determined in the region of <1 μM.[108–110] 2OG binding constants were reported to be in the range of 1–5 μM.[108–110] Kinetic studies with both the catalytic domain of PHD2 and full-length PHD2 have determined K_M values for 2OG ranging from <1 to 60 μM,[109,111–113] with differences probably reflecting experimental conditions. The combined binding and kinetic data are consistent with the proposal that PHD2 activity is unlikely (or less likely) to be limited by Fe(II)/2OG availability in many cells under *normal* conditions, and that the stable PHD2 : Fe(II) : 2OG complex may help to enable an 'ever ready' enzymatic state, primed for O_2-regulated HIF hydroxylation.

The PHDs hydroxylate one or both of two prolyl residues in HIF-α, located within the NODD and CODD.[66–68,80,113] PHD1 and 2 can hydroxylate both NODD and CODD, whereas PHD3 only interacts with and efficiently hydroxylates CODD,[106,112–114] although it does hydroxylate NODD on prolonged incubation.[115] Studies on isolated proteins are supported by cellular work with antibodies selective for hydroxylated NODD and CODD.[116] The reason for the presence of both NODD and CODD hydroxylation sites in higher animals is presently unknown, as hydroxylation of one site is apparently sufficient for targeting HIF to pVHL.[64] Bioinformatics analyses indicate that most lower animals only contain a single PHD2 homologue and one HIF-α hydroxylation site, which is more similar to CODD than NODD.[32,117] Nevertheless, the presence of two conserved prolyl hydroxylation sites in higher animals suggests a differential role for each, possibly enabling context-dependent regulation of sets of HIF target genes. PHD isoform specificity for NODD *vs* CODD is enabled by multiple interactions including with the β2-β3 loop,[106,118,119] a region that is non-conserved amongst the PHDs and has been shown in PHD2 structural studies to be important in substrate binding (see Section 6.2.2). Reported K_M values of PHD2 for CODD and NODD range between ~2 and 40 μM and ~10–130 μM, respectively;[106,109,111–113] the range of differing values is dependent on experimental conditions including the length of the enzyme/substrate constructs used. Overall, the kinetic analyses demonstrate that CODD is preferentially hydroxylated over NODD, in line with cellular studies carried out over O_2 gradients where it was found that CODD hydroxylation is maintained in lower O_2 concentrations than is NODD hydroxylation.[114,116] Surface plasmon resonance analyses indicate more rapid dissociation of PHD2 from

NODD than from CODD, indicating a less stable PHD2 : Fe(II) : 2OG : NODD than PHD2 : Fe(II) : 2OG : CODD complex.[112]

Within the cellular environment many factors have the potential to regulate PHD activity, including relative concentrations of enzymes and (co)substrates. However, if the PHDs are to fulfil a role as O_2-sensing enzymes an essential requirement is that their overall cellular HIF-α prolyl hydroxylation activity changes in response to fluctuations in cellular O_2 concentrations. This requirement has prompted work to connect the biochemical properties of the PHDs with their physiological roles. As described above, the ability of PHD2 to form stable PHD : Fe(II) : 2OG complexes may assist in its role as a hypoxia sensor.[107,108] Early kinetic characterization of the O_2 dependence of the PHDs indicated K_M values in the range of ~70–250 μM,[112,113] while more recent studies of PHD2 have reported even greater O_2 K_M values of >450 μM and 1.7 mM.[109,111] These values are significantly higher than those reported for other 2OG oxygenases, and they are probably greater than atmospheric aqueous O_2 concentrations, indicating that in terms of their kinetic properties the PHDs have the potential to fulfil the requirement to act as O_2 sensors. *In vivo* (at least probably beyond some lung tissues), physiological O_2 concentrations are likely to be significantly lower than the reported K_M values and would vary between tissues/cells/cellular compartments. Investigation of the pre-steady state kinetics of the PHD2 reaction revealed unusually slow reaction with O_2,[120] in contrast to most other similarly studied 2OG oxygenases such as taurine:2OG dioxygenase, a viral collagen prolyl hydroxylase, and deacetoxycephalosporin C synthase.[121–123] This slow reaction may in part be due to stabilization of an Fe-coordinating water molecule at the active site that must dissociate before O_2 can bind (see Section 6.2.2).[109] It is proposed that the apparently slow reaction of PHD2 with O_2 is related to its role as a hypoxia sensor.[120] Importantly PHD activity in cells shows a graded decrease from normoxia to hypoxia, more so than FIH, which retains activity at lower O_2 concentrations (see Section 6.3.3).[116]

It is important to note that O_2 availability is not the only factor that can affect the catalytic activity of the HIF hydroxylases in cells. Nitric oxide and reactive O_2 species (ROS) can also reduce the activity of the HIF hydroxylases,[124,125] possibly by directly affecting the oxidation status of the active site Fe of these enzymes[126] or *via* modulation of other redox sensitive aspects of their structure.[127] Indeed, while the PHDs are relatively resistant to H_2O_2-mediated inactivation, FIH is actually proposed as a potential H_2O_2 sensor.[128] Ascorbate is often used to obtain optimal activity of many 2OG oxygenases *in vitro*, although its effect is enzyme and condition dependent (ascorbate is proposed to prevent scurvy by promoting activity of the procollagen prolyl hydroxylases). Ascorbate promotes activity of purified PHDs *in vitro*,[129] and this molecule is also thought to promote PHD activity in cells;[129] however, the role of ascorbate in PHD catalysis can be substantially fulfilled by other reducing agents[130] and mouse model work reveals ascorbate is not essential for PHD catalysis.[131] The mechanism of action of ascorbate is unknown (and may change depending on context), but it has been proposed to act by maintaining Fe(II) at the active site and/or by completing uncoupled turnover cycles.[132–134] Interestingly, HIF upregulation in tumour cells is proposed to be regulated by (amongst other factors) hypoxic conditions and/or increased ROS.[135–137] Tumour cell redox status (and modulators thereof) may therefore affect HIF hydroxylase activity and hence lead to HIF upregulation. This finding raises the interesting possibility that under some conditions the HIF hydroxylases act as signal integration points for cellular redox status. Finally, some tumours exhibit very substantial upregulation of tricarboxylic acid intermediates (*i.e.* succinate and fumarate)[138] or 2-hydroxyglutarate (derived from reduction of 2OG by a mutated form of isocitrate dehydrogenase).[139,140] Although more work is required to demonstrate their pathophysiological relevance in cancer, it is proposed that elevated levels of these metabolites (possibly coupled with decreased 2OG levels) cause decreased activity of 2OG oxygenases, including the HIF hydroxylases (*e.g.* the inhibitory effect of fumarate on the activity of PHD2).[141–144]

6.2.2 Structural Studies on the HIF Prolyl Hydroxylases

In 2006, the first crystal structures of a HIF prolyl hydroxylase, PHD2, were published in complex with Fe(II) and bicyclic 2OG analogues (Figure 6.4).[107] These structures reveal that PHD2 possesses the DSBH fold that is characteristic of 2OG oxygenases (see Chapter 2), supporting the catalytic triad of His-313, Asp-315 and His-374 that binds Fe(II) at the active site. 2OG (or its mimic) coordinates the Fe *via* its carboxylate and ketone O_2 atoms, and is stabilized by ionic interactions between its carboxylate and Arg-383 of PHD2. A network of five water molecules (not depicted) runs alongside

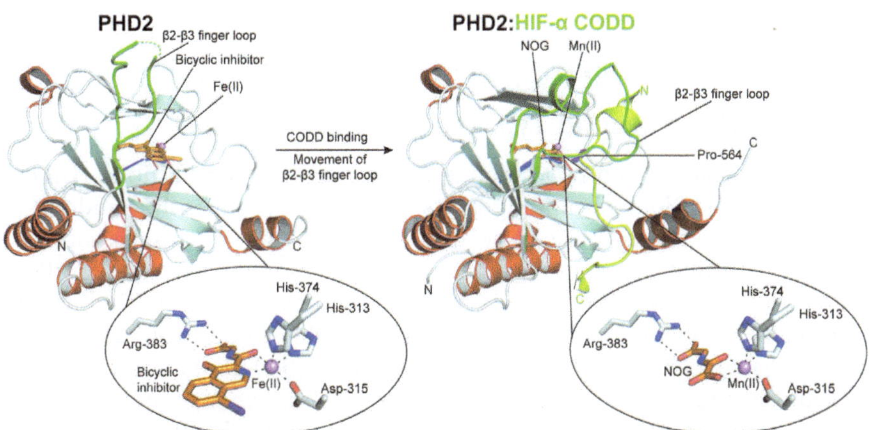

Figure 6.4 Structural overview of the HIF prolyl hydroxylase PHD2. Views from crystal structures of human PHD2 without substrate (left, PDB ID: 2G1M) and in complex with a HIF-1α CODD fragment (right, PDB ID: 3HQR) reveal a conserved mechanism of substrate binding involving conformational changes in the β2-β3 finger loop of PHD2.

the 2OG binding site, extending from the Fe/Asp-315 coordinated water molecule. Notably, the active site is otherwise lined by predominantly hydrophobic residues and appears relatively buried within the overall enzyme structure when compared with many other 2OG oxygenases, including FIH (see Section 6.3.4). These features may rationalize the ability of PHD2 to form a relatively stable Enz : Fe(II) : 2OG complex[108] and/or its slow reaction with O_2.[120]

Structures of PHD2 with 2OG,[107] inhibitors,[145,146] and in complex with Mn(II), *N*-oxalylglycine (NOG, a 2OG analogue) and a CODD peptide substrate[118] have also been reported. The structure of the substrate complex reveals that upon substrate binding, the β2-β3 loop (observed to be flexible in the substrate-free structure[107]) folds to enclose the hydroxylated proline and surrounding residues (the LAPYIP motif), and stabilizes binding of the substrate in the active site (Figure 6.4). The flexible *C*-terminus of PHD2, which plays a role in determining selectivity,[106] has also been shown to interact with substrate. In the substrate structure, the prolyl residue to be hydroxylated is observed in a C4-*endo* conformation; it has been found to preferentially adopt a C4-*exo* conformation upon hydroxylation (as observed in the hydroxylated CODD VCB structures;[69] Figure 6.3), a change that is proposed to promote product release from the PHD2 active site. Interestingly, NOG/2OG appear to coordinate to the active site iron/surrogate metal such that the water-occupied coordination site is not adjacent to the prolyl methylene carbon that becomes hydroxylated.[118] Further, the metal-bound water is positioned to be stabilized by hydrogen bonding. These observations suggest that a rearrangement is required before substrate oxidation can take place and may be a contributing factor in the apparently unusual kinetics of PHD2, including its slow reaction with O_2.[109,120]

6.3 The HIF Asparaginyl Hydroxylase – FIH

6.3.1 FIH is a Second Type of HIF Hydroxylase

FIH was first identified by Mahon *et al.* as a protein that inhibits HIF-mediated transcription *via* binding to its CTAD.[147] Subsequently, Lando *et al.* made the breakthrough discovery that Asn-803 of HIF-1α is hydroxylated,[72] a modification that substantially reduces the tight binding between the HIF-1α CTAD and the transcriptional coactivator proteins CBP (CREB binding protein)/p300, probably by disrupting a hydrophobic interaction. This modification is complementary to those at the HIF-α NODD and CODD regions in that the addition of a single O_2 atom ‘breaks’ rather than ‘makes’ a tight protein : protein interaction. Subsequently, two groups identified FIH as the Fe(II)- and 2OG-dependent HIF-α CTAD Asn hydroxylase,[73,74] but one from a different subfamily to the PHDs; indeed, FIH had already been assigned as a member of the JmjC family of transcriptional regulators,[148] which at the time had not been assigned as 2OG oxygenases. FIH was shown to catalyse hydroxylation at the β-carbon of the HIF-1α CTAD (Figure 6.2) to give the

3*S*-product, rather than the 3*R*-product as observed for the epidermal growth factor-like protein β-hydroxylase.[149,150] Thus, the early studies revealed that FIH catalysis reduces HIF-mediated transcriptional activity under normoxic conditions by blocking the association with CBP/p300; when insufficient O_2 is available, newly synthesized CTAD is no longer (fully) hydroxylated and HIF is transcriptionally active (Figure 6.1). Mouse models and extensive cellular studies have supported a physiologically relevant and non-redundant role for FIH;[151–153] however, subsequent work has revealed that FIH is much more than a simple hydroxylase acting on HIF-α in a 'switch-like' manner. Here we focus on the enzymology of FIH, but we note there are aspects to the role of FIH-catalysed hydroxylation that are beyond our scope (*e.g.* relating to localization[154] and negative regulation by dimerization of CTAD hydroxylated HIF-α).[155]

6.3.2 Alternative Substrates for FIH

Subsequent to its identification as a HIF-α hydroxylase, FIH was found to catalyse the hydroxylation of a diverse range of proteins from the ubiquitous ankyrin repeat domain (ARD) structural family. The first reported ARD substrates for FIH were p105 and IκBα, components of the NFκB signalling system.[156] FIH has since been found to hydroxylate ankyrin repeat sequences from a large number of ARD proteins *in vitro*,[156–167] the majority of which are likely endogenous substrates, at least in some contexts. Identified ARD substrates for FIH include Notch and NFκB transcription factors,[156,157] cytoskeletal ankyrin proteins,[164] RNaseL,[160] ankyrin repeat and SOCS box-containing 4 (ASB4)[159] Tankyrase-2,[163] and apoptosis-stimulating p53 binding protein ASPP2.[167] Proteomic studies suggest that FIH probably interacts with many more ARD proteins, and these are hydroxylated to very different extents (0–80%); complete hydroxylation of ARDs is rare, if present at all.[160,165]

The ankyrin repeat is a structural motif composed of approximately 33 residues, with multiple repeats (3 to >20) common in eukaryotic proteins.[168,169] Despite considerable sequence variation, individual ARDs adopt a very well-characterized stereotypical helix-turn-helix fold, with a β-hairpin loop that connects adjacent repeats.[170] The residue targeted for hydroxylation by FIH occurs near the apex of the β-hairpin loop; this position is occupied by an Asn residue in the majority of eukaryotic ankyrin repeat sequences.[168] FIH displays a preference for Asn hydroxylation (at least in studied cases), but it can catalyse the hydroxylation of other residues at this position including a His residue in Tankyrase-2[163] and Asp residues in human cytoskeletal ankyrin proteins.[164] Hydroxylation of all three residues occurs on the β-carbon to give the 2*S*,3*S* product, analogous with HIF hydroxylation (Figure 6.2).[149,163,164]

Despite the large number of ARD proteins subject to post-translational hydroxylation by FIH, including some of major biomedical importance (*e.g.* Notch,[171,172] IκBα[173,174]), extensive analyses of multiple ARD substrates have yielded few definitive insights as to the functional significance of ARD hydroxylation in a signalling pathway other than HIF. ARD hydroxylation appears to have little influence on the stereotypical folded ankyrin repeat structure (as shown by crystallographic[157,175] and NMR[176] studies), but this modification has been shown to enhance the thermodynamic stability of some ARDs, including artificial ARDs with consensus sequences.[164,175,177] However, given that ARD hydroxylation is rarely, if ever, complete, it would seem that stabilization alone is unlikely to be a primary role of ARD hydroxylation (in contrast to the situation for procollagen prolyl hydroxylation).[178,179]

The ability of multiple ARD proteins to compete with HIF for hydroxylation by FIH has led to the hypothesis that ARD proteins act collectively to regulate the kinetics of the hypoxic response through competition for FIH.[157,161,180] The hydroxylated ARD products bind less well to FIH than the unhydroxylated substrates, thus the kinetics of the hypoxic response may be regulated both by the complement and the hydroxylation status of ARDs accessible to FIH. Such competition has the potential to enable subtle context-dependent modulation of the hypoxic response and, *via* modification of the ARD-pool hydroxylation status, to provide a 'memory' of hypoxic events (ARD proteins have very different lifetimes).[180]

Finally, to further complicate the FIH interaction story, FIH activity is suppressed in macrophages by an X11 protein family member, Mint3/APBA3,[181–183] which inhibits FIH *via* binding of its *N*-terminal domain; as a consequence HIF is activated and glycolysis (the process by which macrophages generate most of their ATP) is upregulated.

6.3.3 Biochemistry of FIH

Crystallographic and spectroscopic studies of FIH suggest that its catalytic mechanism is analogous to that employed by most other 2OG oxygenases.[76,184] This mechanism involves a sequential order of substrate/cosubstrate binding, in which the Fe(II) binds first within the active site (coordinated by His-199, Asp-201 and His-279), followed by 2OG, then substrate, and finally O_2. Substrate binding opens a coordination site on the Fe(II) and stimulates binding/activation of O_2. As for the PHDs, this order of binding has important implications for the role of FIH as an O_2 sensor, minimizing the potential for O_2 activation in the absence of substrate and preventing inactivation of FIH by auto-hydroxylation.[185] In contrast to the PHDs, crystallographic analyses reveal that the observed vacant coordination site is directly adjacent to the methylene that is hydroxylated (at least in the reported FIH:substrate structures, see below), and that the Asp of the His-X-Asp···His Fe(II)-binding motif hydrogen bonds to the substrate.[76,157] Thus, there is a potential for substrate-selective kinetics, including with respect to the O_2 sensitivity of hydroxylation, although as yet there is no evidence for this type of regulation in cells.

The catalytic properties of FIH have been studied extensively *in vitro* using peptide substrates from HIF-1α and, to a lesser extent, ARD proteins.[74,112,162,186] In general, the results imply that FIH may be a more 'typical' 2OG oxygenase than PHD2 with respect to its kinetic profile, *i.e.* FIH forms a less stable enzyme:Fe(II):2OG complex and its activity is less sensitive to

O_2 availability as indicted by lower K_M values (10–240 μM),[112,162,187] consistent with cellular studies showing that FIH activity is retained better in hypoxia than that of the PHDs.[116,188] As for the PHDs, the variation in K_M values probably reflects differences in assay conditions and/or the length of the peptide substrate employed in kinetic studies.[112,187] Notably, FIH has a greater affinity for its (tested) ARD substrates than for the HIF-1α CTAD, and it exhibits a much lower K_M for O_2 with Notch1 as a substrate compared to HIF-1α.[162,187,189] However, quantitatively matching the results of isolated protein experimental observations with those from cell-based experiments is difficult, probably in part due to the promiscuity of FIH.[165] Further, the finding that FIH accepts multiple ARDs, which are likely to be collectively at a much greater concentration than HIF-α in cells, raises the unanswered question of whether there is sufficient ARD-free FIH available to efficiently hydroxylate HIF-α; one possibility is that (some) ARDs can target HIF-α to FIH by binding one of the FIH monomers in the FIH dimer (see 6.3.4).

With regard to its substrate specificity, FIH demonstrates an unusually high level of promiscuity compared to other studied 2OG oxygenases acting on proteins,[190] although work with representatives acting on small molecules has revealed their potential for accepting multiple substrates.[191] There is no strict recognition sequence for hydroxylation by FIH outside of the hydroxylated 'DVNA' motif of HIF-α and, even then, none of these residues is invariant across substrates (not even the Asn).[189] In the context of a 20-mer ankyrin repeat consensus sequence, FIH can catalyse hydroxylation of Asn, Asp, His, Ser, Trp, Ile and Leu residues,[190] although to date only Asn, His and Asp hydroxylations have been identified in cells.

6.3.4 Structural Studies on FIH

Crystallographic studies reveal a homodimeric form of FIH, with each monomer adopting the characteristic DSBH fold and possessing a His-X-Asp...His Fe(II) binding motif.[75–77] Dimerization is mediated by two *C*-terminal helices, and is required for efficient substrate hydroxylation (see also Chapter 2).[75,192] The 2OG binding pocket of FIH is larger than that observed for the PHDs, a difference that can be exploited in the development of selective inhibitors.[193] The mode of 2OG binding also differs from that of the PHDs in that the 5-carboxylate of 2OG is coordinated by Lys-214 from strand IV of the DSBH as opposed to an Arg residue from the VIII strand; this mode of 2OG binding is common to members of the JmjC subfamily of 2OG oxygenases (see Chapter 2) to which FIH belongs.[74]

Substrate-bound structures for FIH have also been reported and reveal that HIF and ARD substrates bind to FIH in a similar overall manner; in the vicinity of the FIH active site the structural organization of the substrate peptide backbones is almost identical for equivalent regions of HIF and Notch peptides.[157] Residues *N*-terminal to the target Asn adopt an elongated conformation and bind in an extended groove in FIH (Site I, Figure 6.5). In the FIH : HIF-1α structure, additional contacts are made with a more *C*-terminal region of the HIF peptide (Site II); these residues form an induced helix and make predominantly hydrophobic interactions on the surface of FIH.[76,157] It is not known whether similar contacts are made between FIH and ankyrin repeats at Site II, as the Notch1 peptide employed in the FIH:Notch structure terminates shortly after the target Asn residue. In contrast to the HIF-α CTADs which are probably disordered,[78,79] ARDs have well-ordered structures that must be unfolded in order to bind productively to FIH[157] (note that ARDs which are not hydroxylated can also bind to FIH).[162,177] As yet there are no structures of an intact ARD protein bound to FIH, hence the molecular

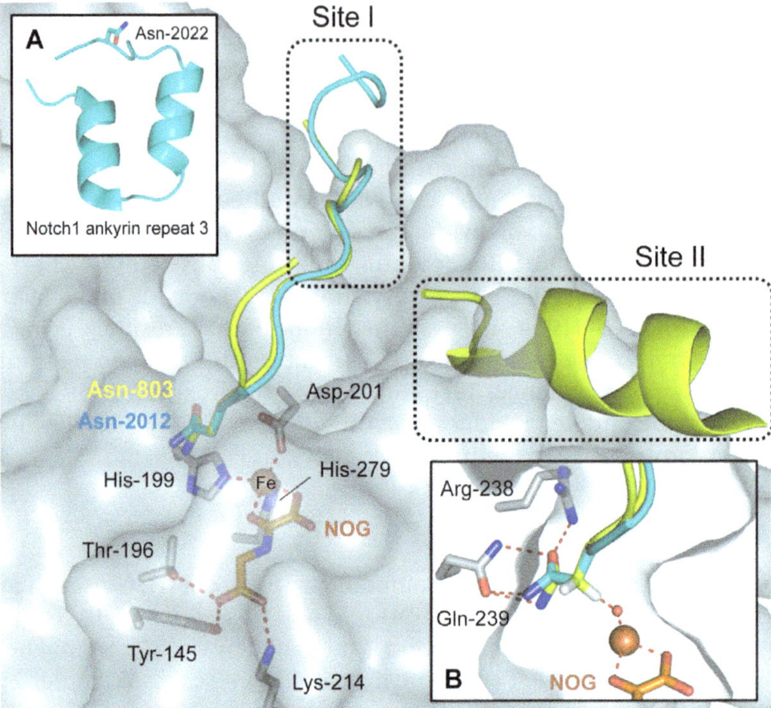

Figure 6.5 Structural insights into substrate binding by FIH. Views from crystal structures of FIH in complex with Fe(II), NOG and peptides from the human HIF-1α CTAD (yellow, PDB ID: 1H2K) or mouse Notch1 ARD (cyan, PDB ID: 3P3P). Superimposition of the peptide backbones reveals similar modes of substrate binding within the active site; the Notch peptide terminates shortly after Asn-2012 and thus does not include residues that would make contacts with FIH at Site II. (A) View from a crystal structure of the human Notch1 ARD showing the stereotypical fold of ankyrin repeat 3 (taken from a crystal structure of the intact ARD, PDB ID: 1YYH). Asn-2022 (equivalent to Asn-2012 in mouse Notch1) is located on a loop between ankyrin repeats 3 and 4, which probably unfold upon binding FIH. (B) FIH residues Arg-238 and Gln-239 form hydrogen bonds that orient the target asparaginyl residues towards the iron (orange ball) within the active site. An iron bound water molecule (small red ball) occupies the proposed position of O_2 binding.

details of ARD binding to FIH are unresolved. Specific interactions outside the immediate vicinity of the active site in part rationalize the different relative efficiencies of ARD hydroxylation (*e.g.* the presence of a His residue in the ARD of Tankyrase-2 contributes to its efficiency as a FIH substrate);[163] although other factors are involved including the stability of the stereotypical ARD fold.[166,177]

Several structures imply that an induced fit involving Trp-296 (and probably other residues) is involved in substrate binding to FIH. In contrast to the PHDs (Figure 6.4), there is no evidence as yet for major loop movements in substrate binding to FIH; the overall structure of FIH is not altered significantly upon substrate binding.[76,157] In general the FIH active site appears more accessible than that of the PHDs, a property that may be related to the promiscuity of FIH substrate binding with respect to ARD substrates, which exhibit considerable sequence variations.

Structures of FIH complexed with either the HIF-1α CTAD or ARD fragments indicate that a precise orientation of the target Asn within the active site of FIH is required for hydroxylation.[76,157] This is achieved by features that include a tight inverse γ-turn in the peptide backbone mediated by a hydrogen bond between the residues flanking the hydroxylated Asn, forming a seven-membered ring[194] that projects the Asn β-carbon toward the Fe(II) (Figure 6.5). Interestingly, the Asp residue of the His-X-Asp...His Fe(II)-binding motif is positioned to hydrogen bond to a backbone amide of the substrate and it can be replaced by a Gly residue while still retaining some activity.[195] FIH residues Arg-238 and, particularly, Gln-239 are also positioned to make important contributions to substrate binding,[196,197] forming hydrogen bonds with the primary amide of the Asn. It should be noted, however, that with ARD substrates (but not, at least efficiently, HIF-α), FIH can catalyse hydroxylation of residues other than Asn, including Asp, His, and (as shown with isolated FIH and ARD peptides) even Ser and Leu residues.[190] Crystallographic analyses have shown how these residues can be accommodated in the FIH active site in a similar, but not identical, manner to the HIF-α CTAD Asn.[190]

6.4 Inhibition of the HIF Hydroxylases

Multiple companies and academic groups have pursued searches for HIF hydroxylase inhibitors. Here we outline general features of these inhibitors and suggest possible areas for future research (although there are many other detailed reviews[198–204]). It is also notable that aspects of the HIF system other than the HIF hydroxylases are potential pharmaceutical targets,[205] from the perspective of either up- or downregulating HIF target genes. Possibilities that have been explored include small molecule approaches for disruption of HIF dimerization,[155,206] DNA binding (HIF : HRE interactions),[207–209] coactivator binding,[210] as well as the blockade of interactions between HIF-α and VHL (for upregulation of HIF target genes).[211,212] However, the major therapeutic focus to date on the HIF system has been on inhibitors of the HIF hydroxylases, predominantly for the treatment of anaemia, although other potential applications include ischaemia-related diseases (*e.g.* coronary heart disease, stroke and diabetic limb ischaemia), gastrointestinal diseases and wound healing. Early studies on inhibition of the PHDs were enabled by pioneering work on inhibition of the procollagen prolyl hydroxylases;[213,214] and although the work on procollagen prolyl hydroxylases has not yet resulted in clinically useful compounds it revealed the potential for small molecule inhibition of the 2OG oxygenases.

Possibly the earliest HIF hydroxylase inhibitor to be examined (although prior to the identification of the HIF hydroxylases) is Co(II), which has been used for the treatment of anaemia,[215–217] and shown to reduce tissue damage in rat models of ischaemia.[218] Co(II) is known to inhibit the PHDs by competition with Fe(II) for binding at the active site, as shown by studies with isolated protein and in cells.[67] However, Co(II) inhibits other 2OG oxygenases,[219] thus whilst it is likely to cause HIF upregulation by PHD inhibition in a physiologically relevant manner, it probably has other effects. The observation that relatively non-specific inhibitors such as $CoCl_2$ (and 2OG competitors, see below) can be used in animals at levels sufficient to induce HIF-α without causing acute toxicity is notable; it also implies that the upregulation of substantial sets of HIF target genes is not intrinsically toxic.[220] At this stage it is unclear whether treatment of specific hypoxia-related diseases *via* PHD inhibition/HIF activation (*e.g.* treating anaemia by upregulation of EPO) is best achieved (from both efficacy, and more importantly, a long-term safety perspective) through the selective upregulation of specific target proteins (*e.g.* EPO) or a presently undefined subset of HIF target genes.

The set of HIF target genes that is expressed is dependent on the context. Thus, the physiological manifestations of HIF hydroxylase inhibition are likely to depend on many factors including, at a cellular level, the sets of HIF and HIF hydroxylase proteins present, the relative contents of iron, 2OG other TCA cycle intermediates, and O_2, the tissues to which inhibitors are targeted, adaptation processes, and many other factors.[12] Despite these variables, however, the current evidence is that small-molecule inhibition of the PHDs is both possible and, at least over the time periods of the clinical trials in progress, not highly toxic. Thus, although the long-term safety of HIF hydroxylase inhibition remains to be determined, the use of PHD inhibitors for applications over short time periods appears to be viable.

Early inhibition studies of the PHDs involved compounds that were initially developed as procollagen prolyl hydroxylase inhibitors.[67] A cell permeable (dimethylated) form of NOG was shown to upregulate HIF-α in both cellular and animal studies, and to promote angiogenesis in the latter.[66,221] Subsequently, other known classes of 2OG oxygenase inhibitors (for the procollagen prolyl hydroxylase and for plant oxygenases, targeted for agrochemical applications, including the gibberellin oxidizing enzymes) have been developed as HIF hydroxylase inhibitors.[111,113,222–224] Most, if not all, of these PHD inhibitors, including the more potent and selective inhibitors identified in subsequent structure–activity relationship and screening studies, are bidentate Fe(II) chelators that compete with 2OG at the PHD active site. Some of

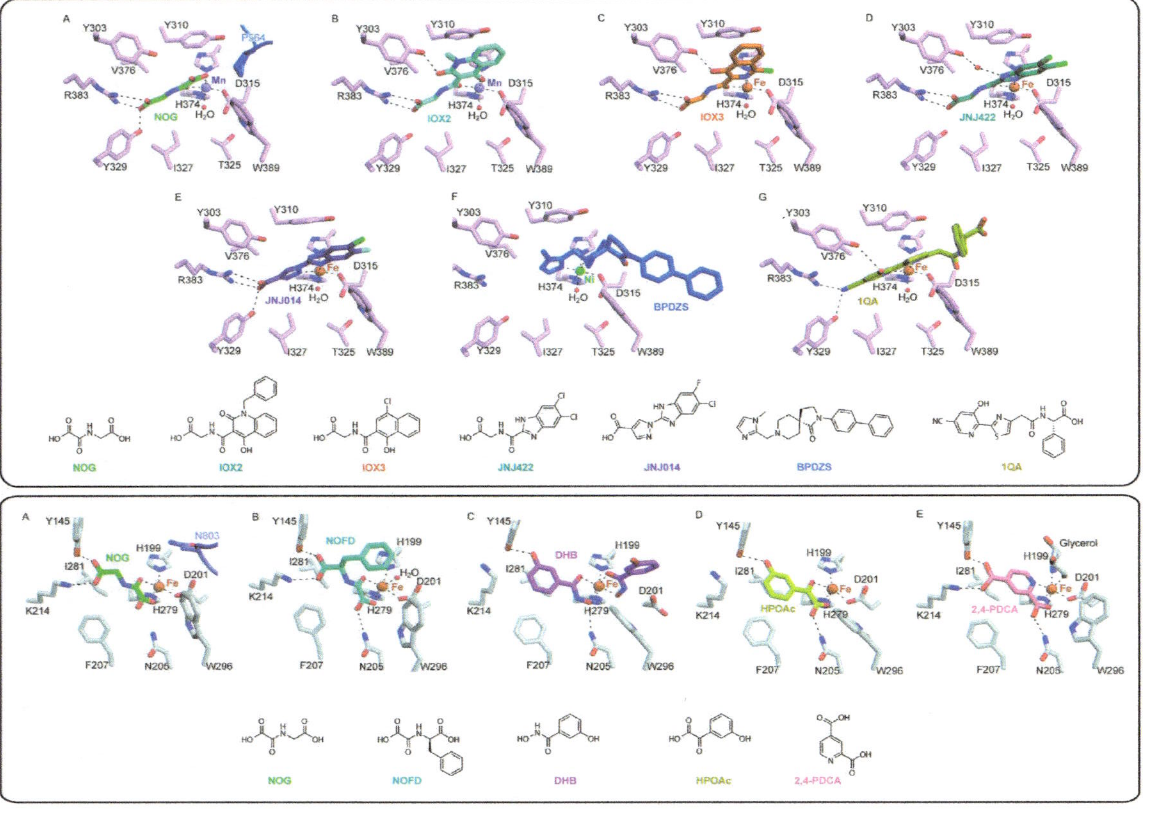

Figure 6.6 Examples of inhibitors of the HIF hydroxylases. Views from crystal structures of PHD2 (upper panel) and FIH (lower panel) in complex with small molecule inhibitors. PHD2 is shown in complex with (A) NOG (PDB ID: 3HQR), (B) IOX2 (PDB ID: 4BQW), (C) IOX3 (PDB ID: 2HBT), (D) JNJ422 (PDB ID: 3OUI), (E) JNJ014 (PDB ID: 3OUH), (F) BPDZS (PDB ID: 4JZR) and (G) 1QA (PDB ID: 4KBZ). FIH is shown in complex with (A) NOG (PDB ID: 1H2K), (B) NOFD (PDB ID: 1YCI), (C) DHB (PDB ID: 2WA4), (D) HPOAc (PDB ID: 2WA3) and (E) 2,4-PDCA (PDB ID: 2W0X).

these inhibitors have also been shown to chelate iron in solution, hence their biological effects may in part be due to iron sequestration. Note that, at least in cellular studies, non-specific iron chelators such as desferrioxamine (used clinically for the treatment of iron overload and related diseases) can cause HIF-α stabilization.[225] The degree to which the inhibitors block HIF-α binding to the PHDs varies; for example, NOG binds in the 2OG-binding pocket and promotes HIF-α binding, whereas (some) more sterically demanding inhibitors block HIF-α binding to the active site. Figure 6.6 shows examples of HIF hydroxylase inhibitors.

To date efforts to devise small molecule based inducers of HIF target genes have principally focused on EPO upregulation *via* PHD inhibition.[226,227] Structural studies on human 2OG oxygenases are also advancing, in part stimulated by pharmaceutical interest in the HIF hydroxylases (see Chapter 2). Crystal structures have been obtained for both PHD2 and FIH in complex with various inhibitors (see Figure 6.6 for representative examples) and have been used in efforts to develop selective inhibitors, as exemplified by early work on the selective inhibition of FIH *versus* the PHDs.

There are few reported examples of small molecules that selectively inhibit PHD isoforms, FIH, HIF1α/HIF2α hydroxylation, or NODD *versus* CODD hydroxylation.[228] The tools to study these inhibitors are now available (including antibodies for NODD/CODD/CTAD hydroxylation status), and are suitable for analyses of isolated components and in cellular/animal contexts. The focus on the PHDs rather than FIH has probably arisen (in part) because without HIF-α stabilization, inhibitors of FIH/CTAD hydroxylation are not likely to be useful. Furthermore, as outlined above, FIH has multiple other substrates, some of physiological relevance (*e.g.* Notch). Nevertheless, the multiple HIF-α isoforms and HIF hydroxylases have different biological roles, some of which are only emerging (*e.g.* the different roles of HIF-1*α* and HIF-2*α* in cancer,[229] and emerging roles for the PHDs in the immune response),[230] thus the development of inhibitors that target different sets of the HIF hydroxylases (including PHD isoform selective inhibitors), and possibly other 2OG oxygenases involved in protein hydroxylation, is of interest for use as functional probes. Such activities will be enabled by advances in functional assignment of human 2OG oxygenases at levels ranging from the biological to the physiological (see Chapters 1, 9, 10, 11).

6.5 PHD and FIH Homologues and Possible Roles in Hypoxia Sensing

6.5.1 Background

The assignment of the roles of the PHDs and FIH in hypoxia signalling raised the question as to whether other 2OG oxygenases play related roles. To date no other 2OG oxygenase has been assigned as playing a major role in hypoxia sensing, although it is very likely that chromatin modifying and other family members help determine the specific sets of genes that are expressed in hypoxically stressed cells. In the following section we summarize links between the HIF hydroxylases and related 2OG oxygenases and we outline recent work on PHD and FIH homologues (ee Chapters 1, 7–11 for more detailed descriptions of the 2OG-dependent demethylases and related enzymes).

6.5.2 PHD Evolution and Ribosomal Protein Prolyl Hydroxylases

The assignment of the PHDs as the most important identified hypoxia sensors in animals coupled to crystallographic analyses led to the identification of related enzymes in human and other organisms. As detailed above, various non-HIF substrates have been suggested for the PHDs, although in most cases the physiological roles of these alternative substrates have yet to be identified (see Section 6.2).[104,105] Nonetheless, the findings regarding alternative PHD substrates support the possibility of oxygenase activities related to those of the PHDs/FIH with signalling roles.

Evidence for the connection of the HIF-PHD-VHL triad in all animals comes from studies on hypoxic sensing in the simplest known animal, *Trichoplax adhaerens*, which contains a HIF system (but probably not a FIH).[32] Moreover, biochemical analysis of the *T. adhaerens* PHD indicates it has similar properties to human PHD2, such as formation of a stable PHD : Fe(II) : 2OG complex. The HIF-PHD-VHL system also occurs in various intermediate and higher animals including *Caenorhabditis elegans*,[67,231–235] *Drosophila melanogaster*,[68,236–238] and Zebrafish (*Danio rerio*);[239,240] thus, the HIF-PHD-VHL system probably extends throughout the vast majority, if not all, of the animal kingdom.[241–243] In contrast, bioinformatic studies imply that while FIH is present in all higher animals, it is only sporadically so in lower animals.[32,244] In simpler eukaryotes, bioinformatic analysis has implied a lack of HIF (or at least transcription factors with recognizable ODDs); however, PHD homologues are apparent in some organisms.[117] The finding that a PHD homologue (PhyA) is present in the social soil amoeba *Dictyostelium dicoideum*[245] is interesting from an evolutionary perspective. Like the PHDs, PhyA catalyses C-4 *trans*-prolyl hydroxylation, but of Skp1, a subunit of the SCF (Skp1-cullin-F box protein complex) class of E3-ubiquitin ligases ($E3^{SCF}$), a modification that has been shown to be important for hypoxia sensing by *Dictyostelium*.[245,246] However, in contrast to what we know for the human PHDs, the Hyp residue in Skp1 undergoes glycosylation, a modification that further biases the prolyl residue to its C-4 *exo* conformation.[247] The overall process is proposed to regulate the assembly and hence activity of the $E3^{SCF}$ ubiquitin ligases, which in turn regulate development, perhaps in a manner regulated by O_2 availability.[248]

Very recently a PHD homologue (PPHD) has been identified in *Pseudomonas* spp. and shown to catalyse the C-4 hydroxylation of a prolyl residue on the Switch1 loop of elongation factor Tu (EF-Tu), one of the most abundant proteins in many bacteria and shown to play a crucial role in ribosome function.[38] Although the biological roles of PPHD-catalysed EF-Tu hydroxylation have yet to be fully defined, deletion of PPHD from *P. aeruginosa* results in upregulation of the iron-chelating siderophore pyocyanin, a known virulence factor in infections. A crystal structure of PPHD from *P. putida* reveals a striking similarity to the catalytic domain of PHD2. The structure of PPHD complexed with EF-Tu reveals that major conformational changes in both PPHD and EF-Tu occur on formation of the enzyme–substrate complex (Figure 6.7). Notably, the β2-β3 loop that folds to enclose the HIF-α substrate of PHD2 plays a similar role in binding the Switch1 loop to PPHD, positioning the EF-Tu and HIF-1α prolyl residues in near identical conformation relative to the active site machinery (Figures 6.4 and 6.7). Notably, relative to its position in isolated EF-Tu, the Switch1 loop of EF-Tu undergoes a major

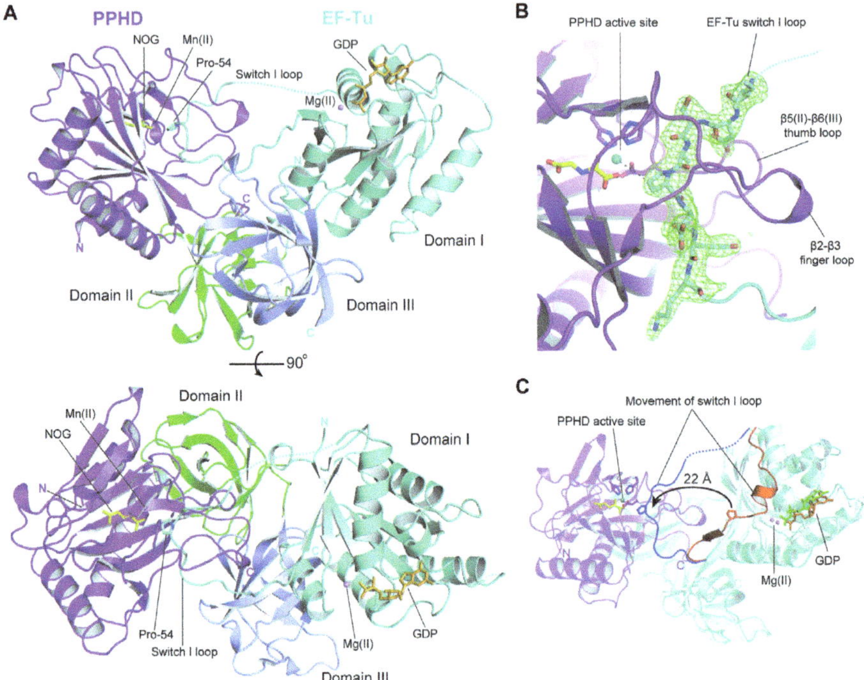

Figure 6.7 Structure of the *Pseudomonas* prolyl hydroxylase, PPHD, in complex with its substrate EF-Tu.[38] (A) Two views from a crystal structure of the PPHD (magenta):EF-Tu (cyan) complex reveal the GDP-bound (inactive) conformation of EF-Tu with its switch I loop present within the active site of PPHD. The EF-Tu Pro-54 is situated for *trans*-4-hydroxylation. (B) Electron density map (green mesh) contoured to 3.0 σ of the EF-Tu switch I loop in the PPHD active site. (C) Superimposition of *P. putida* EF-Tu from the PPHD : EF-Tu complex with its switch I loop (blue) positioned in the PPHD active site and *E. coli* EF-Tu (PDB ID: 1DG1) with its switch I loop (red) positioned adjacent to the nucleotide binding site reveals a ~20 Å movement of the hydroxylated prolyl residue (Hyp-54 *Pseudomonas*/Pro-53 in *E. coli*) based on its position in isolated EF-Tu compared to the PPHD : EF-Tu complex.

conformational change (>20 Å). Because the PPHD:EF-Tu structure is the first of a 2OG-dependent protein hydroxylase in contact with an intact protein, it would seem reasonable to presume that similar major conformational changes occur during protein substrate binding to other members of this enzyme family.

Another prolyl hydroxylase has been found to play a role in translational regulation, but one that catalyses *trans* C-3 rather than C-4 hydroxylations.[35] The OGFOD1 family of 2OG oxygenases is conserved from yeasts and *D. melanogaster* through to humans and catalyses prolyl hydroxylation of the small ribosomal protein 23 (RPS23).[34,35,249] Prior to their assignment as prolyl hydroxylases these enzymes had been shown to play roles in translation in yeast,[250,251] and possibly in the animal hypoxic response.[252] The site of OGFOD1-catalysed hydroxylation (Pro-62 in human RPS23) is the closest point of contact between mRNA and any ribosomal proteins, as predicted by ribosome crystal structures.[253,254] Interestingly, in yeasts and some other lower eukaryotes the OGFOD1 homologues catalyse C-3 and C-4 hydroxylation on the same prolyl residue, but in higher eukaryotes they only appear to catalyse C-4 hydroxylation (Figure 6.2). From a structural perspective OGFOD1 is notable because it has a double DSBH domain architecture with only the *N*-terminal domain having catalytic activity (see Chapter 2). The physiological roles of OGFOD1/OGFOD1 homologues are being elucidated. For example, in *Saccharomyces cerevisiae* the OGFOD1 homologue (TPA1) can play a role in stop codon read-through in a sequence/context dependent manner,[251] although the generality of this role is presently unclear.

In cultured human cells, depletion of OGFOD1 causes cell-type specific induction of stress responses, translational arrest and impaired growth.[249,255] A growth phenotype is also observed in *Drosophila* tissues, associated with both a reduction in cell size and number.[34] Crystallographic analyses on TPA1, the OGFOD1 homologue from *S. cerevisiae*, reveal that the catalytic domain of the OGFOD subfamily contains an active site similar to that of the PHDs,[256] hence achieving selectivity for the PHDs *versus* OGFOD (and other human prolyl hydroxylases) may be difficult.

Finally, the studies on OGFOD1 homologues and related prolyl hydroxylases from *Pseudomonas* and *Dictyostelium* are of interest from an evolutionary perspective because they reveal that the role of prolyl hydroxylation in protein biosynthesis extends from bacteria to humans (although probably not archaea) and encompasses regulation at both transcriptional and translational levels.

6.5.3 Roles for FIH Homologues in Histone Demethylation, RNA Splicing and Ribosomal Hydroxylation

The assignment of FIH as the HIF-α CTAD asparaginyl hydroxylase led to the proposal that many other members of the JmjC protein family are also 2OG oxygenases involved in transcriptional regulation.[74] This prediction was confirmed by work from an independent line of investigation, *i.e.* that many of the JmjC oxygenases are N^{ε}-methyl lysyl histone demethylases.[257,258] The JmjC KDMs, which are ubiquitous and constitute the largest identified KDM family, are described in Chapter 7 (see Kooistra and Helin[259] and Shmakova *et al.*[260] for other reviews). However, it is relevant to note that the histone methylation status and the JmjC KDMs are probably important in regulating HIF gene expression.[46–48] Furthermore, some JmjC KDMs are themselves regulated by HIF (as are some of the HIF hydroxylases).[49–54,90,93–95] The discovery of the 2OG-dependent JmjC KDMs raises the question as to whether they might act in a hypoxic sensing capacity. At present there is no direct evidence for such a role for a KDM, although histone N^{ε}-methylation status can increase in hypoxia in cells.[261–263] In addition, the biochemical properties of at least one JmjC KDM, *i.e.* the relatively slow reaction of the enzyme:Fe(II) : 2OG complex with O_2, suggest that some JmjC KDMs (or indeed nucleic acid modifying 2OG oxygenases such as the ten-eleven translocation (TET) 5-methylcytosine oxygenases[264] or AlkB homologues[265]), have the capacity to act in a hypoxic sensing capacity in animals.[55]

The JmjC family of 2OG oxygenases contains both hydroxylases (including one tRNA hydroxylase)[266] and *N*-methyl demethylases (see Chapters 7, 9, 10). Although the JmjC KDMs have been the subject of intensive investigations, recent work has significantly extended the set of protein hydroxylases related to FIH. The first JmjC protein hydroxylase to be assigned after FIH was JMJD6.[267] This protein was originally assigned as the membrane-associated phosphatidyl serine receptor,[268] but it now seems unlikely that JMJD6 has any direct role in apoptosis.[269] Studies on animals have shown JMJD6 to have an important role in development, including of the cardiovascular system.[269–272] Subsequently, JMJD6 was assigned as a histone H3 demethylase acting on an *N*-methylated argininyl residue in a mechanism analogous to that proposed for the JmjC KDMs.[273] However, JMJD6 was later assigned as a lysyl C-5 hydroxylase,[267] giving products with the 5*S*-stereochemistry rather than the 5*R*-stereochemistry as observed for the procollagen lysyl hydroxylases.[274–276] The assignment of JMJD6 as a hydroxylase appears to be robust and is supported by detailed MS and NMR studies.[267,274]

Hydroxylation substrates/interaction partners of JMJD6 come predominantly from abnormal Arg-Ser (RS)-rich domains of splicing regulatory proteins.[267,277] Given that it interacts with multiple proteins, some of which are hydroxylation substrates, the low selectivity of JMJD6 appears similar to that of FIH. There is evidence that JMJD6 plays a role in mRNA splicing, probably *via* its interaction with splicing regulatory proteins and possibly *via* direct interaction with RNA.[277–281] The tumour suppressor protein p53 is a JMJD6 substrate, although yielding only low levels of activity.[282] JMJD6 can also act as a lysyl hydroxylase of histone H3, although it is unclear if this reaction is of biological relevance.[276] Recent reports of the *N*-methyl arginine demethylase activity of JMJD6 have appeared, including action on RNA helicase A and the oestrogen receptor (ER-α), although these activities have not been definitively shown using isolated proteins.[283,284] There is some evidence that the role of JMJD6 in mRNA splicing in cells may occur in an O_2 regulated manner (as shown by work on vascular endothelial growth factor),[279] although the physiological relevance of these observations in the hypoxic response remains to be determined. Interestingly, the role of JMJD6 in splicing of the ferrochelatase gene is proposed to be regulated by iron availability.[278]

Analyses on JMJD6 are further complicated by changes in its oligomerization status and subcellular localization. Detailed immunolocalization studies have revealed JMJD6 to be distributed throughout the nucleoplasm outside of heterochromatin regions, with sporadic localization in nucleoli.[282,285,286] An alternatively spliced version of JMJD6, lacking its polyserine domain, localizes to the nucleolus.[287] Interestingly, the polyserine domain appears to regulate JMJD6 oligomerization status (at least with isolated JMDJ6), and is also related to its hydroxylation activity and subcellular localization.[288] In a further complication, JMJD6 can auto-hydroxylate several of its own lysyl residues, a property possibly related to its oligomerization status.[289,290] Crystal structures have been reported for JMJD6, but not yet in complex with a substrate.[274,280] These structures, combined with those of other JmjC hydroxylases,[291] support the assignment of its biochemical function as a lysyl C-5 hydroxylase although, as demonstrated by the findings on the promiscuity of FIH (see above), other activities for JMJD6 cannot be ruled out.

Another JmjC oxygenase, JMJD4, has recently been assigned as a lysyl hydroxylase,[292] but one that catalyses hydroxylation at the C-4 rather than the C-5 position of lysyl residues; its activity is thus distinct from both JMJD6 and the collagen lysyl hydroxylases.[274,275] JMJD4 catalyses the hydroxylation of Lys-63 in eukaryotic release factor 1 (eRF1), an important mediator of translation termination in eukaryotes. Lys-63 of eRF1 is located within a highly conserved 'NIKS' motif and its hydroxylation by JMJD4 is required for efficient termination of polypeptide chain elongation. JMJD4 retains substantial activity even under severe hypoxia, suggesting that it is unlikely to act in an O_2-sensing capacity like the HIF hydroxylases. Nevertheless, its identification and functional assignment as a 2OG-dependent hydroxylase highlights the importance of oxygenase-catalysed hydroxylation in regulating fundamental cellular processes such as protein biosynthesis. Furthermore, these results raise the possibility that other as yet uncharacterized JmjC oxygenases, especially those bearing sequence similarity to FIH,[293,294] could potentially function as protein hydroxylases.

Work on an FIH homologue in *Escherichia coli*, YcfD, led to its assignment as a C-3 arginyl hydroxylase acting on the large subunit ribosomal protein L16 (Rpl16).[33] From an enzymology perspective this finding is interesting because other 2OG oxygenases had been shown to act on Arg (VioC)[295,296] and *N*-acylated Arg (clavaminic acid synthase, CAS; see Chapter 1). Amino acid analysis showed the stereochemistry of the YcfD-catalysed hydroxylation to be (2*S*,3*R*), contrasting with the (2*S*,3*S*) stereochemistry of VioC- and CAS-catalysed hydroxylations.[33] A YcfD *E. coli* mutant displays growth impairment under low nutrient conditions. YcfD is homologous to two human JmjC proteins, Myc-induced antigen 53 (MINA53) and NO66, which had been assigned as JmjC KDMs on the basis of cellular studies;[297] however, a subsequent study demonstrated that MINA53 and NO66 are ribosomal oxygenases.[33] Unlike YcfD, MINA53 and NO66 both catalyse C-3 hydroxylation of histidinyl residues, with MINA53 acting on Rpl27a and NO66 acting on Rpl8; in both cases 2*S*,3*S*-products are formed, as for FIH (Figure 6.2).[149] As with the YcfD mutant, reduced expression of MINA53 or NO66 suppresses cell proliferation.[298–300] Structural studies on YcfD, MINA53 and NO66 (see Chapter 3), reveal conservation of their overall folds, including the presence of *C*-terminal hinged helix domains and novel dimerization modes.[291]

6.6 Conclusions

The 2OG-dependent HIF hydroxylases, *i.e.* the PHDs and FIH, have emerged as major regulators of the hypoxic response in animals. Together with work on 2OG oxygenases acting on nucleic acids (see Chapters 8–10) and the JmjC KDMs (see Chapter 6), the collected recent work on PHD and FIH homologues has revealed that 2OG oxygenases probably play major roles at all stages of protein biosynthesis in eukaryotes, including at transcriptional, splicing and translational levels. It is notable that the effects of PHD and FIH catalysed hydroxylations on specific biomacromolecular interactions involved in signalling appear to be very substantial, at least compared to other post-oligomerization modifications catalysed by related oxygenases. Hence, modulation of their activities may be predicted to have substantial physiological effects, consistent with the results of ongoing studies with PHD inhibitors. In some cases there may be 'cross-talk' between the roles of the 2OG oxygenases operating at the different stages in protein biosynthesis, possibly mediated by the acceptance of multiple substrates by some enzymes (*e.g.* as for FIH, JMJD6, and maybe some JmjC KDMs). It now appears likely that 2OG oxygenases other than the PHDs and FIH (*e.g.* some JmjC KDMs) play important roles in the context-dependent regulation of the chronic human hypoxic response, including that mediated by HIF (*i.e.* the major regulatory mechanism for sensing O_2). Whether or not these 2OG oxygenases play physiologically important hypoxic/O_2-sensing roles, however, remains an open question, and one that is of considerable interest from both a basic science and, given the current status of clinical trials with PHD inhibitors, from a medicinal perspective.

Acknowledgements

We thank the Wellcome Trust, the Biotechnology and Biological Sciences Research Council, Cancer Research UK, the Royal Society and the British Heart Foundation for funding our research. RJH acknowledges a William R Miller Junior Research Fellowship, St. Edmund Hall, Oxford. We apologize for incomplete citations due to space constraints; we have focused citations on the roles of 2OG oxygenases in the HIF response. We thank our colleagues and collaborators for discussion and support.

References

1. C. Michiels, *Am. J. Pathol.*, 2004, **164**, 1875-1882.
2. P. Bert, *C. R. Acad. Sci. Paris*, 1882, **94**, 805–807.
3. G. L. Semenza, *Arterioscler., Thromb., Vasc. Biol.*, 2010, **30**, 648–652.
4. J. Cassavaugh and K. M. Lounsbury, *J. Cell. Biochem.*, 2011, **112**, 735–744.
5. G. L. Semenza, *Wiley Interdiscip. Rev.: Syst. Biol. Med.*, 2010, **2**, 336–361.
6. D. R. Mole and P. J. Ratcliffe, *Pediatr. Nephrol.*, 2008, **23**, 681–694.
7. B. F. Palmer and D. J. Clegg, *Mol. Cell. Endocrinol.*, 2014, **397**, 51–58.
8. G. L. Semenza and G. L. Wang, *Mol. Cell. Biol.*, 1992, **12**, 5447–5454.
9. G. L. Wang, B. Jiang, E. A. Rue and G. L. Semenza, *Proc. Natl. Acad. Sci. U.S.A.*, 1995, **92**, 5510–5514.
10. L. E. Huang, Z. Arany, D. M. Livingston and H. F. Bunn, *J. Biol. Chem.*, 1996, **271**, 32253–32259.
11. B. H. Jiang, G. L. Semenza, C. Bauer and H. H. Marti, *Am. J. Physiol.*, 1996, **271**, C1172–C1180.
12. C. J. Schofield and P. J. Ratcliffe, *Biochem. Biophys. Res. Commun.*, 2005, **338**, 617–626.
13. J. D. Webb, M. L. Coleman and C. W. Pugh, *Cell. Mol. Life Sci.*, 2009, **66**, 3539–3554.
14. J. Fandrey and M. Gassmann, *Adv. Exp. Med. Biol.*, 2009, **648**, 197–206.
15. W. G. Kaelin, Jr. and P. J. Ratcliffe, *Mol. Cell*, 2008, **30**, 393–402.
16. G. L. Semenza, B. H. Jiang, S. W. Leung, R. Passantino, J. P. Concordet, P. Maire and A. Giallongo, *J. Biol. Chem.*, 1996, **271**, 32529–32537.
17. S. Bhattacharya, C. L. Michels, M. K. Leung, Z. P. Arany, A. L. Kung and D. M. Livingston, *Genes Dev.*, 1999, **13**, 64–75.
18. Z. Arany, L. E. Huang, R. Eckner, S. Bhattacharya, C. Jiang, M. A. Goldberg, H. F. Bunn and D. M. Livingston, *Proc. Natl. Acad. Sci. U.S.A.*, 1996, **93**, 12969–12973.
19. P. J. Kallio, K. Okamoto, S. O'Brien, P. Carrero, Y. Makino, H. Tanaka and L. Poellinger, *EMBO J.*, 1998, **17**, 6573–6586.
20. W. Gu, X. L. Shi and R. G. Roeder, *Nature*, 1997, **387**, 819–823.
21. S. Reisz-Porszasz, M. R. Probst, B. N. Fukunaga and O. Hankinson, *Mol. Cell. Biol.*, 1994, **14**, 6075–6086.
22. B.-H. Jiang, E. Rue, G. L. Wang, R. Roe and G. L. Semenza, *J. Biol. Chem.*, 1996, **271**, 17771–17778.
23. P. H. Maxwell, C. W. Pugh and P. J. Ratcliffe, *Proc. Natl. Acad. Sci. U.S.A.*, 1993, **90**, 2423–2427.
24. S. Kaluz, M. Kaluzova and E. J. Stanbridge, *Clin. Chim. Acta*, 2008, **395**, 6–13.
25. S. N. Greer, J. L. Metcalf, Y. Wang and M. Ohh, *EMBO J.*, 2012, **31**, 2448–2460.
26. A. Weidemann and R. S. Johnson, *Cell Death Differ.*, 2008, **15**, 621–627.
27. G. L. Semenza, *Cell*, 2012, **148**, 399–408.
28. C. T. Taylor and J. C. McElwain, *Physiology (Bethesda)*, 2010, **25**, 272–279.
29. C. Y. Taabazuing, J. A. Hangasky and M. J. Knapp, *J. Inorg. Biochem.*, 2014, **133**, 63–72.

30. C. Pucciariello and P. Perata, *Plant Signaling Behav.*, 2012, **7**, 813–816.
31. Y. Xu, K. M. Brown, Z. A. Wang, H. van der Wel, C. Teygong, D. Zhang, I. J. Blader and C. M. West, *J. Biol. Chem.*, 2012, **287**, 25098–25110.
32. C. Loenarz, M. L. Coleman, A. Boleininger, B. Schierwater, P. W. Holland, P. J. Ratcliffe and C. J. Schofield, *EMBO Rep.*, 2011, **12**, 63–70.
33. W. Ge, A. Wolf, T. Feng, C. H. Ho, R. Sekirnik, A. Zayer, N. Granatino, M. E. Cockman, C. Loenarz, N. D. Loik, A. P. Hardy, T. D. Claridge, R. B. Hamed, R. Chowdhury, L. Gong, C. V. Robinson, D. C. Trudgian, M. Jiang, M. M. Mackeen, J. S. McCullagh, Y. Gordiyenko, A. Thalhammer, A. Yamamoto, M. Yang, P. Liu-Yi, Z. Zhang, M. Schmidt-Zachmann, B. M. Kessler, P. J. Ratcliffe, G. M. Preston, M. L. Coleman and C. J. Schofield, *Nat. Chem. Biol.*, 2012, **8**, 960–962.
34. M. J. Katz, J. M. Acevedo, C. Loenarz, D. Galagovsky, P. Liu-Yi, M. Perez-Pepe, A. Thalhammer, R. Sekirnik, W. Ge, M. Melani, M. G. Thomas, S. Simonetta, G. L. Boccaccio, C. J. Schofield, M. E. Cockman, P. J. Ratcliffe and P. Wappner, *Proc. Natl. Acad. Sci. U.S.A.*, 2014, **111**, 4025–4030.
35. C. Loenarz, R. Sekirnik, A. Thalhammer, W. Ge, E. Spivakovsky, M. M. Mackeen, M. A. McDonough, M. E. Cockman, B. M. Kessler, P. J. Ratcliffe, A. Wolf and C. J. Schofield, *Proc. Natl. Acad. Sci. U.S.A.*, 2014, **111**, 4019–4024.
36. C. J. Schofield and P. J. Ratcliffe, *Nat. Rev. Mol. Cell Biol.*, 2004, **5**, 343–354.
37. G. L. Semenza, *Annu. Rev. Cell Dev. Biol.*, 1999, **15**, 551–578.
38. J. S. Scotti, I. K. H. Leung, W. Ge, M. A. Bentley, J. Paps, H. B. Kramer, J. Lee, W. Aik, H. Choi, S. M. Paulsen, L. A. H. Bowman, N. D. Loik, S. Horita, C.-h. Ho, N. J. Kershaw, C. M. Tang, T. D. W. Claridge, G. M. Preston, M. A. McDonough and C. J. Schofield, *Proc. Natl. Acad. Sci. U.S.A.*, 2014, **111**, 13331–13336.
39. P. J. Ratcliffe, *J. Clin. Invest.*, 2007, **117**, 862–865.
40. C.-J. Hu, L.-Y. Wang, L. A. Chodosh, B. Keith and M. C. Simon, *Mol. Cell. Biol.*, 2003, **23**, 9361–9374.
41. A. Loboda, A. Jozkowicz and J. Dulak, *Mol. Cells*, 2010, **29**, 435–442.
42. J. W. Kim, I. Tchernyshyov, G. L. Semenza and C. V. Dang, *Cell Metab.*, 2006, **3**, 177–185.
43. P. P. Kapitsinou, Q. Liu, T. L. Unger, J. Rha, O. Davidoff, B. Keith, J. A. Epstein, S. L. Moores, C. L. Erickson-Miller and V. H. Haase, *Blood*, 2010, **116**, 3039–3048.
44. U. Bruning, S. F. Fitzpatrick, T. Frank, M. Birtwistle, C. T. Taylor and A. Cheong, *Cell Mol. Life Sci.*, 2012, **69**, 1319–1329.
45. P. van Uden, N. S. Kenneth and S. Rocha, *Biochem. J.*, 2008, **412**, 477–484.
46. A. J. Krieg, E. B. Rankin, D. Chan, O. Razorenova, S. Fernandez and A. J. Giaccia, *Mol. Cell. Biol.*, 2010, **30**, 344–353.
47. I. Mimura, M. Nangaku, Y. Kanki, S. Tsutsumi, T. Inoue, T. Kohro, S. Yamamoto, T. Fujita, T. Shimamura, J. Suehiro, A. Taguchi, M. Kobayashi, K. Tanimura, T. Inagaki, T. Tanaka, T. Hamakubo, J. Sakai, H. Aburatani, T. Kodama and Y. Wada, *Mol. Cell. Biol.*, 2012, **32**, 3018–3032.
48. W. Luo, R. Chang, J. Zhong, A. Pandey and G. L. Semenza, *Proc. Natl. Acad. Sci. U.S.A.*, 2012, **109**, E3367–E3376.

49. S. Beyer, M. M. Kristensen, K. S. Jensen, J. V. Johansen and P. Staller, *J. Biol. Chem.*, 2008, **283**, 36542–36552.
50. S. Wellmann, M. Bettkober, A. Zelmer, K. Seeger, M. Faigle, H. K. Eltzschig and C. Buhrer, *Biochem. Biophys. Res. Commun.*, 2008, **372**, 892–897.
51. P. J. Pollard, C. Loenarz, D. R. Mole, M. A. McDonough, J. M. Gleadle, C. J. Schofield and P. J. Ratcliffe, *Biochem. J.*, 2008, **416**, 387–394.
52. V. K. Ponnaluri, R. K. Vadlapatla, D. T. Vavilala, D. Pal, A. K. Mitra and M. Mukherji, *Biochem. Biophys. Res. Commun.*, 2011, **415**, 373–377.
53. H. Y. Lee, K. Choi, H. Oh, Y. K. Park and H. Park, *Mol. Cells*, 2014, **37**, 43–50.
54. X. Xia, M. E. Lemieux, W. Li, J. S. Carroll, M. Brown, X. S. Liu and A. L. Kung, *Proc. Natl. Acad. Sci. U.S.A.*, 2009, **106**, 4260–4265.
55. E. M. Sanchez-Fernandez, H. Tarhonskaya, K. Al-Qahtani, R. J. Hopkinson, J. S. McCullagh, C. J. Schofield and E. Flashman, *Biochem. J.*, 2013, **449**, 491–496.
56. P. J. Ratcliffe, *J. Physiol.*, 2013, **591**, 2027–2042.
57. G. L. Semenza, F. Agani, D. Feldser, N. Iyer, L. Kotch, E. Laughner and A. Yu, *Adv. Exp. Med. Biol.*, 2000, **475**, 123–130.
58. S. Salceda and J. Caro, *J. Biol. Chem.*, 1997, **272**, 22642–22647.
59. L. E. Huang, J. Gu, M. Schau and H. F. Bunn, *Proc. Natl. Acad. Sci. U.S.A.*, 1998, **95**, 7987–7992.
60. M. S. Wiesener, H. Turley, W. E. Allen, C. Willam, K. U. Eckardt, K. L. Talks, S. M. Wood, K. C. Gatter, A. L. Harris, C. W. Pugh, P. J. Ratcliffe and P. H. Maxwell, *Blood*, 1998, **92**, 2260–2268.
61. C. M. Robinson and M. Ohh, *FEBS Lett.*, 2014, **588**, 2704–2711.
62. V. Srinivas, L.-P. Zhang, X.-H. Zhu and J. Caro, *Biochem. Biophys. Res. Commun.*, 1999, **260**, 557–561.
63. P. H. Maxwell, M. S. Wiesener, G. W. Chang, S. C. Clifford, E. C. Vaux, M. E. Cockman, C. C. Wykoff, C. W. Pugh, E. R. Maher and P. J. Ratcliffe, *Nature*, 1999, **399**, 271–275.
64. N. Masson, C. Willam, P. H. Maxwell, C. W. Pugh and P. J. Ratcliffe, *EMBO J.*, 2001, **20**, 5197–5206.
65. M. Ivan, K. Kondo, H. Yang, W. Kim, J. Valiando, M. Ohh, A. Salic, J. M. Asara, W. S. Lane and W. G. Kaelin, Jr, *Science*, 2001, **292**, 464–468.
66. P. Jaakkola, D. R. Mole, Y. M. Tian, M. I. Wilson, J. Gielbert, S. J. Gaskell, A. von Kriegsheim, H. F. Hebestreit, M. Mukherji, C. J. Schofield, P. H. Maxwell, C. W. Pugh and P. J. Ratcliffe, *Science*, 2001, **292**, 468–472.
67. A.C. R. Epstein, J. M. Gleadle, L. A. McNeill, K. S. Hewitson, J. O'Rourke, D. R. Mole, M. Mukherji, E. Metzen, M. I. Wilson, A. Dhanda, Y. M. Tian, N. Masson, D. L. Hamilton, P. Jaakkola, R. Barstead, J. Hodgkin, P. H. Maxwell, C. W. Pugh, C. J. Schofield and P. J. Ratcliffe, *Cell*, 2001, **107**, 43–54.
68. R. K. Bruick and S. L. McKnight, *Science*, 2001, **294**, 1337–1340.
69. W. C. Hon, M. I. Wilson, K. Harlos, T. D. W. Claridge, C. J. Schofield, C. W. Pugh, P. H. Maxwell, P. J. Ratcliffe, D. I. Stuart and E. Y. Jones, *Nature*, 2002, **417**, 975–978.

70. C. J. Illingworth, C. Loenarz, C. J. Schofield and C. Domene, *Biochemistry*, 2010, **49**, 6936–6944.
71. J. H. Min, H. F. Yang, M. Ivan, F. Gertler, W. G. Kaelin and N. P. Pavletich, *Science*, 2002, **296**, 1886–1889.
72. D. Lando, D. J. Peet, D. A. Whelan, J. J. Gorman and M. L. Whitelaw, *Science*, 2002, **295**, 858–861.
73. D. Lando, D. J. Peet, J. J. Gorman, D. A. Whelan, M. L. Whitelaw and R. K. Bruick, *Genes Dev.*, 2002, **16**, 1466–1471.
74. K. S. Hewitson, L. A. McNeill, M. V. Riordan, Y. M. Tian, A. N. Bullock, R. W. D. Welford, J. M. Elkins, N. J. Oldham, S. Battacharya, J. Gleadle, P. J. Ratcliffe, C. W. Pugh and C. J. Schofield, *J. Biol. Chem.*, 2002, **277**, 26351–26355.
75. C. E. Dann III, R. K. Bruick and J. Deisenhofer, *Proc. Natl. Acad. Sci. U.S.A.*, 2002, **99**, 15351–15356.
76. J. M. Elkins, K. S. Hewitson, L. A. McNeill, J. F. Seibel, I. Schlemminger, C. W. Pugh, P. J. Ratcliffe and C. J. Schofield, *J. Biol. Chem.*, 2003, **278**, 1802–1806.
77. C. Lee, S. J. Kim, D. G. Jeong, S. M. Lee and S. E. Ryu, *J. Biol. Chem.*, 2003, **278**, 7558–7563.
78. S. A. Dames, M. Martinez-Yamout, R. N. De Guzman, H. J. Dyson and P. E. Wright, *Proc. Natl. Acad. Sci. U.S.A.*, 2002, **99**, 5271–5276.
79. S. J. Freedman, Z.-Y. J. Sun, F. Poy, A. L. Kung, D. M. Livingston, G. Wagner and M. J. Eck, *Proc. Natl. Acad. Sci. U.S.A.*, 2002, **99**, 5367–5372.
80. M. Ivan, K. Kondo, H. F. Yang, W. Kim, J. Valiando, M. Ohh, A. Salic, J. M. Asara, W. S. Lane and W. G. Kaelin, *Science*, 2001, **292**, 464–468.
81. E. Berra, E. Benizri, A. Ginouves, V. Volmat, D. Roux and J. Pouyssegur, *EMBO J.*, 2003, **22**, 4082–4090.
82. F. Oehme, P. Ellinghaus, P. Kolkhof, T. J. Smith, S. Ramakrishnan, J. Hütter, M. Schramm and I. Flamme, *Biochem. Biophys. Res. Commun.*, 2002, **296**, 343–349.
83. P. Koivunen, P. Tiainen, J. Hyvarinen, K. E. Williams, R. Sormunen, S. J. Klaus, K. I. Kivirikko and J. Myllyharju, *J. Biol. Chem.*, 2007, **282**, 30544–30552.
84. G. H. Fong and K. Takeda, *Cell Death Differ.*, 2008, **15**, 635–641.
85. N. A. Smirnova, D. M. Hushpulian, R. E. Speer, I. N. Gaisina, R. R. Ratan and I. G. Gazaryan, *Biochemistry (Mosc)*, 2012, **77**, 1108–1119.
86. J. Myllyharju and P. Koivunen, *Biol. Chem.*, 2013, **394**, 435–448.
87. M. E. Lieb, K. Menzies, M. C. Moschella, R. Ni and M. B. Taubman, *Biochem. Cell Biol.*, 2002, **80**, 421–426.
88. R. J. Appelhoff, Y. M. Tian, R. R. Raval, H. Turley, A. L. Harris, C. W. Pugh, P. J. Ratcliffe and J. M. Gleadle, *J. Biol. Chem.*, 2004, **279**, 38458–38465.
89. C. Willam, P. H. Maxwell, L. Nichols, C. Lygate, Y. M. Tian, W. Bernhardt, M. Wiesener, P. J. Ratcliffe, K. U. Eckardt and C. W. Pugh, *J. Mol. Cell. Cardiol.*, 2006, **41**, 68–77.
90. E. Metzen, U. Berchner-Pfannschmidt, P. Stengel, J. H. Marxsen, I. Stolze, M. Klinger, W. Q. Huang, C. Wotzlaw, T. Hellwig-Burgel, W. Jelkmann, H. Acker and J. Fandrey, *J. Cell Sci.*, 2003, **116**, 1319–1326.

91. K. Takeda, A. Cowan and G. H. Fong, *Circulation*, 2007, **116**, 774–781.
92. J. Myllyharju, *Acta Physiol. (Oxf.)*, 2013, **208**, 148–165.
93. J. H. Marxsen, P. Stengel, K. Doege, P. Heikkinen, T. Jokilehto, T. Wagner, W. Jelkmann, P. Jaakkola and E. Metzen, *Biochem. J.*, 2004, **381**, 761–767.
94. D. P. Stiehl, R. Wirthner, J. Koditz, P. Spielmann, G. Camenisch and R. H. Wenger, *J. Biol. Chem.*, 2006, **281**, 23482–23491.
95. O. Aprelikova, G. V. R. Chandramouli, M. Wood, J. R. Vasselli, J. Riss, J. K. Maranchie, W. M. Linehan and J. C. Barrett, *J. Cell. Biochem.*, 2004, **92**, 491–501.
96. E. P. Cummins, E. Berra, K. M. Comerford, A. Ginouves, K. T. Fitzgerald, F. Seeballuck, C. Godson, J. E. Nielsen, P. Moynagh, J. Pouyssegur and C. T. Taylor, *Proc. Natl. Acad. Sci. U.S.A.*, 2006, **103**, 18154–18159.
97. J. Fu and M. B. Taubman, *J. Biol. Chem.*, 2010, **285**, 8927–8935.
98. J. Xue, X. Li, S. Jiao, Y. Wei, G. Wu and J. Fang, *Gastroenterology*, 2010, **138**, 606–615.
99. A. V. Kuznetsova, J. Meller, P. O. Schnell, J. A. Nash, M. L. Ignacak, Y. Sanchez, J. W. Conaway, R. C. Conaway and M. F. Czyzyk-Krzeska, *Proc. Natl. Acad. Sci. U.S.A.*, 2003, **100**, 2706–2711.
100. L. Xie, K. Xiao, E. J. Whalen, M. T. Forrester, R. S. Freeman, G. Fong, S. P. Gygi, R. J. Lefkowitz and J. S. Stamler, *Sci. Signaling*, 2009, **2**, ra33.
101. W. Luo, H. Hu, R. Chang, J. Zhong, M. Knabel, R. O'Meally, R. N. Cole, A. Pandey and G. L. Semenza, *Cell*, 2011, **145**, 732–744.
102. S. C. Moser, D. Bensaddek, B. Ortmann, J. F. Maure, S. Mudie, J. J. Blow, A. I. Lamond, J. R. Swedlow and S. Rocha, *Dev. Cell*, 2013, **26**, 381–392.
103. W. Luo, B. Lin, Y. Wang, J. Zhong, R. O'Meally, R. N. Cole, A. Pandey, A. Levchenko and G. L. Semenza, *Mol. Biol. Cell*, 2014, **25**, 2788–2796.
104. B. W. Wong, A. Kuchnio, U. Bruning and P. Carmeliet, *Trends Biochem. Sci.*, 2013, **38**, 3–11.
105. K. L. Gorres and R. T. Raines, *Crit. Rev. Biochem. Mol. Biol.*, 2010, **45**, 106–124.
106. E. Flashman, E. A. L. Bagg, R. Chowdhury, J. Mecinovic, C. Loenarz, M. A. McDonough, K. S. Hewitson and C. J. Schofield, *J. Biol. Chem.*, 2008, **283**, 3808–3815.
107. M. A. McDonough, V. Li, E. Flashman, R. Chowdhury, C. Mohr, B. M. Lienard, J. Zondlo, N. J. Oldham, I. J. Clifton, J. Lewis, L. A. McNeill, R. J. Kurzeja, K. S. Hewitson, E. Yang, S. Jordan, R. S. Syed and C. J. Schofield, *Proc. Natl. Acad. Sci. U.S.A.*, 2006, **103**, 9814–9819.
108. L. A. McNeill, E. Flashman, M. R. Buck, K. S. Hewitson, I. J. Clifton, G. Jeschke, T. D. Claridge, D. Ehrismann, N. J. Oldham and C. J. Schofield, *Mol. Biosyst.*, 2005, **1**, 321–324.
109. H. Tarhonskaya, A. Szollossi, I. K. Leung, J. T. Bush, L. Henry, R. Chowdhury, A. Iqbal, T. D. Claridge, C. J. Schofield and E. Flashman, *Biochemistry*, 2014, **53**, 2483–2493.
110. I. K. Leung, E. Flashman, K. K. Yeoh, C. J. Schofield and T. D. Claridge, *J. Med. Chem.*, 2010, **53**, 867–875.

111. J. H. Dao, R. J. Kurzeja, J. M. Morachis, H. Veith, J. Lewis, V. Yu, C. M. Tegley and P. Tagari, *Anal. Biochem.*, 2009, **384**, 213–223.
112. D. Ehrismann, E. Flashman, D. N. Genn, N. Mathioudakis, K. S. Hewitson, P. J. Ratcliffe and C. J. Schofield, *Biochem. J.*, 2007, **401**, 227–234.
113. M. Hirsila, P. Koivunen, V. Gunzler, K. I. Kivirikko and J. Myllyharju, *J. Biol. Chem.*, 2003, **278**, 30772–30780.
114. D. A. Chan, P. D. Sutphin, S. E. Yen and A. J. Giaccia, *Mol. Cell. Biol.*, 2005, **25**, 6415–6426.
115. P. Koivunen, M. Hirsila, K. I. Kivirikko and J. Myllyharju, *J. Biol. Chem.*, 2006, **281**, 28712–28720.
116. Y. M. Tian, K. K. Yeoh, M. K. Lee, T. Eriksson, B. M. Kessler, H. B. Kramer, M. J. Edelmann, C. Willam, C. W. Pugh, C. J. Schofield and P. J. Ratcliffe, *J. Biol. Chem.*, 2011, **286**, 13041–13051.
117. M. S. Taylor, *Gene*, 2001, **275**, 125–132.
118. R. Chowdhury, M. A. McDonough, J. Mecinovic, C. Loenarz, E. Flashman, K. S. Hewitson, C. Domene and C. J. Schofield, *Structure*, 2009, **17**, 981–989.
119. D. Villar, A. Vara-Vega, M. O. Landazuri and L. Del Peso, *Biochem. J.*, 2007, **408**, 231–240.
120. E. Flashman, L. M. Hoffart, R. B. Hamed, J. M. Bollinger, Jr., C. Krebs and C. J. Schofield, *FEBS J.*, 2010, **277**, 4089–4099.
121. L. M. Hoffart, E. W. Barr, R. B. Guyer, J. M. Bollinger, Jr. and C. Krebs, *Proc. Natl. Acad. Sci. U.S.A.*, 2006, **103**, 14738–14743.
122. J. C. Price, E. W. Barr, B. Tirupati, J. M. Bollinger, Jr. and C. Krebs, *Biochemistry*, 2003, **42**, 7497–7508.
123. H. Tarhonskaya, R. Chowdhury, I. K. Leung, N. D. Loik, J. S. McCullagh, T. D. Claridge, C. J. Schofield and E. Flashman, *Biochem. J.*, 2014, **463**, 363–372.
124. Y. Pan, K. D. Mansfield, C. C. Bertozzi, V. Rudenko, D. A. Chan, A. J. Giaccia and M. C. Simon, *Mol. Cell. Biol.*, 2007, **27**, 912–925.
125. E. Metzen, J. Zhou, W. Jelkmann, J. Fandrey and B. Brune, *Mol. Biol. Cell*, 2003, **14**, 3470–3481.
126. D. Gerald, E. Berra, Y. M. Frapart, D. A. Chan, A. J. Giaccia, D. Mansuy, J. Pouyssegur, M. Yaniv and F. Mechta-Grigoriou, *Cell*, 2004, **118**, 781–794.
127. R. Chowdhury, E. Flashman, J. Mecinovic, H. B. Kramer, B. M. Kessler, Y. M. Frapart, J. L. Boucher, I. J. Clifton, M. A. McDonough and C. J. Schofield, *J. Mol. Biol.*, 2011, **410**, 268–279.
128. N. Masson, R. S. Singleton, R. Sekirnik, D. C. Trudgian, L. J. Ambrose, M. X. Miranda, Y. M. Tian, B. M. Kessler, C. J. Schofield and P. J. Ratcliffe, *EMBO Rep.*, 2012, **13**, 251–257.
129. H. J. Knowles, R. R. Raval, A. L. Harris and P. J. Ratcliffe, *Cancer Res.*, 2003, **63**, 1764–1768.
130. E. Flashman, S. L. Davies, K. K. Yeoh and C. J. Schofield, *Biochem. J.*, 2010, **427**, 135–142.
131. K. J. Nytko, N. Maeda, P. Schlafli, P. Spielmann, R. H. Wenger and D. P. Stiehl, *Blood*, 2011, **117**, 5485–5493.
132. L. De Jong, S. P. Albracht and A. Kemp, *Biochim. Biophys. Acta*, 1982, **704**, 326–332.

133. L. De Jong and A. Kemp, *Biochim. Biophys. Acta*, 1984, **787**, 105–111.
134. R. Myllyla, K. Majamaa, V. Gunzler, H. M. Hanauskeabel and K. I. Kivirikko, *J. Biol. Chem.*, 1984, **259**, 5403–5405.
135. N. S. Chandel, D. S. McClintock, C. E. Feliciano, T. M. Wood, J. A. Melendez, A. M. Rodriguez and P. T. Schumacker, *J. Biol. Chem.*, 2000, **275**, 25130–25138.
136. R. D. Guzy and P. T. Schumacker, *Exp. Physiol.*, 2006, **91**, 807–819.
137. P. Gao, H. Zhang, R. Dinavahi, F. Li, Y. Xiang, V. Raman, Z. M. Bhujwalla, D. W. Felsher, L. Cheng, J. Pevsner, L. A. Lee, G. L. Semenza and C. V. Dang, *Cancer Cell*, 2007, **12**, 230–238.
138. P. J. Pollard, J. J. Briere, N. A. Alam, J. Barwell, E. Barclay, N. C. Wortham, T. Hunt, M. Mitchell, S. Olpin, S. J. Moat, I. P. Hargreaves, S. J. Heales, Y. L. Chung, J. R. Griffiths, A. Dalgleish, J. A. McGrath, M. J. Gleeson, S. V. Hodgson, R. Poulsom, P. Rustin and I. P. Tomlinson, *Hum. Mol. Genet.*, 2005, **14**, 2231–2239.
139. L. Dang, D. W. White, S. Gross, B. D. Bennett, M. A. Bittinger, E. M. Driggers, V. R. Fantin, H. G. Jang, S. Jin, M. C. Keenan, K. M. Marks, R. M. Prins, P. S. Ward, K. E. Yen, L. M. Liau, J. D. Rabinowitz, L. C. Cantley, C. B. Thompson, M. G. Vander Heiden and S. M. Su, *Nature*, 2010, **465**, 966.
140. P. S. Ward, J. Patel, D. R. Wise, O. Abdel-Wahab, B. D. Bennett, H. A. Coller, J. R. Cross, V. R. Fantin, C. V. Hedvat, A. E. Perl, J. D. Rabinowitz, M. Carroll, S. M. Su, K. A. Sharp, R. L. Levine and C. B. Thompson, *Cancer Cell*, 2010, **17**, 225–234.
141. K. S. Hewitson, B. M. Lienard, M. A. McDonough, I. J. Clifton, D. Butler, A. S. Soares, N. J. Oldham, L. A. McNeill and C. J. Schofield, *J. Biol. Chem.*, 2007, **282**, 3293–3301.
142. P. Koivunen, M. Hirsila, A. M. Remes, I. E. Hassinen, K. I. Kivirikko and J. Myllyharju, *J. Biol. Chem.*, 2007, **282**, 4524–4532.
143. R. Chowdhury, K. K. Yeoh, Y. M. Tian, L. Hillringhaus, E. A. Bagg, N. R. Rose, I. K. Leung, X. S. Li, E. C. Woon, M. Yang, M. A. McDonough, O. N. King, I. J. Clifton, R. J. Klose, T. D. Claridge, P. J. Ratcliffe, C. J. Schofield and A. Kawamura, *EMBO Rep.*, 2011, **12**, 463–469.
144. W. Xu, H. Yang, Y. Liu, Y. Yang, P. Wang, S. H. Kim, S. Ito, C. Yang, M. T. Xiao, L. X. Liu, W. Q. Jiang, J. Liu, J. Y. Zhang, B. Wang, S. Frye, Y. Zhang, Y. H. Xu, Q. Y. Lei, K. L. Guan, S. M. Zhao and Y. Xiong, *Cancer Cell*, 2011, **19**, 17–30.
145. G. Deng, B. Zhao, Y. Ma, Q. Xu, H. Wang, L. Yang, Q. Zhang, T. B. Guo, W. Zhang, Y. Jiao, X. Cai, J. Zhang, H. Liu, X. Guan, H. Lu, J. Xiang, J. D. Elliott, X. Lin and F. Ren, *Bioorg. Med. Chem.*, 2013, **21**, 6349–6358.
146. R. Chowdhury, J. I. Candela-Lena, M. C. Chan, D. J. Greenald, K. K. Yeoh, Y. M. Tian, M. A. McDonough, A. Tumber, N. R. Rose, A. Conejo-Garcia, M. Demetriades, S. Mathavan, A. Kawamura, M. K. Lee, F. van Eeden, C. W. Pugh, P. J. Ratcliffe and C. J. Schofield, *ACS Chem. Biol.*, 2013, **8**, 1488–1496.
147. P. C. Mahon, K. Hirota and G. L. Semenza, *Genes Dev.*, 2001, **15**, 2675–2686.

148. P. M. Clissold and C. P. Ponting, *Trends Biochem. Sci.*, 2001, **26**, 7–9.
149. L. A. McNeill, K. S. Hewitson, T. D. Claridge, J. r. F. Seibel, L. E. Horsfall and C. J. Schofield, *Biochem. J.*, 2002, **367**, 571–575.
150. J. Stenflo, A. Lundwall and B. Dahlback, *Proc. Natl. Acad. Sci. U.S.A.*, 1987, **84**, 368–372.
151. N. Zhang, Z. Fu, S. Linke, J. Chicher, J. J. Gorman, D. Visk, G. G. Haddad, L. Poellinger, D. J. Peet, F. Powell and R. S. Johnson, *Cell Metab.*, 2010, **11**, 364–378.
152. I. P. Stolze, Y. M. Tian, R. J. Appelhoff, H. Turley, C. C. Wykoff, J. M. Gleadle and P. J. Ratcliffe, *J. Biol. Chem.*, 2004, **279**, 42719–42725.
153. F. Dayan, D. Roux, M. C. Brahimi-Horn, J. Pouyssegur and N. M. Mazure, *Cancer Res.*, 2006, **66**, 3688–3698.
154. M. Lu, J. Zak, S. Chen, L. Sanchez-Pulido, D. T. Severson, J. Endicott, C. P. Ponting, C. J. Schofield and X. Lu, *Cell*, 2014, **157**, 1130–1145.
155. T. H. Scheuermann, Q. Li, H. W. Ma, J. Key, L. Zhang, R. Chen, J. A. Garcia, J. Naidoo, J. Longgood, D. E. Frantz, U. K. Tambar, K. H. Gardner and R. K. Bruick, *Nat. Chem. Biol.*, 2013, **9**, 271–276.
156. M. E. Cockman, D. E. Lancaster, I. P. Stolze, K. S. Hewitson, M. A. McDonough, M. L. Coleman, C. H. Coles, X. Yu, R. T. Hay, S. C. Ley, C. W. Pugh, N. J. Oldham, N. Masson, C. J. Schofield and P. J. Ratcliffe, *Proc. Natl. Acad. Sci. U.S.A.*, 2006, **103**, 14767–14772.
157. M. L. Coleman, M. A. McDonough, K. S. Hewitson, C. Coles, J. Mecinovic, M. Edelmann, K. M. Cook, M. E. Cockman, D. E. Lancaster, B. M. Kessler, N. J. Oldham, P. J. Ratcliffe and C. J. Schofield, *J. Biol. Chem.*, 2007, **282**, 24027–24038.
158. X. Zheng, S. Linke, J. M. Dias, X. Zheng, K. Gradin, T. P. Wallis, B. R. Hamilton, M. Gustafsson, J. L. Ruas, S. Wilkins, R. L. Bilton, K. Brismar, M. L. Whitelaw, T. Pereira, J. J. Gorman, J. Ericson, D. J. Peet, U. Lendahl and L. Poellinger, *Proc. Natl. Acad. Sci. U.S.A.*, 2008, **105**, 3368–3373.
159. J. E. Ferguson, 3rd, Y. Wu, K. Smith, P. Charles, K. Powers, H. Wang and C. Patterson, *Mol. Cell. Biol.*, 2007, **27**, 6407–6419.
160. M. E. Cockman, J. D. Webb, H. B. Kramer, B. M. Kessler and P. J. Ratcliffe, *Mol. Cell. Proteomics*, 2009, **8**, 535–546.
161. J. D. Webb, A. Muranyi, C. W. Pugh, P. J. Ratcliffe and M. L. Coleman, *Biochem. J.*, 2009, **420**, 327–333.
162. S. E. Wilkins, J. Hyvarinen, J. Chicher, J. J. Gorman, D. J. Peet, R. L. Bilton and P. Koivunen, *Int. J. Biochem. Cell Biol.*, 2009, **41**, 1563–1571.
163. M. Yang, R. Chowdhury, W. Ge, R. B. Hamed, M. A. McDonough, T. D. Claridge, B. M. Kessler, M. E. Cockman, P. J. Ratcliffe and C. J. Schofield, *FEBS J.*, 2011, **278**, 1086–1097.
164. M. Yang, W. Ge, R. Chowdhury, T. D. Claridge, H. B. Kramer, B. Schmierer, M. A. McDonough, L. Gong, B. M. Kessler, P. J. Ratcliffe, M. L. Coleman and C. J. Schofield, *J. Biol. Chem.*, 2011, **286**, 7648–7660.
165. R. S. Singleton, D. C. Trudgian, R. Fischer, B. M. Kessler, P. J. Ratcliffe and M. E. Cockman, *J. Biol. Chem.*, 2011, **286**, 33784–33794.

166. S. E. Wilkins, S. Karttunen, R. J. Hampton-Smith, I. Murchland, A. Chapman-Smith and D. J. Peet, *J. Biol. Chem.*, 2012, **287**, 8769–8781.
167. K. Janke, U. Brockmeier, K. Kuhlmann, M. Eisenacher, J. Nolde, H. E. Meyer, H. Mairbaurl and E. Metzen, *J. Cell Sci.*, 2013, **126**, 2629–2640.
168. L. K. Mosavi, T. J. Cammett, D. C. Desrosiers and Z. Y. Peng, *Protein Sci.*, 2004, **13**, 1435–1448.
169. J. Li, A. Mahajan and M. D. Tsai, *Biochemistry*, 2006, **45**, 15168–15178.
170. S. G. Sedgwick and S. J. Smerdon, *Trends Biochem. Sci.*, 1999, **24**, 311–316.
171. I. Espinoza and L. Miele, *Pharmacol. Ther.*, 2013, **139**, 95–110.
172. G. Aquila, M. Pannella, M. B. Morelli, C. Caliceti, C. Fortini, P. Rizzo and R. Ferrari, *Glob. Cardiol. Sci. Pract.*, 2013, **2013**, 364–371.
173. C. Gasparini and M. Feldmann, *Curr. Pharm. Des.*, 2012, **18**, 5735–5745.
174. S. C. Gupta, C. Sundaram, S. Reuter and B. B. Aggarwal, *Biochim. Biophys. Acta*, 2010, **1799**, 775–787.
175. L. Kelly, M. A. McDonough, M. L. Coleman, P. J. Ratcliffe and C. J. Schofield, *Mol. Biosyst.*, 2009, **5**, 52–58.
176. I. L. Devries, R. J. Hampton-Smith, M. M. Mulvihill, V. Alverdi, D. J. Peet and E. A. Komives, *FEBS Lett.*, 2010, **584**, 4725–4730.
177. A. P. Hardy, I. Prokes, L. Kelly, I. D. Campbell and C. J. Schofield, *J. Mol. Biol.*, 2009, **392**, 994–1006.
178. R. A. Berg and D. J. Prockop, *Biochem. Biophys. Res. Commun.*, 1973, **52**, 115–120.
179. L. E. Bretscher, C. L. Jenkins, K. M. Taylor, M. L. DeRider and R. T. Raines, *J. Am. Chem. Soc.*, 2001, **123**, 777–778.
180. B. Schmierer, B. Novak and C. J. Schofield, *BMC Syst. Biol.*, 2010, **4**, 139.
181. H. Tanahashi and T. Tabira, *Biochem. Biophys. Res. Commun.*, 1999, **255**, 663–667.
182. M. Okamoto and T. C. Sudhof, *Eur. J. Cell Biol.*, 1998, **77**, 161–165.
183. T. Sakamoto and M. Seiki, *J. Biol. Chem.*, 2009, **284**, 30350–30359.
184. K. M. Light, J. A. Hangasky, M. J. Knapp and E. I. Solomon, *J. Am. Chem. Soc.*, 2013, **135**, 9665–9674.
185. Y. H. Chen, L. M. Comeaux, S. J. Eyles and M. J. Knapp, *Chem. Commun. (Camb.)*, 2008, 4768–4770.
186. P. Koivunen, M. Hirsilä, V. Günzler, K. I. Kivirikko and J. Myllyharju, *J. Biol. Chem.*, 2004, **279**, 9899–9904.
187. P. Koivunen, M. Hirsila, V. Gunzler, K. I. Kivirikko and J. Myllyharju, *J. Biol. Chem.*, 2004, **279**, 9899–9904.
188. M. B. Pappalardi, D. E. McNulty, J. D. Martin, K. E. Fisher, Y. Jiang, M. C. Burns, H. Zhao, T. Ho, S. Sweitzer, B. Schwartz, R. S. Annan, R. A. Copeland, P. J. Tummino and L. Luo, *Biochem. J.*, 2011, **436**, 363–369.
189. M. E. Cockman, J. D. Webb and P. J. Ratcliffe, *Ann. N. Y. Acad. Sci.*, 2009, **1177**, 9–18.
190. M. Yang, A. P. Hardy, R. Chowdhury, N. D. Loik, J. S. Scotti, J. S. McCullagh, T. D. Claridge, M. A. McDonough, W. Ge and C. J. Schofield, *Angew. Chem., Int. Ed. Engl.*, 2013, **52**, 1700–1704.

191. R. P. Hausinger, *Crit. Rev. Biochem. Mol.*, 2004, **39**, 21–68.
192. D. E. Lancaster, L. A. McNeill, M. A. McDonough, R. T. Aplin, K. S. Hewitson, C. W. Pugh, P. J. Ratcliffe and C. J. Schofield, *Biochem. J.*, 2004, 429–427.
193. M. A. McDonough, L. A. McNeill, M. Tilliet, C. A. Papamicael, Q. Y. Chen, B. Banerji, K. S. Hewitson and C. J. Schofield, *J. Am. Chem. Soc.*, 2005, **127**, 7680–7681.
194. M. Kaewpet, B. Odell, M. A. King, B. Banerji, C. J. Schofield and T. D. Claridge, *Org. Biomol. Chem.*, 2008, **6**, 3476–3485.
195. K. S. Hewitson, S. L. Holmes, D. Ehrismann, A. P. Hardy, R. Chowdhury, C. J. Schofield and M. A. McDonough, *J. Biol. Chem.*, 2008, **283**, 25971–25978.
196. E. Saban, Y. H. Chen, J. A. Hangasky, C. Y. Taabazuing, B. E. Holmes and M. J. Knapp, *Biochemistry*, 2011, **50**, 4733–4740.
197. J. A. Hangasky, G. T. Ivison and M. J. Knapp, *Biochemistry*, 2014, **53**, 5750–5758.
198. S. Nagel, N. P. Talbot, J. Mecinovic, T. G. Smith, A. M. Buchan and C. J. Schofield, *Antioxid. Redox Signaling*, 2010, **12**, 481–501.
199. E. Poon, A. L. Harris and M. Ashcroft, *Expert Rev. Mol. Med.*, 2009, **11**, e26.
200. J. Myllyharju, *Curr. Pharm. Des.*, 2009, **15**, 3878–3885.
201. M. H. Rabinowitz, *J. Med. Chem.*, 2013, **56**, 9369–9402.
202. Y. Hu, J. Liu and H. Huang, *J. Cell. Biochem.*, 2013, **114**, 498–509.
203. E. Muchnik and J. Kaplan, *Expert Opin. Invest. Drugs*, 2011, **20**, 645–656.
204. S. G. Ong and D. J. Hausenloy, *Pharmacol. Ther.*, 2012, **136**, 69–81.
205. I. K. Nordgren and A. Tavassoli, *Chem. Soc. Rev.*, 2011, **40**, 4307–4317.
206. E. Miranda, I. K. Nordgren, A. L. Male, C. E. Lawrence, F. Hoakwie, F. Cuda, W. Court, K. R. Fox, P. A. Townsend, G. K. Packham, S. A. Eccles and A. Tavassoli, *J. Am. Chem. Soc.*, 2013, **135**, 10418–10425.
207. D. T. Jones and A. L. Harris, *Mol. Cancer Ther.*, 2006, **5**, 2193–2202.
208. D. Kong, E. J. Park, A. G. Stephen, M. Calvani, J. H. Cardellina, A. Monks, R. J. Fisher, R. H. Shoemaker and G. Melillo, *Cancer Res.*, 2005, **65**, 9047–9055.
209. N. G. Nickols, C. S. Jacobs, M. E. Farkas and P. B. Dervan, *ACS Chem. Biol.*, 2007, **2**, 561–571.
210. K. M. Cook, S. T. Hilton, J. Mecinovic, W. B. Motherwell, W. D. Figg and C. J. Schofield, *J. Biol. Chem.*, 2009, **284**, 26831–26838.
211. D. L. Buckley, J. L. Gustafson, I. Van Molle, A. G. Roth, H. S. Tae, P. C. Gareiss, W. L. Jorgensen, A. Ciulli and C. M. Crews, *Angew. Chem., Int. Ed. Engl.*, 2012, **51**, 11463–11467.
212. D. L. Buckley, I. Van Molle, P. C. Gareiss, H. S. Tae, J. Michel, D. J. Noblin, W. L. Jorgensen, A. Ciulli and C. M. Crews, *J. Am. Chem. Soc.*, 2012, **134**, 4465–4468.
213. C. J. Cunliffe, T. J. Franklin, N. J. Hales and G. B. Hill, *J. Med. Chem.*, 1992, **35**, 2652–2658.
214. M. Sharir and T. J. Zimmerman, *Curr. Eye Res.*, 1993, **12**, 553–559.

215. E. A. Bowie and P. J. Hurley, *Aust. N. Z. J. Med.*, 1975, **5**, 306–314.
216. H. R. Kasad, G. V. Joglekar, J. H. Balwani and G. S. Mutalik, *Indian Med. J.*, 1964, **58**, 228–230.
217. P. Schleisner, *Acta Med. Scand.*, 1956, **154**, 177–185.
218. M. Matsumoto, Y. Makino, T. Tanaka, H. Tanaka, N. Ishizaka, E. Noiri, T. Fujita and M. Nangaku, *J. Am. Soc. Nephrol.*, 2003, **14**, 1825–1832.
219. R. Sekirnik, N. R. Rose, J. Mecinovic and C. J. Schofield, *Metallomics*, 2010, **2**, 397–399.
220. S. Nagel, M. Papadakis, R. Chen, L. C. Hoyte, K. J. Brooks, D. Gallichan, N. R. Sibson, C. Pugh and A. M. Buchan, *J. Cereb. Blood Flow Metab.*, 2011, **31**, 132–143.
221. M. Milkiewicz, C. W. Pugh and S. Egginton, *J. Physiol.*, 2004, **560**, 21–26.
222. N. C. Warshakoon, S. Wu, A. Boyer, R. Kawamoto, J. Sheville, R. T. Bhatt, S. Renock, K. Xu, M. Pokross, S. Zhou, R. Walter, M. Mekel, A. G. Evdokimov and S. East, *Bioorg. Med. Chem. Lett.*, 2006, **16**, 5616–5620.
223. R. J. Hopkinson, A. Tumber, C. Yapp, R. Chowdhury, W. Aik, K. H. Che, X. S. Li, J. B. L. Kristensen, O. N. F. King, M. C. Chan, K. K. Yeoh, H. Choi, L. J. Walport, C. C. Thinnes, J. T. Bush, C. Lejeune, A. M. Rydzik, N. R. Rose, E. A. Bagg, M. A. McDonough, T. J. Krojer, W. W. Yue, S. S. Ng, L. Olsen, P. E. Brennan, U. Oppermann, S. Muller, R. J. Klose, P. J. Ratcliffe, C. J. Schofield and A. Kawamura, *Chem. Sci.*, 2013, **4**, 3110–3117.
224. O. N. King, X. S. Li, M. Sakurai, A. Kawamura, N. R. Rose, S. S. Ng, A. M. Quinn, G. Rai, B. T. Mott, P. Beswick, R. J. Klose, U. Oppermann, A. Jadhav, T. D. Heightman, D. J. Maloney, C. J. Schofield and A. Simeonov, *PLoS One*, 2010, **5**, e15535.
225. K. J. Woo, T. J. Lee, J. W. Park and T. K. Kwon, *Biochem. Biophys. Res. Commun.*, 2006, **343**, 8–14.
226. M. M. Hsieh, N. S. Linde, A. Wynter, M. Metzger, C. Wong, I. Langsetmo, A. Lin, R. Smith, G. P. Rodgers, R. E. Donahue, S. J. Klaus and J. F. Tisdale, *Blood*, 2007, **110**, 2140–2147.
227. P. Urquilla, A. Fong, S. Oksanen, S. Leigh, E. Turtle, L. Flippin, M. Brenner, E. Muthukrishnan, P. Fourney, A. Lin, D. Yeowell and C. Molineaux, *Blood*, 2004, **104**, 54a.
228. J. K. Murray, C. Balan, A. M. Allgeier, A. Kasparian, V. Viswanadhan, C. Wilde, J. R. Allen, S. C. Yoder, G. Biddlecome, R. W. Hungate and L. P. Miranda, *J. Comb. Chem.*, 2010, **12**, 676–686.
229. R. R. Raval, K. W. Lau, M. G. B. Tran, H. M. Sowter, S. J. Mandriota, J.-L. Li, C. W. Pugh, P. H. Maxwell, A. L. Harris and P. J. Ratcliffe, *Mol. Cell. Biol.*, 2005, **25**, 5675–5686.
230. S. Keely, E. L. Campbell, A. W. Baird, P. M. Hansbro, R. A. Shalwitz, A. Kotsakis, E. N. McNamee, H. K. Eltzschig, D. J. Kominsky and S. P. Colgan, *Mucosal Immunol.*, 2014, **7**, 114–123.
231. H. Jiang, R. Guo and J. A. Powell-Coffman, *Dev. Biol.*, 2001, **235**, 411.
232. C. Shen, D. Nettleton, M. Jiang, S. K. Kim and J. A. Powell-Coffman, *J. Biol. Chem.*, 2005, **280**, 20580–20588.

233. H. Jiang, R. Guo and J. A. Powell-Coffman, *Proc. Natl. Acad. Sci. U.S.A.*, 2001, **98**, 7916–7921.
234. J. Veijola, P. Annunen, P. Koivunen, A. P. Page, T. Pihlajaniemi and K. I. Kivirikko, *Biochem. J.*, 1996, **317**, 721–729.
235. J. Veijola, P. Koivunen, P. Annunen, T. Pihlajaniemi and K. I. Kivirikko, *J. Biol. Chem.*, 1994, **269**, 26746–26753.
236. N. Arquier, P. Vigne, E. Duplan, T. Hsu, P. P. Therond, C. Frelin and G. D'Angelo, *Biochem. J.*, 2006, **393**, 471–480.
237. S. Lavista-Llanos, L. Centanin, M. Irisarri, D. M. Russo, J. M. Gleadle, S. N. Bocca, M. Muzzopappa, P. J. Ratcliffe and P. Wappner, *Mol. Cell. Biol.*, 2002, **22**, 6842–6853.
238. P. Annunen, P. Koivunen and K. I. Kivirikko, *J. Biol. Chem.*, 1999, **274**, 6790–6796.
239. E. van Rooijen, K. Santhakumar, I. Logister, E. Voest, S. Schulte-Merker, R. Giles and F. van Eeden, *Methods Cell Biol.*, 2011, **105**, 163–190.
240. D. A. Rojas, D. A. Perez-Munizaga, L. Centanin, M. Antonelli, P. Wappner, M. L. Allende and A. E. Reyes, *Gene Expression Patterns*, 2007, **7**, 339–345.
241. T. A. Gorr, M. Gassmann and P. Wappner, *J. Insect Physiol.*, 2006, **52**, 349–364.
242. K. T. Rytkonen, T. A. Williams, G. M. Renshaw, C. R. Primmer and M. Nikinmaa, *Mol. Biol. Evol.*, 2011, **28**, 1913–1926.
243. P. H. Maxwell, C. W. Pugh and P. J. Ratcliffe, *Adv. Exp. Med. Biol.*, 2001, **502**, 365–376.
244. R. J. Hampton-Smith and D. J. Peet, *Ann. N. Y. Acad. Sci.*, 2009, **1177**, 19–29.
245. H. van der Wel, A. Ercan and C. M. West, *J. Biol. Chem.*, 2005, **280**, 14645–14655.
246. C. M. West, H. van der Wel and Z. A. Wang, *Development*, 2007, **134**, 3349–3358.
247. P. Teng-umnuay, H. R. Morris, A. Dell, M. Panico, T. Paxton and C. M. West, *J. Biol. Chem.*, 1998, **273**, 18242–18249.
248. Y. Xu, Z. A. Wang, R. S. Green and C. M. West, *BMC Dev. Biol.*, 2012, **12**, 31.
249. R. S. Singleton, P. Liu-Yi, F. Formenti, W. Ge, R. Sekirnik, R. Fischer, J. Adam, P. J. Pollard, A. Wolf, A. Thalhammer, C. Loenarz, E. Flashman, A. Yamamoto, M. L. Coleman, B. M. Kessler, P. Wappner, C. J. Schofield, P. J. Ratcliffe and M. E. Cockman, *Proc. Natl. Acad. Sci. U.S.A.*, 2014, **111**, 4031–4036.
250. B. T. Hughes and P. J. Espenshade, *EMBO J.*, 2008, **27**, 1491–1501.
251. K. M. Keeling, J. Salas-Marco, L. Z. Osherovich and D. M. Bedwell, *Mol. Cell. Biol.*, 2006, **26**, 5237–5248.
252. K. Saito, N. Adachi, H. Koyama and M. Matsushita, *FEBS Lett.*, 2010, **584**, 3340–3347.
253. M. Selmer, C. M. Dunham, F. V. t. Murphy, A. Weixlbaumer, S. Petry, A. C. Kelley, J. R. Weir and V. Ramakrishnan, *Science*, 2006, **313**, 1935–1942.
254. P. Chandramouli, M. Topf, J. F. Menetret, N. Eswar, J. J. Cannone, R. R. Gutell, A. Sali and C. W. Akey, *Structure*, 2008, **16**, 535–548.

255. K. A. Wehner, S. Schutz and P. Sarnow, *Mol. Cell. Biol.*, 2010, **30**, 2006–2016.
256. H. S. Kim, H. L. Kim, K. H. Kim, D. J. Kim, S. J. Lee, J. Y. Yoon, H. J. Yoon, H. Y. Lee, S. B. Park, S. J. Kim, J. Y. Lee and S. W. Suh, *Nucleic Acids Res.*, 2010, **38**, 2099–2110.
257. Y. Tsukada, J. Fang, H. Erdjument-Bromage, M. E. Warren, C. H. Borchers, P. Tempst and Y. Zhang, *Nature*, 2006, **439**, 811–816.
258. S. C. Trewick, P. J. McLaughlin and R. C. Allshire, *EMBO Rep.*, 2005, **6**, 315–320.
259. S. M. Kooistra and K. Helin, *Nat. Rev. Mol. Cell Biol.*, 2012, **13**, 297–311.
260. A. Shmakova, M. Batie, J. Druker and S. Rocha, *Biochem. J.*, 2014, **462**, 385–395.
261. A. B. Johnson, N. Denko and M. C. Barton, *Mutat. Res.*, 2008, **640**, 174–179.
262. X. Zhou, H. Sun, H. Chen, J. Zavadil, T. Kluz, A. Arita and M. Costa, *Cancer Res.*, 2010, **70**, 4214–4221.
263. M. Tausendschon, N. Dehne and B. Brune, *Cytokine*, 2011, **53**, 256–262.
264. J. U. Guo, Y. Su, C. Zhong, G. L. Ming and H. Song, *Cell Cycle*, 2011, **10**, 2662–2668.
265. A. Thalhammer, W. Aik, E. A. L. Bagg and C. J. Schofield, *Drug Discovery Today: Ther. Strategies*, 2012, **9**, e91–e100.
266. A. Noma, R. Ishitani, M. Kato, A. Nagao, O. Nureki and T. Suzuki, *J. Biol. Chem.*, 2010, **285**, 34503–34507.
267. C. J. Webby, A. Wolf, N. Gromak, M. Dreger, H. Kramer, B. Kessler, M. L. Nielsen, C. Schmitz, D. S. Butler, J. R. Yates, 3rd, C. M. Delahunty, P. Hahn, A. Lengeling, M. Mann, N. J. Proudfoot, C. J. Schofield and A. Bottger, *Science*, 2009, **325**, 90–93.
268. V. A. Fadok, D. L. Bratton, D. M. Rose, A. Pearson, R. A. Ezekewitz and P. M. Henson, *Nature*, 2000, **405**, 85–90.
269. J. Bose, A. D. Gruber, L. Helming, S. Schiebe, I. Wegener, M. Hafner, M. Beales, F. Kontgen and A. Lengeling, *J. Biol.*, 2004, **3**, 15.
270. J. R. Hong, G. H. Lin, C. J. Lin, W. P. Wang, C. C. Lee, T. L. Lin and J. L. Wu, *Development*, 2004, **131**, 5417–5427.
271. M. O. Li, M. R. Sarkisian, W. Z. Mehal, P. Rakic and R. A. Flavell, *Science*, 2003, **302**, 1560–1563.
272. Y. Kunisaki, S. Masuko, M. Noda, A. Inayoshi, T. Sanui, M. Harada, T. Sasazuki and Y. Fukui, *Blood*, 2004, **103**, 3362–3364.
273. B. S. Chang, Y. Chen, Y. M. Zhao and R. K. Bruick, *Science*, 2007, **318**, 444–447.
274. M. Mantri, N. D. Loik, R. B. Hamed, T. D. W. Claridge, J. S. O. McCullagh and C. J. Schofield, *ChemBioChem*, 2011, **12**, 531–534.
275. B. Witkop, *Experientia*, 1956, **12**, 372–374.
276. M. Unoki, A. Masuda, N. Dohmae, K. Arita, M. Yoshimatsu, Y. Iwai, Y. Fukui, K. Ueda, R. Hamamoto, M. Shirakawa, H. Sasaki and Y. Nakamura, *J. Biol. Chem.*, 2013, **288**, 6053–6062.
277. A. Heim, C. Grimm, U. Muller, S. Haussler, M. M. Mackeen, J. Merl, S. M. Hauck, B. M. Kessler, C. J. Schofield, A. Wolf and A. Bottger, *Nucleic Acids Res.*, 2014, **42**, 7833–7850.

278. J. Barman-Aksozen, C. Beguin, A. M. Dogar, X. Schneider-Yin and E. I. Minder, *Blood Cells Mol. Dis.*, 2013, **51**, 151–161.
279. J. N. Boeckel, V. Guarani, M. Koyanagi, T. Roexe, A. Lengeling, R. T. Schermuly, P. Gellert, T. Braun, A. Zeiher and S. Dimmeler, *Proc. Natl. Acad. Sci. U.S.A.*, 2011, **108**, 3276–3281.
280. X. Hong, J. Zang, J. White, C. Wang, C. H. Pan, R. Zhao, R. C. Murphy, S. Dai, P. Henson, J. W. Kappler, J. Hagman and G. Zhang, *Proc. Natl. Acad. Sci. U.S.A.*, 2010, **107**, 14568–14572.
281. J. W. Qi, T. T. Qin, L. X. Xu, K. Zhang, G. L. Yang, J. Li, H. Y. Xiao, Z. S. Zhang and L. Y. Li, *Proc. Natl. Acad. Sci. U.S.A.*, 2013, **110**, 13863–13868.
282. F. Wang, L. He, P. Huangyang, J. Liang, W. Si, R. Yan, X. Han, S. Liu, B. Gui, W. Li, D. Miao, C. Jing, Z. Liu, F. Pei, L. Sun and Y. Shang, *PLoS Biol.*, 2014, **12**, e1001819.
283. P. Lawrence, J. S. Conderino and E. Rieder, *Virology*, 2014, **452–453**, 1–11.
284. C. Poulard, J. Rambaud, N. Hussein, L. Corbo and M. Le Romancer, *PLoS One*, 2014, **9**, e87982.
285. P. Hahn, I. Wegener, A. Burrells, J. Bose, A. Wolf, C. Erck, D. Butler, C. J. Schofield, A. Bottger and A. Lengeling, *PLoS One*, 2010, **5**, e13769.
286. P. Cui, B. Qin, N. Liu, G. Pan and D. Pei, *Exp. Cell Res.*, 2004, **293**, 154–163.
287. A. Wolf, M. Mantri, A. Heim, U. Muller, E. Fichter, M. M. Mackeen, L. Schermelleh, G. Dadie, H. Leonhardt, C. Venien-Bryan, B. M. Kessler, C. J. Schofield and A. Bottger, *Biochem. J.*, 2013, **453**, 357–370.
288. G. Han, J. Li, Y. Wang, X. Li, H. Mao, Y. Liu and C. D. Chen, *J. Cell. Biochem.*, 2012, **113**, 1663–1670.
289. M. Mantri, C. J. Webby, N. D. Loik, R. B. Hamed, M. L. Nielsen, M. A. McDonough, J. S. O. McCullagh, A. Bottger, C. J. Schofield and A. Wolf, *MedChemComm*, 2012, **3**, 80–85.
290. M. Mantri, Z. Zhang, M. A. McDonough and C. J. Schofield, *FEBS J.*, 2012, **279**, 1563–1575.
291. R. Chowdhury, R. Sekirnik, N. C. Brissett, T. Krojer, C. H. Ho, S. S. Ng, I. J. Clifton, W. Ge, N. J. Kershaw, G. C. Fox, J. R. Muniz, M. Vollmar, C. Phillips, E. S. Pilka, K. L. Kavanagh, F. von Delft, U. Oppermann, M. A. McDonough, A. J. Doherty and C. J. Schofield, *Nature*, 2014, **510**, 422–426.
292. T. Feng, A. Yamamoto, S. E. Wilkins, E. Sokolova, L. A. Yates, M. Munzel, P. Singh, R. J. Hopkinson, R. Fischer, M. E. Cockman, J. Shelley, D. C. Trudgian, J. Schodel, J. S. McCullagh, W. Ge, B. M. Kessler, R. J. Gilbert, L. Y. Frolova, E. Alkalaeva, P. J. Ratcliffe, C. J. Schofield and M. L. Coleman, *Mol. Cell*, 2014, **53**, 645–654.
293. M. A. McDonough, C. Loenarz, R. Chowdhury, I. J. Clifton and C. J. Schofield, *Curr. Opin. Struct. Biol.*, 2010, **20**, 659–672.
294. R. J. Klose, E. M. Kallin and Y. Zhang, *Nat. Rev. Genet.*, 2006, **7**, 715–727.
295. J. Ju, S. G. Ozanick, B. Shen and M. G. Thomas, *ChemBioChem*, 2004, **5**, 1281–1285.
296. X. Yin and T. M. Zabriskie, *ChemBioChem*, 2004, **5**, 1274–1277.

297. Y. Lu, Q. Chang, Y. Zhang, K. Beezhold, Y. Rojanasakul, H. Zhao, V. Castranova, X. Shi and F. Chen, *Cell Cycle*, 2009, **8**, 2101–2109.
298. C. Suzuki, K. Takahashi, S. Hayama, N. Ishikawa, T. Kato, T. Ito, E. Tsuchiya, Y. Nakamura and Y. Daigo, *Mol. Cancer Ther.*, 2007, **6**, 542–551.
299. K. Teye, M. Tsuneoka, N. Arima, Y. Koda, Y. Nakamura, Y. Ueta, K. Shirouzu and H. Kimura, *Am. J. Pathol.*, 2004, **164**, 205–216.
300. M. Tsuneoka, Y. Koda, M. Soejima, K. Teye and H. Kimura, *J. Biol. Chem.*, 2002, **277**, 35450–35459.

CHAPTER 7

JmjC Lysine Demethylases

XIAODONG CHENG[a] AND RAYMOND C. TRIEVEL*[b]

[a]Department of Biochemistry, Emory University, Atlanta, GA 30322, USA; [b]Department of Biological Chemistry, University of Michigan, Ann Arbor, MI 48109, USA
*E-mail: rtrievel@umich.edu

7.1 Introduction

The Jumonji C (JmjC) proteins are Fe(II)- and 2-oxoglutarate (2OG)-dependent dioxygenases that belong to the cupin superfamily of metalloenzymes. These enzymes possess a conserved JmjC catalytic domain that adopts a canonical β-barrel fold that is conserved throughout the cupin superfamily with the active site Fe(II) typically coordinated by a His–X–Asp/Glu–X_N–His catalytic triad.[1–3] Biochemical studies of numerous JmjC enzymes have demonstrated that they are capable of hydroxylating asparaginyl, histidyl and lysyl side chains in proteins, in addition to catalysing RNA hydroxylation.[4–7] JmjC enzymes possessing lysine demethylase (KDM) activity were first reported in the seminal studies of Zhang and colleagues in 2006,[8] and characterization of this group has rapidly expanded to now include six different human subfamilies that exhibit distinct specificities toward different sites and states of histone lysine methylation (Table 7.1). The reaction mechanism of the JmjC KDMs proceeds through the oxidation of 2OG and Fe(II) by O_2, yielding CO_2, succinate and a highly reactive Fe(IV)-oxo or ferryl intermediate (see Chapter 3).[1] This species oxidizes the methyl group of a methyllysine through a radical-based mechanism, yielding a hydroxymethyl hemiaminal intermediate that decomposes to produce formaldehyde and the demethylated lysine. This mechanism offers chemical

RSC Metallobiology Series No. 3
2-Oxoglutarate-Dependent Oxygenases
Edited by Robert P. Hausinger and Christopher J. Schofield

Published by the Royal Society of Chemistry, www.rsc.org

Table 7.1 JmjC KDMs and their major substrate specificities.

JmjC enzyme	Alternate name(s)	Histone substrate(s)	Reported additional substrates
FBXL11	KDM2A, JHDM1A	H3K36me2, H3K36me1	NF-κB p65 K218, K221
FBXL10	KDM2B, JHDM1B	H3K36me2, H3K36me1	
JMJD1A	KDM3A, JHDM2A	H3K9me2, H3K9me1	
JMJD1B	KDM3B, JHDM2B	H3K9me2, H3K9me1	
JMJD1C	—	—	MDC1 K45
JMJD2A	KDM4A, JHDM3A	H3K9me3 (H3K9me2)[a] H3K36me3 (H3K36me2)[a] H1.4K26me3	
JMJD2B	KDM4B, JHDM3B	H3K9me3 (H3K9me2)[a] H3K36me3 (H3K36me2)[a] H1.4K26me3	
JMJD2C	KDM4C, JHDM3C	H3K9me3 (H3K9me2)[a] H3K36me3 (H3K36me2)[a] H1.4K26me3	Pc2 K191
JMJD2D	KDM4D, JHDM3D	H3K9me3, H3K9me2 H1.4K26me3, H1.4K26me2	
JARID1A	KDM5A, RBP2	H3K4me3, H3K4me2	
JARID1B	KDM5B, PLU1	H3K4me3, H3K4me2	
JARID1C	KDM5C, SMCX	H3K4me3, H3K4me2	
JARID1D	KDM5D, SMCY	H3K4me3, H3K4me2	
UTX	KDM6A	H3K27me3, H3K27me2	
JMJD3	KDM6B	H3K27me3, H3K27me2	
PHF8	—	H3K9me2, H3K9me1, H4K20me1	
PHF2	—	H3K9me2, H3K9me1, H4K20me3	ARID5B K336
KIAA1718	KDM7A, JHDM1D	H3K27me2, H3K27me1	

[a]Parentheses denote substrates that are inefficiently demethylated by certain KDMs.

versatility that enables JmjC KDMs to demethylate mono-, di- and trimethyllysines, in contrast to the LSD1 (KDM1) family of flavin-dependent KDMs that utilize an amine oxidase mechanism, limiting their demethylase activity to mono- and dimethyllysines.[9] This chapter examines the biological functions, mechanism, substrate specificities and structures of the different human families of JmjC KDMs, as well as select homologues from other organisms.

7.2 The FBXL11/FBXL10 Family (also known as the KDM2 Family)

The FBXL11/FBXL10 family represents the first JmjC KDMs to be discovered and comprises two homologues that are also termed JHDM1A and JHDM1B or KDM2A and KDM2B, respectively.[8] Initial characterization of FBXL11 and its related *Saccharomyces cerevisiae* homologue JHD1 by Zhang and coworkers demonstrated that these KDMs preferentially demethylate H3K36me2 (where this nomenclature indicates the histone protein, the target lysine residue, and the degree of methylation of that residue)[10] and, to a lesser extent, H3K36me1 *in vitro*.[8] Overexpression of *FBXL11* in 293T cells decreases the global level of H3K36me2 *in vivo*.[8] Subsequent studies have implicated this KDM in different genomic processes. FBXL11 localizes in nucleoli where it demethylates H3K36me1/me2 to silence ribosomal RNA gene expression under conditions of cellular starvation.[11,12] This KDM has also been shown to be a key regulator of hepatic gluconeogenesis.[13] Short-hairpin RNA (shRNA)-mediated knockdown of FBXL11 stimulated the expression of gluconeogenic genes, including those encoding PEPCK and G6Pase, in hepatic cells and in mice, whereas ectopic overexpression repressed expression of these genes. In addition to histone demethylation, FBXL11 negatively regulates NF-κB signalling by demethylating Lys-218 and Lys-221 in RelA/p65, repressing the expression of NF-κB target genes and diminishing NF-κB-controlled cell growth and proliferation.[14–16] *FBXL11* expression is also regulated *via* NF-κB, indicating that they function in a negative feedback loop.[15] Finally, *FBXL11* is frequently overexpressed in non-small cell lung cancer (NSCLC), and high levels of the corresponding enzyme in NSCLC patients generally correlate with a poor prognosis for recovery.[17,18] Subsequent studies demonstrated that FBXL11 silences the expression of the gene for HDAC3 in NSCLC cell lines, resulting in the upregulation of genes associated with cell cycle progression and proliferation, such as *CDK6*, *NANOS1*, *NEK7* and *RAPH1*, thus promoting tumorigenesis.[17]

FBXL10 is also an H3K36me1/me2-specific demethylase, although its functions are largely distinct from those of FBXL11. FBXL10 plays a pivotal role in recruiting Polycomb Repressive Complex 1 (PRC1) to CpG islands through the enzyme's Cys–X–X–Cys (CXXC) domain (Figure 7.1A) and interacts with the Ring1B E3 ubiquitin ligase subunit to facilitate ubiquitination of H2A Lys-119, forming H2AK119ub.[19–24] This mark is an important modification that facilitates the recruitment of the PRC2 H3K27 methyltransferase complex to chromatin, establishing epigenetically silent Polycomb domains. FBXL10 has also been shown to repress the expression of the gene encoding CDK inhibitor p15 (Ink4B) through mechanisms involving H3K36 demethylation and PRC1 recruitment, thus promoting cell proliferation.[25–27] Correlatively, ectopic expression of *FBXL10* circumvents cellular senescence, whereas knockdown of this KDM results in replicative senescence through p53- and retinoblastoma protein (Rb)-dependent pathways.[25,28] In agreement with these findings, FBXL10 appears to function as a key regulator of stem cell pluripotency in multiple contexts by promoting stem cell renewal and regulating gene expression at early stages of stem cell reprogramming.[19,20,22,23,29–31] Consistent with its roles in promoting cell proliferation, FBXL10 is implicated in the onset or progression of certain haematological malignancies, including acute myeloid leukemia and T-cell lymphomas,[26,28,32] highlighting it as a potential target for chemotherapeutic drug design.

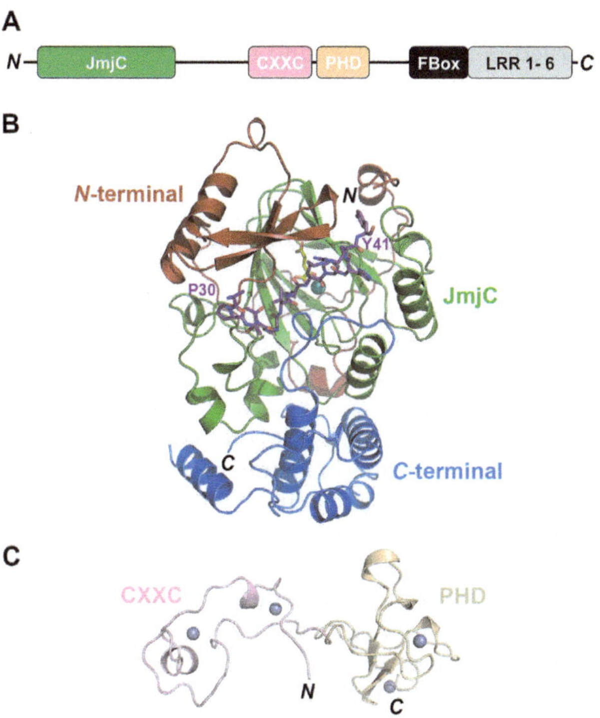

Figure 7.1 Structures of FBXL11 and FBXL10 KDMs. (A) Domain architecture of the FBXL10 KDM family. (B) Crystal structure of the catalytic domain of human FBXL11 bound to an H3K36me2 peptide and 2OG (PDB ID: 4QX7; residues 36–364 and 450–517) with the *N*-terminal region (red) and *C*-terminal region (blue) flanking the JmjC domains (green). The H3K36me2 peptide (residues 29–41) and 2OG are depicted in stick representation with purple and yellow carbon atoms, respectively, and the Ni(II) bound in the Fe(II) coordinate site is shown as a cyan sphere. The terminal residues of the H3K36me2 peptide are labelled for clarity. (C) The tandem CXXC (pink) and PHD (beige) domains of human FBXL10 (PDB ID: 4O64; residues 607–723) are illustrated in ribbon representation. Zn(II) ions are denoted as grey spheres.

The FBXL11 and FBXL10 proteins comprise multiple domains that facilitate their biological functions. The *N*-terminal region of these KDMs harbours the JmjC-containing catalytic domain, whereas the *C*-terminal region includes several domains known to mediate protein:protein or protein:nucleic acid interactions (Figure 7.1A). Crystal structures of the catalytic domain of human FBXL11 in complex with histone H3 peptides bearing different methylated states of H3K36 have offered key insights into its substrate specificity.[33] The FBXL11 catalytic domain consists of a central JmjC domain preceded by an *N*-terminal flanking motif of mixed α-helical and β-sheet structure that is followed by a *C*-terminal domain composed of multiple α-helices (Figure 7.1B). The JmjC domain adopts a canonical β-barrel fold with the 2OG substrate and an inhibitory Ni(II) ion coordinated in the active site within the interior of the barrel. The structure of FBXL11 bound to an H3K36me2 peptide substrate reveals that the peptide adopts a bent U-shaped conformation with sharp kinks in its backbone at Gly-34 and Pro-38. There are several key interactions that are important in conferring H3K36 specificity. Gly-33 and Gly-34 in the H3K36me2 peptide bind in a narrow groove within the histone binding cleft of FBXL11 that cannot accommodate residues with side chains, whereas the pocket that binds Pro-38 stabilizes the sharp turn in the histone H3 peptide backbone. In addition, the side chain of Tyr-41 in the H3K36me2 peptide binds in a pocket on the enzyme's surface through van der Waals interactions and hydrogen bonding. Substitutions of these residues in the H3K36me2 peptide impairs K36me2 demethylation by FBXL11, corroborating the observed interactions in the FBXL11-H3K36me2 complex. Notably, Gly-33, Gly-34, Pro-38 and Tyr-41 in the H3K36 site are not conserved in the sequences flanking the other lysine methylation sites in histones H3 and H4, providing a molecular explanation for the H3K36 specificity of FBXL11 and FBXL10.

In addition to its catalytic domain, the crystal structure of the tandem CXXC Zn motif and PHD domain of human FBXL10 has been determined (Figure 7.1C). The CXXC Zn motif mediates binding of these KDMs to unmodified CpG islands in DNA, promoting rDNA gene silencing by FXBL11 and PRC1 complex recruitment by FBXL10.[12,34–37] Conversely, many PHD domains have been shown to function as chromatin binding modules that can recognize unmodified and methylated forms of H3K4 in addition to other histone modifications.[38] The PHD domain of FBXL10 binds to H3K4me3 and H3K36me2,[39] whereas the PHD domain of FBXL11 does not appear to recognize unmodified or methylated histone H3.[36] Finally, the *C*-terminus of these KDMs is composed of an FBox motif followed by six leucine-rich repeats (LRRs), which are generally categorized as protein:protein interaction modules.[40] In particular, FBox domains frequently mediate interactions with ubiquitin conjugating enzymes,[40] consistent with the association of FBXL10 with the PRC1 subunits that catalyse H2AK119 ubiquitination. Future studies are needed to elucidate the functions of the LRR and FBox domains in mediating recruitment of FBXL11 and FBXL10 to chromatin or other chromatin-modifying complexes.

7.3 The JMJD1 Family (also known as the KDM3 Family)

Shortly after the discovery of the FBXL11/FBXL10 family, Zhang and coworkers reported that JMJD1A and JMJD1B (also known as KDM3A/B and JHDM2A/B) are H3K9me1/me2-specific KDMs.[41] In their initial studies, they demonstrated that JMJD1A physically interacts with the androgen receptor (AR) to upregulate AR target gene expression through the demethylation of H3K9me1 and H3K9me2, modifications that are generally associated with transcriptional repression. Subsequent studies have revealed that this KDM has pivotal roles in the epigenetic regulation of metabolism, meiosis and sex determination.

JMJD1A activates the expression of at least two spermatogenesis-associated genes, encoding transition nuclear protein 1 (Tnp1) and protamine 1 (Prm1) through H3K9me2 demethylation, and is essential to spermatogenesis in a mouse model for male fertility.[42,43] More recent studies have revealed that JMJD1A is a client protein of the chaperone Hsp90 and is localized to the cytoplasm where it contributes to cytoskeletal reorganization during sperm maturation.[44] In addition to its functions in spermatogenesis, JMJD1A plays an important role in sex determination in mice and presumably other mammals by activating the expression of the Y chromosome sex-determining gene *Sry* through demethylation of H3K9me2.[45] Beyond its functions in reproductive biology, several reports have implicated JMJD1A in regulating energy homeostasis by controlling the expression of a host of metabolic genes involved in fat storage and utilization, including *Nr2f2* and *Apoc1*, as well as *Glut4* that regulates glucose transport.[46,47] In agreement with these observations, *Jmjd1a* knockout mice exhibit an obese phenotype as adults with hyperlipidemia and fasting-induced hypothermia that correlates with defects in fatty acid catabolism.[43,46,47]

JMJD1B (also referred to as KDM3B and JHDM2B) also displays H3K9me1/me2 demethylase activity, although its biological functions are not as well characterized as those of JMJD1A. JMJD1B functions as a putative tumour suppressor in colorectal cancer, and its overexpression in breast cancer correlates with an improved prognosis for recovery.[48,49] In contrast, other studies have shown that JMJD1B is frequently upregulated in prostate cancer cells and is implicated in acute lymphoblastic leukaemia by inducing expression of the pro-leukemogenic factor Imo2 through H3K9me1/me2 demethylation.[50,51] In contrast to the defined KDM activities of JMJD1A and JMJD1B, the enzymatic activity of the related homologue JMJD1C remains somewhat controversial. Initial studies of JMJD1C in mouse steroid biosynthesis suggested that it functions as an H3K9me1/me2-specific demethylase and upregulates gene expression.[52] However, subsequent biochemical and genetic studies comparing JMJD1C with JMJD1A and JMJD1B indicated that JMJD1C lacks demethylase activity toward H3K9me1/me2 as well as the other major sites of histone lysine methylation.[53] Consistent with these findings, Watanabe *et al.* have reported that JMJD1C does not exhibit activity toward histone substrates, but instead demethylates Lys-45 in MDC1 during the repair of DNA double-strand breaks (DSBs).[54,55] This activity facilitates interactions with RNF8-RNF168 ubiquitin ligase that result in the ubiquitination of MDC1 and recruitment of the BRCA1-RAP80 complex, illustrating the role of JMJD1C in DSB repair.

The domain structure of JMJD1 KDMs comprises a CH2C4 type Zn finger motif followed by a *C*-terminal JmjC domain (Figure 7.2A). A crystal structure of human JMJD1B bound to the 2OG analogue inhibitor *N*-oxalylglycine (NOG) has been determined, illustrating an unusual architecture for its catalytic domain (Figure 7.2B). The *N*-terminal motif preceding the JmjC domain comprises several α-helices and two three-stranded anti-parallel β-sheets that form β-extension motifs that buttress each side of the central JmjC β-barrel. One of the three-stranded β-sheets from the *N*-terminal region is positioned near the entrance to the active site, implicating it in protein substrate

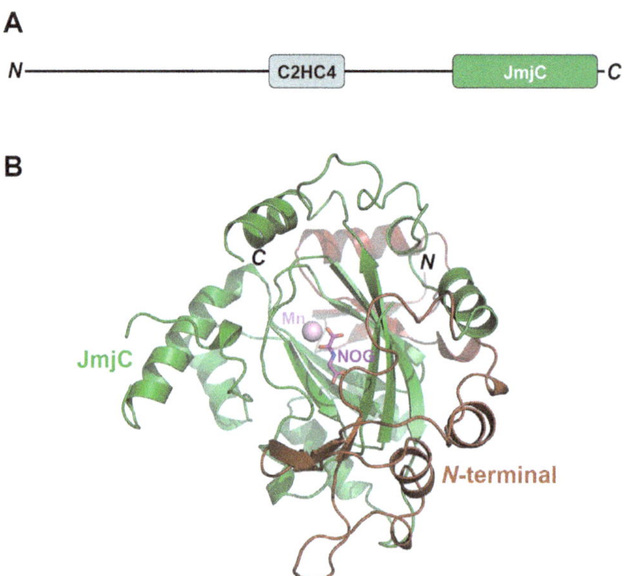

Figure 7.2 Crystal structure of the catalytic domain of JMJD1B. (A) Domain structure of the JMJD1 KDM family. (B) Structure of the catalytic domain of human JMJD1B (PDB ID: 4C8D). The catalytic domain (residues 1380–1720) comprises an *N*-terminal structural motif of mixed α/β topology followed by the JmjC domain that are denoted in red and green, respectively. Bound within the active site are Mn(II) (pink sphere) and the 2OG analogue inhibitor *N*-oxalylglycine (NOG) that is depicted in stick representation with magenta carbon atoms.

recognition. Biochemical studies of JMJD1A by Aburatani and colleagues have revealed that the enzyme dimerizes through interactions between its CH2C4 type Zn finger motif and JmjC domain.[56] This dimerization is essential to its demethylase activity and its ability to processively demethylate H3K9me2 to the unmodified H3K9 state through a hand-off or partial dissociation–reassociation mechanism between the two active sites in the homodimer. These findings substantiate earlier studies demonstrating that truncations or deletions of the C2HC4 Zn finger motif abolish H3K9me1/me2 demethylation.[41] Further structural studies of the JMJD1 KDM family are necessary to understand the mechanisms by which its quaternary structure promotes substrate recognition and facilitates processive demethylation of H3K9me2.

7.4 The JMJD2 Family (also known as the KDM4 Family)

The JMJD2 KDMs were the first family of trimethyllysine-specific demethylases to be identified and are conserved from yeast through mammals.[57–63] Lower organisms such as *S. cerevisiae* and *Caenorhabditis elegans* possess a single JMJD2 enzyme, whereas the genomes of placental mammals encode four homologues termed JMJD2A, JMJD2B, JMJD2C and JMJD2D.[64] In contrast to other JmjC KDM families, the JMJD2 enzymes display differences in their histone methylation site and state specificities. JMJD2A, JMJD2B and JMJD2C exhibit dual selectivity for demethylating trimethylated, and to a lesser extent, dimethylated forms of H3K9 and H3K36, two sites that share no discernable sequence homology. In contrast, JMJD2D is specific for H3K9 and can efficiently demethylate both di- and trimethyllysines.[59] In addition, the JMJD2 KDMs have been reported to catalyse the *in vitro* and *in vivo* demethylation of trimethyl-Lys-26 in the linker histone H1.4 (H1.4K26me3), a methylation site that shares sequence homology with the residues flanking H3K9.[65]

Zhang and colleagues, as well as Shi and coworkers, identified JMJD2A as a trimethyllysine-specific demethylase that selectively demethylates H3K9me3 and H3K36me3 in heterochromatin and euchromatin.[59,60] Since its discovery, JMJD2A has been implicated in numerous biological processes, including cell cycle control, gene regulation, and cellular differentiation and development. Whetstine and coworkers have demonstrated that JMJD2A participates in driving the replication of chromosomal satellite repeats during S phase of the cell cycle.[66] Specifically, demethylation of H3K9me3 by this KDM facilitates displacement of the H3K9me3-binding heterochromatin protein 1γ (HP1γ) from chromatin, resulting in chromatin decondensation that promotes DNA replication within the satellite repeats. Further, overexpression of *JMJD2A* accelerates S phase progression and modulates replication timing, whereas overexpression of the gene encoding HP1γ exhibits the opposite effect. Subsequent studies have revealed that amplification and overexpression of *JMJD2A* in ovarian cancer leads to copy gain of the 1q12h chromosomal region and is statistically associated with a poor prognosis for survival.[67] As with other proteins that regulate cycle progression, JMJD2A degradation is tightly controlled and is subject to ubiquitination by the SKP1-Cul1-FbxL4 ubiquitin ligase, leading to ubiquitin-mediated proteolysis by the proteasome.[68] In addition, ubiquitination of JMJD2A by RNF8 and RNF168 and subsequent proteasomal degradation of this KDM have been shown to be necessary for the binding of p53 binding protein 1 (53BP1) to H4K20me in chromatin, thus recruiting 53BP1 to sites of DNA damage.[69] Further, inhibition of RNF8-dependent ubiquitination of JMJD2A by the peroxisome proliferator-activated receptor γ (PPARγ) and G protein suppressor 2 stimulates the expression of a subset of PPARγ target genes in adipocytes, including hormone-sensitive lipase and adipose triglyceride lipase, illustrating the importance of ubiquitination in regulating the functions of this KDM.[70] In addition to stimulating PPARγ target gene transcription, JMJD2A has also been reported to physically associate with AR and it acts as coactivator of AR-responsive genes.[71] JMJD2A has also been implicated in cardiac gene expression. Mouse knockout and transgenic overexpression studies of *Jmjd2A* have revealed that the KDM upregulates the expression of four-and-a-half LIM domains 1 (FHL1), a mechanotransducing factor in cardiac muscle, through demethylation of H3K9me3, linking it to cardiac hypertrophy.[72] Consistent with these findings, patients suffering from hypertrophic cardiomyopathy display enhanced expression of *JMJD2A*, implicating it as a causative factor in heart disease. Additionally, JMJD2A functions in gene regulatory programmes that control differentiation and development, such as in the neural cell and muscle differentiation and neural crest development.[73–75] Finally, *JMJD2A* overexpression is linked to the incidence of certain cancers, including bladder, breast, colon, endometrial, gastric, lung and ovarian cancer as well as squamous cell carcinoma,[67,76–87] rendering it a putative target for anti-cancer drug design.

The demethylase activity of JMJD2B was first characterized by Jenuwein and colleagues, who demonstrated that it functions to reduce the levels of H3K9me3 in pericentric heterochromatin.[58] Subsequent studies have shown that this KDM participates in multiple genomic processes and that its expression is governed by several transcription factors. Under hypoxic conditions, the hypoxia-inducible factor (HIF)-1α stimulates the expression of JMJD2B, which in turn co-activates a subset of hypoxia-responsive target genes through H3K9me3 demethylation within their promoters, inducing cell proliferation.[88–90] The tumour suppressor p53 also induces expression of JMJD2B, promoting heterochromatic H3K9me3 demethylation during DNA repair.[91–93] Further, studies by Chen *et al.* have demonstrated that JMJD2B activates ataxia telangiectasia mutated (ATM) and ATR (ATM and Rad3-related) signalling pathways in DNA repair and upregulates the expression of genes involved in DNA damage response.[94] In addition to HIF-1α and p53-dependent pathways, JMJD2B has an integral role in oestrogen receptor (ER)-mediated signalling pathways. ERα and JMJD2B exhibit a yin–yang type of relationship wherein the receptor and the KDM drive each other's expression, stimulating breast cancer cell proliferation.[95–98] Moreover, JMJD2B associates with ERα and co-activates the expression of ERα-target genes through H3K9me3 demethylation. JMJD2B also physically associates with the mixed lineage leukaemia 2 (MLL2) H3K4 methyltransferase complex that interacts with ERα.[98] These interactions result in the recruitment of H3K4 methyltransferase and H3K9me3 demethylase activities to ERα-inducible genes, stimulating oestrogen-responsive breast cancer tumorigenesis. In addition to its interplay with ERα, JMJD2B enhances the stability of the AR by blocking its ubiquitin-mediated proteolysis and functions as a coactivator in stimulating the AR target gene expression.[99,100] This KDM also associates with β-catenin and upregulates the expression of β-catenin-inducible genes, promoting cellular proliferation and epithelial–mesenchymal transition in cancer.[101,102] Given its central roles in steroid receptor, β-catenin and HIF signalling pathways, JMJD2B has been implicated in initiation and progression of numerous cancers, including breast, prostate, colorectal and gastric cancer, and has thus been proposed as a target for new chemotherapeutics.[76,89,94,95,97–105]

Similar to JMJD2B, JMJD2C is implicated in steroid receptor-mediated gene activation as well as other genomic processes. Prior to its discovery as a KDM, the gene encoding JMJD2C was initially identified as being amplified in squamous cell carcinoma, particularly in oesophageal cancer.[106,107] Cloos *et al.* were amongst the first to report that JMJD2C possesses H3K9me2/me3 demethylase activity and that shRNA-mediated knockdown of the enzyme diminished cell proliferation, consistent with its pro-proliferative role in tumorigenesis.[57] JMJD2C functions as a co-activator of the expression of HIF-1α responsive genes and is essential to invasive breast cancer growth, consistent with other studies linking this KDM to breast carcinogenesis.[108–111] In addition to breast cancer, JMJD2C physically associates with AR and activates the transcription of AR-responsive genes through H3K9me3 demethylation, promoting prostate cancer cell proliferation and tumorigenesis.[76,112–114] Given that JMJD1A also functions in AR coactivation,[41] JMJD1A and JMJD2C may function in concert to catalyse demethylation of H3K9me3 to an unmodified state. In addition, these KDMs coordinately regulate embryonic development as well as embryonic stem cell (ESC) self-renewal by upregulating the expression of genes involved in ESC maintenance, including *Nanog*, *Tcl1*, *Tcfcp2l1* and *Zfp57*.[115,116] JMJD2C functions in conjunction with JMJD2B to promote ESC self-renewal through PRC2- and Nanog-dependent pathways, respectively.[117] In contrast to its roles in ESC renewal, JMJD2C and JMJD2A are essential to neuronal differentiation by upregulating expression of the brain-derived neurotrophic factor BDNF and repressing the glial fibrillary acidic protein GFAP.[73] Further complicating these findings, Pedersen *et al.* have demonstrated that *Jmjd2c* knockout mice display no overt defects in ESC self-renewal or embryonic development, consistent with mouse knockouts of *Jmjd2b* and *Jmjd2d*.[97,118,119] One potential explanation for these findings is that the JMJD2 KDMs may exhibit a certain degree of functional redundancy, wherein the knockout of one KDM is compensated through the activities of one or more homologues.[118] Finally, beyond its roles in histone demethylation, JMJD2C regulates the relocation of developmental genes between Polycomb bodies and interchromatin granules by demethylating Lys-191 in Polycomb 2 protein (Pc2), implicating this KDM in functions beside chromatin modification.[120]

Similar to other JMJD2 homologues, JMJD2D is implicated in a wide array of nuclear processes. Initial studies of JMJD2D reported that it functions as an AR coactivator through H3K9me3 demethylation,[71] analogous to other JMJD2 homologues.[71,100,113] AR mediates the recruitment of JMJD2D to target genes through physical interactions between the receptor's ligand binding domain and the *C*-terminal region following the KDM's catalytic domain. Characterization of *Jmjd2d* knockout mice has shown that the enzyme also participates in spermatogenesis by promoting H3K9me3 demethylation in developing spermatids but is dispensable for fertility.[119] Poly-ADP ribosylation of JMJD2D regulates gene expression and DNA damage response. Le May *et al.* have reported that Glu-26 and Glu-27 in the JmjN domain of JMJD2D can undergo poly-ADP ribosylation that represses the expression of a subset of retinoic acid receptor (RAR)-responsive genes whose expression is dependent on the RAR coactivator poly-(ADP-ribose) glycohydrolase (PARG).[121] PARG activity, substitutions of Glu-26 and Glu-27 in the JmjN domain, and inhibitors of poly-(ADP-ribose) polymerases (PARPs) reverse these modifications in JMJD2D, resulting in the activation of PARG-dependent RAR target genes. Interestingly, poly-ADP ribosylation of JMJD2D has also been implicated in DNA DSB repair. Khoury-Haddad *et al.* have shown that PARP1 recruits and catalyses the poly-ADP ribosylation of JMJD2D in response to DNA damage.[122] Localization of JMJD2D to sites of DNA damage mediates the association of ATM protein kinase with chromatin, facilitating the phosphorylation of a subset of ATM substrates. Further, JMJD2D elicits the formation of RAD51 and 53BP1 foci when DNA damage is induced by ionizing radiation and participates in non-homologous end joining during DNA DSB repair. Both the recruitment and H3K9me3 demethylase activity of JMJD2D are important in DNA damage response, as active site substitutions in the enzyme and treatment with JMJD2 demethylase inhibitor 8-hydroxyquinoline abrogate DNA repair. Taken together, these findings underscore the importance of JMJD2D in DNA damage response.

The JMJD2 KDMs have emerged as a paradigm for understanding the structures, catalytic mechanism and substrate specificities of the JmjC KDMs. The domain architecture of JMJD2A, JMJD2B and JMJD2C comprises an *N*-terminal catalytic domain composed of JmjN and JmjC domains and a *C*-terminal region harbouring two PHD domains followed by two tandem Tudor domains that mediate chromatin interactions (Figure 7.3A), whereas JMJD2D lacks the *C*-terminal PHD and Tudor domains (Figure 7.3D). Crystal structures of human JMJD2A reveals that its catalytic domain is composed of a JmjN domain, an intervening motif of mixed α/β structure, the JmjC domain, and *C*-terminal motif that is stabilized by a three-Cys/one-His Zn-binding motif (Figure 7.3B).[123] Subsequently determined structures of human JMJD2B, JMJD2C and JMJD2D, as well as the *S. cerevisiae* JMJD2 homologue Rph1, illustrate the overall structural conservation of the catalytic domain within this family.[64,99,124,125]

Structural and biochemical studies have revealed that the JMJD2 KDMs exhibit plasticity in their recognition of the H3K9 and H3K36 methylation sites, with JMJD2A-C accepting both substrates and JMJD2D accepting only the former. Crystal structures of JMJD2A bound to H3K9me3 and H3K36me3 substrate peptides reveal that the peptides adopt different conformations when bound within the enzyme's substrate binding cleft (Figure 7.3B).[126–128] Specifically, the H3K9me3 peptide adopts a kinked conformation with a sharp bend at Thr-11-Gly-12, whereas the H3K36me3 peptide adopts more of an extended conformation that is stabilized in part by interactions with the residues flanking Lys-36 in histone H3. Substitutions of the residues flanking the H3K9 site, including Arg-8, Ser-10, Thr-11, Gly-12 and Gly-13, either impair or abolish demethylation by JMJD2A, as does phosphorylation of Ser-10 and Thr-11.[124,128] The crystal structure of JMJD2D has yielded molecular insights into how its specificity differs from that of other JMJD2 KDMs.[124] The structure of JMJD2D bound to an H3K9me3 substrate peptide reveals that the peptide adopts a bent conformation analogous to that observed in the structure of the JMJD2A-H3K9me3 complex, although there are subtle

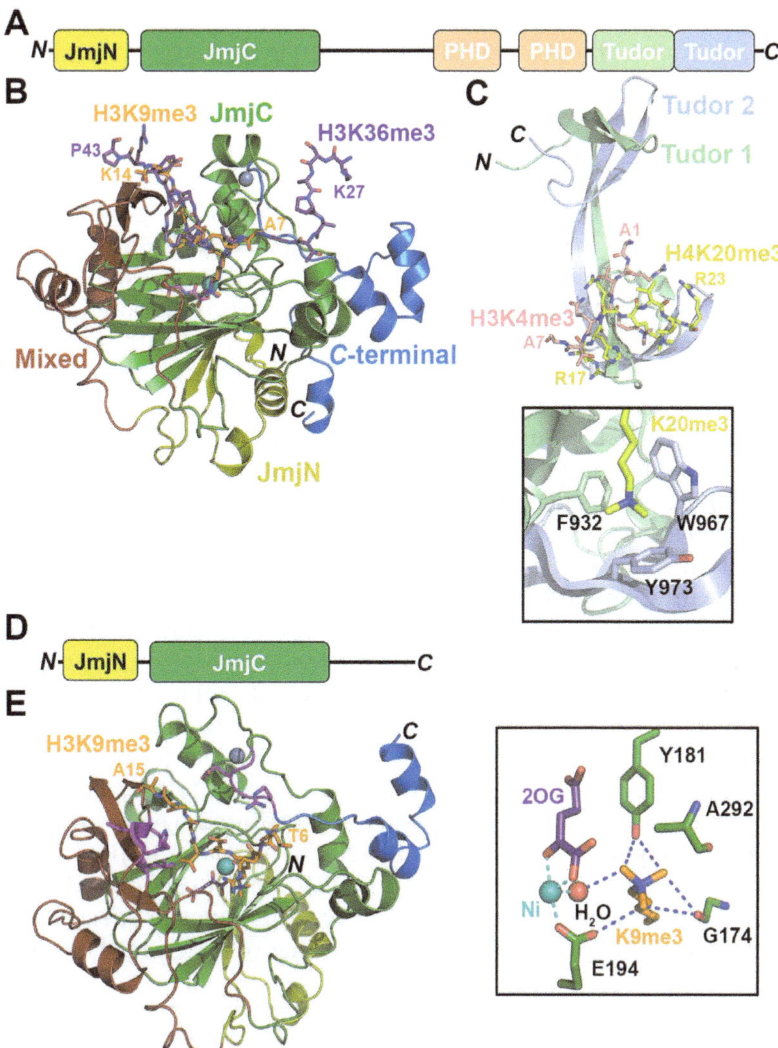

Figure 7.3 Representative structures of the JMJD2 KDM family. (A) Domain architecture of JMJD2A. (B) Crystal structure of the catalytic domain of human JMJD2A (PDB ID: 2OQ6; residues 1–359) bound to an H3K9me3 peptide (residues 7–14, orange carbon atoms) with Zn and Ni ions illustrated as grey and cyan spheres, respectively, and NOG depicted in stick representation with magenta carbon atoms. The different domains and structural motifs composing the catalytic domain are coloured and labelled. This structure was aligned with the coordinates of human JMJD2A bound to an H3K36me3 peptide (PDB ID: 2P5B, H3 residues 26–47, purple carbon atoms) to illustrate the variations in the H3K9me3 and H3K36me3 peptide binding modes (terminal residues of the peptides are labelled for clarity). (C) Structure of the tandem Tudor domains of human JMJD2A (PDB ID: 2GFA) bound to an H3K4me3 peptide. The two Tudor domains (residues 895–1011) are illustrated

yet important differences in how the two KDMs recognize Ser-10 and Thr-11 following Lys-9 in the histone H3 sequence (Figure 7.3E).[124] Docking of the H3K36me3 peptide into the JMJD2D structure through an alignment with a JMJD2A-H3K36me3 peptide complex reveals that the residues in the JMJD2D substrate binding cleft that contact the H3K36me3 peptide diverge from the corresponding conserved amino acids in the substrate binding clefts of JMJD2A, JMJD2B and JMJD2C (shown in magenta in Figure 7.3E). These divergent residues in JMJD2D discriminate against H3K36me3 recognition *via*: (i) steric clashes and electrostatic repulsion with the residues following Lys-36 in histone H3 and (ii) loss of productive hydrogen bonding with the residues preceding Lys-36 in histone H3. In agreement with these observations, Hillringhaus *et al.* have demonstrated that substitution of the amino acids in the substrate binding cleft of JMJD2A by the divergent residues in the cleft of JMJD2D results in impaired H3K36me3 demethylation *in vitro* and *in vivo*, converting JMJD2A into an essentially H3K9me3-specific KDM.[64]

In addition to elucidating their methylation site selectivities, the JMJD2 crystal structures have yielded insights into the molecular basis of their methylation state specificities. The trimethyllysine substrate binds deep within the JmjC domain β-barrel with the trimethylammonium group bound adjacent to the 2OG and Fe(II) binding sites, as illustrated in the JMJD2D-H3K9me3 complex (Figure 7.3E). The trimethyllysyl methyl groups engage in a network of unconventional carbon–oxygen (CH···O) hydrogen bonds that position one of the methyl groups toward the metal centre for hydroxylation and subsequent demethylation.[124,127] The residues that form the trimethyllysine binding pocket are conserved throughout the JMJD2 family, with the exception of Ala-292 in JMJD2D, which is substituted by a Ser in the other JMJD2 homologues. Changing this Ser-288 to an Ala in JMJD2A enables it to efficiently demethylate di- and trimethyllysine substrates, whereas the converse A292S substitution in JMJD2D has the opposite effect, diminishing its activity toward H3K9me2 relative to that of H3K9me3.[123,127] Notably,

in light green and blue and the H3K4me3 peptide (residues 1–10) is depicted with pink carbon atoms. This structure was superimposed with the coordinates of the JMJD2A double Tudor domains bound to an H4K20me3 peptide (PDB ID: 2QQS; H3 residues 16–25, yellow carbon atoms) to illustrate the opposite binding orientations of the H3K4me3 and H4K20me3 peptides (termini labelled for clarity). The inset panel shows the binding of the K20me3 side chain bound within the aromatic cage formed by the two tandem Tudor domains of JMJD2A. (D) Domain architecture of JMJD2D. (E) Crystal structure of the catalytic domain of human JMJD2D bound to an H3K9me3 peptide (PDB ID: 4HON). The JMJD2D catalytic domain (residues 12–342) and H3K9me3 peptide (residues 1–15) are illustrated as in panel A with 2OG depicted with magenta carbon atoms. The inset panel (right) depicts the active site of JMJD2D (green carbon atoms) bound to the K9me3 side chain, 2OG, and the active site Ni ion. Carbon–oxygen (CH···O) hydrogen bonds to the trimethyl lysyl methyl groups are depicted as blue dashed lines, whereas the cyan dashed lines denote coordination to Ni(II).

the methyllysine binding pockets of the UTX, JMJD3 and JARID1 KDMs are structurally conserved with that of JMJD2 KDMs and possess an alanine in the Ser/Ala position, consistent with their selectivity for di- and tri-methyllysines.[129,130] The Ser/Ala position has been proposed to influence the binding of dimethyllysine between catalytically productive and non-productive conformations for demethylation through its propensity to form different patterns of methyl CH···O hydrogen bonds, although further studies are required to substantiate this model.[127,131,132]

Structural and functional studies have also highlighted the mechanisms by which the two tandem Tudor domains in the *C*-terminus of JMJD2A facilitate interactions with chromatin. Interestingly, the double Tudor domains recognize both H3K4me3 and H4K20me3, two methylation sites that share virtually no sequence homology.[133] Crystal structures of the JMJD2A tandem Tudor domains have elucidated the mechanisms by which they recognize the unrelated sequences flanking the H3K4me3 and H4K20me3 sites (Figure 7.3C).[134,135] The double Tudor domains adopt a tightly interwound structure, forming two distinct hybrid Tudor folds. Remarkably, the H3K4me3 and H4K20me3 peptides bind in opposite orientations to the second hybrid Tudor domain with their trimethyllysines bound in an aromatic cage composed of Phe-932, Trp-967 and Tyr-973 through cation-π interactions, as observed in other methyllysine binding domains.[134–136] Interactions with Arg-2 in the H3K4me3 site and Arg-19 in the H4K20me3 site are pivotal to recognition by the second hybrid Tudor domain.[135] Substitutions of the residues in JMJD2A that recognize Arg-2 or Arg-19 in histone H3 result in double Tudor domains that selectively bind either H4K20me3 or H3K4me3, respectively, and largely lose their ability to recognize both methylation sites. Taken together, these studies have provided key insights into the mechanisms underlying chromatin recognition by the tandem Tudor domains of JMJD2A, although the potential functions of the JMJD2 PHD domain in mediating protein:protein or protein:nucleic acids interactions remain largely unexplored. The roles of the PHD domains and how they function in concert with the other domains composing the JMJD2 KDMs remain as topics of future investigation.

7.5 The PHF8/KIAA1718/PHF2 Family (also known as the KDM7 Family)

PHF8 and KIAA1718 (also known as JHDM1D) belong to a small family of JmjC proteins with three members in mice and humans (PHF2, PHF8 and KIAA1718).[137] Mutations in *PHF8* lead to X-linked mental retardation[138] and have been found in affected males with autism spectrum disorders.[139] Knockdown of *KIAA1718* and *PHF8* homologues in zebrafish causes brain defects.[140,141] *PHF8* and *JARID1B* (see next section) are overexpressed in a significant fraction of prostate cancer cases and this effect is associated with high Gleason grade and a poor prognosis in prostate cancer patients, with high statistical significance (*P*-values of 5.7×10^{-8} and 1.0×10^{-7}, respectively), out of 615 epigenetic genes tested (including 32 JmjC demethylases).[51] Importantly, knockdown of *PHF8* in prostate cancer cells[51] or in oesophageal squamous cell carcinoma cells[142] blocks cell proliferation and inhibits cell migration and motility, which may be related to the protein's role as a positive regulator of genes involved in cell adhesion/actin cytoskeleton.[143] PHF8 also functions as a transcriptional co-activator that is specifically recruited by RARα fusions to activate expression of their downstream targets upon ATRA (all-*trans* retinoic acid) treatment in acute promyelocytic leukaemia (APL).[144] While ATRA treatment has been the paradigm of targeted therapy for oncogenic transcription factors, a significant number of APL patients still relapse and become ATRA resistant. Forced expression of *PHF8* resensitizes ATRA-resistant APL cells, whereas its downregulation confers resistance.[144] ATRA sensitivity depends on the PHF8 enzymatic activity as well as CDK1-mediated phosphorylation status of Ser-33 and Ser-84.[144] It was thus suggested that PHF8 dephosphorylation could be pharmacologically manipulated to resurrect ATRA sensitivity to resistant cells.

Both PHF8 and KIAA1718 harbour two domains in their *N*-terminal regions (Figure 7.4A): a PHD domain that binds H3K4me3/me2 and a JmjC domain that demethylates H3K9me2/me1, H3K27me2/me1 or H4K20me1. However, the presence of H3K4me3 on the same peptide as H3K9me2 makes the doubly methylated peptide a significantly better substrate of PHF8.[145–148] In contrast, the presence of H3K4me3 has the opposite effect in that it diminishes the H3K9me2 demethylase activity of KIAA1718 with no adverse effect on its H3K27me2 activity.[145] Differences in substrate specificity between the two enzymes are explained by a bent conformation of PHF8, allowing each of its domains to engage their respective targets, and an extended conformation of KIAA1718, which prevents the access to H3K9me2 by its JmjC domain when its PHD domain engages H3K4me3 (Figure 7.4B). Thus the structural linkage between the PHD domain binding to H3K4me3 and the placement of the catalytic JmjC domains relative to this 'on' epigenetic mark determines which repressive marks are removed by both demethylases. Taken together, the PHF8 and KIAA1718 JmjC domains on their own are promiscuous enzymes; it is the associated PHD domains and linker – a determinant for the relative positioning of the two domains – that are mainly responsible for substrate specificity. In addition, available structures include the catalytic JmjC domain of PHF8 in complex with the small molecule inhibitor Daminozide (PDB ID: 4DO0) and KIAA1718 in complex with E67.[149]

Another structural study on *C. elegans* KIAA1718 suggested that the extended conformation between the PHD and JmjC domains might enable a *trans*-histone peptide-binding mechanism, in which H3K4me3 associated with the PHD domain and the H3K9me2 bound to the JmjC domain could derive from two separate histone H3 molecules of the same nucleosome or two neighbouring nucleosomes.[150] However, this *trans*-binding mechanism can be excluded for human KIAA1718 because the presence of an H3K4me3 *in trans* or *in cis* with H3K9me2 substrate peptide strongly inhibits KIAA1718 activity towards H3K9me2.[145] Nevertheless, the *trans*-binding mechanism is

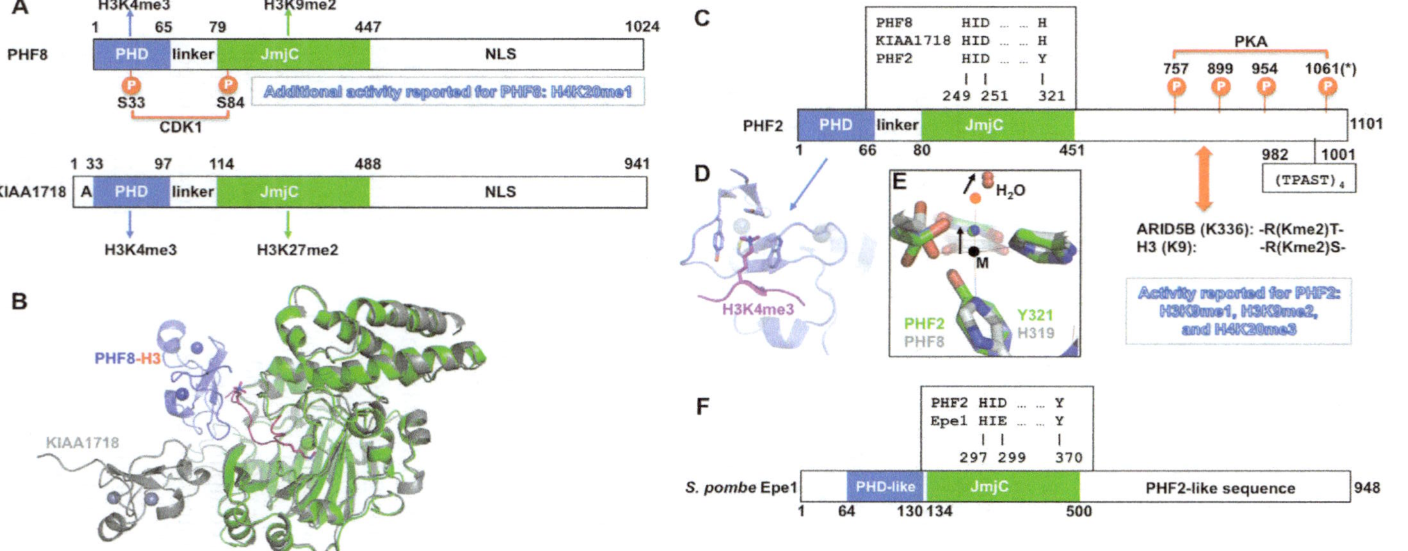

Figure 7.4 Coordinated methyl-lysine erasure between a JmjC and a PHD within the same polypeptide. (A) Schematic representations of PHF8 and KIAA1718. Letter P indicates the CDK1-mediated PHF8 phosphorylation sites.[144] (B) Superimposition of the JmjC domains for PHF8 (coloured) and KIAA1718 (grey) indicates that the PHF8 PHD domain adopts a bent conformation towards the JmjC domain in the presence of H3 substrate binding, whereas the PHD and JmjC domains of KIAA1718 adopt an extended conformation.[145] (C) Schematic representation of PHF2. The iron binding residues (HID...Y/H) of the family members are indicated. Letter P indicates the potential PKA-mediated phosphorylation sites.[155] (D) Structure of the PHF2 PHD domain in complex with H3K4me3 peptide (PDB ID: 3KQI).[154] (E) Superimposition of PHF2–Fe(II) (PDB ID: 3PU8)[153] and PHF8–Fe(II) (PDB ID: 3KV4).[145] The Fe(II) atoms (labelled by letter M) are depicted by small spheres, PHF2 is in colour, whereas PHF8 is in grey. The water molecules (labelled as H_2O) are shown as red small spheres. The arrows indicate the relatively small movements of the metal and metal-bound water molecule between PHF2 and PHF8. One important difference between PHF2 and PHF8 is that Tyr-321 of PHF2 replaces His-319 in PHF8 as one of the ligands. (F) Schematic representation of *S. pombe* Epe1. In addition to the amino acid changes in the Fe(II) coordination, the PHD-like domain loses Cys and His residues important for Zn(II) binding.[153]

an attractive model for PHF8 if the flexible loop between the PHD and JmjC enables the enzyme to adopt an extended conformation to allow binding of two peptides simultaneously. The *trans*-binding mechanism could explain the finding that PHF8 also functions *in vivo* as an H4K20me1 demethylase while its PHD domain interacts with H3K4me3/me2 in the context of the nucleosome.[141,151] One has to explain, however, why PHF8 is only active on monomethylated Lys-20 of histone H4, whereas it is active on di- and mono-methylated Lys-9 and Lys-27 of histone H3. One possibility is that only H4K20me1 coexists with H3K4me3/me2 *in vivo*. The amount of H4K20me1, as well as the PHF8 protein level, is tightly regulated during the cell cycle.[152]

PHF2 has the same domain architecture as that of PHF8 and KIAA1718 (Figure 7.4C) with its PHD domain binding H3K4me3 (Figure 7.4D) at submicromolar affinity,[153,154] but its JmjC domain is enzymatically inactive *in vitro*.[153,155] In other structurally examined JmjC domains, two His and one Asp or Glu [*i.e.* the Hx(D/E)...H motif] bind to the Fe(II) (Figure 7.4E). The isolated domain structures of PHD and JmjC of PHF2 have been characterized.[153,154] Human or mouse PHF2 and *Schizosaccharomyces pombe* Epe1 have a tyrosine at the position corresponding to the distal iron-binding His (Tyr-321 and Tyr-370, respectively). However, the Y321H substitution does not render PHF2 an active demethylase on histone peptides, despite the metal binding site in PHF2 closely resembling the Fe(II) sites in other JmjC domains examined (Figure 7.4E).[153] Other regulatory factors must be required for the reported enzymatic activity of PHF2 *in vivo*.

The aforementioned CDK1-mediated PHF8 phosphorylation sites are located on surfaces of the *N*-terminal PHD (Ser-33) and JmjC (Ser-84) domains, respectively. Notably, PHF2 becomes an active H3K9me2 demethylase through PKA-mediated phosphorylation, with four potential phosphorylation sites located in its *C*-terminal half (Figure 7.4C).[155] The phosphorylated PHF2 associates with ARID5B, a DNA-binding protein, and induces demethylation of H3K9me2 as well as methylated ARID5B at Lys-336, both sharing the Arg–Lys–(Thr/Ser) sequence.[155,156] It is worth noting that, in a separate study, PHF2 was reported to demethylate H3K9me1 *in vivo*, as detected by immunostaining of cells expressing green fluorescent protein-tagged PHF2 with anti-H3K9me antibodies.[154] Furthermore, PHF2 has also been linked to demethylation of H4K20me3, in a NF-κB-dependent fashion,[157] unlike the PHF8-mediated demethylation of H4K20me1. *Phf2* null mice have partial neonatal death and growth retardation and exhibit less adipose tissue and reduced adipocyte numbers compared with control littermates, and the tamoxifen-induced conditional knockout of *Phf2* results in impaired adipogenesis in stromal vascular cells.[158]

S. pombe Epe1 was proposed to be a putative histone demethylase that could act by oxidative demethylation.[159] However, recombinant Epe1 purified from Sf9 cells lacks histone KDM activity,[8] whereas functional characterization *in vivo* suggested that Epe1 is involved in changes in methylation patterns of H3K4 and H3K9 in fission yeast.[160] This raises the question of whether Epe1, which shares significant sequence homology with PHF2

(Figure 7.4F),[153] including the above-mentioned tyrosine at the corresponding iron-binding position and one of the phosphorylation sites (Ser-757 of PHF2 and Ser-761 of Epe1), could become an active histone demethylase upon phosphorylation.

7.6 The JARID1 Family (also known as the KDM5 Family)

There are four JARID (JmjC, AT-rich interactive domain) family proteins in mammalian cells: JARID1A, B, C and D (Figure 7.5), all of which demethylate H3K4me3. JARID1A (also known as RBP2) was originally identified as an Rb-binding protein[161] and the interaction between these proteins may enhance the tumour-suppressing activity of Rb.[162] Expression of JARID1B (also known as PLU-1) is restricted to the testis in normal tissues,[163,164] but is upregulated in prostate cancer,[51,165] breast cancer[163,166,167] and melanomas.[168] Significantly, JARID1B physically associates with AR and regulates its function in human prostate cancer cells.[165] JARID1C (also known as SMCX) and JARID1D (also known as SMCY) are located on sex chromosomes (the latter names derive from Selected Mouse cDNA on the X and Y, respectively). Mutations in *JARID1C* are associated with adult kidney cancer (clear cell renal cell carcinoma)[169] and X-linked mental retardation.[170–173] Reducing the levels of JARID1C in primary neurons reverses the downregulation of key neuronal genes caused by mutant Huntington expression.[174]

In addition to being upregulated in advanced and metastatic prostate tumours[165] and being required for continuous growth of melanoma cells,[168] JARID1B is required for embryonic survival, contributes to cell proliferation in the mammary gland and in ER + breast cancer cells,[175] and is highly expressed in human mammary tumours and breast cancer cell lines, but not in normal adult breast tissue.[163] Knockdown of *JARID1B* leads to upregulation of tumour suppressor genes including *BRCA1*.[166] Downregulation of *JARID1B* in breast cancer cells decreases tumour formation potential in mouse syngeneic or xenograft models.[175] A screen against a small library of ~15 000 small molecules using Flag-tagged JARID1B identified 2-(4-methylphenyl)-1,2-benzisothiazol-3(2H)-one as a JARID1B inhibitor with an IC_{50} of ~3 μM *in vitro*.[176] JARID1C acts as a transcriptional corepressor downstream of a number of important signalling pathways and is dysregulated in human cancers, including TGF-β/Smad3[177] and tumour suppressor von Hippel–Lindau.[178] Inter-individual variation in *JARID1C* expression correlates with disease-free survival in primary breast tumours.[179]

Like the PHF2-ARID5B complex, the JARID JmjC family proteins (including Little imaginal discs 2, Lid2, in *S. pombe*) also interact with DNA, although in this case it is *via* an ARID DNA-binding domain within the same polypeptide (Figure 7.5). The ARID domain present in JARID1A and JARID1B binds to CG-rich sequences, CCGCCC[180] and GCAC(A/C),[181] respectively. The ARID–DNA interaction is required for JARID1A/RBP2 demethylase activity

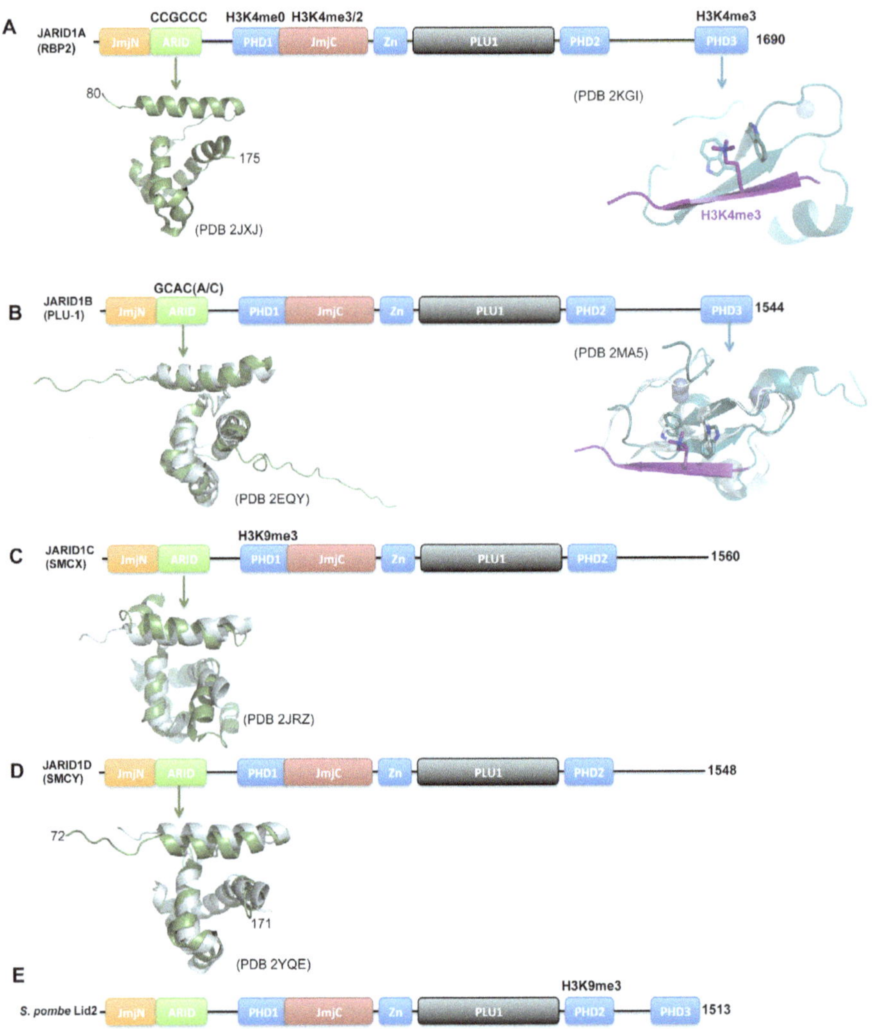

Figure 7.5 Schematic representations of human JARID1/KDM5 family members (A–D) and *S. pombe* Lid2 (E). (A–D) NMR structures have been determined for the ARID domain in the absence of DNA.[180,188] For JARID1B, 1C and 1D, the ARID domain structure (in light green) is superimposed individually with that of JARID1A (in grey) (panels B–D). JARID1A PHD3 structure (panel A, in light blue) was determined in complex with histone H3K4me3 peptide (in magenta).[184] The NMR structure of JARID1B PHD3 (panel B, in light blue), in the absence of histone peptide, is superimposed with that of JARID1A PHD3-H3 complex.

in cells.[180] In addition, JARID contains several PHD domains surrounding its JmjC domain that demethylates H3K4me3[172,182] and at least one of them binds H3K9me3 (PHD1 in JARID1C/SMCX[172] or PHD2 in Lid2, Figure 7.5).[183] Mutation or deletion of this PHD domain impairs the demethylase activity on H3K4me3.[172,183]

In JARID1A, PHD3 binds H3K4me3/me2 and PHD1 interacts with H3K4me0,[184] the substrate and product of its JmjC domain. The substrate H3K4me3/me2 binding by PHD3 may serve as a 'boundary factor' to protect H3K4me3 from JARID1A-mediated demethylation.[184] On the other hand, the binding of product H3K4me0 by PHD1 may serve as a 'seed' to propagate H3K4me0 formation by JARID1A; other histone-methylating enzymes also contain domains both to synthesize and bind a specific histone mark and thereby propagate it, for example, mammalian G9a/GLP (for H3K9me2/me1)[185] and *S. pombe* Clr4 (for H3K9me3).[186] One discrepancy in the literature is that JARID1A PHD1 was reported to bind H3K4me0 in pull-down assays,[184] whereas JARID1C PHD1 was thought to be an H3K9me3 binder;[172] the domains share 67% sequence identity or 87% similarity.

The JARID family is unique in that the catalytic JmjC domain has an atypical insertion of ARID and PHD1 domains not found in other JmjC-containing histone KDMs (Figure 7.5). However, the presence of a PHD domain immediately followed by the catalytic JmjC domain has a similar domain organization as the PHF8 family (Figure 7.4), suggesting chromatin cross-talk between a repressing H3K9me3 bound by PHD1 (in the case of JARID1C) and removal of activating methyl groups of H3K4me3 by the JmjC of JARID1. The *N*-terminal fragment of KDM5B/JARID1B (residues 1–769), containing both the PHD and the catalytic JmjC domain, demonstrated that the presence of H3K9 methylation does not affect the catalytic activity of JARID1, regardless of whether the methylated H3K9 peptide exists in *cis* or *trans*.[187] Of course, the interpretation of this observation assumes that, like the PHD1 domain of JARID1C,[172] the corresponding PHD domain in KDM5B binds to methylated H3K9. As mentioned above, the corresponding PHD1 in JARID1A was reported to bind H3K4me0,[184] the product of the demethylation reaction.

Currently, no structural information is available on the catalytic domains of JARID1/KDM5 subfamily enzymes. However, JARID1A PHD3 in complex with H3K4me3 peptide[184] and NMR structures of the ARID domains of all four JARID1/KDM5 family members are available (Fig. 7.5).[180,188] Interestingly, a JmjC demethylase in rice, JMJ703, has a JARID1-like activity on H3K4me3, but contains a canonical overall folding of JMJD2 proteins with no insertion within the catalytic domain.[189]

7.7 The UTX/JMJD3 Family (also known as the KDM6 Family)

Like JARID1C/SMCX and JARID1D/SMCY, encoded by two closely related genes located on sex chromosomes X and Y, UTX and UTY are encoded by two closely related Ubiquitously Transcribed Tetratricopeptide Repeat genes on sex chromosome X and Y, respectively. Several groups in 2007 identified UTX and JMJD3 (the third family member) as JmjC histone demethylases of H3K27me2/me3,[190–194] whereas no corresponding enzymatic activity has been reported for UTY *in vitro*[190,194] or *in vivo*.[195] However, UTY contributes to male-specific antigens, which can lead to the sex-specific tissue-transplantation rejection response,[196] and maintains demethylase-independent functions during mouse embryonic development.[195]

Since the discovery of its enzymatic demethylase activity, UTX has been proven to be essential during normal development, as it is required for cellular reprogramming,[197] embryonic development[195,198–201] and tissue-specific differentiation.[202] UTX is a member of the MLL2 H3K4 methyltransferase complex,[203] suggesting a dynamic interplay between H3K4 methylation (an 'on' mark) and demethylation of H3K27me2/me3 (an 'off' mark) during transcriptional gene activation, and its demethylase activity is linked to regulation of *HOX*[190,192] and *Rb* gene networks.[204,205] Importantly, loss-of-function mutations of both *MLL2* and *UTX* have been found in Kabuki syndrome patients,[206–209] a hereditary disorder of mental retardation, and both gene defects contribute to cancer pathogenesis.[210–214] In addition, both UTX and JMJD3 were shown to interact with the SWI/SNF remodelling complex, independent of their demethylase activities,[215] to regulate chromatin accessibility and thereby enable transcriptional regulation.

UTX and UTY contain a tetratricopeptide repeat (TPR) in the *N*-terminal half that might be important for protein:protein interactions and the JmjC domain in the *C*-terminal half (Figure 7.6A). JMJD3 lacks the TPR domain but shares extensive homology with UTX and UTY, both within and outside the JmjC domain (Figure 7.6D). UTX and UTY share 88% sequence similarity (83% identity) in humans, and 82% sequence similarity in mouse. Within the catalytic JmjC domains, the similarity is even more significant, at 98% and 97% for human and mouse UTX/Y pairs, respectively, including the invariant residues for Fe(II), 2OG and substrate H3 peptide binding; thus, it is an intriguing puzzle why UTY is inactive. However, a study suggested that UTY possesses KDM activity but, at least when tested *in vitro* in recombinant form, at a substantially lower level than for UTX and JMJD3.[216] This is, at least in part, due to substitution of an isoleucine in UTX and JMJD3 for Pro-1214 in UTY (Figure 7.6H–J), potentially affecting substrate peptide binding.

The structure of a *C*-terminal fragment of human UTX (residues 880–1401), including the catalytic JmjC domain followed by a helical domain and a Zn-binding domain (Figure 7.6B and C), has been determined in the presence and absence of histone H3K27me3 peptide.[217] The most notable finding of the structural study is that the *C*-terminal Zn-binding domain is involved in recognizing the portion of the histone H3 peptide (residues 17–23) *N*-terminal to the target Lys (H3K27),[217] excluding the near-cognate histone H3K9 whose immediately neighbouring residues (Ala–Arg–Lys–Ser) share identity with H3K27 (Figure 7.6C).

Multiple studies have highlighted UTX and JMJD3 as important players in cancer biology. The most selective reported small molecule compound against JMJD3 and UTX, GSK-J1 (Figure 7.6E), was discovered based on the structural insights of the mouse and human JMJD3 proteins (Figure 7.6F, G) and shown to have an IC_{50} of 60 nM for inhibiting JMJD3.[218] GSK-J1 is active against both JMJD3 and UTX, but it is inactive against most members

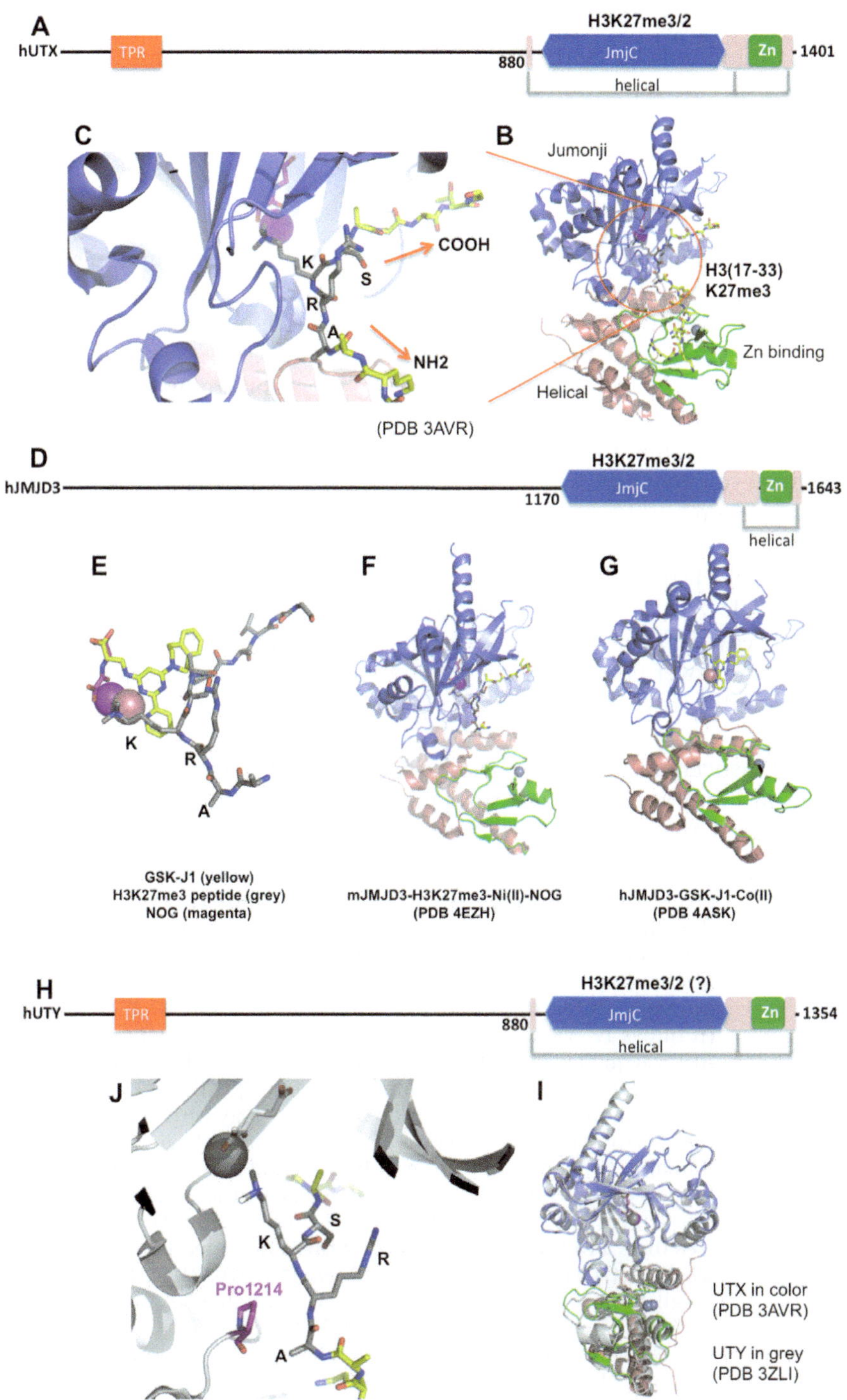
A
hUTX
TPR
H3K27me3/2
880
JmjC
Zn
1401
helical
C
K
S
R
A
COOH
NH2
B
Jumonji
H3(17-33)
K27me3
Zn binding
Helical
(PDB 3AVR)
D
hJMJD3
H3K27me3/2
1170
JmjC
Zn
1643
helical
E
F
G
K
R
A
GSK-J1 (yellow)
H3K27me3 peptide (grey)
NOG (magenta)
mJMJD3-H3K27me3-Ni(II)-NOG
(PDB 4EZH)
hJMJD3-GSK-J1-Co(II)
(PDB 4ASK)
H
hUTY
TPR
H3K27me3/2 (?)
880
JmjC
Zn
1354
helical
J
I
S
K
R
Pro1214
A
UTX in color
(PDB 3AVR)
UTY in grey
(PDB 3ZLI)

of a panel of JmjC demethylases. Remarkably, the cell-penetrating derivative GSK-J4 inhibited the JMJD3-induced loss of nuclear H3K27me3 levels and allowed specific inhibition of H3K27 demethylation at promoter regions of JMJD3 and UTX target genes.[218]

7.8 Conclusion

The past decade has witnessed tremendous advances in our understanding of the biological roles, catalytic mechanisms and molecular basis for substrate recognition of the JmjC KDMs. Although substantial progress has been made, many important questions regarding this essential class of enzymes remain to be explored. Among the more intriguing questions is to what extent the JmjC KDMs demethylate non-histone substrates and participate in signalling pathways beyond chromatin modifications. There is growing evidence that protein lysine methylation is a dynamic modification that occurs on hundreds of proteins, and numerous non-histone proteins are known to be substrates of lysine methyltransferases.[219,220] Indeed, FBXL11, JMJD1C and JMJD2C, have been reported to demethylate non-histone proteins (Table 7.1),[15,55,120] suggesting that other substrates of the JmjC KDMs await discovery. Thus, ongoing investigations into JmjC KDM specificity will probably uncover new substrates for these enzymes, further expanding our understanding of their functions in signal transduction and other biological pathways.

Other aspects of the JmjC KDMs also merit future exploration. As illustrated in numerous examples through this chapter, JmjC KDMs can form complexes with a myriad of nuclear proteins, including steroid receptors,

Figure 7.6 Structures of mammalian UTX, UTY and JMJD3. (A) UTX catalytic domain is located in the *C*-terminal region, containing the catalytic JmjC domain (blue), a helical domain (brown) and a Zn-binding domain (green) important for H3K27me3 peptide specificity. (B) Structure of human UTX (residues 886–1395) in complex with H3K27me3 peptide (residues 17–33 in yellow). (C) Enlarged active site of UTX showing the ARKS sequence shared by H3K9 and H3K27. The directions for the *N*- and *C*-termini of H3 peptide are labelled. (D) JMJD3 lacks the *N*-terminal TPR domain but shares extensive homology with UTX and UTY in the *C*-terminal catalytic region (JmjC domain, helical and Zn binding domain). (E) Superimposition of human and mouse JMJD3 structures (panels F and G) reveals that GSK-J1 chelates the metal ion and occupies part of both the 2OG and peptide binding sites. (F) Mouse JMJD3 (residues 1157–1638) in complex with H3K27me3 peptide (residues 24–34). (G) Human JMJD3 (residues 1177–1638) in complex with small molecule inhibitor GSK-J1. (H) Human UTX and UTY share 88% sequence similarity overall and 98% similarity in their respective *C*-terminal catalytic region. (I) The *C*-terminal domain structure of human UTY (in grey; PDB ID: 3ZLI) is highly conserved with that of human UTX (in colour; PDB ID: 3AVR). (J) Model of human UTY in complex with H3K27me3 peptide (taken from human UTX structure) indicates that Pro-1214 is involved in peptide substrate binding.

transcription factors and other chromatin-modifying enzymes. Although crystal structures of different domains in JmjC KDMs have been determined, structural and functional studies of these enzymes bound to other nuclear proteins and complexes have not been investigated extensively, thus limiting our understanding of how these enzymes function in the context of different nuclear complexes. Correlatively, many JmjC KDMs comprise one or more domains known to facilitate protein:protein and protein:nucleic acid interactions that may function as platforms for the assembly of multi-subunit complexes, although many of these domains and their interacting partners have not been thoroughly characterized.

Another related aspect of JmjC KDM biology is the question of functional redundancy. In humans, JmjC KDM families possess two or more homologues (Table 7.1) and in certain contexts, the functions of these isoforms appear to be redundant. However, as highlighted in this chapter, many JmjC KDMs participate in genomic processes that are cell lineage or cell cycle-specific and are not shared with their closely related homologues. These findings imply that a division of labour exists within each JmjC KDM family wherein each homologue is expressed and may have specific functions during differentiation and development or in somatic cell types. However, the degree of functional redundancy *versus* functional specificity within the various families of these enzymes remains an open question. Future studies of the JmjC KDMs hold the key to unlocking the answers to these questions and will undoubtedly hold some unexpected surprises regarding the biological functions of this important family of enzymes.

Acknowledgements

The work in the Cheng Laboratory was supported by grants GM049245-21 and DK094346-01 from the National Institutes of Health. X.C. is a Georgia Research Alliance Eminent Scholar. Research in the Trievel Laboratory is supported by grants CHE-1213484 from the National Science Foundation and AI111182-01 from the National Institutes of Health.

References

1. L. J. Walport, R. J. Hopkinson and C. J. Schofield, *Curr. Opin. Chem. Biol.*, 2012, **16**, 525–534.
2. R. P. Hausinger, *Crit. Rev. Biochem. Mol. Biol.*, 2004, **39**, 21–68.
3. W. Aik, M. A. McDonough, A. Thalhammer, R. Chowdhury and C. J. Schofield, *Curr. Opin. Struct. Biol.*, 2012, **22**, 691–700.
4. A. Noma, R. Ishitani, M. Kato, A. Nagao, O. Nureki and T. Suzuki, *J. Biol. Chem.*, 2010, **285**, 34503–34507.
5. J. M. Elkins, K. S. Hewitson, L. A. McNeill, J. F. Seibel, I. Schlemminger, C. W. Pugh, P. J. Ratcliffe and C. J. Schofield, *J. Biol. Chem.*, 2003, **278**, 1802–1806.

6. C. J. Webby, A. Wolf, N. Gromak, M. Dreger, H. Kramer, B. Kessler, M. L. Nielsen, C. Schmitz, D. S. Butler, J. R. Yates, 3rd, C. M. Delahunty, P. Hahn, A. Lengeling, M. Mann, N. J. Proudfoot, C. J. Schofield and A. Bottger, *Science*, 2009, **325**, 90–93.
7. W. Ge, A. Wolf, T. Feng, C. H. Ho, R. Sekirnik, A. Zayer, N. Granatino, M. E. Cockman, C. Loenarz, N. D. Loik, A. P. Hardy, T. D. Claridge, R. B. Hamed, R. Chowdhury, L. Gong, C. V. Robinson, D. C. Trudgian, M. Jiang, M. M. Mackeen, J. S. McCullagh, Y. Gordiyenko, A. Thalhammer, A. Yamamoto, M. Yang, P. Liu-Yi, Z. Zhang, M. Schmidt-Zachmann, B. M. Kessler, P. J. Ratcliffe, G. M. Preston, M. L. Coleman and C. J. Schofield, *Nat. Chem. Biol.*, 2012, **8**, 960–962.
8. Y. Tsukada, J. Fang, H. Erdjument-Bromage, M. E. Warren, C. H. Borchers, P. Tempst and Y. Zhang, *Nature*, 2006, **439**, 811–816.
9. Y. Shi, F. Lan, C. Matson, P. Mulligan, J. R. Whetstine, P. A. Cole, R. A. Casero and Y. Shi, *Cell*, 2004, **119**, 941–953.
10. B. M. Turner, *Nat. Struct. Mol. Biol.*, 2005, **12**, 110–112.
11. Y. Tanaka, K. Okamoto, K. Teye, T. Umata, N. Yamagiwa, Y. Suto, Y. Zhang and M. Tsuneoka, *EMBO J.*, 2010, **29**, 1510–1522.
12. Y. Tanaka, T. Umata, K. Okamoto, C. Obuse and M. Tsuneoka, *Cell Struct. Funct.*, 2014, **39**, 79–92.
13. D. Pan, C. Mao, T. Zou, A. Y. Yao, M. P. Cooper, V. Boyartchuk and Y. X. Wang, *PLoS Genet.*, 2012, **8**, e1002761.
14. T. Lu, M. W. Jackson, A. D. Singhi, E. S. Kandel, M. Yang, Y. Zhang, A. V. Gudkov and G. R. Stark, *Proc. Natl. Acad. Sci. U. S. A.*, 2009, **106**, 16339–16344.
15. T. Lu, M. W. Jackson, B. Wang, M. Yang, M. R. Chance, M. Miyagi, A. V. Gudkov and G. R. Stark, *Proc. Natl. Acad. Sci. U. S. A.*, 2010, **107**, 46–51.
16. T. Lu, M. Yang, D. B. Huang, H. Wei, G. H. Ozer, G. Ghosh and G. R. Stark, *Proc. Natl. Acad. Sci. U. S. A.*, 2013, **110**, 13510–13515.
17. S. S. Dhar, H. Alam, N. Li, K. W. Wagner, J. Chung, Y. W. Ahn and M. G. Lee, *J. Biol. Chem.*, 2014, **289**, 7483–7496.
18. K. W. Wagner, H. Alam, S. S. Dhar, U. Giri, N. Li, Y. Wei, D. Giri, T. Cascone, J. H. Kim, Y. Ye, A. S. Multani, C. H. Chan, B. Erez, B. Saigal, J. Chung, H. K. Lin, X. Wu, M. C. Hung, J. V. Heymach and M. G. Lee, *J. Clin. Invest.*, 2013, **123**, 5231–5246.
19. M. J. Barrero and J. C. Izpisua Belmonte, *Nat. Cell Biol.*, 2013, **15**, 348–350.
20. N. P. Blackledge, A. M. Farcas, T. Kondo, H. W. King, J. F. McGouran, L. L. Hanssen, S. Ito, S. Cooper, K. Kondo, Y. Koseki, T. Ishikura, H. K. Long, T. W. Sheahan, N. Brockdorff, B. M. Kessler, H. Koseki and R. J. Klose, *Cell*, 2014, **157**, 1445–1459.
21. A. M. Farcas, N. P. Blackledge, I. Sudbery, H. K. Long, J. F. McGouran, N. R. Rose, S. Lee, D. Sims, A. Cerase, T. W. Sheahan, H. Koseki, N. Brockdorff, C. P. Ponting, B. M. Kessler and R. J. Klose, *Elife*, 2012, **1**, e00205.
22. J. He, L. Shen, M. Wan, O. Taranova, H. Wu and Y. Zhang, *Nat. Cell Biol.*, 2013, **15**, 373–384.

23. F. Kottakis, P. Foltopoulou, I. Sanidas, P. Keller, A. Wronski, B. T. Dake, S. A. Ezell, Z. Shen, S. P. Naber, P. W. Hinds, E. McNiel, C. Kuperwasser and P. N. Tsichlis, *Cancer Res.*, 2014, **74**, 3935–3946.
24. X. Wu, J. V. Johansen and K. Helin, *Mol. Cell*, 2013, **49**, 1134–1146.
25. J. He, E. M. Kallin, Y. Tsukada and Y. Zhang, *Nat. Struct. Mol. Biol.*, 2008, **15**, 1169–1175.
26. J. He, A. T. Nguyen and Y. Zhang, *Blood*, 2011, **117**, 3869–3880.
27. N. Popov and J. Gil, *Epigenetics*, 2010, **5**, 685–690.
28. R. Pfau, A. Tzatsos, S. C. Kampranis, O. B. Serebrennikova, S. E. Bear and P. N. Tsichlis, *Proc. Natl. Acad. Sci. U. S. A.*, 2008, **105**, 1907–1912.
29. T. Konuma, S. Nakamura, S. Miyagi, M. Negishi, T. Chiba, H. Oguro, J. Yuan, M. Mochizuki-Kashio, H. Ichikawa, H. Miyoshi, M. Vidal and A. Iwama, *Exp. Hematol.*, 2011, **39**, 697–709 e695.
30. G. Liang, J. He and Y. Zhang, *Nat. Cell Biol.*, 2012, **14**, 457–466.
31. T. Wang, K. Chen, X. Zeng, J. Yang, Y. Wu, X. Shi, B. Qin, L. Zeng, M. A. Esteban, G. Pan and D. Pei, *Cell Stem Cell*, 2011, **9**, 575–587.
32. S. Nakamura, L. Tan, Y. Nagata, T. Takemura, A. Asahina, D. Yokota, T. Yagyu, K. Shibata, S. Fujisawa and K. Ohnishi, *Mol. Carcinog.*, 2013, **52**, 57–69.
33. Z. Cheng, P. Cheung, A. J. Kuo, E. T. Yukl, C. M. Wilmot, O. Gozani and D. J. Patel, *Genes Dev.*, 2014, **28**, 1758–1771.
34. N. P. Blackledge, J. P. Thomson and P. J. Skene, *Cold Spring Harbor. Perspect. Biol.*, 2013, **5**, a018648.
35. N. P. Blackledge, J. C. Zhou, M. Y. Tolstorukov, A. M. Farcas, P. J. Park and R. J. Klose, *Mol. Cell*, 2010, **38**, 179–190.
36. J. C. Zhou, N. P. Blackledge, A. M. Farcas and R. J. Klose, *Mol. Cell. Biol.*, 2012, **32**, 479–489.
37. H. K. Long, N. P. Blackledge and R. J. Klose, *Biochem. Soc. Trans.*, 2013, **41**, 727–740.
38. C. A. Musselman and T. G. Kutateladze, *Mol. Interventions*, 2009, **9**, 314–323.
39. A. Janzer, K. Stamm, A. Becker, A. Zimmer, R. Buettner and J. Kirfel, *J. Biol. Chem.*, 2012, **287**, 30984–30992.
40. B. Kobe and A. V. Kajava, *Curr. Opin. Struct. Biol.*, 2001, **11**, 725–732.
41. K. Yamane, C. Toumazou, Y. Tsukada, H. Erdjument-Bromage, P. Tempst, J. Wong and Y. Zhang, *Cell*, 2006, **125**, 483–495.
42. Y. Okada, G. Scott, M. K. Ray, Y. Mishina and Y. Zhang, *Nature*, 2007, **450**, 119–123.
43. Y. Okada, K. Tateishi and Y. Zhang, *J. Androl.*, 2010, **31**, 75–78.
44. I. Kasioulis, H. M. Syred, P. Tate, A. Finch, J. Shaw, A. Seawright, M. Fuszard, C. H. Botting, S. Shirran, I. R. Adams, I. J. Jackson, V. van Heyningen and P. L. Yeyati, *Mol. Biol. Cell*, 2014, **25**, 1216–1233.
45. S. Kuroki, S. Matoba, M. Akiyoshi, Y. Matsumura, H. Miyachi, N. Mise, K. Abe, A. Ogura, D. Wilhelm, P. Koopman, M. Nozaki, Y. Kanai, Y. Shinkai and M. Tachibana, *Science*, 2013, **341**, 1106–1109.

46. T. Inagaki, M. Tachibana, K. Magoori, H. Kudo, T. Tanaka, M. Okamura, M. Naito, T. Kodama, Y. Shinkai and J. Sakai, *Genes Cells*, 2009, **14**, 991–1001.
47. K. Tateishi, Y. Okada, E. M. Kallin and Y. Zhang, *Nature*, 2009, **458**, 757–761.
48. Y. Liu, P. Zheng, T. Ji, X. Liu, S. Yao, X. Cheng, Y. Li, L. Chen, Z. Xiao, J. Zhou and J. Li, *Gut*, 2013, **62**, 571–581.
49. E. Paolicchi, F. Crea, W. L. Farrar, J. E. Green and R. Danesi, *Crit. Rev. Oncol. Hematol.*, 2013, **86**, 97–103.
50. J. Y. Kim, K. B. Kim, G. H. Eom, N. Choe, H. J. Kee, H. J. Son, S. T. Oh, D. W. Kim, J. H. Pak, H. J. Baek, H. Kook, Y. Hahn, D. Chakravarti and S. B. Seo, *Mol. Cell. Biol.*, 2012, **32**, 2917–2933.
51. M. Bjorkman, P. Ostling, V. Harma, J. Virtanen, J. P. Mpindi, J. Rantala, T. Mirtti, T. Vesterinen, M. Lundin, A. Sankila, A. Rannikko, E. Kaivanto, P. Kohonen, O. Kallioniemi and M. Nees, *Oncogene*, 2012, **31**, 3444–3456.
52. S. M. Kim, J. Y. Kim, N. W. Choe, I. H. Cho, J. R. Kim, D. W. Kim, J. E. Seol, S. E. Lee, H. Kook, K. I. Nam, Y. Y. Bhak and S. B. Seo, *Nucleic Acids Res.*, 2010, **38**, 6389–6403.
53. M. Brauchle, Z. Yao, R. Arora, S. Thigale, I. Clay, B. Inverardi, J. Fletcher, P. Taslimi, M. G. Acker, B. Gerrits, J. Voshol, A. Bauer, D. Schubeler, T. Bouwmeester and H. Ruffner, *Plos One*, 2013, **8**, e60549.
54. J. Lu and M. J. Matunis, *Nat. Struct. Mol. Biol.*, 2013, **20**, 1346–1348.
55. S. Watanabe, K. Watanabe, V. Akimov, J. Bartkova, B. Blagoev, J. Lukas and J. Bartek, *Nat. Struct. Mol. Biol.*, 2013, **20**, 1425–1433.
56. S. Goda, T. Isagawa, Y. Chikaoka, T. Kawamura and H. Aburatani, *J. Biol. Chem.*, 2013, **288**, 36948–36956.
57. P. A. Cloos, J. Christensen, K. Agger, A. Maiolica, J. Rappsilber, T. Antal, K. H. Hansen and K. Helin, *Nature*, 2006, **442**, 307–311.
58. B. D. Fodor, S. Kubicek, M. Yonezawa, R. J. O'Sullivan, R. Sengupta, L. Perez-Burgos, S. Opravil, K. Mechtler, G. Schotta and T. Jenuwein, *Genes Dev.*, 2006, **20**, 1557–1562.
59. J. R. Whetstine, A. Nottke, F. Lan, M. Huarte, S. Smolikov, Z. Chen, E. Spooner, E. Li, G. Zhang, M. Colaiacovo and Y. Shi, *Cell*, 2006, **125**, 467–481.
60. R. J. Klose, K. Yamane, Y. Bae, D. Zhang, H. Erdjument-Bromage, P. Tempst, J. Wong and Y. Zhang, *Nature*, 2006, **442**, 312–316.
61. T. Kim and S. Buratowski, *J. Biol. Chem.*, 2007, **282**, 20827–20835.
62. R. J. Klose, K. E. Gardner, G. Liang, H. Erdjument-Bromage, P. Tempst and Y. Zhang, *Mol. Cell. Biol.*, 2007, **27**, 3951–3961.
63. S. Tu, E. M. Bulloch, L. Yang, C. Ren, W. C. Huang, P. H. Hsu, C. H. Chen, C. L. Liao, H. M. Yu, W. S. Lo, M. A. Freitas and M. D. Tsai, *J. Biol. Chem.*, 2007, **282**, 14262–14271.
64. L. Hillringhaus, W. W. Yue, N. R. Rose, S. S. Ng, C. Gileadi, C. Loenarz, S. H. Bello, J. E. Bray, C. J. Schofield and U. Oppermann, *J. Biol. Chem.*, 2011, **286**, 41616–41625.

65. P. Trojer, J. Zhang, M. Yonezawa, A. Schmidt, H. Zheng, T. Jenuwein and D. Reinberg, *J. Biol. Chem.*, 2009, **284**, 8395–8405.
66. J. C. Black, A. Allen, C. Van Rechem, E. Forbes, M. Longworth, K. Tschop, C. Rinehart, J. Quiton, R. Walsh, A. Smallwood, N. J. Dyson and J. R. Whetstine, *Mol. Cell*, 2010, **40**, 736–748.
67. J. C. Black, A. L. Manning, C. Van Rechem, J. Kim, B. Ladd, J. Cho, C. M. Pineda, N. Murphy, D. L. Daniels, C. Montagna, P. W. Lewis, K. Glass, C. D. Allis, N. J. Dyson, G. Getz and J. R. Whetstine, *Cell*, 2013, **154**, 541–555.
68. C. Van Rechem, J. C. Black, T. Abbas, A. Allen, C. A. Rinehart, G. C. Yuan, A. Dutta and J. R. Whetstine, *J. Biol. Chem.*, 2011, **286**, 30462–30470.
69. F. A. Mallette, F. Mattiroli, G. Cui, L. C. Young, M. J. Hendzel, G. Mer, T. K. Sixma and S. Richard, *EMBO J.*, 2012, **31**, 1865–1878.
70. M. D. Cardamone, B. Tanasa, M. Chan, C. T. Cederquist, J. Andricovich, M. G. Rosenfeld and V. Perissi, *Cell Rep.*, 2014, **8**, 163–176.
71. S. Shin and R. Janknecht, *Biochem. Biophys. Res. Commun.*, 2007, **359**, 742–746.
72. Q. J. Zhang, H. Z. Chen, L. Wang, D. P. Liu, J. A. Hill and Z. P. Liu, *J. Clin. Invest.*, 2011, **121**, 2447–2456.
73. A. Cascante, S. Klum, M. Biswas, B. Antolin-Fontes, F. Barnabe-Heider and O. Hermanson, *J. Mol. Biol.*, 2014, **426**, 3467–3477.
74. P. H. Strobl-Mazzulla, T. Sauka-Spengler and M. Bronner-Fraser, *Dev. Cell*, 2010, **19**, 460–468.
75. L. Verrier, F. Escaffit, C. Chailleux, D. Trouche and M. Vandromme, *PLoS Genet.*, 2011, **7**, e1001390.
76. W. L. Berry and R. Janknecht, *Cancer Res.*, 2013, **73**, 2936–2942.
77. W. L. Berry, S. Shin, S. A. Lightfoot and R. Janknecht, *Int. J. Oncol.*, 2012, **41**, 1701–1706.
78. C. E. Hu, Y. C. Liu, H. D. Zhang and G. J. Huang, *Biochem. Biophys. Res. Commun.*, 2014, **449**, 1–7.
79. T. D. Kim, S. Shin, W. L. Berry, S. Oh and R. Janknecht, *J. Cell. Biochem.*, 2012, **113**, 1368–1376.
80. B. X. Li, M. C. Zhang, C. L. Luo, P. Yang, H. Li, H. M. Xu, H. F. Xu, Y. W. Shen, A. M. Xue and Z. Q. Zhao, *J. Exp. Clin. Cancer Res.*, 2011, **30**, 90.
81. L. L. Li, A. M. Xue, B. X. Li, Y. W. Shen, Y. H. Li, C. L. Luo, M. C. Zhang, J. Q. Jiang, Z. D. Xu, J. H. Xie and Z. Q. Zhao, *Breast Cancer Res.*, 2014, **16**, R56.
82. H. L. Wang, M. M. Liu, X. Ma, L. Fang, Z. F. Zhang, T. F. Song, J. Y. Gao, Y. Kuang, J. Jiang, L. Li, Y. Y. Wang and P. L. Li, *Asian Pac. J. Cancer Prev.*, 2014, **15**, 3051–3056.
83. L. C. Young and M. J. Hendzel, *Biochem. Cell Biol.*, 2013, **91**, 369–377.
84. M. Kogure, M. Takawa, H. S. Cho, G. Toyokawa, K. Hayashi, T. Tsunoda, T. Kobayashi, Y. Daigo, M. Sugiyama, Y. Atomi, Y. Nakamura and R. Hamamoto, *Cancer Lett.*, 2013, **336**, 76–84.
85. F. A. Mallette and S. Richard, *Cell Rep.*, 2012, **2**, 1233–1243.
86. B. X. Li, C. L. Luo, H. Li, P. Yang, M. C. Zhang, H. M. Xu, H. F. Xu, Y. W. Shen, A. M. Xue and Z. Q. Zhao, *Exp. Ther. Med.*, 2012, **4**, 755–761.

87. X. Ding, H. Pan, J. Li, Q. Zhong, X. Chen, S. M. Dry and C. Y. Wang, *Sci. Signaling*, 2013, **6**, ra28 1–13.
88. S. Beyer, M. M. Kristensen, K. S. Jensen, J. V. Johansen and P. Staller, *J. Biol. Chem.*, 2008, **283**, 36542–36552.
89. L. Fu, L. Chen, J. Yang, T. Ye, Y. Chen and J. Fang, *Carcinogenesis*, 2012.
90. P. J. Pollard, C. Loenarz, D. R. Mole, M. A. McDonough, J. M. Gleadle, C. J. Schofield and P. J. Ratcliffe, *Biochem. J.*, 2008, **416**, 387–394.
91. B. L. Adamsen, K. L. Kravik, O. P. Clausen and P. M. De Angelis, *Int. J. Oncol.*, 2007, **31**, 1491–1500.
92. H. Zheng, L. Chen, W. J. Pledger, J. Fang and J. Chen, *Oncogene*, 2014, **33**, 734–744.
93. R. B. Slee, C. M. Steiner, B. S. Herbert, G. H. Vance, R. J. Hickey, T. Schwarz, S. Christan, M. Radovich, B. P. Schneider, D. Schindelhauer and B. R. Grimes, *Oncogene*, 2012, **31**, 3244–3253.
94. L. Chen, L. Fu, X. Kong, J. Xu, Z. Wang, X. Ma, Y. Akiyama, Y. Chen and J. Fang, *Br. J. Cancer*, 2014, **110**, 1014–1026.
95. J. Yang, A. M. Jubb, L. Pike, F. M. Buffa, H. Turley, D. Baban, R. Leek, K. C. Gatter, J. Ragoussis and A. L. Harris, *Cancer Res.*, 2010, **70**, 6456–6466.
96. L. Gaughan, J. Stockley, K. Coffey, D. O'Neill, D. L. Jones, M. Wade, J. Wright, M. Moore, S. Tse, L. Rogerson and C. N. Robson, *Nucleic Acids Res.*, 2013, **41**, 6892–6904.
97. M. Kawazu, K. Saso, K. I. Tong, T. McQuire, K. Goto, D. O. Son, A. Wakeham, M. Miyagishi, T. W. Mak and H. Okada, *Plos One*, 2011, **6**, e17830.
98. L. Shi, L. Sun, Q. Li, J. Liang, W. Yu, X. Yi, X. Yang, Y. Li, X. Han, Y. Zhang, C. Xuan, Z. Yao and Y. Shang, *Proc. Natl. Acad. Sci. U. S. A.*, 2011, **108**, 7541–7546.
99. C. H. Chu, L. Y. Wang, K. C. Hsu, C. C. Chen, H. H. Cheng, S. M. Wang, C. M. Wu, T. J. Chen, L. T. Li, R. Liu, C. L. Hung, J. M. Yang, H. J. Kung and W. C. Wang, *J. Med. Chem.*, 2014, **57**, 5975–5985.
100. K. Coffey, L. Rogerson, C. Ryan-Munden, D. Alkharaif, J. Stockley, R. Heer, K. Sahadevan, D. O'Neill, D. Jones, S. Darby, P. Staller, A. Mantilla, L. Gaughan and C. N. Robson, *Nucleic Acids Res.*, 2013, **41**, 4433–4446.
101. L. Zhao, W. Li, W. Zang, Z. Liu, X. Xu, H. Yu, Q. Yang and J. Jia, *Clin. Cancer Res.*, 2013, **19**, 6419–6429.
102. W. L. Berry, T. D. Kim and R. Janknecht, *Int. J. Oncol.*, 2014, **44**, 1341–1348.
103. J. G. Kim, J. M. Yi, S. J. Park, J. S. Kim, T. G. Son, K. Yang, M. A. Yoo and K. Heo, *Biochim. Biophys. Acta*, 2012, **1819**, 1200–1207.
104. W. Li, L. Zhao, W. Zang, Z. Liu, L. Chen, T. Liu, D. Xu and J. Jia, *Biochem. Biophys. Res. Commun.*, 2011, **416**, 372–378.
105. G. Toyokawa, H. S. Cho, Y. Iwai, M. Yoshimatsu, M. Takawa, S. Hayami, K. Maejima, N. Shimizu, H. Tanaka, T. Tsunoda, H. I. Field, J. D. Kelly, D. E. Neal, B. A. Ponder, Y. Maehara, Y. Nakamura and R. Hamamoto, *Cancer Prev. Res.*, 2011, **4**, 2051–2061.
106. Z. Q. Yang, I. Imoto, A. Pimkhaokham, Y. Shimada, K. Sasaki, M. Oka and J. Inazawa, *Jpn. J. Cancer Res.*, 2001, **92**, 423–428.

107. Z. Q. Yang, I. Imoto, Y. Fukuda, A. Pimkhaokham, Y. Shimada, M. Imamura, S. Sugano, Y. Nakamura and J. Inazawa, *Cancer Res.*, 2000, **60**, 4735–4739.
108. B. Berdel, K. Nieminen, Y. Soini, M. Tengstrom, M. Malinen, V. M. Kosma, J. J. Palvimo and A. Mannermaa, *BMC Cancer*, 2012, **12**, 516.
109. G. Liu, A. Bollig-Fischer, B. Kreike, M. J. van de Vijver, J. Abrams, S. P. Ethier and Z. Q. Yang, *Oncogene*, 2009, **28**, 4491–4500.
110. W. Luo, R. Chang, J. Zhong, A. Pandey and G. L. Semenza, *Proc. Natl. Acad. Sci. U. S. A.*, 2012, **109**, E3367–E3376.
111. J. Wu, S. Liu, G. Liu, A. Dombkowski, J. Abrams, R. Martin-Trevino, M. S. Wicha, S. P. Ethier and Z. Q. Yang, *Oncogene*, 2012, **31**, 333–341.
112. E. Metzger, N. Yin, M. Wissmann, N. Kunowska, K. Fischer, N. Friedrichs, D. Patnaik, J. M. Higgins, N. Potier, K. H. Scheidtmann, R. Buettner and R. Schule, *Nat. Cell Biol.*, 2008, **10**, 53–60.
113. M. Wissmann, N. Yin, J. M. Muller, H. Greschik, B. D. Fodor, T. Jenuwein, C. Vogler, R. Schneider, T. Gunther, R. Buettner, E. Metzger and R. Schule, *Nat. Cell Biol.*, 2007, **9**, 347–353.
114. F. Crea, L. Sun, A. Mai, Y. T. Chiang, W. L. Farrar, R. Danesi and C. D. Helgason, *Mol. Cancer*, 2012, **11**, 52.
115. Y. H. Loh, W. Zhang, X. Chen, J. George and H. H. Ng, *Genes Dev.*, 2007, **21**, 2545–2557.
116. J. Wang, M. Zhang, Y. Zhang, Z. Kou, Z. Han, D. Y. Chen, Q. Y. Sun and S. Gao, *Biol. Reprod.*, 2010, **82**, 105–111.
117. P. P. Das, Z. Shao, S. Beyaz, E. Apostolou, L. Pinello, A. De Los Angeles, K. O'Brien, J. M. Atsma, Y. Fujiwara, M. Nguyen, D. Ljuboja, G. Guo, A. Woo, G. C. Yuan, T. Onder, G. Daley, K. Hochedlinger, J. Kim and S. H. Orkin, *Mol. Cell*, 2014, **53**, 32–48.
118. M. T. Pedersen, K. Agger, A. Laugesen, J. V. Johansen, P. A. Cloos, J. Christensen and K. Helin, *Mol. Cell. Biol.*, 2014, **34**, 1031–1045.
119. N. Iwamori, M. Zhao, M. L. Meistrich and M. M. Matzuk, *Biol. Reprod.*, 2011, **84**, 1225–1234.
120. L. Yang, C. Lin, W. Liu, J. Zhang, K. A. Ohgi, J. D. Grinstein, P. C. Dorrestein and M. G. Rosenfeld, *Cell*, 2011, **147**, 773–788.
121. N. Le May, I. Iltis, J. C. Ame, A. Zhovmer, D. Biard, J. M. Egly, V. Schreiber and F. Coin, *Mol. Cell*, 2012, **48**, 785–798.
122. H. Khoury-Haddad, N. Guttmann-Raviv, I. Ipenberg, D. Huggins, A. D. Jeyasekharan and N. Ayoub, *Proc. Natl. Acad. Sci. U. S. A.*, 2014, **111**, E728–E737.
123. Z. Chen, J. Zang, J. Whetstine, X. Hong, F. Davrazou, T. G. Kutateladze, M. Simpson, Q. Mao, C. H. Pan, S. Dai, J. Hagman, K. Hansen, Y. Shi and G. Zhang, *Cell*, 2006, **125**, 691–702.
124. S. Krishnan and R. C. Trievel, *Structure*, 2013, **21**, 98–108.
125. Y. Chang, J. Wu, X. J. Tong, J. Q. Zhou and J. Ding, *Biochem. J.*, 2011, **433**, 295–302.
126. Z. Chen, J. Zang, J. Kappler, X. Hong, F. Crawford, Q. Wang, F. Lan, C. Jiang, J. Whetstine, S. Dai, K. Hansen, Y. Shi and G. Zhang, *Proc. Natl. Acad. Sci. U. S. A.*, 2007, **104**, 10818–10823.

127. J. F. Couture, E. Collazo, P. A. Ortiz-Tello, J. S. Brunzelle and R. C. Trievel, *Nat. Struct. Mol. Biol.*, 2007, **14**, 689–695.
128. S. S. Ng, K. L. Kavanagh, M. A. McDonough, D. Butler, E. S. Pilka, B. M. Lienard, J. E. Bray, P. Savitsky, O. Gileadi, F. von Delft, N. R. Rose, J. Offer, J. C. Scheinost, T. Borowski, M. Sundstrom, C. J. Schofield and U. Oppermann, *Nature*, 2007, **448**, 87–91.
129. E. Metzger and R. Schule, *Nat. Struct. Mol. Biol.*, 2007, **14**, 252–254.
130. T. Swigut and J. Wysocka, *Cell*, 2007, **131**, 29–32.
131. R. Marmorstein and R. C. Trievel, *Biochim. Biophys. Acta*, 2009, **1789**, 58–68.
132. J. R. Wilson, *Nat. Struct. Mol. Biol.*, 2007, **14**, 682–684.
133. J. Kim, J. Daniel, A. Espejo, A. Lake, M. Krishna, L. Xia, Y. Zhang and M. T. Bedford, *EMBO Rep.*, 2006, **7**, 397–403.
134. Y. Huang, J. Fang, M. T. Bedford, Y. Zhang and R. M. Xu, *Science*, 2006, **312**, 748–751.
135. J. Lee, J. R. Thompson, M. V. Botuyan and G. Mer, *Nat. Struct. Mol. Biol.*, 2008, **15**, 109–111.
136. S. D. Taverna, H. Li, A. J. Ruthenburg, C. D. Allis and D. J. Patel, *Nat. Struct. Mol. Biol.*, 2007, **14**, 1025–1040.
137. R. J. Klose, E. M. Kallin and Y. Zhang, *Nat. Rev. Genet.*, 2006, **7**, 715–727.
138. C. Loenarz, W. Ge, M. L. Coleman, N. R. Rose, C. D. Cooper, R. J. Klose, P. J. Ratcliffe and C. J. Schofield, *Hum. Mol. Genet.*, 2010, **19**, 217–222.
139. C. Nava, F. Lamari, D. Heron, C. Mignot, A. Rastetter, B. Keren, D. Cohen, A. Faudet, D. Bouteiller, M. Gilleron, A. Jacquette, S. Whalen, A. Afenjar, D. Perisse, C. Laurent, C. Dupuits, C. Gautier, M. Gerard, G. Huguet, S. Caillet, B. Leheup, M. Leboyer, C. Gillberg, R. Delorme, T. Bourgeron, A. Brice and C. Depienne, *Transl. Psychiatry*, 2012, **2**, e179.
140. Y. Tsukada, T. Ishitani and K. I. Nakayama, *Genes Dev.*, 2010, **24**, 432–437.
141. H. H. Qi, M. Sarkissian, G. Q. Hu, Z. Wang, A. Bhattacharjee, D. B. Gordon, M. Gonzales, F. Lan, P. P. Ongusaha, M. Huarte, N. K. Yaghi, H. Lim, B. A. Garcia, L. Brizuela, K. Zhao, T. M. Roberts and Y. Shi, *Nature*, 2010, **466**, 503–507.
142. X. Sun, J. J. Qiu, S. Zhu, B. Cao, L. Sun, S. Li, P. Li, S. Zhang and S. Dong, *PLoS One*, 2013, **8**, e77353.
143. E. Asensio-Juan, C. Gallego and M. A. Martinez-Balbas, *Nucleic Acids Res.*, 2012, **40**, 9429–9440.
144. M. F. Arteaga, J. H. Mikesch, J. Qiu, J. Christensen, K. Helin, S. C. Kogan, S. Dong and C. W. So, *Cancer Cell*, 2013, **23**, 376–389.
145. J. R. Horton, A. K. Upadhyay, H. H. Qi, X. Zhang, Y. Shi and X. Cheng, *Nat. Struct. Mol. Biol.*, 2010, **17**, 38–43.
146. W. Feng, M. Yonezawa, J. Ye, T. Jenuwein and I. Grummt, *Nat. Struct. Mol. Biol.*, 2010, **17**, 445–450.
147. D. Kleine-Kohlbrecher, J. Christensen, J. Vandamme, I. Abarrategui, M. Bak, N. Tommerup, X. Shi, O. Gozani, J. Rappsilber, A. E. Salcini and K. Helin, *Mol. Cell*, 2010, **38**, 165–178.
148. K. Fortschegger, P. de Graaf, N. S. Outchkourov, F. M. van Schaik, H. T. Timmers and R. Shiekhattar, *Mol. Cell. Biol.*, 2010, **30**, 3286–3298.

149. A. K. Upadhyay, D. Rotili, J. W. Han, R. Hu, Y. Chang, D. Labella, X. Zhang, Y. S. Yoon, A. Mai and X. Cheng, *J. Mol. Biol.*, 2012, **416**, 319–327.
150. Y. Yang, L. Hu, P. Wang, H. Hou, Y. Lin, Y. Liu, Z. Li, R. Gong, X. Feng, L. Zhou, W. Zhang, Y. Dong, H. Yang, H. Lin, Y. Wang, C. D. Chen and Y. Xu, *Cell Res.*, 2010, **20**, 886–898.
151. W. Liu, B. Tanasa, O. V. Tyurina, T. Y. Zhou, R. Gassmann, W. T. Liu, K. A. Ohgi, C. Benner, I. Garcia-Bassets, A. K. Aggarwal, A. Desai, P. C. Dorrestein, C. K. Glass and M. G. Rosenfeld, *Nature*, 2010, **466**, 508–512.
152. H. J. Lim, N. V. Dimova, M. K. Tan, F. D. Sigoillot, R. W. King and Y. Shi, *Mol. Cell. Biol.*, 2013, **33**, 4166–4180.
153. J. R. Horton, A. K. Upadhyay, H. Hashimoto, X. Zhang and X. Cheng, *J. Mol. Biol.*, 2011, **406**, 1–8.
154. H. Wen, J. Li, T. Song, M. Lu, P. Y. Kan, M. G. Lee, B. Sha and X. Shi, *J. Biol. Chem.*, 2010, **285**, 9322–9326.
155. A. Baba, F. Ohtake, Y. Okuno, K. Yokota, M. Okada, Y. Imai, M. Ni, C. A. Meyer, K. Igarashi, J. Kanno, M. Brown and S. Kato, *Nat. Cell Biol.*, 2011, **13**, 668–675.
156. K. Hata, R. Takashima, K. Amano, K. Ono, M. Nakanishi, M. Yoshida, M. Wakabayashi, A. Matsuda, Y. Maeda, Y. Suzuki, S. Sugano, R. H. Whitson, R. Nishimura and T. Yoneda, *Nat. Commun.*, 2013, **4**, 2850.
157. J. D. Stender, G. Pascual, W. Liu, M. U. Kaikkonen, K. Do, N. J. Spann, M. Boutros, N. Perrimon, M. G. Rosenfeld and C. K. Glass, *Mol. Cell*, 2012, **48**, 28–38.
158. Y. Okuno, F. Ohtake, K. Igarashi, J. Kanno, T. Matsumoto, I. Takada, S. Kato and Y. Imai, *Diabetes*, 2013, **62**, 1426–1434.
159. S. C. Trewick, P. J. McLaughlin and R. C. Allshire, *EMBO Rep.*, 2005, **6**, 315–320.
160. N. Ayoub, K. Noma, S. Isaac, T. Kahan, S. I. Grewal and A. Cohen, *Mol. Cell. Biol.*, 2003, **23**, 4356–4370.
161. D. Defeo-Jones, P. S. Huang, R. E. Jones, K. M. Haskell, G. A. Vuocolo, M. G. Hanobik, H. E. Huber and A. Oliff, *Nature*, 1991, **352**, 251–254.
162. E. V. Benevolenskaya, H. L. Murray, P. Branton, R. A. Young and W. G. Kaelin, Jr, *Mol. Cell*, 2005, **18**, 623–635.
163. P. J. Lu, K. Sundquist, D. Baeckstrom, R. Poulsom, A. Hanby, S. Meier-Ewert, T. Jones, M. Mitchell, P. Pitha-Rowe, P. Freemont and J. Taylor-Papadimitriou, *J. Biol. Chem.*, 1999, **274**, 15633–15645.
164. A. Barrett, B. Madsen, J. Copier, P. J. Lu, L. Cooper, A. G. Scibetta, J. Burchell and J. Taylor-Papadimitriou, *Int. J. Cancer*, 2002, **101**, 581–588.
165. Y. Xiang, Z. Zhu, G. Han, X. Ye, B. Xu, Z. Peng, Y. Ma, Y. Yu, H. Lin, A. P. Chen and C. D. Chen, *Proc. Natl. Acad. Sci. U. S. A.*, 2007, **104**, 19226–19231.
166. K. Yamane, K. Tateishi, R. J. Klose, J. Fang, L. A. Fabrizio, H. Erdjument-Bromage, J. Taylor-Papadimitriou, P. Tempst and Y. Zhang, *Mol. Cell*, 2007, **25**, 801–812.
167. A. Barrett, S. Santangelo, K. Tan, S. Catchpole, K. Roberts, B. Spencer-Dene, D. Hall, A. Scibetta, J. Burchell, E. Verdin, P. Freemont and J. Taylor-Papadimitriou, *Int. J. Cancer*, 2007, **121**, 265–275.

168. A. Roesch, M. Fukunaga-Kalabis, E. C. Schmidt, S. E. Zabierowski, P. A. Brafford, A. Vultur, D. Basu, P. Gimotty, T. Vogt and M. Herlyn, *Cell*, 2010, **141**, 583–594.
169. G. L. Dalgliesh, K. Furge, C. Greenman, L. Chen, G. Bignell, A. Butler, H. Davies, S. Edkins, C. Hardy, C. Latimer, J. Teague, J. Andrews, S. Barthorpe, D. Beare, G. Buck, P. J. Campbell, S. Forbes, M. Jia, D. Jones, H. Knott, C. Y. Kok, K. W. Lau, C. Leroy, M. L. Lin, D. J. McBride, M. Maddison, S. Maguire, K. McLay, A. Menzies, T. Mironenko, L. Mulderrig, L. Mudie, S. O'Meara, E. Pleasance, A. Rajasingham, R. Shepherd, R. Smith, L. Stebbings, P. Stephens, G. Tang, P. S. Tarpey, K. Turrell, K. J. Dykema, S. K. Khoo, D. Petillo, B. Wondergem, J. Anema, R. J. Kahnoski, B. T. Teh, M. R. Stratton and P. A. Futreal, *Nature*, 2010, **463**, 360–363.
170. L. R. Jensen, M. Amende, U. Gurok, B. Moser, V. Gimmel, A. Tzschach, A. R. Janecke, G. Tariverdian, J. Chelly, J. P. Fryns, H. Van Esch, T. Kleefstra, B. Hamel, C. Moraine, J. Gecz, G. Turner, R. Reinhardt, V. M. Kalscheuer, H. H. Ropers and S. Lenzner, *Am. J. Hum.Genet.*, 2005, **76**, 227–236.
171. A. Tzschach, S. Lenzner, B. Moser, R. Reinhardt, J. Chelly, J. P. Fryns, T. Kleefstra, M. Raynaud, G. Turner, H. H. Ropers, A. Kuss and L. R. Jensen, *Hum. Mutat.*, 2006, **27**, 389.
172. S. Iwase, F. Lan, P. Bayliss, L. de la Torre-Ubieta, M. Huarte, H. H. Qi, J. R. Whetstine, A. Bonni, T. M. Roberts and Y. Shi, *Cell*, 2007, **128**, 1077–1088.
173. M. Tahiliani, P. Mei, R. Fang, T. Leonor, M. Rutenberg, F. Shimizu, J. Li, A. Rao and Y. Shi, *Nature*, 2007, **447**, 601–605.
174. M. Vashishtha, C. W. Ng, F. Yildirim, T. A. Gipson, I. H. Kratter, L. Bodai, W. Song, A. Lau, A. Labadorf, A. Vogel-Ciernia, J. Troncosco, C. A. Ross, G. P. Bates, D. Krainc, G. Sadri-Vakili, S. Finkbeiner, J. L. Marsh, D. E. Housman, E. Fraenkel and L. M. Thompson, *Proc. Natl. Acad. Sci. U. S. A.*, 2013, **110**, E3027–E3036.
175. S. Catchpole, B. Spencer-Dene, D. Hall, S. Santangelo, I. Rosewell, M. Guenatri, R. Beatson, A. G. Scibetta, J. M. Burchell and J. Taylor-Papadimitriou, *Int. J. Oncol.*, 2011, **38**, 1267–1277.
176. J. Sayegh, J. Cao, M. R. Zou, A. Morales, L. P. Blair, M. Norcia, D. Hoyer, A. J. Tackett, J. S. Merkel and Q. Yan, *J. Biol. Chem.*, 2013, **288**, 9408–9417.
177. T. D. Kim, S. Shin and R. Janknecht, *Biochem. Biophys. Res. Commun.*, 2008, **366**, 563–567.
178. X. Niu, T. Zhang, L. Liao, L. Zhou, D. J. Lindner, M. Zhou, B. Rini, Q. Yan and H. Yang, *Oncogene*, 2012, **31**, 776–786.
179. N. Patani, W. G. Jiang, R. F. Newbold and K. Mokbel, *Anticancer Res.*, 2011, **31**, 4115–4125.
180. S. Tu, Y. C. Teng, C. Yuan, Y. T. Wu, M. Y. Chan, A. N. Cheng, P. H. Lin, L. J. Juan and M. D. Tsai, *Nat. Struct. Mol. Biol.*, 2008, **15**, 419–421.
181. A. G. Scibetta, S. Santangelo, J. Coleman, D. Hall, T. Chaplin, J. Copier, S. Catchpole, J. Burchell and J. Taylor-Papadimitriou, *Mol. Cell. Biol.*, 2007, **27**, 7220–7235.

182. R. J. Klose, Q. Yan, Z. Tothova, K. Yamane, H. Erdjument-Bromage, P. Tempst, D. G. Gilliland, Y. Zhang and W. G. Kaelin, Jr, *Cell*, 2007, **128**, 889–900.
183. F. Li, M. Huarte, M. Zaratiegui, M. W. Vaughn, Y. Shi, R. Martienssen and W. Z. Cande, *Cell*, 2008, **135**, 272–283.
184. G. G. Wang, J. Song, Z. Wang, H. L. Dormann, F. Casadio, H. Li, J. L. Luo, D. J. Patel and C. D. Allis, *Nature*, 2009, **459**, 847–851.
185. R. E. Collins, J. P. Northrop, J. R. Horton, D. Y. Lee, X. Zhang, M. R. Stallcup and X. Cheng, *Nat. Struct. Mol. Biol.*, 2008, **15**, 245–250.
186. K. Zhang, K. Mosch, W. Fischle and S. I. Grewal, *Nat. Struct. Mol. Biol.*, 2008, **15**, 381–388.
187. L. H. Kristensen, A. L. Nielsen, C. Helgstrand, M. Lees, P. Cloos, J. S. Kastrup, K. Helin, L. Olsen and M. Gajhede, *FEBS J.*, 2012, **279**, 1905–1914.
188. C. Koehler, S. Bishop, E. F. Dowler, P. Schmieder, A. Diehl, H. Oschkinat and L. J. Ball, *Biomol. NMR Assignments*, 2008, **2**, 9–11.
189. Q. Chen, X. Chen, Q. Wang, F. Zhang, Z. Lou, Q. Zhang and D. X. Zhou, *PLoS Genet.*, 2013, **9**, e1003239.
190. F. Lan, P. E. Bayliss, J. L. Rinn, J. R. Whetstine, J. K. Wang, S. Chen, S. Iwase, R. Alpatov, I. Issaeva, E. Canaani, T. M. Roberts, H. Y. Chang and Y. Shi, *Nature*, 2007, **449**, 689–694.
191. K. Agger, P. A. Cloos, J. Christensen, D. Pasini, S. Rose, J. Rappsilber, I. Issaeva, E. Canaani, A. E. Salcini and K. Helin, *Nature*, 2007, **449**, 731–734.
192. M. G. Lee, R. Villa, P. Trojer, J. Norman, K. P. Yan, D. Reinberg, L. Di Croce and R. Shiekhattar, *Science*, 2007, **318**, 447–450.
193. F. De Santa, M. G. Totaro, E. Prosperini, S. Notarbartolo, G. Testa and G. Natoli, *Cell*, 2007, **130**, 1083–1094.
194. S. Hong, Y. W. Cho, L. R. Yu, H. Yu, T. D. Veenstra and K. Ge, *Proc. Natl. Acad. Sci. U. S. A.*, 2007, **104**, 18439–18444.
195. K. B. Shpargel, T. Sengoku, S. Yokoyama and T. Magnuson, *PLoS Genet.*, 2012, **8**, e1002964.
196. A. Greenfield, D. Scott, D. Pennisi, I. Ehrmann, P. Ellis, L. Cooper, E. Simpson and P. Koopman, *Nat. Genet.*, 1996, **14**, 474–478.
197. A. A. Mansour, O. Gafni, L. Weinberger, A. Zviran, M. Ayyash, Y. Rais, V. Krupalnik, M. Zerbib, D. Amann-Zalcenstein, I. Maza, S. Geula, S. Viukov, L. Holtzman, A. Pribluda, E. Canaani, S. Horn-Saban, I. Amit, N. Novershtern and J. H. Hanna, *Nature*, 2012, **488**, 409–413.
198. S. Lee, J. W. Lee and S. K. Lee, *Dev. Cell*, 2012, **22**, 25–37.
199. G. G. Welstead, M. P. Creyghton, S. Bilodeau, A. W. Cheng, S. Markoulaki, R. A. Young and R. Jaenisch, *Proc. Natl. Acad. Sci. U. S. A.*, 2012, **109**, 13004–13009.
200. C. Wang, J. E. Lee, Y. W. Cho, Y. Xiao, Q. Jin, C. Liu and K. Ge, *Proc. Natl. Acad. Sci. U. S. A.*, 2012, **109**, 15324–15329.
201. S. Thieme, T. Gyarfas, C. Richter, G. Ozhan, J. Fu, D. Alexopoulou, M. H. Muders, I. Michalk, C. Jakob, A. Dahl, B. Klink, J. Bandola, M. Bachmann, E. Schrock, F. Buchholz, A. F. Stewart, G. Weidinger, K. Anastassiadis and S. Brenner, *Blood*, 2013, **121**, 2462–2473.

202. J. Van der Meulen, F. Speleman and P. Van Vlierberghe, *Epigenetics*, 2014, **9**, 658–668.
203. I. Issaeva, Y. Zonis, T. Rozovskaia, K. Orlovsky, C. M. Croce, T. Nakamura, A. Mazo, L. Eisenbach and E. Canaani, *Mol. Cell. Biol.*, 2007, **27**, 1889–1903.
204. J. K. Wang, M. C. Tsai, G. Poulin, A. S. Adler, S. Chen, H. Liu, Y. Shi and H. Y. Chang, *Genes Dev.*, 2010, **24**, 327–332.
205. M. Terashima, A. Ishimura, M. Yoshida, Y. Suzuki, S. Sugano and T. Suzuki, *Biochem. Biophys. Res. Commun.*, 2010, **399**, 238–244.
206. S. B. Ng, A. W. Bigham, K. J. Buckingham, M. C. Hannibal, M. J. McMillin, H. I. Gildersleeve, A. E. Beck, H. K. Tabor, G. M. Cooper, H. C. Mefford, C. Lee, E. H. Turner, J. D. Smith, M. J. Rieder, K. Yoshiura, N. Matsumoto, T. Ohta, N. Niikawa, D. A. Nickerson, M. J. Bamshad and J. Shendure, *Nat. Genet.*, 2010, **42**, 790–793.
207. M. C. Hannibal, K. J. Buckingham, S. B. Ng, J. E. Ming, A. E. Beck, M. J. McMillin, H. I. Gildersleeve, A. W. Bigham, H. K. Tabor, H. C. Mefford, J. Cook, K. Yoshiura, T. Matsumoto, N. Matsumoto, N. Miyake, H. Tonoki, K. Naritomi, T. Kaname, T. Nagai, H. Ohashi, K. Kurosawa, J. W. Hou, T. Ohta, D. Liang, A. Sudo, C. A. Morris, S. Banka, G. C. Black, J. Clayton-Smith, D. A. Nickerson, E. H. Zackai, T. H. Shaikh, D. Donnai, N. Niikawa, J. Shendure and M. J. Bamshad, *Am. J. Med. Genet. Part A*, 2011, **155A**, 1511–1516.
208. D. Lederer, B. Grisart, M. C. Digilio, V. Benoit, M. Crespin, S. C. Ghariani, I. Maystadt, B. Dallapiccola and C. Verellen-Dumoulin, *Am. J. Hum. Genet.*, 2012, **90**, 119–124.
209. N. Miyake, S. Mizuno, N. Okamoto, H. Ohashi, M. Shiina, K. Ogata, Y. Tsurusaki, M. Nakashima, H. Saitsu, N. Niikawa and N. Matsumoto, *Hum. Mutat.*, 2013, **34**, 108–110.
210. G. van Haaften, G. L. Dalgliesh, H. Davies, L. Chen, G. Bignell, C. Greenman, S. Edkins, C. Hardy, S. O'Meara, J. Teague, A. Butler, J. Hinton, C. Latimer, J. Andrews, S. Barthorpe, D. Beare, G. Buck, P. J. Campbell, J. Cole, S. Forbes, M. Jia, D. Jones, C. Y. Kok, C. Leroy, M. L. Lin, D. J. McBride, M. Maddison, S. Maquire, K. McLay, A. Menzies, T. Mironenko, L. Mulderrig, L. Mudie, E. Pleasance, R. Shepherd, R. Smith, L. Stebbings, P. Stephens, G. Tang, P. S. Tarpey, R. Turner, K. Turrell, J. Varian, S. West, S. Widaa, P. Wray, V. P. Collins, K. Ichimura, S. Law, J. Wong, S. T. Yuen, S. Y. Leung, G. Tonon, R. A. DePinho, Y. T. Tai, K. C. Anderson, R. J. Kahnoski, A. Massie, S. K. Khoo, B. T. Teh, M. R. Stratton and P. A. Futreal, *Nat. Genet.*, 2009, **41**, 521–523.
211. M. Tumino, M. Licciardello, G. Sorge, M. C. Cutrupi, F. Di Benedetto, L. Amoroso, R. Catania, M. Pennisi, S. D'Amico and A. Di Cataldo, *Am. J. Med. Genet. Part A*, 2010, **152A**, 1536–1539.
212. J. H. Merks, H. N. Caron and R. C. Hennekam, *Am. J. Med. Genet. Part A*, 2005, **134A**, 132–143.
213. S. Scherer, U. Theile, V. Beyer, R. Ferrari, C. Kreck and M. Rister, *Am. J. Med. Genet. Part A*, 2003, **122A**, 76–79.

214. O. Ijichi, K. Kawakami, Y. Matsuda, N. Ikarimoto, K. Miyata, H. Takamatsu and M. Tokunaga, *Acta Paediatr. Jpn.*, 1996, **38**, 66–68.
215. S. A. Miller, S. E. Mohn and A. S. Weinmann, *Mol. Cell*, 2010, **40**, 594–605.
216. L. J. Walport, R. J. Hopkinson, M. Vollmar, S. K. Madden, C. Gileadi, U. Oppermann, C. J. Schofield and C. Johansson, *J. Biol. Chem.*, 2014, **289**, 18302–18303.
217. T. Sengoku and S. Yokoyama, *Genes Dev.*, 2011, **25**, 2266–2277.
218. L. Kruidenier, C. W. Chung, Z. Cheng, J. Liddle, K. Che, G. Joberty, M. Bantscheff, C. Bountra, A. Bridges, H. Diallo, D. Eberhard, S. Hutchinson, E. Jones, R. Katso, M. Leveridge, P. K. Mander, J. Mosley, C. Ramirez-Molina, P. Rowland, C. J. Schofield, R. J. Sheppard, J. E. Smith, C. Swales, R. Tanner, P. Thomas, A. Tumber, G. Drewes, U. Oppermann, D. J. Patel, K. Lee and D. M. Wilson, *Nature*, 2012, **488**, 404–408.
219. X. Zhang, H. Wen and X. Shi, *Acta Biochim. Biophys. Sin. (Shanghai)*, 2012, **44**, 14–27.
220. K. E. Moore and O. Gozani, *Biochim. Biophys. Acta*, 2014, **1839**, 1395–1403.

CHAPTER 8

AlkB and Its Homologues – DNA Repair and Beyond

TINA A. MÜLLER[a] AND ROBERT P. HAUSINGER*[a,b]

[a]Department of Microbiology and Molecular Genetics, Michigan State University, East Lansing, MI 48824, USA; [b]Department of Biochemistry and Molecular Biology, Michigan State University, East Lansing, MI 48824, USA
*E-mail: hausinge@msu.edu

8.1 Introduction

DNA is constantly exposed to both endogenous alkylating agents like the metabolic methyl donor *S*-adenosylmethionine and exogenous chemical species such as found in tobacco smoke or cancer treatment drugs. The resultant alkylated base adducts can be toxic and mutagenic; thus, organisms have evolved multiple DNA repair pathways to remove these lesions from the genome.[1–6] The most efficient chemical approach to repair an alkylated DNA base is direct dealkylation to restore the unalkylated base without breaking the sugar-phosphate backbone. One enzyme that uses this process for DNA repair is AlkB. The expression of the gene encoding this protein in *Escherichia coli* was found to be induced by methyl methanesulfonate (MMS) as part of the adaptive response to alkylation damage.[7,8] Despite knowing that AlkB functions to repair alkylated lesions, the mechanism of this reaction was unclear for 20 years until computational studies predicted it to be a member of the diverse group of Fe(II)/2-oxoglutarate (2OG)-dependent dioxygenases.[9,10] The enzyme was shown to be an oxidative demethylase that hydroxylates *N*-alkyl groups on DNA bases such as 1-methyladenine (1meA) or 3-methylcytosine

RSC Metallobiology Series No. 3
2-Oxoglutarate-Dependent Oxygenases
Edited by Robert P. Hausinger and Christopher J. Schofield

Published by the Royal Society of Chemistry, www.rsc.org

(3meC), producing unstable intermediates that decompose to restore the native bases while releasing formaldehyde (Figure 8.1).[11,12]

Homologues of *E. coli* AlkB have been found in other bacteria, viruses, metazoa and plants, but not archaea.[13–18] (Note that some bacterial proteins are also named AlkB, but possess dinuclear iron sites that hydroxylate alkanes; they are unrelated to the dioxygenase proteins of interest here). Humans have nine AlkB homologues (here referred to by the general name Alkbh, although the human proteins are correctly denoted ALKBH) 1–8 and FTO (sometimes referred to as Alkbh9) (Figure 8.2). Two of the human homologues repair lesions in alkylated DNA in cell lines,[19,20] whereas the other representatives are involved in a variety of roles including mRNA or tRNA hydroxylation and acting on protein substrates.[21–24] A few studies have described AlkB-like proteins found in organisms other than *E. coli* and mammals (*e.g.* other bacteria, the protozoan *Trypanosoma brucei* and the plant *Arabidopsis thaliana*), but these homologues also appear to be involved in DNA repair, thus confirming the importance of oxidative demethylation to protect the organism from cytotoxic and mutagenic damage.[15–17,25–28]

This chapter focuses first on the archetypal AlkB from *E. coli*, describing its role in DNA repair, examining its reaction mechanism, and detailing its biophysical characterization. Second, we discuss the mammalian homologues Alkbh2 and Alkbh3 that also function in repair of alkylated DNA. Next, we briefly describe the RNA-demethylases Alkbh5 and FTO (see also Chapter 9), the tRNA modifying enzyme Alkbh8 (see Chapter 10), and involvement of Alkbh4 and Alkbh7 in several cellular functions. Finally, we summarize what is known about Alkbh1 and Alkbh6 where their enzyme functions remain uncertain.

8.2 Repair of Alkylated DNA (and RNA) in *E. coli*: AlkB

AlkB is one of four enzymes that function in *E. coli*'s adaptive response to alkylation damage. When exposed to alkylating agents, the dual-acting DNA demethylase/transcriptional activator Ada increases expression of its own gene as well as *alkA*, *alkB* and *aidB*.[7,8,29] Ada exhibits both methyl phosphotriester methyltransferase activity, which repairs alkylation damage to the DNA backbone, and 4-methylthymine and 6-methylguanine methyltransferase activities, that shifts the methyl group from these lesions to catalytic Cys residues of the protein; AlkA is a 3-methyladenosine glycosylase, creating an abasic (AP or apurinic/apyrimidinic) site in the DNA; AlkB is an Fe(II)/2OG-dependent dioxygenase that hydroxylates 1meA and 3meC which then decompose to the native base and formaldehyde with concomitant formation of succinate and CO_2 (Figure 8.1); and the role of AidB remains unknown.[30] It was shown that *alkB* mutants are defective in processing *N*-methylation damage in single-stranded (ss) DNA and later studies confirmed this specificity with the purified enzyme.[31,32]

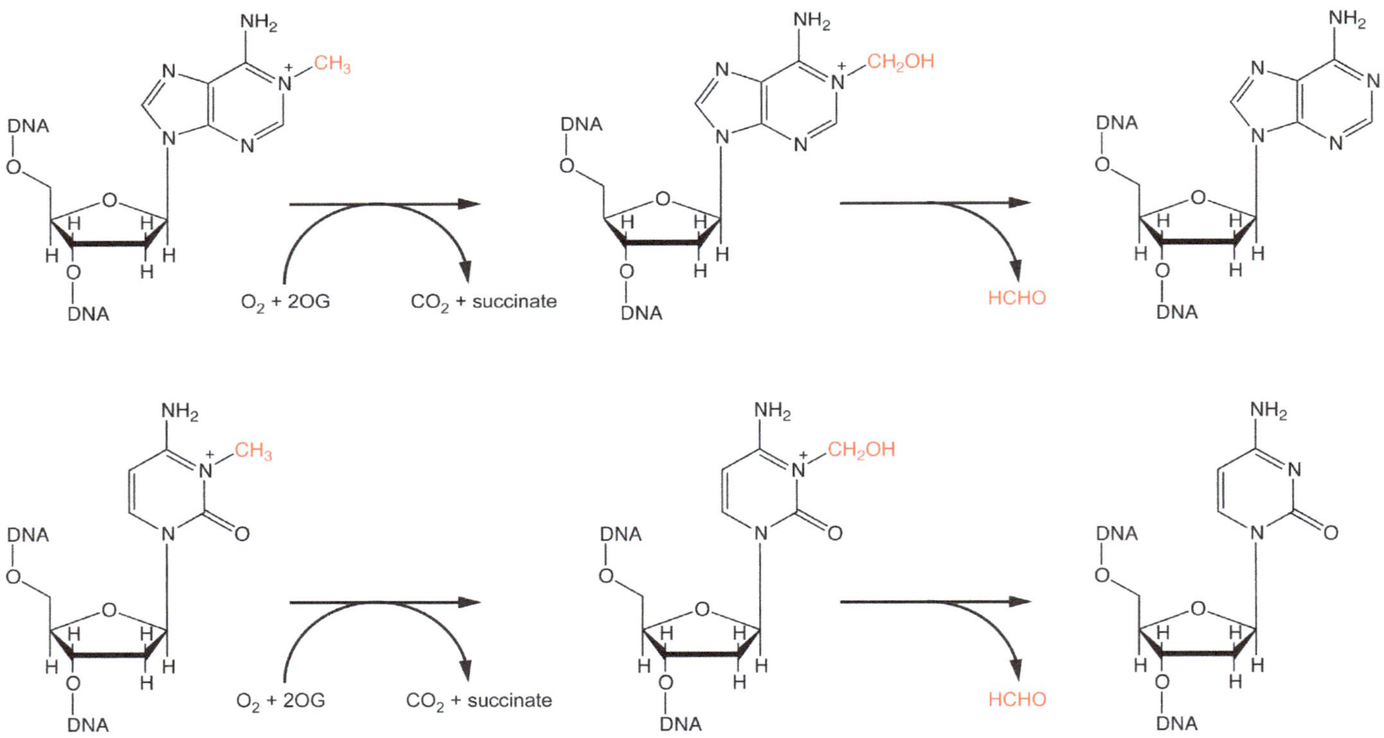

Figure 8.1 Oxidative demethylation reactions catalysed by *E. coli* AlkB using DNA (or RNA) containing 1meA (top) and 3meC (bottom) lesions.

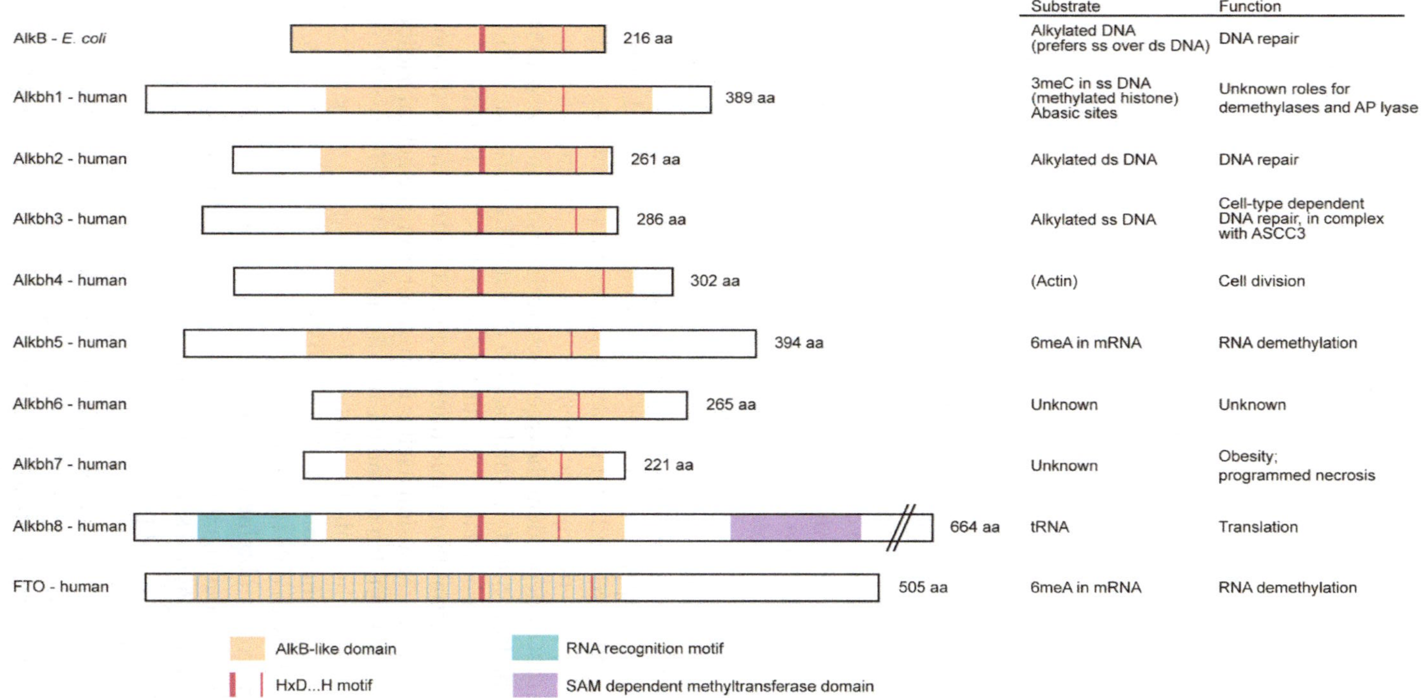

Figure 8.2 Overview of *E. coli* AlkB and its nine human homologues including a schematic representation of the sequence organization of the proteins, their preferred substrates, and cellular roles, when known. The AlkB-like domains for homologues 1–8 are shown, as identified by the multiple sequence alignment published by Kurowski.[13] For FTO (Alkbh9), the demethylase domain was assigned according to the region listed in the UniProtKB/Swiss-Prot protein database. The overall lower sequence identity is indicated by the hatching pattern.

AlkB possesses robust hydroxylation activity and shows broad substrate specificity. It acts on the *N*-alkylated bases that disrupt double-stranded (ds) DNA, such as the above-mentioned 1meA and 3meC, along with 3-methylthymine, 1-methylguanine and 6-methyladenine.[11,12,31,33–36] The AlkB reaction which releases formaldehyde is reminiscent of the *N*-demethylation catalysed by histone demethylases (see Chapter 7). AlkB's substrate range further includes exocyclic substrates such as N^2-alkylguanine or N^4-alkylcytosine, where the protonated form is preferentially turned over according to molecular dynamics studies.[34,35,37,38] Likewise, etheno DNA lesions, which are by-products of lipid peroxidation, are repaired.[39,40] It has also been reported that AlkB hydroxylates 1,N^6-ethanoA, a highly damaging lesion in mammalian cells produced by the antitumour agent 1,3-bis (chloroethyl)-1-nitrosourea; however, AlkB does not restore the native base, rather it hydroxylates the C8 position leading to the non-toxic and only weakly mutagenic N^6 adduct.[41] While AlkB hydroxylates a wide variety of base adducts, the lesions are not repaired at equal rates – the dioxygenase shows a preference for 1meA and 3meC and, at least with isolated components, for ssDNA over dsDNA.[33,34,41–43] It is noteworthy that the overall catalytic efficiency is similar when using DNA samples containing 1meA and 3meC, but the K_m and k_{cat} vary widely for such substrates, a phenomenon called k_{cat}–K_m compensation.[40] The minimal substrate for hydroxylation is 1-methyl-dAMP(5′), while nucleosides such as 1meA and 3meC only stimulate 2OG turnover, but are not demethylated, indicating that AlkB requires a phosphate 5′ of the damaged base.[44,45] AlkB also readily acts on *N*-alkylated RNA, as shown using chemically treated mRNA and tRNA.[11,31,46–48]

A large number of crystal structures of AlkB have been solved with various substrates, products and metals at the active site (see two examples in Figure 8.3).[40,49–52] The core enzyme contains the double-stranded β-helix fold typical for Fe(II)/2OG-dependent dioxygenases; residues His-131, Asp-133 and His-187 bind the iron cofactor, which is also chelated by 2OG that is further stabilized at its C5 carboxylate by a salt bridge to Arg-204. The substrate is recognized by a structurally flexible nucleotide recognition lid located in the *N*-terminal region and the lesion itself is flipped out from the helical backbone and sandwiched between Trp-69 and the iron ligand His-131 (Figure 8.3A, right panel). It is interesting how binding of the different substrates to the active site explains AlkB's preference for ssDNA. The ssDNA and dsDNA substrates bind almost identically, with only the lesion-containing strand interacting with the protein.[49,52] Structures obtained with duplex DNA show the DNA is kinked, thereby pressing the damaged base into the active site and stacking the two flanking nucleotides (Figure 8.3B, right panel).[49] AlkB does not possess a 'finger residue' to fill the vacant position in dsDNA, unlike Alkbh2 (see below) or DNA glycosylases.[49] The complementary strand must be accommodated, an energetically unfavorable situation, and this requirement makes flipping out the base more costly for a ds-substrate than for a ss-substrate. The interactions between AlkB and substrate do not discriminate between a deoxy- and ribonucleotide backbone, consistent with the activity measurements showing high

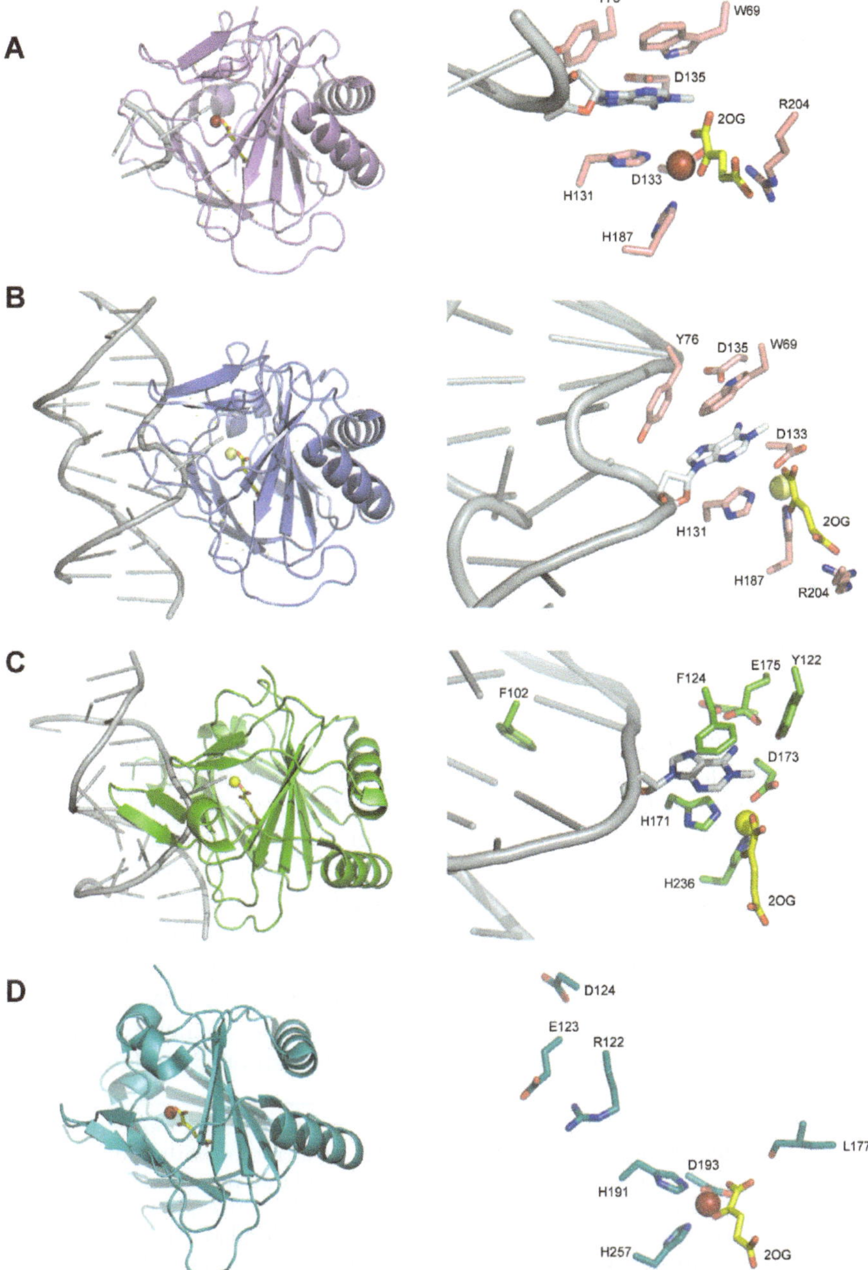

Figure 8.3 Views of crystal structures for *E. coli* AlkB and human homologues Alkbh2 and Alkbh3. (A) The left panel shows a cartoon view of AlkB bound to the trinucleotide T-1meA-T (PDB ID: 2FD8). AlkB is coloured in light purple and the DNA is shown in grey. The right panel illustrates a close-up view of the active site residues (light purple) and the

turnover numbers with both DNA and RNA.[52] In contrast, Asp-135 is involved in catalysis and responsible for the preference of 1meA and 3meC over the alkylated bases.[51]

Of additional structural interest, comparison of AlkB apoprotein with the holo-enzyme containing bound substrate indicated little movement between the two states.[52] The authors speculated that Tyr-76 rotates upon binding of the methylated base to close the active site pocket. It is noteworthy that O_2 diffusion into the active site might be gated because open and closed states of a putative oxygen binding tunnel were found in different crystal structures;[51] however, self-hydroxylation of Trp-178 can take place in the absence of the primary substrate, leading to the generation of a blue chromophore, indicating that O_2 access is not completely blocked.[53] In contrast to the small structural changes due to substrate binding found by using X-ray analysis, NMR structures of AlkB showed that the apoprotein is highly dynamic whereas the protein becomes globular and forms a more stable hydrophobic core upon binding of Fe(II), 2OG and DNA.[54] Earlier NMR structures also suggested differences in the dynamic properties of the AlkB-2OG and AlkB-succinate complexes and indicated that cosubstrate decarboxylation leads to an enhanced dynamic state.[55]

Oxidative demethylation of alkylated DNA by AlkB is thought to occur according to the canonical reaction mechanism of Fe(II)/2OG-dependent dioxygenases, most clearly detailed in studies of the taurine-degrading enzyme TauD (see Chapter 3).[4,5,56] That is, the substrate is hydroxylated by a multi-step mechanism including an Fe(IV)-oxo species to form an unstable intermediate that decomposes to products. In order to visualize the predicted intermediates, anaerobic crystals of AlkB with bound substrates were exposed to air and *in crystallo* turnover was investigated.[57] For example, this treatment of DNA containing the 3-methylthymine lesion captured the theoretically predicted hemiaminal base intermediate, confirming the proposed reaction mechanism. Of interest, the 2OG cofactor was observed in the 'off-line' position, *i.e.* its C1 carboxylate is opposite of His-187, resulting in an oxygen binding site that is opposite of His-131 and away from the primary substrate.[51] The oxygen binding position was further examined by using quantum mechanics/molecular mechanics methods,[58] and two oxygen

trinucleotide is shown as a ribbon with 1meA in stick mode. (B) On the left, AlkB (PDB ID: 3BI3) is depicted as a cartoon with bound ds DNA and the lesion flipped out of the helix and into the active site. The protein is coloured in purple. On the right, the active site is shown using the same colouring as in A. The DNA is visualized as a ribbon with 1meA in stick mode. (C) The left side depicts a cartoon view of Alkbh2 (PDB ID: 3BUC) with bound ds DNA and the lesion pointed into the active site. Alkbh2 is in green. The panel on the right shows the active site residues in green and the DNA in cartoon view with the damaged base in stick mode. (D) Crystal structure of Alkbh3 (PDB ID: 2IUW) with the cartoon view on the left and the active site on the right. The protein is in teal. In all panels, the DNA is in grey, 2OG is in yellow and the metal is the coloured sphere (red for iron or yellow for manganese substitution).

binding positions are proposed. The favoured oxygen binding site is suggested to be opposite of His-131 and the authors posit a mechanism in which an Arg residue helps the Fe(IV)-oxo intermediate to isomerize prior to hydrogen abstraction so that it is directed toward the substrate.

8.3 Repair of Alkylated DNA in Mammals: Alkbh2 and Alkbh3

Four of the nine mammalian homologues have been shown to possess measurable DNA demethylation activity, namely Alkbh1-3 and FTO.[20,46,59,60] Alkbh1 hydroxylates 3meC in ssDNA at low rates; however, despite several attempts from our lab as well as others, this protein has not been shown to be involved in DNA demethylation *in vivo* and *alkbh1*$^{-/-}$ mouse fibroblasts are not more sensitive to alkylating agents than wild-type cells.[61] FTO possesses low levels of 3-methylthymine and 3-methyluracil demethylase activity, but has now been classified as a 6-methyladenine demethylase acting on RNA (see Section 8.4 and Chapter 9).[21] In contrast, Alkbh2 and Alkbh3 are *bona fide* demethylases of alkylated DNA.

Alkbh2 and Alkbh3 were the first mammalian homologues shown to possess activity analogous to AlkB. Each of these enzymes demethylates DNA bases *in vitro* and complements an *alkB*$^{-}$ *E. coli* strain.[20,46] The two proteins efficiently catalyse the repair reaction at rates that approach the diffusion limit, but they show different substrate specificities and kinetic parameters.[33,46] Alkbh2 acts on both dsDNA and ssDNA, preferring the former, whereas Alkbh3 only demethylates ssDNA and ssRNA. Like AlkB, Alkbh2 hydroxylates ethenoA as well as ethenoC.[62,63] The latter substrate is converted at rates similar to alkylated lesions, but the activity can be blocked by the base excision repair enzyme alkyladenine DNA glycosylase, which binds to this type of adduct without initiating repair. This result indicates an overlap of the direct DNA repair and the base excision repair pathways.[62] As expected for an Fe(II)/2OG-dependent dioxygenase, the activity is fully dependent on Fe(II), mutating the iron ligands results in a catalytically inactive variant,[33] and nickel ions inhibit Alkbh2 activity by competing for the metal binding site.[64,65]

Numerous crystal structures have been solved for Alkbh2, but only one for Alkbh3 (Figure 8.3, panels C and D). Both enzymes have the typical jelly-roll motif of Fe(II)/2OG-dependent dioxygenases at the core, the predicted iron-binding ligands, and the expected binding of 2OG (when co-crystallized) at the active site.[49,66] Five residues of Alkbh2 (Phe-124, His-171, Tyr-122, Glu-175 and Asp-174) interact with the 1meA lesion in contrast to the two flanking amino acids of AlkB (Figure 8.3A, B and C). Also unlike AlkB, Alkbh2 has two loops interacting with the opposite strand, explaining its substrate preference for dsDNA. While both enzymes employ a base-flipping mechanism, Alkbh2 uses a finger residue, Phe-102, to intercalate into the DNA, thus filling the space vacated by the flipped out lesion.[49] It is also noteworthy that only alkylated substrates are positioned correctly in the active site to be hydroxylated, ensuring that native bases are not unnecessarily modified.[50] DNA repair proteins need to scan the genome in search of the damaged base. In order to gain insights into Alkbh2's search mode, it was crystallized with different duplex DNAs. Alkbh2 detects the damaged base due to the weakened base pairing caused by the methyl group at the N1 or N3 position for A and C, respectively.[50,67]

Alkbh3 was crystallized in the presence of Fe(II) and 2OG without substrate, but comparison to AlkB and Alkbh2 still reveals interesting features (Figure 8.3D).[66] Two β-strands, β4 and β5, form a hairpin and a lid over the active site, equivalent to the nucleotide recognition lid in AlkB. Alkbh3's active site is more polar than the hydrophobic core in either AlkB or Alkbh2, which might explain the narrower substrate specificity of this mammalian homologue. Arg-122 is predicted to intercalate into DNA, analogous to Phe-102 in Alkbh2, indicating that Alkbh3 uses a different mechanism to flip out the lesion than that used by AlkB (Figure 8.3D, right panel). Only a few of the DNA backbone–protein interactions are conserved between AlkB and Alkbh3, and the authors speculate that the DNA is bound differently in these two proteins. Alkbh2 and Alkbh3 exhibit different preferences for their substrates, attributed to a short stretch of amino acids. When Arg-122/Glu-123/Asp-124 of Alkbh3 is replaced by the corresponding Val-Phe-Gly residues of Alkbh2, the former enzyme gains activity with dsDNA; however, the single E123F variant results in a similar change in a different study.[68,69] In the Alkbh3 crystal structure, approximately 69% of Leu-177 is oxidized; of interest, substitution of this residue with Ala, Asn or Gln renders the protein inactive. The authors speculated that the modified residue acts as a buffer stop to prevent the pyrimidine substrates from penetrating too deeply into the active site. Alternatively, it could impair 2OG binding, thereby preventing uncoupled turnover.

In order to study the *in vivo* role of Alkbh2 and Alkbh3, *Alkbh2*$^{-/-}$ and *Alkbh3*$^{-/-}$ mice were created and analysed.[19] Both types of null mice (and the double mutants) are viable, shown to be generated at Mendelian ratios, and found to be phenotypically indistinguishable from wild-type animals. DNA extracted from livers of mice deficient in Alkbh2 shows 1meA accumulation, unlike DNA of *alkbh3*$^{-/-}$ mice.[63] Growth of *alkbh2*$^{-/-}$ mouse embryonic fibroblasts (MEFs) is slightly affected in the presence of MMS, these cells are more sensitive to MMS treatment, and they show a greater mutation rate.[19,63,70] Localization studies corroborate these findings, *i.e.* Alkbh2 is found exclusively in the nucleus.[71] A more detailed analysis showed that Alkbh2 is in the nucleolus and, upon MMS exposure, it redistributes to the nucleoplasm.[72] Alkbh2 was found to interact with several nucleolar proteins, the DNA binding factor Ku70/80 and rDNA gene regions, while its overexpression stimulated the transcriptional activity of rDNA promoters, suggesting that this homologue plays an important role in demethylation of rDNA genes and thereby functions in transcription. Taken together, these observations demonstrate that mammalian alkylation repair relies primarily on Alkbh2 and suggest that Alkbh2 and Alkbh3 have distinct roles in the body.[72]

Whereas Alkbh2 was established early on as the main *in vivo* DNA repair demethylase, Alkbh3's role has remained unclear. Just recently, Alkbh3 was found to be in a complex with activating signal cointegrator complex (ASCC) 3 helicase. ASCC3 is a 3′–5′ DNA helicase and its unwinding activity could reasonably promote Alkbh3-mediated alkylation repair of dsDNA.[73] ASCC3 together with Alkbh3 were shown to prevent accumulation of 3meC in specific cancer cells, whereas loss of the dioxygenase causes sensitivity to alkylation damage. This finding suggests that Alkbh3 acts as a demethylase in a cell-type specific manner.[73] Indeed, Alkbh3 and Alkbh2 overexpression have been associated with cancer and either homologue has been found in a wide variety of cancers such as non-small cell lung cancer, rectal cancer, paediatric brain tumours, bladder cancer, as well as prostate cancer. It has been speculated that overexpression of these proteins contributes to cancer cell line survival, making these homologues an interesting target in cancer therapy.[74–80] At least one study showed that downregulation of Alkbh2 increases the sensitivity of cells to cisplatin, whereas Alkbh2 is downregulated in gastric cancer and its overexpression inhibits cell proliferation.[81] Surprisingly, to the best of our knowledge no studies have reported on regulation of either homologue upon exposure to alkylating agents. It would be interesting to study whether such chemicals, used for instance in cancer therapy, affect expression of Alkbh2 or Alkbh3.

8.4 AlkB Homologues Involved in Cellular Processes Other than DNA Repair

In recent years it was revealed that several AlkB homologues are involved in cellular processes other than DNA repair of alkylation damage. Alkbh5 and FTO remove methylation from the N^6 position of adenine in mRNA (see Chapter 9), whereas Alkbh8 acts on tRNA (see Chapter 10). Two other AlkB representatives function in cellular processes that are independent of nucleic acids: Alkbh4 is involved in cell division and Alkbh7 plays a role in obesity.

An initial study examining the functional role of Alkbh4 showed that it interacts with the transcriptional co-activator/histone acetyltransferase p300, other transcription factors, and unknown proteins associated with nucleosomes; the authors speculated that Alkbh4 might have a role in transcriptional activation.[82] In contrast, a more recent publication demonstrates that Alkbh4 localizes to the contractile ring and mid-body structure formed during cell division.[24] Alkbh4 interacts with a specifically post-translationally modified actin variant, actin K84me1, and depletion of the enzyme leads to impaired cell division and increased polynucleation and apoptosis.[24] Because the critical interaction between actin and non-muscle myosin II depends on the methylation status of the former, the authors speculated that Alkbh4 mediates cleavage furrow formation by demethylating actin and thereby allowing the cognate myosin to move forward on the actin filament. It is important to point out that Alkbh4 deletion in mice was embryonically lethal, in sharp contrast to the Alkbh1-3, Alkbh5, Alkbh7 and FTO depleted animals which are viable and show only a modest phenotype.[19,22,24,61,83,84] Despite the evidence from these cellular studies, no *in vitro* substrate has been found for Alkbh4 and it still remains to be determined whether Alkbh4 is a true protein or nucleic acid demethylase or possesses another activity.

Mammalian Alkbh7 also has a function beyond DNA repair. One study demonstrated that the protein does not possess 1meA or 3meC demethylase activity, and its depletion in HeLa cells led to a surprising increased resistance to alkylating DNA damaging agents and H_2O_2.[85] Interestingly, complementation with the gene encoding wild-type protein or a 2OG binding-deficient mutant each rescue the phenotype, suggesting that 2OG binding is not required for Alkbh7's *in vivo* function. Alkylating agents can induce programmed necrosis that is characterized by poly-ADP-ribose polymerase (PARP) hyperactivation leading to NAD and ATP depletion, mitochondrial dysfunction, reactive oxygen species formation and cell death. While cells depleted of Alkbh7 react like wild-type cells to the stress, the absence of this protein enables the cell to recover quickly and re-establish normal PARP and NAD levels. The authors therefore speculated that Alkbh7 depletion suppresses damage-induced necrosis.[85] In agreement with these findings, Alkbh7 is localized to the mitochondria[85–87] and Alkbh7-depleted cells maintain their mitochondrial function after exposure to alkylating agents.[85]

An *Alkbh7*$^{-/-}$ mouse was created by deletion of exons 2–4 encompassing the region encoding the Fe(II)/2OG dioxygenase domain.[86] These Alkbh7 null animals have an obese phenotype, which is more pronounced in males. The severity of the symptoms increases when the animals are on a high-fat diet due to production of more adipose tissue. Detailed analysis revealed higher levels of short-chain fatty acids, indicating that *Alkbh7*$^{-/-}$ mice are impaired in fatty acid oxidation. In conclusion, Alkbh7 is a mitochondrial protein and has been implicated in two different cellular processes – fatty acid metabolism and cell necrosis. Similar to the situation for Alkbh4, no *in vitro* activity has been associated with either role connected to Alkbh7.

8.5 AlkB Homologues with Unknown Functions: Alkbh1 and Alkbh6

No functional roles are yet associated with the last two members of the mammalian AlkB homologues, Alkbh1 and Alkbh6. Several studies have examined aspects of Alkbh1 (described below), whereas information about Alkbh6 is limited to knowledge that it is highly expressed in testis and pancreas and is localized throughout the cell.[71] No *in vitro* activity has been found for this protein and its function still remains to be determined.

Alkbh1 is the mammalian homologue with the greatest sequence similarity to AlkB,[13] yet only trace levels of DNA demethylase activity are observed.[59] Two different genotypes of Alkbh1 deficient mice have been created; one has a deletion of exon 3 and the other of exon 6.[61,83,88] Both types of *Alkbh1*$^{-/-}$ mice exhibit an unusual sex-ratio distortion in favour of males. Furthermore, *Alkbh1*$^{-/-}$ embryos show intra-uterine growth retardation, the pups are smaller, and they are born at a non-Mendelian ratio.

The cellular function of Alkbh1 remains ambiguous, at best. This homologue of AlkB has been reported as being localized to the nucleus or the mitochondria, or detected diffusely throughout the cell, depending on the study.[59,71,88,89] Likewise, expression of Alkbh1 is reported to be elevated in a wide variety of tissues, including trophoblast lineages of the developing placenta, muscle and heart, or spleen.[59,71,88] It is therefore not surprising that roles hypothesized for this homologue are also highly diverse. Alkbh1 was proposed to be a 'histone dioxygenase' functioning in neural differentiation of murine stem cells,[89] whereas another study found the expression of non-repeat piRNA clusters is upregulated in *Alkbh1*- and *Tzfp*-deficient murine testes, consistent with a role in transposon control in these cells.[90] *In vitro*, Alkbh1 exhibits lyase activity, separate from its putative hydroxylase activity, at AP sites in DNA, and is even capable of introducing dsDNA breaks when two AP sites are in close proximity on the opposing DNA strands.[91] This chemistry is independent of 2OG, oxygen, Fe(II) or a functional Fe(II) binding site. Interestingly, Alkbh1 covalently binds to the 5′ product of cleavage, forming a protein–DNA adduct.[92] The AP lyase activity, but not the demethylase activity, was also reported for the *Schizosaccharomyces pombe* homologue.[93] AP endonuclease activity is known to be associated with base excision repair and class switch recombination, but Alkbh1 is dispensable for both of these processes.[61] To conclude, despite many efforts, the function and enzymatic activity of Alkbh1 remains to be determined.

8.6 Conclusions

AlkB and its homologues have been studied extensively, but fundamental questions remain to be addressed for several of these proteins. A clear role in the repair of alkylation-damaged DNA has been demonstrated for the proteins from several bacteria, a plant and a protozoan, along with mammalian Alkbh2 and Alkbh3.[11,19,20,46] An enlarged picture of AlkB function has emerged through studies of other homologues that play additional and diverse cellular roles. For example, Alkbh5 and FTO are demethylases of RNA, although the biological roles of this chemistry are still unclear (see also Chapter 9).[21,22] Alkbh8 catalyses a hydroxylation step during the synthesis of a modified tRNA (see also Chapter 10).[94] Alkbh1 is an AP lyase in addition to having reported DNA and histone demethylase activities.[89,91,93] Alkbh1-deficient mice have shown a peculiar phenotype of a distorted sex ratio[61,83,88] and this homologue has been implicated in placental trophoblast lineage differentiation as well as transposon control in pachytene spermatocytes,[88–90] but overall there remains no clear connection to a robust enzymatic activity. Similarly, the cellular roles of Alkbh4 in cell division and Alkbh7 in programmed necrosis and obesity have been studied, but no *in vitro* activity has been found.[24,86] In conclusion, AlkB and its homologues play many important cellular roles, yet we only partially understand the extent of these functions and the diversity of the enzymatic reactions catalysed.

Acknowledgements

AlkB and Alkbh1 studies in the author's laboratory were supported by the National Institutes of Health (R21AI79430 to T.A.M. and GM063584 to R.P.H.).

References

1. P. Ø. Falnes, A. Klungland and I. Alseth, *Neuroscience*, 2007, **145**, 1222–1232.
2. B. Sedgwick, P. A. Bates, J. Paik, S. C. Jacobs and T. Lindahl, *DNA Repair (Amst)*, 2007, **6**, 429–442.
3. C.-G. Yang, K. Garcia and C. He, *ChemBioChem*, 2009, **10**, 417–423.
4. C. Yi and C. He, *Cold Spring Harb. Perspect. Biol.*, 2013, **5**, a012575.
5. C. Yi, C. G. Yang and C. He, *Acc. Chem. Res.*, 2009, **42**, 519–529.
6. D. Fu, J. A. Calvo and L. D. Samson, *Nat. Rev. Cancer*, 2012, **12**, 104–120.
7. H. Kataoka, Y. Yamamoto and M. Sekiguchi, *J. Bacteriol.*, 1983, **153**, 1301–1307.
8. H. Kondo, Y. Nakabeppu, H. Kataoka, S. Kuhara, S. Kawabata and M. Sekiguchi, *J. Biol. Chem.*, 1986, **261**, 15772–15777.
9. L. Aravind and E. V. Koonin, *Genome Biol.*, 2001, **2**, RESEARCH0007.
10. J. M. Simmons, T. A. Müller and R. P. Hausinger, *Dalton Trans.*, 2008, 5132–5142.
11. P. Ø. Falnes, R. F. Johansen and E. Seeberg, *Nature*, 2002, **419**, 178–182.
12. S. C. Trewick, T. F. Henshaw, R. P. Hausinger, T. Lindahl and B. Sedgwick, *Nature*, 2002, **419**, 174–178.
13. M. A. Kurowski, A. S. Bhagwat, G. Papaj and J. M. Bujnicki, *BMC Genomics*, 2003, **4**, 48.
14. E. van den Born, A. Bekkelund, M. N. Moen, M. V. Omelchenko, A. Klungland and P. Ø. Falnes, *Nucleic Acids Res.*, 2009, **37**, 7124–7136.
15. E. van den Born, M. V. Omelchenko, A. Bekkelund, V. Leihne, E. V. Koonin, V. V. Dolja and P. Ø. Falnes, *Nucleic Acids Res.*, 2008, **36**, 5451–5461.
16. M. S. Bratlie and F. Drabløs, *BMC Genomics*, 2005, **6**, 1.
17. L. V. Mello and D. J. Rigden, *FEBS Lett.*, 2012, **586**, 3908–3913.
18. D. Mielecki, D. L. Zugaj, A. Muszewska, J. Piwowarski, A. Chojnacka, M. Mielecki, J. Nieminuszczy, M. Grynberg and E. Grzesiuk, *PLoS One*, 2012, **7**, e30588.
19. J. Ringvoll, L. M. Nordstrand, C. B. Vågbø, V. Talstad, K. Reite, P. A. Aas, K. H. Lauritzen, N. B. Liabakk, A. Bjørk, R. W. Doughty, P. Ø. Falnes, H. E. Krokan and A. Klungland, *EMBO J.*, 2006, **25**, 2189–2198.
20. T. Duncan, S. C. Trewick, P. Koivisto, P. A. Bates, T. Lindahl and B. Sedgwick, *Proc. Natl. Acad. Sci. U. S. A.*, 2002, **99**, 16660–16665.

21. G. Jia, Y. Fu, X. Zhao, Q. Dai, G. Zheng, Y. Yang, C. Yi, T. Lindahl, T. Pan, Y. G. Yang and C. He, *Nat. Chem. Biol.*, 2011, **7**, 885–887.
22. G. Zheng, J. A. Dahl, Y. Niu, P. Fedorcsak, C. M. Huang, C. J. Li, C. B. Vågbø, Y. Shi, W. L. Wang, S. H. Song, Z. Lu, R. P. Bosmans, Q. Dai, Y. J. Hao, X. Yang, W. M. Zhao, W. M. Tong, X. J. Wang, F. Bogdan, K. Furu, Y. Fu, G. Jia, X. Zhao, J. Liu, H. E. Krokan, A. Klungland, Y. G. Yang and C. He, *Mol. Cell*, 2013, **49**, 18–29.
23. E. van den Born, C. B. Vågbø, L. Songe-Moller, V. Leihne, G. F. Lien, G. Leszczynska, A. Malkiewicz, H. E. Krokan, F. Kirpekar, A. Klungland and P. Ø. Falnes, *Nat. Commun.*, 2011, **2**, 172.
24. M. M. Li, A. Nilsen, Y. Shi, M. Fusser, Y. H. Ding, Y. Fu, B. Liu, Y. Niu, Y. S. Wu, C. M. Huang, M. Olofsson, K. X. Jin, Y. Lv, X. Z. Xu, C. He, M. Q. Dong, J. M. Rendtlew Danielsen, A. Klungland and Y. G. Yang, *Nat. Commun.*, 2013, **4**, 1832.
25. D. Colombi and S. L. Gomes, *J. Bacteriol.*, 1997, **179**, 3139–3145.
26. T. J. Meza, M. N. Moen, C. B. Vågbø, H. E. Krokan, A. Klungland, P. E. Grini and P. Ø. Falnes, *Nucleic Acids Res.*, 2012, **40**, 6620–6631.
27. J. M. Simmons, D. J. Koslowsky and R. P. Hausinger, *Exp. Parasitol.*, 2012, **131**, 92–100.
28. D. Mielecki, S. Saumaa, M. Wrzesinski, A. M. Maciejewska, K. Zuchniewicz, A. Sikora, J. Piwowarski, J. Nieminuszczy, M. Kivisaar and E. Grzesiuk, *PLoS One*, 2013, **8**, e76198.
29. I. Teo, B. Sedgwick, M. W. Kilpatrick, T. V. McCarthy and T. Lindahl, *Cell*, 1986, **45**, 315–324.
30. P. Landini and M. R. Volkert, *J. Bacteriol.*, 2000, **182**, 6543–6549.
31. P. Ø. Falnes, M. Bjørås, P. A. Aas, O. Sundheim and E. Seeberg, *Nucleic Acids Res.*, 2004, **32**, 3456–3461.
32. S. Dinglay, S. C. Trewick, T. Lindahl and B. Sedgwick, *Genes Dev.*, 2000, **14**, 2097–2105.
33. D. H. Lee, S. G. Jin, S. Cai, Y. Chen, G. P. Pfeifer and T. R. O'Connor, *J. Biol. Chem.*, 2005, **280**, 39448–39459.
34. D. Li, J. C. Delaney, C. M. Page, X. Yang, A. S. Chen, C. Wong, C. L. Drennan and J. M. Essigmann, *J. Am. Chem. Soc.*, 2012, **134**, 8896–8901.
35. A. M. Maciejewska, J. Poznanski, Z. Kaczmarska, B. Krowisz, J. Nieminuszczy, A. Polkowska-Nowakowska, E. Grzesiuk and J. T. Kusmierek, *J. Biol. Chem.*, 2013, **288**, 432–441.
36. P. Koivisto, P. Robins, T. Lindahl and B. Sedgwick, *J. Biol. Chem.*, 2004, **279**, 40470–40474.
37. J. C. Delaney, L. Smeester, C. Wong, L. E. Frick, K. Taghizadeh, J. S. Wishnok, C. L. Drennan, L. D. Samson and J. M. Essigmann, *Nat. Struct. Mol. Biol.*, 2005, **12**, 855–860.
38. Y. Mishina, C. G. Yang and C. He, *J. Am. Chem. Soc.*, 2005, **127**, 14594–14595.
39. A. M. Maciejewska, B. Sokolowska, A. Nowicki and J. T. Kusmierek, *Mutagenesis*, 2011, **26**, 401–406.
40. B. Yu and J. F. Hunt, *Proc. Natl. Acad. Sci. U.S. A.*, 2009, **106**, 14315–14320.

41. L. E. Frick, J. C. Delaney, C. Wong, C. L. Drennan and J. M. Essigmann, *Proc. Natl. Acad. Sci. U. S. A.*, 2007, **104**, 755–760.
42. D. Li, B. I. Fedeles, N. Shrivastav, J. C. Delaney, X. Yang, C. Wong, C. L. Drennan and J. M. Essigmann, *Chem. Res. Toxicol.*, 2013, **26**, 1182–1187.
43. D. Li, J. C. Delaney, C. M. Page, A. S. Chen, C. Wong, C. L. Drennan and J. M. Essigmann, *J. Nucleic Acids*, 2010, **2010**, 369434.
44. P. Koivisto, T. Duncan, T. Lindahl and B. Sedgwick, *J. Biol. Chem.*, 2003, **278**, 44348–44354.
45. R. W. Welford, I. Schlemminger, L. A. McNeill, K. S. Hewitson and C. J. Schofield, *J. Biol. Chem.*, 2003, **278**, 10157–10161.
46. P. A. Aas, M. Otterlei, P. Ø. Falnes, C. B. Vågbø, F. Skorpen, M. Akbari, O. Sundheim, M. Bjørås, G. Slupphaug, E. Seeberg and H. E. Krokan, *Nature*, 2003, **421**, 859–863.
47. C. B. Vågbø, E. K. Svaasand, P. A. Aas and H. E. Krokan, *DNA Repair (Amst)*, 2013, **12**, 188–195.
48. R. Ougland, C. M. Zhang, A. Liiv, R. F. Johansen, E. Seeberg, Y. M. Hou, J. Remme and P. Ø. Falnes, *Mol. Cell*, 2004, **16**, 107–116.
49. C. G. Yang, C. Yi, E. M. Duguid, C. T. Sullivan, X. Jian, P. A. Rice and C. He, *Nature*, 2008, **452**, 961–965.
50. C. Yi, B. Chen, B. Qi, W. Zhang, G. Jia, L. Zhang, C. J. Li, A. R. Dinner, C. G. Yang and C. He, *Nat. Struct. Mol. Biol.*, 2012, **19**, 671–676.
51. B. Yu, W. C. Edstrom, J. Benach, Y. Hamuro, P. C. Weber, B. R. Gibney and J. F. Hunt, *Nature*, 2006, **439**, 879–884.
52. P. J. Holland and T. Hollis, *PLoS One*, 2010, **5**, e8680.
53. T. F. Henshaw, M. Feig and R. P. Hausinger, *J. Inorg. Biochem.*, 2004, **98**, 856–861.
54. B. Bleijlevens, T. Shivarattan, K. S. van den Boom, A. de Haan, G. van der Zwan, P. J. Simpson and S. J. Matthews, *Biochemistry*, 2012, **51**, 3334–3341.
55. B. Bleijlevens, T. Shivarattan, E. Flashman, Y. Yang, P. J. Simpson, P. Koivisto, B. Sedgwick, C. J. Schofield and S. J. Matthews, *EMBO Rep.*, 2008.
56. P. K. Grzyska, E. H. Appelman, R. P. Hausinger and D. A. Proshlyakov, *Proc. Natl. Acad. Sci. U. S. A.*, 2010, **107**, 3982–3987.
57. C. Yi, G. Jia, G. Hou, Q. Dai, W. Zhang, G. Zheng, X. Jian, C. G. Yang, Q. Cui and C. He, *Nature*, 2010, **468**, 330–333.
58. M. G. Quesne, R. Latifi, L. E. Gonzalez-Ovalle, D. Kumar and S. P. de Visser, *Chemistry*, 2014, **20**, 435–446.
59. M. P. Westbye, E. Feyzi, P. A. Aas, C. B. Vågbø, V. A. Talstad, B. Kavli, L. Hagen, O. Sundheim, M. Akbari, N. B. Liabakk, G. Slupphaug, M. Otterlei and H. E. Krokan, *J. Biol. Chem.*, 2008, **283**, 25046–25056.
60. T. Gerken, C. A. Girard, Y. C. Tung, C. J. Webby, V. Saudek, K. S. Hewitson, G. S. Yeo, M. A. McDonough, S. Cunliffe, L. A. McNeill, J. Galvanovskis, P. Rorsman, P. Robins, X. Prieur, A. P. Coll, M. Ma, Z. Jovanovic, I. S. Farooqi, B. Sedgwick, I. Barroso, T. Lindahl, C. P. Ponting, F. M. Ashcroft, S. O'Rahilly and C. J. Schofield, *Science*, 2007, **318**, 1469–1472.
61. T. A. Müller, K. Yu, R. P. Hausinger and K. Meek, *PLoS One*, 2013, **8**, e67403.

62. D. Fu and L. D. Samson, *DNA Repair (Amst)*, 2012, **11**, 46–52.
63. J. Ringvoll, M. N. Moen, L. M. Nordstrand, L. B. Meira, B. Pang, A. Bekkelund, P. C. Dedon, S. Bjelland, L. D. Samson, P. Ø. Falnes and A. Klungland, *Cancer Res.*, 2008, **68**, 4142–4149.
64. N. C. Giri, H. Sun, H. Chen, M. Costa and M. J. Maroney, *Biochemistry*, 2011, **50**, 5067–5076.
65. H. Chen, N. C. Giri, R. Zhang, K. Yamane, Y. Zhang, M. Maroney and M. Costa, *J. Biol. Chem.*, 2010, **285**, 7374–7383.
66. O. Sundheim, C. B. Vågbø, M. Bjørås, M. M. Sousa, V. Talstad, P. A. Aas, F. Drabløs, H. E. Krokan, J. A. Tainer and G. Slupphaug, *EMBO J.*, 2006, **25**, 3389–3397.
67. L. Lu, C. Yi, X. Jian, G. Zheng and C. He, *Nucleic Acids Res.*, 2010, **38**, 4415–4425.
68. B. Chen, H. Liu, X. Sun and C. G. Yang, *Mol. Biosyst.*, 2010, **6**, 2143–2149.
69. V. T. Monsen, O. Sundheim, P. A. Aas, M. P. Westbye, M. M. Sousa, G. Slupphaug and H. E. Krokan, *Nucleic Acids Res.*, 2010, **38**, 6447–6455.
70. S. L. Nay, D. H. Lee, S. E. Bates and T. R. O'Connor, *DNA Repair (Amst)*, 2012, **11**, 502–510.
71. K. Tsujikawa, K. Koike, K. Kitae, A. Shinkawa, H. Arima, T. Suzuki, M. Tsuchiya, Y. Makino, T. Furukawa, N. Konishi and H. Yamamoto, *J. Cell. Mol. Med.*, 2007, **11**, 1105–1116.
72. P. Li, S. Gao, L. Wang, F. Yu, J. Li, C. Wang, J. Li and J. Wong, *Cell Rep.*, 2013, **4**, 817–829.
73. S. Dango, N. Mosammaparast, M. E. Sowa, L. J. Xiong, F. Wu, K. Park, M. Rubin, S. Gygi, J. W. Harper and Y. Shi, *Mol. Cell*, 2011, **44**, 373–384.
74. S. Y. Choi, J. H. Jang and K. R. Kim, *Clin. Exp. Med.*, 2011, **11**, 219–226.
75. M. Tasaki, K. Shimada, H. Kimura, K. Tsujikawa and N. Konishi, *Br. J. Cancer*, 2011, **104**, 700–706.
76. N. Konishi, M. Nakamura, E. Ishida, K. Shimada, E. Mitsui, R. Yoshikawa, H. Yamamoto and K. Tsujikawa, *Clin. Cancer Res.*, 2005, **11**, 5090–5097.
77. V. Cetica, L. Genitori, L. Giunti, M. Sanzo, G. Bernini, M. Massimino and I. Sardi, *J. Neuro-Oncol.*, 2009, **94**, 195–201.
78. T. C. Johannessen, L. Prestegarden, A. Grudic, M. E. Hegi, B. B. Tysnes and R. Bjerkvig, *Neuro-Oncology*, 2013, **15**, 269–278.
79. S. Y. Lee, S. K. Luk, C. P. Chuang, S. P. Yip, S. S. To and Y. M. Yung, *Br. J. Cancer*, 2010, **103**, 362–369.
80. S. S. Wu, W. Xu, S. Liu, B. Chen, X. L. Wang, Y. Wang, S. F. Liu and J. Q. Wu, *Acta Pharmacol. Sin.*, 2011, **32**, 393–398.
81. W. Gao, L. Li, P. Xu, J. Fang, S. Xiao and S. Chen, *J. Gastroenterol. Hepatol.*, 2011, **26**, 577–584.
82. L. G. Bjørnstad, T. J. Meza, M. Otterlei, S. M. Olafsrud, L. A. Meza-Zepeda and P. Ø. Falnes, *PLoS One*, 2012, **7**, e49045.
83. L. M. Nordstrand, J. Svärd, E. Larsen, A. Nilsen, R. Ougland, K. Furu, G. F. Lien, T. Rognes, S. H. Namekawa, J. T. Lee and A. Klungland, *PLoS One*, 2010, **5**, e13827.

84. J. Fischer, L. Koch, C. Emmerling, J. Vierkotten, T. Peters, J. C. Bruning and U. Ruther, *Nature*, 2009, **458**, 894–898.
85. D. Fu, J. J. Jordan and L. D. Samson, *Genes Dev.*, 2013, **27**, 1089–1100.
86. A. Solberg, A. B. Robertson, J. M. Aronsen, Ø. Rognmo, I. Sjaastad, U. Wisløff and A. Klungland, *J. Mol. Cell. Biol.*, 2013, **5**, 194–203.
87. H. W. Rhee, P. Zou, N. D. Udeshi, J. D. Martell, V. K. Mootha, S. A. Carr and A. Y. Ting, *Science*, 2013, **339**, 1328–1331.
88. Z. Pan, S. Sikandar, M. Witherspoon, D. Dizon, T. Nguyen, K. Benirschke, C. Wiley, P. Vrana and S. M. Lipkin, *Dev. Dyn.*, 2008, **237**, 316–327.
89. R. Ougland, D. Lando, I. Jonson, J. A. Dahl, M. N. Moen, L. M. Nordstrand, T. Rognes, J. T. Lee, A. Klungland, T. Kouzarides and E. Larsen, *Stem Cells*, 2012, **30**, 2672–2682.
90. L. M. Nordstrand, K. Furu, J. Paulsen, T. Rognes and A. Klungland, *Nucleic Acids Res.*, 2012, **40**, 10950–10963.
91. T. A. Müller, K. Meek and R. P. Hausinger, *DNA Repair (Amst)*, 2010, **9**, 58–65.
92. T. A. Müller, M. M. Andrzejak and R. P. Hausinger, *Biochem. J.*, 2013, **452**, 509–518.
93. H. Korvald, P. Ø. Falnes, J. K. Laerdahl, M. Bjørås and I. Alseth, *DNA Repair (Amst)*, 2012, **11**, 453–462.
94. Y. Fu, Q. Dai, W. Zhang, J. Ren, T. Pan and C. He, *Angew. Chem., Int. Ed. Engl.*, 2010, **49**, 8885–8888.

CHAPTER 9

RNA Demethylation by FTO and ALKBH5

GUANQUN ZHENG[a] AND CHUAN HE*[a]

[a]Department of Chemistry, Institute for Biophysical Dynamics, University of Chicago, 929 East 57th Street, Chicago, IL, USA
*E-mail: chuanhe@uchicago.edu

9.1 Introduction

Cellular DNA can be methylated by endogenous and exogenous methylating agents. The resulting DNA methylation adducts can be cytotoxic and/or mutagenic, and must be promptly repaired. One prominent repair mechanism is direct oxidative demethylation mediated through the AlkB family proteins (see Chapter 8). AlkB, characterized as a DNA repair enzyme in *Escherichia coli*, uses a non-haem mononuclear Fe(II) and the cosubstrates 2-oxoglutarate (2OG) and dioxygen to perform a unique oxidative dealkylation repair of N^1-methyladenine and N^3-methylcytosine lesions on genomic DNA.[1,2] The AlkB family of Fe(II)/2OG-dependent dioxygenases is widespread from bacteria to humans.[3,4] To date, nine human homologues of AlkB have been identified and termed ALKBH1-ALKBH8 and FTO (fat mass and obesity-associated). Proteins of this family catalyse a wide range of biological oxidations, representing an important class of hydroxylases inside cells.[1–3,5–12]

While methylation can be present as a form of nucleic acid damage, it is also well known for its biological signalling and regulatory roles. Reversible methylations on DNA and histones have been well accepted as key processes to regulate gene expression. For instance, DNA methylation in the form of 5-methylcytosine (5mC) carries another layer of heritable information

RSC Metallobiology Series No. 3
2-Oxoglutarate-Dependent Oxygenases
Edited by Robert P. Hausinger and Christopher J. Schofield

Published by the Royal Society of Chemistry, www.rsc.org

independent of genomic sequence, and is deeply involved in gene regulation, genetic imprinting, development and diseases.[13,14] Recent discoveries of the sequential oxidized derivatives of 5mC in mammals, *i.e.* 5-hydroxymethylcytosine (5hmC), 5-formylcytosine (5fC) and 5-carboxylcytosine (5caC) (see Chapter 11), not only reveal the complexity of the demethylation processes, but also provide the possibility of additional regulatory DNA modifications besides 5mC.[15–18] Histone methylation is another type of epigenetic marker that mainly occurs on lysine or arginine residues of histones following translation.[19–22] Information on enzymes responsible for methylation and those catalysing oxidative demethylation (Chapter 7) has emerged, and the antagonized activities were shown to establish a dynamic balance of histone methylation where the regulatory significance has been well established.[23–25]

In contrast to the well-known DNA and histone methylations, reversible chemical modifications (*e.g.* methylation) on RNA remained as unexplored territory for a long time. RNA carries genetic information from DNA to protein in the central dogma.[26,27] Cellular RNAs contain more than a hundred structurally distinct post-transcriptional modifications at thousands of sites, serving versatile coding, structural and catalytic functions.[28,29] N^6-Methyladenosine (m^6A) is one of the most interesting RNA modifications because it presents as the most abundant internal modification in the messenger RNA (mRNA) of all higher eukaryotes.[30,31] Each mammalian mRNA contains ~3 m^6A modifications within the consensus sequences of Pu[G> A]m^6AC[A/C/U].[32–34] The m^6A modification is post-transcriptionally installed by a multicomponent m^6A methyltransferase (MT) complex yet to be fully characterized;[35] the core complex has been recently revealed to contain a METTL3-METTL14 heterocomplex with participation of the splicing regulator WTAP.[36] Although essential to cell viability and development,[37–39] the exact role of m^6A in biological systems was poorly understood. We had envisioned that m^6A could be formed reversibly and, analogous to reversible DNA and histone methylations, it might play regulatory roles with biological significance. Our efforts in searching for the reversible RNA modification were rewarded with the first-time discovery of the reversible RNA modification as well as the first two RNA demethylases, FTO and ALKBH5, both of which are AlkB human homologues.[40,41]

9.2 FTO

The *Fto* (fat mass and obesity associated) gene was first discovered in a fused-toe mutant mouse as a result of a genomic deletion of six genes, which included *Fto*.[42] It is a 9-exon gene located on human chromosome 16 and mouse chromosome 8. Later, in 2007, three independent genome-wide-association studies, designed to search for type 2 diabetes susceptibility genes, revealed a common variant of *FTO* that generates a predisposition to obesity.[43–45] Homozygous variants of *FTO* affect an estimated one billion people in the world.[46] *FTO* is ubiquitously expressed,[6,47] with high levels found in the hypothalamus within the brain, a region that controls energy homeostasis.[6]

Subsequent studies further revealed the association of *FTO* to the control of food intake,[48–51] albeit with no correlation to energy expenditure.[52,53]

Mutants in a mouse model further confirmed the importance of FTO in mammalian energy homeostasis. The loss of *Fto* by replacing its exons 2 and 3 with a neomycin resistance cassette results in postnatal growth retardation and a significant reduction in adipose tissue and lean body mass.[54] Missense mutations yielding I367F and R316Q in mouse FTO lead to leanness and growth retardation, respectively.[55,56] Overexpression of *Fto* in mice gives rise to increased food intake and can result in obesity in a dose-dependent manner.[57]

FTO belongs to the AlkB family of non-haem Fe(II)/2OG-dependent dioxygenases that typically catalyse demethylations/hydroxylations. FTO was initially found to oxidatively demethylate *N*^{3}-methylthymine (m^{3}T) in single-stranded DNA (ssDNA)[6,58] and *N*^{3}-methyluracil (m^{3}U) in single-stranded RNA (ssRNA)[58] *in vitro*. However, the observed activity is extremely low compared with that of other AlkB proteins, not to mention the lack of physiological relevance of these two modifications *in vivo*, suggesting other functions for FTO. Indeed, we discovered that FTO is capable of efficiently demethylating the ubiquitous m^{6}A in RNA both *in vitro* and inside cells. Knockdown of FTO leads to increased amounts of m^{6}A in mRNA, whereas overexpression of FTO results in decreased amounts of m^{6}A in mRNA isolated from HeLa and 293FT human cells, demonstrating m^{6}A in RNA as a main physiologically relevant substrate of FTO.[40]

Like other AlkB family proteins, FTO uses a mononuclear Fe(II) site, as well as 2OG as a cosubstrate, to activate the dioxygen molecule for the oxidative demethylation of m^{6}A. The catalysis involves two reaction phases (Figure 9.1A). Firstly, the dioxygen is activated in the enzyme active site to form a putative Fe(IV)-oxo intermediate, accompanied by the conversion of 2OG to succinate and CO_2. Secondly, the Fe(IV)-oxo species may abstract an H atom followed by hydroxyl radical rebound to hydroxylate the methyl group. During the process of FTO-mediated oxidative demethylation, m^{6}A can be oxidized to *N*^{6}-hydroxymethyladenosine (hm^{6}A), and hm^{6}A could be further oxidized to *N*^{6}-formyladenosine (f^{6}A) (Figure 9.1B). Both hm^{6}A and f^{6}A fragment in water to afford the final demethylated product.[59,60]

The oxidation intermediates hm^{6}A and f^{6}A are relatively stable, and can be detected and confirmed *in vitro* and *in vivo*.[59] Both hm^{6}A and f^{6}A have half-lives of ~3 hours under physiological conditions, and are detected in mRNA isolated from human cells and mouse tissues.[59] Reminiscent of the DNA demethylation intermediates 5hmC, 5fC and 5caC in mammalian cells,[18] these RNA oxidation derivatives may also play potential biological roles.

FTO prefers single-stranded to double-stranded substrates.[58] Crystallographic comparison of the FTO structure with those of other AlkB proteins revealed that FTO possesses an extra loop that competes with the unmethylated strand of the DNA/RNA duplex for binding to FTO.[61] In addition to the conserved catalytic AlkB-like domain, FTO contains a unique *C*-terminal domain with a novel fold comprised mainly of α-helices, which is also required for the

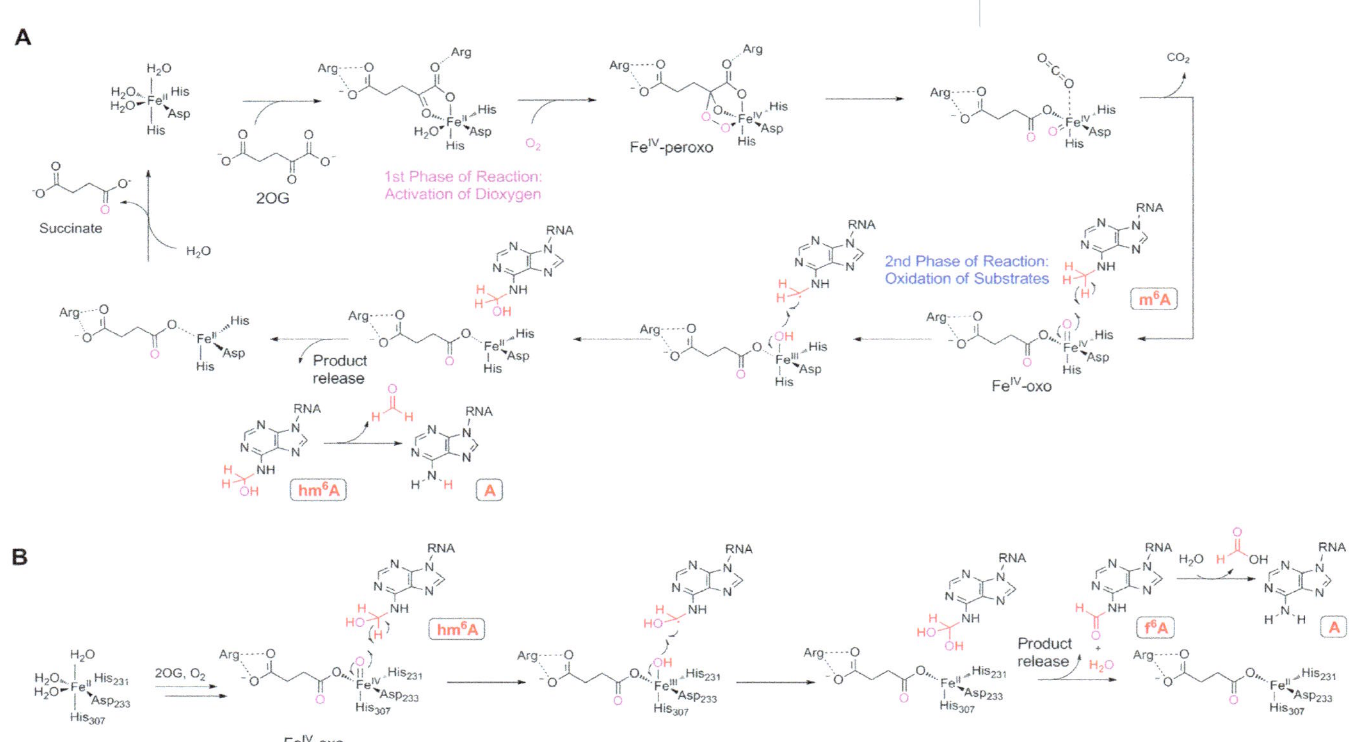

Figure 9.1 Oxidative demethylation of m^{6}A. (A) Proposed mechanism of demethylase-mediated oxygen activation and oxidative demethylation of m^{6}A, which occurs in both FTO and ALKBH5. (B) Oxidative conversion of hm^{6}A in RNA by FTO to yield intermediate f^{6}A and final product A.

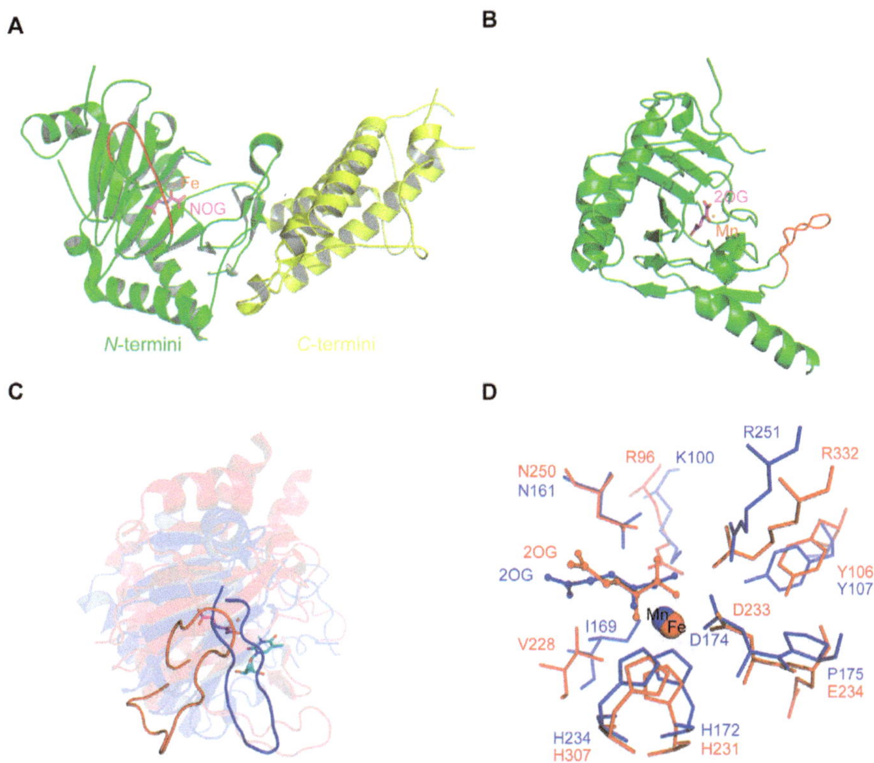

Figure 9.2 Views from crystal structures of m^{6}A RNA demethylases. (A) Crystal structure of FTO (PDB ID: 3LFM). The catalytic AlkB-like domain (residues 32–326) and *C*-terminal domain (residues 327–498) of FTO are coloured in green and yellow, respectively. The loop for single-stranded substrate recognition is highlighted in red. Fe(II) is shown in orange and *N*-oxalylglycine (NOG, a 2OG homologue) is shown in magenta. (B) Crystal structure of zebrafish ALKBH5 (fALKBH5, PDB ID: 4NPL). The single-stranded substrate recognition loop is highlighted in red. Mn, substituting for Fe(II), is labelled in orange and 2OG is labelled in magenta. (C) Structure alignment of FTO and fALKBH5 with the recognition loop highlighted. FTO and fALKBH5 are coloured red and blue, respectively. Fe(II) is labelled in orange, NOG in magenta, and m^{3}T in cyan for illustration. (D) Superimposition of residues of the catalytic cores in FTO and hALKBH5. Structures of FTO and fALKBH5 are coloured red and blue, respectively.

catalytic activity (Figure 9.2A).[61] In spite of these findings, a crystal structure of m^{6}A-bound FTO complex is yet to be obtained. Moreover, existing tools (crosslinking and immunoprecipitation, or CLIP,[62] and photoactivatable-ribonucleoside-enhanced crosslinking and immunoprecipitation, PAR-CLIP)[63] have so far failed to identify RNA substrates of FTO (unpublished data from our work and other labs), suggesting labile interactions between FTO and its substrates. A crystal structure with bound substrates would be very useful in understanding the basis of substrate selectivity. Innovative approaches that can trap and identify bound substrates of FTO are highly desired in order to understand the chemical basis of its exact biological functions.

FTO is present in the cell nucleus in a dot-like manner.[6,40] It partially co-localizes with nuclear speckles where mRNA methylation and splicing take place.[40] Immunofluorescence analysis indicates that FTO is recruited to the nuclear speckles to process nascently transcribed mRNA.[40] Recent analysis comparing the m^{6}A-IP (immunoprecipitation)-enriched mRNA from wild-type and *Fto*$^{-/-}$ mouse brain showed that FTO specifically demethylates a functionally unique subset of mRNA involved in neuronal function.[64] Interestingly, the protein levels of these targets significantly attenuate whereas their mRNA levels remain unaltered, suggesting a potential role of m^{6}A in affecting mRNA translation.[64] Very recent work showed that proteins that specifically recognize m^{6}A modification can dramatically affect the mRNA stability, therefore impacting protein production through mRNA methylation.[65] Small-molecule inhibitors of human FTO have been developed with the hope of their use as therapeutics of obesity and diabetes diseases in the future.[66,67]

9.3 ALKBH5

The discovery of FTO as the first m^{6}A mRNA demethylase raised the possibility of the existence of other demethylases. Mammalian homologues of FTO were biochemically tested towards m^{6}A-containing oligonucleotides. These experiments revealed that recombinant ALKBH5 also exhibits efficient m^{6}A demethylation activity.[41] Cell-based assays confirmed this activity. An ~9% increase of m^{6}A level in total mRNA was observed upon the knockdown of ALKBH5 for 48 hours, whereas an ~29% decrease of m^{6}A level in mRNA resulted from the overexpression of ALKBH5 for 24 hours.[41] ALKBH5 has been characterized as an Fe(II)/2OG-dependent dioxygenase that catalyses the decarboxylation of 2OG.[68] It was annotated as a potential mRNA-binding protein based on photocrosslinking-based mRNA-bound proteomics.[69,70] However, the exact role of ALKBH5 was largely unknown until our discovery that ALKBH5 is an RNA demethylase. The possibility that ALKBH5 also works on m^{6}A in other RNA species, such as ribosomal and long non-coding RNA, cannot be excluded.

ALKBH5 prefers single-stranded over double-stranded substrates, consistent with its role as an RNA demethylase.[41] It exhibits a higher demethylation activity towards m^{6}A-containing consensus sequences over non-consensus sequences.[41] ALKBH5 adopts the conserved catalytic AlkB-like domain as seen for FTO (Figure 9.1A). However, in contrast to FTO, neither hm^{6}A nor f^{6}A can be detected during the oxidative demethylation of m^{6}A by ALKBH5. Because both hm^{6}A and f^{6}A have half lives of ~3 hours,[59] ALKBH5 may catalyse decomposition of hm^{6}A, thus avoiding further oxidation of hm^{6}A to f^{6}A.

ALKBH5 exhibits high sequence conservation among different vertebrates.[3,71] Recently, the crystal structure of the apoprotein form of zebrafish

ALKBH5 (fALKBH5) has been solved (Figure 9.2B).[71] fALKBH5 shares over 70% identity with human ALKBH5 and exhibits the same biochemical activity. With a conserved double-stranded β-helix catalytic core as found in other AlkB proteins, fALKBH possesses a loop different from that of FTO, which also interferes with the complementary strand of a potential duplex substrate (Figure 9.2B and C).[71] Although the demethylation process of ALKBH5 is proposed to be slightly different from that of FTO based on the different reaction intermediates, the active sites of these two proteins are quite conserved except for three residues. Lys-100, Ile-169 and Pro-175 in fALKBH5 correspond to Arg-96, Val-228 and Glu-234 in FTO (Figure 9.2D).[71] Interestingly, these three residues are highly conserved for ALKBH5 among species.[71] It will be interesting to test whether these differences explain the absence of intermediates in the ALKBH5-catalysed demethylation of m^6A.

Previous studies showed that ALKBH5 could be regulated by protein arginine methyltransferase 7 (PRMT7) under genotoxic stresses or by hypoxia-inducible factor 1α (HIF-1α) upon hypoxia stimulations.[68,72] To investigate how ALKBH5 could impact cells as an m^6A RNA demethylase, a series of experiments was performed in HeLa cells as a model cell line. ALKBH5 was found to be a nuclear protein that co-localizes with nuclear speckles where its demethylation activity influences mRNA processing factor assembly. When ALKBH5 was knocked down, the signals of phosphorylated SC35 (serine/arginine-rich splicing factor 2) disappeared; the signals can be recovered by complementation with wild-type ALKBH5, but not with a demethylation-inactive mutant ALKBH5.[41] Further studies revealed that ALKBH5 deficiency increases the rate of nascent RNA synthesis as well as the rate of nuclear polyadenylated RNA export. Complementary expression of wild-type *ALKBH5* could alleviate the pronounced RNA export phenomena, but the inactive *ALKBH5* mutant did not have the same effect, suggesting a role of demethylation activity in RNA metabolism.[41]

In mice, ALKBH5 is ubiquitously expressed with the highest mRNA level detected in testes.[41] The knockout of *Alkbh5* results in increased m^6A levels in mRNA isolated from the mouse organs compared with those from wild-type litter mates, which strongly supported that m^6A on mRNA is a physiologically-relevant substrate for ALKBH5.[41] *Alkbh5*-deficient mice exhibit impaired male fertility as a result of compromised spermatogenesis. Apoptosis of pachytene- and metaphase-stage spermatocytes and aberrant spermiogenesis account for the generation of a low number of spermatozoa with poor quality, which explains the impaired fertility of male mice.[41]

9.4 Conclusions and Perspectives

We have discovered two RNA demethylases that are capable of demethylating m^6A in RNA both *in vitro* and *in vivo*. The discovery of RNA demethylases has opened up a rapidly emerging field of reversible m^6A RNA modification in biological regulation. The topology of human, mouse and yeast m^6A RNA methylomes have been mapped by antibody-based affinity purification followed by

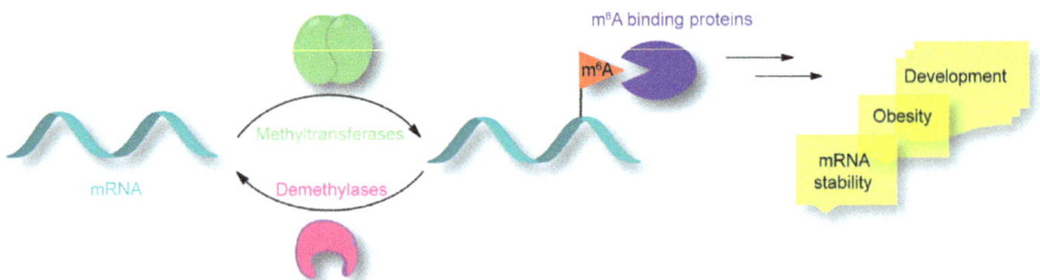

Figure 9.3 A potential mechanistic model for m^{6}A-dependent gene regulation. The m^{6}A modifications on mRNA or other RNAs are dynamically regulated by methyltransferases and demethylases. The selective m^{6}A-binding proteins recognize the modification and recruit additional complexes to perform downstream biological functions.

high-throughput sequencing.[33,34,73] m^{6}A sites are shown to be highly conserved across cell types, tissues and stimuli. Functional studies revealed that m^{6}A regulates RNA processing and mammalian circadian clockwork.[74] The loss of methyltransferase activity not only leads to apoptosis in HeLa cells,[75] but also significantly impairs development in *Arabidopsis*[38] and in *Drosophila*.[39]

FTO and ALKBH5 both co-localize with nuclear speckles where methyltransferases reside,[36] yet they have distinct biological functions. It is tempting to envisage that the dynamics of m^{6}A in mRNA can be temporally and spatially regulated by enzymes with antagonizing activities during mRNA processing. The m^{6}A fraction at specific adenosine sites on mRNA can be recognized as regulatory information by downstream proteins, the m^{6}A reader proteins (Figure 9.3). The recent characterization of the first reader protein, YTHDF2, revealed an m^{6}A-dependent mRNA degradation mechanism,[65] which provided strong evidence for this functional scenario. In addition, the newly discovered intermediates during FTO demethylation might have their own biological roles through recruiting or repelling certain reader proteins, which would add more complexity to this regulation based on the m^{6}A modification.

The growing acknowledgement of the roles of m^{6}A in RNA fills a gap in the regulation of gene expression control by reversible RNA modifications. Regulation through reversible RNA modifications provides a rapid response mechanism to various stimuli and signals compared with those acting on DNA and proteins. Many questions remain to be answered. For example, the regulatory mechanisms of FTO and ALKBH5 at the molecular level and their precise regulatory functions remain to be solved. The roles of m^{6}A in various biological processes need to be further elucidated. Answers to these questions are critical to advance the research on RNA epigenetics.

Acknowledgements

We thank our colleagues for discussions. This study was supported by National Institutes of Health (GM071440).

References

1. S. C. Trewick, T. F. Henshaw, R. P. Hausinger, T. Lindahl and B. Sedgwick, *Nature*, 2002, **419**, 174–178.
2. P. O. Falnes, R. F. Johansen and E. Seeberg, *Nature*, 2002, **419**, 178–182.
3. M. A. Kurowski, A. S. Bhagwat, G. Papaj and J. M. Bujnicki, *BMC Genomics*, 2003, **4**, 48.
4. M. S. Bratlie and F. Drablos, *BMC Genomics*, 2005, **6**, 1.
5. J. Ringvoll, L. M. Nordstrand, C. B. Vagbo, V. Talstad, K. Reite, P. A. Aas, K. H. Lauritzen, N. B. Liabakk, A. Bjork, R. W. Doughty, P. O. Falnes, H. E. Krokan and A. Klungland, *EMBO J.*, 2006, **25**, 2189–2198.
6. T. Gerken, C. A. Girard, Y. C. L. Tung, C. J. Webby, V. Saudek, K. S. Hewitson, G. S. H. Yeo, M. A. McDonough, S. Cunliffe, L. A. McNeill, J. Galvanovskis, P. Rorsman, P. Robins, X. Prieur, A. P. Coll, M. Ma, Z. Jovanovic, I. S. Farooqi, B. Sedgwick, I. Barroso, T. Lindahl, C. P. Ponting, F. M. Ashcroft, S. O'Rahilly and C. J. Schofield, *Science*, 2007, **318**, 1469–1472.
7. L. Sanchez-Pulido and M. A. Andrade-Navarro, *BMC Biochem.*, 2007, **8**, 23.
8. S. Dango, N. Mosammaparast, M. E. Sowa, L. J. Xiong, F. Wu, K. Park, M. Rubin, S. Gygi, J. W. Harper and Y. Shi, *Mol. Cell*, 2011, **44**, 373–384.
9. Y. Fu, Q. Dai, W. Zhang, J. Ren, T. Pan and C. He, *Angew. Chem., Int. Ed.*, 2010, **49**, 8885–8888.
10. E. van den Born, C. B. Vågbø, L. Songe-Møller, V. Leihne, G. F. Lien, G. Leszczynska, A. Malkiewicz, H. E. Krokan, F. Kirpekar, A. Klungland and P. Ø. Falnes, *Nat. Commun.*, 2011, **2**, 172.
11. T. Duncan, S. C. Trewick, P. Koivisto, P. A. Bates, T. Lindahl and B. Sedgwick, *Proc. Natl. Acad. Sci. U.S.A.*, 2002, **99**, 16660–16665.
12. R. Ougland, D. Lando, I. Jonson, J. A. Dahl, M. N. Moen, L. M. Nordstrand, T. Rognes, J. T. Lee, A. Klungland, T. Kouzarides and E. Larsen, *Stem Cells*, 2012, **30**, 2672–2682.
13. P. A. Jones, *Nat. Rev. Genet.*, 2012, **13**, 484–492.
14. Z. D. Smith and A. Meissner, *Nat. Rev. Genet.*, 2013, **14**, 204–220.
15. M. Tahiliani, K. P. Koh, Y. H. Shen, W. A. Pastor, H. Bandukwala, Y. Brudno, S. Agarwal, L. M. Iyer, D. R. Liu, L. Aravind and A. Rao, *Science*, 2009, **324**, 930–935.
16. S. Ito, L. Shen, Q. Dai, S. C. Wu, L. B. Collins, J. A. Swenberg, C. He and Y. Zhang, *Science*, 2011, **333**, 1300–1303.
17. Y. F. He, B. Z. Li, Z. Li, P. Liu, Y. Wang, Q. Y. Tang, J. P. Ding, Y. Y. Jia, Z. C. Chen, L. Li, Y. Sun, X. X. Li, Q. Dai, C. X. Song, K. L. Zhang, C. He and G. L. Xu, *Science*, 2011, **333**, 1303–1307.
18. C. X. Song and C. He, *Trends Biochem. Sci.*, 2013, **38**, 480–484.
19. A. J. Bannister and T. Kouzarides, *Nature*, 2005, **436**, 1103–1106.
20. D. Chen, H. Ma, H. Hong, S. S. Koh, S. M. Huang, B. T. Schurter, D. W. Aswad and M. R. Stallcup, *Science*, 1999, **284**, 2174–2177.
21. H. B. Wang, Z. Q. Huang, L. Xia, Q. Feng, H. Erdjument-Bromage, B. D. Strahl, S. D. Briggs, C. D. Allis, J. M. Wong, P. Tempst and Y. Zhang, *Science*, 2001, **293**, 853–857.

22. B. D. Strahl, S. D. Briggs, C. J. Brame, J. A. Caldwell, S. S. Koh, H. Ma, R. G. Cook, J. Shabanowitz, D. F. Hunt, M. R. Stallcup and C. D. Allis, *Curr. Biol.*, 2001, **11**, 996–1000.
23. R. J. Klose and Y. Zhang, *Nat. Rev. Mol. Cell Biol.*, 2007, **8**, 307–318.
24. C. Martin and Y. Zhang, *Nat. Rev. Mol. Cell Biol.*, 2005, **6**, 838–849.
25. E. L. Greer and Y. Shi, *Nat. Rev. Genet.*, 2012, **13**, 343–357.
26. F. Crick, *Nature*, 1970, **227**, 561–563.
27. M. Nirenberg, *Trends Biochem. Sci.*, 2004, **29**, 46–54.
28. C. Q. Yi and T. Pan, *Acc. Chem. Res.*, 2011, **44**, 1380–1388.
29. M. A. Machnicka, K. Milanowska, O. Osman Oglou, E. Purta, M. Kurkowska, A. Olchowik, W. Januszewski, S. Kalinowski, S. Dunin-Horkawicz, K. M. Rother, M. Helm, J. M. Bujnicki and H. Grosjean, *Nucleic Acids Res.*, 2013, **41**, D262–D267.
30. M. T. Tuck, *Int. J. Biochem. Cell Biol.*, 1992, **24**, 379–386.
31. G. Jia, Y. Fu and C. He, *Trends Genet.*, 2013, **29**, 108–115.
32. J. E. Harper, S. M. Miceli, R. J. Roberts and J. L. Manley, *Nucleic Acids Res.*, 1990, **18**, 5735–5741.
33. D. Dominissini, S. Moshitch-Moshkovitz, S. Schwartz, M. Salmon-Divon, L. Ungar, S. Osenberg, K. Cesarkas, J. Jacob-Hirsch, N. Amariglio, M. Kupiec, R. Sorek and G. Rechavi, *Nature*, 2012, **485**, 201–206.
34. K. D. Meyer, Y. Saletore, P. Zumbo, O. Elemento, C. E. Mason and S. R. Jaffrey, *Cell*, 2012.
35. Z. Bodi, J. D. Button, D. Grierson and R. G. Fray, *Nucleic Acids Res.*, 2010, **38**, 5327–5335.
36. J. Liu, Y. Yue, D. Han, X. Wang, Y. Fu, L. Zhang, G. Jia, M. Yu, Z. Lu, X. Deng, Q. Dai, W. Chen and C. He, *Nat. Chem. Biol.*, 2013, DOI: 10.1038/nchembio.1432.
37. M. J. Clancy, M. E. Shambaugh, C. S. Timpte and J. A. Bokar, *Nucleic Acids Res.*, 2002, **30**, 4509–4518.
38. S. L. Zhong, H. Y. Li, Z. Bodi, J. Button, L. Vespa, M. Herzog and R. G. Fray, *Plant Cell*, 2008, **20**, 1278–1288.
39. C. F. Hongay and T. L. Orr-Weaver, *Proc. Natl. Acad. Sci. U.S.A.*, 2011, **108**, 14855–14860.
40. G. Jia, Y. Fu, X. Zhao, Q. Dai, G. Zheng, Y. Yang, C. Yi, T. Lindahl, T. Pan, Y. G. Yang and C. He, *Nat. Chem. Biol.*, 2011, **7**, 885–887.
41. G. Zheng, J. A. Dahl, Y. Niu, P. Fedorcsak, C. M. Huang, C. J. Li, C. B. Vagbo, Y. Shi, W. L. Wang, S. H. Song, Z. Lu, R. P. Bosmans, Q. Dai, Y. J. Hao, X. Yang, W. M. Zhao, W. M. Tong, X. J. Wang, F. Bogdan, K. Furu, Y. Fu, G. Jia, X. Zhao, J. Liu, H. E. Krokan, A. Klungland, Y. G. Yang and C. He, *Mol. Cell*, 2013, **49**, 18–29.
42. T. Peters, K. Ausmeier and U. Ruther, *Mamm. Genome*, 1999, **10**, 983–986.
43. C. Dina, D. Meyre, S. Gallina, E. Durand, A. Korner, P. Jacobson, L. M. S. Carlsson, W. Kiess, V. Vatin, C. Lecoeur, J. Delplanque, E. Vaillant, F. Pattou, J. Ruiz, J. Weill, C. Levy-Marchal, F. Horber, N. Potoczna, S. Hercberg, C. Le Stunff, P. Bougneres, P. Kovacs, M. Marre, B. Balkau, S. Cauchi, J. C. Chevre and P. Froguel, *Nat. Genet.*, 2007, **39**, 724–726.

44. T. M. Frayling, N. J. Timpson, M. N. Weedon, E. Zeggini, R. M. Freathy, C. M. Lindgren, J. R. B. Perry, K. S. Elliott, H. Lango, N. W. Rayner, B. Shields, L. W. Harries, J. C. Barrett, S. Ellard, C. J. Groves, B. Knight, A. M. Patch, A. R. Ness, S. Ebrahim, D. A. Lawlor, S. M. Ring, Y. Ben-Shlomo, M. R. Jarvelin, U. Sovio, A. J. Bennett, D. Melzer, L. Ferrucci, R. J. F. Loos, I. Barroso, N. J. Wareham, F. Karpe, K. R. Owen, L. R. Cardon, M. Walker, G. A. Hitman, C. N. A. Palmer, A. S. F. Doney, A. D. Morris, G. D. Smith, A. T. Hattersley, M. I. McCarthy and W. T. C. Control, *Science*, 2007, **316**, 889–894.
45. A. Scuteri, S. Sanna, W. M. Chen, M. Uda, G. Albai, J. Strait, S. Najjar, R. Nagaraja, M. Orru, G. Usala, M. Dei, S. Lai, A. Maschio, F. Busonero, A. Mulas, G. B. Ehret, A. A. Fink, A. B. Weder, R. S. Cooper, P. Galan, A. Chakravarti, D. Schlessinger, A. Cao, E. Lakatta and G. R. Abecasis, *PloS Genet.*, 2007, **3**, 1200–1210.
46. G. S. Yeo, *J. Neuroendocrinol.*, 2012, **24**, 393–394.
47. J. S. McTaggart, S. Lee, M. Iberl, C. Church, R. D. Cox and F. M. Ashcroft, *PLoS One*, 2011, **6**.
48. J. Wardle, C. Llewellyn, S. Sanderson and R. Plomin, *Int. J. Obes.*, 2009, **33**, 42–45.
49. J. Wardle, S. Carnell, C. M. A. Haworth, I. S. Farooqi, S. O'Rahilly and R. Plomin, *J. Clin. Endocrinol. Metab.*, 2008, **93**, 3640–3643.
50. N. J. Timpson, P. M. Emmett, T. M. Frayling, I. Rogers, A. T. Hattersley, M. I. McCarthy and G. D. Smith, *Am. J. Clin. Nutr.*, 2008, **88**, 971–978.
51. J. E. Cecil, R. Tavendale, P. Watt, M. M. Hetherington and C. N. Palmer, *N. Engl. J. Med.*, 2008, **359**, 2558–2566.
52. J. R. Speakman, K. A. Rance and A. M. Johnstone, *Obesity (Silver Spring)*, 2008, **16**, 1961–1965.
53. A. Haupt, C. Thamer, H. Staiger, O. Tschritter, K. Kirchhoff, F. Machicao, H. U. Häring, N. Stefan and A. Fritsche, *Exp. Clin. Endocrinol. Diabetes*, 2009, **117**, 194–197.
54. J. Fischer, L. Koch, C. Emmerling, J. Vierkotten, T. Peters, J. C. Bruning and U. Rüther, *Nature*, 2009, **458**, 894–898.
55. C. Church, S. Lee, E. A. Bagg, J. S. McTaggart, R. Deacon, T. Gerken, A. Lee, L. Moir, J. Mecinovic, M. M. Quwailid, C. J. Schofield, F. M. Ashcroft and R. D. Cox, *PLoS Genet.*, 2009, **5**, e1000599.
56. S. Boissel, O. Reish, K. Proulx, H. Kawagoe-Takaki, B. Sedgwick, G. S. Yeo, D. Meyre, C. Golzio, F. Molinari, N. Kadhom, H. C. Etchevers, V. Saudek, I. S. Farooqi, P. Froguel, T. Lindahl, S. O'Rahilly, A. Munnich and L. Colleaux, *Am. J. Hum. Genet.*, 2009, **85**, 106–111.
57. C. Church, L. Moir, F. McMurray, C. Girard, G. T. Banks, L. Teboul, S. Wells, J. C. Bruning, P. M. Nolan, F. M. Ashcroft and R. D. Cox, *Nat. Genet.*, 2010, **42**, 1086–1092.
58. G. F. Jia, C. G. Yang, S. D. Yang, X. Jian, C. Q. Yi, Z. Q. Zhou and C. He, *FEBS Lett.*, 2008, **582**, 3313–3319.
59. Y. Fu, G. F. Jia, X. Q. Pang, R. N. Wang, X. Wang, C. J. Li, S. Smemo, Q. Dai, K. A. Bailey, M. A. Nobrega, K. L. Han, Q. Cui and C. He, *Nat. Commun.*, 2013, **4**, 1798.

60. Y. Fu and C. He, *Curr. Opin. Chem. Biol.*, 2012, **16**, 516–524.
61. Z. F. Han, T. H. Niu, J. B. Chang, X. G. Lei, M. Y. Zhao, Q. Wang, W. Cheng, J. J. Wang, Y. Feng and J. J. Chai, *Nature*, 2010, **464**, 1205–1209.
62. J. Ule, K. B. Jensen, M. Ruggiu, A. Mele, A. Ule and R. B. Darnell, *Science*, 2003, **302**, 1212–1215.
63. M. Hafner, M. Landthaler, L. Burger, M. Khorshid, J. Hausser, P. Berninger, A. Rothballer, M. Ascano, Jr., A. C. Jungkamp, M. Munschauer, A. Ulrich, G. S. Wardle, S. Dewell, M. Zavolan and T. Tuschl, *Cell*, 2010, **141**, 129–141.
64. M. E. Hess, S. Hess, K. D. Meyer, L. A. Verhagen, L. Koch, H. S. Bronneke, M. O. Dietrich, S. D. Jordan, Y. Saletore, O. Elemento, B. F. Belgardt, T. Franz, T. L. Horvath, U. Ruther, S. R. Jaffrey, P. Kloppenburg and J. C. Bruning, *Nat. Neurosci.*, 2013, **16**, 1042–1048.
65. X. Wang, Z. Lu, A. Gomez, G. C. Hon, Y. Yue, D. Han, Y. Fu, M. Parisien, Q. Dai, G. Jia, B. Ren, T. Pan and C. He, *Nature*, 2013, DOI: 10.1038/nature12730.
66. W. Aik, M. Demetriades, M. K. Hamdan, E. A. Bagg, K. K. Yeoh, C. Lejeune, Z. Zhang, M. A. McDonough and C. J. Schofield, *J. Med. Chem.*, 2013, **56**, 3680–3688.
67. B. Chen, F. Ye, L. Yu, G. Jia, X. Huang, X. Zhang, S. Peng, K. Chen, M. Wang, S. Gong, R. Zhang, J. Yin, H. Li, Y. Yang, H. Liu, J. Zhang, H. Zhang, A. Zhang, H. Jiang, C. Luo and C. G. Yang, *J. Am. Chem. Soc.*, 2012, **134**, 17963–17971.
68. A. Thalhammer, Z. Bencokova, R. Poole, C. Loenarz, J. Adam, L. O'Flaherty, J. Schodel, D. Mole, K. Giaslakiotis, C. J. Schofield, E. M. Hammond, P. J. Ratcliffe and P. J. Pollard, *PLoS One*, 2011, **6**, e16210.
69. A. G. Baltz, M. Munschauer, B. Schwanhausser, A. Vasile, Y. Murakawa, M. Schueler, N. Youngs, D. Penfold-Brown, K. Drew, M. Milek, E. Wyler, R. Bonneau, M. Selbach, C. Dieterich and M. Landthaler, *Mol. Cell*, 2012, **46**, 674–690.
70. A. Castello, B. Fischer, K. Eichelbaum, R. Horos, B. M. Beckmann, C. Strein, N. E. Davey, D. T. Humphreys, T. Preiss, L. M. Steinmetz, J. Krijgsveld and M. W. Hentze, *Cell*, 2012, **149**, 1393–1406.
71. W. Chen, L. Zhang, G. Zheng, Y. Fu, Q. Ji, F. Liu, H. Chen and C. He, *FEBS Lett.*, **588**, 892–898.
72. V. Karkhanis, L. Wang, S. Tae, Y. J. Hu, A. N. Imbalzano and S. Sif, *J. Biol. Chem.*, 2012, **287**, 29801–29814.
73. S. Schwartz, S. D. Agarwala, M. R. Mumbach, M. Jovanovic, P. Mertins, A. Shishkin, Y. Tabach, T. S. Mikkelsen, R. Satija, G. Ruvkun, S. A. Carr, E. S. Lander, G. R. Fink and A. Regev, *Cell*, 2013, **155**, 1409–1421.
74. M. H. Hastings, *Cell*, 2013, **155**, 740–741.
75. J. A. Bokar, M. E. Shambaugh, D. Polayes, A. G. Matera and F. M. Rottman, *RNA*, 1997, **3**, 1233–1247.

CHAPTER 10

Role of ALKBH8 in the Synthesis of Wobble Uridine Modifications in tRNA

PÅL Ø. FALNES*[a] AND ANGELA YEUAN YEN HO[a]

[a]Department of Biosciences, University of Oslo, PO Box 1066 Blindern, 0316 Oslo, Norway
*E-mail: pal.falnes@ibv.uio.no

10.1 *Escherichia coli* AlkB and Its Homologues – A Brief History

10.1.1 Discovery of the AlkB Function

Spurious methylations and other alkylations represent a serious threat to the integrity of the genome, and all organisms possess systems dedicated to repairing such damage.[1] In the 1980s, several alkylation repair enzymes were discovered through the isolation and complementation of *Escherichia coli* mutants that were hypersensitive to methylating agents. The affected gene in one of these mutants, discovered in 1982, was denoted *alkB*,[2] however it would take almost two decades before the function of the encoded AlkB protein was unravelled. The discovery of the AlkB function was primarily instigated by a bioinformatics study published in 2001 that placed *E. coli* AlkB and its sequence homologues in the superfamily of 2-oxoglutarate (2OG)- and Fe(II)-dependent dioxygenases.[3] This assignment spurred researchers to

RSC Metallobiology Series No. 3
2-Oxoglutarate-Dependent Oxygenases
Edited by Robert P. Hausinger and Christopher J. Schofield

Published by the Royal Society of Chemistry, www.rsc.org

revisit the AlkB mystery, now armed with the two putative cofactors 2OG and Fe(II). Indeed, it soon became clear that AlkB is a 2OG- and Fe(II)-dependent dioxygenase that catalyses the hydroxylation of the deleterious methyl group found on certain DNA lesions such as 1-methyladenine and 3-methylcytosine.[4,5] The resulting hydroxymethyl moiety is then spontaneously released as formaldehyde, regenerating the lesion-free base. Additional information on AlkB is described in Chapter 8.

10.1.2 Mammalian AlkB Homologues

In the wake of the initial sequencing of the human genome, bioinformatics analysis soon revealed that humans have eight putative homologues of AlkB, now denoted ALKBH1–8 (according to HUGO nomenclature).[6–8] Later, the obesity-associated protein FTO was shown to be a more distant member of the ALKBH family, and it is now alternatively referred to as ALKBH9.[9] The two first ALKBH proteins to be characterized *in vitro* were ALKBH2 and ALKBH3, which show repair activities similar to *E. coli* AlkB.[6,7] Remarkably, ALKBH3 and *E. coli* AlkB show robust enzymatic activity on RNA as well as on DNA, suggesting a possible significance of AlkB-mediated RNA repair.[6,10] When other ALKBHs were tested and found to lack DNA/RNA repair activity, however, it became clear that several of these enzymes are likely to have functions different from that of the bacterial AlkB protein. The first ALKBH protein to be assigned a function beyond repair was ALKBH8, a bifunctional enzyme involved in generating certain tRNA modifications and the primary focus of the present chapter. It was later demonstrated that FTO and ALKBH5 also have roles in RNA modification (see Chapter 9), as these enzymes are responsible for demethylating N^6-methyladenosine (m^6A) modifications in mRNA.[11,12] Finally, ALKBH proteins have been implicated in protein demethylation, as genetic ablation of ALKBH1 and ALKBH4 leads to increased methylation levels at specific Lys residues in histone H2A and actin, respectively.[13,14]

10.2 Formation of Modified Wobble Uridines in Eukaryotic tRNAs

10.2.1 Wobble Uridine Modifications

tRNAs in all kingdoms of life are subject to extensive post-transcriptional modification and a tRNA typically contains several different modified nucleosides. Approximately 100 different tRNA modifications have been reported, with methylation, introduced in a single reaction by a dedicated methyltransferase (MTase) representing the most common type of modification.[15] However, many tRNAs also contain more complex 'hypermodifications', usually generated in a multistep process involving several different enzymes. The wobble nucleoside in the tRNA anticodon loop base-pairs with the third nucleobase in the matching codon in mRNA, and modification at this position is particularly frequent.[16,17] Wobble modifications play important roles both in assuring efficiency and accuracy of translation, and wobble uridines are almost invariably modified.

Eukaryotic wobble uridine-containing tRNAs carry the modified nucleosides ncm^{5}U or mcm^{5}U as well as further modified forms of these species (Figure 10.1; see legend for chemical structures and full names). In mammals, further modifications of mcm^{5}U and ncm^{5}U include 2-thiolation, 2′-*O*-ribose methylation and hydroxylation, yielding the modifications mcm^{5}s^{2}U, ncm^{5}s^{2}U, mcm^{5}Um, (*S*)-mchm^{5}U and (*R*)-mchm^{5}U, respectively.

10.2.2 Enzymes Responsible for Wobble Uridine Modification

The current understanding of the formation of eukaryotic wobble uridine modifications primarily originates from studies of the budding yeast *Saccharomyces cerevisiae*, but the enzymatic pathways appear to be conserved in other organisms such as animals and plants.

In *S. cerevisiae* it was shown that formation of mcm^{5}U and ncm^{5}U requires the so-called Elongator complex, consisting of the six subunits Elp1–Elp6 (Figure 10.2).[18] The phenotypes of Elp-deficient yeast cells suggested a role for the Elongator complex in various processes, including protein acetylation, transcriptional regulation and exocytosis.[19] Interestingly, the phenotypes of Elp mutants can be reversed by over-expression of certain wobble uridine containing tRNAs, indicating that the primary function of Elongator is wobble uridine modification.[20,21] The presence of both a *S*-adenosylmethionine (SAM)-radical domain and an acetyltransferase domain in Elp3 indicates an enzymatic function for the Elongator complex, but the direct involvement of Elongator in the biochemical reactions leading to uridine modification remains to be demonstrated.[22,23]

In yeast, the 2-thiolation found on wobble uridines is introduced by a thiouridylase constituted by the Ctu1 and Ctu2 proteins (also referred to as Tuc1 and Tuc2) (Figure 10.2).[24,25] The same mechanism appears to operate in mammals, which have convincing putative orthologues of these proteins. The final methylation step leading to the formation of mcm^{5}U from its precursor cm^{5}U was shown in *S. cerevisiae* to be performed by the MTase Trm9, the activity of which depends on the accessory protein Trm112, a required partner for several yeast MTases.[26–28] As described in detail below, mammalian ALKBH8 represents the orthologue of yeast Trm9; ALKBH8 also carries an oxygenase activity which catalyses the hydroxylation of mcm^{5}U into (*S*)-mchm^{5}U specifically in tRNA$^{Gly(UCC)}$ (Figure 10.2).

10.3 Unravelling the Function of Mammalian ALKBH8

Unlike the other human ALKBH proteins, ALKBH8 contains additional domains besides its defining AlkB domain. The AlkB domain is flanked by an *N*-terminal RNA recognition motif (RRM), and a SAM-dependent MTase

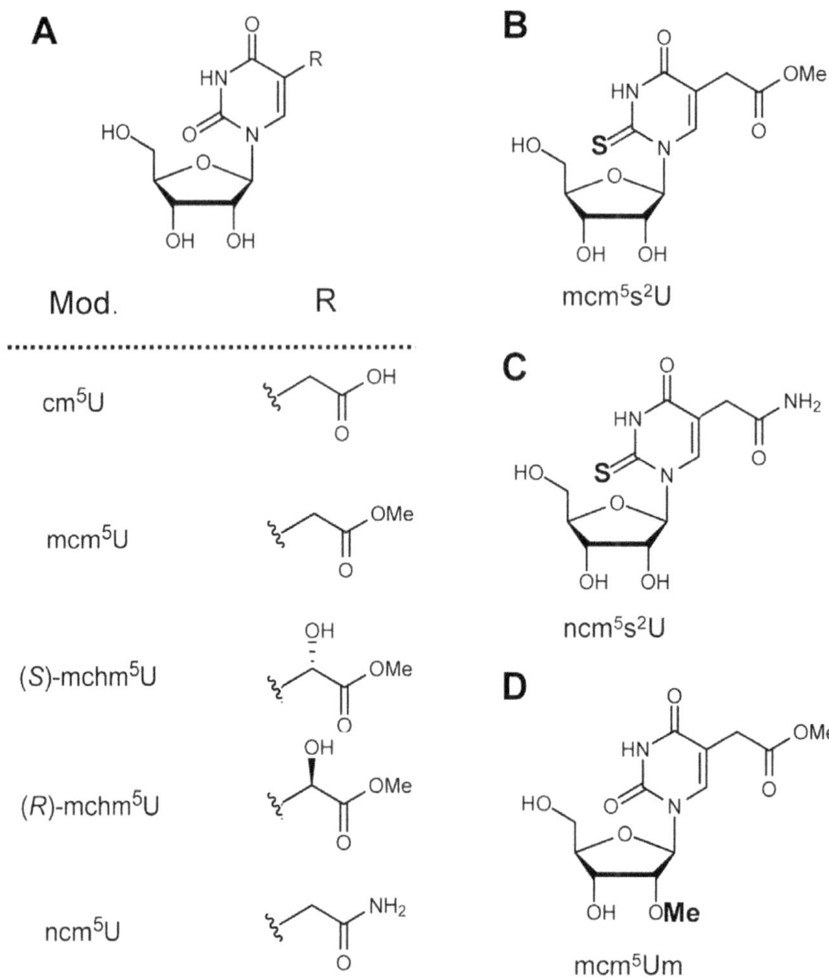

Figure 10.1 Modified uridine nucleosides found at the wobble position of mammalian tRNAs. (A) 5-carboxymethyluridine (cm^5U), 5-methoxycarbonylmethyluridine (mcm^5U), 5-(*S*)-[methoxycarbonyl-(hydroxy)methyl]uridine [(*S*)-$mchm^5U$], 5-(*R*)-[methoxycarbonyl-(hydroxy)methyl]uridine [(*R*)-$mchm^5U$], and 5-carbamoylmethyluridine (ncm^5U). (B) 5-methoxycarbonyl-2-thiouridine (mcm^5s^2U), resulting from 2-thiolation of mcm^5U. (C) 5-carbamoylmethyluridine-2-thiouridine (ncm^5s^2U), resulting from 2-thiolation of ncm^5U. (D) 5-methoxycarbonyl-2′-*O*-methyluridine (mcm^5Um), resulting from 2′-*O*-ribose methylation of mcm^5U. Note that cm^5U is not found in mature tRNAs, but serves as a precursor for the synthesis of mcm^5U (and, possibly, ncm^5U).

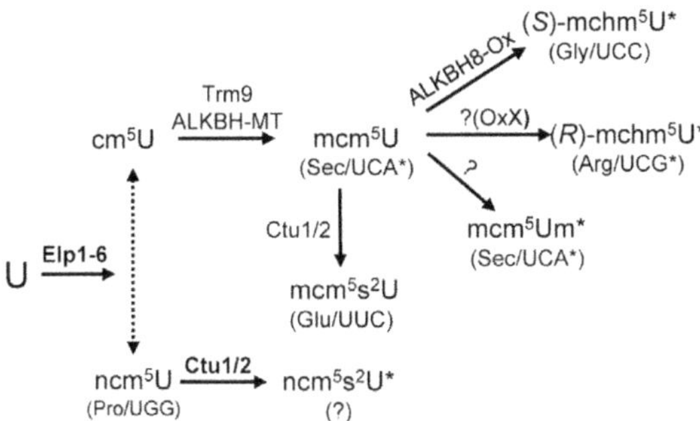

Figure 10.2 Pathways of wobble uridine modifications in mammals. For each modification, a tRNA carrying the modification is indicated (residue/anticodon in parenthesis), except for ncm^{5}s^{2}U, where such information is not available. Asterisks indicate modifications and tRNA species that are present in mammals, but absent in the yeast *S. cerevisiae*. OxX indicates an as yet unidentified oxygenase/hydroxylase predicted to catalyse the formation of (*R*)-mchm^{5}U from mcm^{5}U. Ctu1/2 indicate the mammalian homologues of the yeast Ctu1/Ctu2 enzymes, which presumably catalyse the indicated reactions (but this has not yet been experimentally demonstrated).

domain which displays strong sequence similarity to the *S. cerevisiae* tRNA MTase Trm9 (Figure 10.3). This fusion of protein domains clearly suggested that ALKBH8 may be involved in RNA modification.

10.3.1 The ALKBH8 MTase

Consistent with a role for ALKBH8 in RNA modification, two independent studies showed that the MTase portion of human ALKBH8 interacts strongly with the human TRM112 protein, and this complex was found to possess an enzymatic activity identical to that of the yeast Trm9/Trm112 complex (Figure 10.4).[29,30] Moreover, analysis of the wobble uridine modification pattern in *ALKBH8* knockout mice and in mammalian cells where *ALKBH8* was knocked down by siRNAs showed that ALKBH8 catalyses the corresponding MTase reaction also *in vivo*.[29,30] Indeed, *ALKBH8* knockout mice are completely devoid of mcm^{5}U (and its further modified derivatives), and show an accumulation of cm^{5}U, indicating that ALKBH8 represents the major Trm9 orthologue in mammalian cells.[30] Furthermore, the tRNA modification pattern of *ALKBH8* knockout mice yields additional knowledge on the formation of wobble uridines in mammals.[30] Firstly, in these mice the tRNA species that normally contain wobble mcm^{5}U not only show the expected accumulation of cm^{5}U, but also of ncm^{5}U, indicating that the synthesis of these two nucleosides is interconnected. Secondly, the tRNAs that normally contain mcm^{5}s^{2}U or mcm^{5}Um instead contain cm^{5}U in the *ALKBH8* knockout mice, but not its 2-thiolated or the 2′-*O*-ribose methylated derivatives (cm^{5}s^{2}U or cm^{5}Um, respectively), indicating that these additional modifications depend on the presence of the full mcm^{5}U modification.

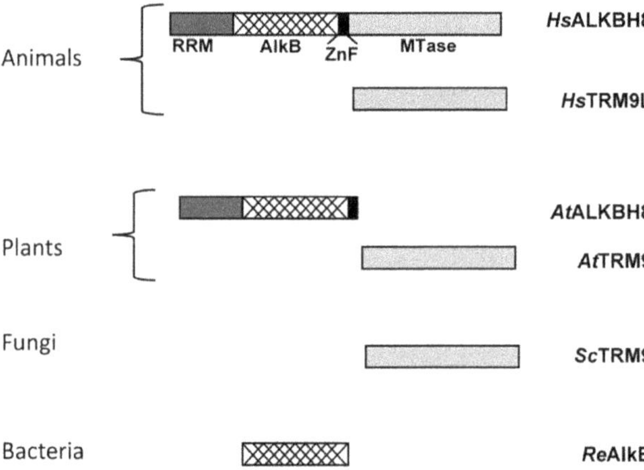

Figure 10.3 Domain architectures of ALKBH8 homologues in different organisms. *Homo sapiens* (*Hs*), *Arabidopsis thaliana* (*At*), *Saccharomyces cerevisiae* (*Sc*) and *Rhizobium etli* (*Re*) are indicated as examples of the ALKBH8/Trm9-containing members of the organism groups mammals, plants, fungi and bacteria, respectively. AlkB domain, crosshatched; RNA recognition motif (RRM), dark grey; Cys-rich zinc finger (ZnF), black; Trm9-like MTase domain, light grey. Domain sizes are not drawn to scale.

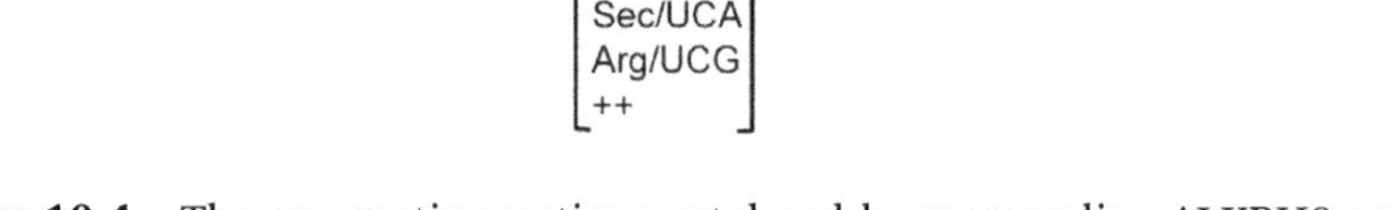

Figure 10.4 The enzymatic reactions catalysed by mammalian ALKBH8 on wobble uridines. The MTase domain of ALKBH8 (ALKBH8-MT) catalyses, together with its partner TRM112, the transfer of a methyl group from the methyl donor *S*-adenosylmethionine (SAM) to wobble position cm^5U in various tRNA species, yielding mcm^5U and *S*-adenosylhomocysteine (SAH). The oxygenase portion of ALKBH8 (ALKBH8-Ox), *i.e.* the RRM, AlkB and ZnF regions, catalyses the stereospecific hydroxylation of wobble mcm^5U in $tRNA^{Gly(UCC)}$ to give (*S*)-$mchm^5U$, in a 2OG-dependent dioxygenase reaction.

10.3.2 The ALKBH8 Hydroxylase

The domain architecture of ALKBH8 suggested the presence of both MTase and demethylase activities, yielding the interesting possibility that ALKBH8 acts as a molecular switch that either introduces or removes a regulatory methyl group, depending on a specific signal. However, the alternative possibility existed that the end result of the ALKBH8 oxygenase reaction is not demethylation, but rather the introduction of a stable hydroxyl moiety. Actually, a previously published finding suggested that the latter may be true; in 1988, it was shown that $tRNA^{Gly(UCC)}$ from silkworm (*Bombyx mori*) contains the nucleoside (*S*)-$mchm^5U$, *i.e.* a hydroxylated version of mcm^5U, in the wobble position.[31] Indeed, a biochemical study reported that ALKBH8, when incubated with an artificial substrate corresponding to the anticodon stem-loop of human $tRNA^{Gly(UCC)}$ with mcm^5U at the wobble position, catalyses the conversion of wobble mcm^5U into (*S*)-$mchm^5U$ *in vitro*.[32] A parallel study investigated the presence of the (*S*)-$mchm^5U$ modification in mammals as well as the role of ALKBH8 in its biogenesis.[33] Firstly, the presence in mammalian tRNA of (*S*)-$mchm^5U$ and its diastereomer (*R*)-$mchm^5U$ was investigated. As observed in silkworm, $tRNA^{Gly(UCC)}$ carries (*S*)-$mchm^5U$ at the wobble position while, interestingly, $tRNA^{Arg(UCG)}$ was shown to have wobble (*R*)-$mchm^5U$.[33] Secondly, the role of ALKBH8 in the formation of the two $mchm^5U$ diastereomers was studied by using three different gene-targeted mouse models: the *ALKBH8* knockout mouse, as well as two knock-in mice where the gene knockout had been complemented by *ALKBH8*-derived transgenes containing inactivating mutations in the regions encoding the oxygenase or the MTase domain, denoted KI(MT^+) and KI(Ox^+), respectively. While tRNA from the *ALKBH8* knockout mice and the KI(Ox^+) mice are devoid of both $mchm^5U$ diastereomers, the KI(MT^+) mice have (*R*)-$mchm^5U$, but not (*S*)-$mchm^5U$, indicating that the MTase activity of ALKBH8 is required for the formation of both diastereomers, whereas only the formation of (*S*)-$mchm^5U$ requires the oxygenase.[33] These findings were corroborated with biochemical experiments showing that the incubation of cm^5U-containing tRNA from *ALKBH8* knockout mice with the ALKBH8 and its cofactors yields (*S*)-$mchm^5U$ (Figure 10.4). In addition to unravelling the function of the ALKBH8 oxygenase, results of these experiments predict the existence of another oxygenase, denoted OxX, responsible for the formation of (*R*)-$mchm^5U$ in $tRNA^{Arg(UCG)}$.

10.4 The Structure of Human ALKBH8

A crystal structure at 3.0 Å resolution was recently reported for part of human ALKBH8 (see Chapter 2), *i.e.* a protein construct containing the RRM and AlkB domains, but lacking the MTase domain.[34] The AlkB domain in ALKBH8 is followed by a cluster of three conserved Cys residues (Cys-341, Cys-343, Cys-349) which, together with a histidine residue (His-242), constitute a zinc finger (ZnF) that contributes to the overall stability of the protein (Figure 10.3). Previously reported structures of AlkB domains have revealed the presence of a subdomain of particular importance for nucleic acid substrate recognition, the so-called 'nucleotide recognition lid' (NRL).[35] Although the ALKBH8 oxygenase shows much narrower substrate specificity than ALKBHs involved in DNA repair, the region corresponding to the NRL is more disordered than in the structures of the repair proteins.[34] It was therefore suggested that the binding of the tRNA substrate induces a disorder-to-order transition, thereby providing the high specificity in substrate recognition. The RRM moiety of ALKBH8 is, based on its similarity with other structurally characterized RRMs, proposed to contain a pyrimidine-binding cavity involved in substrate recognition, and it was also shown that the *N*-terminal basic alpha-helix of ALKBH8 contributes to RNA binding, thereby constituting a non-canonical part of the RRM.[34]

10.5 Organismal Distribution of ALKBH8

Apparently, all animals have an ALKBH8 homologue with a RRM/AlkB/ZnF/MTase domain organization, *i.e.* identical to that of mammalian ALKBH8, suggesting that the ALKBH8 homologues from animals all catalyse the same reactions (Figure 10.3). Also, the reported presence of the (*S*)-mchm^{5}U modification both in silkworm and in the nematode *Caenorhabditis elegans* indicates that ALKBH8 function is conserved in animals.[31,33] In addition, the genomes of vertebrates encode a protein with substantial similarity to yeast Trm9 (and, thus, to the MTase portion of ALKBH8), denoted TRM9L (TRM9-Like) (Figure 10.3). The enzymatic activity of TRM9L has not yet been uncovered, but it appears to be different from that of the yeast Trm9 and the ALKBH8 MTase. This notion is supported by the observation that genetic ablation of *ALKBH8* completely abolishes the formation of mcm^{5}U in a wide range of mouse tissues, indicating that TRM9L does not function as a back-up.

In plants, such as *Arabidopsis thaliana*, the two enzymatic activities of human ALKBH8 are shared between two different proteins (Figure 10.3). The protein encoded by the *AT1G36310* gene in *A. thaliana* is a sequence homologue of yeast Trm9 and the ALKBH8 MTase, and was indeed shown to catalyse the methylation of cm^{5}U into mcm^{5}U, in a reaction that depends on a plant Trm112 homologue.[36] A protein corresponding to the RRM/AlkB/ZnF moieties of human ALKBH8 is encoded by the *AT1G31600* gene, inactivation of which has been shown to block the formation of (*S*)-mchm^{5}U from mcm^{5}U.[36]

While the Trm9-like MTase function of ALKBH8 appears to be conserved in all eukaryotes, the ALKBH8 oxygenase shows a much more limited distribution. Proteins with similarity to this oxygenase are, for example, absent in the yeasts *S. cerevisiae* and *Schizosaccaromyces pombe*, and these organisms have accordingly been shown to lack the (*S*)-mchm^{5}U modification.[33]

The genomes of some α-proteobacteria encode proteins with high sequence similarities to the AlkB domains of eukaryotic ALKBH8s (~35% sequence identity to human ALKBH8), but they are devoid of the RRM, ZnF and MTase moieties (Figure 10.3).[37] This close similarity may suggest a role in tRNA modification for these bacterial ALKBH8 homologues. However, only one such protein, from *Rhizobium etli*, has been functionally characterized, and this enzyme displayed repair activity on etheno lesions in DNA.[37] Thus, further investigations are required to establish whether these bacterial enzymes are involved in nucleic acid modification or repair.

10.6 Biological and Regulatory Significance of ALKBH8

Arguably, many RNA modifications may be regarded as expansions of the nucleoside repertoire available for building RNAs and merely serve the purpose of improving RNA function, *e.g.* by increasing stability or enhancing interactions with partner molecules. Such modifications will typically represent static entities that are not subject to regulation, but recent findings indicate that certain RNA modifications do indeed play regulatory roles. In particular, the recent demonstrations of ALKBH5 and FTO as demethylases regulating the level of the modified base m^6A in mRNA, thereby modulating mRNA stability, strongly support the notion that some RNA modifications serve regulatory purposes.[11,12,38]

10.6.1 Potential Regulation of Translation by Wobble Uridine Modifications

It was demonstrated several years ago that stress can lead to alterations in tRNA modification patterns.[39] This finding was systematically addressed in a recent study where the levels of 23 tRNA modifications in *S. cerevisiae* were measured as cells were exposed to various types of chemical stressors.[40] It was reported that such exposure typically leads to considerable changes in the level of several modified tRNA nucleosides and that different chemicals cause distinct alterations in the tRNA modification patterns. On this basis, the authors proposed that these changes reflect a cellular 'reprogramming' of tRNA modifications and that such reprogramming may play an important role in regulating the protein expression at the level of translation.

Indeed, yeast Trm9, the counterpart of the mammalian ALKBH8 MTase, has been suggested to be responsible for selectively upregulating the translation of DNA repair proteins in response to DNA damage. It was shown that the *Trm9* mutant is hypersensitive towards the DNA-damaging methylating agent methylmethane sulfonate (MMS) and that some repair-associated proteins particularly rich in the Arg codon AGA (which requires the mcm^5U-containing $tRNA^{Arg(UCU)}$ for decoding) are inefficiently translated in the *Trm9* mutant.[41] Moreover, reporter constructs containing runs of ten consecutive, identical codons that require mcm^5U-modified tRNAs for decoding are also less efficiently translated in the *Trm9* mutant. In agreement with these results, Elongator mutants of *S. pombe*, which are deficient in mcm^5U/ncm^5U

modification, were shown to be hypersensitive towards oxidative stress, and this phenotype was shown to be due to inefficient decoding of the Lys codon AAA by the hypomodified $tRNA^{Lys(UUU)}$.[42]

When codon-run reporter constructs such as those described above were tested in cells from ALKBH8-deficient mice, their translation appeared to be similar to those in wild-type cells. Taken together with the lack of overt phenotypes in these mice, these results suggest that the modifications introduced by ALKBH8 are not strictly required for efficient protein synthesis, but rather enhance its accuracy and/or efficacy.

10.6.2 Selenoproteins

The human genome has ~25 genes encoding selenoproteins, *i.e.* proteins that contain the non-canonical amino acid residue selenocysteine, the so-called 21st amino acid.[43] The insertion of selenocysteine during translation is mediated by $tRNA^{Sec(UCA)}$, a specialized tRNA that recognizes the codon UGA, normally serving as stop codon.[44] Interestingly, $tRNA^{Sec(UCA)}$ carries the modified nucleoside mcm^5U at the wobble position.[44] Two subpopulations of $tRNA^{Sec(UCA)}$ exist, one which has wobble mcm^5U and another with the 2′-*O*-ribose methylated derivative mcm^5Um.[45] A large body of work from the Hatfield group has suggested that such ribose methylation plays a regulatory role in selenoprotein expression.[44] It has been demonstrated that certain conditions, *e.g.* stress or addition of selenium, that lead to increased expression of some selenoproteins, *e.g.* glutathione peroxidase 1 (Gpx1), also cause the $tRNA^{Sec(UCA)}$ population to shift towards the mcm^5Um-containing form.[45] Conversely, a substitution of a nearby position (A37G) which abolishes the 2′-*O*-ribose methylation of wobble mcm^5U in $tRNA^{Sec(UCA)}$ was shown to suppress the expression of stress-inducible proteins such as Gpx1.[46] In mice deficient in ALKBH8, $tRNA^{Sec(UCA)}$ carries the wobble modifications ncm^5U and cm^5U, *i.e.* 2′-*O*-ribose methylation is lacking.[30] Indeed, these mice show a reduction in Gpx1 levels, whereas the expression of thioredoxin reductase 1, a constitutively expressed selenoprotein not correlated with mcm^5Um levels, is unaltered. These results indicate that the ALKBH8 function is required for efficient expression of at least a subset of the selenoproteome. However, given the modest (~50%) reduction in Gpx1 levels, the data do not strongly support the notion that mcm^5U ribose methylation is a potent regulator of stress-inducible selenoproteins. Actually, a study of the plant *A. thaliana* suggests that ribose methylation of mcm^5U may not represent an important regulatory cue.[36] *A. thaliana*, which does not have selenoproteins (or $tRNA^{Sec(UCA)}$), normally lacks mcm^5U, but does harbour the further modifications (*S*)-$mchm^5U$ and mcm^5s^2U. However, plants deficient in the ALKBH8 oxygenase showed an accumulation of both mcm^5U and mcm^5Um, suggesting that mcm^5U not subjected to further hydroxyl- or 2-thio-modification may instead be prone to ribose methylation by a non-specific mechanism.[36]

10.6.3 Cancer

One study has suggested a role for ALKBH8 in cancer progression. siRNA-mediated knockdown of *ALKBH8* was found to induce apoptosis in urothelial carcinoma cell lines and also to inhibit the growth of human bladder cancer xenografts in nude mice.[47] As the study was performed before the ALKBH8 function was unravelled, tRNA modification status was not investigated and it is somewhat difficult to connect the obtained results with the ALKBH8 function. However, as rapidly growing cancer cells are likely to depend on efficient protein translation, it is not unexpected that reducing the activity of a tRNA modification enzyme will affect tumour growth.

The human TRM9L protein (also known as KIAA1456/C8orf79) was initially reported to be downregulated in colorectal cancers, suggesting its possible role as a tumour repressor.[48] These findings were followed up by a study that investigated the effect of re-expressing *TRM9L* in a *TRM9L*-deficient colon cancer cell line.[49] Re-expression of *TRM9L* inhibits the ability of the cells to form tumours in mice and also causes increased sensitivity towards hypoxia. Interestingly, the inhibitory effect on tumour growth is reduced when using *TRM9L* mutants carrying putatively inactivating substitutions in the MTase domain, indicating that the enzymatic activity of TRM9L is required.

10.7 Conclusion and Future Perspectives

Recent *in vitro* and *in vivo* studies have firmly established mammalian ALKBH8 as a bifunctional tRNA modification enzyme holding two distinct biochemical activities. ALKBH8 has a MTase activity that targets several different wobble uridine-containing tRNAs and a specialized oxygenase (AlkB) activity that modifies $tRNA^{Gly(UCC)}$. However, it remains to be investigated whether ALKBH8 is subject to regulation and thus plays a role in modulating protein translation. Furthermore, it will be of great interest to unravel the enzymatic activity of bacterial ALKBH8 proteins and mammalian TRM9L. Finally, the molecular mechanisms underlying the potential involvement of ALKBH8 and TRM9L in cancer require further investigation.

Acknowledgements

The work from our group on the ALKBH8 function in mammals and plants represents close collaborations with the groups of Arne Klungland and Paul Grini, respectively, and their contributions are gratefully acknowledged. Our research on ALKBH8 has been supported by grants from the Research Council of Norway, the Polish–Norwegian Research Fund and the Norwegian Cancer Society.

References

1. B. Sedgwick, *Nat. Rev. Mol. Cell Biol.*, 2004, **5**, 148.
2. H. Kataoka, Y. Yamamoto and M. Sekiguchi, *J. Bacteriol.*, 1983, **153**, 1301.
3. L. Aravind and E. V. Koonin, *Genome Biol.*, 2001, **2**, RESEARCH0007.
4. P. O. Falnes, R. F. Johansen and E. Seeberg, *Nature*, 2002, **419**, 178.
5. S. C. Trewick, T. F. Henshaw, R. P. Hausinger, T. Lindahl and B. Sedgwick, *Nature*, 2002, **419**, 174.
6. P. A. Aas, M. Otterlei, P. O. Falnes, C. B. Vagbo, F. Skorpen, M. Akbari, O. Sundheim, M. Bjoras, G. Slupphaug, E. Seeberg and H. E. Krokan, *Nature*, 2003, **421**, 859.
7. T. Duncan, S. C. Trewick, P. Koivisto, P. A. Bates, T. Lindahl and B. Sedgwick, *Proc. Natl. Acad. Sci. U.S.A.*, 2002, **99**, 16660.
8. M. A. Kurowski, A. S. Bhagwat, G. Papaj and J. M. Bujnicki, *BMC Genomics*, 2003, **4**, 48.
9. T. Gerken, C. A. Girard, Y. C. Tung, C. J. Webby, V. Saudek, K. S. Hewitson, G. S. Yeo, M. A. McDonough, S. Cunliffe, L. A. McNeill, J. Galvanovskis, P. Rorsman, P. Robins, X. Prieur, A. P. Coll, M. Ma, Z. Jovanovic, I. S. Farooqi, B. Sedgwick, I. Barroso, T. Lindahl, C. P. Ponting, F. M. Ashcroft, S. O'Rahilly and C. J. Schofield, *Science*, 2007, **318**, 1469.
10. R. Ougland, C. M. Zhang, A. Liiv, R. F. Johansen, E. Seeberg, Y. M. Hou, J. Remme and P. O. Falnes, *Mol. Cell*, 2004, **16**, 107.
11. G. Jia, Y. Fu, X. Zhao, Q. Dai, G. Zheng, Y. Yang, C. Yi, T. Lindahl, T. Pan, Y. G. Yang and C. He, *Nat. Chem. Biol.*, 2011, **7**, 885.
12. G. Zheng, J. A. Dahl, Y. Niu, P. Fedorcsak, C. M. Huang, C. J. Li, C. B. Vagbo, Y. Shi, W. L. Wang, S. H. Song, Z. Lu, R. P. Bosmans, Q. Dai, Y. J. Hao, X. Yang, W. M. Zhao, W. M. Tong, X. J. Wang, F. Bogdan, K. Furu, Y. Fu, G. Jia, X. Zhao, J. Liu, H. E. Krokan, A. Klungland, Y. G. Yang and C. He, *Mol. Cell*, 2013, **49**, 18.
13. M. M. Li, A. Nilsen, Y. Shi, M. Fusser, Y. H. Ding, Y. Fu, B. Liu, Y. Niu, Y. S. Wu, C. M. Huang, M. Olofsson, K. X. Jin, Y. Lv, X. Z. Xu, C. He, M. Q. Dong, J. M. Rendtlew Danielsen, A. Klungland and Y. G. Yang, *Nat. Commun.*, 2013, **4**, 1832.
14. R. Ougland, D. Lando, I. Jonson, J. A. Dahl, M. N. Moen, L. M. Nordstrand, T. Rognes, J. T. Lee, A. Klungland, T. Kouzarides and E. Larsen, *Stem Cells*, 2012, **30**, 2672.
15. W. A. Cantara, P. F. Crain, J. Rozenski, J. A. McCloskey, K. A. Harris, X. Zhang, F. A. Vendeix, D. Fabris and P. F. Agris, *Nucleic Acids Res.*, 2011, **39**, D195–D201.
16. P. F. Agris, F. A. Vendeix and W. D. Graham, *J. Mol. Biol.*, 2007, **366**, 1.
17. E. M. Gustilo, F. A. Vendeix and P. F. Agris, *Curr. Opin. Microbiol.*, 2008, **11**, 134.
18. B. Huang, M. J. Johansson and A. S. Bystrom, *RNA*, 2005, **11**, 424.
19. J. Q. Svejstrup, *Curr. Opin. Cell Biol.*, 2007, **19**, 331.
20. C. Chen, B. Huang, M. Eliasson, P. Ryden and A. S. Bystrom, *PLoS. Genet.*, 2011, **7**, e1002258.

21. A. Esberg, B. Huang, M. J. Johansson and A. S. Bystrom, *Mol. Cell*, 2006, **24**, 139.
22. C. Paraskevopoulou, S. A. Fairhurst, D. J. Lowe, P. Brick and S. Onesti, *Mol. Microbiol.*, 2006, **59**, 795.
23. B. O. Wittschieben, G. Otero, B. T. de, J. Fellows, H. Erdjument-Bromage, R. Ohba, Y. Li, C. D. Allis, P. Tempst and J. Q. Svejstrup, *Mol. Cell*, 1999, **4**, 123.
24. G. R. Bjork, B. Huang, O. P. Persson and A. S. Bystrom, *RNA*, 2007, **13**, 1245.
25. M. Dewez, F. Bauer, M. Dieu, M. Raes, J. Vandenhaute and D. Hermand, *Proc. Natl. Acad. Sci. U.S.A.*, 2008, **105**, 5459.
26. P. Studte, S. Zink, D. Jablonowski, C. Bar, H. T. von der, M. F. Tuite and R. Schaffrath, *Mol. Microbiol.*, 2008, **69**, 1266.
27. H. R. Kalhor and S. Clarke, *Mol. Cell. Biol.*, 2003, **23**, 9283.
28. D. Liger, L. Mora, N. Lazar, S. Figaro, J. Henri, N. Scrima, R. H. Buckingham, T. H. van, V. Heurgue-Hamard and M. Graille, *Nucleic Acids Res.*, 2011, **39**, 6249.
29. D. Fu, J. A. Brophy, C. T. Chan, K. A. Atmore, U. Begley, R. S. Paules, P. C. Dedon, T. J. Begley and L. D. Samson, *Mol. Cell. Biol.*, 2010, **30**, 2449.
30. L. Songe-Moller, E. van den Born, V. Leihne, C. B. Vagbo, T. Kristoffersen, H. E. Krokan, F. Kirpekar, P. O. Falnes and A. Klungland, *Mol. Cell. Biol.*, 2010, **30**, 1814.
31. M. Kawakami, S. Takemura, T. Kondo, T. Fukami and T. Goto, *J. Biochem.*, 1988, **104**, 108.
32. Y. Fu, Q. Dai, W. Zhang, J. Ren, T. Pan and C. He, *Angew. Chem., Int. Ed. Engl.*, 2010, **49**, 8885.
33. E. van den Born, C. B. Vagbo, L. Songe-Moller, V. Leihne, G. F. Lien, G. Leszczynska, A. Malkiewicz, H. E. Krokan, F. Kirpekar, A. Klungland and P. O. Falnes, *Nat. Commun.*, 2011, **2**, 172.
34. C. Pastore, I. Topalidou, F. Forouhar, A. C. Yan, M. Levy and J. F. Hunt, *J. Biol. Chem.*, 2012, **287**, 2130.
35. B. Yu, W. C. Edstrom, J. Benach, Y. Hamuro, P. C. Weber, B. R. Gibney and J. F. Hunt, *Nature*, 2006, **439**, 879.
36. V. Leihne, F. Kirpekar, C. B. Vagbo, E. van den Born, H. E. Krokan, P. E. Grini, T. J. Meza and P. O. Falnes, *Nucleic Acids Res.*, 2011, **39**, 7688.
37. E. van den Born, A. Bekkelund, M. N. Moen, M. V. Omelchenko, A. Klungland and P. O. Falnes, *Nucleic Acids Res.*, 2009, **21**, 7124.
38. X. Wang, Z. Lu, A. Gomez, G. C. Hon, Y. Yue, D. Han, Y. Fu, M. Parisien, Q. Dai, G. Jia, B. Ren, T. Pan and C. He, *Nature*, 2014, **505**, 117.
39. M. Buck and B. N. Ames, *Cell*, 1984, **36**, 523.
40. C. T. Chan, M. Dyavaiah, M. S. DeMott, K. Taghizadeh, P. C. Dedon and T. J. Begley, *PLoS. Genet.*, 2010, **6**, e1001247.
41. U. Begley, M. Dyavaiah, A. Patil, J. P. Rooney, D. DiRenzo, C. M. Young, D. S. Conklin, R. S. Zitomer and T. J. Begley, *Mol. Cell*, 2007, **28**, 860.
42. J. Fernandez-Vazquez, I. Vargas-Perez, M. Sanso, K. Buhne, M. Carmona, E. Paulo, D. Hermand, M. Rodriguez-Gabriel, J. Ayte, S. Leidel and E. Hidalgo, *PLoS Genet.*, 2013, **9**, e1003647.

43. J. Lu and A. Holmgren, *J. Biol. Chem.*, 2009, **284**, 723.
44. D. L. Hatfield, B. A. Carlson, X. M. Xu, H. Mix and V. N. Gladyshev, *Prog. Nucleic Acid Res. Mol. Biol.*, 2006, **81**, 97.
45. A. M. Diamond, I. S. Choi, P. F. Crain, T. Hashizume, S. C. Pomerantz, R. Cruz, C. J. Steer, K. E. Hill, R. F. Burk and J. A. McCloskey, *J. Biol. Chem.*, 1993, **268**, 14215.
46. B. A. Carlson, X. M. Xu, V. N. Gladyshev and D. L. Hatfield, *J. Biol. Chem.*, 2005, **280**, 5542.
47. K. Shimada, M. Nakamura, S. Anai, V. M. De, M. Tanaka, K. Tsujikawa, Y. Ouji and N. Konishi, *Cancer Res.*, 2009, **69**, 3157.
48. J. M. Flanagan, S. Healey, J. Young, V. Whitehall, D. A. Trott, R. F. Newbold and G. Chenevix-Trench, *Genes, Chromosomes Cancer*, 2004, **40**, 247.
49. U. Begley, M. S. Sosa, A. Avivar-Valderas, A. Patil, L. Endres, Y. Estrada, C. T. Chan, D. Su, P. C. Dedon, J. A. Aguirre-Ghiso and T. Begley, *EMBO Mol. Med.*, 2013, **5**, 366.

CHAPTER 11

The TET/JBP Family of Nucleic Acid Base-Modifying 2-Oxoglutarate and Iron-Dependent Dioxygenases

L. ARAVIND*[a], DAPENG ZHANG[a], AND LAKSHMINARAYAN M. IYER[a]

[a]National Center for Biotechnology Information, National Library of Medicine, National Institutes of Health, Bethesda, MD 20894, USA
*E-mail: aravind@ncbi.nlm.nih.gov

11.1 Introduction

11.1.1 Modification of Bases in Nucleic Acids

Catalytic modification of the four standard bases in DNA and RNA is observed across all three superkingdoms of life, namely bacteria, archaea and eukaryotes, and also in several viruses.[1–3] Of these modifications, several altered bases in rRNAs and tRNAs are highly conserved and might even show a near universal occurrence in cellular organisms, pointing to their presence in the last universal common ancestor of all life.[1,2,4] The changed bases in RNAs appear to be critical for mediating specific interactions required for robust decoding of genetic information, which is central to the process of translation. On the other hand modifications of DNA bases

RSC Metallobiology Series No. 3
2-Oxoglutarate-Dependent Oxygenases
Edited by Robert P. Hausinger and Christopher J. Schofield

Published by the Royal Society of Chemistry, www.rsc.org

are more sporadically distributed, although some of them do occur across vast phylogenetic distances.[3,5] Moreover, drastic modifications of bases (hypermodified bases) are relatively uncommon in DNA, most likely due to selective constraints imposed by double-helical pairing that precludes modification of certain positions. DNA base modifications first came to light in studies on prokaryotic restriction-modification (R-M) systems, which comprise a major defence mechanism that discriminates cellular from invasive DNA and degrades the latter.[6–8] In these systems, self DNA is distinguished from non-self DNA by methylation of cytosine at the 5th position of the pyrimidine ring (5mC), of adenine on the NH_2 group bound to the 6 position of the purine ring (N6mA), or less frequently of cytosine at the NH_2 group linked to the 4 position of the ring (N4mC). Subsequently, it was found that bacteriophages counter these restriction mechanisms using their own modified bases, which include 5-hydroxymethyl cytosine (5hmC), 5-hydroxymethyl uracil (5hmU), and their more complex derivatives such as hypermodified thymines and glycosylated derivatives of 5hmC.[3,9–11]

5mC and N6mA modifications are also observed in eukaryotic DNA.[3,12,13] In particular, 5mC is an important epigenetic mark in animals and plants, *i.e.* it provides an additional layer of information over and beyond the genetic information coded by the four standard bases, a discovery that has sparked enormous interest. Developmental studies in mammalian systems revealed two points where 5mC epigenetic marks are almost entirely erased in both the male and female genomes.[14–16] First, when the primordial germ cells, which eventually give rise to the gametes, are set aside in the developing organism, and second, shortly after the gametes fuse to form the zygote. Another unusual DNA modification that came to light in eukaryotes was base J, which was discovered in euglenozoans, including human parasites of the genera *Trypanosoma* and *Leishmania*.[3,17] Base J is a hypermodified thymine (Figure 11.1), β-D-glucosyl-hydroxymethyluracil, which for a time was also believed to function as an epigenetic mark in organisms possessing it. Base J is synthesized in DNA *via* a two-step reaction from thymine: oxidation of its methyl group at the 5 position to yield 5hmU, followed by conjugation of glucose to the newly generated hydroxyl group. The discovery of modified bases in eukaryotic DNA inspired researchers to identify the enzymes catalysing these modifications and to further understand their biological significance. Methyltransferases catalysing *de novo* methylation of DNA for establishment of epigenetic marks and those involved in their subsequent maintenance over cell divisions were characterized first and found to be closely related to their counterparts in bacterial R-M systems.[12,18] This finding was followed by identification of thymine hydroxylases catalysing the first step of base J biosynthesis.[19,20] These belong to a class of 2-oxoglutarate (2OG) and iron-dependent dioxygenases (2OGFeDOs),[21–23] which has opened a door for unravelling the enigma of the erasure of epigenetic methyl marks in eukaryotes such as mammals.

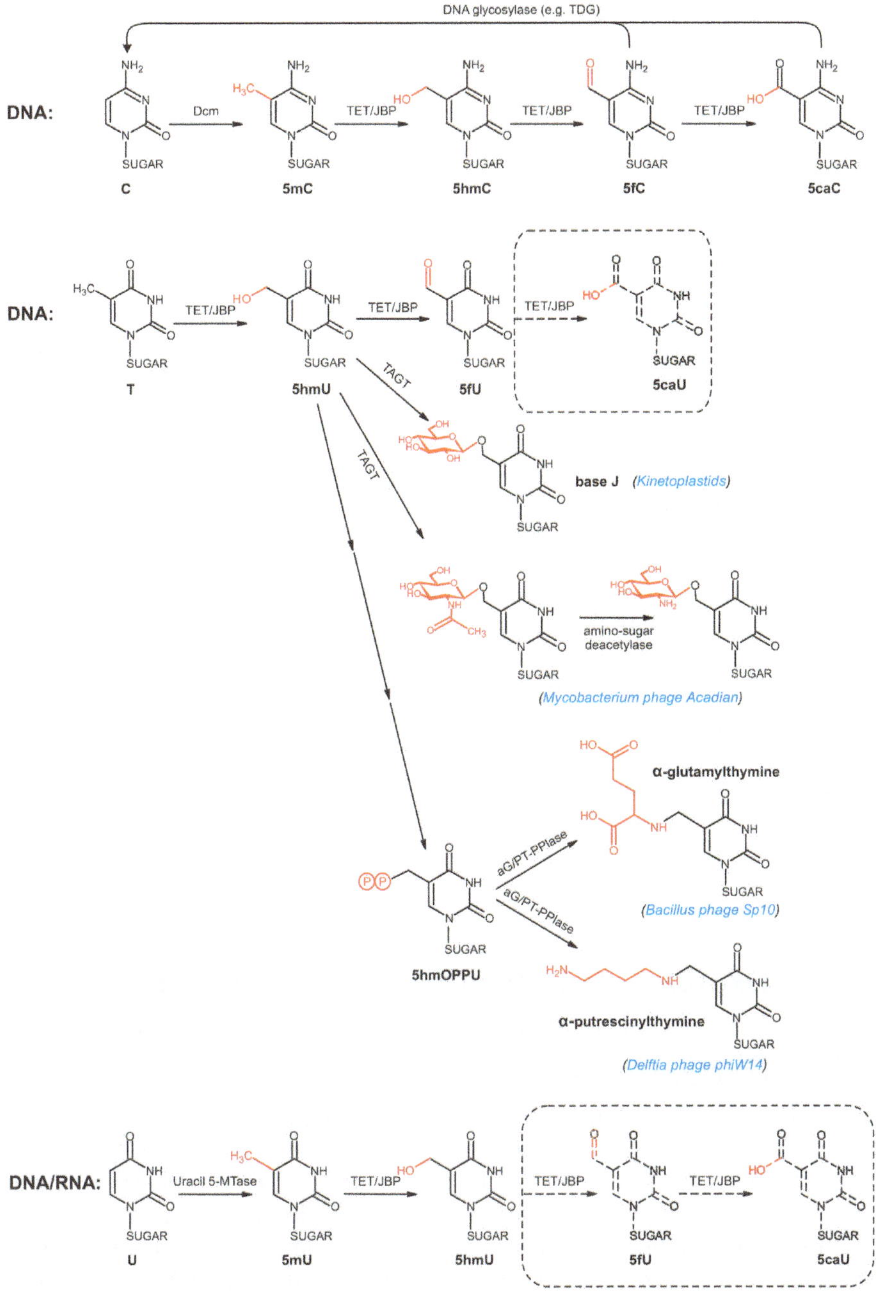

Figure 11.1 Major DNA/RNA nucleic base modification pathways catalysed by TET/JBP and other enzymes. Chemical groups added during the biochemical reactions are highlighted in red in the ensuing products. The enzymes catalysing the reactions are indicated above the reaction arrows. Predicted modifications are shown in dotted chemical forms and boxes. Abbreviations: aG/PT-PPlase, α glutamyl/α putrescinyl thymine pyrophosphorylase; Dcm, DNA cytosine methyltransferase; MTase, methyltransferase; TAGT, TET/JBP-associated glycosyltransferase.

11.1.2 Discovery of the TET/JBP Family of Dioxygenases

The first members of the TET/JBP family to be characterized were the JBP1 and JBP2 proteins from *Trypanosoma* and *Leishmania*, which catalyse the hydroxylation of the methyl group in thymine forming 5hmU as the first step in the synthesis of base J, which is characteristic of these organisms (Figure 11.1; see Chapter 12 for details).[19,20,24] Sequence analysis of the JBP hydroxylase domains revealed that they are members of the 2OGFeDOs, whose previously undetected representatives are present in numerous eukaryotes, bacteria and viruses.[25] Among the sequences identified were the metazoan TET proteins (Tet1, Tet2 and Tet3; oncogenes in humans), whose domain architectures closely paralleled the DNMT1 DNA methyltransferases, with an *N*-terminal DNA-binding Cys–X–X–Cys (CXXC) domain combined with a *C*-terminal 2OGFeDO catalytic domain. This observation led to the prediction that the TET proteins would oxidatively modify 5mC to generate its oxidized derivatives.[25] Follow-up experimental studies showed that indeed the Tet proteins are 2OGFeDOs that successively generate 5hmC, 5-formyl cytosine (5fC) and 5-carboxycytosine (5caC) *in situ* from the 5mC in DNA.[26,27] The presence of 5hmC had been reported earlier in mammalian DNA, but there had been debate regarding whether those results were partly or entirely an artefact of non-biological oxidation.[28] Identification of the catalytic activity of the TET proteins established beyond doubt that 5hmC is indeed an endogenous modification with potential biological significance both as a novel epigenetic mark and as an intermediate in 5mC demethylation.[29]

In this chapter we review the TET/JBP family in terms of their characterized catalytic activity, structural features, evolutionary diversity and potential biological functions.

11.2 Oxidative Modifications of Nucleic Acid Bases Catalysed by the TET/JBP Family of Dioxygenases

To date several distinct members of the TET/JBP family from diverse eukaryotes have been biochemically characterized and found to generate a number of distinct oxidized bases (Figure 11.1). The archetypal TET proteins from Metazoa catalyse three successive oxidations of 5mC in DNA generating 5hmC, 5fC and 5caC (top line of Figure 11.1).[26,27] This order of action correlates with the concentrations of these bases detected in mammalian DNA, with 5hmC being the most common followed by 5fC and 5caC.[26] Outside metazoans, multiple members of the TET/JBP family encoded by transposon-linked lineage-specific gene expansions in fungi have also been characterized.[30] At least four members of the expansion from the mushroom *Coprinopsis cinerea* have been shown to possess 2OGFeDO activity that generates 5hmC or its more oxidized derivatives from 5mC in DNA (Anjana Rao, personal communication).[31] One of the best characterized of these proteins, CC1G_05589 (CcTet), generates 5fC as the dominant product at neutral pH, unlike the metazoan enzymes. This preference for generating 5fC is even more pronounced at acidic pH, with very little 5caC generated under these conditions.[31] However, investigation of *Coprinopsis cinerea* DNA indicates that in cells the predominant form of the oxidized base is 5caC rather than 5fC, suggesting that one or more of the other TET/JBP family members encoded by this organism might drive further conversion of the lower oxidized forms to 5caC (Anjana Rao, personal communication). A study of one of the eight members of the TET/JBP family from the early-branching eukaryote *Naegleria*,[5] a heterolobosean, has shown that it modifies 5mC serially, generating all three oxidized derivatives.[32] While the *Naegleria* enzyme converts 5mC to 5hmC faster than it converts the latter to 5fC, incubation for an hour yields predominantly 5caC with much lower fractions of 5fC.[32] All TET proteins studied to date use DNA with either 5mC, 5hmC or 5fC as substrates for further oxidation.

Interestingly, a recent study suggests that the mammalian TET proteins operate not just on cytosine but also thymine (Figure 11.1, second line), converting it to 5hmU – a reaction very similar to that catalysed by the euglenozoan JBPs.[33] The original studies on base J biosynthesis suggested that the JBPs might only convert thymine to 5hmU, which is then glycosylated.[19,20,24] However, in trypanosomatids with a knockout of the glycosyltransferase that conjugates glucose to 5hmU it was seen that the JBPs serially convert 5hmU to 5fU and also perhaps 5caU at much lower levels.[34] These observations suggest that different members of the TET/JBP family might have a closer similarity in their activities than was previously believed. Interestingly, two glycosyltransferases (GREB1 and GREB1L) related to the one involved in base J synthesis have been identified in metazoans such as humans.[5] It remains to be seen whether they might be able to use any of the 5-hydroxymethylpyrimidines that are generated by the TET proteins as substrates. The bacteriophage and bacterial members of the TET/JBP family have not yet been biochemically characterized. However, the genomic contexts of certain bacteriophage versions of genes, which are closest to those encoding euglenozoan JBPs, show strong linkages to genes for glycosyltransferases that are closely related to those involved in base J synthesis.[5] Hence, the corresponding enzymes are likely to catalyse the synthesis of J and related bases in these viruses (Figure 11.1). Other bacteriophage versions are linked to enzymes involved in the biosynthesis of hypermodified thymines, such as α-putrescinylthymine and α-glutamylthymine.[3,5,9,11] Here again the corresponding TET/JBP enzymes are likely to generate 5hmU as the first step in the biosynthesis of these hypermodified thymines (Figure 11.1). In certain phages, such as the cyanophage MED4-117, the TET/JBP-coding gene is linked to a gene encoding uracil-5-methyltransferase.[5] Here, the TET/JBP enzyme might generate 5hmU from T after methyltransferase-catalysed methylation of uracil, which is incorporated in place of thymine, a phenomenon observed in certain bacteriophages.[3] In stramenopile and chlorophyte algae there is a distinct subfamily of predicted RNA-modifying TET/JBP enzymes, a subset of which (from stramenopile algae) is also fused to the same family of uracil-5-methyltransferases.[5,25] This correlation suggests that one of the RNA modifications they generate is 5hmU from uracil (Figure 11.1).

11.3 Structure, Domain Architectures and Genomic Context of the TET/JBP Family of Dioxygenases

11.3.1 Structure of the Catalytic Domain

The catalytic domain of the Tet/JBP family displays a double-stranded β-helix (DSBH) fold (Figure 11.2),[35–37] the common denominator of a large class of evolutionarily related domains that includes: (i) cyclic nucleotide-binding domains; (ii) sugar-binding domains (*e.g.* the AraC sugar-binding domain); (iii) sugar isomerases; (iv) non-haem Fe-dependent dioxygenases that do not use 2OG; and (v) two superfamilies of dioxygenases that use 2OG as a cosubstrate, namely the Jumonji-related and the classical 2OGFeDOs. Dioxygenases of the DSBH fold incorporate both oxygen atoms from molecular oxygen into a wide range of substrates, including peptides, nucleic acids and small molecules.[35–37] The classical 2OGFeDOs and Jumonji-related proteins are distinct from other dioxygenases of this fold in utilizing the cosubstrate 2OG, which they oxidize to CO_2 plus succinate, containing one of the oxygen atoms, while incorporating the second oxygen atom into a distinct substrate molecule. The conserved DSBH core shared by all the above-stated domains contains eight β-strands (Figure 11.2); in classical 2OGFeDOs this core is elaborated further by two additional *N*-terminal strands (strands-1N and 2N; Figure 11.2), which in the structure stack on either side of the core DSBH, and are each linked to helical segments that pack against one of the exposed faces of the DSBH β-sheet (Figure 11.2).[35] In all 2OG-utilizing DSBH fold enzymes the second strand of the core DSBH domain bears a conserved motif (typically His-x-Asp) and the seventh strand bears a conserved His, which together chelate Fe(II).[21,23,35–37] Unique to the classical 2OGFeDOs is a conserved Arg on the eighth strand that binds the 2-oxoacid cofactor *via* a salt bridge.[35] In reactions catalysed by the TET/JBP enzymes the target base is directed away from the rest of the nucleic acid chain and taken into the dioxygenase active site, lying at the aperture of the interior cavity of the DSBH (Figure 11.2).[32,38] Here the target base appears to be positioned *via* a cation-π interaction with the catalytic Fe(II). Thus, in the case of DNA-modifying TET/JBPs, this base is flipped out of the DNA double helix with accompanying distortion of the helical axis (Figure 11.2).

The crystal structures of the TET/JBP enzymes (represented by human TET2 and *Naegleria* Tet1) reveal that members of this family possess characteristic structural features that distinguish them from the rest of the 2OGFeDO superfamily.[5,32,38] First, they have an *N*-terminal insert of a β-hairpin-like element (between strand-1N and the helix downstream of it; Figure 11.2), which is also found in another major family of nucleic acid-modifying 2OGFeDOs, the AlkB family.[39,40] In the TET/JBP family, this insert is of critical importance because it is inserted deep into the helical axis of DNA, close to the minor groove, and causes the target base to be flipped out into the active site.[32,38]

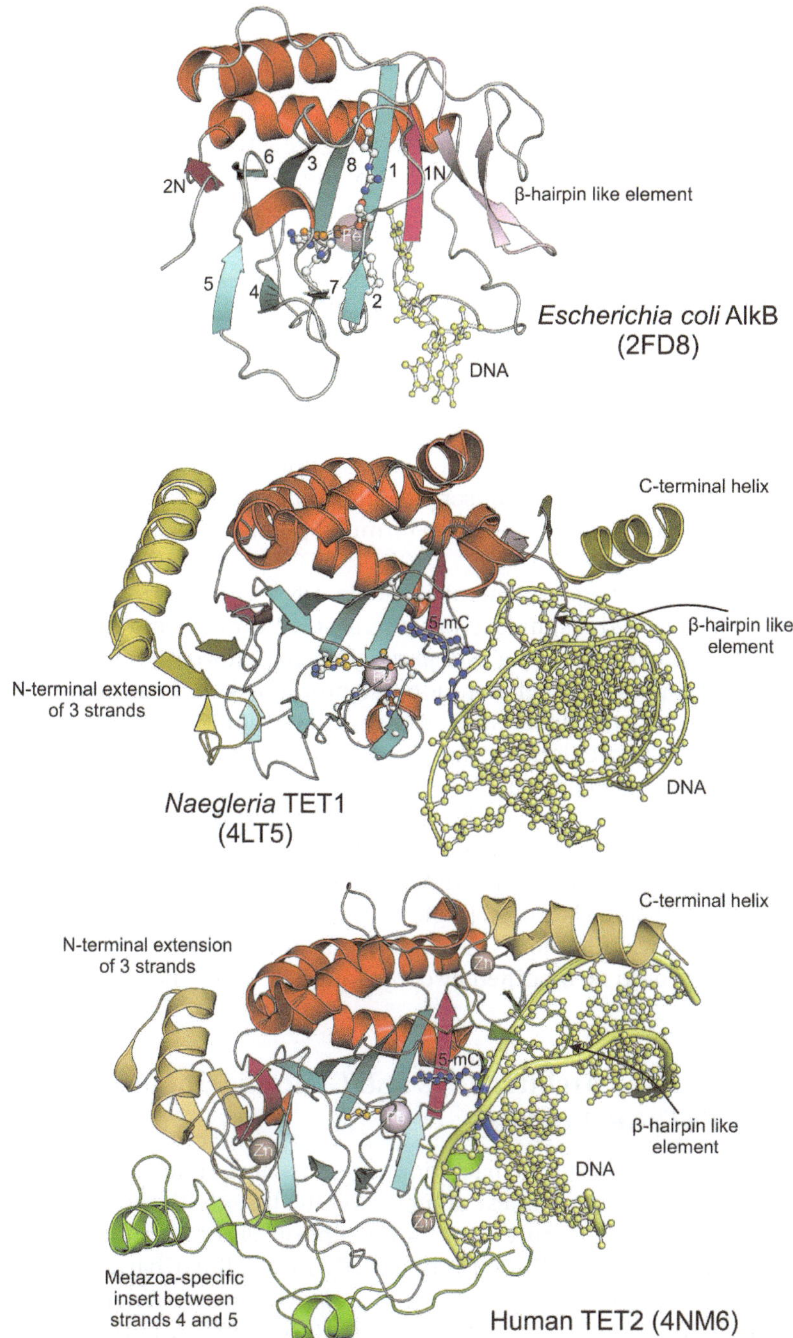

Figure 11.2 Cartoon representation of the structures of three members of the 2OGFeDO superfamily including bacterial AlkB (Protein Databank, PDB, identifier: 2FD8), *Naegleria* TET1 (4LT5) and human TET2 (4NM6). The conserved core of the DSBH fold is shown in red for α-helices and light blue for β-strands (numbered 1–8). The 2OGFeDO specific strands 1N and 2N are coloured bright pink. Additional structural features specific to each of the proteins are also highlighted.

Second, a subset of the TET/JBP proteins, *e.g.* the fungal and metazoan TETs, but not the JBPs, *Naegleria* Tet1, or the bacterial and viral TET/JBPs, have a further insert upstream of strand-1N[5] (Figure 11.2). In the structure this insert is positioned on the opposite side of the above insert and forms a second DNA-binding interface in those versions possessing it (Figure 11.2).[38] In metazoan versions this insert and the previous one show a distinct sequence signature with conserved Cys residues, which along with conserved residues elsewhere in the protein chelate Zn ions to stabilize the insert.[25] Third, the TET/JBP enzymes are set apart from other 2OGFeDOs by an *N*-terminal extension of three strands augmenting the DSBH, two of which pack with conserved strand-2N.[5,32,38] In metazoan TETs the loops between the strands of the *N*-terminal extension are stabilized by conserved Zn(II)-chelating Cys residues (Figure 11.2). Finally, the catalytic domain of the TET/JBP family is also distinguished by a conserved *C*-terminal helix which contacts the DNA adjacent to the DNA-binding *N*-terminal β-hairpin-like element (Figure 11.2).[25,32,38]

While bacteriophage, eukaryote-specific DNA virus and bacterial TET/JBP proteins show the simplest structures, structural elaboration is seen in the predicted RNA-modifying, *Naegleria*-type, euglenozoan, fungal-type and metazoan versions.[5] This elaboration has proceeded *via* both further development of pre-existing inserts of the ancestral versions as well as emergence of new inserts, particularly in the fungal-type and metazoan versions. The metazoan proteins acquired 12 distinct Cys or His residues that chelate a total of three Zn ions.[25,38] Additionally, the metazoan versions possess a unique, giant insert (~250–600 amino acids) between strand-4 and strand-5, which is largely unstructured save for a few helical and β-hairpin elements at its extreme *N*- and *C*-termini (Figure 11.2).[38] This unstructured region has potential SUMOylation motifs suggesting that at least some of the metazoan TET proteins could be modified by conjugation of SUMO polypeptides.[25]

11.3.2 Domain Architectures and Genomic Context of the TET/JBP Proteins

The TET/JBP proteins from bacteriophages, bacteria and large eukaryotic DNA viruses show the simplest domain architectures with more-or-less just the catalytic domain (Figure 11.3). An exception is the version from *Legionella drancourtii* (gi: 493927700), the strictly intracellular bacterial parasite of *Acanthamoeba*, which possesses a long *N*-terminal extension that might be required for its trafficking into the host cell (Figure 11.3).[5] However, most bacteriophage and bacterial versions show conserved gene-neighbourhood linkages to several distinct sets of genes (Figure 11.3B):[5] (i) those coding for glycosyltransferases and in some cases other sugar-modifying enzymes such as acetylated amino-sugar deacetylases (*e.g.* the *Persicivirga* phage P12024L base J biosynthesis related glycosyltransferase, gi: 399528505, Figures 11.1 and 11.3B); (ii) enzymes such as the 5-hydroxymethylpyrimidine

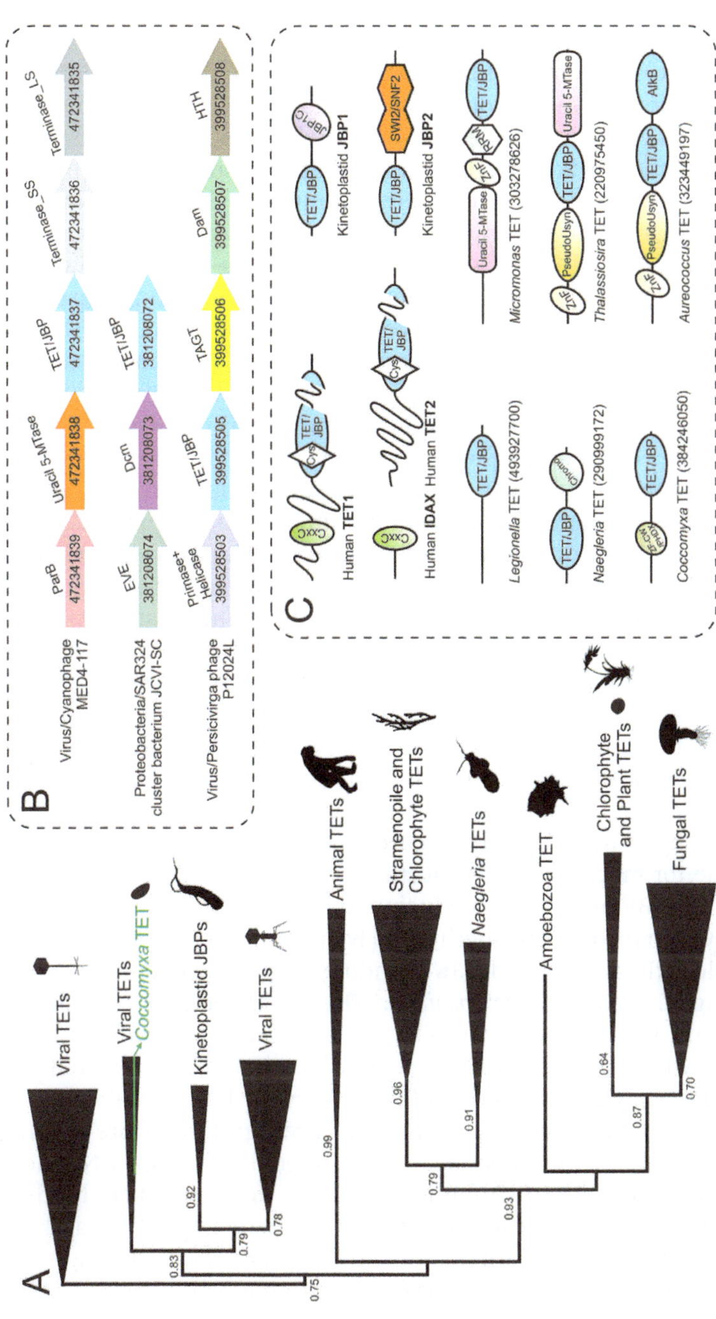

Figure 11.3 Evolutionary relationships of the TET/JBP clades and multiple origins of the eukaryotic versions from bacteriophages (A), genomic context (B) and domain architectures (C) of the representative TET/JBP proteins. Tree nodes supporting values greater than 75% are shown. For abbreviations refer to the legend for Figure 11.1 and Dam, DNA adenine methyltransferase; Dcm, EVE, a 5hmC-specific DNA-binding domain; HTH, helix-turn-helix; PseudoUsyn, pseudouridine synthase; Terminase_SS and Terminase_LS, small and large subunits of the terminase; ZnF, zinc finger; and ZF-CW/PHDX, CW/PHDX type of zinc finger.

phosphorylating P-loop kinases and α-glutamyl and α-putrescinyl thymine pyrophosphorylases and other accessory proteins involved in hypermodified thymine biosynthesis;[5,9,11] (iii) uracil-5-methyltransferases; (iv) enzymes involved in other DNA modifications such as N6mA-generating methyltransferases; and (v) cytosine-5-methyltransferases which could generate 5mC for further modification by the product of the associated TET/JBP genes (*e.g.* SAR324 cluster bacterium; gi: 281208072). These associations show that the TET/JBP enzymes of bacteriophages and bacteria are part of a biosynthetic network for the generation of a range of modified bases in DNA from 5-hydroxymethylpyrimidine to glycosylated and hypermodified thymines. In several bacteriophages these gene neighbourhoods occur sandwiched between the genes for a ParB nuclease implicated in chromosome partitioning and the terminase genes involved in packaging the phage head with DNA (Figure 11.3B).[5] This observation suggests that these DNA modifications might play a key role in viral DNA packaging, consistent with previous observations regarding the need for such modifications in this process.[9,11,41,42]

The eukaryotic DNA-modifying versions of the TET/JBP proteins often show more complex domain architectures that typically combine the 2OGFeDO domain with other domains typical of chromatin proteins (Figure 11.3C).[5,25] One example is the CXXC domain, a Zn-dependent DNA-binding domain that recognizes the CpG sequence typically in its unmethylated form, but certain versions also tolerate methylation or hydroxymethylation;[43,44] an *N*-terminal fusion to this domain is seen in the metazoan TET proteins. Another example is the JBP1C domain, a DNA-binding domain with the helix-turn-helix fold, which is fused to the *C*-terminus of the TET/JBP domain in the kinetoplastid JBP1 proteins;[25] this domain binds DNA-containing base J.[45] A third example is the SWI2/SNF2-type superfamily-II helicase domain fused to the *C*-terminus of the catalytic domain in the kinetoplastid JBP2;[20,44] this component is likely to catalyse ATP-hydrolysis-dependent chromatin remodelling, to make the DNA accessible for modification. A fourth example found in at least one of the *Naegleria* TET/JBP proteins is the fusion to a *C*-terminal chromodomain, which is involved in recognizing methylated lysines on histones H3 and H4.[25] A final example is the methylated H3K4-binding PHDX/Zf-CW domain in the chlorophyte alga *Coccomyxa subellipsoidea*.[5] These architectures suggest that the predominant theme is the combination of the catalytic domain with other domains which probably recognize modified DNA or methylated histones (Figure 11.3C). Thus, the action of the eukaryotic DNA-modifying TET/JBP proteins is likely to be closely coupled with recognition of pre-existing epigenetic marks on DNA and histones. The predicted eukaryotic RNA-modifying members of the TET/JBP family are fused to several RNA-binding domains (*e.g.* the RNA-recognition-motif/RRM, the zinc-finger/ZnF, and the cysteinyl tRNA synthetase *C*-terminal domains) and/or RNA-modifying enzymatic domains (pseudouridine synthase, uracil-5-methyltransferase, AlkB family 2OGFeDO, and decarboxylase domains) (Figure 11.3C).[25] These architectures suggest that the modifications catalysed by these enzymes are likely to be coupled with other RNA modifications catalysed by linked enzymes.

11.4 Evolutionary History and Adaptations of the TET/JBP Family of Dioxygenases

Within the 2OGFeDO superfamily, the TET/JBP family appears to be most closely related to the AlkB demethylase family (Chapter 8)[39] and another 2OGFeDO family that is often found in gene neighbourhoods linked to the AlkB family.[44,46] The TET/JBP family shares with these groups not just the preference for modifying bases in nucleic acids, but also the conserved structural element that helps direct the target base into the active site cavity for modification. Structural features and phylogenetic analysis strongly suggest that the TET/JBP family emerged first in bacteriophages (Figure 11.3A).[5] Recent comparative genomic analysis has shown that the caudate bacteriophages infecting most major bacterial lineages have evolved numerous distinct systems for generation of modified bases in DNA.[5] This finding suggests that there has been selection for diversity of DNA modification in these phages, thereby making them fertile 'nurseries' for the innovation of novel modifications.[47] In biological terms these modifications appear to be required for packaging of viral DNA.[9,11,41,42] In this capacity the modified bases possibly function as epigenetic signals that allow for the marking of a genome's length of DNA for packaging into the empty capsid by the phage terminase. In addition, the modifications appear to help evade attacks by host restriction systems; this feature might account for the great variability in the phage DNA modification systems.[3,6] From bacteriophages, the genes encoding TET/JBP proteins have been transferred to several bacterial genomes; in some cases these genes are still associated with integrated prophages or their remnants.[5] While some of the bacterial versions might modify the cellular genomic DNA, in at least certain cases, like the *Legionella drancourtii* version, they appear to have been recruited as effectors, which might modify the DNA of the host cells in which they reside. Likewise, TET/JBP-encoding genes also appear to have been transferred from bacteriophages to eukaryotic viruses of the nucleo-cytoplasmic large DNA virus (NCLDV) class, in particular those belonging to the mimivirus-phycodnavirus clade.[5] These NCLDV versions might help protect viral DNA from the restriction attack some of these viruses are known to launch on their hosts and/or they might be deployed as regulatory factors to influence host or viral gene expression.[48]

Eukaryotes appear to have acquired their genes encoding TET/JBP proteins on at least three independent occasions *via* lateral transfer from bacteriophages.[5] In the case of the euglenozoan JBPs the transferred element appears to have included the genes for both the JBP and the glycosyltransferase which catalyses the subsequent step in base J biosynthesis. Thus, both JBP and the base J-generating glycosyltransferase are most closely related to the cognate enzymes of phages of the *Persicivirga* phage P12024L type.[5] In basidiomycete fungi the TET/JBP family has undergone massive expansions (10 to over 200 paralogues). Genomic analysis revealed that many of these copies are linked to novel DNA transposons belonging to three families, namely Kyakuja, Dileera and Zisupton.[30] The expansions of genes encoding the TET/JBPs are mirrored by expansions of the associated transposons, suggesting their linkage. Moreover, these expansion events appear to coincide with speciation events in fungi.[30] Interestingly, the majority of both the TET/JBP proteins and the linked transposase domains are predicted to be active proteins. This is in contrast to several other previously studied transposons expansions, where most copies are inactive.[30] Hence, these transposons are possibly at least partially 'domesticated' and might play a role in genomic plasticity, which in turn could be regulated by the associated TET/JBP-coding genes. These genes encoding fungal-type TET/JBP proteins, while predominantly present in basidiomycete fungi such as mushrooms, have also been laterally transferred to ascomycete and glomeromycete fungi on occasions, and also to certain chlorophytes and plants.[30] The fission yeast *Schizosaccharomyces* possesses a single catalytically inactive member of the TET/JBP family Cif1, which is believed to regulate, independently of dioxygenase activity, the induction of an epigenetically transmitted prion state required for resistance against protein misfolding stress.[49]

Unlike in fungi, metazoans as a rule typically display only a single copy of a TET/JBP-coding gene; however, in both types of organisms the presence of these genes is strongly correlated with the presence of a *bona fide* DNA 5C methyltransferase.[30] Early in vertebrate evolution this single copy appears to have undergone a triplication and the resulting paralogues were then conserved throughout vertebrate evolution.[25] In the vertebrate TET2 protein, the *N*-terminal CXXC domain was separated from the rest of the protein by a chromosomal inversion, which converted it into a separate gene, CXXC4/IDAX.[50] However, this separated CXXC domain physically interacts with TET2 and regulates its activity.

11.5 Functional Implications of the TET/JBP Family

Outside of kinetoplastids and vertebrates, there has been little directed exploration using experiments to assess the biological significance of the TET/JBP family. Base J generated by kinetoplastid JBPs was shown to block elongation of the polycistronic kinetoplastid transcripts beyond the transcription stop sites (see Chapter 12 for details).[51–53]

11.5.1 Biological Roles of the Vertebrate TET Proteins

The functions of the vertebrate TET proteins are not entirely understood but enormous progress has been made in the past 5 years. *Tet1*$^{-/-}$ mice have low birth weight, but otherwise they are viable and fertile.[54] In placental mammals Tet1 negatively regulates key transcription factors of the trophectoderm such as Cdx2, Eomes and Elf5, and it suppresses differentiation towards the extra-embryonic lineage, consistent with the knockout results.[55] While there is evidence for Tet1 regulating gene expression and differentiation in murine embryonic stem (ES) cells, the evidence from the knockout studies suggests that by itself it is not an essential player in ES cell pluripotency.[54] *Tet2*$^{-/-}$ mice are fertile, but they show haematological abnormalities. These mutant animals have more haematopoietic stem cells (HSCs) than normal mice, and their HSCs show self-renewal and proliferation in culture and in serial transfers of progenitor cells in mice.[56] At least one strain of Tet2-deficient mice develops a condition similar to human chronic myelomonocytic leukaemia (CMML) typified by a gross abundance of monocytes. TET2 is a major player in human myeloid malignancies – approximately one-fifth of all and over half of the human CMML samples show somatic deletions or loss-of-function mutations in the TET2-encoding gene.[57–59] Moreover, gain-of-function mutations in genes encoding isocitrate dehydrogenase (IDH) 1 and IDH2 are frequently found in glioblastomas and myeloid leukaemias.[60] As a result these enzymes generate an excess of 2-hydroxyglutarate, which inhibits the TETs among other 2OG-utilizing enzymes such as the Jumonji-C related lysine demethylases. These observations, along with a typically low level of 5hmC in cancer cells (*e.g.* its loss in melanoma cells) point to a key role for TET in tumorigenesis and progression.[61]

Tet3$^{-/-}$ mice die at or soon after birth, pointing to a role for Tet3 in embryonic development.[62] Immediately after fertilization the male pronucleus loses most of its staining for 5mC and shows an increase in 5hmC staining as assessed with specific antibodies. This effect is abrogated by siRNA targeting of Tet3 and is also not observed in fertilizations using Tet3-negative oocytes.[62] The placental mammal-specific PGC7/Stella/Dppa3 protein, with a distinctive CXCXXC motif containing a DNA-binding domain,[44] protects maternal DNA, and the few imprinted loci in the paternal genome, from Tet3-mediated oxidation.[63,64] In *PGC7* knockout mice, both the paternal and maternal genomes become hydroxymethylated by Tet3. This result suggests that the Tet3-Pgc7 antagonism is the primary control apparatus for the epigenetic 5mC mark during zygote formation, especially in the male pronucleus. Double knockout of *Tet1* and *Tet2* results in mice that are born alive, but die within a few days of birth.[65] Given the greater severity of the double mutant than the single mutants there could be a back-up role between Tet1 and Tet2. In primordial germ cells of placental mammals a genome-wide loss of 5mC staining has been observed (at approximate day E11.5 in mice). Tet3 is not prominent at this stage, but Tet1 and Tet2 are present, suggesting that they have overlapping roles in oxidation of 5mC marks at this point in germ cell development. The *Tet1/2/3* triple-knockout mouse ES cells resulted in impaired differentiation with resultant poorly differentiated embryoid bodies and teratomas.[66] This result suggests that the vertebrate TETs back up each other to a certain extent, but at least one of them is necessary for ES cells to form differentiated cells.

11.5.2 Vertebrate TET Proteins, Demethylation and Epigenetics

The removal of the 5mC methyl mark, both locally and in genome-wide events like those observed after zygote formation and primordial germ cell development, has been a major conundrum for over two decades.[67] The results noted in Section 11.5.1 suggest that the TET proteins are indeed major players in this process, at least in metazoans. However, there are several details that have only recently become apparent. In the event following zygote formation, when much of the 5mC mark as assessed by specific antibody staining is lost,[15,16] a lower fraction of the mark is removed as assessed by bisulfite sequencing (which does not distinguish 5mC and 5hmC).[62] This finding indicates that rather than complete demethylation, much of the 5mC mark is converted to 5hmC, with additional parts to 5fC and 5caC, which would score as demethylation by the bisulfite sequencing method. Hence, the major 'demethylation' event in early mammalian development appears to be primarily the 5mC-oxidizing action of Tet3.[62] However, other lines of evidence suggest that oxidized methylcytosines are intermediates in complete demethylation, *i.e.* restoration of the cytosine. The DNA glycosylase TDG is specific to 5fC and 5caC and can remove these bases leaving abasic lesions that are then repaired using the base-excision-repair (BER) machinery (Figure 11.1).[26] Indeed, components of the BER system that act downstream of the abasic sites created by DNA-glycosylases, namely APE1, XRCC1 and PARP1, have been observed in the context of both post-zygotic and primordial germ cell demethylation.[68,69] There have also been early reports of an unknown DNA glycosylase that is specific to 5hmC.[70] Thus, in addition to removal of 5fC and 5caC, even 5hmC could trigger BER. Consistent with this notion, a study aimed at systematically detecting 5hmC-DNA-binding proteins recovered the DNA glycosylases Neil1 and Mbd4 along with the DNA-repair proteins HELLS and RecqL (both DNA helicases) as 5hmC DNA binders.[71] Furthermore, this study also identified C3ORF37 as a 5hmC DNA binder.[71] This protein is a member of the SRAP superfamily of autopeptidases, which are key players in bacterial SOS response systems.[72] The SRAP domain is predicted to sense 5hmC and initiate a DNA repair response that could facilitate its removal from DNA, similar to the SOS response initiated by DNA damage in bacteria.[72] Thus, the emerging evidence strongly favours a role for TET-mediated oxidation during demethylation in metazoans. It is possible that such a role might be more general for the TET proteins including those from certain fungi, algae and *Naegleria*.

Other lines of evidence suggest that the oxidized 5mC marks might function as epigenetic marks in their own right. In ES cells, 5hmC (rather than 5mC) is enriched near the transcription start sites (TSS) of genes, and the so-called 'bivalent' promoters with both H3K4 and H3K27 trimethylation are specifically enriched for both Tet1 and 5hmC.[73–75] Genes with bivalent promoters tend to be expressed at low levels in ES cells but are upregulated upon differentiation,[76] suggesting a possible role for 5hmC in genes primed for future expression. The incidence of 5hmC also sharply rises near enhancers, along with H3K27 acetylation, concomitant with the first signs of cell differentiation.[77] However, 5hmC is depleted at the actual binding site.[78] In mammalian brain tissues, where there was an early report of 5hmC enrichment in Purkinje neurons, several studies point to potential epigenetic roles for 5hmC.[29] In contrast to the ES cells, hardly any 5hmC was detected at the TSSs in the mature hippocampus and cerebellum.[73,75] However, in neural tissue 5hmC in gene bodies is significantly correlated with elevated expression of genes activated during development, resulting in the building up of high levels of 5hmC. Curiously, the 5hmC mark is significantly depleted on sex chromosomes in neural tissues.[73,75] A study on the action of Tet3 in the brain has revealed that Tet3-driven accumulation of 5hmC in the infralimbic prefrontal cortex possibly facilitates behavioral adaptation in mice.[79] In terms of the action of the oxidized methylcytosines, some reports have noted altered binding of methylated DNA-recognition factors containing the MBD/TAM domain, with 5hmC inhibiting the DNA binding of some of these proteins.[80] Chromatin structure is also clearly altered by 5hmC in DNA because neurons enriched in 5hmC and TET1 overexpressing cells show enlarged or decondensed nuclei.[27,29] Finally, certain transcription factors like Zhx1 and Zhx2 specifically bind 5hmC-containing DNA,[71] whereas 5caC in the CpG dinucleotide enhances binding of the bHLH domain transcription factor dimer Tcf3-Ascl1.[81] Despite these findings, considerably more research remains to be done to obtain a more complete understanding of the roles of the TET/JBP proteins in epigenetics.

11.6 Concluding Remarks

Discovery of the JBP proteins,[20] leading to the definition of the TET/JBP family in 2009,[25] has inspired an enormous amount of research into these proteins and the modifications generated by them.[82] Yet, there remain several major lacunae in our understanding. In large part these gaps in knowledge stem from the fact that most of the research has been on mammalian and kinetoplastid systems, which only represent a small part of the diversity of this family.[5,82] Key areas of future research likely to yield rich dividends can be enumerated: (i) Deciphering the complete system of proteins involved in demethylation, especially the coupling of TET/JBP catalysis with BER; in this regard the role of the SRAP domain proteins is likely to be of great interest.[71,72] (ii) Working out the relationship between the modified bases generated by the TET/JBP proteins and the observed biological phenomena in which they have been implicated.[82] (iii) Better understanding of the biochemistry and biology of the non-mammalian versions, especially those from fungi, algae, *Naegleria* and bacteriophages.[5,32] (iv) The role of the TET/JBP in RNA modification in chlorophytes and stramenopiles, including the identification of the substrate RNAs and the types of modifications catalysed by the multiple TET/JBP enzymes coded by their genomes.[25] (v) Deciphering the base hypermodification pathways in which the TET/JBP enzymes create the initial 5-hydroxymethyl pyrimidines and the potential epigenetic roles of the hypermodified bases in eukaryotes that possess them.[5] (vi) Understanding the roles of the catalytically inactive versions such as *S. pombe* Cif1.[30,49] Given these outstanding issues we expect the TET/JBP enzymes to be among the most interesting members of the 2OGFeDO superfamily from the viewpoint of future research.

Acknowledgements

Research by the authors is supported by the Intramural Research Program of the National Library of Medicine, NIH, Department of Health and Human Services, USA.

References

1. A. Czerwoniec, S. Dunin-Horkawicz, E. Purta, K. H. Kaminska, J. M. Kasprzak, J. M. Bujnicki, H. Grosjean and K. Rother, *Nucleic Acids Res.*, 2009, **37**, D118–D121.
2. H. Grosjean, *DNA and RNA modification enzymes : structure, mechanism, function, and evolution*, Landes Bioscience, Austin, Tex., 2009.
3. J. H. Gommers-Ampt and P. Borst, *FASEB J.*, 1995, **9**, 1034–1042.
4. V. Anantharaman, E. V. Koonin and L. Aravind, *Nucleic Acids Res.*, 2002, **30**, 1427–1464.
5. L. M. Iyer, D. Zhang, A. M. Burroughs and L. Aravind, *Nucleic Acids Res.*, 2013, **41**, 7635–7655.
6. T. A. Bickle and D. H. Kruger, *Microbiol. Rev.*, 1993, **57**, 434–450.
7. R. J. Roberts, M. Belfort, T. Bestor, A. S. Bhagwat, T. A. Bickle, J. Bitinaite, R. M. Blumenthal, S. Degtyarev, D. T. Dryden, K. Dybvig, K. Firman, E. S. Gromova, R. I. Gumport, S. E. Halford, S. Hattman, J. Heitman, D. P. Hornby, A. Janulaitis, A. Jeltsch, J. Josephsen, A. Kiss, T. R. Klaenhammer, I. Kobayashi, H. Kong, D. H. Kruger, S. Lacks, M. G. Marinus, M. Miyahara, R. D. Morgan, N. E. Murray, V. Nagaraja, A. Piekarowicz, A. Pingoud, E. Raleigh, D. N. Rao, N. Reich, V. E. Repin, E. U. Selker, P. C. Shaw, D. C. Stein, B. L. Stoddard, W. Szybalski, T. A. Trautner, J. L. Van Etten, J. M. Vitor, G. G. Wilson and S. Y. Xu, *Nucleic Acids Res.*, 2003, **31**, 1805–1812.
8. R. J. Roberts, T. Vincze, J. Posfai and D. Macelis, *Nucleic Acids Res.*, 2010, **38**, D234–D236.
9. H. Witmer and C. Wiatr, *J. Virol.*, 1985, **53**, 522–527.
10. K. L. Maltman, J. Neuhard and R. A. Warren, *Biochemistry*, 1981, **20**, 3586–3591.
11. R. A. Warren, *Annu. Rev. Microbiol.*, 1980, **34**, 137–158.
12. T. H. Bestor, *Philos. Trans. R. Soc. London, Ser. B*, 1990, **326**, 179–187.
13. T. H. Bestor, *Hum. Mol. Genet.*, 2000, **9**, 2395–2402.
14. P. Hajkova, K. Ancelin, T. Waldmann, N. Lacoste, U. C. Lange, F. Cesari, C. Lee, G. Almouzni, R. Schneider and M. A. Surani, *Nature*, 2008, **452**, 877–881.
15. J. Oswald, S. Engemann, N. Lane, W. Mayer, A. Olek, R. Fundele, W. Dean, W. Reik and J. Walter, *Curr. Biol.*, 2000, **10**, 475–478.
16. W. Mayer, A. Niveleau, J. Walter, R. Fundele and T. Haaf, *Nature*, 2000, **403**, 501–502.
17. P. Borst and R. Sabatini, *Annu. Rev. Microbiol.*, 2008, **62**, 235–251.
18. M. G. Goll and T. H. Bestor, *Annu. Rev. Biochem.*, 2005, **74**, 481–514.

19. L. J. Cliffe, T. N. Siegel, M. Marshall, G. A. Cross and R. Sabatini, *Nucleic Acids Res.*, 2010, **38**, 3923–3935.
20. Z. Yu, P. A. Genest, B. T. Riet, K. Sweeney, C. Dipaolo, R. Kieft, E. Christodoulou, A. Perrakis, J. M. Simmons, R. P. Hausinger, H. G. van Luenen, D. J. Rigden, R. Sabatini and P. Borst, *Nucleic Acids Res.*, 2007.
21. K. S. Hewitson, L. A. McNeill, J. M. Elkins and C. J. Schofield, *Biochem. Soc. Trans.*, 2003, **31**, 510–515.
22. J. N. Barlow, J. E. Baldwin, I. J. Clifton, E. Gibson, C. M. Hensgens, J. Hajdu, T. Hara, A. Hassan, P. John, M. D. Lloyd, P. L. Roach, A. Prescott, J. K. Robinson, Z. H. Zhang and C. J. Schofield, *Biochem. Soc. Trans.*, 1997, **25**, 86–90.
23. A. G. Prescott, *J. Exp. Bot.*, 1993, **44**, 849–861.
24. S. Vainio, P. A. Genest, B. Ter Riet, H. van Luenen and P. Borst, *Mol. Biochem. Parasitol.*, 2009, **164**, 157–161.
25. L. M. Iyer, M. Tahiliani, A. Rao and L. Aravind, *Cell Cycle*, 2009, **8**, 1698–1710.
26. Y. F. He, B. Z. Li, Z. Li, P. Liu, Y. Wang, Q. Tang, J. Ding, Y. Jia, Z. Chen, L. Li, Y. Sun, X. Li, Q. Dai, C. X. Song, K. Zhang, C. He and G. L. Xu, *Science*, 2011, **333**, 1303–1307.
27. M. Tahiliani, K. P. Koh, Y. Shen, W. A. Pastor, H. Bandukwala, Y. Brudno, S. Agarwal, L. M. Iyer, D. R. Liu, L. Aravind and A. Rao, *Science*, 2009, **324**, 930–935.
28. N. W. Penn, R. Suwalski, C. O'Riley, K. Bojanowski and R. Yura, *Biochem. J.*, 1972, **126**, 781–790.
29. S. Kriaucionis and N. Heintz, *Science*, 2009, **324**, 929–930.
30. L. M. Iyer, D. Zhang, R. F. de Souza, P. J. Pukkila, A. Rao and L. Aravind, *Proc. Natl. Acad. Sci. U. S. A.*, 2014, **111**, 1676–1683.
31. L. Zhang, W. Chen, L. M. Iyer, J. Hu, G. Wang, Y. Fu, M. Yu, Q. Dai, L. Aravind and C. He, *J. Am. Chem. Soc.*, 2014, **136**, 4801–4804.
32. H. Hashimoto, J. E. Pais, X. Zhang, L. Saleh, Z. Q. Fu, N. Dai, I. R. Correa, Jr., Y. Zheng and X. Cheng, *Nature*, 2014, **506**, 391–395.
33. T. Pfaffeneder, F. Spada, M. Wagner, C. Brandmayr, S. K. Laube, D. Eisen, M. Truss, J. Steinbacher, B. Hackner, O. Kotljarova, D. Schuermann, S. Michalakis, O. Kosmatchev, S. Schiesser, B. Steigenberger, N. Raddaoui, G. Kashiwazaki, U. Muller, C. G. Spruijt, M. Vermeulen, H. Leonhardt, P. Schar, M. Muller and T. Carell, *Nat. Chem. Biol.*, 2014, **10**, 574–581.
34. W. Bullard, J. Lopes da Rosa-Spiegler, S. Liu, Y. Wang and R. Sabatini, *J. Biol. Chem.*, 2014, **289**, 20273–20282.
35. L. M. Iyer, S. Abhiman, R. F. de Souza and L. Aravind, *Nucleic Acids Res.*, 2010, **38**, 5261–5279.
36. W. Aik, M. A. McDonough, A. Thalhammer, R. Chowdhury and C. J. Schofield, *Curr. Opin. Struct. Biol.*, 2012, **22**, 691–700.
37. J. M. Dunwell, A. Purvis and S. Khuri, *Phytochemistry*, 2004, **65**, 7–17.
38. L. Hu, Z. Li, J. Cheng, Q. Rao, W. Gong, M. Liu, Y. G. Shi, J. Zhu, P. Wang and Y. Xu, *Cell*, 2013, **155**, 1545–1555.

39. L. Aravind and E. V. Koonin, *Genome Biol.*, 2001, **2**, RESEARCH0007.
40. B. Yu, W. C. Edstrom, J. Benach, Y. Hamuro, P. C. Weber, B. R. Gibney and J. F. Hunt, *Nature*, 2006, **439**, 879–884.
41. A. Droge and P. Tavares, *J. Mol. Biol.*, 2000, **296**, 103–115.
42. M. B. Lobocka, D. J. Rose, G. Plunkett, 3rd., M. Rusin, A. Samojedny, H. Lehnherr, M. B. Yarmolinsky and F. R. Blattner, *J. Bacteriol.*, 2004, **186**, 7032–7068.
43. C. Frauer, A. Rottach, D. Meilinger, S. Bultmann, K. Fellinger, S. Hasenoder, M. Wang, W. Qin, J. Soding, F. Spada and H. Leonhardt, *PLoS One*, 2011, **6**, e16627.
44. L. M. Iyer, S. Abhiman and L. Aravind, *Prog. Mol. Biol. Transl. Sci.*, 2011, **101**, 25–104.
45. T. Heidebrecht, E. Christodoulou, M. J. Chalmers, S. Jan, B. Ter Riet, R. K. Grover, R. P. Joosten, D. Littler, H. van Luenen, P. R. Griffin, P. Wentworth, Jr., P. Borst and A. Perrakis, *Nucleic Acids Res.*, 2011, **39**, 5715–5728.
46. L. V. Mello and D. J. Rigden, *FEBS Lett.*, 2012, **586**, 3908–3913.
47. L. Aravind, A. M. Burroughs, D. Zhang and L. M. Iyer, *Cold Spring Harbor Perspect. Biol.*, 2014, **6**, a016063.
48. I. V. Agarkova, D. D. Dunigan and J. L. Van Etten, *J. Virol.*, 2006, **80**, 8114–8123.
49. P. B. Beauregard, R. Guerin, C. Turcotte, S. Lindquist and L. A. Rokeach, *J. Cell Sci.*, 2009, **122**, 1342–1351.
50. M. Ko, J. An, H. S. Bandukwala, L. Chavez, T. Aijo, W. A. Pastor, M. F. Segal, H. Li, K. P. Koh, H. Lahdesmaki, P. G. Hogan, L. Aravind and A. Rao, *Nature*, 2013, **497**, 122–126.
51. D. Ekanayake and R. Sabatini, *Eukaryotic Cell*, 2011, **10**, 1465–1472.
52. D. Reynolds, L. Cliffe, K. U. Forstner, C. C. Hon, T. N. Siegel and R. Sabatini, *Nucleic Acids Res.*, 2014.
53. H. G. van Luenen, C. Farris, S. Jan, P. A. Genest, P. Tripathi, A. Velds, R. M. Kerkhoven, M. Nieuwland, A. Haydock, G. Ramasamy, S. Vainio, T. Heidebrecht, A. Perrakis, L. Pagie, B. van Steensel, P. J. Myler and P. Borst, *Cell*, 2012, **150**, 909–921.
54. M. M. Dawlaty, K. Ganz, B. E. Powell, Y. C. Hu, S. Markoulaki, A. W. Cheng, Q. Gao, J. Kim, S. W. Choi, D. C. Page and R. Jaenisch, *Cell Stem Cell*, 2011, **9**, 166–175.
55. K. Williams, J. Christensen, M. T. Pedersen, J. V. Johansen, P. A. Cloos, J. Rappsilber and K. Helin, *Nature*, 2011, **473**, 343–348.
56. M. Ko, H. S. Bandukwala, J. An, E. D. Lamperti, E. C. Thompson, R. Hastie, A. Tsangaratou, K. Rajewsky, S. B. Koralov and A. Rao, *Proc. Natl. Acad. Sci. U. S. A.*, 2011, **108**, 14566–14571.
57. M. Ko, Y. Huang, A. M. Jankowska, U. J. Pape, M. Tahiliani, H. S. Bandukwala, J. An, E. D. Lamperti, K. P. Koh, R. Ganetzky, X. S. Liu, L. Aravind, S. Agarwal, J. P. Maciejewski and A. Rao, *Nature*, 2010, **468**, 839–843.
58. M. Ko and A. Rao, *Blood*, 2011, **118**, 4501–4503.
59. U. Bacher, C. Haferlach, S. Schnittger, A. Kohlmann, W. Kern and T. Haferlach, *Ann. Hematol.*, 2010, **89**, 643–652.

60. A. P. Im, A. R. Sehgal, M. P. Carroll, B. D. Smith, A. Tefferi, D. E. Johnson and M. Boyiadzis, *Leukemia*, 2014.
61. C. G. Lian, Y. Xu, C. Ceol, F. Wu, A. Larson, K. Dresser, W. Xu, L. Tan, Y. Hu, Q. Zhan, C. W. Lee, D. Hu, B. Q. Lian, S. Kleffel, Y. Yang, J. Neiswender, A. J. Khorasani, R. Fang, C. Lezcano, L. M. Duncan, R. A. Scolyer, J. F. Thompson, H. Kakavand, Y. Houvras, L. I. Zon, M. C. Mihm, Jr., U. B. Kaiser, T. Schatton, B. A. Woda, G. F. Murphy and Y. G. Shi, *Cell*, 2012, **150**, 1135–1146.
62. T. P. Gu, F. Guo, H. Yang, H. P. Wu, G. F. Xu, W. Liu, Z. G. Xie, L. Shi, X. He, S. G. Jin, K. Iqbal, Y. G. Shi, Z. Deng, P. E. Szabo, G. P. Pfeifer, J. Li and G. L. Xu, *Nature*, 2011, **477**, 606–610.
63. T. Nakamura, Y. J. Liu, H. Nakashima, H. Umehara, K. Inoue, S. Matoba, M. Tachibana, A. Ogura, Y. Shinkai and T. Nakano, *Nature*, 2012, **486**, 415–419.
64. M. Wossidlo, T. Nakamura, K. Lepikhov, C. J. Marques, V. Zakhartchenko, M. Boiani, J. Arand, T. Nakano, W. Reik and J. Walter, *Nat. Commun.*, 2011, **2**, 241.
65. M. M. Dawlaty, A. Breiling, T. Le, G. Raddatz, M. I. Barrasa, A. W. Cheng, Q. Gao, B. E. Powell, Z. Li, M. Xu, K. F. Faull, F. Lyko and R. Jaenisch, *Dev. Cell*, 2013, **24**, 310–323.
66. M. M. Dawlaty, A. Breiling, T. Le, M. I. Barrasa, G. Raddatz, Q. Gao, B. E. Powell, A. W. Cheng, K. F. Faull, F. Lyko and R. Jaenisch, *Dev. Cell*, 2014, **29**, 102–111.
67. S. K. Ooi and T. H. Bestor, *Cell*, 2008, **133**, 1145–1148.
68. P. Hajkova, S. J. Jeffries, C. Lee, N. Miller, S. P. Jackson and M. A. Surani, *Science*, 2010, **329**, 78–82.
69. M. Wossidlo, J. Arand, V. Sebastiano, K. Lepikhov, M. Boiani, R. Reinhardt, H. Scholer and J. Walter, *EMBO J.*, 2010, **29**, 1877–1888.
70. S. V. Cannon, A. Cummings and G. W. Teebor, *Biochem. Biophys. Res. Commun.*, 1988, **151**, 1173–1179.
71. C. G. Spruijt, F. Gnerlich, A. H. Smits, T. Pfaffeneder, P. W. Jansen, C. Bauer, M. Munzel, M. Wagner, M. Muller, F. Khan, H. C. Eberl, A. Mensinga, A. B. Brinkman, K. Lephikov, U. Muller, J. Walter, R. Boelens, H. van Ingen, H. Leonhardt, T. Carell and M. Vermeulen, *Cell*, 2013, **152**, 1146–1159.
72. L. Aravind, S. Anand and L. M. Iyer, *Biol. Direct.*, 2013, **8**, 20.
73. K. E. Szulwach, X. Li, Y. Li, C. X. Song, H. Wu, Q. Dai, H. Irier, A. K. Upadhyay, M. Gearing, A. I. Levey, A. Vasanthakumar, L. A. Godley, Q. Chang, X. Cheng, C. He and P. Jin, *Nat. Neurosci.*, 2011, **14**, 1607–1616.
74. W. A. Pastor, U. J. Pape, Y. Huang, H. R. Henderson, R. Lister, M. Ko, E. M. McLoughlin, Y. Brudno, S. Mahapatra, P. Kapranov, M. Tahiliani, G. Q. Daley, X. S. Liu, J. R. Ecker, P. M. Milos, S. Agarwal and A. Rao, *Nature*, 2011, **473**, 394–397.
75. C. X. Song, K. E. Szulwach, Y. Fu, Q. Dai, C. Yi, X. Li, Y. Li, C. H. Chen, W. Zhang, X. Jian, J. Wang, L. Zhang, T. J. Looney, B. Zhang, L. A. Godley, L. M. Hicks, B. T. Lahn, P. Jin and C. He, *Nat. Biotechnol.*, 2011, **29**, 68–72.

76. B. E. Bernstein, T. S. Mikkelsen, X. Xie, M. Kamal, D. J. Huebert, J. Cuff, B. Fry, A. Meissner, M. Wernig, K. Plath, R. Jaenisch, A. Wagschal, R. Feil, S. L. Schreiber and E. S. Lander, *Cell*, 2006, **125**, 315–326.
77. A. A. Serandour, S. Avner, F. Oger, M. Bizot, F. Percevault, C. Lucchetti-Miganeh, G. Palierne, C. Gheeraert, F. Barloy-Hubler, C. L. Peron, T. Madigou, E. Durand, P. Froguel, B. Staels, P. Lefebvre, R. Metivier, J. Eeckhoute and G. Salbert, *Nucleic Acids Res.*, 2012, **40**, 8255–8265.
78. M. Yu, G. C. Hon, K. E. Szulwach, C. X. Song, P. Jin, B. Ren and C. He, *Nat. Protoc.*, 2012, **7**, 2159–2170.
79. X. Li, W. Wei, Q. Y. Zhao, J. Widagdo, D. Baker-Andresen, C. R. Flavell, A. D'Alessio, Y. Zhang and T. W. Bredy, *Proc. Natl. Acad. Sci. U. S. A.*, 2014, **111**, 7120–7125.
80. H. Hashimoto, Y. Liu, A. K. Upadhyay, Y. Chang, S. B. Howerton, P. M. Vertino, X. Zhang and X. Cheng, *Nucleic Acids Res.*, 2012, **40**, 4841–4849.
81. J. P. Golla, J. Zhao, I. K. Mann, S. K. Sayeed, A. Mandal, R. B. Rose and C. Vinson, *Biochem. Biophys. Res. Commun.*, 2014, **449**, 248–255.
82. W. A. Pastor, L. Aravind and A. Rao, *Nat. Rev. Mol. Cell Biol.*, 2013, **14**, 341–356.

CHAPTER 12

2-Oxoglutarate-Dependent Hydroxylases Involved in DNA Base J (β-D-Glucopyranosyloxymethyluracil) Synthesis

DAVID REYNOLDS[a], LAURA CLIFFE[a], AND ROBERT SABATINI*[a]

[a]Department of Biochemistry and Molecular Biology, University of Georgia, Davison Life Sciences Building, 120 Green Street, Athens, GA 30602-7229, USA
*E-mail: rsabatini@bmb.uga.edu

12.1 Introduction to Base J Localization and Function

Base J (β-D-glucopyranosyloxymethyluracil) is an evolutionarily conserved nuclear DNA modification consisting of an *O*-linked glucosylation of thymidine (Figure 12.1). This DNA modification is found within members of the kinetoplastid family, a group of early-diverged unicellular eukaryotes including *Trypanosoma brucei*, *T. cruzi* and *Leishmania* spp. Base J was initially discovered in the African trypanosome *T. brucei*, the causative agent of human African trypanosomiasis (sleeping sickness).[1] The key regulatory step of base J synthesis is the initial hydroxylation of thymidine at specific sites within the genome by Fe(II)/2-oxoglutarate (2OG)-dependent dioxygenases (JBP1

RSC Metallobiology Series No. 3
2-Oxoglutarate-Dependent Oxygenases
Edited by Robert P. Hausinger and Christopher J. Schofield

Published by the Royal Society of Chemistry, www.rsc.org

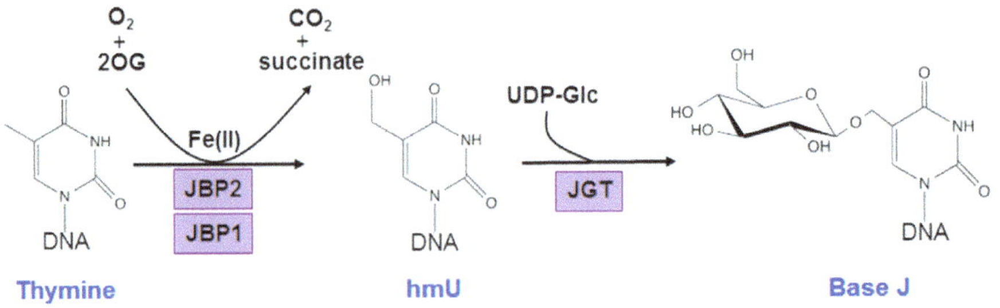

Figure 12.1 The biosynthesis of base J by a two-step modification of a specific thymidine base in the DNA. Base J is synthesized in DNA by a two-step mechanism involving thymidine hydroxylation and glycosylation. JBP1 and JBP2 are members of the Fe(II)/2OG dioxygenase family that utilize 2OG and O_2 as cosubstrates to hydroxylate thymidine bases in dsDNA, releasing succinate and CO_2 as by-products. The intermediate hmU is then glycosylated by a base J-associated glucosyltransferase (JGT) forming base J. UDP-glucose (UDP-Glc) is the activated sugar donor.

and JBP2). In this chapter we highlight our current knowledge of the structure and function of these two dioxygenases, focusing on how they regulate base J synthesis and how this regulation has been pivotal in unravelling the function of this unusual epigenetic mark.

The early history of base J discovery, biosynthesis and function have been reviewed in detail.[2,3] Therefore, we begin by briefly discussing recent advances in J localization and function, focusing primarily on *T. brucei*, the organism in which the majority of the work on J biosynthesis has been carried out. We also briefly describe work performed in two related kinetoplastids, *T. cruzi* and *L. tarentolae*, which has helped to define the function of the modified base in the regulation of gene expression.

12.1.1 Base J Localization

Base J localizes to repetitive DNA sequences, namely telomeric repeats, as well as the 5S rRNA repeats, mini-exon repeats and centromeric DNA repeats.[2,3] Specific to *T. brucei*, base J localizes to specialized subtelomeric sites, termed expression sites (ESs) (Figure 12.2), which are important for immune evasion in the mammalian host. J is also enriched at other repetitive sequences in the *T. brucei* genome including the 50-bp and 70-bp repeats within ESs and the 177-bp repeats in minichromosomes.[4]

Technological advances in DNA sequencing have enabled the genome-wide characterization of J localization by anti-base J immunoprecipitation (IP) and high-throughput sequencing. This analysis led to the discovery of a minor fraction of base J at sites of RNA polymerase II (Pol II) transcription initiation and termination (Figure 12.2) in *T. brucei*, *T. cruzi*, *L. major* and *L. tarentolae*.[5–7] The resolution of J localization has been further improved using single-molecule, real-time (SMRT) DNA sequencing, which allows for strand-specific, base-resolution detection of modified bases in DNA.[8] The localization of J

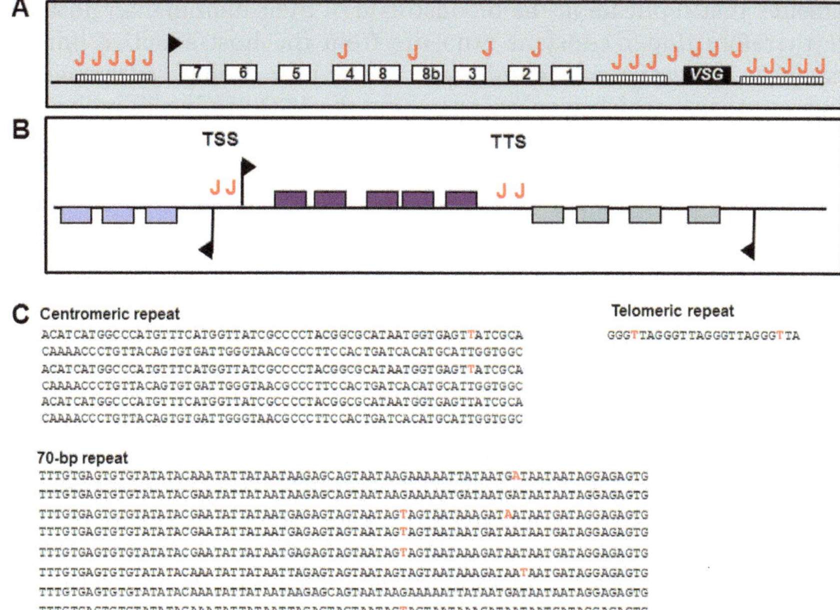

Figure 12.2 Genomic localization of base J. (A) Base J is found within the silent telomeric VSG expression sites in *T. brucei.* The presence of J was determined by immunoprecipitation of J-containing DNA fragments by using an antibody against the modified base followed by a combination of DNA hybridization with various probes or high-throughput DNA sequencing approaches. The hatched boxes represent the 50 bp upstream of the promoter, 70 bp upstream of the *VSG* gene, and telomeric DNA repeats. (B) Base J is also found at internal sites in the genome, including RNA polymerase II transcription start sites (TSS) and transcription termination sites (TTS). (C) Strand-specific, base-resolution detection of base J within repetitive DNA sequences as identified by SMRT DNA sequencing. A portion of one DNA strand for each repeat is indicated. Thymidines that are modified to base J are highlighted in red. Highlighted A's indicate the modified T is located on the opposing DNA strand.

determined by SMRT sequencing of the *T. brucei* genome closely matches the J-IP sequencing analyses, thus localizing the modified base at regions involved in transcription initiation, elongation and termination as well as repetitive DNA sequences, including telomeric and centromeric DNA repeats (Figure 12.2 and unpublished data). Although no conserved DNA motif associated with J synthesis was identified, SMRT sequencing results closely match previously identified DNA strand and sequence bias of J within the telomeric repeat.[9]

12.1.2 Base J Function

Base J was initially described in *T. brucei* based on its localization within the variant surface glycoprotein (*VSG*) gene ESs, which are crucial for parasite growth in the mammalian host.[10] African trypanosomes are extracellular parasites that replicate in the bloodstream of their mammalian host, and are therefore under constant exposure from the host adaptive immune response. *T. brucei* evades elimination by the host through antigenic variation, a process which involves switching the VSG produced by the parasite.[11] The *T. brucei* genome contains over 1000 *VSG* genes, but only one of these genes is expressed at a time. Periodic switching of the actively expressed *VSG* gene maintains a population of trypanosomes capable of avoiding antibody recognition, thus avoiding the adaptive immune system of the host. Monoallelic *VSG* expression is achieved through regulated transcription from ESs. Although there are 15 ESs, only one is active at a time, whilst the remainder are silenced. Interestingly, J is found in the silent, but not the active, ES.[12] Furthermore, transcriptional activation of a silent ES results in the loss of J from the site, while the previously active ES is modified to contain J. The presence of J only in mammalian, but not insect stage, parasites,[13] as well as its association with silent *VSG* gene expression sites, supports the hypothesis that J plays a role in the regulation of antigenic variation in *T. brucei*. However, to date little experimental evidence exists to support this hypothesis.

Unlike *T. brucei*, base J is found in all developmental stages of *T. cruzi* and *Leishmania* spp.[14] This finding, along with the apparent essential nature of J in kinetoplastids that do not undergo antigenic variation,[6,15] indicates a broad functional role for the modified base. Although the reduction of base J has not revealed any telomeric defects, the identification of J at Pol II initiation and termination sites suggests the modification is involved in regulating gene expression. Indeed, the loss of base J from these chromosome-internal regions in *T. cruzi* and *L. tarentolae* alters Pol II transcription and leads to global gene expression changes.[6,7]

Reduction of J in *T. cruzi* upon deleting the gene encoding one of the enzymes involved in J synthesis results in chromatin changes at Pol II initiation sites including decreased nucleosome abundance and increased histone acetylation, a mark associated with active chromatin.[16] Pol II recruitment to initiation sites and transcription rates both increase upon J loss, resulting in global changes in gene expression and parasite virulence.[6,16] Similar decreases of base J in *L. tarentolae* also lead to genome-wide changes in gene expression, possibly due in part to altered transcription initiation.[7] However, the loss of base J at transcription termination sites clearly leads to a termination defect, with subsequent read-through transcription and the generation of antisense RNAs. Thus, base J represents a novel epigenetic modification of kinetoplastid DNA involved in regulating transcription and gene expression.

Unique for eukaryotes, transcription in kinetoplastids is polycistronic, *i.e.* multiple genes are arranged in long units that are co-transcribed. Individual mRNAs are processed from the pre-mRNA through the addition of a mini-exon sequence, called the spliced leader, and polyadenylation.[17] Unlike bacterial operons, however, kinetoplastid polycistronic gene clusters do not contain functionally related genes. While the loss of base J clearly affects gene expression in *T. cruzi* and *L. tarentolae*, how and whether kinetoplastids employ

J to fine-tune Pol II transcription remains unknown. Because the two dioxygenases JBP1 and JBP2 appear to be the key regulators of J synthesis, understanding how they govern the synthesis and localization of J in the genome is important in fully understanding the role of J in controlling kinetoplastid gene expression.

12.2 The Two-Step Biosynthesis Pathway

J is synthesized in a two-step pathway (Figure 12.1). The first step involves the oxidation of specific thymidine residues in DNA by a thymidine hydroxylase (TH, JBP1 or JBP2), forming the intermediate base hydroxymethyluracil (hmU). This intermediate is converted into base J by addition of a glucose molecule by a glucosyltransferase (GT). Once the two-step mechanism of J synthesis was defined it became clear that the enzymes involved are unique to the J synthesis pathway. The J-specific GT is unusual in that it operates in the nucleus, unlike the predominantly cytoplasmic localization of other known GTs. Thymine hydroxylases that oxidize the free base in the pyrimidine salvage pathway are well characterized, but THs that oxidize the base in DNA were unknown prior to the discovery of base J. Following characterization of the JBP dioxygenases, homology-based searches resulted in the identification of the ten-eleven translocation (TET) oxygenases in mammalian cells.[18] The TETs hydroxylate 5-methylcytosine (5meC) in DNA by a mechanism analogous to the JBPs (see Chapter 11).[19] Based on their sequence similarity, TH and TET proteins have been grouped together in the TET/JBP subfamily of dioxygenases (see Chapter 11).[20,21]

There is an abundance of evidence in support of the two-step synthesis pathway. The specific localization of J within the genome indicates that thymidine residues are modified in DNA, rather than at the nucleotide level followed by incorporation during DNA replication. This result is in contrast to the related glucosyl-hydroxymethylcytosine (hmC) base in T-even bacteriophages, where synthesis of the modified hmC at the nucleotide level results in glucosylation of up to 90% of the thymidine in the genome.[22] In addition, the detection of hmU in the DNA of bloodstream form trypanosomes by post-labelling and thin layer chromatographic analysis,[23] and by mass spectrometry (unpublished data), indicates it is a freely available intermediate. Finally, growth of trypanosomes in media containing hmU allows cells to bypass the first step of the synthesis pathway.[5,24] This circumvention results in the synthesis of J at random sites within the genome, implying that the J-specific GT is non-specific for DNA sequence and is able to glycosylate hmU present anywhere in the genome. Furthermore, it follows that the regulation of J biosynthesis and localization occurs primarily at the level of the TH enzymes. While these data provide indirect support for the two-step pathway, unambiguous support is provided by recent detailed *in vitro* and *in vivo* characterization of the enzymes that catalyse each step in J synthesis.

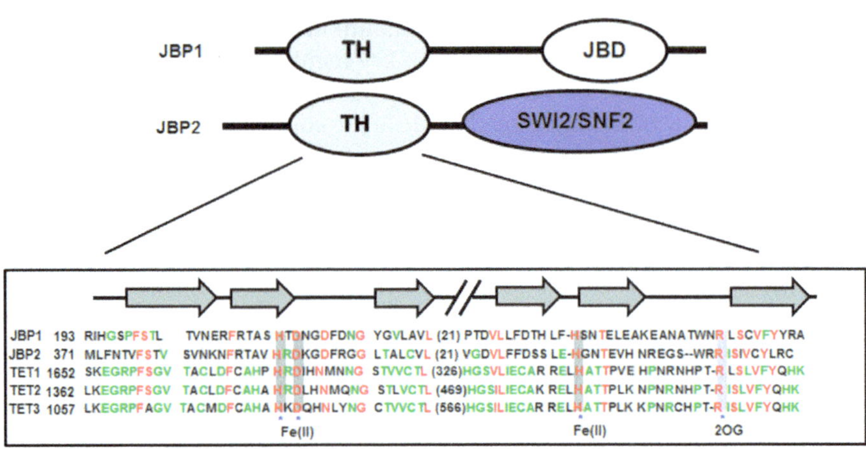

Figure 12.3 Functional domains of JBP1 and JBP2. The white oval within the C-terminus of JBP1 represents the 20 kDa minimal J-binding domain (JBD). The blue oval at the C-terminus of JBP2 represents the domain homologous to the SWI2/SNF2 family of chromatin remodelling ATPases. The region shared between JBP1 and JBP2 at the *N*-terminus is indicated by the grey oval (TH). Within this region is the ~70-residue motif which is related to the double-stranded β-helix domain of the members of the Fe(II)/2OG-dependent dioxygenase family. For the dioxygenase domain, a multiple sequence alignment of selected JBP/TET family proteins (*T. brucei* JBP1/2 and human TET1–3) is shown. The key residues conserved in members of the JBP/TET family predicted to be involved in iron (HXD···H) and 2OG binding (R) are indicated. Sequences that constitute the conserved β-helix fold are shown above the alignment (adapted from Shen and Zhang[52]). Numbers represent the amino acid numbers. TH, thymidine hydroxylase.

12.2.1 Characterization of the Two Dioxygenases in J Biosynthesis

JBP1 was initially discovered based on its ability to bind J-containing DNA substrates.[25] The ability of JBP1 to bind base J is dependent on the *C*-terminal domain (Figure 12.3). *In silico* screening of the *T. brucei* genome led to the identification of JBP2, based on its homology to the *N*-terminus of JBP1.[26] Unlike JBP1, JBP2 is unable to bind base J; instead, JBP2 contains a domain at its *C*-terminus homologous to the SWI2/SNF2 family of chromatin remodelling ATPases. The importance of both JBP1 and JBP2 for J synthesis has been confirmed *in vivo*, through gene deletion studies in bloodstream form *T. brucei*. The loss of either JBP1 or JBP2 results in a 20- and 8-fold reduction in J levels in the genome, respectively.[27,28] Re-expression of the *JBP* gene in the corresponding knockout cell line rescued J levels. The simultaneous deletion of genes encoding both JBP1 and JBP2 generated a trypanosome cell line unable to synthesize base J.[29] JBP null trypanosomes fed hmU can convert this base to J, illustrating that the JBP enzymes are critical for the first step of J biosynthesis.

These results suggest JBP1 and JBP2 are directly involved in thymidine oxidation and their related *N*-terminus regions might contain a TH domain. Upon close examination it is found that the conserved *N*-terminus shares weak homology with enzymes of the Fe(II)/2-oxoglutarate (2OG)-dependent dioxygenase superfamily,[30] where hydroxylation is driven by the oxidative decarboxylation of 2OG to form succinate and CO_2. These enzymes catalyse the oxidation of a wide variety of substrates using ferrous iron as cofactor and 2OG plus molecular oxygen as cosubstrates. Enzymes in the dioxygenase superfamily are typically identified on a structural level by the presence of a jelly-roll or double-stranded β-helix fold (see Chapter 2), which contains four key conserved residues involved in the binding of Fe(II) and 2OG that are essential for catalytic activity (Figure 12.3).[31,32] Substitution of any of the four conserved residues by alanine abolishes the ability of both JBP1 and JBP2 to stimulate *de novo* J synthesis *in vivo*.[29,30,33] This loss of activity is not due to the inability of the mutant JBP to enter the nucleus or, in the case of JBP1, to bind J-DNA.

To unambiguously characterize the JBPs as Fe(II)/2OG-dependent dioxygenases we developed an *in vitro* TH assay, using recombinant JBP1 produced in *E. coli*. This assay demonstrated that JBP1 stimulates the hydroxylation of thymidine specifically in the context of dsDNA and, as expected with this family of enzymes, was dependent on Fe(II), 2OG and O_2.[34] Under anaerobic conditions, the addition of Fe(II) to JBP1 and 2OG results in the formation of a broad absorption spectrum centered at 530 nm, attributed to metal chelation by 2OG that is bound to JBP, a spectroscopic signature of Fe(II)/2OG-dependent dioxygenases. The *N*-terminal TH domain of JBP1 is sufficient for full activity and mutation of residues involved in coordinating Fe(II) inhibit iron binding and the formation of hmU.[34] Thymidine oxidation is inhibited both *in vitro* and *in vivo* by using previously identified 2OG-dependent dioxygenase inhibitors. For example dimethyloxalylglycine (DMOG) is taken up by cells and undergoes ester hydrolysis to form *N*-oxalylglycine, a well-characterized inhibitor of these enzymes.[35] *In vitro* TH reactions performed in the presence of *N*-oxalylglycine demonstrate a significant reduction in thymidine-to-hmU conversion. Moreover, the growth of *T. brucei* in the presence of DMOG results in a complete loss of detectable J.[34] The inhibition of *in vivo* J synthesis by DMOG was recapitulated in *T. cruzi* and *L. major*.

12.2.2 Identification of the Glucosyl Transferase

The culturing of insect-stage *T. brucei* or the bloodstream-form J null (JBP1 and JBP2 KO) in media containing hmU results in the synthesis of base J randomly throughout the genome.[24] Therefore, the GT functions regardless of the DNA sequence context and is expressed in both trypanosome life stages. The significantly higher levels of J stimulated by hmU addition to bloodstream J-null cells compared to insect-stage cells suggests downregulation of the GT upon differentiation along with the JBPs. While attempts to purify the GT from nuclear extracts have failed,[2,36] a recent computational screen identified a gene encoding a GT-like sequence in the kinetoplastid genome as a strong candidate for the base J-associated GT. Iyer *et al.* identified gene sequences related to GT-A/fringe-like glycosyltransferases with operonic association with TET/JBP-like encoding genes that are predicted to modify DNA or RNA in certain phages.[21] The operonic association with TET/JBP enzymes suggests that these TET/JBP-associated glycosyltransferases could glycosylate hydroxylated bases generated by the former enzyme. We have recently confirmed that the GT homologue in kinetoplastids is the base J-associated GT, which we now refer to as JGT (unpublished data). We have found that recombinant JGT whose gene was cloned from *T. brucei* and *L. major* utilizes UDP-Glc to transfer glucose to hmU in the context of dsDNA. The deletion of both alleles of *JGT* from *T. brucei* generates a cell line that completely lacks base J and re-expression of *JGT* restores J synthesis. These studies confirm the identity of the J-specific GT and the two-step J synthesis model. As discussed below, the identification of the GT provides a useful tool in understanding the regulation of DNA modification by the two dioxygenases.

12.3 Regulation of J Synthesis by Two Thymidine Hydroxylases

12.3.1 JBP Structure

The genomes of all J-containing organisms encode two distinct thymidine 2OG-dependent dioxygenases, JBP1 and JBP2. Although both proteins are capable of hydroxylating thymidine, the *C*-terminal domain differs significantly (Figure 12.3), which is strongly suggestive of a differential role for each in the J biosynthesis pathway. JBP1 has a J-DNA binding domain in the *C*-terminal half of the protein that is essential for function *in vivo*. Both gel shift and fluorescence anisotropy measurements reveal that JBP1 binds to J only in the context of double-stranded DNA.[37,38] Optimal DNA binding requires at least five base pairs flanking J, where recognition is dependent upon J itself and the base immediately 5′ of J.[39] JBP1 does not make any sequence-specific contacts with the bases surrounding the modified base, but rather it recognizes J when presented in any sequence context. However, it appears that interactions with the base immediately 5′ of J is essential for the proper orientation of the glucose moiety of the modified base. Analysis of JBP1 binding to various modified DNA substrates indicates that the phosphoryl oxygen of the base upstream of J locks the glucose moiety, *via* hydrogen bonds of the essential 2- and 3-hydroxyl groups, into an ‘edge on’ conformation necessary for optimal JBP1 binding.[38] A crystal structure of the 160-residue DNA binding domain reveals a novel ‘helical bouquet’ fold where a single Asp residue is essential for recognition of base J in DNA and JBP1 function *in vivo*.[40] Recent fluorescent polarization measurements indicate that JBP1 binds J-DNA in a two-step reaction where the protein undergoes a conformational change upon DNA binding.[41] Presumably, this conformational change allows the TH domain to come into proximity with the DNA.

JBP2 does not bind the modified base directly, but is able to bind chromatin in a base J-independent manner.[26] Re-expression of *JBP2* in bloodstream-form cells that lack JBP1 and JBP2 leads to *de novo* J synthesis at specific sites of the genome.[5] We believe that interaction of JBP2 with its DNA substrate is driven by the *C*-terminal SWI2/SNF2 domain. Mutation of key residues within the ATPase region of the SWI2/SNF2 domain eliminates JBP2 function,[26] but whether ATP hydrolysis is required for JBP2 to bind to/remodel chromatin structure to potentiate oxidase activity is unknown. Moreover, it is unknown whether recognition of chromatin by JBP2 is driven at the level of DNA sequence or structure, or potentially *via* interactions with histone variants that co-localize with J.[5,42]

Base J synthesis is developmentally regulated in *T. brucei*; it is detectable in bloodstream-form, but not insect-stage, parasite DNA. Characterization of different life cycle stages reveals that developmental regulation of J synthesis is controlled at both steps. Insect-stage cells downregulate the production of both dioxygenases and the GT enzyme. However, hmU feeding or ectoptic expression of JBP2 in insect-stage trypanosomes results in J synthesis.[24,29] This result implies that the developmental regulation of J synthesis is governed primarily by the dioxygenases.

The results from ectopic expression of *JBPs* in insect-stage trypanosomes allow the proposition of a model for J synthesis where JBP2 is the key regulator of site-specific *de novo* J synthesis.[3,27] According to this model the role of JBP1 is to amplify and maintain the levels of the modified base. However, this model indicating a separation of function was modified upon realization that both JBP1 and JBP2 have *de novo* J synthesis capability in the bloodstream form trypanosome. Re-expression of each *JBP* in the JBP1/JBP2 KO cells stimulates high levels of base J.[5] Interestingly, the localization of J stimulated by each JBP is different. JBP2 stimulates J at high levels within the telomeres whereas JBP1 stimulates *de novo* synthesis primarily at genome internal sites. Optimal J synthesis (level and localization) occurs only upon repression of both JBPs. Whilst both dioxygenases are able to stimulate *de novo* synthesis of J, the inability of JBP2 to bind J may be essential for maintaining *de novo* J synthesis. Analysis of JBP1 concentrations in the wild-type trypanosome nucleus indicates there are 30–60 fold more J residues in the genome than molecules of JBP1 per cell.[43] The large number of high-affinity binding sites would restrict JBP1 function to specific regions within the genome, thus explaining the inability of JBP1 to stimulate *de novo* J synthesis in a telomere fragmentation assay.[27] In wild-type cells, telomeric cleavage results in the growth of a new telomere that contains J; however, in a JBP2 KO cell line, the new telomere lacks base J despite the presence of endogenous JBP1. Presumably, the remaining large number of high-affinity JBP1 binding sites precludes any interactions with the newly generated telomeric array. In a wild-type cell, therefore, we believe that JBP2 provides specific basal J-DNA for high-affinity JBP1 binding, in turn directing the localization of JBP1-stimulated J synthesis.

The proposed functional separation of each JBP, related to chromatin substrate preference and ability to bind base J, may help explain the evolutionary conservation of two dioxygenase enzymes in the biosynthesis pathway.

12.3.2 Replication-Independent Oxidation

Based upon its ability to bind base J, it was proposed that one key function of JBP1 is to maintain J following DNA replication.[25,26] SMRT-sequencing of J within the trypanosome genome indicates sequences are primarily hemi-modified (Figure 12.2 and unpublished data). This result indicates there is no strict co-replicative maintenance of J and that its genomic localization occurs at the level of *de novo* J synthesis. This conclusion is supported by the finding that JBP stimulation of *de novo* J formation is replication independent.[5]

To date, no enzyme capable of removing the modified base has been identified.[44] This finding, in combination with hmU feeding experiments, has led to the proposal that J is lost *via* passive dilution during replication, rather than by active removal.[26,28] However, dynamic regulation of J is still possible by regulating hmU formation and its conversion to J. In combination with thymine-to-uracil (U) conversion in the pyrimidine salvage pathway, the recently identified activity of TET proteins has provided insight into how the JBPs could regulate hmU levels. In the thymine salvage pathway, the Fe(II)/2OG-dependent dioxygenase thymine hydroxylase converts the base to hmU, formyluracil (fU) and carboxyuracil (caU) through three successive oxidation reactions.[45] Thymine-to-uracil conversion is completed by isoorotate decarboxylase-mediated decarboxylation of caU.[45] Similarly, TET-mediated oxidation of 5mC to hmC, fC and caC (Chapter 11) has led to the possibility that these cytosine analogues may play a role in dynamic regulation of 5mC.[46–48] The similarity in chemistry between thymine-to-uracil conversion, TET oxidation of 5mC and oxidative modification of thymidine in trypanosomes suggests the possibility that JBP enzymes can mediate iterative oxidation of thymidine (Figure 12.4). Preliminary (unpublished) data indicate they do. We have detected low levels of fU in the genome of wild-type trypanosomes. In addition, during the *in vitro* JBP1 TH assay, hmU is rapidly generated and then lost over time. Inducible ablation of GT mRNA *in vivo* by using RNAi leads to a similar increase and decrease of hmU. While further work is needed, we believe JBP iterative oxidation could provide a key regulatory step. For example, regulated interactions between GT and the dioxygenases could allow for the conversion of hmU to fU/caU, leading to thymine. In the thymine salvage pathway the conversion is completed by decarboxylation, and an isoorotate decarboxylase is present in the trypanosome genome. The possibility that JBP enzymes mediate iterative oxidation of thymidine has important consequences on understanding the role of DNA modification in trypanosomes. All of the studies of J function to date have utilized parasites with reduced levels of both J and its intermediate hmU through deletion of

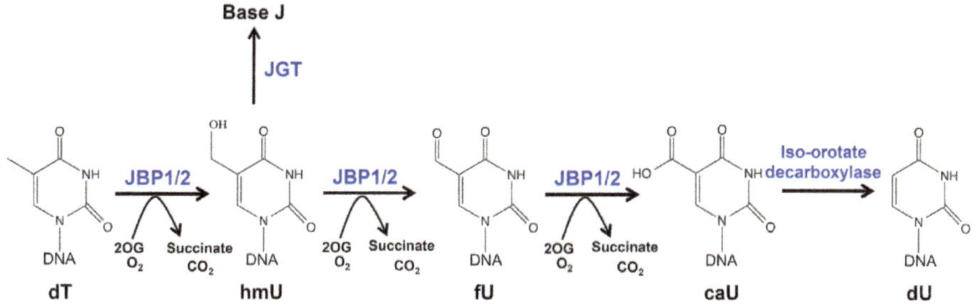

Figure 12.4 Proposed mechanism of iterative oxidation of thymidine initiated by the JBP enzymes. Similar to thymine hydroxylase activity in fungi and the human TET enzymes, the JBP proteins can potentially catalyse sequential conversion of thymine in DNA to 5-hydroxymethyluracil (5-hmU), 5-formyluracil (5-fU) and 5-carboxyuracil (5-caU). Each consecutive oxidation reaction would require O_2 and 2OG and release CO_2 and succinate. 5-caU would then be converted to uracil (dU) by an iso-orotate decarboxylase that is encoded in the kinetoplastid genome. Uracil would presumably be removed by base excision repair.

the dioxygenases. Future studies of the GT KO cell lines will allow us to distinguish between the roles of J and hmU (and potentially fU and caU).

12.4 Regulation of Thymine Oxidation by Metabolism and Host–Parasite Interactions

As discussed in other chapters in this book, the Fe(II)/2OG-dependent dioxygenase enzyme family encompasses a large group of enzymes that catalyse the hydroxylation of a diverse variety of substrates, including but not limited to DNA, protein, RNA and lipid. Most dioxygenases utilize Fe(II) as a cofactor and 2OG and O_2 as cosubstrates. Succinate and carbon dioxide are released as by-products. As expected, succinate is known to inhibit this class of enzymes by product inhibition. In addition, the requirement of O_2 for enzyme activity has led to the suggestion that dioxygenases function as direct O_2 sensors. It is noteworthy that kinetoplastid parasites experience large environmental differences during their life cycle progression between the insect vector and mammalian hosts. In *T. brucei* for example, these differences include, among others, a change in temperature from 37 °C in the mammalian host to 27 °C in the insect vector, as well as a change in the availability of glucose, the preferred energy source of the parasite. Within the various hosts, these pathogens are also exposed to changing O_2 conditions. Oxygen concentrations range from 0 to 21% within human host tissues, depending upon the proximity to blood vessels and the O_2 consumptive activity of the cell. For example, *T. cruzi* experiences varying O_2 concentrations, from high levels on the skin to low levels within various tissues (*i.e.* gut and muscle) of the human host and insect vector. Accordingly, these organisms are capable of tightly regulating their gene expression, enabling them to adapt to a diverse array of environmental conditions including O_2 concentration. The potential role of dioxygenases as O_2 and metabolic sensors has exciting implications for JBP1/2 as key epigenetic regulators of gene expression in response to the parasite environment and different host niches.

12.4.1 JBP as Oxygen Sensors

We have shown that *T. cruzi* decreases its J levels after the loss of JBP1, resulting in increased Pol II recruitment and transcription initiation. The resultant changes in gene expression have direct effects on the host–parasite relationship, as indicated by enhanced mammalian cell invasion and delayed egress.[6] Growth of *T. cruzi* in physiologically relevant low O_2 tensions (hypoxia) leads to reduced levels of J, presumably through inhibition of both JBP1/2 activities. This reduced-J state resulted in similar phenotypic changes in parasite virulence as previously characterized in the JBP1 KO cells. These results clearly demonstrate an O_2 dependence for epigenetic regulation of virulence gene expression. Conclusive demonstration of the oxygen-sensing capabilities of the JBPs, however, will require measuring the K_m for O_2 coupled to appropriate *in vivo* studies. Future studies are also needed to better describe the mechanistic link between hypoxic signalling and JBP function, exploring the parasite response to different levels of O_2, and examining the corresponding changes in J synthesis, transcription and gene expression profiles.

12.4.2 JBP Regulation by Parasite Metabolism

In addition to O_2 concentrations, the metabolic state of the cell may be important in regulating JBP1/2 activity. As previously discussed, base J is developmentally regulated in *T. brucei* through downregulation of the biosynthesis machinery. This downregulation may not fully explain the loss of J synthesis however, because JBP1 and JBP2 are not fully active when over-expressed in insect-stage cells compared to similar expression levels in bloodstream-form cells.[5,29] While it is not clear to what extent this expression is affected by the reduced levels of the GT, hmU feeding leads to significant levels of J in insect-stage parasites,[24] suggesting an additional mechanism exists to reduce JBP function in this life stage.

In the bloodstream of the mammalian host, trypanosomes utilize free glucose as their major carbon source. During differentiation to the insect form, however, parasites undergo a metabolic shift and utilize amino acids available in the insect vector as their primary carbon source. As a consequence, rather than pyruvate as the major metabolic end product, insect-stage parasites produce high levels of succinate.[49–51] Succinate is known to inhibit some 2OG-dependent dioxygenases and, consistent with this, we have shown succinate significantly inhibits JBP1 activity *in vitro*.[34] Accordingly, upon re-expression of *JBP1* to achieve identical levels of the enzyme, the ~800-fold decrease in succinate/2OG ratio in bloodstream- *versus* insect-stage parasites corresponds to an ~15-fold increase in JBP1 activity.[5] Thus, the high levels of succinate may inhibit the activity of any low level of JBP1/2 expressed in the insect-stage parasite as well as help explain the origins of developmental regulation of J synthesis in *T. brucei.* Adaptation to life without J in the insect stage may have necessitated the development of new mechanisms to regulate chromatin and gene expression, which might explain the apparent lack of a phenotype of the bloodstream-form *T. brucei* upon loss of J.

12.5 Conclusions and Future Goals

Since the initial discovery of base J in 1993, significant progress has been made in elucidating its synthesis pathway. Two distinct thymidine 2OG-dependent dioxygenases, JBP1 and JBP2, have been identified that perform the initial oxidation of thymidine in DNA. The characterization of the THs has been crucial in developing our understanding of the function of this important regulatory epigenetic mark.

It is clear, however, that despite our recent progress, further work remains to fully understand the mechanism of JBP function and the biological role of base J. For example, it is unclear how the two dioxygenases work together to regulate the genomic distribution of J. What is the basis of the apparent chromatin substrate specificity? What aspects of chromatin do they recognize and bind, and does this occur at the level of DNA sequence or structure? Are there JBP-associated proteins that are involved in these interactions? How do the *C*-terminal domains regulate thymidine oxidase activity in the *N*-terminal domain? Finally, are the dioxygenases able to stimulate iterative oxidation of thymidine and formation of additional analogues in the kinetoplastid genome?

Thus, continued analysis of the THs will continue to shed light on the biological function of base J and its role in the parasite during an infection of a mammalian host. The recent identification of the GT in the synthesis pathway provides an additional tool to study J function.

References

1. J. H. Gommers-Ampt, F. Van Leeuwen, A. L. de Beer, J. F. Vliegenthart, M. Dizdaroglu, J. A. Kowalak, P. F. Crain and P. Borst, *Cell*, 1993, **75**, 1129–1136.
2. P. Borst and R. Sabatini, *Annu. Rev. Microbiol.*, 2008, **62**, 235–251.
3. R. Sabatini, L. Cliffe, L. Vainio, P. Borst. Enzymatic Formation of the Hypermodified DNA base J. In: Grosjean H, ed. *DNA and RNA Modification Enzymes: Comparative Structure, Mechanism, Function, Cellular Interactions and Evolution.* Texas: Landes Biosciences, 2009:120–131.
4. F. van Leeuwen, R. Kieft, M. Cross and P. Borst, *Mol. Biochem. Parasitol.*, 2000, **109**, 133–145.
5. L. J. Cliffe, T. N. Siegel, M. Marshall, G. A. Cross and R. Sabatini, *Nucleic Acids Res.*, 2010, **38**, 3923–3935.

6. D. K. Ekanayake, T. Minning, B. Weatherly, K. Gunasekera, D. Nilsson, R. Tarleton, T. Ochsenreiter and R. Sabatini, *Mol. Cell. Biol.*, 2011, **31**, 1690–1700.
7. H. G. van Luenen, C. Farris, S. Jan, P. A. Genest, P. Tripathi, A. Velds, R. M. Kerkhoven, M. Nieuwland, A. Haydock, G. Ramasamy, S. Vainio, T. Heidebrecht, A. Perrakis, L. Pagie, B. van Steensel, P. J. Myler and P. Borst, *Cell*, 2012, **150**, 909–921.
8. B. A. Flusberg, D. R. Webster, J. H. Lee, K. J. Travers, E. C. Olivares, T. A. Clark, J. Korlach and S. W. Turner, *Nat. Methods*, 2010, **7**, 461–465.
9. F. van Leeuwen, M. de Kort, G. A. van der Marel, J. H. van Boom and P. Borst, *Anal. Biochem.*, 1998, **258**, 223–229.
10. A. Bernards, N. van Harten-Loosbroek and P. Borst, *Nucleic Acids Res.*, 1984, **12**, 4153–4170.
11. E. Pays, *Trends Parasitol.*, 2005, **21**, 517–520.
12. F. van Leeuwen, E. R. Wijsman, R. Kieft, G. A. van der Marel, J. H. van Boom and P. Borst, *Genes Dev.*, 1997, **11**, 3232–3241.
13. J. Gommers-Ampt, J. Lutgerink and P. Borst, *Nucleic Acids Res.*, 1991, **19**, 1745–1751.
14. F. van Leeuwen, M. C. Taylor, A. Mondragon, H. Moreau, W. Gibson, R. Kieft and P. Borst, *Proc. Natl. Acad. Sci.*, 1998, **95**, 2366–2371.
15. P. A. Genest, B. ter Riet, C. Dumas, B. Papadopoulou, H. G. van Luenen and P. Borst, *Nucleic Acids Res.*, 2005, **33**, 1699–1709.
16. D. Ekanayake and R. Sabatini, *Eukaryotic Cell*, 2011, **10**, 1465–1472.
17. S. Martinez-Calvillo, J. C. Vizuet-de-Rueda, L. E. Florencio-Martinez, R. G. Manning-Cela and E. E. Figueroa-Angulo, *J. Biomed. Biotechnol.*, 2010, 525241.
18. L. M. Iyer, M. Tahiliani, A. Rao and L. Aravind, *Cell Cycle*, 2009, **8**, 1698–1710.
19. M. Tahiliani, K. P. Koh, Y. Shen, W. A. Pastor, H. Bandukwala, Y. Brudno, S. Agarwal, L. M. Iyer, D. R. Liu, L. Aravind and A. Rao, *Science*, 2009, **324**, 930–935.
20. W. A. Pastor, L. Aravind and A. Rao, *Nat. Rev. Mol. Cell. Biol.*, 2013, **14**, 341–356.
21. L. M. Iyer, D. Zhang, A. Maxwell Burroughs and L. Aravind, *Nucleic Acids Res.*, 2013, **41**, 7635–7655.
22. J. H. Gommers-Ampt and P. Borst, *FASEB J.*, 1995, **9**, 1034–1042.
23. J. H. Gommers-Ampt, A. J. Teixeira, G. van de Werken, W. J. van Dijk and P. Borst, *Nucleic Acids Res.*, 1993, **21**, 2039–2043.
24. F. van Leeuwen, R. Kieft, M. Cross and P. Borst, *Mol. Cell. Biol.*, 1998, **18**, 5643–5651.
25. M. Cross, R. Kieft, R. Sabatini, M. Wilm, M. de Kort, G. van der Marel, J. van Boom, F. van Leeuwen and P. Borst, *EMBO J.*, 1999, **18**, 6573–6581.
26. C. DiPaolo, R. Kieft, M. Cross and R. Sabatini, *Mol. Cell*, 2005, **17**, 441–451.
27. R. Kieft, V. Brand, D. K. Ekanayake, K. Sweeney, C. DiPaolo, W. S. Reznikoff and R. Sabatini, *Mol. Biochem. Parasitol.*, 2007, **156**, 24–31.
28. M. Cross, R. Kieft, R. Sabatini, A. Dirks-Mulder, I. Chaves and P. Borst, *Mol. Microbiol.*, 2002, **46**, 37–47.

29. L. J. Cliffe, R. Kieft, T. Southern, S. R. Birkeland, M. Marshall, K. Sweeney and R. Sabatini, *Nucleic Acids Res.*, 2009.
30. Z. Yu, P. A. Genest, B. ter Riet, K. Sweeney, C. DiPaolo, R. Kieft, E. Christodoulou, A. Perrakis, J. M. Simmons, R. P. Hausinger, H. G. van Luenen, D. J. Rigden, R. Sabatini and P. Borst, *Nucleic Acids Res.*, 2007, **35**, 2107–2115.
31. C. J. Schofield and Z. Zhang, *Curr. Opin. Struct. Biol.*, 1999, **9**, 722–731.
32. R. P. Hausinger, *Crit. Rev. Biochem. Mol. Biol.*, 2004, **39**, 21–68.
33. S. Vainio, P. A. Genest, B. ter Riet, H. van Luenen and P. Borst, *Mol. Biochem. Parasitol.*, 2009, **164**, 157–161.
34. L. J. Cliffe, G. Hirsch, J. Wang, D. Ekanayake, W. Bullard, M. Hu, Y. Wang and R. Sabatini, *J. Biol. Chem.*, 2012, **287**, 19886–19895.
35. R. Nathan, M. A. M. Rose, O. N. F. King, A. Kawamura and C. J. Schofield, *Chem. Soc. Rev.*, 2011, **40**, 4364–4397.
36. S. Ulbert. *DNA Repair and Antigenic Variation in Trypanosoma Brucei.* Amsterdam: University of Amsterdam. 2003.
37. R. Sabatini, N. Meeuwenoord, J. H. van Boom and P. Borst, *J. Biol. Chem.*, 2002, **277**, 958–966.
38. R. K. Grover, S. J. Pond, Q. Cui, P. Subramaniam, D. A. Case, D. P. Millar and P. Wentworth Jr, *Angew. Chem., Int. Ed. Engl.*, 2007, **46**, 2839–2843.
39. R. Sabatini, N. Meeuwenoord, J. H. van Boom and P. Borst, *J. Biol. Chem.*, 2002, **277**, 28150–28156.
40. T. Heidebrecht, E. Christodoulou, M. J. Chalmers, S. Jan, B. Ter Riet, R. K. Grover, R. P. Joosten, D. Littler, H. van Luenen, P. R. Griffin, P. Wentworth Jr., P. Borst and A. Perrakis, *Nucleic Acids Res.*, 2011, **39**, 5715–5728.
41. T. Heidebrecht, A. Fish, E. von Castelmur, K. A. Johnson, G. Zaccai, P. Borst and A. Perrakis, *J. Am. Chem. Soc.*, 2012, **134**, 13357–13365.
42. T. N. Siegel, D. R. Hekstra, L. E. Kemp, L. M. Figueiredo, J. E. Lowell, D. Fenyo, X. Wang, S. Dewell and G. A. Cross, *Genes Dev.*, 2009, **23**, 1063–1076.
43. C. B. Toaldo, R. Kieft, A. Dirks-Mulder, R. Sabatini, H. G. van Luenen and P. Borst, *Mol. Biochem. Parasitol.*, 2005, **143**, 111–115.
44. S. Ulbert, L. Eide, E. Seeberg and P. Borst, *DNA Repair (Amst)*, 2004, **3**, 145–154.
45. J. A. Smiley, M. Kundracik, D. A. Landfried, V. R. Barnes Sr. and A. A. Axhemi, *Biochim. Biophys. Acta*, 2005, **1723**, 256–264.
46. J. U. Guo, Y. Su, C. Zhong, G. L. Ming and H. Song, *Cell*, 2011, **145**, 423–434.
47. K. Williams, J. Christensen, M. T. Pedersen, J. V. Johansen, P. A. Cloos, J. Rappsilber and K. Helin, *Nature*, 2011, **473**, 343–348.
48. G. Ficz, M. R. Branco, S. Seisenberger, F. Santos, F. Krueger, T. A. Hore, C. J. Marques, S. Andrews and W. Reik, *Nature*, 2011, **473**, 398–402.
49. A. G. Tielens and J. J. van Hellemond, *Trends Parasitol.*, 2009, **25**, 482–490.
50. K. W. van Grinsven, J. Van Den Abbeele, P. Van den Bossche, J. J. van Hellemond and A. G. Tielens, *Eukaryotic Cell*, 2009, **8**, 1307–1311.
51. J. Cazzulo. Energy Metabolism in *Trypanosoma cruzi*. In: Avila JL, Harris JR, eds. *Intracellular Parasites*: Springer US, 1992:235–257.
52. L. Shen and Y. Zhang, *Curr Opin. Cell Biol.*, 2013, **25**, 289–296.

CHAPTER 13

Dioxygenases of Carnitine Biosynthesis: 6-N-Trimethyllysine and γ-Butyrobetaine Hydroxylases

FRÉDÉRIC M. VAZ*[a] AND NAOMI VAN VLIES[a]

[a]Laboratory Genetic Metabolic Diseases, Departments of Paediatrics and Clinical Chemistry, Emma Children's Hospital, Academic Medical Center, 1105 AZ, Amsterdam, The Netherlands
*E-mail: f.m.vaz@amc.nl

13.1 Introduction to Carnitine Biosynthesis and Metabolism

13.1.1 Carnitine Occurrence and Function

Carnitine (3-hydroxy-4-*N*-trimethylaminobutyrate) is a small, water-soluble molecule which is probably present in all animal species, as well as many micro-organisms and plants. In mammals, carnitine plays an indispensable role in intermediary metabolism, where it is involved in fatty acid metabolism.[1] Prokaryotes can use carnitine both as a carbon and nitrogen source for aerobic growth and some bacteria use carnitine as an osmoprotectant.[2,3]

RSC Metallobiology Series No. 3
2-Oxoglutarate-Dependent Oxygenases
Edited by Robert P. Hausinger and Christopher J. Schofield

Published by the Royal Society of Chemistry, www.rsc.org

Plants contain relatively little carnitine,[4] and its role in their metabolism is poorly characterized.

In mammals, the main function of carnitine is to catalyse the transport of activated fatty acids across organellar membranes. The acyl groups of activated fatty acids (*i.e.* acyl-coenzyme A (CoA) thioesters) are reversibly transesterified to the 3-hydroxy group of carnitine by the action of carnitine acyltransferases, yielding *O*-acyl-carnitine esters (Figure 13.1A). The esters can be shuttled across membranes by specific transporters and reconverted to acyl-CoA thioesters for use in metabolism. This reversible acylation also enables carnitine to be used as an 'acyl sink' to ensure enough free CoA is available for other metabolic reactions and the detoxification of organic/fatty acids by exporting them as *O*-acyl-carnitine ester derivatives from the cell (and, eventually, the body). Carnitine is particularly important for mitochondrial β-oxidation as long-chain fatty acids can only enter the mitochondrion as acyl-carnitines, making carnitine essential for cellular energy generation in many tissues that rely mainly on energy from fatty acids (Figure 13.1).[1]

13.1.2 Carnitine Sources and Biosynthesis

Humans obtain most of their carnitine through their diet, primarily *via* the consumption of meat, poultry, fish and dairy products, which contain considerable amounts of carnitine.[5] Due to their (very) low carnitine contents, plants do not significantly contribute to carnitine intake. Apart from dietary intake, humans acquire carnitine *via de novo* synthesis from the essential amino acids lysine and methionine in the liver, kidney and brain. Protein-derived 6-*N*-trimethyllysine (TML) is sequentially converted into carnitine *via* the intermediates 3-hydroxy-6-*N*-trimethyllysine (HTML), 4 *N*-trimethylaminobutyraldehyde (TMABA) and 4 *N*-trimethylaminobutyrate (γ-butyrobetaine, γBB) (Figure 13.1B).[1] In most mammals, only the liver and kidney are capable of carnitine synthesis. Other cells/tissues therefore depend on carnitine import *via* active uptake from the blood. This transport system is also involved in the renal tubular reabsorption and intestinal absorption of carnitine. Taken together, carnitine homeostasis is maintained by dietary intake, a modest rate of endogenous synthesis, and efficient tubular reabsorption of carnitine by the kidney.

In general, prokaryotes can interconvert γBB and carnitine, and they often can completely catabolize both compounds,[2,3] but there are no reports describing the biosynthesis of carnitine from TML as occurs in fungi and mammals.

As can be seen in Figure 13.1B, the first and the last steps of carnitine biosynthesis are performed by 2-oxoglutarate (2OG)-dependent Fe(II)-containing non-haem dioxygenases, namely 6-*N*-trimethyllysine hydroxylase (TMLH) and γ-butyrobetaine hydroxylase (γBBH). These two enzymes are discussed below, with emphasis on the role of these enzymes in mammalian metabolism.

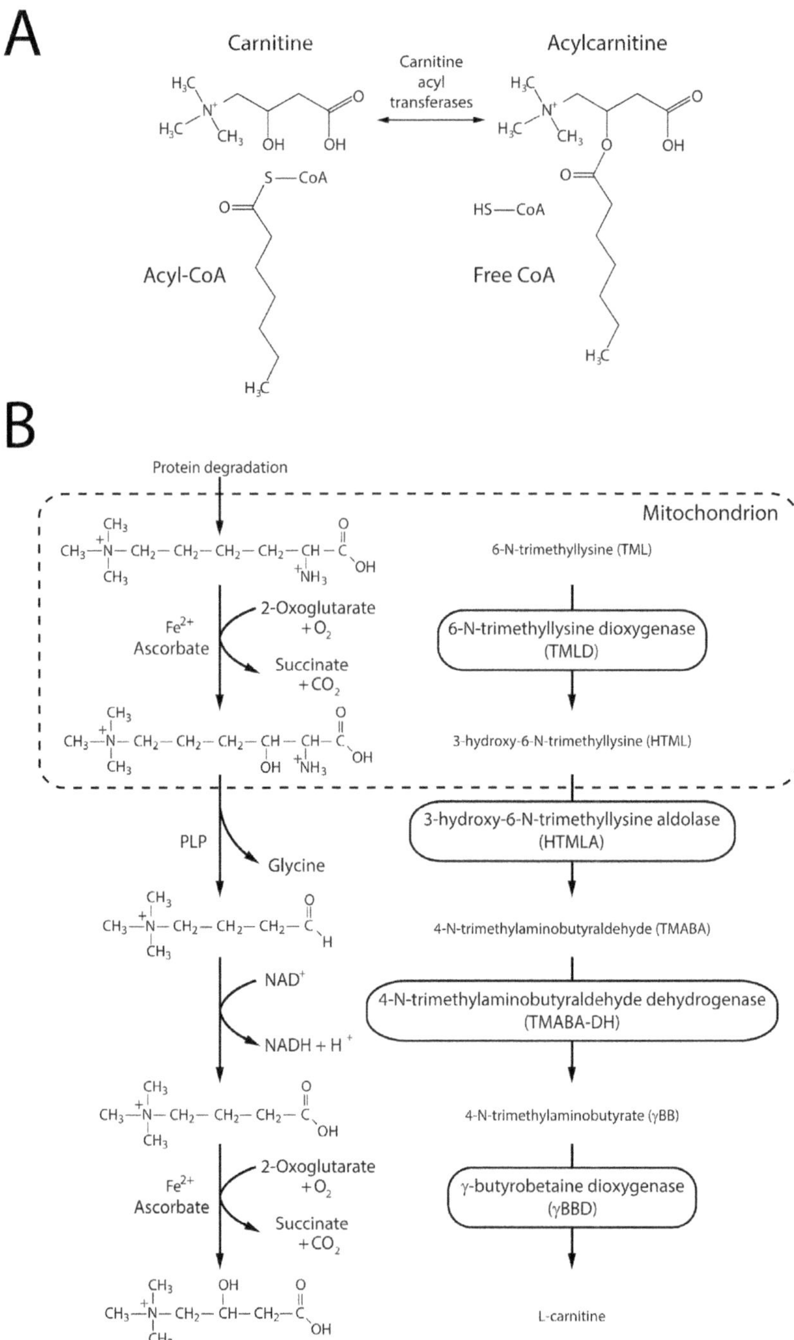

Figure 13.1 Carnitine metabolism and biosynthesis. (A) The reaction catalysed by carnitine acyl-transferases. The acyl moiety of acyl-CoA is transesterified to the 3-hydroxy group of carnitine by the action of these

13.2 Roles of Trimethyllysine and γ-Butyrobetaine Hydroxylases in Carnitine Biosynthesis

TMLH and γBBH are both Fe(II)- and 2OG-dependent dioxygenases, which hydroxylate their substrate *via* a ferryl intermediate (Fe(IV)=O) (see Chapter 3). The activity of TMLH and γBBH is promoted by the presence of ascorbate (vitamin C), one action of which is likely to maintain the iron in the ferrous state. These enzymes are closely related to each other in sequence and, although both enzymes contain the 2-His-1-carboxylate motif characteristic of 2OG-dependent dioxygenases, they bear little sequence similarity to most other members of this enzyme family. Higher organisms typically contain two genes encoding homologues of TMLH/γBBH, suggesting that these enzymes belong to a distinct subclass of 2OG-dependent dioxygenases that is conserved through evolution.

Although some bacterial protein sequences are tentatively assigned as corresponding to TMLH, no experimental evidence exists that shows this enzyme activity is present in these species. The low number of sequences in the TrEMBL and Swiss-Prot protein databases for TMLH (15 sequences, 1 prokaryotic) and γBBH (107 sequences, including many of prokaryotic origin), suggests that carnitine synthesis from TML is probably present only in multi-organellar organisms and not present in prokaryotes. We speculate that γBBH is used by prokaryotes to enable the use of γBB as a carbon and nitrogen source and that the product carnitine is not involved in the oxidation of fatty acids as in eukaryotes.

13.2.1 Trimethyllysine Hydroxylase (E.C. 1.14.11.8)

TMLH is the first enzyme of carnitine biosynthesis and catalyses the stereospecific hydroxylation of TML to give HTML, most likely forming the L-*erythro* isomer with respect to the configuration of the carbon atoms bearing the 2-amino and 3-hydroxy groups.[6] In the fungus *Neurospora crassa*, free lysine is trimethylated by a specific methylase that uses *S*-adenosylmethionine as methyl donor to yield TML.[7] In contrast, in mammals the primary substrate TML is obtained by hydrolysis of TML-containing proteins, including calmodulin, myosin, actin, cytochrome *c* and histones,[1,8–10] whereas there is no evidence for direct methylation of TML. It is believed that the availability of TML, which is determined by the extent of peptide-linked lysine methylation

enzymes, which results in the formation of an acyl-carnitine ester. (B) Carnitine biosynthesis pathway in mammals. Carnitine is synthesized in four enzymatic steps. After release of 6-*N*-trimethyllysine (TML) by lysosomal protein degradation, this compound is hydroxylated by TML hydroxylase (TMLH) producing 3-hydroxy-6-*N*-trimethyllysine (HTML). HTML is cleaved by HTML aldolase (HTMLA) into 4-trimethylaminobutyraldehyde (TMABA) and glycine. Subsequently, TMABA is oxidized by TMABA-dehydrogenase (TMABA-DH) to form 4-*N*-trimethylaminobutyrate (γ-butyrobetaine, γBB). Finally, γBB is metabolized by γ-butyrobetaine hydroxylase (γBBH), yielding L-carnitine.

and the rate of protein turnover, limits the rate of carnitine biosynthesis.[11,12] TMLH activity is, therefore, not rate-limiting for carnitine biosynthesis, at least under non-stressed conditions.

13.2.1.1 *TMLH Occurrence in Different Species*

The measurement of TMLH enzyme activity has only been published for rat,[13] human,[14–16] mouse[17] and the fungus *N. crassa*.[13] Enzymatically, most is known about rat and *N. crassa* TMLH. Genetically, most is known about human TMLH and the last part of this section briefly discusses human *TMLHE* deficiency and the recent discovery that this may be a risk factor for developing autism spectrum disorders.

13.2.1.2 Rattus norvegicus/Mus musculus

The first characterization of TMLH catalytic activity was performed by Hulse *et al.*,[18] and later by Sachan and Hoppel,[19] using rat liver and kidney, respectively. TMLH was found to require 2OG, Fe(II) and O_2 and its activity is stimulated by ascorbate. Inclusion of calcium ions produced a two-fold increase in specific activity, but other divalent cations (Mg^{2+}, Mn^{2+} or Zn^{2+}) inhibited TMLH activity.[19,20] Ascorbate is unlikely to be an actual cofactor, but it probably assists in maintaining the iron in the ferrous state; other reducing agents (*e.g.* dithiothreitol or 3-mercaptoethanol) are also effective, but ascorbate stimulates the hydroxylation reaction to the greatest extent. TMLH activity was initially often measured by using radiolabelled TML and counting the radioactivity of the product HTML after its isolation from the incubation medium by ion exchange chromatography.[15,21,22] Davis has described an alternative enzyme assay using unlabelled TML followed by derivatization of the isolated HTML with *o*-phthalaldehyde and detection by HPLC.[23] More recently, enzyme assays have been developed that measure the produced HTML using tandem mass spectrometry and deuterium-labelled internal standards.[17] Using this latter assay it is even possible to simultaneously measure TMLH and γBBH activity, *i.e.* both dioxygenases of carnitine biosynthesis.

TMLH, like many 2OG-dependent dioxygenases, is sensitive to oxidation, but the enzyme activity can be preserved during isolation and storage at −80 °C by the presence of 2 mM ascorbate, 5 mM dithiothreitol and 10 g l^{-1} glycerol. Using these conditions, the rat enzyme can be purified to apparent homogeneity as a homodimer of ~87 kDa or a denatured monomer of 43 kDa.[24] The rat enzyme has a broad pH optimum between 6.5 and 7.5 at 37 °C,[20,24] exhibits K_m values of 1.1 mM for TML and 109 μM for 2OG, and requires 54 μM Fe(II) for half-maximal activity for the stated assay conditions. Relatively little is known about the enzyme properties of mouse TMLH, but one paper reported K_m values of 164 μM for TML and 605 μM for 2OG using the mouse liver enzyme.[17] In the rat, TMLH activity is greatest in the kidney, present in liver and heart, and low in muscle and brain.[17,20]

Subcellular fractionation experiments using differential centrifugation pointed to a mitochondrial localization for TMLH,[19] as did earlier experiments performed using rat liver mitochondria.[18] A mitochondrial localization was confirmed by Nycodenz density gradient analysis of rat kidney samples, and evidence was found that TMLH is produced as a 49 kDa precursor that is processed to a mature protein of 43 kDa upon mitochondrial import.[24] More recently, progressive membrane solubilization with digitonin and protease protection experiments show that rat TMLH is located in the mitochondrial matrix.[25] This result means that TML must enter the mitochondrion and that the product of the TMLH reaction, HTML, must be exported back to the cytosol where the remainder of the carnitine biosynthesis takes place.[25] Using the purified rat protein, mass spectrometric (MS) peptide fingerprinting was performed to identify the corresponding cDNA/gene in rat, mouse and man,[24] and this study showed the TMLH protein in all three organisms is encoded by a gene located on the X-chromosome, now called *TMLHE*. A recent paper showed that *TMLHE* mRNA expression is low after birth, but it increases to adult values a week post-partum.[26] Other studies in mouse have shown that *TMLHE* expression is regulated by PPARα,[27–29] suggesting that carnitine biosynthesis is activated upon increased fatty acid oxidation.

13.2.1.2 Neurospora crassa

In the 1970s, Broquist and coworkers characterized the carnitine biosynthesis pathway in the filamentous fungus *N. crassa* and showed that it proceeds *via* the same intermediate metabolites as it does in mammals.[30,31] In the microorganism the TML is produced by direct methylation of free lysine,[7] whereas in mammals protein-bound lysine is trimethylated and only becomes available for carnitine synthesis after hydrolysis of the TML-containing proteins.[8] The first direct characterization of the *N. crassa* enzyme activity was performed by Sachan and Broquist,[13] who showed that the *N. crassa* TMLH is also a 2OG-dependent dioxygenase that requires Fe(II) and is stimulated by ascorbate. These researchers uncovered another difference with the mammalian carnitine biosynthesis system; *N. crassa* TMLH is cytosolic and not mitochondrial. In 2002, the corresponding cDNA/gene was identified based on homology with genes encoding known TMLH proteins, and heterologous expression confirms that the protein product exhibits TMLH activity.[32] This recombinant enzyme was kinetically characterized (K_m TML: 0.33 mM, K_m 2OG: 133 μM)[32] and shown to have properties similar to the mammalian enzymes.

13.2.1.4 Homo sapiens

In 2001, the gene encoding human TMLH, *TMLHE*, was identified based on sequence homology to the rat cDNA, which was in turn obtained after purification of the corresponding kidney enzyme followed by MALDI-TOF MS peptide analysis.[24] The product of the human *TMLHE* cDNA consists of a protein of 421 amino acids with a predicted molecular mass of 49 kDa. When the human TMLH cDNA is expressed in COS cells the recombinant human protein has the same apparent molecular mass as the purified rat enzyme (43 kDa). This result suggests that, like rat TMLH, the human enzyme is produced as a 49 kDa precursor that is cleaved upon mitochondrial import resulting in a mature protein of 43 kDa.[24] No kinetic information is available for the human enzyme.

TMLHE is located on the X chromosome and maps to the extreme end of Xq28. It is alternatively spliced into at least five mRNA products where only the most abundant transcript (*TMLHE*-a) encodes a protein with TMLH activity.[33,34] Expression of the other transcripts may be a way to negatively regulate the TMLH activity/expression.[33] In 2011, Beaudet and coworkers identified a male autism patient with a deletion in the *TMLHE* gene.[35] Further studies showed that *TMLHE* deficiency is very common in males (1/350) and results in increased substrate concentration (TML) and decreased product levels (HTML and γBB) in plasma and urine.[36] This newly identified inborn error of metabolism is almost always non-symptomatic, but it does render the affected individuals auxotrophic for carnitine. Based on a comparison of the frequency of *TMLHE* deficiency in controls and simplex/multiplex autism spectrum disorders, it was concluded that *TMHLE* deficiency probably is a weak risk factor for disorders and is neither necessary nor sufficient to cause the symptoms.[36] A recent independent study also found mutations in *TMLHE* which, although they did not reach statistical significance, were more frequently found in autistic patients.[37]

13.2.2 γ-Butyrobetaine Hydroxylase (E.C. 1.14.11.1)

γBBH is the better characterized enzyme of the two animal carnitine biosynthesis dioxygenases. The hydroxylation of γBB is stereospecific and in all investigated species results in the production of L-carnitine.[38] Like TMLH, γBBH requires ascorbate, 2OG, Fe(II) and O_2,[39] and millimolar concentrations of zinc ions inhibit γBBH activity.[17] The enzyme is cytosolic and differentially expressed in mammalian tissues. γBBH activity has been found in liver, kidney, brain, testis[17] and epididymis,[40,41] but not in other tissues. γBB is readily hydroxylated to carnitine in extracts of human, cat, cow, hamster, rabbit, pig and rhesus monkey kidneys, and the activities in these tissues exceed or equal the γBBH activity in liver. In contrast, γBBH activity is not present, or is very low, in cebus monkey, sheep, dog, guinea pig, mouse and rat kidney; in these species γBBH activity is predominantly in the liver.[17,42–48] The reason for this species-dependent difference in kidney/liver expression of BBH is not clear, but the differential expression of this enzyme dictates which organs can or cannot synthesize carnitine. The human brain is unique as it contains detectable γBBH activity,[15] making human the only mammal known to date that can synthesize carnitine *in cerebro*. The human enzyme has a great capacity and is not rate limiting for carnitine synthesis as almost all administered γBB is converted into carnitine.[49,50]

13.2.2.1 Measurement of γBBH

In the 1970s and 1980s, γBBH activity was usually measured radiochemically using labelled γBB.[39,45,51] The enzyme activity can also be determined by measuring the γBB-dependent release of [^{14}C]-CO_2 that is produced from the decarboxylation of [1-^{14}C]-2OG to succinate. This method, however, requires the measurement of γBB-independent activity because the mitochondrial 2OG dehydrogenase complex also produces CO_2 from 2OG. Alternatively, γBBH activity can be measured by using a two-step procedure in which the carnitine produced from unlabelled γBB is measured in a radioisotopic assay.[52,53] A disadvantage of this assay is that the endogenous carnitine content of tissue homogenates also needs to be determined. More recently, a γBBH enzyme assay has been described that measures the produced carnitine using tandem MS and deuterium-labelled internal standards.[17] Another assay uses an alternative substrate, (3*S*)-3-fluoro-4-(trimethylammonio)butanoate, which releases fluoride ion when hydroxylated; the fluoride can be measured by using tert-butyldimethylsilyl-protected fluorescein.[54]

Studies in the 1960s using rat liver were the first to show that γBB is the substrate of an oxygenase that produces carnitine.[55] The *Pseudomonas* and rat γBBH proteins are both relatively easy to isolate and hence remain the best characterized. These and the human enzyme are discussed below.

13.2.2.2 Pseudomonas *sp. AK 1*

A strain of *Pseudomonas aeruginosa* that uses γBB as the sole source of carbon and nitrogen was isolated and named *Pseudomonas* sp. AK 1.[56] The authors showed that, when grown on γBB, carnitine is produced.[57] The partially purified *Pseudomonas* γBBH is stimulated considerably by 2OG and requires molecular oxygen, Fe(II) for activity and ascorbate stimulated its activity.[56,58] The enzyme was purified by different groups,[59–62] and kinetically characterized (K_m γBB: 2.4 mM, K_m 2OG: 0.5 mM).[59,61] *Pseudomonas* sp. AK 1 γBBH has a subunit mass of ~39 kDa and a native molecular mass of ~90 kDa, suggesting the enzyme is a homodimer.[60] In 1993, Ruetschi *et al.* determined the complete sequence of *Pseudomonas* sp. AK 1 γBBH by Edman degradation and noted very little sequence identity to other known dioxygenases.[62] Recently, a Chinese group reported the cloning and heterologous expression of γBBH from *Pseudomonas* sp. L-1.[63] The amino acid sequence is nearly identical to that reported for *Pseudomonas* sp. AK 1 in 1993.[62]

13.2.2.3 Rattus norvegicus

Rat liver γBBH was purified by several groups and kinetically characterized with variable results for the K_m values (K_m γBB: 50–80 μM, K_m 2OG: 0.04–0.5 mM).[39,64–68] The enzyme has a broad pH optimum from 6.4 to 8.0 and has the same cofactor requirements as *Pseudomonas* γBBH.[39] The rat cDNA was identified using the bacterial and human protein sequences, and is expressed mainly in liver. Low expression of higher molecular weight mRNAs is also detected in testis and epididymis,[69] corroborating the observed γBBH enzyme activity in rat testis.[17,41] Galland and coworkers investigated the expression of rat γBBH in the liver during development.[69] The corresponding mRNA appears after weaning and reaches maximal values at the adult stage, a finding that is corroborated at the mRNA level by a recent study.[26] The mRNA expression data are in agreement with enzymatic data of two groups who showed that γBBH activity in liver homogenates increases from low values in the foetus to adult values after a week post-partum.[26,70] In rat, γBBH expression is influenced by the thyroid status, with hypothyroidism resulting in low γBBH expression and hyperthyroidism in high γBBH expression.[71] Several studies show that rodent γBBH, like TMLH in these organisms, is regulated by PPARα, confirming the auxiliary role of carnitine biosynthesis in fatty acid metabolism.[27,29,72,73] Mouse γBBH was recently characterized and showed kinetic parameters similar to that of rat (K_m γBB: 40 µM, K_m 2OG: 63 µM).[17]

13.2.2.4 Homo sapiens

In man, γBBH activity is found in kidney, liver and brain.[14,15] Also, placenta and foetal kidney, liver, spinal cord and brain contain γBBH activity, suggesting that carnitine biosynthesis may supply carnitine for placental and foetal metabolism.[74] The human enzyme was first isolated in the 1980s from kidney and subsequently kinetically characterized by several groups with variable results for the K_m values (K_m γBB: 12–200 µM, K_m 2OG: 0.1–0.3 mM).[44,54,75–77] In 1998, the purified rat liver enzyme was *N*-terminally sequenced and the resulting sequence was used to screen the human expressed sequence tag database, resulting in the identification of the human cDNA.[78] The sequence contains an open reading frame of 1161 base pairs, encodes a protein of 44.7 kDa, and the corresponding γBBH gene, named *BBOX1*, is localized on 11q14-15. Northern blot analysis confirms that the human enzyme is expressed in kidney, liver and brain.[78] The heterologously expressed human enzyme was purified and crystal structures of it have been elucidated.[77,79] The recombinant enzyme was kinetically characterized and, of special interest, the K_m of γBB is lower than that measured using γBBH purified from human kidney (~27 µM *versus* 110–200 µM) for unknown reasons. The crystal structure (see Chapter 2) of human γBBH confirms that the enzyme functions as a homodimer.[77,79]

In the late 1980s, 3-(2,2,2-trimethylhydrazinium)propionate (THP), also known as Mildronate®, was identified as a non-competitive enzyme inhibitor of γBBH.[66] Further study of the structure and reaction mechanism of human γBBH revealed that THP is not solely an inhibitor, but it actually is hydroxylated which leads to fragmentation of THP into multiple reaction products.[79,80] Studies showed that THP not only inhibits γBBH but also inhibits carnitine transport across the plasma membrane and hence leads to tissue and eventually urinary carnitine loss in addition to carnitine biosynthesis inhibition.[81,82] Its cardioprotective effect (mainly tested in rats) was proposed to be based on lowering of the carnitine levels in the heart, which results in inhibition of fatty acid oxidation, decreased levels of harmful long-chain acyl-carnitines, and conservation of ATP.[83,84] In addition, it has been shown that THP interferes with carnitine/acyl-carnitine transport into the mitochondria, which also contributes to the decrease in fatty acid utilization.[85] When healthy volunteers were treated with 1 g/day of THP for 4 weeks, plasma carnitine levels decreased by 20%. This decrease is very moderate as compared to levels observed in mice and rats on similar doses. The difference in efficacy of THP to deplete carnitine stores between rodents and humans is possibly related to their diet. While rodent diets are low in carnitine, the volunteers in this study consumed a diet containing meat and therefore could partly compensate for the renal carnitine loss caused by the THP supplementation.[86] Although the decrease in plasma carnitine level at a dose of 1 g of THP per day is relatively moderate, this carnitine depletion is adequate to confer cardioprotective effects.[87] More recently, inhibitors of γBBH with much lower IC_{50} values, as compared to THP, have been described.[88] To date, it is still unclear whether these inhibitors also affect cellular carnitine uptake and mitochondrial carnitine transport and whether they have the same clinical benefit as THP.

13.3 Concluding Remarks

Much fundamental enzymological research on the two dioxygenases of carnitine biosynthesis, TMLH and γBBH, was performed at the end of the 20th century. The subsequent identification of the coding genes and the elucidation of the crystal structure of human γBBH have further increased our knowledge of these proteins; however, our understanding of the regulation of these enzymes is just beginning to unfold. Given the recent discovery that TMLH deficiency is a very frequent inborn error of metabolism and probably related to autism spectrum disorders, carnitine biosynthesis research in general and the function of TMLH (and γBBH) specifically is entering a new and interesting phase.

References

1. F. M. Vaz and R. J. Wanders, *Biochem. J.*, 2002, **361**, 417–429.
2. H. P. Kleber, *FEMS Microbiol. Lett.*, 1997, **147**, 1–9.
3. C. J. Rebouche and H. Seim, *Annu. Rev. Nutr.*, 1998, **18**, 39–61.
4. R. A. Panter and J. B. Mudd, *FEBS Lett.*, 1969, **5**, 169–170.
5. C. J. Rebouche, *FASEB J.*, 1992, **6**, 3379–3386.
6. R. F. Novak, T. J. Swift and C. L. Hoppel, *Biochem. J.*, 1980, **188**, 521–527.
7. P. R. Borum and H. P. Broquist, *J. Biol. Chem.*, 1977, **252**, 5651–5655.
8. W. A. Dunn and S. Englard, *J. Biol. Chem.*, 1981, **256**, 12437–12444.
9. W. A. Dunn, G. Rettura, E. Seifter and S. Englard, *J. Biol. Chem.*, 1984, **259**, 10764–10770.
10. J. LaBadie, W. A. Dunn and N. N. Aronson, Jr, *Biochem. J.*, 1976, **160**, 85–95.

11. C. J. Rebouche, L. J. Lehman and L. Olson, *J. Nutr.*, 1986, **116**, 751–759.
12. A. T. Davis and C. L. Hoppel, *J. Nutr.*, 1986, **116**, 760–767.
13. D. S. Sachan and H. P. Broquist, *Biochem. Biophys. Res. Commun.*, 1980, **96**, 870–875.
14. C. J. Rebouche and A. G. Engel, *J. Biol. Chem.*, 1980, **255**, 8700–8705.
15. C. J. Rebouche and A. G. Engel, *Biochim. Biophys. Acta*, 1980, **630**, 22–29.
16. C. J. Rebouche and A. G. Engel, *Clin. Chim. Acta*, 1980, **106**, 295–300.
17. N. van Vlies, R. J. Wanders and F. M. Vaz, *Anal. Biochem.*, 2006, **354**, 132–139.
18. J. D. Hulse, S. R. Ellis and L. M. Henderson, *J. Biol. Chem.*, 1978, **253**, 1654–1659.
19. D. S. Sachan and C. L. Hoppel, *Biochem. J.*, 1980, **188**, 529–534.
20. R. Stein and S. Englard, *Arch. Biochem. Biophys.*, 1982, **217**, 324–331.
21. L. M. Henderson, P. J. Nelson and L. Henderson, *Fed. Proc.*, 1982, **41**, 2843–2847.
22. R. Stein and S. Englard, *Anal. Biochem.*, 1981, **116**, 230–236.
23. A. T. Davis, *J. Chromatogr.*, 1987, **422**, 253–256.
24. F. M. Vaz, R. Ofman, K. Westinga, J. W. Back and R. J. Wanders, *J. Biol. Chem.*, 2001, **276**, 33512–33517.
25. N. van Vlies, R. Ofman, R. J. Wanders and F. M. Vaz, *FEBS J.*, 2007, **274**, 5845–5851.
26. B. Ling, C. Aziz and J. Alcorn, *Nutr. Metab.*, 2012, **9**, 66.
27. A. Koch, B. Konig, G. I. Stangl and K. Eder, *Exp. Biol. Med.*, 2008, **233**, 356–365.
28. J. Gloerich, D. M. van den Brink, J. P. Ruiter, N. van Vlies, F. M. Vaz, R. J. Wanders and S. Ferdinandusse, *J. Lipid Res.*, 2007, **48**, 77–85.
29. G. Wen, H. Kuhne, C. Rauer, R. Ringseis and K. Eder, *Biochem. Pharmacol.*, 2011, **82**, 175–183.
30. D. W. Horne and H. P. Broquist, *J. Biol. Chem.*, 1973, **248**, 2170–2175.
31. R. A. Kaufman and H. P. Broquist, *J. Biol. Chem.*, 1977, **252**, 7437–7439.
32. J. H. Swiegers, F. M. Vaz, I. S. Pretorius, R. J. Wanders and F. F. Bauer, *FEMS Microbiol. Lett.*, 2002, **210**, 19–23.
33. J. Monfregola, A. Cevenini, A. Terracciano, N. van Vlies, S. Arbucci, R. J. Wanders, M. D'Urso, F. M. Vaz and M. V. Ursini, *J. Cell. Physiol.*, 2005, **204**, 839–847.
34. J. Monfregola, G. Napolitano, I. Conte, A. Cevenini, C. Migliaccio, M. D'Urso and M. V. Ursini, *Gene*, 2007, **395**, 86–97.
35. P. B. Celestino-Soper, C. A. Shaw, S. J. Sanders, J. Li, M. T. Murtha, A. G. Ercan-Sencicek, L. Davis, S. Thomson, T. Gambin, A. C. Chinault, Z. Ou, J. R. German, A. Milosavljevic, J. S. Sutcliffe, E. H. Cook, Jr., P. Stankiewicz, M. W. State and A. L. Beaudet, *Hum. Mol. Genet.*, 2011, **20**, 4360–4370.
36. P. B. Celestino-Soper, S. Violante, E. L. Crawford, R. Luo, A. C. Lionel, E. Delaby, G. Cai, B. Sadikovic, K. Lee, C. Lo, K. Gao, R. E. Person, T. J. Moss, J. R. German, N. Huang, M. Shinawi, D. Treadwell-Deering, P. Szatmari, W. Roberts, B. Fernandez, R. J. Schroer, R. E. Stevenson, J. D. Buxbaum, C. Betancur, S. W. Scherer, S. J. Sanders, D. H. Geschwind, J. S. Sutcliffe,

M. E. Hurles, R. J. Wanders, C. A. Shaw, S. M. Leal, E. H. Cook, Jr., R. P. Goin-Kochel, F. M. Vaz and A. L. Beaudet, *Proc. Natl. Acad. Sci.*, 2012, **109**, 7974–7981.
37. C. Nava, F. Lamari, D. Heron, C. Mignot, A. Rastetter, B. Keren, D. Cohen, A. Faudet, D. Bouteiller, M. Gilleron, A. Jacquette, S. Whalen, A. Afenjar, D. Perisse, C. Laurent, C. Dupuits, C. Gautier, M. Gerard, G. Huguet, S. Caillet, B. Leheup, M. Leboyer, C. Gillberg, R. Delorme, T. Bourgeron, A. Brice and C. Depienne, *Transl. Psychiatry*, 2012, **2**, e179.
38. S. Englard, J. S. Blanchard and C. F. Midelfort, *Biochemistry*, 1985, **24**, 1110–1116.
39. G. Lindstedt and S. Lindstedt, *J. Biol. Chem.*, 1970, **245**, 4178–4186.
40. V. Tanphaichitr and H. P. Broquist, *J. Biol. Chem.*, 1973, **248**, 2176–2181.
41. A. L. Carter, T. O. Abney, H. Braver and A. H. Chuang, *Biol. Reprod.*, 1987, **37**, 68–72.
42. S. Englard and H. H. Carnicero, *Arch. Biochem. Biophys.*, 1978, **190**, 361–364.
43. J. D. Erfle, *Biochem. Biophys. Res. Commun.*, 1975, **64**, 553–557.
44. G. Lindstedt, S. Lindstedt and I. Nordin, *Scand. J. Clin. Lab. Invest.*, 1982, **42**, 477–485.
45. S. Englard, L. J. Horwitz and J. T. Mills, *J. Lipid Res.*, 1978, **19**, 1057–1063.
46. A. Kondo, J. S. Blanchard and S. Englard, *Arch. Biochem. Biophys.*, 1981, **212**, 338–346.
47. G. Cederblad, J. Holm, G. Lindstedt, S. Lindstedt, I. Nordin and T. Schersten, *FEBS Lett.*, 1979, **98**, 57–60.
48. M. Fischer, J. Keller, F. Hirche, H. Kluge, R. Ringseis and K. Eder, *Comp. Biochem. Physiol. A*, 2009, **153**, 324–331.
49. A. L. Olson and C. J. Rebouche, *J. Nutr.*, 1987, **117**, 1024–1031.
50. C. J. Rebouche, E. P. Bosch, C. A. Chenard, K. J. Schabold and S. E. Nelson, *J. Nutr.*, 1989, **119**, 1907–1913.
51. A. L. Carter and F. W. Stratman, *FEBS Lett.*, 1980, **111**, 112–114.
52. A. Sandor, P. E. Minkler, S. T. Ingalls and C. L. Hoppel, *Clin. Chim. Acta*, 1988, **176**, 17–27.
53. F. M. Vaz, S. van Gool, R. Ofman, L. Ijlst and R. J. Wanders, *Adv. Exp. Med. Biol.*, 1999, **466**, 117–124.
54. A. M. Rydzik, I. K. Leung, G. T. Kochan, A. Thalhammer, U. Oppermann, T. D. Claridge and C. J. Schofield, *ChemBioChem*, 2012, **13**, 1559–1563.
55. G. Lindstedt and S. Lindstedt, *Biochem. Biophys. Res. Commun.*, 1962, **7**, 394–397.
56. G. Lindstedt, S. Lindstedt, T. Midtvedt and M. Tofft, *Biochemistry*, 1967, **6**, 1262–1270.
57. G. Lindstedt, S. Lindstedt, T. Midtvedt and M. Tofft, *J. Bacteriol.*, 1970, **101**, 1094–1095.
58. G. Lindstedt, S. Lindstedt, B. Olander and M. Tofft, *Biochim. Biophys. Acta*, 1968, **158**, 503–505.
59. G. Lindstedt, S. Lindstedt and M. Tofft, *Biochemistry*, 1970, **9**, 4336–4342.
60. G. Lindstedt, S. Lindstedt and I. Nordin, *Biochemistry*, 1977, **16**, 2181–2188.

61. S. F. Ng, H. M. Hanauske-Abel and S. Englard, *J. Biol. Chem.*, 1991, **266**, 1526–1533.
62. U. Ruetschi, I. Nordin, B. Odelhog, H. Jornvall and S. Lindstedt, *Eur. J. Biochem.*, 1993, **213**, 1075–1080.
63. X. Lu, P. Zhang, Q. Li, H. Liu, X. Lin and X. Ma, *Wei sheng wu xue bao = Acta Microbiologica Sinica*, 2012, **52**, 602–610.
64. G. Lindstedt, *Biochemistry*, 1967, **6**, 1271–1282.
65. R. S. Wehbie, N. S. Punekar and H. A. Lardy, *Biochemistry*, 1988, **27**, 2222–2228.
66. B. Z. Simkhovich, Z. V. Shutenko, D. V. Meirena, K. B. Khagi, R. J. Mezapuke, T. N. Molodchina, I. J. Kalvins and E. Lukevics, *Biochem. Pharmacol.*, 1988, **37**, 195–202.
67. H. Noel, R. Parvin and S. V. Pande, *Biochem. J.*, 1984, **220**, 701–706.
68. S. Galland, F. Le Borgne, D. Guyonnet, P. Clouet and J. Demarquoy, *Mol. Cell. Biochem.*, 1998, **178**, 163–168.
69. S. Galland, F. Le Borgne, F. Bouchard, B. Georges, P. Clouet, F. Grand-Jean and J. Demarquoy, *Biochim. Biophys. Acta*, 1999, **1441**, 85–92.
70. P. Hahn, *Life Sci.*, 1981, **28**, 1057–1060.
71. S. Galland, B. Georges, F. Le Borgne, G. Conductier, J. V. Dias and J. Demarquoy, *Cell. Mol. Life Sci.*, 2002, **59**, 540–545.
72. A. Gutgesell, R. Ringseis, C. Brandsch, G. I. Stangl, F. Hirche and K. Eder, *Metab.: Clin. Exp.*, 2009, **58**, 226–232.
73. N. van Vlies, S. Ferdinandusse, M. Turkenburg, R. J. Wanders and F. M. Vaz, *Biochim. Biophys. Acta*, 2007, **1767**, 1134–1142.
74. N. A. Oey, N. van Vlies, F. A. Wijburg, R. J. Wanders, T. Attie-Bitach and F. M. Vaz, *Placenta*, 2006, **27**, 841–846.
75. E. Holme, S. Linstedt and I. Nordin, *Biochem. Biophys. Res. Commun.*, 1982, **107**, 518–524.
76. S. Lindstedt and I. Nordin, *Biochem. J.*, 1984, **223**, 119–127.
77. K. Tars, J. Rumnieks, A. Zeltins, A. Kazaks, S. Kotelovica, A. Leonciks, J. Sharipo, A. Viksna, J. Kuka, E. Liepinsh and M. Dambrova, *Biochem. Biophys. Res. Commun.*, 2010, **398**, 634–639.
78. F. M. Vaz, S. van Gool, R. Ofman, L. Ijlst and R. J. Wanders, *Biochem. Biophys. Res. Commun.*, 1998, **250**, 506–510.
79. I. K. Leung, T. J. Krojer, G. T. Kochan, L. Henry, F. von Delft, T. D. Claridge, U. Oppermann, M. A. McDonough and C. J. Schofield, *Chem. Biol.*, 2010, **17**, 1316–1324.
80. L. Henry, I. K. Leung, T. D. Claridge and C. J. Schofield, *Bioorg. Med. Chem. Lett.*, 2012, **22**, 4975–4978.
81. B. Georges, F. Le Borgne, S. Galland, M. Isoir, D. Ecosse, F. Grand-Jean and J. Demarquoy, *Biochem. Pharmacol.*, 2000, **59**, 1357–1363.
82. M. Spaniol, H. Brooks, L. Auer, A. Zimmermann, M. Solioz, B. Stieger and S. Krahenbuhl, *Eur. J. Biochem.*, 2001, **268**, 1876–1887.
83. Y. Hayashi, T. Kirimoto, N. Asaka, M. Nakano, K. Tajima, H. Miyake and N. Matsuura, *Eur. J. Pharmacol.*, 2000, **395**, 217–224.

84. Y. Hayashi, K. Tajima, T. Kirimoto, H. Miyake and N. Matsuura, *Pharmacology*, 2000, **61**, 238–243.
85. M. Tsoko, F. Beauseigneur, J. Gresti, I. Niot, J. Demarquoy, J. Boichot, J. Bezard, L. Rochette and P. Clouet, *Biochem. Pharmacol.*, 1995, **49**, 1403–1410.
86. E. Liepinsh, I. Konrade, E. Skapare, O. Pugovics, S. Grinberga, J. Kuka, I. Kalvinsh and M. Dambrova, *J. Pharm. Pharmacol.*, 2011, **63**, 1195–1201.
87. M. Dambrova, E. Liepinsh and I. Kalvinsh, *Trends Cardiovasc. Med.*, 2002, **12**, 275–279.
88. K. Tars, J. Leitans, A. Kazaks, D. Zelencova, E. Liepinsh, J. Kuka, M. Makrecka, D. Lola, V. Andrianovs, D. Gustina, S. Grinberga, E. Liepinsh, I. Kalvinsh, M. Dambrova, E. Loza and O. Pugovics, *J. Med. Chem.*, 2014, **57**, 2213–2236.

CHAPTER 14

Phytanoyl-CoA Hydroxylase: A 2-Oxoglutarate-Dependent Dioxygenase Crucial for Fatty Acid Alpha-Oxidation in Humans

RONALD J. A. WANDERS*[a], SACHA FERDINANDUSSE[a], MEREL S. EBBERINK[a], AND HANS R. WATERHAM[a]

[a]Laboratory Genetic Metabolic Diseases, Departments of Paediatrics, Emma Children's Hospital, and Clinical Chemistry, Academic Medical Center, University of Amsterdam, Meibergdreef 9, 1105 AZ, Amsterdam, the Netherlands
*E-mail: r.j.wanders@amc.uva.nl

14.1 Introduction

Hansen and coworkers were the first to identify phytanic acid (3,7,11,14-tetramethylhexadecanoic acid) in butterfat, which was soon followed by its identification in a variety of other food sources.[1,2]

In vivo phytanic acid occurs as a racemic mixture (3*R*,7*R*,11*R*,15)-tetramethylhexadecanoic acid, and (3*S*,7*R*,11*R*,15)-tetramethylhexadecanoic acid. Quantitative assessment of phytanic acid revealed high levels in milk, butter, cheese and meat from cows and sheep, while in meat from pigs and poultry

RSC Metallobiology Series No. 3
2-Oxoglutarate-Dependent Oxygenases
Edited by Robert P. Hausinger and Christopher J. Schofield

Published by the Royal Society of Chemistry, www.rsc.org

hardly any phytanic acid could be detected. Furthermore, almost no phytanic acid was detected in vegetables, while some species of fish and fish oils were found to contain large amounts.

Subsequent work revealed that phytanic acid is present in some non-ruminants such as rats, pigs and humans. The amount of phytanic acid usually ranges from about 0.01–0.3% of the total fatty acid pool, but it can exceed the 10% mark in milk from cows fed on ensilage, the fermented grass that is used as winter feed for cattle. High amounts of phytanic acid have also been detected in Antarctic krill (1.4%), the plankton that forms the basis of the oceanic food chain, which accounts for the presence of phytanic acid in, for example, molluscs, fish oil, whale oil and whale milk.[3]

Early on, it was hypothesized that (E)-phytol (3,7,11,15-tetrahexadec-2-en-1-ol), a constituent of the chlorophyll molecule, is the direct precursor of phytanic acid. Because almost all photosynthetic organisms use chlorophyll, phytol is abundantly present in nature. As a constituent of the large quantities of grasses consumed by ruminant animals, chlorophyll is consumed in massive amounts, from which the phytol moiety is indeed released and converted into phytanic acid. To show that phytol could be converted into phytanic acid, feeding studies were performed in which animals were fed phytol-supplemented diets. Feeding of these diets resulted in the accumulation of phytanic acid in rats and rabbits, with levels reaching as high as 50% of total fatty acids.[4,5] Although it was originally thought otherwise, humans are virtually unable to release phytol from the chlorophyll molecule. In contrast to humans, animals with a ruminant digestive system liberate phytol from chlorophyll by the action of their associated bacteria. The implication is that the major source of phytanic acid for humans is not chlorophyll-bound phytol but rather phytanic acid itself as present in food products.

Interest in phytanic acid was greatly stimulated by the discovery of Klenk and Kahlke of phytanic acid accumulation in plasma and tissues from patients affected by Refsum disease.[6] Inspired by this discovery, much work has been done since then on the metabolism of phytanic acid - a process called α-oxidation. The enzymology of the α-oxidation pathway and the enzymatic and molecular defects in Refsum disease have since been identified. In this chapter we describe the current state of knowledge regarding phytanic acid metabolism and the enzymatic and molecular basis of Refsum disease at the level of phytanoyl-coenzyme A (CoA) 2-hydroxylase.

14.2 Metabolism of Phytanic Acid and the Role of Phytanoyl-CoA 2-Hydroxylase

The discovery of elevated phytanic acid levels in patients suffering from Refsum disease in the early 1960s was soon followed by studies aimed to resolve the pathway of oxidation of phytanic acid and to pinpoint the enzymatic defects in Refsum disease. To this end, Steinberg *et al.* administered radiolabelled [U–$^{14}C_{20}$]-phytol and [U–$^{14}C_{20}$]-phytanic acid to mice,[7] followed by fatty acid analysis of the livers. This effort revealed the presence of labelled pristanic acid (2,6,10,14-tetramethylpentadecanoic acid), which is one carbon atom shorter than phytanic acid (Figure 14.1). In addition, a number of other metabolites, including 4,8,12-trimethyltridecanoic acid and 4,8-dimethylnonanoic acid, were observed, representing the expected products of pristanic acid β-oxidation. These studies provided convincing evidence in favour of a metabolic pathway that would first involve the oxidative decarboxylation of phytanic acid to pristanic acid, followed by repeated cycles of β-oxidation (Figure 14.1). Nevertheless, the enzymology of the α-oxidation pathway catalysing the oxidative decarboxylation of phytanic acid long remained an enigma, which explains why the true enzyme defect in Refsum disease was only discovered in 1997.[8]

Following the discovery of α-oxidation as the preferred pathway of oxidation of phytanic acid, research focused on the resolution of the α-oxidation pathway, including the identity of the enzymatic reactions and the enzymes involved. These studies were performed mainly with rat liver homogenates and isolated rat liver mitochondria. As reviewed earlier in detail,[9] these studies had led to erroneous conclusions, including the claim that α-oxidation occurs in mitochondria with the first step catalysed by a cytochrome P450 type of enzyme. The underlying basis for this misconception is now entirely clear and is the consequence of the inadequacy of the incubation conditions adopted by the various authors. Most importantly, the reaction media lacked certain essential cofactors, notably 2-oxoglutarate (2OG), and contained the wrong substrate, *i.e.* phytanic acid instead of phytanoyl-CoA.

Several seminal observations have revolutionized our way of thinking about the mechanism of phytanic acid α-oxidation through the years. First, Poulos *et al.*[10] reported the accumulation of phytanic acid in plasma of Zellweger patients and the deficient oxidation of phytanic acid in Zellweger fibroblasts, suggesting that phytanic acid oxidation would be peroxisomal rather than mitochondrial. Second, rates of phytanic acid oxidation in intact cells, including hepatocytes[11] and cultured human skin fibroblasts,[12] were > 50-fold higher compared to homogenates, suggesting that essential components required for α-oxidation were missing in the experiments with homogenates and/or isolated mitochondria.[9] Third, Poulos *et al.*[13] showed that formic acid is the major product of α-oxidation rather than CO_2. Finally, and perhaps most importantly, Watkins *et al.*[14] discovered that phytanic acid first had to be activated to phytanoyl-CoA before α-oxidation could take place. Taken together, these findings prompted a thorough reinvestigation of the α-oxidation pathway by different groups. Starting from the concept that phytanoyl-CoA is the most plausible substrate for α-oxidation and that peroxisomes are the most likely site of phytanic acid α-oxidation, Mihalik *et al.*[15] succeeded in identifying a novel enzyme, *i.e.* phytanoyl-CoA 2-hydroxylase, which converts phytanoyl-CoA into 2-hydroxyphytanoyl-CoA. It was found that formation of 2-hydroxyphytanoyl-CoA is strictly dependent on the presence of 2OG and Fe(II), and less so on ascorbate, indicating that the enzyme

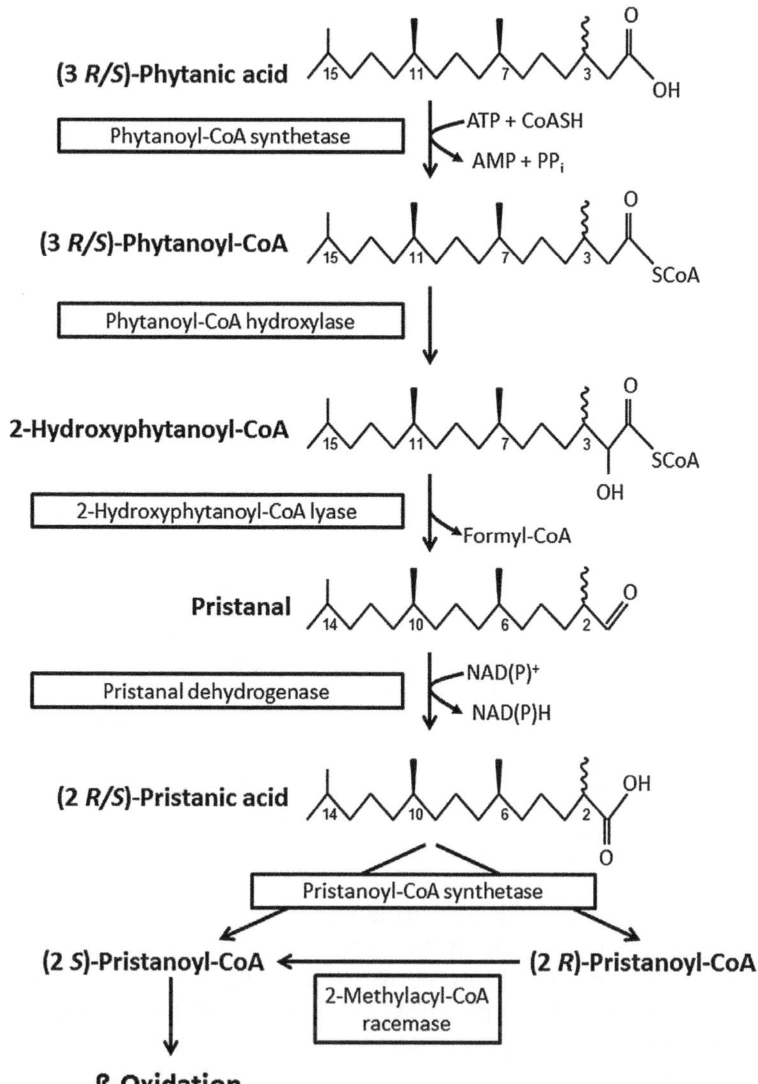

Figure 14.1 Enzymology of the phytanic acid α-oxidation pathway. Phytanic acid occurs as a racemic mixture of two stereoisomers, (3*R*,7*R*,11*R*,15)- and (3*S*,7*R*,11*R*,15)-tetramethylhexadecanoic acid, under *in vivo* conditions. After activation to the corresponding CoA thioesters, both (3*S*)-phytanoyl-CoA and (3*R*)-phytanoyl-CoA undergo α-oxidation to produce the two stereoisomers of pristanic acid, (2*S*,6*R*,10*R*,14)- and (2*R*,6*R*,10*R*,14)-tetramethylpentadecanoic acid. After further activation, (2*S*)-pristanoyl-CoA can undergo direct β-oxidation whereas (2*R*)-pristanoyl-CoA first requires conversion into (2*S*)-pristanoyl-CoA because the first enzyme involved in peroxisomal β-oxidation only reacts with (2*S*)-acyl-CoAs and not with (2*R*)-acyl-CoAs. The enzyme 2-methylacyl-CoA racemase, localized in both peroxisomes and mitochondria, catalyses the interconversion between (2*R*)- and (2*S*)-pristanoyl-CoA. For further details see text.

belongs to the group of 2OG-dependent dioxygenases. Subsequent studies in rat liver[16] and human liver[17] essentially confirmed the findings of Mihalik *et al.*[15] and have set the stage for clarifying the remainder of the α-oxidation pathway as described below.

14.2.1 Phytanoyl-CoA Synthetase

The first obligatory step in the degradation of phytanic acid is its activation to phytanoyl-CoA (Figure 14.1). As reviewed in detail elsewhere,[18] several acyl-CoA synthetases localized in different subcellular compartments, including the mitochondrial outer membrane, mitochondria-associated membranes, endoplasmic reticulum (ER) and the cytosol,[19] have been shown to activate phytanic acid. The exact role of these different synthetases in the degradation of phytanic acid has remained obscure.

14.2.2 Phytanoyl-CoA 2-Hydroxylase

Phytanoyl-CoA 2-hydroxylase catalyses the second step of the α-oxidation pathway (see Figure 14.1), and has been purified from different sources. Furthermore, the encoding genes have been identified in several species, including man, rat and mouse.[20] The amino acid sequences of the human, rat and mouse enzymes have revealed marked homologies and contain clear PTS2 sequences present in all three proteins, *i.e.* RLQIVLGHL in the human hydroxylase and RLQVLLGHL in the rat and mouse hydroxylases.[21] These PTS2 sequences are recognized by a specific cycling receptor, named PEX7, which escorts these proteins from their site of biosynthesis in the cytosol to the peroxisomal membrane, after which they are imported into the peroxisome.[22–26]

The mature hydroxylase as purified from rat liver peroxisomes has a smaller molecular weight than the nascent product due to cleavage of the PTS2 domain after import into peroxisomes.[21] Recent studies have identified the enzyme catalysing the cleavage of this 30 amino acid peptide from the phytanoyl-CoA hydroxylase.[27] The same enzyme, named Tysnd1 (trypsin domain containing 1), also catalyses the intra-peroxisomal cleavage of other peroxisomal proteins including acyl-CoA oxidase 1, D-bifunctional protein and sterol-carrier-protein X. Interestingly a mouse model in which the gene coding for this cysteine endopeptidase (*Tynsd1*) was disrupted, has recently been described.[28] As a consequence of the deficiency of Tysnd1, the activity of multiple enzymes including phytanoyl-CoA 2-hydroxylase is reduced, which explains why phytanic acid metabolism is grossly impaired in the *Tysnd1*$^{(-/-)}$ mouse.[28]

In the amino acid sequences of 2OG-dependent hydroxylases, a conserved iron-binding motif (His–X–Asp/Glu...His) has been identified (see Chapter 1). On the basis of site-directed mutagenesis studies, His-175 was proposed as the first residue in this sequence for the human hydroxylase whereas His-264 was identified as the most likely candidate for the second histidine. A second motif present in many 2OG-dependent hydroxylases (Arg–X–Ser) is often involved in binding 2OG; this motif is not present in phytanoyl-CoA 2-hydroxylases, but Arg-275 was identified as the arginine which binds the 5-carboxy group of 2OG in the human enzyme.[29] The latter structural study also confirmed the metal ligands as His-175, Asp-177 and His-264.

At least *in vitro*, phytanoyl-CoA 2-hydroxylase has the capacity to catalyse the hydroxylation of other substrates besides phytanoyl-CoA, including 3-methyl-branched acyl-CoAs and straight-chain acyl-CoA esters. Interestingly, phytanoyl-CoA 2-hydroxylase accepts both 3*S*- as well as 3*R*-phytanoyl-CoA as substrates, generating products with the same relative stereochemistry, *i.e.* 2*R*,3*S*-hydroxyphytanoyl-CoA and 2*S*,3*R*-hydroxyphytanoyl-CoA, respectively[16] (see Figure 14.1). Apparently, the stereochemistry of the methyl group at C-3 determines the stereochemistry of the hydroxylated product at C-2. *In vitro* studies by Mukherji *et al.*[30] have shown that the mature 13.2 kDa form of sterol-carrier-protein (SCP2) greatly stimulates phytanoyl-CoA hydroxylase activity.

14.2.3 2-Hydroxyphytanoyl-CoA Lyase

2-Hydroxyphytanoyl-CoA lyase, now renamed as 2-hydroxyacyl-CoA lyase (HACL1), catalyses the cleavage of 2-hydroxyphytanoyl-CoA to pristanal and formyl-CoA (Figure 14.1).[31,32] The protein was purified by Foulon *et al.*[31] and is a thiamine diphosphate-dependent enzyme that also requires Mg(II) for optimal activity. The human HACL1 protein contains a non-canonical PTS1 sequence (SNM). A thiamine diphosphate-binding motif, typical for decarboxylases/transketolases, is present in the *C*-terminal half of the protein. HACL1 is a homotetramer composed of 60 kDa subunits and catalyses the cleavage of a range of other 2-hydroxyacyl-CoA species.[31]

14.2.4 Pristanal Dehydrogenase

Pristanal, synthesized in the HACL1 reaction, is converted into pristanic acid by a dehydrogenase (Figure 14.1). Differential and density gradient centrifugation experiments have shown that peroxisomes contain pristanal dehydrogenase activity in addition to mitochondria and microsomes.[33] The true identity of the aldehyde dehydrogenase catalysing this reaction has remained unresolved so far. Initially, the fatty aldehyde dehydrogenase (FALDH) encoded by *ALDH3A2* was supposed to catalyse this reaction.[34] The finding that phytanic acid α-oxidation is completely normal in cells from patients with Sjögren Larsson syndrome, caused by mutations in *ALDH3A2*, argue against this conclusion. Recent data, however, have brought FALDH back into the arena again as a possible candidate. Indeed, evidence indicates that an alternative splice product of *ALDH3A2* targets to peroxisomes. It must be concluded, however, that at this moment the identity of the pristanal dehydrogenase involved in phytanic acid α-oxidation remains uncertain (see Wanders and Komen[35] and Wanders *et al.*[36] for more detailed discussion).

14.2.5 Pristanic Acid Activation

Just like phytanic acid, pristanic acid activation can occur at different subcellular locations, including peroxisomes, mitochondria and the ER.[37] As discussed in more detail elsewhere, it has not been established with certainty which synthetase actually catalyses the activation of pristanic acid that is derived from pristanal. A likely candidate is the enzyme ACSVL1 encoded by *SLC27A2*, which has also been shown to activate phytanic acid.[38] This enzyme was first purified by Hashimoto *et al.* from rat liver,[39] and subsequently cloned.[40] This work was followed up by that of Steinberg *et al.*,[38] who identified the human ACSVL1, and showed that the enzyme is localized to the ER membrane and peroxisomes. Because of its subcellular location, with its catalytic site exposed to the peroxisomal interior, it has been suggested that peroxisomal ACSVL1 catalyses the activation of pristanic acid as produced from pristanal within the peroxisome (see Figure 4C in Jansen *et al.*[33]).

14.3 Disorders of Phytanic Acid Metabolism

The disorders of fatty acid α-oxidation can be subdivided into two classes depending upon whether the deficiency is caused by a primary deficiency of one of the enzymes involved in α-oxidation, due to mutations in the encoding genes, or is the secondary consequence of a defect in peroxisome formation which causes α-oxidation to be deficient.

14.3.1 Primary Disorders of Peroxisomal Alpha-Oxidation

Phytanoyl-CoA 2-hydroxylase deficiency is the only primary disorder of fatty acid α-oxidation and is caused by mutations in the gene coding for phytanoyl-CoA 2-hydroxylase. So far, no deficiencies of any of the other four enzymes involved in α-oxidation have been identified in humans. The outcome associated with phytanoyl-CoA hydroxylase deficiency is Refsum disease. Although initially described as an entity characterized by four cardinal features including retinitis pigmentosa, cerebellar ataxia, polyneuropathy and elevated cerebrospinal fluid protein, it is now clear that the phenotypic spectrum of Refsum disease is very diverse, with night blindness in childhood as the most common abnormality observed in all patients. Less constant features include: cerebellar ataxia, polyneuropathy, sensory neural hearing loss, ichthyosis, skeletal malformations and cardiac abnormalities. Most features progress with age. Wierzbicki *et al.*[41] studied the cumulative incidence of the following features over decades: retinitis pigmentosa (15/15), anosmia (14/15), neuropathy (11/15), deafness (10/15).

14.3.2 Secondary Disorders of Peroxisomal Alpha-Oxidation

Peroxisomal α-oxidation is also deficient in disorders affecting peroxisome biogenesis. This is true for the Zellweger spectrum disorders, in which the formation of peroxisomes is disturbed leading to the absence (or marked deficiency) of morphologically identifiable peroxisomes. Moreover, α-oxidation is also deficient in rhizomelic chondrodysplasia punctata (RCDP) type 1. In contrast to the generalized defect in peroxisome biogenesis for Zellweger spectrum disorders, only one branch of the peroxisome biogenesis machinery is affected in RCDP type 1. The RCDP type 1 disorder is caused by mutations in the *PEX7* gene, encoding the PTS2-receptor, which plays an indispensable role in the proper transport across the peroxisomal membrane of a subset of peroxisomal matrix proteins equipped with a PTS2-signal. Phytanoyl-CoA hydroxylase is one of the PTS2-proteins and is deficient in RCDP type 1 as a consequence of the loss of function of the PTS2-receptor.[22]

14.4 Molecular Basis of Phytanoyl-CoA 2-Hydroxylase Deficiency

At present, 30 different mutations have been reported in *PHYH* encoding phytanoyl-CoA 2-hydroxylase, including 17 missense mutations, 5 deletions, 2 insertions and 6 splice site mutations (see Table 14.1), along with one non-disease-causing variant (636A > G). Mutations are distributed across the entire *PHYH* gene, although exons 6 and 7 harbour most mutations. Most of these mutations lead to proteins with substitutions clustering either around the 2OG-binding pocket or the Fe(II)-binding site as shown by crystallographic analysis of human phytanoyl-CoA 2-hydroxylase.[42] There are missense mutations reported in codons for each of the three individual amino acids that form the first portion of the iron-binding His-X-Asp motif (amino acid positions 175–177). These include: 524A > G (His-175 → Arg), 526C > A (Gln-176 → Lys) and 530A > G (Asp-177 → Gly). Mukherji *et al.*[43] showed that an amino acid substitution of Arg-275 results in impaired 2OG binding, whereas an amino acid substitution of Gln-176, Gly-204 or Asn-269 causes partial uncoupling of 2OG decarboxylation. The 530A > G (Asp-177 → Gly) and 734 G > A (Arg-245 → Gln) mutations are located on one allele. Subsequent studies have shown that only the 530A > G (Asp-177 → Gly) mutation gives rise to an impaired phytanoyl-CoA 2-hydroxylase enzyme activity, whereas expression of the 734 G > A (Arg-245 → Gln) results in low normal enzyme activity.[44] Another interesting mutation is 85C > T (Pro-29 → Ser) for which the protein substitution is located outside the active domain; this variant protein shows normal enzyme activity *in vitro*. This substitution is situated next to the cleavage point of Tysnd1, which possibly hinders proper cleavage of the PTS2 sequence from the mature protein.[43]

Most of the splice site, deletion and insertion mutations lead to a truncated phytanoyl-CoA 2-hydroxylase protein, if stably expressed at all. Jansen *et al.*[44] studied the effects of two splice site mutations that lead to skipping

Table 14.1 Mutations in the *PHYH* gene[a] and their consequences.

Mutation			
Nucleotide	**Amino acids**	**Affected protein domain**	**References**
85C > T	Pro-29 → Ser	PTS2 sequence	Mukherji *et al.*[47]
135-1G > C	Tyr-46_Arg-82 del[b]	Fe(II) and 2OG binding site	Jansen *et al.*[44]
135-2A > G	Tyr-46_Arg-82 del[b]	Fe(II) and 2OG binding site	Jansen *et al.*[48]
164delT	Leu-55 fsX12[c]	Fe(II) and 2OG binding site	Jansen *et al.*[20]
247A > T	Asn-83 → Tyr	—	Jansen *et al.*[20]
	Glu-86 fsX26[c]	Fe(II) and 2OG binding site	Jansen *et al.*[20]
375_376delGG	Glu-126 fsX1[c]	Fe(II) and 2OG binding site	Jansen *et al.*[20]
415_678del[d]	Ile-139_Glu-226 del[b]	Fe(II) and 2OG binding site	Chahal *et al.*[45]
457delG	Ala-152 fsX5[c]	Fe(II) and 2OG binding site	Jansen *et al.*[20]
497-2A > G	Ala-166 fsX3[c]	Fe(II) and 2OG binding site	Jansen *et al.*[20]
517C > T	Pro-173 → Ser	Fe(II) binding site, β-strand core	Jansen *et al.*[20]
524A > G	His-175 → Arg	Fe(II) binding site	Jansen *et al.*[20]
526C > A	Gln-176 → Lys	Fe(II) binding site	Jansen *et al.*[47]
530A > G	Asp-177 → Gly	Fe(II) binding site	Jansen *et al.*[20]
576_577insGCC	192_193 ins Ala	2OG binding site, β-strand core	Jansen *et al.*[20]
577T > C	Trp-193 → Arg	2OG binding site, β-strand core	Jansen *et al.*[20]
589 G > C	Glu-197 → Gln	2OG binding site, β-strand core	Jansen *et al.*[20]
595A > T	Ile-199 → Phe	2OG binding site, β-strand core	Jansen *et al.*[20]
610 G > A	Gly-204 → Ser	2OG binding site	Mukherji *et al.*[47]
658C > T	His-220 → Tyr	Fe(II) binding site	Jansen *et al.*[20]
678+2T > G	Ala-166 fsX3[c]	Fe(II) and 2OG binding site	Jansen *et al.*[20]
678+5 G > T	Ala-166 fsX3[c]	Fe(II) and 2OG binding site	Jansen *et al.*[20]
679-1G > T	Ala-166 fsX3[c]	Fe(II) and 2OG binding site	Jansen *et al.*[20]
683_684insG	Gly-228 fsX2[c]	2OG binding site	Jansen *et al.*[20]
734 G > A	Arg-245 → Gln	β-strand core	Jansen *et al.*[20]
770T > C	Phe-257 → Ser	2OG binding site, β-strand core	Jansen *et al.*[20]
805A > C	Asn-269 → His	2OG binding site	Jansen *et al.*[20] and Mukherji *et al.*[47]
823C > T	Arg-275 → Trp	2OG binding site	Mihalik *et al.*[49]
824 G > A	Arg-275 → Gln	2OG binding site	Jansen *et al.*[20]
830C > A	Ala-277 → Glu	β-strand core	Kohlschutter *et al.*[50]

[a]Reference sequence of *PHYH*: GenBank accession number NM_006214.3. Nucleotide numbering starting at the first adenine of the translation initiation codon ATG.
[b]Mutation causes an in-frame deletion (del) which results in no or reduced enzyme activity.
[c]Mutation causes a frameshift (fs) which leads to a truncated PHYH protein or instable *PHYH* mRNA.
[d]Observed in *PHYH* mRNA.

of exon 3, which causes an in-frame deletion of 37 amino acids (Tyr-46_Arg-82 del). When expressed in *S. cerevisiae* this variant protein is detectable by Western blot analysis, but completely lacks enzymatic activity. Fibroblasts of a patient homozygous for an in-frame deletion (415_678del) showed reduced α-oxidation of phytanic acid when compared to control fibroblasts.[45]

In summary, much has been learned in recent years about the pathway of phytanic acid oxidation and the inborn errors involved with this pathway. It is remarkable that so far out of the five enzymes participating in phytanic acid α-oxidation only the deficiency of phytanoyl-CoA 2-hydroxylase has been described. It may be that the clinical signs and symptoms of patients with genetic defects in any of the other enzymes show a different phenotype. Surely, in these days of exome sequencing we will soon find out whether deficiencies in any of these enzymes exist. The availability of a mouse mutant, in which the gene coding for phytanoyl-CoA 2-hydroxylase has been disrupted, opens the way to obtain information about the pathophysiology of the phytanic acid α-oxidation pathway and possible modes of treatment.[46]

Acknowledgements

Work described in this chapter was financially supported by ITN Project PERoxisome formation, FUnction, MEtabolism (acronym PERFUME) FP7 Call for Proposals: FP7-PEOPLE-2012-ITN Proposal No. 316723. The authors gratefully acknowledge Mrs Maddy Festen for expert preparation of the manuscript and Mr Jos Ruiter for artwork.

References

1. R. P. Hansen and F. B. Shorland, *Biochem. J.*, 1951, **50**, 207.
2. R. P. Hansen, F. B. Shorland and N. J. COOKE, *Biochem. J.*, 1952, **52**, 203.
3. R. P. Hansen, *N. Z. J. Sci.*, 1980, **23**, 259.
4. C. E. Mize, J. Avigan, J. H. Baxter, H. M. Fales and D. Steinberg, *J. Lipid Res.*, 1966, 7, 692.
5. D. Steinberg, J. Avigan, C. E. Mize, J. H. Baxter, J. Cammermeyer, H. M. Fales and P. F. Highet, *J. Lipid Res.*, 1966, **7**, 684.
6. S. Ferdinandusse and S. M. Houten, *Biochim. Biophys. Acta*, 2006, **1763**, 1427.
7. C. E. Mize, D. Steinberg, J. Avigan and H. M. Fales, *Biochem. Biophys. Res. Commun.*, 1966, **25**, 359.
8. G. A. Jansen, R. J. A. Wanders, P. A. Watkins and S. J. Mihalik, *N. Engl. J. Med.*, 1997, **337**, 133.
9. R. J. A. Wanders, G. A. Jansen and M. D. Lloyd, *Biochim. Biophys. Acta*, 2003, **1631**, 119.
10. A. Poulos, *Lipids*, 1995, **30**, 1.
11. S. Huang, P. P. Van Veldhoven, F. Vanhoutte, G. Parmentier, H. J. Eyssen and G. P. Mannaerts, *Arch. Biochem. Biophys.*, 1992, **296**, 214.

12. R. J. A. Wanders and C. W. T. van Roermund, *Biochim. Biophys. Acta*, 1993, **1167**, 345.
13. A. Poulos, P. Sharp, H. Singh, D. W. Johnson, W. F. Carey and C. Easton, *Biochem. J.*, 1993, **292**, 457.
14. P. A. Watkins, A. E. Howard and S. J. Mihalik, *Biochim. Biophys. Acta*, 1994, **1214**, 288.
15. S. J. Mihalik, A. M. Rainville and P. A. Watkins, *Eur. J. Biochem.*, 1995, **232**, 545.
16. K. Croes, M. Casteels, E. de Hoffmann, G. P. Mannaerts and P. P. Van Veldhoven, *Eur. J. Biochem.*, 1996, **240**, 674.
17. G. A. Jansen, S. J. Mihalik, P. A. Watkins, H. W. Moser, C. Jakobs, S. Denis and R. J. A. Wanders, *Biochem. Biophys. Res. Commun.*, 1996, **229**, 205.
18. R. J. A. Wanders, J. C. Komen and S. Ferdinandusse, *Biochim. Biophys. Acta*, 2011, **1811**, 498.
19. T. M. Lewin, J. H. Kim, D. A. Granger, J. E. Vance and R. A. Coleman, *J. Biol. Chem.*, 2001, **276**, 24674.
20. G. A. Jansen, H. R. Waterham and R. J. A. Wanders, *Hum. Mutat.*, 2004, **23**, 209.
21. G. A. Jansen, R. Ofman, S. Denis, S. Ferdinandusse, E. M. Hogenhout, C. Jakobs and R. J. A. Wanders, *J. Lipid. Res.*, 1999, **40**, 2244.
22. H. R. Waterham and M. S. Ebberink, *Biochim. Biophys. Acta*, 2012, **1822**, 1430.
23. F. D. Mast, A. Fagarasanu, B. Knoblach and R. A. Rachubinski, *Physiology. (Bethesda.)*, 2010, **25**, 347.
24. I. J. Lodhi and C. F. Semenkovich, *Cell Metab.*, 2014.
25. Y. Fujiki, Y. Yagita and T. Matsuzaki, *Biochim. Biophys. Acta*, 2012, **1822**, 1337.
26. H. W. Platta, S. Hagen, C. Reidick and R. Erdmann, *Biochimie*, 2014, **98**, 16.
27. I. V. Kurochkin, Y. Mizuno, A. Konagaya, Y. Sakaki, C. Schonbach and Y. Okazaki, *EMBO J.*, 2007, **26**, 835.
28. Y. Mizuno, Y. Ninomiya, Y. Nakachi, M. Iseki, H. Iwasa, M. Akita, T. Tsukui, N. Shimozawa, C. Ito, K. Toshimori, M. Nishimukai, H. Hara, R. Maeba, T. Okazaki, A. N. Alodaib, A. M. Al, M. Jacob, F. S. Alkuraya, Y. Horai, M. Watanabe, H. Motegi, S. Wakana, T. Noda, I. V. Kurochkin, Y. Mizuno, C. Schonbach and Y. Okazaki, *PLoS Genet.*, 2013, **9**, e1003286.
29. M. A. McDonough, K. L. Kavanagh, D. Butler, T. Searls, U. Oppermann and C. J. Schofield, *J. Biol. Chem.*, 2005, **280**, 41101.
30. M. Mukherji, N. J. Kershaw, C. J. Schofield, A. S. Wierzbicki and M. D. Lloyd, *Chem. Biol.*, 2002, **9**, 597.
31. V. Foulon, V. D. Antonenkov, K. Croes, E. Waelkens, G. P. Mannaerts, P. P. Van Veldhoven and M. Casteels, *Proc. Natl. Acad. Sci. U. S. A.*, 1999, **96**, 10039.
32. G. A. Jansen, N. M. Verhoeven, S. Denis, G. J. Romeijn, C. Jakobs, H. J. ten Brink and R. J. A. Wanders, *Biochim. Biophys. Acta*, 1999, **1440**, 176.
33. G. A. Jansen, D. M. van den Brink, R. Ofman, O. Draghici, G. Dacremont and R. J. A. Wanders, *Biochem. Biophys. Res. Commun.*, 2001, **283**, 674.

34. N. M. Verhoeven, C. Jakobs, G. Carney, M. P. Somers, R. J. A. Wanders and W. B. Rizzo, *FEBS Lett.*, 1998, **429**, 225.
35. R. J. A. Wanders and J. C. Komen, *Biochem. Soc. Trans.*, 2007, **35**, 865.
36. P. P. Van Veldhoven, *J. Lipid Res.*, 2010, **51**, 2863.
37. R. J. A. Wanders, S. Denis, C. W. T. van Roermund, C. Jakobs and H. J. ten Brink, *Biochim. Biophys. Acta*, 1992, **1125**, 274.
38. S. J. Steinberg, S. J. Wang, D. G. Kim, S. J. Mihalik and P. A. Watkins, *Biochem. Biophys. Res. Commun.*, 1999, **257**, 615.
39. Y. Uchida, N. Kondo, T. Orii and T. Hashimoto, *J. Biochem. (Tokyo)*, 1996, **119**, 565.
40. A. Uchiyama, T. Aoyama, K. Kamijo, Y. Uchida, N. Kondo, T. Orii and T. Hashimoto, *J. Biol. Chem.*, 1996, **271**, 30360.
41. A. S. Wierzbicki, M. D. Lloyd, C. J. Schofield, M. D. Feher and F. B. Gibberd, *J. Neurochem.*, 2002, **80**, 727.
42. C. J. Schofield and M. A. McDonough, *Biochem. Soc. Trans.*, 2007, **35**, 870.
43. M. Mukherji, N. J. Kershaw, C. H. MacKinnon, I. J. Clifton, A. S. Wierzbicki, C. J. Schofield and M. D. Lloyd, *Chem. Commun.*, 2001, 972.
44. G. A. Jansen, E. M. Hogenhout, S. Ferdinandusse, H. R. Waterham, R. Ofman, C. Jakobs, O. H. Skjeldal and R. J. A. Wanders, *Hum. Mol. Genet.*, 2000, **9**, 1195.
45. A. Chahal, M. Khan, S. G. Pai, E. Barbosa and I. Singh, *FEBS Lett.*, 1998, **429**, 119.
46. S. Ferdinandusse, A. W. Zomer, J. C. Komen, C. E. Van den brink, M. Thanos, F. P. Hamers, R. J. A. Wanders, P. T. Van Der Saag, B. T. Poll-Thé and P. M. Brites, *Proc. Natl. Acad. Sci. U. S. A.*, 2008, **105**, 17712.
47. M. Mukherji, W. Chien, N. J. Kershaw, I. J. Clifton, C. J. Schofield, A. S. Wierzbicki and M. D. Lloyd, *Hum. Mol. Genet.*, 2001, **10**, 1971.
48. G. A. Jansen, S. Ferdinandusse, E. M. Hogenhout, N. M. Verhoeven, C. Jakobs and R. J. A. Wanders, *Adv. Exp. Med. Biol.*, 1999, **466**, 371.
49. S. J. Mihalik, J. C. Morrell, D. Kim, K. A. Sacksteder, P. A. Watkins and S. J. Gould, *Nat. Genet.*, 1997, **17**, 185.
50. A. Kohlschutter, R. Santer, Z. Lukacs, C. Altenburg, M. J. Kemper and K. Ruther, *J. Child Neurol.*, 2012, **27**, 654.

CHAPTER 15

Role of 2-Oxoglutarate-Dependent Oxygenases in Flavonoid Metabolism

STEFAN MARTENS*[a] AND ULRICH MATERN[b]

[a]Department of Food Quality and Nutrition Department, IASMA Research and Innovation Centre, Fondazione Edmund Mach (FEM), Via E. Mach 1, 38010 San Michele all'Adige (TN), Italy; [b]Institut für Pharmazeutische Biologie und Biotechnologie, FB Pharmazie, Philipps-Universität Marburg, Deutschhausstrasse 17a, D-35032 Marburg, Germany
*E-mail: stefan.martens@fmach.it

15.1 Introduction

15.1.1 Physiological Significance and Use of Flavonoids

Flavonoids represent a group of secondary products that are ubiquitous in spermatophytic plants and which fulfil a multitude of physiological and ecological roles.[1,2] Their functions extend even to the nutritional and medicinal use of plant flavonoid extracts due to long-established beneficial properties, including inhibition of cell proliferation, antimutagenic, antimicrobial, anti-inflammatory and antihypertensive activities.[3,4] Some of these effects, at least, might be assigned to radical scavenging. Flavonoids also fulfill an important role in the propagation of the producing plant; the accumulation of flavonoids, as well as other secondary metabolites, was probably an essential prerequisite for the vegetal land invasion and adaptation to biotic and abiotic stresses. The numerous functions of flavonoids help plants to cope

RSC Metallobiology Series No. 3
2-Oxoglutarate-Dependent Oxygenases
Edited by Robert P. Hausinger and Christopher J. Schofield

Published by the Royal Society of Chemistry, www.rsc.org

with stress by shielding from UV radiation or modulating auxin transport and regulating fertility. The flavonoid-based pigmentation of flowers attracts pollinators and the inducing activity of flavonoids on nodulation genes in bacteria, as well as their role as phytoalexins in the defence of plants against pathogens, are considered essential factors in the survival of land-grown plants.[1,5,6] This proposal is supported by studies on the individual patterns of flavonoids accumulating in different plant species.

Flavonoids are subclassified into nine groups including flavanones, dihydroflavonols, flavan-3,4-ols, anthocyanidins, isoflavones, flavones, flavonols, flavan-3-ols and proanthocyanidins (PAs) according to the oxidation state, the substitution pattern of ring C (see Section 15.3.1.1) and the degree of oligomerization.[6] The overall structural diversity is reflected by the more than 10 000 flavonoids that have been structurally defined to date. The specific sets of flavonoids present in individual plants/plant species have probably evolved according to the physiological and ecological needs to meet adapting and conflicting demands to various environmental pressures under which plant communities grow.[2,7,8] The expansion in chemodiversity of flavonoid subgroups is achieved by enzymatic modifications of the flavonoid core structure, including hydroxylations, glycosylations, acylations, prenylations and methylations.[1]

15.2 Modes of Action of Oxygenases and Conserved Motifs

The branched pathway to flavonoids has been completely unravelled and the biochemistry and genetics of individual steps have been extensively investigated in many plant species and tissues. Accordingly, this pathway represents one of the best studied chapters in plant secondary metabolism (Figure 15.1).[1,7,9] The oxygenases participate primarily in the later stages of flavonoid biosynthesis and fall into two categories. One type is made of mixed-function, cytochrome P450-dependent monooxygenases (CYPs) that use one atom of molecular oxygen to oxidize their substrate, while the second atom ends up in water. CYPs are haem-thiolate membrane-associated proteins generally bound to the cytoplasmic surface of the endoplasmic reticulum and require NADPH for enzyme activity.[10] The second group includes dioxygenases that use both atoms of molecular oxygen for the oxidation of substrate(s). These enzymes are commonly soluble non-haem, iron-containing cytosolic dioxygenases that require ferrous iron as cofactor and, for example in the case of flavonoid biosynthesis, 2-oxoglutarate (2OG) as a cosubstrate for catalytic activity.[11] A particularly intriguing aspect is the oxygen- and iron-sensing roles ascribed to some 2OG-dependent dioxygenases (2ODDs) in animals (Chapter 6), which link with a Fe-sensing function for a 2ODD in *Arabidopsis thaliana*.[12] In the latter instance, Fe deficiency triggers the synthesis of two 2ODDs, one of which catalyses the *ortho*-hydroxylation of feruloyl-CoA that is pivotal to coumarin biosynthesis. These observations are compatible with

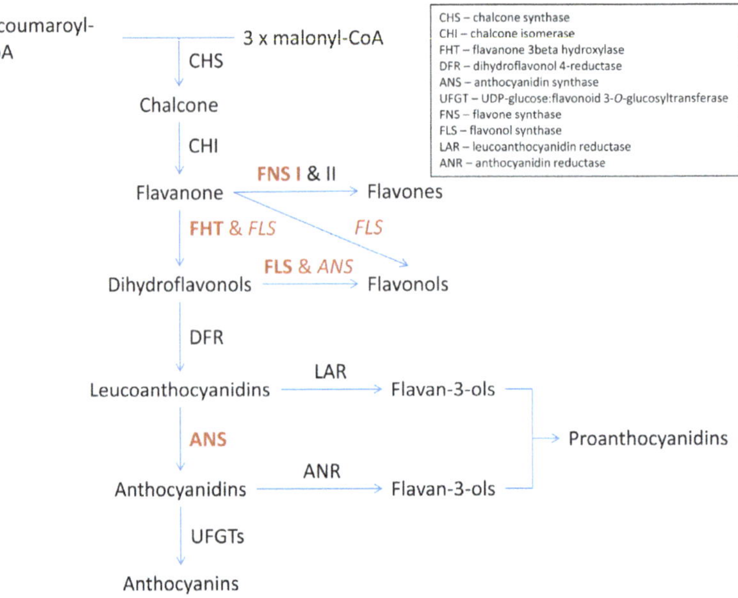

Figure 15.1 2-Oxoglutarate-dependent dioxygenases (2ODDs) involved in the main branches of flavonoid biosynthesis. 2ODDs are marked in red, main steps are shown in bold, and 2ODDs with side activities found *in vitro* and *in vivo* are in italics.

the proposal that within the 2ODD superfamily (E.C. 1.14.11.-), which is the second largest in the plant genome, the evolution and diversification of a particular subfamily is partly responsible for the diversity and complexity of specialized metabolites in land plants.[11]

The catalytic process of 2ODDs has been reviewed elsewhere in detail (see Chapter 3)[13–15] and only some major aspects are addressed here. These enzymes possess a core double-stranded β-helix, or a jelly-roll fold, which is composed of two sets of four antiparallel β-strands forming a distorted barrel-like structure that binds Fe(II) and 2OG on its inside (Chapter 2).[15,16] The cofactor Fe(II) is coordinated to a conserved His-x-Asp/Glu…His motif as well as the 2-oxo and carboxylate moieties of 2OG, which is held in position by electrostatic and hydrogen bonding of its C5 carboxylate to a Lys or Arg residue. Binding of substrate commonly follows that of 2OG and the ternary oxygenase:2OG:substrate complex then binds oxygen.[15] The decarboxylation of 2OG produces succinate and a reactive ferryl species [Fe(IV)=O ←→ Fe(III)–O•][17] that carries out the two-electron oxidation leading to hydroxylation, desaturation, epoxidation, demethylation or cyclization of the substrate.[13] Apparently, the flexibility of substrate and cosubstrate coordination to the catalytic Fe(II) is greater in 2ODDs as compared to CYPs where oxygen binds to one side of the haem ring, the other haem face is coordinated by a Cys ligand, and protein side chains interact with substrate. Furthermore, the topology of 2OG binding, possibly associated with an induced spatial adaption of the active site, affects the specificity of subsequent substrate binding. Overall, the more flexible coordination chemistry of 2ODDs is proposed to facilitate the evolution of new catalytic activities.[14]

15.3 2OG-Dependent Oxygenases Involved in Flavonoid Biosynthesis

Despite the fact that all 2ODDs associated with flavonoid biosynthesis depend on Fe(II) and 2OG as cofactor and cosubstrate, respectively, and use the electrons gained from decarboxylation of 2OG to generate succinate and the reactive enzyme-ferryl species, their substrate specificities and catalytic function differ considerably. Six 2ODDs involved in the flavonoid pathway have been identified (four are shown in Figure 15.1) and three of these occur widely distributed in spermatophytes: anthocyanidin synthase (ANS; synonymous with leucoanthocyanidin synthase, LDOX),[18] flavanone 3β-hydroxylase (FHT or F3H)[19] and flavonol synthase (FLS).[20] Flavone synthase I (FNS I), however, appears to be confined mainly to species of the Apiaceae family.[21,22] Two other 2ODDs, catalysing the 6-hydroxylation of partially methylated flavonols and demethylation of polymethoxylated flavones, were reported only recently from a single plant species.[23,24]

15.3.1 Hydroxylation – Hydroxylases

15.3.1.1 Hydroxylation of Ring C

The flavonoid pathway uses *p*-coumaroyl-CoA, provided by the general phenylpropanoid pathway, as a starter substrate which is condensed by chalcone synthase (CHS) with three units of malonyl-CoA to yield the intermediate tetraketide naringenin chalcone (Figure 15.1).[9] Intramolecular and stereospecific cyclization by Michael-type nucleophilic attack on the 2′-hydroxyl on the α,β-unsaturated double bond, catalysed by chalcone isomerase (CHI),[25] transforms the chalcone to the flavanone (2*S*)-naringenin containing a central γ-pyrone ring C (Figure 15.2). Flavanones are substrates of FHT/F3H (E.C. 1.14.11.9), which is the first 2ODD involved in the unbranched upstream sequence of reactions en route to flavonols/anthocyanidins and/or flavan-3-ols/PA biosynthesis (Figure 15.1). FHT stereospecifically catalyses the 3β-hydroxylation of (2*S*)-flavanones, such as naringenin (NAR) or eriodictyol (ERI) (Figure 15.2), to give the (2*R*,3*R*)-dihydroflavonols (DHFs) dihydrokaempferol (DHK) and dihydroquercetin (DHQ), respectively.[26,27] FHT activity was first demonstrated in crude extracts of *Matthiola incana* flower buds and illuminated parsley cells.[19,20] In *Arabidopsis* the gene encoding FHT was pinpointed to the *tt6* locus, and the loss-of-function mutant line does not accumulate wild-type amounts of flavonols (kaempferol, Km, or quercetin, Qu, and derivatives). Nevertheless, a recent careful analysis of 'leaky' mutant FHT phenotype alleles[28] revealed that the lack of FHT activity *in vivo* can be partially compensated by FLS and ANS, two 2ODDs required later in the flavonoid pathway (Figure 15.1; see Sections 15.3.2.2 and 15.3.3).

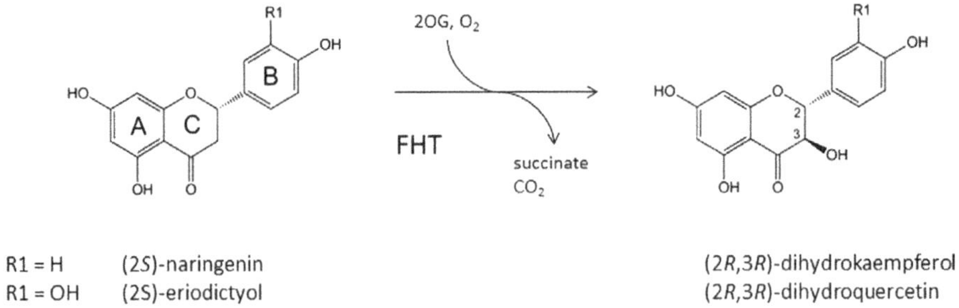
R1
OH
B
HO
O
A
C
OH
O
2OG, O_2
FHT
succinate
CO_2
R1
OH
HO
O
2
3
OH
OH
O
R1 = H (2*S*)-naringenin
R1 = OH (2S)-eriodictyol
(2*R*,3*R*)-dihydrokaempferol
(2*R*,3*R*)-dihydroquercetin

Figure 15.2 Reaction scheme for flavanone 3β-hydroxylase.

In coloured tissues the modification of anthocyanin accumulation can be easily tracked.[29] Antisense expression to block the expression of the gene encoding FHT in orange carnation (*Dianthus caryophyllus*) results in flower colour modifications up to the complete loss of anthocyanin pigmentation.[30] Furthermore, the effect of antisense or RNAi experiments can be mimicked by applying a structural analogue of 2OG, prohexadion-Ca, to the plant. This strategy revealed several uncommon flavonoids in treated apple leaves. The new compounds were assumed to be responsible, or at least involved in, the lower pathogen incidence of fire blight and other plant diseases. Inhibition of FHT activity, which is essential at an early stage of flavonoid biosynthesis, led to a rechannelling of intermediates from flavanones towards 3-deoxyflavonoids and to the accumulation of phenolic acids.[31]

15.3.1.2 *Hydroxylation of Ring A*

Hydroxylation of the aromatic ring A of flavanones at carbon 6 or 8 (Figure 15.3) has been reported, but these types of flavonoids have restricted occurrence.[9] Notably, both CYP and 2ODD enzymes are involved here. One CYP each was identified from soybean and basil with very narrow substrate specificities. The soybean (*Glycine max*) flavonoid 6-hydroxylase (F6H) was proposed to selectively catalyse the hydroxylation of 6-substituted isoflavonoids, whereas the enzyme from basil (*Ocimum basilicum*) was attributed to the formation of 6-substituted flavones.[24,32] In contrast, *Chrysosplenium americanum* expresses a 2ODD that was purified to near homogeneity and shown to catalyse the 6-hydroxylation of partially methylated flavonols. The monomeric enzyme exhibits strict specificity for carbon 6 of flavonols possessing a 7-methoxyl group.[23] Its cDNA was cloned and functionally expressed in *Escherichia coli*. The recombinant F6H showed substrate preference for 3,7,4′-trimethylquercetin as had been found also for the native enzyme (Figure 15.3) and is involved in biosynthesis of polymethoxylated flavonols in *Chrysosplenium*.[33]

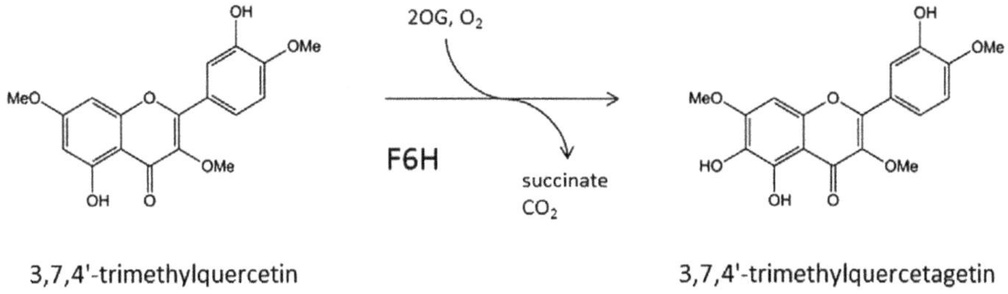

3,7,4'-trimethylquercetin → ($2OG$, O_2; F6H; succinate, CO_2) → 3,7,4'-trimethylquercetagetin

Figure 15.3 Reaction scheme for flavonoid 6-hydroxylase.

15.3.2 Desaturation – Desaturases

15.3.2.1 Desaturation of Flavanones

The oxidation of (2*S*)-NAR to apigenin (Ap) (Figure 15.4) was initially observed with extracts from irradiated parsley cell cultures and ascribed to a soluble FNS I (E.C. 1.14.11.22).[20,34] The mode of 2,3-desaturation appears to be extraordinary because the direct abstraction of the vicinal hydrogen atoms was postulated rather than a sequential hydroxylation and subsequent elimination of water. This assumption was based on the fact that *in vitro* synthetic 2-hydroxylated flavanone was not accepted as an intermediate.[21] Although FNS I was among the first enzymes of flavonoid biosynthesis to be characterized at the biochemical level, cloning of its cDNA and functional expression were not accomplished until two and a half decades later, using young parsley leaves.[35] Subsequently, the gene encoding FNS I was characterized from other Apiaceae species.[22] It should be noted, however, that another kind of enzyme, FNS II, also oxidizes flavanones to flavones in other plants.[36] FNS II was identified as a microsomal CYP class enzyme.[7] The capacity for catalysing one particular reaction by two very different classes of oxygenases resembles that observed for flavonoid 6-hydroxylation relying on a 2ODD in *Chrysosplenium* or a CYP in soybean and basil (see Section 15.3.1.2).

For a long time FNS I had been considered to be restricted solely to species of the Apiaceae. Meanwhile, however, a cDNA putatively encoding FNS I was cloned from rice and functionally expressed in *E. coli*.[37] Furthermore, in horsetail (*Equisetum arvense*, Equisetaceae) FNS activity was demonstrated to be dependent on 2OG, suggesting the involvement of a 2ODD.[38] Very recently a cDNA encoding a putative FNS I was isolated from the liverwort *Plagiochasma appendicalatum* (Aytoniaceae) and functionally characterized by expression in *E. coli*. The recombinant enzyme was found to have high FNS I activity and converted NAR to Ap, but 2-hydroxynaringenin was identified as a second product.[39] 2-Hydroxynaringenin was found to slowly dehydrate spontaneously to Ap. Nevertheless, the *Plagiochasma* FNS I was unable to catalyse this dehydration,[39] a result that is fully compatible with the previous mechanistic proposal.[21] Moreover, significant FLS side activity of this FNS I was detected by conversion of DHK to Km (Figures 15.1 and 15.5), whereas the

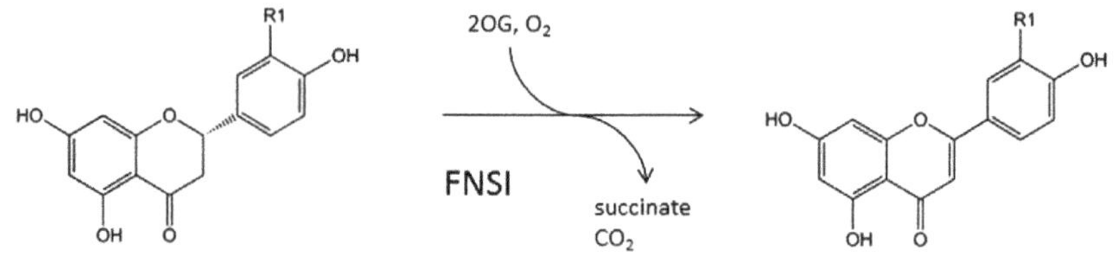

R1 = H (2*S*)-naringenin apigenin
R1 = OH (2*S*)-eriodictyol luteolin

Figure 15.4 Reaction scheme for flavone synthase I.

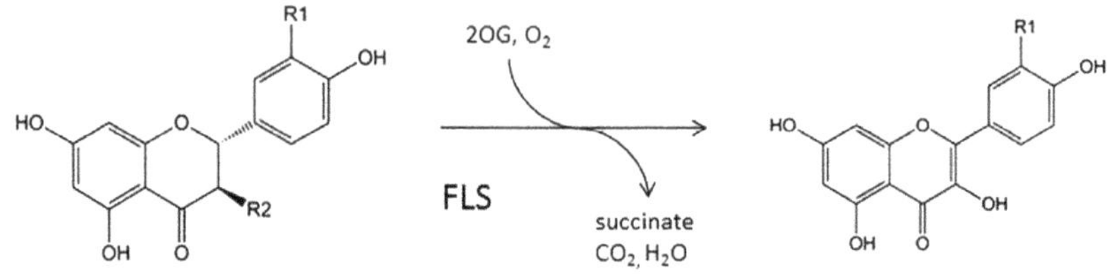

R1 = H; R2 = OH (2*R*,3*R*)-dihydrokaempferol kaempferol

R1 = OH; R2 = OH (2*R*,3*R*)-dihydroquercetin quercetin
R1 = OH; R2 = H (2*S*)-eriodictyol

Figure 15.5 Reaction scheme for flavonol synthase.

respective 3′-hydroxylated flavanone/flavanol ERI or DHQ were not accepted as substrates in these assays, indicating a hitherto unknown and relatively narrow substrate specificity. Homology modelling and mutational analysis attributed Tyr-240 (corresponding to Val-235 in ANS from *A. thaliana* or Pro-221 in FNS I from *Petroselinium crispum*) and Leu-311 (corresponding to Phe-304 in ANS or Phe-292 in FNS I from the same two sources), in close proximity to the active site (Figure 15.6), to be most relevant for flavanone 2-hydroxylation or FLS side activity. In particular, an efficient π-stacking interaction of Tyr-240 with the A-ring of NAR was assumed to be essential for 2-hydroxyflavanone formation, while the L311F variant increased the FLS side activity.[39]

The marginal structural changes in the FNS I sequence affecting the FLS side activity and the occurrence of genes encoding FNS I and observation of these activities in distant plant taxons suggest a close evolutionary relationship of these 2ODDs. Accordingly, recent molecular studies of Apiaceae 2ODDs strongly suggest the evolution of FNS I from FHT by gene duplication and gain of function,[40] in line also with the proposed flexibility of substrate and cosubstrate binding.[14] It is thus likely that FNS I arose more than once during the evolution of land plants from recruitment of a suitable 2ODD.

```
              ---------+---------+---------+---------+---------+---------+---------+---------+
                      10        20        30        40        50        60        70        80
Ara_tha_ANS   MVAVERVESLAKSGIIS--IPKEYIRPKEELESINDVFLEEKKEDGPQVPTIDLKNIESDDEKIRENCIEELKKASLDWG    78
Ara_tha_FLS1  -MEVERVQDISSSSLLTEAIPLEFIRSEKEQPAIT-----TFRGPTPAIPVVDLS--DPDEESVRR---AVVK-ASEEWG    68
Pet_cri_FNSI  -MAPTTITALAKEKTLN----LDFVRDEDERPKVA------YNQFSNEIPIISLAGLDDDSDGRRPEICRKIVKACEDWG    69
Ara_tha_FHT   -MAPGTLTELAGESKLN----SKFVRDEDERPKVA------YNVFSDEIPVISLAGIDD-VDGKRGEICRQIVEACENWG    68

              ---------+---------+---------+---------+---------+---------+---------+---------+
                      90       100       110       120       130       140       150       160
Ara_tha_ANS   VMHLINHGIPADLMERVKKAGEEFFSLSVEEKEKYANDQATGKIQGYGSKLANNASGQLEWEDYFFHLAYPEEKRDLSIW   158
Ara_tha_FLS1  LFQVVNHGIPTELIRRLQDVGRKFFELPSSEKESVAKPEDSKDIEGYGTKLQKDPEGKKAWVDHLFHRIWPPSCVNYRFW   148
Pet_cri_FNSI  IFQVVDHGIDSGLISEMTRLSREFFALPAEEK--LEYDTTGGKRGGFTISTVLQGDDAMDWREFVTYFSYPINARDYSRW   147
Ara_tha_FHT   IFQVVDHGVDTNLVADMTRLARDFFALPPEDK--LRFDMSGGKKGGFIVSSHLQGEAVQDWREIVTYFSYPVRNRDYSRW   146

              ---------+---------+---------+---------+---------+---------+---------+---------+
                     170       180       190       200       210       220       230   ↓   240
Ara_tha_ANS   PKTPSDYIEATSEYAKCLRLLATKVFKALSVGLGLEPDRLEKEVGGLEELLLQMKINYYPKCPQPELALGVEAHTDVSAL   238
Ara_tha_FLS1  PKNPPEYREVNEEYAVHVKKLSETLLGILSDGLGLKRDALKEGLGG-EMAEYMMKINYYPPCPRPDLALGVPAHTDLSGI   227
Pet_cri_FNSI  PKKPEGWRSTTEVYSEKLMVLGAKLLEVLSEAMGLEKGDLTKACVD---MEQKVLINYYPTCPQPDLTLGVRRHTDPGTI   224
Ara_tha_FHT   PNKPEGWVKVTEEYSERLMSLACKLLEVLSEAMGLEKESLTNACVD---MDQKIVVNYYPKCPQPDLTLGLKRHTDPGTI   223

              ---------+---------+---------+---------+---------+---------+---------+---------+
                     250       260       270       280       290       300   ↓   310       320
Ara_tha_ANS   TFILHNMVPGLQLFYEG--KWVTAKCVPDSIVMHIGDTLEILSNGKYKSILHRGLVNKEKVRISWAVFCEPPKDKIVLKP   316
Ara_tha_FLS1  TLLVPNEVPGLQVFKDD--HWFDAEYIPSAVIVHIGDQILRLSNGRYKNVLHRTTVDKEKTRMSWPVFLEPPREKIVG-P   304
Pet_cri_FNSI  TILLQDMVGGLQATRDGGKTWITVQPVEGAFVVNLGDHGHYLSNGRFRNADHQAVVNSTSSRLSIATFQNPAQNAIVY-P   303
Ara_tha_FHT   TLLLQDQVGGLQATRDNGKTWITVQPVEGAFVVNLGDHGHFLSNGRFKNADHQAVVNSNSSRLSIATFQNPAPDATVY-P   302

              ---------+---------+---------+---------+---------+---------+---
                     330       340       350       360       370       380
Ara_tha_ANS   LPEMVSV---ESPAKFPPRTFAQHIEHKLFGKEQEELVSEKND.                                        357
Ara_tha_FLS1  LPELTGD---DNPPKFKPFAFKDYSYRKLNKLPLD.                                                337
Pet_cri_FNSI  LKIREGEKAILDEAITYAEMYKKCMTKHIEVATRKKLAKEKRLQDEKAKLEMKSKSADENLA.                     366
Ara_tha_FHT   LKVREGEKAILEEPITFAEMYKRKMGRDLELARLKKLAKEERDHKEVA------KPVDQIFA                      358
```

Figure 15.6 Alignment of 2ODDs involved in flavonoid biosynthesis showing conserved residues. Functionally relevant and conserved amino acids are marked in bold, iron and 2OG binding sites are underlined, and arrows indicate the sites of mutagenesis in *Plagiochasma appendicalatum*.[39]

15.3.2.2 Desaturation of Dihydroflavonols

The formation of flavonols from flavanones or DHFs by the activity of FLS (E.C. 1.14.11.23) (Figure 15.1) might be considered as a branch from the major flavonoid pathway to anthocyanidins and PAs, which relies on the NADPH-dependent reduction of DHFs to leucoanthocyanidins by the action of dihydroflavonol 4-reductase (DFR) (Figure 15.1).[6] Flavonols are found as pigments in some yellow flowers or affect the coloration as co-pigments of anthocyanins. They are essential for pollen tube growth in some plants and shield the tissues from UV irradiation and oxidative damage.[1,41,42]

FLS catalyses *in planta* the carbon 2/3 desaturation of genuine (2*R*,3*R*)-DHFs, *i.e.* DHK or DHQ, to the flavonols Km or Qu (Figure 15.5). This activity was first described with enzyme extracts from irradiated parsley cells and later from various other plant species.[6,20] FLS-encoding cDNA was initially cloned from *Petunia hybrida*, where the locus *Fl* was found to control flavonol synthesis in flowers, and greatly reduced flavonol contents were reported for *flfl* mutant lines.[41] The impact of flavonols as co-pigments on coloration was shown in various plants, such as soybean, petunia or tobacco. A single deletion in the soybean gene led to a truncated, inactive FLS protein and substantially reduced flavonol contents in petals; the colour of these flowers changed from purple to magenta, while the anthocyanin content remained unchanged.[42] Altered flower coloration was also found in transgenic antisense petunia plants, which was assigned to a co-pigmentation effect producing deeper red flowers than a non-transgenic control. However, in tobacco the same approach resulted in the production of red instead of light pink flowers, which is due to a decrease in flavonols and correlates with increases in anthocyanin production.[41]

Recombinantly produced FLSs from *Arabidopsis* (AtFLS1) or *Citrus unshiu* surprisingly converted *in vitro* (2*S*)-NAR to Km, thus exhibiting both FHT and FLS activities (Figure 15.1). Trace amounts of DHK detected in these incubations suggest that a sequential formation of DHK and Km (Figure 15.1) had occurred. Control assays employing the unnatural substrate (2*R*)-NAR, however, revealed DHK as the only product which was identified as the (−)-trans isomer.[43,44] Meanwhile, such FHT side activity of FLS was demonstrated also for the FLS activities from other plant species and raised the possibility of an alternative access to flavonol synthesis *in vivo*, *i.e.* in *FHT* mutants.[45–47]

The *Arabidopsis* genome contains five genes with high sequence similarity to *AtFLS1*.[48] AtFLS1 predominantly determines the flavonol levels in *Arabidopsis*, and the residual four FLS-like polypeptides initially were considered non-functional. However, the *Atfls1* mutant still accumulates significant amounts of Km and Qu derivatives and AtFLS3 was identified later as a second active FLS.[28,47,49]

15.3.2.3 Oxidation to Anthocyanidin

Anthocyanins are most conspicuous due to their characteristic red, blue or purple colour decorating and distinguishing plants, but also serve essential functions in plant reproduction by recruiting pollinators and seed dispersal.[1,5] The *in planta* function was proven by complementation of mutants, such as maize A2 or *Arabidopsis* tds4, with homologous and heterologous genes and restoring anthocyanidin or PA production in aleurone or seeds, respectively.[18,50] Anthocyanidin biosynthesis requires colourless (2*R*,3*S*,4*S*)-*cis*-leucoanthocyanidins (flavan-3,4-*cis*-diols) as substrates, which are oxidized to coloured anthocyanidins by the action of ANS/LDOX) (E.C. 1.14.11.19) (Figure 15.1). Although several cDNAs and genes encoding putative ANS proteins have been isolated from various plant species coincidently with FHT and FLS genes,[6] the demonstration of *in vitro* activity was first accomplished with *Perilla frutescens* extracts using leucoanthocyanidin as substrate and followed by acid treatment.[51] Later, the activity was also reported from *Spinacia oleracea*, *Phytolacca americana* and *Gerbera* hybrids.[52,53]

The conversion of leucoanthocyanidin to anthocyanidin most likely proceeds by initial C3 oxidation (*cis*-hydroxylation) followed by dehydration of the geminal diol to give the C2/C3 enol, which on further dehydration releases the anthocyanidin (Figure 15.7).[54] Subsequently, the aromatic C nucleus is stabilized through glucosylation at the 3-hydroxyl by the common UDP-glucose:flavonoid 3-O-glucosyltransferase (Figure 15.1). In *Arabidopsis*, as in most other plants, ANS is encoded by a single gene and several respective mutants (tt11, tt17, tt18, tt19 and tds4) have been described.[55,56] Although ANS was designated to catalyse *in situ* the conversion of leucoanthocyanidin to anthocyanidin, *Arabidopsis* ANS was the first flavonoid-committed 2ODD proven *in vitro* as a non-specific enzyme exhibiting typical FLS activities, *i.e.* the oxidation of the natural FLS substrates *trans*-DHFs to flavonols (Figure 15.1) or the formation of *cis*-DHF from (2*R*)-NAR.[17,45,57,58] Furthermore, leucoanthocyanidins, the natural substrates of ANS, were converted *in vitro* predominantly to the respective *cis*-DHFs or flavonols and only to a minor extent to the anticipated anthocyanidins.[17,58] This may be explained by the proposed catalytic mechanism, because dehydration of the initial oxidation product from leucoanthocyanidin (C3 geminal diol) can lead also to a C3/C4 enol, which on further dehydration releases DHF. Moreover, leucoanthocyanidins

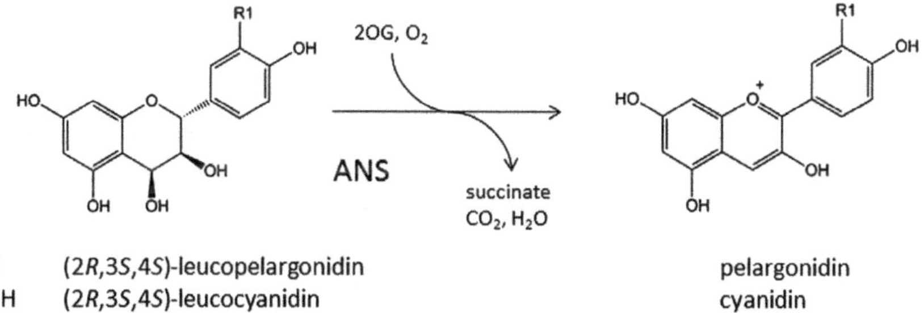

Figure 15.7 Reaction scheme for anthocyanidin synthase.

epimerize readily at C4, such that both *trans*- and *cis*-epimers are present in enzyme assays. Both C2/C3-enols and C3/C4-enols probably arise by favoured antiperiplanar *trans*-elimination of water, and thus it was suggested that *trans*-leucoanthocyanidins may preferentially lead to C3/C4-enols and DHFs, whereas *cis*-leucoanthocyanidins more readily end up in the C2/C3-enol and leucoanthocyanidins.[54]

The ambiguous role of ANS activity was recently documented also *in planta* with *Arabidopsis fls1* mutant lines. These lines still accumulate about 5% flavonols as compared to the wild-type plants, whereas only trace amounts of flavonols could be detected in the *Arabidopsis* double mutant lines *fls*/*ans*. These results provide strong evidence for FLS activity of ANS *in planta*, which may contribute a significant proportion of the flavonol contents in *Arabidopsis* and possibly also other plants.[47,59] The bi-/multifunctionality of ANS was further confirmed by *in vitro* studies employing (+)-catechin as a potential substrate and recombinant ANS from *Gerbera* hybrids.[53] This compound constitutes an essential monomer in the composition of PAs and lacks a 3-hydroxyl. Nevertheless, (+)-catechin was converted in these assays to an unnatural 4-4 dimer, cyanidin and traces of Qu.

15.3.3 Demethylation – Demethylases

Demethylation reactions appear to be common in plant secondary metabolism, but solid biochemical or genetic studies are scarce. In a few instances, such as nicotine and codamine biosynthesis, *N*- or *O*-demethylations were reported to be catalysed by CYP monooxygenases.[24] However, in the course of biosynthesis of the 6-substituted flavone nevadensin in basil the 7-*O*-demethylation of gardenin B (Figure 15.8) was recently described and ascribed to a 2ODD activity. The activity was documented with crude enzyme extracts of trichomes in the presence of the essential cofactors (Fe(II) and 2OG) and represents the first example of a demethylation 2ODD in this pathway, while P450-directed assays with membrane fractions in the presence of NADPH were unsuccessful.[24] Nevertheless, the corresponding gene coding for a flavonoid 7-*O*-demethylase has not yet been identified.

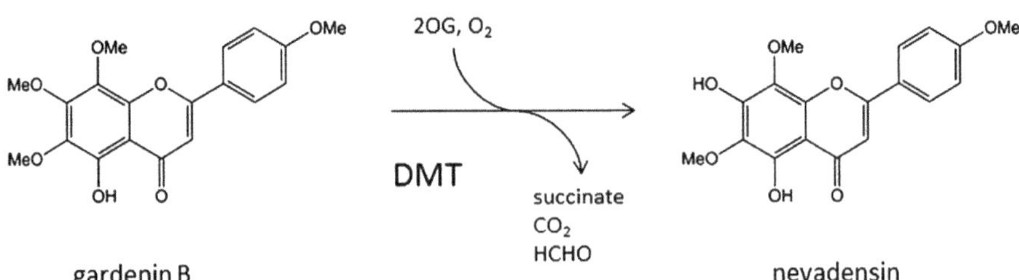

Figure 15.8 Reaction scheme for flavone demethylase.

15.4 Genetic Identification of Residues Relevant for ANS Activity

2ODDs constitute a large superfamily of oxygenases showing quite low overall sequence similarity. Nevertheless, the polypeptide sequence of the catalytic core in the carboxy terminus and the spatial tertiary structure are conserved.[11,60] The sandwich structure of the double-stranded β-helix (Chapter 2)[15,16] contains and protects the essential Fe(II)- and 2OG-binding sites,[14] which require at least two conserved His and one Asp residues (His-232–X–Asp-234–X_n–His-288 in AtANS) for Fe(II) coordination (Figure 15.9)[14,61] as well as a tripeptide (Arg-298–X–Ser-300 in AtANS, eight to ten residues downstream of the third conserved His) for C5-carboxyl binding of 2OG (Figure 15.9).[26]

Point mutations of codons for either His or the Asp residue in AtFLS1 completely abolished the activity, corroborating the importance of correct coordination of Fe(II) to a 2-His-1-carboxylate catalytic triad.[62] Site-directed mutagenesis was also employed to examine the 2OG binding site. While substitution of the Arg residue completely ablated the activity of AtFLS1, approximately 50% of the wild-type activity was retained after altering the Ser residue.[26] This demonstrates that Ser-300 is not absolutely essential for AtANS activity, but might be required for sustaining high enzyme activity.[62] Further variants focused on residues that may severely affect polypeptide folding. Two strictly conserved Pro residues in motifs 2 (Pro-162 AtANS) and 3 (Pro-218 AtANS) as well as Gly (Gly-78 AtANS) and His residues (His-85 AtANS) in motif 1 (Figure 15.9) were exchanged and shown to modulate rather than block the ANS activity.

15.5 Evolution

Plants recruit a multitude of activities from enzyme superfamilies to produce the numerous specialized metabolites needed for establishing themselves in a highly competitive or even hostile environment. Many of these metabolites probably serve more than one function, *e.g.* flavonoids support pollination by attracting insects, but also possess ultraviolet protecting and radical

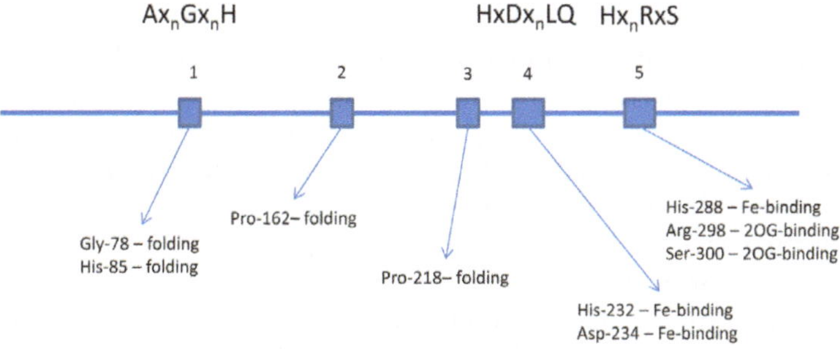

Figure 15.9 Functionally important 2ODD structural motifs.

scavenging qualities. The increasing insight into the enzymology of specialized metabolite formation gained in recent years has also brought into focus the role of 2ODDs. While the biochemical significance of these oxygenases in secondary plant metabolism is not in question, the phylogeny of the 2ODD superfamily of genes remained unclear for a long time, and a first comprehensive genome-wide analysis of six model plants was published only recently.[11,63] The number of putative 2ODDs (>100) in the genomes of *Arabidopsis* and *Oryza* is equivalent to the number of CYPs and plant family 1 glycosyltransferases in these species.[10,64] Furthermore, it should be noted that the flexible coordination chemistry of 2ODDs might favour the evolution of new catalytic activities.[14] The diversification of plant 2ODD genes was suggested, therefore, to contribute significantly to the structural variety of specialized metabolites through oxidative modifications in concert with CYP activities and other enzyme classes.[11] A high degree of identity had been observed early on when cloning the first genes encoding 2ODDs, and this observation led to the assumption of their evolution from a common ancestral dioxygenase. The bi- or multifunctionality of bacterial and fungal 2ODDs, *e.g.* as in clavulanic acid biosynthesis, further supported the idea that new substrate specificities of 2ODDs might evolve easily. This assumption is compatible with the results derived later from plants for flavonoid 2ODDs. The process of evolution of these genes may be explained by tandem gene duplication followed by mutation, especially in the coding region, generating two genes with new but related specificities.[14,40,65]

Polypeptide alignments and phylogenetic analyses ranked the plant 2ODD superfamily into three large clusters, designated as classes DOXA (9 clades), DOXB (7 clades) and DOXC (57 clades).[11] Most of the 2ODDs from land plants belong to the latter class, while just a few were reported from *Chlamydomonas*, and the numbers of 2ODDs assigned to the DOXC class differ considerably across the plant species. This led to the suggestion that the DOXC class of 2ODDs is partly responsible for the diversity of specialized metabolites in land plants.[11] Gene expansion thus occurred after the split from the common ancestor of land plants and was followed by large-scale duplication of these 2ODD-encoding genes. The diversification of genes within the DOXC class may be considered as a consequence of segmented or whole genome duplication which is common to other plant genes.

Flavonoid-committed 2ODDs have been assigned to the two phylogenetically distant clades DOXC28 and DOXC47.[11] Clade DOXC28 includes FHTs from various angiosperm and gymnosperm plants, but also FNS I, which has so far been reported only from some Apiaceae species or rice[6] and more recently from the liverwort *P. appendicalatum*.[39] Nevertheless, lower plant genomes, such as *Selaginella* or *Physcomitrella*, generally do not contain FHT orthologues, although DHF-derived flavonoids were observed.[66] The inconsistency suggests that in these plants a 2ODD of a different clade or a non-2ODD enzyme or multifunctional enzyme, such as ANS, might be involved in DHF formation.[11,47] The distinct clade DOXC47 accommodates both ANS and FLS. These enzymes are known to hydroxylate flavanol or leucoanthocyanidin at carbon-3 in the α- orientation, which is in contrast to the C-3β hydroxylation catalysed by FHT[11] and the substrate specificity of FNSI for flavanon- or dihydroflavon-3β-ol.

The low substrate specificity of ANS and FLS enzymes characterized so far may reflect an incomplete evolution, which does not exclude the existence of isoforms with different substrate specificities in a given plant species.[44,58] Overall, another mode of subclassification of flavonoid-2ODDs might be feasible by taking into account the stringency of FHT and FNS I for (2*S*)-NAR as substrate. The category of 2ODDs with a narrow substrate specificity (FHT, FNS I, F6H) clearly differs from that of the group with a broad substrate specificity (FLS, ANS).[6] This kind of subclassification is in full agreement with the phylogenetic results described above.[11]

15.6 Concluding Remarks

For a long time the 2ODDs of the highly branched flavonoid pathway were considered to exhibit narrow substrate and product (stereo)specificities. This assumption, however, had to be revised during the last decade, because at least the 'late' enzymes, FLS and ANS, have been shown to catalyse more than one reaction both *in vitro* and *in vivo*.[6,43,44] Bi- or even multifunctional enzymes are probably essential features of the complex metabolic grid of plant-specialized metabolites. Multifunctionality may also be a prerequisite to readily develop related genes with new functions including by derivatization of existing metabolites, which might provide an advantage for habitat adaption.[67] The recent *in vitro* and *in vivo* studies seem to support such a scenario for the complex involvement of 2ODDs in flavonoid biosynthesis. The use of sophisticated analytical equipment has documented the formation of trace amounts of apparent 'side' products. Taking the promiscuity of 2ODDs into consideration may be important when performing metabolic engineering experiments with the aim of modulating single steps of the flavonoid pathway and/or flavonoid patterns in plants or microorganisms. Side activities should be monitored to control an undesired flux of intermediates in the transgenic organism. The latter approach provides in the long run an advantage to the *in vitro* use of the rather labile oxygenases for an extended period of time.

Acknowledgements

One author (S.M.) gratefully acknowledges the contribution by the ADP 2011–14 project funded by the Autonomous Province of Trento.

References

1. V. Cheynier, G. Comte, K. M. Davies, V. Lattanzio and S. Martens, *Plant Physiol. Biochem.*, 2013, **72**, 1.
2. K. Saito, K. Yonekura-Sakakibara, R. Nakabayashi, Y. Higashi, M. Yamazaki, T. Tohge and A. R. Fernie, *Plant Physiol. Biochem.*, 2013, **72**, 21.

3. D. Vauzour, A. Rodriguez-Mateos, G. Corona, M. J. Oruna-Concha and J. P. E. Spencer, *Nutrients*, 2010, **2**, 1106.
4. D. Del Rio, A. Rodriguez-Mateos, J. P. E. Spencer, M. Tognolini, G. Borges and A. Crozier, *Antioxid. Redox Signaling*, 2013, **18**, 1818.
5. B. Winkel-Shirley, *Plant Physiol.*, 2001, **126**, 485.
6. S. Martens, A. Preuss and U. Matern, *Phytochemistry*, 2010, **71**, 1040.
7. S. Martens and A. Mithöfer, *Phytochemistry*, 2005, **66**, 2399.
8. N. C. Veitch and R. J. Grayer, *Nat. Prod. Rep.*, 2011, **28**, 1626.
9. K. M. Davies, K. E. Schwinn, *Flavonoids – Chemistry, Biochemistry and Applications*, ed. O.M. Andersen and K.R. Markham, Taylor & Francis, Boca Raton, 2006, 143–218.
10. D. Nelson and D. Werck-Reichhart, *Plant J.*, 2011, **66**, 194.
11. Y. Kawai, E. Ono and M. Mizutani, *Plant J.*, 2014, **78**, 328.
12. G. Vigani, P. Morandini and I. Murgia, *Front. Plant Sci.*, 2013, **4**, 169.
13. T. D. Bugg, *Tetrahedron*, 2003, **59**, 7075.
14. M. Mantri, Z. Zhang, M. A. McDonough and C. J. Schofield, *FEBS J.*, 2012, **279**, 1563.
15. W. Aik, M. A. McDonough, A. Thalhammer, R. Chowdhury and C. J. Schofield, *Curr. Opin. Struct. Biol.*, 2012, **22**, 691.
16. I. J. Clifton, M. A. McDonough, D. Ehrismann, N. J. Kershaw, N. Granatino and C. J. Schofield, *J. Inorg. Biochem.*, 2006, **100**, 644.
17. J. J. Turnbull, M. J. Nagle, J. F. Seibel, R. W. Welford, G. H. Grant and C. J. Schofield, *Bioorg. Med. Chem. Lett.*, 2003, **13**, 3853.
18. A. Menssen, S. Höhmann, W. Martin, P. S. Schnable, P. A. Peterson, H. Saedler and A. Gierl, *EMBO J.*, 1990, **9**, 3051.
19. G. Forkmann, *Planta*, 1980, **148**, 157.
20. L. Britsch, W. Heller and H. Grisebach, *Z. Naturforsch.*, 1981, **36C**, 742.
21. L. Britsch, *Arch. Biochem. Biophys.*, 1990, **282**, 152.
22. Y. Gebhardt, S. Witte, G. Forkmann, R. Lukacin, U. Matern and S. Martens, *Phytochemistry*, 2005, **66**, 1273.
23. D. Anzellotti and R. K. Ibrahim, *Arch. Biochem. Biophys.*, 2000, **382**, 161.
24. A. Berim and D. R. Gang, *J. Biol. Chem.*, 2013, **288**, 1795.
25. J. M. Jez, M. E. Bowman, R. A. Dixon and J. P. Noel, *Nat. Struct. Biol.*, 2000, **7**, 786.
26. R. Lukacin, I. Gröning, U. Pieper and U. Matern, *Eur. J. Biochem.*, 2000, **267**, 853.
27. R. Lukacin, C. Urbanke, I. Gröning and U. Matern, *FEBS Lett.*, 2000, **467**, 353.
28. D. K. Owens, K. C. Crosby, J. Runac, B. A. Howard and B. S. J. Winkel, *Plant Physiol. Biochem.*, 2008, **46**, 833.
29. S. Martens, J. Knott, C. A. Seitz, L. Janvari, S.-N. Yu and G. Forkmann, *Biochem. Eng. J.*, 2003, **14**, 227.
30. A. Zuker, T. Tzfira, H. Ben-Meir, M. Ovadis, E. Shklarman, H. Itzhaki, G. Forkmann, S. Martens, I. Neta-Sharir, D. Weiss and A. Vainstein, *Mol. Breed.*, 2002, **9**, 33.

31. S. Roemmelt, N. Zimmermann, W. Rademacher and D. Treutter, *Phytochemistry*, 2003, **64**, 709.
32. A. O. Latunde-Dada, F. Cabello-Hurtado, N. Czittrich, L. Didierjean, C. Schopfer, N. Hertkorn, D. Werck-Reichhart and J. Ebel, *J. Biol. Chem.*, 2001, **276**, 1688.
33. D. Anzellotti and R. K. Ibrahim, *BMC Plant Biol.*, 2004, **4**, 20.
34. A. Sutter, J. Poulton and H. Grisebach, *Arch. Biochem. Biophys.*, 1975, **170**, 547.
35. S. Martens, G. Forkmann, U. Matern and R. Lukačin, *Phytochemistry*, 2001, **58**, 43.
36. G. Kochs, R. Welle and H. Grisebach, *Planta*, 1987, **171**, 519.
37. Y. J. Lee, J. H. Kim, B. G. Kim, Y. Lim and J. H. Ahn, *BMB Rep.*, 2008, **41**, 68.
38. M. Bredebach, U. Matern and S. Martens, *Phytochemistry*, 2011, **72**, 557.
39. X. J. Han, Y. F. Wu, S. Gao, H. N. Yu, R. X. Xu, H. X. Lou and A. X. Cheng, *FEBS Lett.*, 2014, **588**, 2307.
40. Y. H. Gebhardt, S. Witte, H. Steuber, U. Matern and S. Martens, *Plant Physiol.*, 2007, **144**, 1442.
41. T. A. Holton, F. Brugliera and Y. Tanaka, *Plant J.*, 1993, **4**, 1003.
42. R. Takahashi, S. M. Githiri, K. Hatayama, E. G. Dubouzet, N. Shimada, T. Aoki, S. Ayabe, T. Iwashina, K. Toda and H. Matsumura, *Plant Mol. Biol.*, 2007, **63**, 125.
43. A. G. Prescott, N. P. Stamford, G. Wheeler and J. L. Firmin, *Phytochemistry*, 2002, **60**, 589.
44. R. Lukacin, F. Wellmann, L. Britsch, S. Martens and U. Matern, *Phytochemistry*, 2003, **62**, 287.
45. J. J. Turnbull, J. Nakajima, R. W. Welford, M. Yamazaki, K. Saito and C. J. Schofield, *J. Biol. Chem.*, 2004, **279**, 1206.
46. J. R. Almeida, E. D'Amico, A. Preuss, F. Carbone, C. H. de Vos, B. Deiml, F. Mourgues, G. Perrotta, T. C. Fischer, A. G. Bovy, S. Martens and C. Rosati, *Arch. Biochem. Biophys.*, 2007, **465**, 61.
47. A. Preuss, R. Stracke, B. Weisshaar, A. Hillebrecht, U. Matern and S. Martens, *FEBS Lett.*, 2009, **583**, 1981.
48. M. K. Pelletier, J. R. Murrell and B. W. Shirley, *Plant Physiol.*, 1997, **113**, 1437.
49. D. K. Owens, A. B. Alerding, K. C. Crosby, A. B. Bandara, J. H. Westwood and B. S. J. Winkel, *Plant Physiol.*, 2008, **147**, 1046.
50. A. M. Reddy, V. S. Reddy, B. E. Scheffler, U. Wienand and A. R. Reddy, *Metab. Eng.*, 2007, **9**, 95.
51. K. Saito, M. Kobayashi, Z. Gong, Y. Tanaka and M. Yamazaki, *Plant J.*, 1999, **17**, 181.
52. S. Shimada, Y. T. Inoue and M. Sakuta, *Plant J.*, 2005, **44**, 950.
53. F. Wellmann, M. Griesser, W. Schwab, S. Martens, W. Eisenreich, U. Matern and R. Lukacin, *FEBS Lett.*, 2006, **580**, 1642.
54. R. C. Wilmouth, J. J. Turnbull, R. W. D. Welford, I. J. Clifton, A. G. Prescott and C. J. Schofield, *Structure*, 2002, **10**, 93.
55. S. Abrahams, E. Lee, A. R. Walker, G. J. Tanner, P. J. Larkin and A. R. Ashton, *Plant J.*, 2003, **35**, 624.

56. I. Appelhagen, O. Jahns, L. Bartelniewoehner, M. Sagasser, B. Weisshaar and R. Stracke, *Gene*, 2011, **484**, 61.
57. R. W. Welford, J. J. Turnbull, T. D. Claridge, A. G. Prescott and C. J. Schofield, *Chem. Commun.*, 2001, **18**, 1828.
58. J. J. Turnbull, W. J. Sobey, R. T. Aplin, A. Hassan, J. L. Firmin, C. J. Schofield and A. G. Prescott, *Chem. Commun.*, 2000, **24**, 2473.
59. R. Stracke, R. C. De Vos, L. Bartelniewoehner, H. Ishihara, M. Sagasser, S. Martens and B. Weisshaar, *Planta*, 2009, **229**, 427.
60. M. Punta, P. C. Coggill, R. Y. Eberhardt, J. Mistry, J. Tate, C. Boursnell, N. Pang, K. Forslund, G. Ceric, J. Clements, A. Heger, L. Holm, E. L. L. Sonnhammer, S. R. Eddy, A. Bateman and R. D. Finn, *Nucl. Acids Res.*, 2012, **40**, D290.
61. E. L. Hegg and L. Que, Jr., *Eur. J. Biochem.*, 1997, **250**, 625.
62. C. S. Chua, D. Biermann, K. S. Goo and T. S. Sim, *Phytochemistry*, 2008, **69**, 66.
63. R. P. Hausinger, *Crit. Rev. Biochem. Mol. Biol.*, 2004, **39**, 21.
64. L. Caputi, M. Malnoy, V. Goremykin, S. Nikiforova and S. Martens, *Plant J.*, 2012, **69**, 1030.
65. E. DeCarolis and V. de Luca, *Phytochemistry*, 1994, **36**, 1093.
66. H. A. Stafford, *Plant Physiol.*, 1999, **96**, 680.
67. W. Schwab, *Phytochemistry*, 2003, **62**, 837.

CHAPTER 16

Gibberellin Metabolism

PETER HEDDEN*[a] AND ANDREW L. PHILLIPS[a]

[a]Rothamsted Research, Harpenden, Hertfordshire, AL5 2JQ, UK
*E-mail: peter.hedden@rothamsted.ac.uk

16.1 Introduction

The gibberellins (GAs) comprise a large group of diterpenoid tetracyclic carboxylic acids, some members of which function as growth hormones in higher plants. They are also produced in lower vascular plants, where they have a role in sporulation, but not growth,[1] as well as in some species of fungi and bacteria, in which they may function to modify the behaviour of their plant hosts.[2] The biosynthetic pathways to the biologically active compounds in higher plants are summarized in Scheme 16.1. These compounds are formed from the hydrocarbon *ent*-kaurene, which is oxidized initially by cytochrome P450 monooxygenases to GA_{12} and GA_{53}, which are then further converted to the biologically active hormones GA_4 and GA_1, respectively, by 2-oxoglutarate-dependent dioxygenases (ODDs). These conversions are catalysed by two families of ODDs, the GA 20-oxidases (GA20ox), which are responsible for the removal of C-20 with formation of a C-4 to C-10 lactone, and the GA 3-oxidases (GA3ox), which introduce a 3β-hydroxyl group. Further families of ODDs, known as GA 2-oxidases (GA2ox), introduce a 2β-hydroxyl group as an inactivating reaction and in some cases oxidize C-2 further to the ketone (Scheme 16.2).

The first indication that ODDs are involved in GA biosynthesis was obtained from studies with cell-free homogenates of *Cucurbita maxima* (pumpkin) endosperm.[3] In this system the conversion of the precursor GA_{12}-aldehyde

RSC Metallobiology Series No. 3
2-Oxoglutarate-Dependent Oxygenases
Edited by Robert P. Hausinger and Christopher J. Schofield

Published by the Royal Society of Chemistry, www.rsc.org

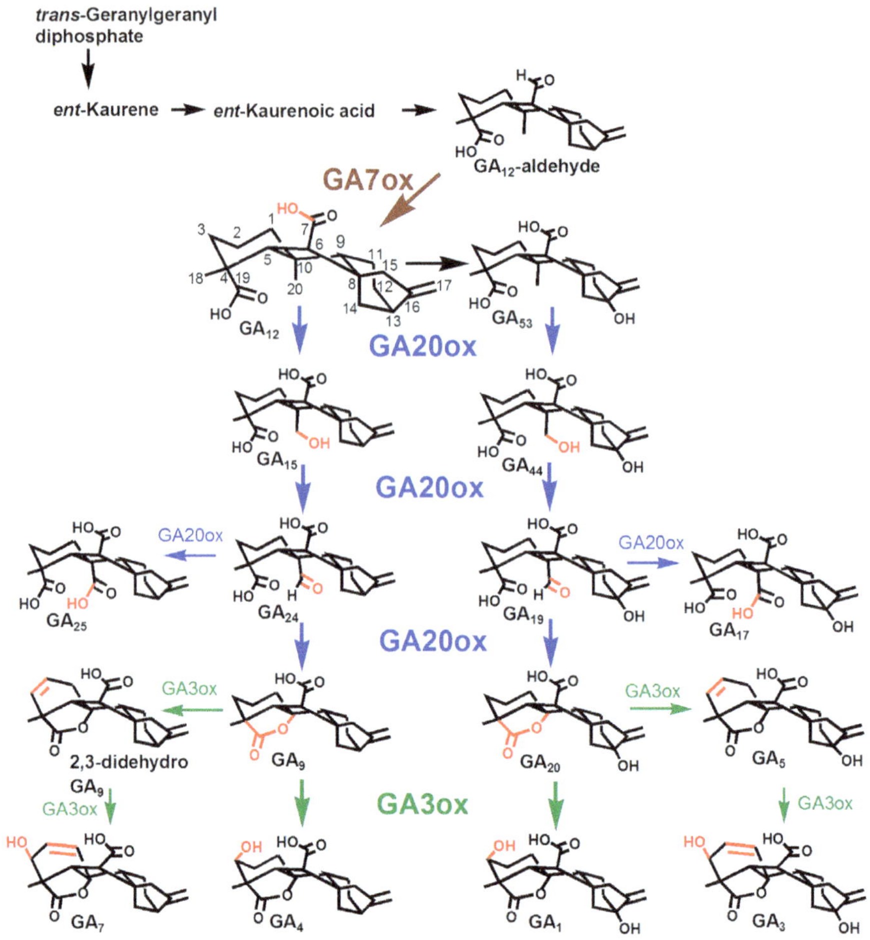

Scheme 16.1 Biosynthetic pathways from *trans*-geranylgeranyl diphosphate to the biologically active gibberellins GA_1, GA_3, GA_4 and GA_7 indicating reactions catalysed by the 2-oxoglutarate-dependent dioxygenases: GA7ox, GA 7-oxidase; GA20ox, GA 20-oxidase; GA3ox, GA 3-oxidase. Carbon numbers are marked on GA_{12}. The molecular changes introduced by each reaction are shown in red.

to the 2β,3β-dihydroxylated C_{20}-GA GA_{43} (not shown) was found to have an absolute requirement for 2-oxoglutarate (2OG) and Fe(II). Other 2-oxo acids and transition metal cations are ineffective and, in fact, the latter are inhibitory. The reactions required for the observed conversion comprise sequential oxidation at C-7, C-20, C-3 and C-2, suggesting potentially four different ODDs acting on GAs are present in the homogenate, although it was found subsequently that a single multifunctional enzyme introduces both the 2β- and 3β-hydroxyl groups in this tissue.[4] The presence of ODD-type GA 7-oxidases (GA7ox) has been confirmed in *C. maxima*[5,6] and *Cucumis sativus*

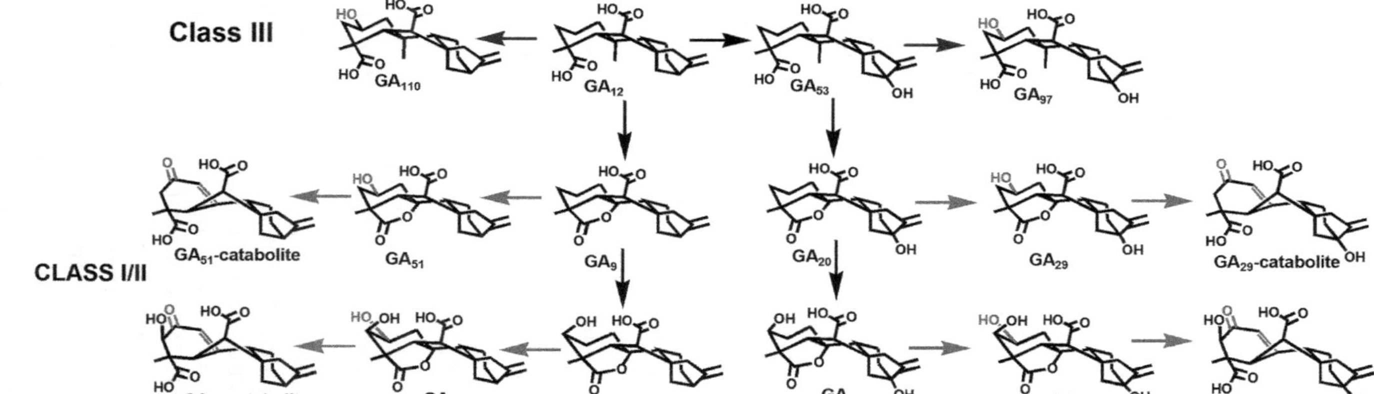

Scheme 16.2 Reactions catalysed by the GA 2-oxidases (GA2ox) that result in the formation of biologically inactive products. Class I and II enzymes act mainly on C_{19}-GAs, converting them to 2β-hydroxylated products and in some cases oxidizing them further to the GA-catabolites. Class III enzymes act mainly on the C_{20} 10-methyl precursors. Some enzymes from each class act on both C_{19}- and C_{20}-GAs. The molecular changes introduced by each reaction are shown in grey.

(cucumber),[7] but so far evidence that they are present in species outside the Cucurbitaceae is sparse.

In contrast to the enzymes catalysing the early steps of GA biosynthesis that are encoded by a single copy or a limited number of genes, the ODDs of GA biosynthesis form gene families, members of which are differentially regulated and differ in their expression domains, although there may be considerable redundancy.[8] A phylogenetic tree for GA ODDs in the dicot species *Arabidopsis thaliana* (Arabidopsis) and *C. sativus*, and the monocot rice (*Oryza sativa*) is presented in Figure 16.1. In the gene/enzyme nomenclature, equivalent numbers in unrelated species does not denote orthology, except within plant families; rather, the number has usually been assigned in order of gene discovery and it is not possible to relate the different clade members across plant families, indicating that they are due to separate, relatively recent gene duplications.[9] The genes encoding ODD enzymes acting on GAs are major sites of regulation for GA biosynthesis in higher plants, their expression responding to developmental stage, hormonal status and the environment. Given the importance of GAs in determining plant form and in setting a number of developmental switches, the mechanisms underlying the regulation of GA metabolism is currently an active avenue of research.

This review describes the four classes of ODD enzymes currently known to be involved in GA biosynthesis, focusing on their biochemical functions. Plant tissues for which GAs are essential growth regulators produce the biologically active compounds GA_1 and GA_4, with, in some cases, minor amounts of other GAs such as GA_3 and GA_5 that are unsaturated in the A ring. However, developing seeds of many species produce very high amounts of GAs with, in some cases, a large array of different structures. The very high rates of GA biosynthesis and the presence in seeds of ODDS with relatively low regiospecificity could account for the formation of these unusual structures.

16.2 Gibberellin 7-Oxidases

GA7ox enzymes oxidize the 6-aldehyde group of GA_{12}-aldehyde to a carboxylic acid group in GA_{12} (Scheme 16.1). This reaction is also catalysed by a multifunctional cytochrome P450 (*ent*-kaurenoic acid oxidase; KAO), which converts *ent*-kaurenoic acid to GA_{12} in three steps *via* 7β-hydroxy-*ent*-kaurenoic acid and GA_{12}-aldehyde.[14] While KAO is present in all species for which a genome sequence is available, GA7ox has so far been found in the Cucurbitaceae, with one report of its presence in potato (*Solanum tuberosum*).[15] Recombinant GA7ox enzymes from the Cucurbitaceae are multifunctional: in addition to its 7-oxidase activity the pumpkin enzyme (CmGA7ox1) converts GA_{12} to GA_{14} (not shown) by 3β-hydroxylation, a putative 15-hydroxy GA_{12} (not shown),[16] and two unidentified minor products. The cucumber genome contains two *GA7ox* genes, one of which (*CsGA7ox1*) is genetically and functionally closely related to the pumpkin gene.[7] While both recombinant enzymes possess GA_{12}-aldehyde 7-oxidase activity,

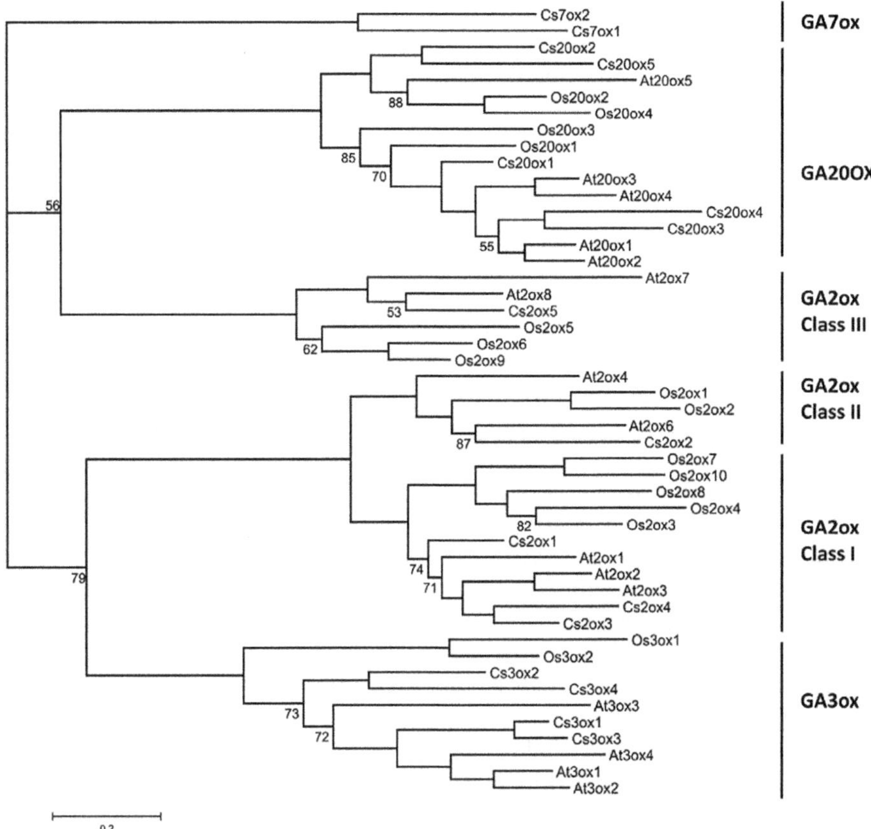

Figure 16.1 Phylogenetic relationships between gibberellin ODDs from *Arabidopsis thaliana* (At), *Cucumis sativus* (Cs) and *Oryza sativa* (Os). Polypeptide sequences were aligned in MUSCLE[10] within Geneious (Biomatters Ltd) and unaligned regions removed. Model selection and optimization were carried out in TOPALi v2.5[11] (Milne *et al.*, 2004) and the tree was generated by using MrBayes[12] using the WAG + I + G substitution model with 200 000 generations and 40% burnin. The unrooted tree was drawn using MEGA6.[13] Only branch support values of less than 95% are shown; the scale bar indicates number of substitutions per site.

CsGA7ox1 also shows 3β-hydroxylase and 15α-hydroxylase activities with respect to GA_{12}. The second enzyme (CsGA7ox2), which is more divergent from the pumpkin GA7ox, converts GA_{12} to GA_{111} by 12α-hydroxylation.[7] The physiological function of the GA7ox enzymes is unclear; the presence of KAO and the apparent limited species distribution of the GA7ox species would suggest they do not play a major role in the production of biologically active GAs, although this has not been tested by experiments with mutants lacking GA7ox activity. The pumpkin gene (*CmGA7ox1*) is highly expressed in developing seeds,[5] for which a role of GA is not established, but the gene is also expressed in vegetative tissues,[17] which suggests a potential involvement in growth regulation. Indeed ectopic expression of *CmGA7ox1* in Arabidopsis, which does not contain an ODD-type GA7ox, results in growth stimulation, indicating that 7-oxidase activity by KAO may be limiting for GA production.[18] However, it was not possible to detect an increase in GA levels, possibly because the analysis did not have sufficient spatial resolution to detect effects on specific tissues.

16.3 Gibberellin 20-Oxidases

The multifunctional GA20ox enzymes catalyse the oxidative removal of C-20 in the conversion of C_{20}- to C_{19}-GAs (Schemes 16.1 and 16.3). The initial substrates are normally GA_{12} and its 13-hydroxylated analogue GA_{53}, which contain a methyl group (C-20) on C-10. The C-10 methyl group is oxidized first to the alcohol and then the aldehyde, from which C-20 is cleaved to produce the C_{19} product with formation of a C-4 to C-10 lactone. The intermediates and products in this reaction sequence from GA_{12} are GA_{15}, GA_{24} and GA_9, while from GA_{53}, they are GA_{44}, GA_{19} and GA_{20}. The alcohol intermediates GA_{15} and GA_{44} readily form δ-lactones with the C-4α-carboxyl group (C-19) under acidic conditions and are always isolated from plant tissues in this form, but the open lactone form (20-alcohol and unesterified C-4α carboxyl) is required for further oxidation by the GA20ox.[19] In the oxidation of the alcohol to the aldehyde, the 20-*pro*-R hydrogen is abstracted stereospecifically.[20] Typically, *in planta* the intermediates accumulate to higher amounts than the final product, with the aldehyde often at the highest concentration, indicative of C-20 cleavage limiting C_{19}-GA formation. Analysis of the enzyme kinetics for each reaction with the recombinant wheat (*Triticum aestivum*) enzyme TaGA20ox1 revealed very low K_m values for GA_{12} and GA_{53}, at around 60 nM, but much larger values for the alcohol and aldehyde intermediates.[21] However, in contrast to the situation *in vivo*, the aldehyde intermediates do not normally accumulate appreciably when C-10-methyl GAs are incubated with GA20ox enzymes *in vitro*, in which case oxidation of the alcohol appears to be limiting.[21] Alcohol oxidation *in planta* is apparently boosted by an enzyme that catalyses only the alcohol to aldehyde conversion and accepts the lactone form as a substrate.[20] The nature of this enzyme is unknown, but it has some characteristics of an ODD.[22,23]

The mechanism for C-20 cleavage from the aldehyde is not fully understood. It does not involve oxidation to the carboxylic acid, which can occur as a minor side reaction of GA20ox activity, but the carboxyl product is not converted to C_{19}-GAs.[24] In the formation of C_{19}-GAs, C-20 is lost as CO_2 indicating that two rounds of oxidation are required,[25] but no intermediate has been detected and it may remain bound at the enzyme active site.[26] Ward *et al.*[19] provided evidence for the involvement of a C-10 radical that could react with the 19-carboxyl group to form the γ-lactone. It was proposed that initial oxidation of the aldehyde produces a C-20 acyl radical that forms an enzyme-bound intermediate.[19,26] A second round of

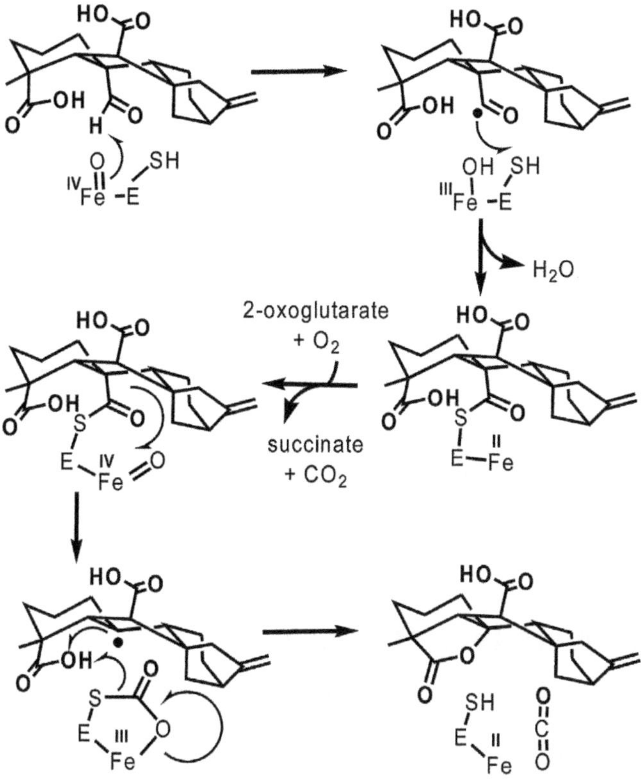

Scheme 16.3 A proposed mechanism for loss of C-20 in the formation of C_{19}-GAs (GA_9) from the C_{20}-GA aldehyde intermediate (GA_{24}) involving a hypothetical enzyme-bound intermediate. E, enzyme.

oxidation would result in cleavage of the C-10 to C-20 bond, with transient formation of the C-10 radical, followed by release of the lactone product and CO_2. This proposed mechanism is shown in Scheme 16.3, which includes a hypothetical intermediate that is bound to the enzyme through a thioester linkage, although evidence for such an intermediate is currently lacking. An alternative fate for the C-20 acyl radical is to combine with the Fe-bound hydroxyl to form the 20-carboxyl. While this is a relatively minor reaction of most 20-oxidases, a GA20ox from pumpkin seed (CmGA20ox1) produces the C-10-carboxyl as the sole end-product.[27] Due to its very high abundance, it was possible to purify this GA20ox enzyme from pumpkin endosperm,[27] thus paving the way for its cDNA to be cloned by antibody screening of an expression library.[28] This approach allowed for the first cloning of an ODD cDNA in GA biosynthesis. Examples of functionally abnormal GA20ox enzymes are rare, but one such enzyme from Arabidopsis, AtGA20ox5, was found *in vitro* to lack the capacity for C-20 cleavage, accumulating the aldehyde (GA_{24}) when incubated with GA_{12}.[29]

In most higher plant species that have been investigated, GA20ox activity is likely to be limiting for bioactive GA production, *e.g.* such limitation has been demonstrated by over-expression in Arabidopsis.[30,31] This species contains five *GA20ox* genes,[32,33] including *AtGA20ox5*, which encodes a functionally abnormal GA20ox enzyme as described above, while rice contains four *GA20ox* genes (Figure 16.1).[34] There is partial gene redundancy, particularly in vegetative tissues, as demonstrated by silencing of the most highly expressed gene in either species, which causes a semi-dwarf phenotype as opposed to the severe dwarfism displayed in mutants of the single-copy genes for the enzymes involved in early steps in the pathway (reviewed by Ross *et al.*).[35] In Arabidopsis, loss of *AtGA20ox1* reduces internode length and stem height, but has only a minor effect on fertility, even though GAs are necessary for normal reproductive development,[32] indicating greater gene redundancy in reproductive tissues. This dwarf trait due to mutations in *AtGA20ox1* has been found in natural Arabidopsis populations,[36] while similar traits in rice and barley (*Hordeum vulgare*), due to mutations in the cereal *GA20ox2*, have been selected in breeding for lodging resistance and increased yield.[37–40]

16.4 Gibberellin 3-Oxidases

GA3ox enzymes catalyse the introduction of a 3β-hydroxyl group in the final step in the biosynthesis of the major biologically active GAs. The normal substrates for this enzyme are GA_9 and its 13-hydroxylated analogue, GA_{20}, which are converted to GA_4 and GA_1, respectively (Scheme 16.1). These enzymes from vegetative organs of dicotyledonous species, such as the Arabidopsis enzyme AtGA3ox1[41] and the pea (*Pisum sativum*) enzyme PsGA3ox1[42,43] are regio- and stereospecific, producing only the 3β-hydroxylated products. Both enzymes show a preference for the non-13-hydroxylated substrate GA_9 over GA_{20}, with about a 10-fold difference in K_m values. However, GA3ox enzymes from rice[44] and wheat[21] exhibit lower regiospecificity, possessing, in addition to 3β-hydroxylase activity, low levels of 2β-hydroxylase, 2,3-desaturase and, in the case of the wheat enzyme, even 13-hydroxylase activity. This last case involves hydroxylation at a remote site on the molecule compared to the normal location and would require the substrate to bind to the active site in a very different orientation. The cereal enzymes convert the 2,3-dehydro substrate GA_5 to GA_3, requiring movement of the double bond to the 1,2 position, whereas the Arabidopsis GA3ox1 converts GA_5 to GA_6 by epoxidation (not shown).[41] This minor side reaction from GA_{20} to GA_3 *via* GA_5 is characteristic of cereals, having been shown also in maize,[45] and accounts for the low levels of GA_3 found in vegetative tissues of cereals.[46] Formation of 1,2-dehydro GAs *via* 2,3-unsaturated intermediates has also been described for developing seeds of *Marah macrocarpus* and apple (*Malus pumila*), in which the reaction involves initial removal of the 1β-H, movement of the double bond and hydroxylation on C-3β.[47] Work with recombinant enzymes indicated that, in contrast to the situation in cereals, the formation of the

1,2-dehydro GAs GA_7 and GA_3 in *M. macrocarpus* seeds requires the action of two GA3ox enzymes, one of which (MmGA3ox2) introduces the 2,3-double bond and the second (MmGA3ox1) adds the 3β-hydroxyl.[48] Both enzymes and a third GA3ox (MmGA3ox3) catalyse hydroxylation of substrates with a saturated A-ring at the 3β position, this being the sole activity for MmGA3ox1 and MmGA3ox3 with these substrates. However, MmGA3ox2 produces a similar product profile to that found for the wheat enzyme TaGA3ox2, including 13-hydroxylation. Although most GA3ox enzymes that have been characterized are specific for C_{19}-GA substrates, developing seeds of some species contain 3β-hydroxylated C_{20}-GAs, indicative of an early 3β-hydroxylation step. Indeed recombinant GA3-oxidases expressed from developing pumpkin seed cDNAs act preferentially on C_{20}-GAs,[4,16] while three GA3-oxidases from cucumber act on both C_{19}- and C_{20}-GAs.[7]

The *GA3ox* genes form relatively small families, with four members in Arabidopsis[49] and only two in rice[34] (Figure 16.1). These genes have quite distinct expression patterns, but the partially GA-deficient phenotypes of mutants imply movement of GA between tissues or organs.[50] Two of the Arabidopsis enzymes, AtGA3ox1 and AtGA3ox2, are involved in the development of vegetative organs,[51] while all four enzymes have a role in reproductive development.[50] In contrast, a single rice paralogue, OsGA3ox2, is responsible for GA biosynthesis throughout the plant, while OsGA3ox1 is present in specific GA-rich reproductive tissues, the tapetum and scutellum epithelium, where it may enhance GA production for export to other tissues.[52] It has been shown for a number of species that a single GA3ox paralogue has a major influence on stem height such that loss of this enzyme through mutation results in semi-dwarfism. Such dwarf lines include the *ga4* (*atga3ox1*) mutant of Arabidopsis,[53] *dwarf-1* maize,[45] *d18* (*osga3ox2*) rice[44] and *le* (psga3ox1) pea,[42,43] this last mutation being responsible for the stem height trait used by Mendel in his experiments on the nature of inheritance. Due to the expression of other *GA3ox* orthologues in reproductive organs, these mutants are fertile and have desirable traits when in crop species.

16.5 Gibberellin 2-Oxidases

Formation of 2β-hydroxylated GAs by GA2ox is one of the main mechanisms for GA deactivation that allows precise control of GA concentrations in plant tissues. On the basis of biochemical function, GA2ox representatives comprise two main groups of ODDs, one containing enzymes that act on C_{19}-GAs[54] and a second group acting on C_{20}-GAs[55] (Scheme 16.2). This grouping correlates to some extent with their phylogenetic relationship (Figure 16.1), with Class I and II enzymes utilizing C_{19}-GAs, and class III enzymes oxidizing C_{20}-GAs.[56,57] However, this relationship is not absolute, because a cucumber enzyme belonging to Class I acts on both forms of GAs.[7] Classes I and II are subclades of the C_{19}-GA2ox enzymes,[56–58] which accept the bioactive 3β-hydroxyGAs and their non-3β-hydroxylated precursors as substrates, the latter commonly being the preferred substrate *in vitro*.[54] The products from the 2β-hydroxylation of both types of C_{19}-GA substrate are found in plant tissues.[46] Some C_{19}-GA2ox enzymes oxidize 2β-hydroxyGAs further to the 2-ketones, known as GA-catabolites, in which the C-4 to C-10 lactone is cleaved to form a carboxyl group on C-4 with a double bond between C-10 and C-1, C-5 or C-9 (shown in Scheme 16.2 for C-1,10). This rearrangement probably occurs during isolation or as a consequence of derivatization prior to GC-MS analysis. While the ability of C_{19}-GA2ox enzymes to perform both or just one oxidation step may relate to whether they are in Class I or II, respectively,[57] this has not been rigorously tested and there is evidence to the contrary.[7] The ability to perform the second oxidation step may in fact be a consequence of high levels of enzyme activity and, because this is normally tested *in vitro* with recombinant protein, it is unwise to extrapolate too far to the *in vivo* situation; the functional distinction between Classes I and II is therefore unclear. GA-catabolites are normally present in organs, such as developing seeds and fruit, in which GAs are highly abundant,[59,60] indicating that their formation is associated with high rates of GA metabolism. The substrate specificity of the C_{20}-GA 2-oxidases has not been comprehensively tested: the Arabidopsis enzymes GA2ox7 and 8 are active against GA_{12} and GA_{53}, converting them to the endogenous products GA_{110} and GA_{97}, respectively.[55] 2β-Hydroxylated forms of other C_{20}-GA intermediates have been detected in plant tissues,[61] but their biosynthetic origin is unclear.

The *GA2ox* gene families are relatively large in size compared to other gene families coding for the ODDs of GA biosynthesis. For example, seven and ten *GA2ox* genes have been identified in Arabidopsis and rice, respectively,[62] although the genes encoding C_{20}-GA2ox proteins (Class III) may be considered to constitute a separate phylogenetic clade (Figure 16.1). An additional Arabidopsis *GA2ox* sequence identified is a pseudogene, while the biochemical functions of all the rice gene products have not been confirmed. Although the *GA2ox* genes differ substantially in expression patterns and regulation, there is considerable gene redundancy and multiple genes normally need to be silenced in order to modify GA content and growth.[63] There are, however, exceptions, a notable one being the *slender* (*sln*) mutant of pea, which contains a mutation in *PsGA2ox1*.[64,65] Loss of *PsGA2ox1*, which is highly expressed in developing seeds, results in accumulation of the precursor GA_{20} in mature pea seeds, which on germination convert it to GA_1 by 3β-hydroxylation resulting in a slender (overgrowth) phenotype in the seedling.[66] This finding exemplifies an important function of GA2ox during late seed development when GA deactivation serves to prevent the accumulation of bioactive GA that could induce premature germination.

16.6 Gibberellin Biosynthesis in Other Organisms

The lycophyte (spike moss) *Selaginella moellendorffii* contains single functional *GA20ox* and *GA3ox* paralogues, with each corresponding protein preferring non-13-hydroxylated substrates, consistent with this organism producing GA_4;[67] however, there is no indication that *S. moellendorffii* contains *GA2ox* genes.[68] GA-related compounds that function as antheridiogens have been identified in a number of fern species,[46] and their biosynthesis was shown recently to involve ODD-type GA20ox and GA3ox enzymes.[69] GAs were first isolated from the rice-pathogenic fungus *Gibberella fujikuroi* (now reclassified as *Fusarium fujikuroi*), from which they acquired their name. This species produces mainly GA_3 (gibberellic acid) and is used for the production of this product for commercial use. The enzymes for GA_3 biosynthesis in *F. fujikuroi* are encoded by a cluster of seven genes, four of which encode cytochrome P450 monooxygenases, including those responsible for 3β-hydroxylation and cleavage of C-20, reactions that are carried out by ODDs in higher plants.[33] The fifth oxidase, a desaturase that introduces a double bond at the C-1,2 position in the conversion of GA_4 to GA_7, has characteristics of an ODD, although it has little homology with known ODDs apart from putative Fe- and 2OG-binding residues.[70] The desaturase has a large number of homologues in fungi, although their functions are unknown. In contrast to the A-ring desaturation reactions in plants, which involves oxidation on the β face,[47] α-H atoms are lost in C-1,2-desaturation in the fungus.[71] Moreover the desaturase forms 2α-hydroxy GAs as by-products, consistent with action on the α face of the GA molecule.[72]

16.7 Regulation of Gene Expression

The genes encoding ODDs are major sites of regulation in GA biosynthesis, responding to developmental and environmental signals. The molecular mechanisms by which these genes are regulated is an active area of research that has been reviewed in some detail recently,[8] and it is appropriate to give only a brief summary here. Within the gene families the different paralogues show distinct spatial expression patterns, with a particular paralogue often having a major role for GA production in a specific tissue. The highest rates of expression of *GA20ox* and *GA3ox* correspond to sites that accumulate relatively high GA concentrations, such as expanding leaves, anthers and developing seeds, while *GA2ox* expression is greatest in fully expanded organs.[73] An important mechanism by which GA concentrations are regulated in GA-responsive tissues involves the DELLA component of the GA signalling pathway, which upregulates expression of certain *GA20ox* and *GA3ox* paralogues and downregulates *GA2ox* genes.[74,75] As GAs act by destabilizing DELLA proteins, this mechanism serves to establish GA homeostasis. The DELLA proteins function as transcriptional regulators by interacting with transcription factors, either to inhibit or promote their activity.[76] Several transcription factors have been shown to regulate expression of particular genes encoding ODD enzymes to allow for GA homeostasis, *i.e.* they regulate expression in a GA-dependent manner (reviewed by Hedden and Thomas)[8], while the Arabidopsis INDETERMINATE DOMAIN protein GAF1 promotes expression of several feedback regulated genes in association with the DELLA protein GAI.[77] ODD genes are also regulated by other classes of hormones, the primary example being auxin, which upregulates expression of *GA20ox* or *GA3ox*, depending on species, and downregulates *GA2ox* expression.[78,79] Thus, growth stimulation by auxin (*e.g.* in stems, roots and fruit) is in part mediated by GA.

Plant growth and development are strongly influenced by light signals, for which the GA-biosynthetic pathway is a major target. Germination of positively photoblastic (light-sensitive) seeds, such as those of Arabidopsis, is stimulated by red light. In the dark the transcription factor PIL5 suppresses expression of Arabidopsis *GA3ox1* and *GA3ox2* and stimulates *GA2ox2* expression, while in red light PIL5 is degraded, allowing the synthesis of GA_4, which promotes germination. Growth reduction on exposure of dark-grown seedlings to light (de-etiolation) is accompanied by rapid upregulation of *GA2ox* genes, and a slower downregulation of *GA20ox* and *GA3ox* expression, the process involving both phytochrome and cryptochrome light receptors.[80,81] Changes in day length promote GA biosynthesis in species in which flowering is induced by both photoperiod and GAs.[82,83] Responses to the change in photoperiod include stimulation of *GA20ox* expression in the leaves, which produce GA as a mobile signal that induces floral differentiation at the shoot apex.[82] In some species floral differentiation is accompanied by a reduction of *GA2ox* expression below the shoot apex that allows entry of GA without deactivation.[84,85]

Reduced growth with an associated decrease in GA content is a typical response to stress, allowing the plant to concentrate its resource on withstanding the adverse conditions. It has been shown for several abiotic stresses that the treatment results in increased expression of specific *GA2ox* genes.[86] In the case of Arabidopsis subject to salt stress, upregulation of *AtGA2ox7*, which encodes a C_{20}-GA2ox, is mediated by the AP2-type transcription factor DDF1,[87] while the related transcription factor CBF1 mediated cold-induced expression of *AtGA2ox3*.[88]

16.8 Gibberellin ODDs as Targets in Crop Improvement

The GA-biosynthetic pathway has been a prime target in the introduction of agronomically beneficial traits into crop species. An important example is the incorporation of *semidwarf1* (*sd1*) mutant alleles into rice to prevent plant lodging, which was a major factor in the green revolution.[89] As described above, *SD1* is a *GA20ox* (*OsGA20ox2*), loss of which reduces stem height without affecting fertility,[37–39] while the *sdw1* and *denso* dwarfing mutations in barley result in reduced expression of the barley orthologue.[40] In legumes, where GA3ox rather than GA20ox activity limits GA_1 biosynthesis, natural mutations in *GA3ox1* result in dwarfism in pea (*le*)[42,43] and alfalfa (*Medicago sativa*; *dmf1*).[90] GA3ox is a target for the acylcyclohexanedione-type growth retardants, such as prohexadione-calcium and trinexapac-ethyl, which are believed to compete with 2OG at the enzyme active site.[91,92] The older growth retardant daminozide is proposed to have a similar mode of action.[93] The acylcyclohexanediones also inhibit the GA2oxs,[94] and in some circumstances when GA content is determined more by rates of deactivation than synthesis, application of these inhibitors stimulates growth.[95]

Although not yet exploited in agriculture, the introduction of genes for ectopic *ODD* expression has been tested extensively (reviewed by Phillips).[96] Expression of *GA20ox* increases GA content and growth in numerous species, including tomato,[97] *Solanum nigrum*[98] and hybrid aspen (*Populus tremula* x *P. tremuloides*);[99] this last case resulted in increased xylem fibre length with improved paper-making quality. Increased biomass and fibre production were also demonstrated in tobacco by either *GA20ox* overexpression[100] or silencing of *GA2ox* genes,[101] the latter producing the greater effect. Constitutive expression of the fungal GA desaturase gene in *S. nigrum* and *Nicotiana sylvestris* produced very substantial growth stimulation by promoting the synthesis of 1,2-unsaturated GAs, which are resistant to inactivation by GA2ox.[70] In terms of introduction of dwarfism, which is an important agricultural trait for many crops, there are numerous examples of the ectopic expression of *GA2ox* in commercially valuable species, including rice,[102] wheat,[103] turf grass (*Paspalum notatum*)[104] and hybrid poplar (*Populus tremula* × *P. alba*).[105] Dwarfism has also been achieved by silencing biosynthetic genes, particularly *GA20ox* genes, such as in apple[106] and in rice, in which silencing each of three *GA20ox* paralogues produces dwarfism, indicating partial redundancy of these genes with respect to stem growth.[107–109]

Acknowledgements

The authors thank Dr Alison Huttly for assistance with the phylogenetic analysis. They are supported at Rothamsted Research by the 20:20 Wheat® programme, funded by the Biotechnology and Biological Sciences Research Council of the United Kingdom.

References

1. K. Aya, Y. Hiwatashi, M. Kojima, H. Sakakibara, M. Ueguchi-Tanaka, M. Hasebe and M. Matsuoka, *Nature Commun.*, 2011, **2**.
2. X. Hou, L. Y. C. Lee, K. Xia, Y. Yen and H. Yu, *Dev. Cell*, 2010, **19**, 884–894.
3. P. Hedden and J. E. Graebe, *J. Plant Growth Regul.*, 1982, **1**, 105–116.
4. T. Lange, S. Robatzek and A. Frisse, *Plant Cell*, 1997, **9**, 1459–1467.
5. T. Lange, *Proc. Natl. Acad. Sci. U. S. A.*, 1997, **94**, 6553–6558.
6. T. Lange, A. Schweimer, D. A. Ward, P. Hedden and J. E. Graebe, *Planta*, 1994, **195**, 98–107.
7. M. J. P. Lange, A. Liebrandt, L. Arnold, S. M. Chmielewska, A. Felsberger, E. Freier, M. Heuer, D. Zur and T. Lange, *Phytochemistry*, 2013, **90**, 62–69.
8. P. Hedden and S. G. Thomas, *Biochem. J.*, 2012, **444**, 11–25.
9. F. Han and B. Zhu, *Gene*, 2011, **473**, 23–35.

10. R. C. Edgar, *Nucleic Acids Res.*, 2004, **32**, 1792–1797.
11. I. Milne, F. Wright, G. Rowe, D. F. Marshall, D. Husmeier and G. McGuire, *Bioinformatics*, 2004, **20**, 1806–1807.
12. F. Ronquist, M. Teslenko, P. van der Mark, D. L. Ayres, A. Darling, S. Hohna, B. Larget, L. Liu, M. A. Suchard and J. P. Huelsenbeck, *Syst. Biol.*, 2012, **61**, 539–542.
13. K. Tamura, G. Stecher, D. Peterson, A. Filipski and S. Kumar, *Mol. Biol. Evol.*, 2013, **30**, 2725–2729.
14. C. A. Helliwell, P. M. Chandler, A. Poole, E. S. Dennis and W. J. Peacock, *Proc. Natl. Acad. Sci. U. S. A.*, 2001, **98**, 2065–2070.
15. K. R. Fixen, S. C. Thomas and C. B. S. Tong, *J. Plant Growth Regul.*, 2012, **31**, 342–350.
16. A. Frisse, M. J. Pimenta and T. Lange, *Plant Physiol.*, 2003, **131**, 1220–1227.
17. T. Lange, J. Kappler, A. Fischer, A. Frisse, T. Padeffke, S. Schmidtke and M. J. P. Lange, *Plant Physiol.*, 2005, **139**, 213–223.
18. A. Radi, T. Lange, T. Niki, M. Koshioka and M. J. P. Lange, *Plant Physiol.*, 2006, **140**, 528–536.
19. J. L. Ward, P. Gaskin, R. G. S. Brown, G. S. Jackson, P. Hedden, A. L. Phillips, C. L. Willis and M. H. Beale, *J. Chem. Soc. Perkin*, 2002, **1**, 232–241.
20. J. L. Ward, G. J. Jackson, M. H. Beale, P. Gaskin, P. Hedden, L. N. Mander, A. L. Phillips, H. Seto, M. Talon, C. L. Willis, T. M. Wilson and J. A. D. Zeevaart, *Chem. Commun.*, 1997, 13–14.
21. N. E. J. Appleford, D. J. Evans, J. R. Lenton, P. Gaskin, S. J. Croker, K. Devos, A. L. Phillips and P. Hedden, *Planta*, 2006, **223**, 568–582.
22. S. J. Gilmour, J. A. D. Zeevaart, L. Schwenen and J. E. Graebe, *Plant Physiol.*, 1986, **82**, 190–195.
23. E. Grosselindemann, M. J. Lewis, P. Hedden and J. E. Graebe, *Planta*, 1992, **188**, 252–257.
24. Y. Kamiya and J. E. Graebe, *Phytochemistry*, 1983, **22**, 681–689.
25. Y. Kamiya, N. Takahashi and J. E. Graebe, *Planta*, 1986, **169**, 524–528.
26. P. Hedden, *Physiol. Plant.*, 1997, **101**, 709–719.
27. T. Lange, *Planta*, 1994, **195**, 108–115.
28. T. Lange, P. Hedden and J. E. Graebe, *Proc. Natl. Acad. Sci. U. S. A.*, 1994, **91**, 8552–8556.
29. A.R. G. Plackett, S. J. Powers, N. Fernandez-Garcia, T. Urbanova, Y. Takebayashi, M. Seo, Y. Jikumaru, R. Benlloch, O. Nilsson, O. Ruiz-Rivero, A. L. Phillips, Z. A. Wilson, S. G. Thomas and P. Hedden, *Plant Cell*, 2012, **24**, 941–960.
30. J. P. Coles, A. L. Phillips, S. J. Croker, R. GarciaLepe, M. J. Lewis and P. Hedden, *Plant J.*, 1999, **17**, 547–556.
31. S. S. Huang, A. S. Raman, J. E. Ream, H. Fujiwara, R. E. Cerny and S. M. Brown, *Plant Physiol.*, 1998, **118**, 773–781.
32. I. Rieu, O. Ruiz-Rivero, N. Fernandez-Garcia, J. Griffiths, S. J. Powers, F. Gong, T. Linhartova, S. Eriksson, O. Nilsson, S. G. Thomas, A. L. Phillips and P. Hedden, *Plant J.*, 2008, **53**, 488–504.

33. P. Hedden, A. L. Phillips, M. C. Rojas, E. Carrera and B. Tudzynski, *J. Plant Growth Regul.*, 2002, **20**, 319–331.
34. T. Sakamoto, K. Miura, H. Itoh, T. Tatsumi, M. Ueguchi-Tanaka, K. Ishiyama, M. Kobayashi, G. K. Agrawal, S. Takeda, K. Abe, A. Miyao, H. Hirochika, H. Kitano, M. Ashikari and M. Matsuoka, *Plant Physiol.*, 2004, **134**, 1642–1653.
35. J. J. Ross, I. C. Murfet and J. B. Reid, *Physiol. Plant.*, 1997, **100**, 550–560.
36. L. Barboza, S. Effgen, C. Alonso-Blanco, R. Kooke, J. J. B. Keurentjes, M. Koornneef and R. Alcazar, *Proc. Natl. Acad. Sci. U. S. A.*, 2013, **110**, 15818–15823.
37. L. Monna, N. Kitazawa, R. Yoshino, J. Suzuki, H. Masuda, Y. Maehara, M. Tanji, M. Sato, S. Nasu and Y. Minobe, *DNA Res.*, 2002, **9**, 11–17.
38. A. Sasaki, M. Ashikari, M. Ueguchi-Tanaka, H. Itoh, A. Nishimura, D. Swapan, K. Ishiyama, T. Saito, M. Kobayashi, G. S. Khush, H. Kitano and M. Matsuoka, *Nature*, 2002, **416**, 701–702.
39. W. Spielmeyer, M. H. Ellis and P. M. Chandler, *Proc. Natl. Acad. Sci. U. S. A.*, 2002, **99**, 9043–9048.
40. Q. J. Jia, X. Q. Zhang, S. Westcott, S. Broughton, M. Cakir, J. M. Yang, R. Lance and C. D. Li, *Theor. Appl. Genet.*, 2011, **122**, 1451–1460.
41. J. Williams, A. L. Phillips, P. Gaskin and P. Hedden, *Plant Physiol.*, 1998, **117**, 559–563.
42. D. R. Lester, J. J. Ross, P. J. Davies and J. B. Reid, *Plant Cell*, 1997, **9**, 1435–1443.
43. D. N. Martin, W. M. Proebsting and P. Hedden, *Proc. Natl. Acad. Sci. U. S. A.*, 1997, **94**, 8907–8911.
44. H. Itoh, M. Ueguchi-Tanaka, N. Sentoku, H. Kitano, M. Matsuoka and M. Kobayashi, *Proc. Natl. Acad. Sci. U. S. A.*, 2001, **98**, 8909–8914.
45. C. R. Spray, M. Kobayashi, Y. Suzuki, B. O. Phinney, P. Gaskin and J. Macmillan, *Proc. Natl. Acad. Sci. U. S. A.*, 1996, **93**, 10515–10518.
46. J. MacMillan, *J. Plant Growth Regul.*, 2002, **20**, 387–442.
47. K. S. Albone, P. Gaskin, J. Macmillan, B. O. Phinney and C. L. Willis, *Plant Physiol.*, 1990, **94**, 132–142.
48. D. A. Ward, J. MacMillan, F. Gong, A. L. Phillips and P. Hedden, *Phytochemistry*, 2010, **71**, 2010–2018.
49. P. Hedden and A. L. Phillips, *Trends Plant Sci.*, 2000, **5**, 523–530.
50. J. H. Hu, M. G. Mitchum, N. Barnaby, B. T. Ayele, M. Ogawa, E. Nam, W. C. Lai, A. Hanada, J. M. Alonso, J. R. Ecker, S. M. Swain, S. Yamaguchi, Y. Kamiya and T. P. Sun, *Plant Cell*, 2008, **20**, 320–336.
51. M. G. Mitchum, S. Yamaguchi, A. Hanada, A. Kuwahara, Y. Yoshioka, T. Kato, S. Tabata, Y. Kamiya and T. P. Sun, *Plant J.*, 2006, **45**, 804–818.
52. M. Kaneko, H. Itoh, Y. Inukai, T. Sakamoto, M. Ueguchi-Tanaka, M. Ashikari and M. Matsuoka, *Plant J.*, 2003, **35**, 104–115.
53. H. H. Chiang, I. Hwang and H. M. Goodman, *Plant Cell*, 1995, **7**, 195–201.
54. S. G. Thomas, A. L. Phillips and P. Hedden, *Proc. Natl. Acad. Sci. U. S. A.*, 1999, **96**, 4698–4703.

55. F. M. Schomburg, C. M. Bizzell, D. J. Lee, J. A. D. Zeevaart and R. M. Amasino, *Plant Cell*, 2003, **15**, 151–163.
56. D. J. Lee and J. A. D. Zeevaart, *Plant Physiol.*, 2005, **138**, 243–254.
57. J. C. Serrani, R. Sanjuan, O. Ruiz-Rivero, M. Fos and J. L. Garcia-Martinez, *Plant Physiol.*, 2007, **145**, 246–257.
58. R. C. Elliott, J. J. Ross, J. L. Smith, D. R. Lester and J. B. Reid, *J. Plant Growth Regul.*, 2001, **20**, 87–94.
59. J. L. Garcia-martinez, C. Santes, S. J. Croker and P. Hedden, *Planta*, 1991, **184**, 53–60.
60. V. M. Sponsel and J. Macmillan, *Planta*, 1980, **150**, 46–52.
61. L. N. Mander, D. J. Owen, S. J. Croker, P. Gaskin, P. Hedden, M. J. Lewis, M. Talon, D. A. Gage, J. A. D. Zeevaart, M. L. Brenner and C. X. Sheng, *Phytochemistry*, 1996, **43**, 23–28.
62. S. F. Lo, S. Y. Yang, K. T. Chen, Y. L. Hsing, J. A. D. Zeevaart, L. J. Chen and S. M. Yu, *Plant Cell*, 2008, **20**, 2603–2618.
63. I. Rieu, S. Eriksson, S. J. Powers, F. Gong, J. Griffiths, L. Woolley, R. Benlloch, O. Nilsson, S. G. Thomas, P. Hedden and A. L. Phillips, *Plant Cell*, 2008, **20**, 2420–2436.
64. D. N. Martin, W. M. Proebsting and P. Hedden, *Plant Physiol.*, 1999, **121**, 775–781.
65. D. R. Lester, J. J. Ross, J. J. Smith, R. C. Elliott and J. B. Reid, *Plant J.*, 1999, **19**, 65–73.
66. J. J. Ross, J. B. Reid, S. M. Swain, O. Hasan, A. T. Poole, P. Hedden and C. L. Willis, *Plant J.*, 1995, **7**, 513–523.
67. K. Hirano, M. Nakajima, K. Asano, T. Nishiyama, H. Sakakibara, M. Kojima, E. Katoh, H. Xiang, T. Tanahashi, M. Hasebe, J. A. Banks, M. Ashikari, H. Kitano, M. Ueguchi-Tanaka and M. Matsuoka, *Plant Cell*, 2007, **19**, 3058–3079.
68. A. Shimada, M. Ueguchi-Tanaka, T. Nakatsu, M. Nakajima, Y. Naoe, H. Ohmiya, H. Kato and M. Matsuoka, *Nature*, 2008, **456**, 520–U544.
69. J. Tanaka, K. Yano, K. Aya, K. Horano, S. Takehara, E. Koketsu, R. L. Ordonio, S.-H. Park, M. Nakajima, M. Ueguchi-Tanaka and M. Matsuoka, *Science*, 2014, **346**, 469–473.
70. A. Bhattacharya, S. Kourmpetli, D. A. Ward, S. G. Thomas, F. Gong, S. J. Powers, E. Carrera, B. Taylor, F. N. D. Gonzalez, B. Tudzynski, A. L. Phillips, M. R. Davey and P. Hedden, *Plant Physiol.*, 2012, **160**, 837–845.
71. R. Evans, J. R. Hanson and A. F. White, *J. Chem. Soc. C*, 1970, 2601.
72. B. Tudzynski, M. Mihlan, M. C. Rojas, P. Linnemannstons, P. Gaskin and P. Hedden, *J. Biol. Chem.*, 2003, **278**, 28635–28643.
73. J. J. Ross, S. E. Davidson, C. M. Wolbang, E. Bayly-Stark, J. J. Smith and J. B. Reid, *Funct. Plant Biol.*, 2003, **30**, 83–89.
74. D. E. Weston, R. C. Elliott, D. R. Lester, C. Rameau, J. B. Reid, I. C. Murfet and J. J. Ross, *Plant Physiol.*, 2008, **147**, 199–205.
75. R. Zentella, Z.-L. Zhang, M. Park, S. G. Thomas, A. Endo, K. Murase, C. M. Fleet, Y. Jikumaru, E. Nambara, Y. Kamiya and T.-P. Sun, *Plant Cell*, 2007, **19**, 3037–3057.

76. J. M. Daviere and P. Achard, *Development*, 2013, **140**, 1147–1151.
77. J. Fukazawa, H. Teramura, S. Murakoshi, K. Nasuno, N. Nishida, T. Ito, M. Yoshida, Y. Kamiya, S. Yamaguchi and Y. Takahashi, *Plant Cell,* 2014, **26**, 2920–2938.
78. J. B. Reid, S. E. Davidson and J. J. Ross, *Plant Signaling Behav.*, 2011, **6**, 406–408.
79. M. Frigerio, D. Alabadi, J. Perez-Gomez, L. Garcia-Carcel, A. L. Phillips, P. Hedden and M. A. Blazquez, *Plant Physiol.*, 2006, **142**, 553–563.
80. F. Hirose, N. Inagaki, A. Hanada, S. Yamaguchi, Y. Kamiya, A. Miyao, H. Hirochika and M. Takano, *Plant Cell Physiol.*, 2012, **53**, 1570–1582.
81. X.-Y. Zhao, X.-H. Yu, X.-M. Liu and C.-T. Lin, *J Integrative Plant Biol.*, 2007, **49**, 21–27.
82. R. W. King, *Russ. J. Plant Physiol.*, 2012, **59**, 479–490.
83. E. Mutasa-Gottgens and P. Hedden, *J. Exp. Bot.*, 2009, **60**, 1979–1989.
84. R. W. King, L. N. Mander, T. Asp, C. P. MacMillan, C. A. Blundell and L. T. Evans, *Mol. Plant*, 2008, **1**, 295–307.
85. T. Sakamoto, M. Kobayashi, H. Itoh, A. Tagiri, T. Kayano, H. Tanaka, S. Iwahori and M. Matsuoka, *Plant Physiol.*, 2001, **125**, 1508–1516.
86. E. H. Colebrook, S. G. Thomas, A. L. Phillips and P. Hedden, *J. Exp. Biol.*, 2014, **217**, 67–75.
87. H. Magome, S. Yamaguchi, A. Hanada, Y. Kamiya and K. Oda, *Plant J.*, 2008, **56**, 613–626.
88. P. Achard, F. Gong, S. Cheminant, M. Alioua, P. Hedden and P. Genschik, *Plant Cell*, 2008, **20**, 2117–2129.
89. P. Hedden, *Trends Genet.*, 2003, **19**, 5–9.
90. A. Dalmadi, P. Kalo, J. Jakab, A. Saskoi, T. Petrovics, G. Deak and G. B. Kiss, *Plant Cell Rep.*, 2008, **27**, 1271–1279.
91. R. G. S. Brown, L. Yan, M. H. Beale and P. Hedden, *Phytochemistry*, 1998, **47**, 679–687.
92. W. Rademacher, *Annu. Rev. Plant Physiol. Plant Mol. Biol.*, 2000, **51**, 501–531.
93. R. G. S. Brown, H. Kawaide, Y. Y. Yang, W. Rademacher and Y. Kamiya, *Physiol. Plant.*, 1997, **101**, 309–313.
94. D. L. Griggs, P. Hedden, K. E. Templesmith and W. Rademacher, *Phytochemistry*, 1991, **30**, 2513–2517.
95. J. F. Martinez-Garcia and J. L. Garcia-Martinez, *Planta*, 1992, **188**, 245–251.
96. A. L. Phillips, in *Plant Hormones Biosynthesis, Signal Transduction, Action!*, ed. P. J. Davies, Kluwer Academic Publishers, Dordrecht, Revised 3rd edn edn., 2010, pp. 582–609.
97. N. Garcia-Hurtado, E. Carrera, O. Ruiz-Rivero, M. P. Lopez-Gresa, P. Hedden, F. Gong and J. L. Garcia-Martinez, *J. Exp. Bot.*, 2012, **63**, 5803–5813.
98. A. Bhattacharya, D. A. Ward, P. Hedden, A. L. Phillips, J. B. Power and M. R. Davey, *Plant Cell Rep.*, 2012, **31**, 945–953.
99. M. E. Eriksson, M. Israelsson, O. Olsson and T. Moritz, *Nat. Biotechnol.*, 2000, **18**, 784–788.

100. S. Biemelt, H. Tschiersch and U. Sonnewald, *Plant Physiol.*, 2004, **135**, 254–265.
101. J. Dayan, M. Schwarzkopf, A. Avni and R. Aloni, *Plant Biotech. J.*, 2010, **8**, 425–435.
102. T. Sakamoto, Y. Morinaka, K. Ishiyama, M. Kobayashi, H. Itoh, T. Kayano, S. Iwahori, M. Matsuoka and H. Tanaka, *Nat. Biotechnol.*, 2003, **21**, 909–913.
103. N. E. J. Appleford, M. D. Wilkinson, Q. Ma, D. J. Evans, M. C. Stone, S. P. Pearce, S. J. Powers, S. G. Thomas, H. D. Jones, A. L. Phillips, P. Hedden and J. R. Lenton, *J. Exp. Bot.*, 2007, **58**, 3213–3226.
104. M. Agharkar, P. Lomba, F. Altpeter, H. Zhang, K. Kenworthy and T. Lange, *Plant Biotech. J.*, 2007, **5**, 791–801.
105. A. A. Elias, V. B. Busov, K. R. Kosola, C. Ma, E. Etherington, O. Shevchenko, H. Gandhi, D. W. Pearce, S. B. Rood and S. H. Strauss, *Plant Physiol.*, 2012, **160**, 1130–1144.
106. S. M. Bulley, F. M. Wilson, P. Hedden, A. L. Phillips, S. J. Croker and D. J. James, *Plant Biotech. J.*, 2005, **3**, 215–223.
107. T. Oikawa, M. Koshioka, K. Kojima, H. Yoshida and M. Kawata, *Plant Mol. Biol.*, 2004, **55**, 687–700.
108. F. Qiao, Q. Yang, C. L. Wang, Y. L. Fan, X. F. Wu and K. J. Zhao, *Euphytica*, 2007, **158**, 35–45.
109. X. Qin, J. H. Liu, W. S. Zhao, X. J. Chen, Z. J. Guo and Y. L. Peng, *Mol. Plant-Microbe Interact.*, 2013, **26**, 227–239.

CHAPTER 17

2-Oxoglutarate-Dependent Oxygenases of Cephalosporin Synthesis

INGER ANDERSSON*[a] AND KARIN VALEGÅRD[a]

[a]Department of Cell and Molecular Biology, Uppsala University, Box 596, S-751 24, Uppsala, Sweden
*E-mail: inger@xray.bmc.uu.se

17.1 Introduction

Several steps in the biosynthesis of cephalosporins are catalysed by mononuclear Fe(II)-dependent oxygenases that utilize 2-oxoglutarate (2OG) as a cosubstrate (Figure 17.1). Cephalosporins and the closely related penicillins are among the medicinally most important antibiotics in use. These β-lactam antibiotics are secondary metabolites produced by filamentous fungi and unicellular bacteria. Their biosynthesis pathways include unconventional and 'difficult' steps beyond the reach of current synthetic chemistry, therefore most β-lactam-type antibiotics are produced from fermentation-derived materials. In nature, these reactions are catalysed by metal-dependent oxygenases that exploit the oxidative capacity of molecular oxygen. This chapter is concerned with the structure and function of the 2OG-dependent oxygenases of the cephalosporin/cephamycin C biosynthetic pathway.

RSC Metallobiology Series No. 3
2-Oxoglutarate-Dependent Oxygenases
Edited by Robert P. Hausinger and Christopher J. Schofield

Published by the Royal Society of Chemistry, www.rsc.org

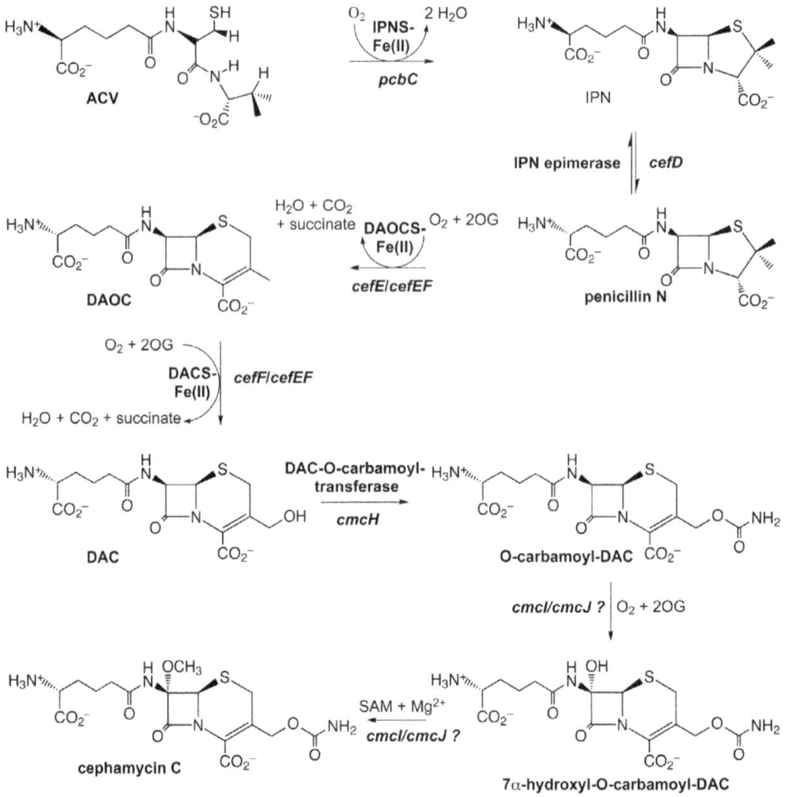

Figure 17.1 The cephamycin biosynthetic pathway. Biosynthesis of naturally occurring penicillins and cephalosporins begins with a condensation of three amino acids (L-Cys, L-Val and the unusual L-α-aminoadipic acid) to form the linear tripeptide δ-(L-α-aminoadipyl)-L-cysteinyl-D-valine (ACV), catalysed by ACV synthetase (not shown). The ring closure of ACV to form the bicyclic isopenicillin N (IPN) is catalysed by IPN synthase (IPNS).[22] Cephalosporin biosynthesis begins with the isomerization of the L-α-aminoadipoyl side chain of IPN by IPN epimerase to form penicillin N. Expansion of the five-membered thiazolidine ring of penicillin N, catalysed by deacetoxycephalosporin C synthase (DAOCS) in prokaryotes,[19] forms the six-membered dihydrothiazine ring common to all cephalosporins. The resulting compound, deacetoxycephalosporin C (DAOC), is hydroxylated at the C-3′ position by deacetylcephalosporin C (DAC) synthase (DACS) to form DAC. In the cephamycin-specific biosynthetic pathway of actinomycetes the C-3′ hydroxyl group of DAC is carbamoylated by DAC-O-carbamoyltransferase,[14] and the C-7 position is methoxylated in two steps: hydroxylation, followed by methylation of the hydroxylated intermediate to form cephamycin C (the order of C-3′-carbamoylation and C-7 methoxylation may be arbitrary).[12,13]

17.2 Overview of Cephalosporin/Cephamycin C Biosynthesis

17.2.1 Biosynthesis of the Bicyclic β-Lactam Core

The step common to all penicillin and cephalosporin producers is the remarkable cyclization of the unusual tripeptide, δ-(L-α-aminoadipoyl)-L-cysteinyl-D-valine (ACV), leading to the formation of the bicyclic nucleus of penicillins in a single step, using dioxygen and a mononuclear ferrous centre (Figure 17.1).[1,2] The reaction is catalysed by isopenicillin N (IPN) synthase (IPNS, see Chapter 19). IPNS is related to the 2OG-dependent oxygenases, but does not use 2OG in catalysis. Instead IPNS uses the complete four-electron oxidizing power of dioxygen to form the bicyclic β-lactam core in a single step. In organisms producing cephalosporins, such as the eukaryotic *Acremonium* or the prokaryotic *Streptomyces* spp., the product of this reaction is processed by an epimerase that transforms the L-α-aminoadipoyl side chain of IPN to give the D-configured side chain of penicillin N.

17.2.2 Discovery of Cephalosporin Antibiotics

Cephalosporin-producing microorganisms were first isolated at a sewage outlet in Sardinia by Giuseppe Brotzu in 1945.[3,4] Brotzu was intrigued by the 'self-purification' properties of sewage systems and suspected the presence of microorganisms producing antibiotics. The isolated strain of *Cephalosporium acremonium* (now *Acremonium crysogenum*) secretes material active against both Gram-positive and Gram-negative bacteria. Subsequent studies led by Abraham in Oxford showed that this strain produces a number of anti-bacterial compounds, some of which are resistant to hydrolysis by β-lactamases. These chemical studies and X-ray crystallographic investigations by Hodgkin and coworkers showed that the new compound contains a six-membered dihydrothiazine ring fused to the β-lactam ring instead of the five-membered penam ring of penicillins.[5,6] The isolation of the cephalosporin C nucleus, 7-aminocephalosporanic acid (7-ACA), enabled the modification of the side groups and eventually paved the way for the production of semisynthetic cephalosporins with improved pharmacokinetic properties, lowered toxicity and increased resistance to penicillinases.[4]

17.2.3 Cephalosporin Biosynthesis

The committed step in cephalosporin biosynthesis is the expansion of the five-membered thiazolidine ring of the penicillin nucleus into the six-membered dihydrothiazine ring of the cephalosporin nucleus (Figure 17.1).[1] The reaction requires dioxygen, and is catalysed by a 2OG-dependent oxygenase, deacetoxycephalosporin C synthase (DAOCS). In prokaryotic cephalosporin producers, the product of this reaction, DAOC, is hydroxylated at the C-3′ position to form deacetylcephalosporin C (DAC) as catalysed by a second 2OG-dependent oxygenase, deacetylcephalosporin C synthase (DACS).[7] In eukaryotic cephalosporin producers, the reaction is catalysed by a bifunctional enzyme, deacetoxy/deacetylcephalosporin C synthase (DAOC/DACS) that catalyses both the ring expansion and the following 3′-hydroxylation reaction.[8] The prokaryotic and eukaryotic enzymes are closely related by sequence, suggesting transfer of genes from bacteria to fungi.[9] In eukaryotic fungi DAC is further acetylated to give cephalosporin C, which is the terminal reaction in some cephalosporin-producing fungi.

17.2.4 Biosynthesis of Cephamycin C

Prokaryotic organisms such as *Streptomyces* and *Nocardia* convert DAC to O-carbamoyl-DAC (Figure 17.1). This compound is subsequently converted to the 7-methoxy-cephalosporin, cephamycin C, which is less vulnerable to β-lactam hydrolysing enzymes. The 7α-methoxyl group is derived from molecular oxygen[10] and methionine,[11] and the system can use O-carbamoyl-DAC or cephalosporin C as substrates.[12] Methoxylation was shown to occur in two steps, hydroxylation at C-7, followed by methylation of the hydroxylated intermediate.[13] Two genes (*cmcI* and *cmcJ*) in *Amycolatopsis lactamdurans* (previously *Nocardia lactamdurans*) were subsequently identified to encode enzymes responsible for the C-7-methoxylation reaction.[14,15] The *cmcI* and *cmcJ* genes are also present in *S. clavuligerus*.[16] Studies with recombinant and native CmcI and CmcJ led to the proposal that the 7α-methoxylation is performed by the two-component system,[14,15] using ferrous iron, 2OG and *S*-adenosyl-methionine (SAM). This set of substrates indicates that CmcI/CmcJ may involve the third 2OG-dependent ferrous enzyme in the cephamycin C pathway (Figure 17.1). The role of the individual components is not clear, but CmcI has frequently been annotated as the hydroxylase and CmcJ as the methyltransferase. As indicated below, this assignment is probably mistaken.

A crystal structure of *S. clavuligerus* CmcI[17] reveals the polypeptide chain folds into a *C*-terminal Rossmann domain that binds SAM in a fashion similar to the common binding mode of this cofactor in SAM-dependent methyltransferases.[18] This observation suggested CmcI is a methyltransferase that catalyses the transfer of a methyl group from SAM to the 7α-hydroxycephalosporin in the second step of the cephamycin biosynthesis. A magnesium ion is bound at a site close to the SAM binding site and is ligated by residues Asp-160, Glu-186 and Asp-187. Docking of the putative substrate, 7α-hydroxy-O-carbamoyl-DAC, places the 7-hydroxy group of the β-lactam ring as a ligand to the Mg with its A-side facing the methyl group of SAM at a distance that would allow methylation of the hydroxyl group. There was no evidence of an active site with the features of a 2OG-dependent oxygenase in the CmcI enzyme. CmcJ features a HXD motif common to iron binding sites in 2OG-dependent oxygenases in its sequence, indicating that the hydroxylation reaction may be catalysed by the CmcJ protein. Another possible explanation would be that methoxylation is performed by a CmcI/J complex and that the active site is not complete until the complex is formed. In fact, the fold of the *N*-terminal domain of CmcI makes it unlikely that CmcI would be stable as a monomer.[17] It has not yet been possible to produce soluble recombinant CmcJ, so the question of whether CmcJ performs this task alone or in a complex with CmcI requires further work.

17.3 DAOCS Structure, Function and Mechanism

17.3.1 Overall Fold and Arrangement of the Active Site

The crystal structure of *Streptomyces clavuligerus* DAOCS[19] was the first for a 2OG-dependent oxygenase. The protein folds into a conserved double-stranded β-helix (DSBH, also named a jelly-roll fold or Greek key motif) fold with flanking helices (Figure 17.2).[20] The DSBH barrel is a common motif in 2OG-dependent oxygenases; further variations of this motif are found in the extended family of 2OG-dependent oxygenases (see Chapter 2).[21] The barrel provides three iron binding ligands, two His residues and an Asp residue that are part of a conserved HXD/E...H motif.[19,22] The three protein residues constitute one face of an octahedral Fe(II) centre forming a two-His, one-carboxylate facial triad iron-binding motif.[22–24] In the unliganded holoenzyme with Fe(II) bound, the remaining three coordination sites opposite the facial triad are occupied by solvent. These three sites are available for binding of exogenous ligands, such as cofactors, inhibitors and substrates, including dioxygen. This is in contrast to haem enzymes, which have the four equatorial ligand-binding sites occupied by the porphyrin ring and one axial site occupied by a variable amino acid residue that affixes the metallo-porphyrin in the active site. Thus, in the haem enzymes only a single site is available to bind a single exogenous reactant at a time, an arrangement that bars the use of a cosubstrate such as 2OG. In this respect, the 2OG-dependent oxygenases offer a much more versatile platform for catalysis. However, this flexibility may come at the price of a more uncontrollable catalysis often, at least when working with isolated enzyme, leading to oxidative damage.[25–27]

17.3.2 Monomer-to-Trimer Transition and the Role of the *C*-Terminus

Recombinant DAOCS occurs in solution predominately as a monomer, which is in equilibrium with a trimeric form.[19,28] Crystal packing favours the trimeric form, which is stabilized by an intersubunit contact between the flexible *C*-terminal arm (residues 308–311) of one monomer and the active site of the neighbouring monomer in a cyclic fashion.[19,28] In particular, Lys-310 from one subunit interacts with the active-site pocket of the neighbouring subunit.

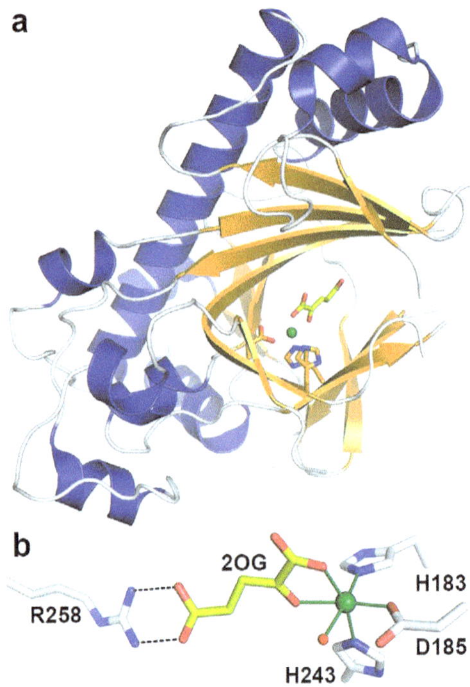

Figure 17.2 Structure of DAOCS. (a) Ribbon diagram with beta sheets in yellow, and helices in blue. (b) Active site in the complex with Fe(II) and 2OG (PDB code: 1rxg).[19] The iron is displayed as a green sphere, active-site residues are depicted as sticks and 2OG is in yellow. The figure was generated using Pymol.[80]

The trimeric oligomerization in this crystal form fixes the *C*-terminal arm and partially blocks access to the active site. Addition of Fe(II) or 2OG shifts the equilibrium toward the monomeric form that appears to be the active form in solution.[28] The soaking of Fe(II) or Fe(II) and 2OG into crystals of the DAOCS apoprotein leaves the crystal intact, but causes disorder of residues 309–311 and results in increased access to the active site. It has been suggested that the trimeric structure may have some functional significance, for example in protecting the *C*-terminal arm (rich in lysines and arginines) against proteases in the apoprotein or resting form of DAOCS.[19] There is also evidence that the *C*-terminus is involved in substrate binding and catalysis (see Section 17.3.4).

17.3.3 Extension of the *N*-Terminus: Breaking of the Trimer Interaction Observed in Crystals of DAOCS Apoprotein

Modification of DAOCS by creating a gene fusion that attaches a His-tag to its *N*-terminal Met-1 residue[29] breaks the trimer formation observed in crystals of DAOCS apoprotein.[19,28] It appears that the *N*-terminal His-tag acts as a spacer hindering the close contact of the monomers required to form the trimer.[29] As expected, the modification does not change the overall protein fold, but some significant changes are observed in the *C*-terminal region. Residues 268–299 rotate by 16° toward the active site on a hinge consisting of Pro-267 and Asn-268 and the *C*-terminal residues 300–311 become disordered. The altered crystal packing resulting from the introduction of the *N*-terminal His-tag in turn influences the binding of ligands to the enzyme (see Section 17.3.5.3). It is possible that conformational changes of the *C*-terminus are involved in catalysis, perhaps by the folding to an α-helix as observed in IPNS or in the plant enzyme anthocyanidin synthase.[22,30]

17.3.4 Influence of Substitutions in the *C*-Terminal Region on Catalysis

The sequence and length of the *C*-terminus are important in determining substrate selectivity and ring expansion activity. Substitutions of the *C*-terminal residues 304–308 were found to influence substrate selectivity.[31–40] In several cases, alterations to non-polar residues were found to alter selectivity from penicillin N, which has a polar side chain, towards clinically useful penicillins with hydrophobic side chains. Truncation of the *C*-terminus[33,36] also alters the selectivity from penicillin N towards penicillin G, whereas shortening the polypeptide chain from the *C*-terminus by up to six residues diminishes activity or increases the rate of uncoupling of substrate conversion from 2OG decarboxylation.[19,32,41] These results suggest that the *C*-terminal arm may be involved in catalysis, *e.g.* by assisting the binding of the substrates and products during catalysis or by isolating reactive intermediates.

17.3.5 Ligand Binding to DAOCS

17.3.5.1 Iron Binding Site and the Binding of 2OG to the Ferrous Ion

In DAOCS, the ferrous ion is ligated by His-183, Asp-185, His-243 and three solvent molecules in an almost perfect octahedral geometry.[19] Comparison of the apoprotein and Fe(II)-bound holoenzyme structures shows that there are very few active-site differences. The position of the iron in the ternary DAOCS–Fe(II)–2OG complex superimposes with the position of the iron in the binary enzyme–iron complex. The cosubstrate binds in a bidentate manner, replacing two solvent ligands to the iron. The 1-carboxylate of 2OG binds *trans* to His-243, and the 2-carbonyl group binds *trans* to Asp-185. The latter site is equivalent to the position where nitric oxide, a dioxygen analogue, binds in IPNS.[42] The stereochemistry of the cosubstrate binding is similar to that proposed from chemical and mutagenesis studies for prolyl 4-hydroxylase,[43] and spectroscopic studies of clavaminate synthase (CAS).[44] The 5-carboxylate of the cosubstrate forms a salt bridge with Arg-258 and is hydrogen bonded to Ser-260 (Figure 17.2). The importance of this interaction has been demonstrated by substitutions of Arg-258 that reduce activity and alter cosubstrate selectivity.[45,46] Residues corresponding to Arg-258 and Ser-260 are conserved within the DAOCS subfamily of enzymes. The only accessible site at the iron in the 2OG complex of DAOCS is located in a hydrophobic pocket lined by residues Ile-192, Val-262, Phe-225, Phe-264 and the methyl group of Thr-190. This site is occupied by a solvent molecule in the structure[19] (Figure 17.2) and has been proposed to be the binding site for dioxygen in DAOCS and related 2-oxoacid-dependent ferrous enzymes.

17.3.5.2 Binding of Succinate and Uncoupled Turnover

Information on the 2OG-dependent oxygenases[2,47–49] indicates that the binding mode and coordination of 2OG to the active-site metal follow a similar pattern in DAOCS; however, the sequence of events following this step appears to be more diverse. Structural studies on crystals of DAOCS soaked with succinate and bicarbonate have shown the co-product binds with the proximal carboxylate ligated to the iron in a monodentate manner.[41,50] The coordination geometry of the iron is octahedral in both cases, but the binding mode of succinate and the character of the ligands differ. A crystal structure of a complex with succinate and carbon dioxide bound to the iron was obtained from a deletion mutant of DAOCS (truncating the *C*-terminus at residue 307) and soaking with succinate and bicarbonate.[41] In this complex, carbon dioxide appears to bind at a position *trans* to His-243 as expected if CO_2 were produced by oxidative decarboxylation of 2OG, while succinate binds with its carboxylate to the site originally occupied by the carbonyl group of 2OG. Yet the distal carboxylate of succinate does not interact with Arg-258/Ser-260 (the R(K)XS motif) as seen in the complex with 2OG. Instead, the distal carboxylate interacts with Arg-162 and points towards solution as if it were about to dissociate from the active site. In contrast, the soaking of succinate and bicarbonate into crystals of the full-length DAOCS[50] results in octahedral coordination geometry of the iron, but here a second water molecule takes up the expected position of carbon dioxide. Succinate is bound in a position that would be expected if it were produced by oxidative decarboxylation of 2OG, with its proximal carboxylate group ligated monodentate to iron and the distal carboxylate stabilized by interactions with the RXS motif (Arg-258 and Ser-260). Presumably the differences in binding of succinate may arise from slight differences, *e.g.* in the length of the *C*-terminus, differences in crystal packing and local differences in pH in the crystal.

Uncoupled turnover of 2OG in the absence of the prime substrate to yield succinate and carbon dioxide has been detected for many 2OG-dependent oxygenases[43,51–53] and may produce succinate bound in an orientation as suggested by the succinate complexes with DAOCS. This reaction further suggests that a reactive species can form with oxygen without the binding of the prime substrate (see further below). Ascorbate may be required as an alternative oxygen acceptor in these cases.[54]

17.3.5.3 Binding of Prime Substrates and Products

Crystal structures of DAOCS in complex with the alternative substrates penicillin G and ampicillin and the product DAOC[29,50] show these compounds bind at the iron in a mode that overlaps with the binding sites of 2OG and succinate (Figure 17.3). The side chain of the penicillin substrates bind in a pocket of mainly hydrophobic character and, as predicted, the carbonyl and carboxylate groups of the bicyclic nucleus are stabilized by electrostatic/hydrogen bonding interactions with the guanidinium groups of Arg-160 and Arg-162[28,55] and by the binding to well-ordered water molecules. As predicted,[56] the substrates coordinate their sulphur atom directly to the pentacoordinate iron *trans* to His-243 at the position expected to be occupied by carbon dioxide after oxidative decarboxylation of 2OG (Figure 17.3). Presumably, a ferryl group (*i.e.* Fe(IV)–oxo) would occupy the vacant position liberated by the water molecule observed at the sixth position *trans* to His-183 in the 2OG complex.[50] The β-methyl (*pro*-S) group to be inserted into the cephem ring is placed such that a ferryl oxygen at the vacant sixth site would be in a position to abstract a hydrogen atom from the methyl group. The natural cephalosporin product (DAOC) binds in the same pocket as the penicillin substrates (Figure 17.3c) with the sulphur atom directly coordinated to the iron.

Ligand binding in crystals of the His-tagged enzyme[29] point to iron coordination according to the same topological principles as in the trimeric structure, but with a certain flexibility in the binding of the side chain of the ligand. The alternative substrate ampicillin ligates to the iron *via* the sulphur atom *trans* to His-243 with an iron–sulphur distance of 2 Å (PDB entry 1w2n) as in the trimeric model. One solvent molecule occupies the fifth position *trans* to

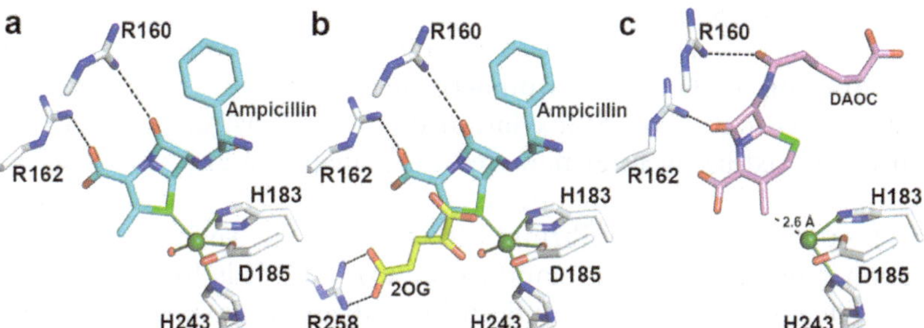

Figure 17.3 Ligand complexes with DAOCS. (a) Fe(II) and ampicillin (PDB code: 1w2n),[29] (b) the same complex as in (a) showing the overlapping 2OG position (PDB code: 1rxg)[19,29] and (c) Fe(II) and DAOC (PDB code: 1w2o).[29] Amino acids within 3 Å of the ligand and the ferrous ion are shown. One water molecule is bound at a fifth site on the iron in the ampicillin complex. The binding orientation of ampicillin and DAOC to DAOCS is similar to that of ACV[42] and isopenicillin N to IPNS.[78] The figure was generated using Pymol.[80]

Asp-185. The penicillin nucleus is rotated by about 70° relative to its position in the trimeric complex and the 2-amino-phenylacetyl side chain overlaps with the position occupied by residues 304–306 of the *C*-terminal arm of a neighbouring molecule in the trimeric form. This observation indicates that trimeric arrangement may disturb substrate binding and that ligand binding may be a molecular trigger for the trimer-to-monomer transition. Binding of DAOC to the monomeric His-tagged enzyme (PDB entry 1w2o) is similar to its binding in the trimeric form, but the product is shifted to a position further away from the iron. The sulphur of the cephalosporin nucleus is positioned at an iron–sulphur distance of 5 Å and the methyl group is located at a distance of 2.6 Å from the iron. The carbonyl oxygen atom of the α-aminoadipoyl side chain of DAOC is within hydrogen bonding distance to the guanidinium group of Arg-160.

Binding of the prime substrate/product depends on the presence in the soaking solution of 2OG/succinate (bound to the iron in a 50/50 mixture overlapping with the prime substrate/product). Soaking of prime substrates or products in the absence of 2OG (or succinate) results in binding at the 2OG/succinate site such that the C3 carboxylate of the thiazolidine ring ligates the iron and the carbonyl oxygen atom of the α-aminoadipoyl side chain interacts with the RXS motif (PDB entry 1uof). This binding mode was deemed to be unproductive.[50]

17.3.6 Mechanistic Implications of the Structures for Ring Expansion

17.3.6.1 Suggested Mechanism of DAOCS-Catalysed Ring Expansion Based on the Crystal Structures

Based on the various ligand complexes with cosubstrate/co-product and prime substrates/product, a mechanism was proposed for the ring expansion catalysed by DAOCS,[50] with the main features summarized in Figure 17.4. Binding of 2OG to the iron activates the ferrous ion for oxygen binding,[19] and oxidative decarboxylation of the cosubstrate results in formation of the oxidizing intermediates plus succinate and carbon dioxide as by-products. The oxidizing species is assumed to be a planar persuccinic acid Fe(II) species in equilibrium with a more reactive ferryl Fe(IV)=O form. Carbon dioxide is a weak ligand to the iron and it is likely to leave easily, while succinate remains bound and stabilizes the oxidizing iron species. The planar persuccinic acid species has been described as a 'booby trap' oxidizing species ready to react when the substrate replaces the succinate at the iron,[50] a process termed negative catalysis.[57,58] Density functional theory (DFT) calculations combined with *ab initio* quantum molecular dynamics and simulated annealing, based on the DAOCS crystal structures,[50,59] support the energetic accessibility of this mechanism. Binding of the cosubstrate was shown to enhance the reactivity of the ferrous ion, while the reactivity of the oxidizing intermediate is turned down by the

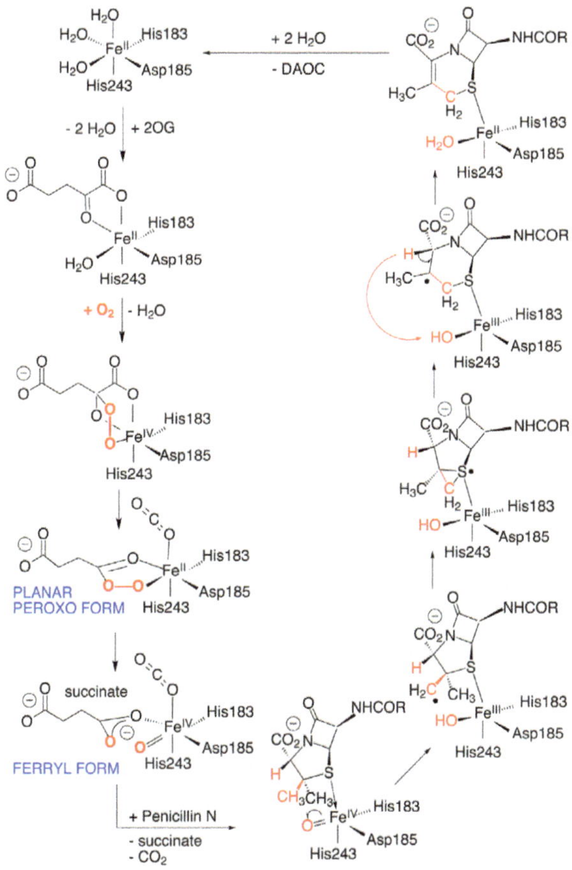

Figure 17.4 Mechanism for the ring expansion reaction catalysed by DAOCS as proposed by Valegård *et al.* and based on the mode of penicillin and cephalosporin binding.[50] In the early steps of the reaction, one of the oxygen atoms of dioxygen is incorporated into succinate while the second oxygen atom remains on the iron. The oxidizing species is formed by generation of a planar persuccinic acid species that is in equilibrium with the more reactive ferryl, Fe(IV)=O, form. This species can remove two hydrogen atoms from the five-membered thiazolidine ring to form the six-membered dihydrothiazine ring of the cephalosporin product.

negative charge of the reaction product, succinate. Binding of the prime substrate expels succinate and allows the penicillin substrate to bind *via* its sulphur atom to the iron. This brings the requisite β-methyl group of the penicillin core within easy reacting distance to the expected position of the ferryl oxygen. Ring expansion, possibly involving radical formation,[60–62] and transfer of two hydrogen atoms to the oxygen results in the formation of water and the cephalosporin product. The binding of the penicillin sulphur atom to the iron may allow electron transfer to the oxygen directly or *via* the oxidized iron during catalysis.

17.3.6.2 Comparison with Other 2OG-Dependent Oxygenases

The structural data show the 2OG cosubstrate/succinate co-product overlaps with the binding site of penicillin/cephalosporin in the enzyme, indicating these compounds cannot bind simultaneously at the iron without a substantial rearrangement of the active site.[50] This finding was unexpected in the light of structural studies on other 2OG-dependent enzymes, *e.g.* CAS,[63,64] alkylsulfatase,[65] taurine/2OG dioxygenase (TauD),[66,67] the hypoxia-inducible transcription factor (HIF) prolyl 4-hydroxylase 2,[68] and the DNA/RNA repair enzyme AlkB,[69] which show simultaneous binding of cosubstrate/substrate in crystal structures. In most of these structures the ferrous ion is five-coordinate, but six-coordinate iron is also observed.[69] Based on this structural evidence combined with data from spectroscopic studies[70,71] and earlier models,[72] a mechanism was proposed for 2OG oxygenases where the prime substrate binds proximal to the iron, but is not directly ligated to the iron.[47,63,73] Binding of the substrate expels the water molecule from the sixth site and the metal centre is converted from octahedral to pentacoordinate square pyramidal geometry. This change in coordination is proposed to serve as a conformational trigger for oxygen binding at the vacated site, leading to decarboxylation of the cosubstrate, formation of an oxidizing species, and oxidation of the substrate in these enzymes.[70,71,74] The evidence for this mechanism is incomplete; crystal structures show 2OG binds to the iron in two different positions, implying initial oxygen binding either *trans* to the His in the HXD motif (proximal site)[19,64,69,75] or opposite to the distal His (corresponding to His-243 in DAOCS).[30,63,65,66] Structures of the human HIF prolyl 4-hydroxylase and an algal prolyl 4-hydroxylase also show the binding of inhibitors compatible with oxygen binding *trans* to the proximal His.[68,76] To achieve the correct geometry for substrate oxidation, in some cases the possibility of rearrangements of the iron coordination with respect to the ligation of 2OG, and/or the position of the oxidizing species, has been raised. Alternative binding of 2OG in crystal structures of CAS indicates that rearrangement of 2OG binding may be possible.[64] Rearrangement of the ferryl species prior to oxidation of the prime substrate has been proposed as a mechanism to achieve the correct catalytic ligation around the iron.[64,66,69]

The available crystallographic evidence indicates that DAOCS may behave differently to most other 2OG oxygenases, leading to the proposal that the oxidizing iron species has to be formed before the binding of the prime substrate.[50] Steady-state kinetic measurements on DAOCS,[29,50,77] showing substrate inhibition and interference between cosubstrate and prime substrate, is consistent with this mechanism. Some steady-state kinetic studies on prolyl 4-hydroxylase[51,54] show that the binding of dioxygen may precede that of the peptide substrate, in line with the proposition that prolyl 4-hydroxylase may follow the same mechanism as proposed for DAOCS.

Studies on prolyl 4-hydroxylase and many other 2OG-dependent oxygenases[43,51] have shown that 2OG oxidation can take place without the prime substrate. In some cases such uncoupled turnover can lead to self-hydroxylation, *e.g.* as for TfdA[26] and TauD.[27] Thus, generating the oxidizing species in the absence of the prime substrate poses the problem of how to store a reactive species awaiting the prime substrate so as to avoid inactivation or self-destruction. DFT calculations on DAOCS[50,59] have provided support for the presence of a relatively stable persuccinic acid intermediate in equilibrium with a more reactive ferryl species. These results indicate that the 'tamed' peroxo intermediate may be stored transiently to contain the reactivity of the oxidizing species awaiting the encounter with the penicillin substrate (this differs from the conventional mechanism, see Chapter 3).

Comparison of penicillin binding in DAOCS and the closely related IPNS[78] shows that the topology of binding of the prime substrate is similar in these enzymes. The thiolate of ACV and the thiazolidine ring of the penicillin substrates coordinate directly to the catalytic iron and the α-aminoadipoyl side chains occupy similar positions. Thus, in this respect, DAOCS seems to be more closely related to IPNS than to other 2OG-dependent oxygenases. Enzymes belonging to the extended family of non-haem iron oxygenases, to which DAOCS belongs, catalyse a remarkable variety of reactions.[79] Several enzymes, including the bifunctional DAOC/DACS, catalyse multiple reactions within the same active site. Given this variation, it may be that different 2OG-dependent oxygenases employ distinct mechanisms to generate the oxidizing species. Haem structures provide a solid container for highly charged iron species, and excess charge can be distributed safely. Mononuclear ferrous enzymes have to cope with the reactive species in different ways; the iron is directly ligated to the protein without any protective container. This coordination gives rise to rich chemistry and many side reactions. The loose substrate and product specificities on a versatile catalytic platform provide advantages for such systems in evolutionary terms. The differences observed in the mechanism of the various 2OG-dependent enzymes reviewed here are a manifestation of this versatility.

Note added in proof: Recent biochemical studies[81] suggest that the ring expansion of penicillin N catalysed by DAOCS may follow the consensus mechanism proposed for other 2OG oxygenases.

References

1. J. E. Baldwin & C. J. Schofield., *The Chemistry of β-Lactams*, ed. M. I. Page, Blackie, London, 1992, pp. 1–78.
2. C. J. Schofield, J. E. Baldwin, M. F. Byford, I. J. Clifton, C. Hensgens and P. L. Roach, *Curr. Opin. Struct. Biol.*, 1997, **7**, 857–864.
3. G. Brotzu., *Lavori Istituto Igiene Cagliari*, 1948, 1–11.
4. E. P. Abraham, *Drugs*, 1987, **34**, Suppl. 2, 1–14.
5. E. P. Abraham and G. G. F. Newton, *Biochem. J.*, 1961, **79**, 377–393.
6. D. C. Hodgkin and E. N. Maslen, *Biochem. J.*, 1961, **79**, 393–402.
7. S. E. Jensen, D. W. Westlake and S. Wolfe, *J. Antibiot.*, 1985, **38**, 263–265.
8. A. Scheidegger, M. T. Küenzi and J. Nüesch, *J. Antibiot.*, 1984, **37**, 522–531.
9. Y. Aharonowitz, G. Cohen and J. F. Martin, *Annu. Rev. Microbiol.*, 1992, **46**, 461–495.
10. J. O'Sullivan, R. T. Aplin, C. M. Stevens and E. P. Abraham, *Biochem. J.*, 1979, **179**, 47–52.

11. J. G. Whitney, D. R. Brannon, J. A. Mabe and K. J. Wicker, *Antimicrob. Agents Chemother.*, 1972, **1**, 247–251.
12. J. O'Sullivan and E. P. Abraham, *Biochem. J.*, 1980, **186**, 613–616.
13. J. D. Hood, A. Elson, M. L. Gilpin and A. G. Brown, *J. Chem. Soc., Chem. Commun.*, 1983, 1187–1188.
14. J. J. R. Coque, F. J. Enguita, J. F. Martin and P. Liras, *J. Bacteriol.*, 1995, **177**, 2230–2235.
15. F. J. Enguita, P. Liras, A. L. Leitao and J. F. Martin, *J. Biol. Chem.*, 1996, **271**, 33225–33230.
16. D. C. Alexander and S. E. Jensen, *J. Bacteriol.*, 1998, **180**, 4068–4079.
17. L. M. Öster, D. R. Lester, A. C. Terwisscha van Scheltinga, M. Svenda, M. van Lun, C. Généreux and I. Andersson, *J. Mol. Biol.*, 2006, **358**, 546–558.
18. J. L. Martin and F. McMillan, *Curr. Opin. Struct. Biol.*, 2002, **12**, 783–793.
19. K. Valegård, A. C. Terwisscha van Scheltinga, M. D. Lloyd, T. Hara, S. Ramaswamy, A. Perrakis, A. Thompson, H.-J. Lee, J. E. Baldwin, C. J. Schofield, J. Hajdu and I. Andersson, *Nature*, 1998, **394**, 805–809.
20. S. C. Harrison, A. J. Olson, C. E. Schutt, F. K. Winkler and G. Bricogne, *Nature*, 1978, **276**, 368–373.
21. I. J. Clifton, M. A. McDonough, D. Ehrismann, N. J. Kershaw, N. Granatino and C. J. Schofield, *J. Inorg. Biochem.*, 2006, **100**, 644–669.
22. P. L. Roach, I. J. Clifton, V. Fülöp, K. Harlos, G. J. Barton, J. Hajdu, I. Andersson, C. J. Schofield and J. E. Baldwin, *Nature*, 1995, **375**, 700–704.
23. E. L. Hegg and L. Que, *Eur. J. Biochem.*, 1997, **250**, 625–629.
24. L. Que, *Nat. Struct. Mol. Biol.*, 2000, **7**, 182–184.
25. J. N. Barlow, Z. H. Zhang, P. John, J. E. Baldwin and C. J. Schofield, *Biochemistry*, 1997, **36**, 3563–3569.
26. A. Liu, R. Y. N. Ho, L. Que, Jr., M. J. Ryle, B. S. Phinney and R. P. Hausinger, *J. Am. Chem. Soc.*, 2001, **123**, 5126–5127.
27. M. J. Ryle, A. Liu, R. B. Muthukumaran, R. Y. N. Ho, K. D. Koehntop, J. McCracken, L. Que, Jr. and R. P. Hausinger, *Biochemistry*, 2003, **42**, 1854–1862.
28. M. D. Lloyd, H.-J. Lee, K. Harlos, Z.-H. Zhang, J. E. Baldwin, C. J. Schofield, J. M. Charnock, C. D. Garner, T. Hara, A. C. Terwisscha van Scheltinga, K. Valegård, J. A. C. Viklund, J. Hajdu, I. Andersson, Å. Danielsson and R. Bhikhabhai, *J. Mol. Biol.*, 1999, **287**, 943–960.
29. L. M. Öster, A. C. Terwisscha van Scheltinga, K. Valegård, A. MacKenzie Hose, A. Dubus, J. Hajdu and I. Andersson, *J. Mol. Biol.*, 2004, **343**, 157–171.
30. R. C. Wilmouth, J. J. Turnbull, R. W. Welford, I. J. Clifton, A. G. Prescott and C. J. Schofield, *Structure*, 2002, **10**, 93–103.
31. H. S. Chin, J. Sim and T. S. Sim, *Biochem. Biophys. Res. Commun.*, 2001, **287**, 507–513.
32. H. S. Chin and T. S. Sim, *Biochem. Biophys. Res. Commun.*, 2002, **295**, 55–61.
33. H.-J. Lee, C. J. Schofield and M. D. Lloyd, *Biochem. Biophys. Res. Commun.*, 2002, **292**, 66–70.
34. C.-L. Wei, Y.-B. Yang, W.-C. Wang, W.-C. Liu, J.-S. Hsu and Y.-C. Tsai, *Appl. Environ. Microbiol.*, 2003, **69**, 2306–2312.
35. H. S. Chin, K. S. Goo and T. S. Sim, *Appl. Environ. Microbiol.*, 2004, **70**, 607–609.

36. M. D. Lloyd, S. J. Lipscomb, K. S. Hewitson, C. M. H. Hensgens, J. E. Baldwin and C. J. Schofield, *J. Biol. Chem.*, 2004, **279**, 15420–15426.
37. X. B. Wu, K. Q. Fan, Q. H. Wang and K. Q. Yang, *FEMS Microbiol. Lett.*, 2005, **246**, 103–110.
38. K. S. Goo, C. S. Chua and T.-S. Sim, *Proteins*, 2008, **70**, 739–747.
39. X. B. Wu, X. Y. Tian, J. J. Ji, W. B. Wu, K. Q. Fan and K. Q. Yang, *Biotechnol. Lett.*, 2011, **33**, 805–812.
40. J. Ji, K. Fan, X. Tian, Y. Zhang and K. Yang, *Appl. Environ. Microbiol.*, 2012, **78**, 7809–7812.
41. H.-J. Lee, M. D. Lloyd, K. Harlos, I. J. Clifton, J. E. Baldwin and C. J. Schofield, *J. Mol. Biol.*, 2001, **308**, 937–948.
42. P. L. Roach, I. J. Clifton, C. M. H. Hensgens, N. Shibata, C. J. Schofield, J. Hajdu and J. E. Baldwin, *Nature*, 1997, **387**, 827–830.
43. J. Myllyharju and K. I. Kivirikko, *EMBO J.*, 1997, **16**, 1173–1180.
44. E. G. Pavel, J. Zhou, R. W. Busby, M. Gunsior, C. A. Townsend and E. I. Solomon, *J. Am. Chem. Soc.*, 1998, **120**, 743–753.
45. H.-J. Lee, M. D. Lloyd, I. J. Clifton, K. Harlos, A. Dubus, J. E. Baldwin, J.-M. Frère and C. J. Schofield, *J. Biol. Chem.*, 2001, **276**, 18290–18295.
46. H.-J. Lee, Y.-F. Dai, C.-Y. Shiau, C. J. Schofield and M. D. Lloyd, *Eur. J. Biochem.*, 2003, **270**, 1301–1307.
47. R. P. Hausinger, *Crit. Rev. Biochem. Mol. Biol.*, 2004, **39**, 21–68.
48. V. Purpero and G. R. Moran, *J. Biol. Inorg. Chem.*, 2007, **12**, 587–601.
49. M. A. McDonough, C. Loenarz, R. Chowdhury, I. J. Clifton and C. J. Schofield, *Curr. Opin. Struct. Biol.*, 2010, **20**, 659–672.
50. K. Valegård, A. C. Terwisscha van Scheltinga, A. Dubus, G. Ranghino, L. M. Öster, J. Hajdu and I. Andersson, *Nat. Struct. Mol. Biol.*, 2004, **11**, 95–101.
51. L. Tuderman, R. Myllylä and K. I. Kivirikko, *Eur. J. Biochem.*, 1977, **80**, 341–348.
52. U. Puistola, T. M. Turpeenniemi-Hujanen, R. Myllylä and K. I. Kivirikko, *Biochim. Biophys. Acta*, 1980, **611**, 51–60.
53. D. F. Counts, G. J. Cardinale and S. Udenfriend, *Proc. Natl. Acad. Sci. U. S. A.*, 1978, **75**, 2145–2149.
54. R. Myllylä, K. Majamaa, V. Günzler, H. M. Hanauske-Abel and K. I. Kivirikko, *J. Biol. Chem.*, 1984, **259**, 5403–5405.
55. S. J. Lipscomb, H. J. Lee, M. Mukherji, J. E. Baldwin, C. J. Schofield and M. D. Lloyd, *Eur. J. Biochem.*, 2002, **269**, 2735–2739.
56. J. E. Baldwin, R. M. Adlington, N. P. Crouch, C. J. Schofield, N. J. Turner and R. T. Aplin, *Tetrahedron*, 1991, **47**, 9881–9900.
57. J. Retey, *Angew. Chem., Int. Ed.*, 1990, **29**, 355–361.
58. A. Szöke, W. G. Scott and J. Hajdu, *FEBS Lett.*, 2003, **553**, 18–20.
59. G. Cicero, C. Carbonera, K. Valegård, J. Hajdu, I. Andersson and G. Ranghino, *Int. J. Quant. Chem.*, 2007, **107**, 1514–1522.
60. C. P. Pang, R. L. White, E. P. Abraham, D. H. G. Crout, M. Lutstorf, P. J. Morgan and A. E. Derome, *Biochem. J.*, 1984, **222**, 777–788.
61. C. A. Townsend, A. B. Theis, A. S. Neese, E. B. Barrabee and D. Poland, *J. Am. Chem. Soc.*, 1985, **107**, 4760–4767.

62. J. E. Baldwin, R. M. Adlington, T. W. Kang, E. Lee and C. J. Schofield, *Tetrahedron*, 1988, **44**, 5953–5957.
63. Z. Zhang, J. Ren, D. K. Stammers, J. E. Baldwin, K. Harlos and C. J. Schofield, *Nat. Struct. Mol. Biol.*, 2000, **7**, 127–133.
64. Z. Zhang, J.-S. Ren, K. Harlos, C. H. McKinnon, I. J. Clifton and C. J. Schofield, *FEBS Lett.*, 2002, **517**, 7–12.
65. I. Müller, A. Kahnert, T. Pape, G. M. Sheldrick, W. Meyer-Klaucke, T. Dierks, M. Kertesz and I. Usón, *Biochemistry*, 2004, **43**, 3075–3088.
66. J. M. Elkins, M. J. Ryle, I. J. Clifton, J. C. Dunning Hotopp, J. S. Lloyd, N. I. Burzlaff, J. E. Baldwin, R. P. Hausinger and P. L. Roach, *Biochemistry*, 2002, **41**, 5185–5192.
67. J. R. O'Brien, D. J. Schuller, V. S. Yang, B. D. Dillard and W. N. Lanzilotta, *Biochemistry*, 2003, **42**, 5547–5554.
68. M. A. McDonough, V. Li, E. Flashman, R. Chowdhury, C. Mohr, B. M. Lienard, J. Zondlo, N. J. Oldham, I. J. Clifton, J. Lewis, L. A. McNeill, R. J. Kurzeja, K. S. Hewitson, E. Yang, S. Jordan, R. S. Syed and C. J. Schofield, *Proc. Natl. Acad. Sci. U. S. A.*, 2006, **103**, 9814–9819.
69. B. Y. Yu, W. C. Edstrom, Y. Benach, Y. Hamuro, P. C. Weber, B. R. Gibney and J. F. Hurt, *Nature*, 2006, **439**, 879–884.
70. E. I. Solomon, T. C. Brunold, M. I. Davis, J. N. Kemsley, S.-K. Lee, N. Lehnert, F. Neese, A. J. Skulan, Y.-S. Yang and J. Zhou, *Chem. Rev.*, 2000, **100**, 235–349.
71. J. Zhou, W. L. Kelly, B. O. Bachmann, M. Gunsior, C. A. Townsend and E. I. Solomon, *J. Am. Chem. Soc.*, 2001, **123**, 7388–7398.
72. H. M. Hanauske-Abel and V. A. Günzler, *J. Theor. Biol.*, 1982, **94**, 421–455.
73. M. Costas, M. P. Mehn, M. P. Jensen and L. Que, Jr., *Chem. Rev.*, 2004, **104**, 939–966.
74. J. C. Price, E. W. Barr, B. Tirupati, J. M. Bollinger, Jr. and C. Krebs, *Biochemistry*, 2003, **42**, 7497–7508.
75. I. J. Clifton, L. X. Doan, M. C. Sleeman, M. Topf, H. Suzuki, R. C. Wilmouth and C. J. Schofield, *J. Biol. Chem.*, 2003, **278**, 20843–20850.
76. M. K. Koski, R. Hieta, C. Böllner, K. I. Kivirikko, J. Myllyharju and R. K. Wierenga, *J. Biol. Chem.*, 2007, **282**, 37112–37123.
77. A. Dubus, M. D. Lloyd, H.-J. Lee, C. J. Schofield, J. E. Baldwin and J.-M. Frère, *Cell. Mol. Life Sci.*, 2001, **58**, 835–843.
78. N. I. Burzlaff, P. J. Rutledge, I. J. Clifton, C. M. H. Hensgens, M. Pickford, R. M. Adlington, P. L. Roach and J. E. Baldwin, *Nature*, 1999, **401**, 721–724.
79. A. G. Prescott, *J. Exp. Bot.*, 1993, **44**, 849–861.
80. *The PyMOL Molecular Graphics System*, Version 1.3 Schrödinger, LLC.
81. H. Tarhonskaya, A. Szöllössi, I. K. H. Leung, J. T. Bush, L. Henry, R. Chowdhury, A. Iqbal, T. D. W. Claridge, C. J. Schofield and E. Flashman, *Biochemistry*, 2014, **53**, 2483–2493.

CHAPTER 18

Recent Advances in the Structural and Mechanistic Biology of Non-Haem Fe(II), 2-Oxoglutarate and O_2-Dependent Halogenases

JANET L. SMITH*[a,b] AND DHEERAJ KHARE[a]

[a]Life Sciences Institute University of Michigan, 210 Washtenaw Avenue, Ann Arbor, MI 48109, USA; [b]Department of Biological Chemistry, University of Michigan, 210 Washtenaw Avenue, Ann Arbor, MI 48109, USA
*E-mail: janetsmith@umich.edu

18.1 Introduction

Halogenated natural products are widespread in nature, with more than 4000 such compounds documented.[1–3] These structurally diverse molecules exhibit medically valuable properties and include important antibiotics such as chloramphenicol and vancomycin. The sources of many halogenated natural products are multi-enzyme assembly lines, including polyketide synthase (PKS), non-ribosomal peptide synthetase (NRPS) and hybrid pathways, where the halogenase is a tailoring enzyme. Some mega-synthases employ transient halogenation to provide a potent leaving group for a downstream reaction.[4,5] Nature employs a number of halogenation strategies that vary in the activating cosubstrate and cofactor.[3,6–10]

RSC Metallobiology Series No. 3
2-Oxoglutarate-Dependent Oxygenases
Edited by Robert P. Hausinger and Christopher J. Schofield

Published by the Royal Society of Chemistry, www.rsc.org

Haloperoxidases use a hydrogen peroxide cosubstrate and a haem or vanadate cofactor.[11] *S*-adenosyl-L-methionine-dependent halogenases act on this substrate and subsequently cannibalize 5′-halo-5′-deoxyadenosine to create an organohalogen.[12] FAD-dependent halogenases incorporate halogens into aromatic groups using a dioxygen cosubstrate and $FADH_2$ cofactor.[13] The Fe(II)-, 2-oxoglutarate (2OG)- and O_2-dependent halogenases, the subjects of this review, employ a non-haem Fe(II) cofactor with 2OG and O_2 cosubstrates in halogenation of CH_2 and CH_3 centres. They chlorinate carbon centres of pathway intermediates that are tethered to peptidyl carrier protein (PCP) or acyl carrier protein (ACP) domains through a thioester link to a phosphopantetheinyl (PPant) cofactor.

18.2 Catalysis by Fe(II)/2OG/O_2-Dependent Halogenases

The Fe(II)/2OG/O_2-dependent halogenases are part of a fascinating enzyme family known as the Fe(II)/2OG-dependent oxygenases, which employ a mononuclear non-haem iron centre to activate molecular oxygen for a variety of reactions, including hydroxylation, halogenation, cyclization, isomerization, desaturation and stereoinversion.[14,15] The Fe(II)/2OG-dependent oxygenases have in common a highly reactive ferryl (Fe(IV)=O) intermediate, resulting from decarboxylative oxidation of 2OG to succinate and CO_2.[16] The hydroxylases and halogenases have analogous reaction mechanisms (Scheme 18.1), which have been probed in numerous biochemical and spectroscopic studies.[17] Insights have also come from crystal structures of three halogenases,[18–20] and of multiple hydroxylases such as TauD.[21] The non-haem ferrous iron is always coordinated by two histidine side chains and two oxygen atoms of the cosubstrate 2OG (Scheme 18.1). A fifth ligand position is occupied by a protein carboxylate oxygen (aspartate or glutamate) in the hydroxylases and by a halide ion in the halogenases. Water binds at the sixth position in the absence of substrate. The substrate is not an iron ligand at any point in the catalytic cycle, however substrate binding triggers H_2O release from the iron centre followed by O_2 binding, which initiates catalysis. In the fully assembled metal centre, O_2 quickly reacts with 2OG to form succinate and release CO_2, resulting in a characteristic five-coordinate Fe(IV)=O state.[16,22] The highly reactive ferryl centre abstracts a hydrogen atom from the substrate to form a ferric hydroxylate and a substrate free radical.[23,24] At this point the hydroxylase and halogenase mechanisms diverge. In the hydroxylases, the substrate free radical is resolved by rebound from the ferric hydroxyl,[25,26] whereas the halogen atom is transferred to the substrate in the halogenases (Scheme 18.1).[27]

Spectroscopic evidence for Cl-Fe(IV)=O and Br-Fe(IV)=O intermediates came from studies of the halogenases SyrB2 and CytC3, which demonstrated that chlorination proceeds through a ferryl intermediate that cleaves a C–H

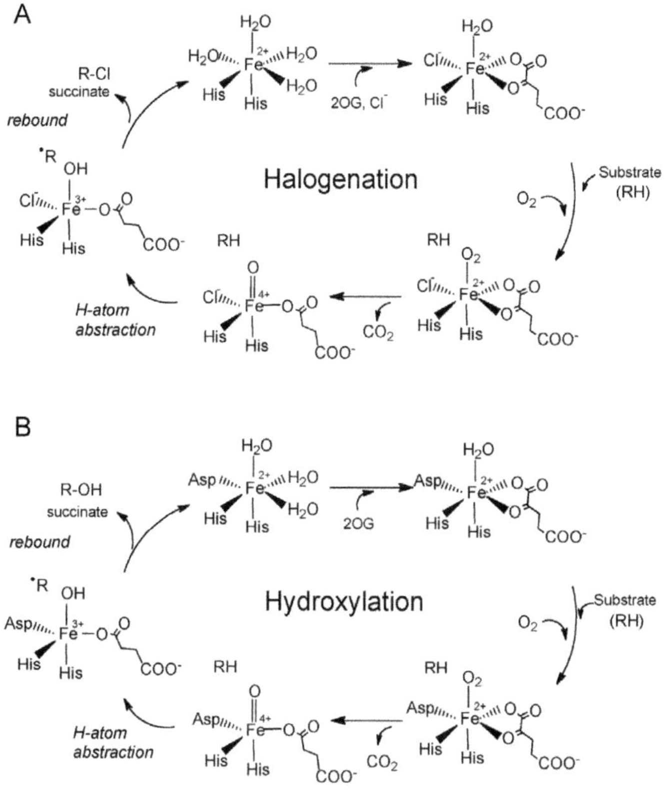

Scheme 18.1 Fe(II)/2OG/O_2-dependent halogenase and hydroxylase reaction mechanisms. The halogenase (A) and hydroxylase (B) mechanisms are similar in the early steps in the catalytic cycle. The ferrous iron is coordinated by two histidine side chains, and by a halide ion in the halogenases and by a carboxylate side chain (aspartate or glutamate) in the hydroxylases. The remainder of the octahedral coordination sphere is occupied by water molecules in the resting enzyme. The cosubstrate 2OG is a bidentate iron ligand, displacing two waters. 2OG binding stimulates substrate (here 'RH') binding, which induces release of the remaining water ligand. The chemical steps begin when O_2 replaces the last water ligand. Iron-activated dioxygen reacts with 2OG, leading to CO_2 release and creating the reactive Fe(IV)=O species. The ferryl group removes a hydrogen atom from the substrate, creating the critical substrate radical and Fe(III)–OH. The final chemical step resolves the substrate radical in a rebound mechanism, by halogen transfer in the halogenases and by hydroxyl transfer in the hydroxylases.

bond.[22,28,29] The reactive Fe(IV)=O species was characterized using synchrotron-based nuclear resonance vibrational spectroscopy, clearly demonstrating an initial step of H-atom abstraction, followed by halogenation.[30] The resting state of the enzyme is regenerated after dissociation of the halogenated product and succinate.

18.3 Biological Range of Fe(II)/2OG/O_2-Dependent Halogenases

All identified Fe(II)/2OG/O_2-dependent halogenases exist within natural product biosynthetic pathways (Figures 18.1 and 18.2). Three characterized halogenases, CmaB, SyrB2 and BarB1/BarB2, perform halogenation reactions in NRPS pathways and act on single amino-acid building blocks tethered to carrier proteins. CmaB, the first-described Fe(II)/2OG/O_2-dependent halogenase, performs a cryptic chlorination step at the γ-position of L-*allo*-isoleucine (Figure 18.2).[5] A second enzyme, CmaC, uses the activated chlorocarbon to form the cyclopropane of coronatine, the pathway product. SyrB2 catalyses the chlorination and bromination of the methyl group of an L-Thr building block in the production of the

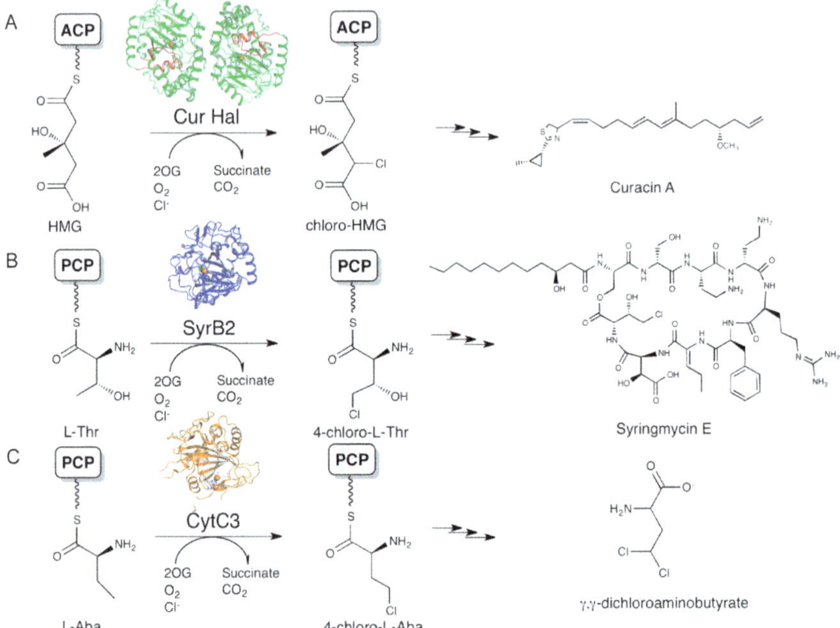

Figure 18.1 Halogenation reactions of structurally characterized Fe(II)/2OG/O_2-dependent enzymes. Three Fe(II)/2OG/O_2-dependent halogenase reactions are shown with the 3D structures of the enzymes. All are from natural product biosynthetic pathways in which the substrate is tethered to the phosphopantetheine (PPant) cofactor of a carrier protein (acyl carrier protein, ACP; peptidyl carrier protein, PCP). The halogenation reactions are shown for (A) Cur Hal in the curacin A pathway (PDB 3NNF);[20] (B) SyrB2 in the syringomycin E pathway (PDB 2FCU);[18] and (C) CytC3, which directly forms the antibiotic γ,γ-dichloroaminobutyrate (PDB 3GJB).[19] The Jam Hal from the jamaicamide pathway[4] catalyses an identical reaction to Cur Hal and is not shown. The natural dimer of Cur Hal is shown; SyrB2 and CytC3 are monomers.

phytotoxin syringomycin E by *Pseudomonas syringae*.[31,32] Similarly, BarB1 and BarB2 from the marine cyanobacterium *Moorea producens* (formerly *Lyngbya majuscula*)[33] have been implicated in the triple chlorination of the unactivated methyl group of an L-leucine building block in barbamide biosynthesis.[27,34] CytC3, a close relative of SyrB2 (58% sequence identity), chlorinates L-2-aminobutyric acid (L-Aba) in the biosynthesis of the antibiotic γ,γ-dichloroaminobutyrate by a soil *Streptomyces*.[35] The *Streptomyces* halogenase Thr3 catalyses the production of 4-chlorothreonine and can replace SyrB2 in the biosynthesis of syringomycin.[36] The kutzneride biosynthetic pathway possesses two Fe(II)/2OG/O_2-dependent halogenases, KtzD and KthP.[37,38]

Chlorination of non-amino-acid substrates was first demonstrated in the biosynthesis of curacin A, a natural product with antimitotic properties isolated from the marine cyanobacterium *M. producens*.[33,39–42] Unlike the stand-alone halogenases of the NRPS pathways, Cur Hal is embedded within a larger PKS polypeptide (CurA) and acts on an ACP-tethered, β-branched diketide intermediate. Cur Hal specifically chlorinates 3-hydroxy-3-methylglutaryl-ACP (HMG-ACP) at carbon-4. This cryptic chlorination step leads to formation of the unique cyclopropane ring in curacin A. Following dehydration of chloro-HMG-ACP by CurE ECH_1 and decarboxylation by CurF ECH_2, chloride is eliminated by the CurF ER, an NADPH-dependent cyclopropanase.[4] Interestingly, the jamaicamide pathway catalyses halogenation, dehydration and decarboxylation reactions using catalytic domains highly similar to those in the curacin pathway, but produces vinyl chloride and not cyclopropane.[4,43] The curacin (Cur) and jamaicamide (Jam) halogenases are 92% identical. Finally, HctB, the most recently characterized

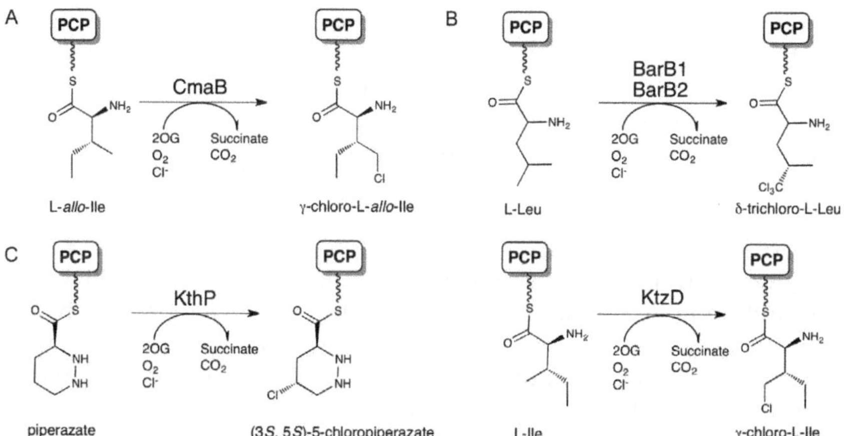

Figure 18.2 Halogenation reactions of Fe(II)/2OG/O_2-dependent enzymes. Halogenation reactions are shown for (A) CmaB from the coronatine pathway;[5] (B) BarB1 and BarB2 from the barbamide pathway;[27,34] and (C) KthP and KtzD from the kutzneride pathway.[37,38]

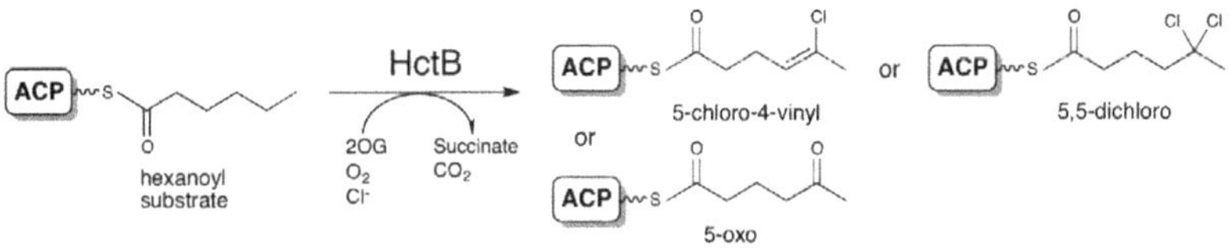

Figure 18.3 Reactions catalysed by the Fe(II)/2OG/O_2-dependent halogenase HctB. The hectochlorin biosynthetic pathway includes an intriguing halogenase that has been shown to produce vinyl chloride and oxygenation products in addition to 5,5-dichlorohexanoyl-ACP.[44]

Fe(II)/2OG/O_2-dependent halogenase, is from the hectochlorin pathway of the same biological source as the Cur and Jam pathways. HctB is predominantly a fatty acyl-ACP halogenase, but also catalyses oxygenation reactions (Figure 18.3).[44]

18.4 Three-Dimensional Structures of Fe(II), 2OG and O_2-Dependent Halogenases

Crystal structures of three Fe(II)/2OG/O_2-dependent halogenases have provided important insights into the catalytic mechanism.[18–20] The overall halogenase reactions catalysed by these enzymes are highlighted in Figure 18.1, and their structures are shown in Figure 18.4. Despite low sequence identity (<16%), SyrB2, CytC3 and Cur Hal share a common 'cupin' or double-stranded β-helix (DSBH) core with other Fe(II)/2OG/O_2-dependent oxygenases (see Chapter 2).[45–47] Proteins of the DSBH superfamily are functionally diverse and characterized by a barrel-like 'cup' consisting of a major antiparallel β-sheet and a minor β-sheet and surrounded by helices. The DSBH 'barrel' houses the active site, which in the Fe(II)/2OG/O_2-dependent oxygenases consists of the iron atom, 2OG, and two histidine ligands. In the halogenases, chloride is an iron ligand, whereas the hydroxylases have a carboxylate ligand from Asp or Glu in this position, as seen in TauD (Figure 18.4).[21]

The first crystal structure of a Fe(II)/2OG/O_2-dependent halogenase was for SyrB2.[18] Three high-resolution structures (1.6–1.8 Å) were reported for complexes with iron and 2OG; with iron, 2OG and chloride; and with iron, 2OG and bromide. These structures established the geometry of the halogenase active site, including the iron ligands His-116 and His-235, and the halide coordination site at the position occupied by a protein carboxylate in the hydroxylases (Figure 18.4B and C). As with other family members, 2OG is a bidentate iron ligand anchored in the active site *via* hydrogen bonding with a conserved arginine side chain. Two 2.2 Å crystal structures of CytC3 as the free enzyme and with bound iron and 2OG[19] revealed an open active site pocket for the halogenase. The active form of CytC3 is a monomer,[35] however it crystallized as a dimer in which disordering of 40 amino acids (residues

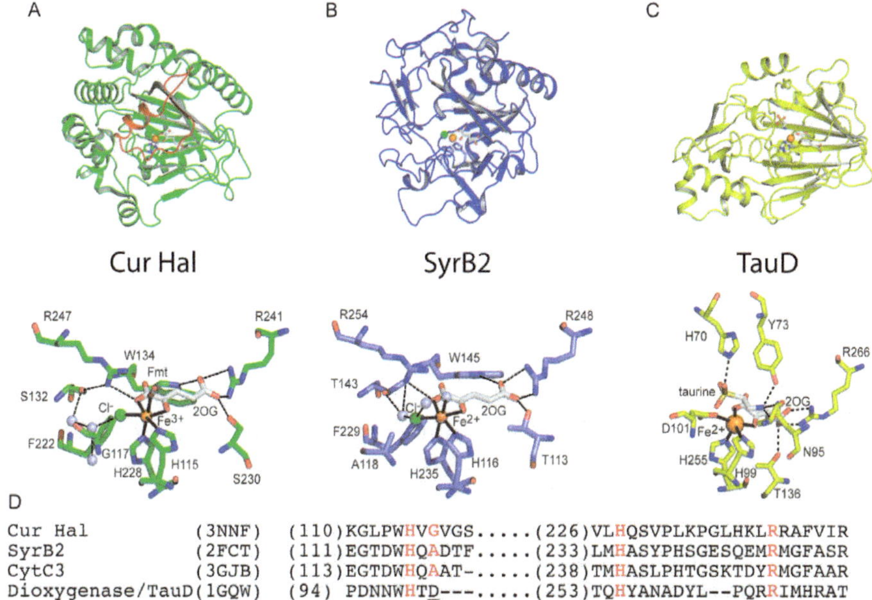

Figure 18.4 Structures of Fe(II)/2OG/O_2-dependent halogenases. Overall structure and active site detail for the halogenases. (A) Cur Hal (green, PDB 3NNF),[20] (B) SyrB2 (blue, PDB 2FCU)[18] and (C) the hydroxylase TauD (yellow, PDB 1GQW).[21] Identical views are shown for the three proteins, all of which have bound iron (orange) and 2OG (white C). The halogenases bind chloride (green) at the coordination site occupied by the Asp-101 carboxylate in the TauD hydroxylase. The positions analogous to TauD Asp-101 are Cur Hal Gly-117 and SyrB2 Ala-118. Cur Hal has formate from the crystallization buffer (Fmt, white C) in the sixth ligand position, where O_2 binds. Product taurine (white C) is bound to TauD. All active sites include an ionic interaction of 2OG with a conserved arginine side chain (Cur Hal Arg-241, SyrB2 Arg-248, TauD Arg-266). (D) Structure-based alignment of key regions of the Cur Hal, SyrB2, CytC3 and TauD sequences showing the two histidine ligands, the aspartate ligand (TauD) and the 2OG-binding arginine.

178–219) exposed a surface that was subsequently buried in the dimer interface. CytC3 and SyrB2 are closely related (58% amino acid sequence identity), and SyrB2 can be used as a basis for modelling a complete monomeric CytC3. Although the artifactual CytC3 dimer trapped by crystallization resulted in shifts to some active site residues relative to SyrB2, the two CytC3 structures taken together revealed no protein conformational differences associated with iron and 2OG binding.

In contrast to SyrB2 and CytC3, the Cur Hal domain exists as a dimer in solution and in the four reported crystal structures,[20] consistent with the dimeric architecture of the PKS protein from which it was excised. The crystal structures of Cur Hal captured the protein in two different conformations (Figure 18.5) and in four different ligand states. A 'closed' form at 2.2 Å resolution had Fe(III), 2OG and Cl^- bound in the active site (Figure 18.4A). A formate ion occupied the sixth coordination position, where dioxygen binds during the catalytic cycle. An 'open' form at 2.6 Å resolution contained no active site ligands. A 15 amino-acid lid (residues 45–60) comprising helix α3 and strand β2 (highlighted in red) covers the active site in the closed form and is disordered in the open form (Figure 18.5A). Through structures of two additional ligand states (iron only at 2.9 Å, formate only at 2.7 Å), 2OG binding was determined to trigger the switch from the open to the closed form. A series of small conformational changes, á la Rube Goldberg, begin when 2OG engages conserved Arg-241 and progressively close the active site by ordering the lid. In the opposite direction, when Arg-241 is not engaged in an ionic interaction, it swings out of the active site, disrupting strands β2 and β3 of the central β-sheet, resulting in complete disordering of strand β2 as well as the preceding helix α3 (Figure 18.5B). Remarkably, iron binding has no effect on the conformational change, as both the open and closed forms were observed for the iron-bound and iron-free halogenase. In addition to the dramatic motions of the active site lid, smaller secondary changes alter access to the substrate-binding site. These motions serve to block access to the substrate site when 2OG is not present, thereby suggesting a rationale for the Cur Hal structural contortions, which have not been seen in any other Fe(II)/2OG-dependent oxygenase. The unique substrate for Cur Hal has a carboxylate group adjacent to the halogenation site. As the substrate is not an iron ligand, its carboxylate must be kept away from the active site until other ligands have bound. Perhaps the open form with its blocked substrate site serves to exclude HMG-ACP from the active site before the iron is coordinated by 2OG and Cl^- (Figure 18.5).

The chloride ligands in both SyrB2 and Cur Hal occupy a generally hydrophobic pocket and also interact with a water molecule (Figure 18.4). Iron in Cur Hal was assumed to be in the ferric state as the protein was crystallized in aerobic conditions, whereas both SyrB2 and CytC3 contain ferrous iron as they were crystallized under anaerobic conditions. The Fe–Cl coordination distance in SyrB2 (2.4 Å) is slightly longer than in Cur Hal (2.3 Å). Both distances are consistent with X-ray absorption fine structure data for the chloro-ferryl intermediate, which resulted in a chloride ligand at 2.3 Å from the iron and an oxygen ligand at 1.7 Å.[48] Taken together, the structures demonstrate the robustness of the DSBH fold. Both fold and active site details are largely unchanged by changes in the iron oxidation state (ferrous *vs* ferric) or by presence or absence of iron.

Unfortunately, none of the crystal structures of Fe(II)/2OG/O_2-dependent halogenases includes a substrate, substrate analogue or carrier protein. Thus details of substrate binding have been deduced from models of substrate complexes. For example, the SyrB2 substrate, tethered to the PPant arm of the carrier protein (SyrB1), was modelled and subjected to error-corrected density functional theory (DFT) calculations, which suggested that the hydroxylation and hydrogen abstraction steps are less sensitive to substrate positioning than is the halogenation step.[49]

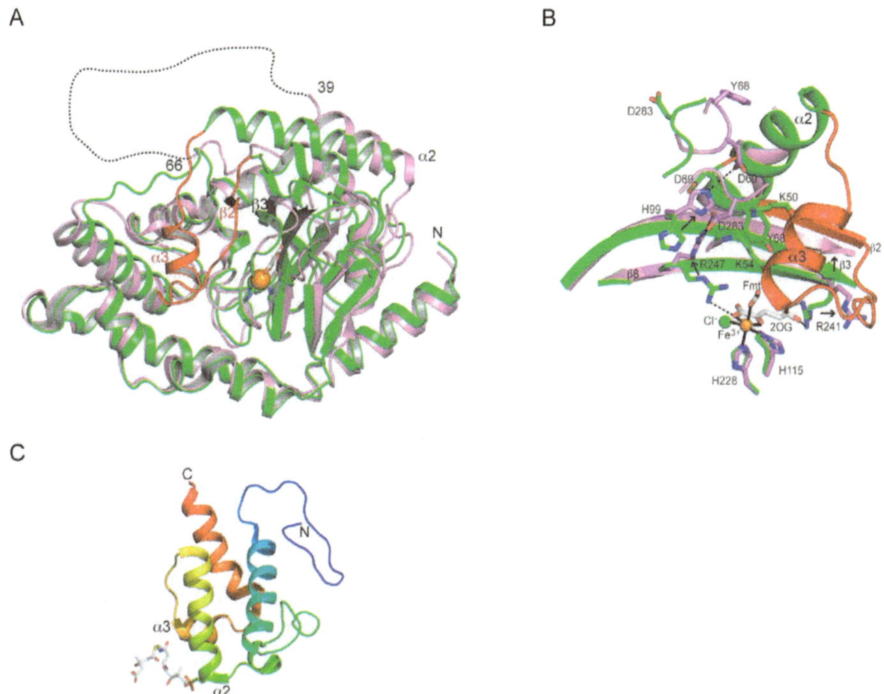

Figure 18.5 Cur Hal dynamics. (A) Superposition of 'open' (pink) and 'closed' (green) forms of Cur Hal. The active site lid (red in the closed form) is disordered in the open form (dotted line). (B) Rearrangements in Cur Hal active site upon 2OG binding. Lys-50, Lys-54 and Tyr-68 in the active site lid are implicated in substrate binding in site-directed mutagenesis experiments.[20] (C) Solution NMR structure of the ACP domain that delivers substrate to Cur Hal (PDB 2LIW).[50] Amino acids in helices α2 and α3 are implicated in Cur Hal interactions.

Cur Hal works specifically on S-HMG-ACP to produce 4-Cl-HMG.[4] The enzyme–substrate interaction was probed by site-directed mutagenesis coupled with a mass spectrometry-based assay.[20] Interestingly, three amino acids in the active site lid, Lys-50, Lys-54 and Tyr-68, were implicated in substrate recognition (Figure 18.5B). Solution NMR structures of holo-ACP and HMG-ACP showed the expected 4-helix bundle architecture (Figure 18.5C), and mutational analysis highlighted the role of residues from ACP helices α2 and α3 near the PPant attachment site, as forming specific interactions with Cur Hal.[50]

18.5 Halogenation *versus* Hydroxylation

The presence of a chloride ligand in the halogenases in the position of a protein carboxylate ligand in the hydroxylases is insufficient to explain the partitioning between halogenation and hydroxylation as both reactions have been shown to proceed through the ferryl intermediate. Several computational and DFT calculations have focused on SyrB2 to investigate the mechanism of substrate halogenation *versus* hydroxylation.[51] Carbon dioxide generated in the 2OG decarboxylation step was proposed to remove the hydroxyl group (from the hydroxo iron complex) to form bicarbonate *via* an OH trapping mechanism, which would prevent formation of the hydroxylated product.[52] Alternatively, it was suggested that protonation of the hydroxyl intermediate might prevent the halogenase from acting as a hydroxylase.[53]

The similarity of the halogenation and hydroxylation mechanisms is reflected in experimental observations of catalytic plasticity. In SyrB2, substrate positioning was shown to control the partitioning between halogenation and hydroxylation. The catalytic outcome switches to hydroxylation when the native substrate, L-Thr, a C4 amino acid, is replaced by L-norvaline, a C5 amino acid.[54] Additional studies of selective halogenation by SyrB2 led to the conclusion that rebound will occur from whichever iron ligand (chloride or hydroxide) is nearest the substrate radical.[55] A detailed thermochemical analysis of possible reaction mechanisms revealed that the strengths of the Fe–OH and Fe–Cl bonds in the radical intermediates are key to the regioselectivity switch between hydroxylation and halogenation.[56] Recently, wild-type SyrB2 was shown to mediate aliphatic azidation and nitration reactions on an unactivated carbon substrate in the presence of N_3^- and NO_3^- anions.[57] The theme of catalytic promiscuity is most dramatic for HctB, which catalyses halogenation, hydroxylation and oxidation reactions on the substrate hexanoyl-ACP in the presence of chloride.[44] The structure of the iron centre was shown to undergo changes in response to chloride binding.[58]

In contrast to variations of the substrates leading to alterations in the reaction outcome, enzyme engineering has not succeeded in interconverting halogenases and hydroxylases. Efforts to convert SyrB2 to a hydroxylase by substituting conserved Ala-118 with Asp and Glu in the chloride site resulted in inactive enzymes. Both variants bound Fe(II), however no chlorination or hydroxylation products were detected.[18] Similarly, efforts to engineer the TauD hydroxylase to perform chlorination resulted in a protein with no chlorination or hydroxylation activity,[59] clearly demonstrating the importance of the coordination site for natural reactions. DFT calculations were interpreted to suggest that lack of chlorination in mutant hydroxylases was due to poor binding of chlorine in the active site and that the variant halogenases do not hydroxylate because of the larger energetic barrier, *i.e.* the Cl transfer is preferred as the Fe(III)–Cl bond is weaker than the Fe(III)–OH bond.[51]

18.6 Conclusions

The Fe(II)/2OG/O_2-dependent halogenases have a well-understood catalytic mechanism and highly similar active site structures despite their rather low overall sequence identities. The most important remaining question is how the structure of the substrate and its position within the active site dictates whether the free radical generated by removal of hydrogen by the activated ferryl species is resolved by halide or hydroxide rebound. The recently characterized HctB, which produces both halogenated and oxygenated products, is a promising candidate to address this question. The enzymes show a remarkable tailoring to their natural substrates. This is especially striking in Cur Hal, where rather complex conformational changes protect the iron centre from erroneous coordination by a substrate carboxyl group. Structures of informative complexes with substrates would be highly valuable in understanding substrate and catalytic selectivity, and perhaps in adapting the halogenases as synthetic tools.

References

1. G. W. Gribble, *Chemosphere*, 2003, **52**, 289–297.
2. C. Wagner, M. El Omari and G. M. Konig, *J. Nat. Prod.*, 2009, **72**, 540–553.
3. F. H. Vaillancourt, E. Yeh, D. A. Vosburg, S. Garneau-Tsodikova and C. T. Walsh, *Chem. Rev.*, 2006, **106**, 3364–3378.
4. L. Gu, B. Wang, A. Kulkarni, T. W. Geders, R. V. Grindberg, L. Gerwick, K. Hakansson, P. Wipf, J. L. Smith, W. H. Gerwick and D. H. Sherman, *Nature*, 2009, **459**, 731–735.
5. F. H. Vaillancourt, E. Yeh, D. A. Vosburg, S. E. O'Connor and C. T. Walsh, *Nature*, 2005, **436**, 1191–1194.
6. L. C. Blasiak and C. L. Drennan, *Acc. Chem. Res.*, 2009, **42**, 147–155.
7. C. S. Neumann, D. G. Fujimori and C. T. Walsh, *Chem. Biol.*, 2008, **15**, 99–109.
8. D. G. Fujimori and C. T. Walsh, *Curr. Opin. Chem. Biol.*, 2007, **11**, 553–560.
9. A. Butler and M. Sandy, *Nature*, 2009, **460**, 848–854.
10. C. D. Murphy, *Nat. Prod. Rep.*, 2006, **23**, 147–152.
11. J. M. Winter and B. S. Moore, *J. Biol. Chem.*, 2009, **284**, 18577–18581.
12. A. S. Eustaquio, F. Pojer, J. P. Noel and B. S. Moore, *Nat. Chem. Biol.*, 2008, **4**, 69–74.
13. K. H. van Pee and E. P. Patallo, *Appl. Microbiol. Biotechnol.*, 2006, **70**, 631–641.
14. R. P. Hausinger, *Crit. Rev. Biochem. Mol. Biol.*, 2004, **39**, 21–68.
15. W. C. Chang, Y. Guo, C. Wang, S. E. Butch, A. C. Rosenzweig, A. K. Boal, C. Krebs and J. M. Bollinger, Jr., *Science*, 2014, **343**, 1140–1144.
16. J. C. Price, E. W. Barr, B. Tirupati, J. M. Bollinger, Jr. and C. Krebs, *Biochemistry*, 2003, **42**, 7497–7508.
17. C. Krebs, D. Galonic Fujimori, C. T. Walsh and J. M. Bollinger, Jr., *Acc. Chem. Res.*, 2007, **40**, 484–492.
18. L. C. Blasiak, F. H. Vaillancourt, C. T. Walsh and C. L. Drennan, *Nature*, 2006, **440**, 368–371.
19. C. Wong, D. G. Fujimori, C. T. Walsh and C. L. Drennan, *J. Am. Chem. Soc.*, 2009, **131**, 4872–4879.
20. D. Khare, B. Wang, L. Gu, J. Razelun, D. H. Sherman, W. H. Gerwick, K. Hakansson and J. L. Smith, *Proc. Natl. Acad. Sci. U. S. A.*, 2010, **107**, 14099–14104.

21. J. M. Elkins, M. J. Ryle, I. J. Clifton, J. C. Dunning Hotopp, J. S. Lloyd, N. I. Burzlaff, J. E. Baldwin, R. P. Hausinger and P. L. Roach, *Biochemistry*, 2002, **41**, 5185–5192.
22. D. G. Fujimori, E. W. Barr, M. L. Matthews, G. M. Koch, J. R. Yonce, C. T. Walsh, J. M. Bollinger, Jr., C. Krebs and P. J. Riggs-Gelasco, *J. Am. Chem. Soc.*, 2007, **129**, 13408–13409.
23. L. M. Hoffart, E. W. Barr, R. B. Guyer, J. M. Bollinger, Jr. and C. Krebs, *Proc. Natl. Acad. Sci. U. S. A.*, 2006, **103**, 14738–14743.
24. J. C. Price, E. W. Barr, T. E. Glass, C. Krebs and J. M. Bollinger, Jr., *J. Am. Chem. Soc.*, 2003, **125**, 13008–13009.
25. J. C. Price, E. W. Barr, L. M. Hoffart, C. Krebs and J. M. Bollinger, Jr., *Biochemistry*, 2005, **44**, 8138–8147.
26. M. Costas, M. P. Mehn, M. P. Jensen and L. Que, Jr., *Chem. Rev.*, 2004, **104**, 939–986.
27. D. P. Galonic, F. H. Vaillancourt and C. T. Walsh, *J. Am. Chem. Soc.*, 2006, **128**, 3900–3901.
28. D. P. Galonic, E. W. Barr, C. T. Walsh, J. M. Bollinger, Jr. and C. Krebs, *Nat. Chem. Biol.*, 2007, **3**, 113–116.
29. E. Flashman and C. J. Schofield, *Nat. Chem. Biol.*, 2007, **3**, 86–87.
30. S. D. Wong, M. Srnec, M. L. Matthews, L. V. Liu, Y. Kwak, K. Park, C. B. Bell, 3rd., E. E. Alp, J. Zhao, Y. Yoda, S. Kitao, M. Seto, C. Krebs, J. M. Bollinger, Jr. and E. I. Solomon, *Nature*, 2013, **499**, 320–323.
31. F. H. Vaillancourt, J. Yin and C. T. Walsh, *Proc. Natl. Acad. Sci. U. S. A.*, 2005, **102**, 10111–10116.
32. F. H. Vaillancourt, D. A. Vosburg and C. T. Walsh, *ChemBioChem*, 2006, **7**, 748–752.
33. N. Engene, E. C. Rottacker, J. Kastovsky, T. Byrum, H. Choi, M. H. Ellisman, J. Komarek and W. H. Gerwick, *Int. J. Syst. Evol. Microbiol.*, 2012, **62**, 1171–1178.
34. Z. Chang, P. Flatt, W. H. Gerwick, V. A. Nguyen, C. L. Willis and D. H. Sherman, *Gene*, 2002, **296**, 235–247.
35. M. Ueki, D. P. Galonic, F. H. Vaillancourt, S. Garneau-Tsodikova, E. Yeh, D. A. Vosburg, F. C. Schroeder, H. Osada and C. T. Walsh, *Chem. Biol.*, 2006, **13**, 1183–1191.
36. M. R. Fullone, A. Paiardini, R. Miele, S. Marsango, D. C. Gross, S. Omura, E. Ros-Herrera, M. C. Bonaccorsi di Patti, A. Lagana, S. Pascarella and I. Grgurina, *FEBS J.*, 2012, **279**, 4269–4282.
37. W. Jiang, J. R. Heemstra, Jr., R. R. Forseth, C. S. Neumann, S. Manaviazar, F. C. Schroeder, K. J. Hale and C. T. Walsh, *Biochemistry*, 2011, **50**, 6063–6072.
38. C. S. Neumann and C. T. Walsh, *J. Am. Chem. Soc.*, 2008, **130**, 14022–14023.
39. Z. Chang, N. Sitachitta, J. V. Rossi, M. A. Roberts, P. M. Flatt, J. Jia, D. H. Sherman and W. H. Gerwick, *J. Nat. Prod.*, 2004, **67**, 1356–1367.
40. P. Verdier-Pinard, N. Sitachitta, J. V. Rossi, D. L. Sackett, W. H. Gerwick and E. Hamel, *Arch. Biochem. Biophys.*, 1999, **370**, 51–58.
41. A. V. Blokhin, H. D. Yoo, R. S. Geralds, D. G. Nagle, W. H. Gerwick and E. Hamel, *Mol. Pharmacol.*, 1995, **48**, 523–531.

42. W. H. Gerwick, Philip J. Proteau, Dale G. Nagle, Ernest Hamel, Andrei Blokhin and Doris L. Slate,*J. Org. Chem.*, 1994, **59**, 1243–1245.
43. D. J. Edwards, B. L. Marquez, L. M. Nogle, K. McPhail, D. E. Goeger, M. A. Roberts and W. H. Gerwick, *Chem. Biol.*, 2004, **11**, 817–833.
44. S. M. Pratter, J. Ivkovic, R. Birner-Gruenberger, R. Breinbauer, K. Zangger and G. D. Straganz, *ChemBioChem*, 2014, **15**, 567–574.
45. I. J. Clifton, M. A. McDonough, D. Ehrismann, N. J. Kershaw, N. Granatino and C. J. Schofield,*J. Inorg. Biochem.*, 2006, **100**, 644–669.
46. J. M. Dunwell, A. Culham, C. E. Carter, C. R. Sosa-Aguirre and P. W. Goodenough, *Trends Biochem. Sci.*, 2001, **26**, 740–746.
47. G. D. Straganz and B. Nidetzky, *ChemBioChem*, 2006, **7**, 1536–1548.
48. M. L. Matthews, C. M. Krest, E. W. Barr, F. H. Vaillancourt, C. T. Walsh, M. T. Green, C. Krebs and J. M. Bollinger, *Biochemistry*, 2009, **48**, 4331–4343.
49. H. J. Kulik and C. L. Drennan,*J. Biol. Chem.*, 2013, **288**, 11233–11241.
50. A. Busche, D. Gottstein, C. Hein, N. Ripin, I. Pader, P. Tufar, E. B. Eisman, L. Gu, C. T. Walsh, D. H. Sherman, F. Lohr, P. Guntert and V. Dotsch, *ACS Chem. Biol.*, 2012, **7**, 378–386.
51. H. J. Kulik, L. C. Blasiak, N. Marzari and C. L. Drennan,*J. Am. Chem. Soc.*, 2009, **131**, 14426–14433.
52. S. P. de Visser and R. Latifi,*J. Phys. Chem. B*, 2009, **113**, 12–14.
53. S. Pandian, M. A. Vincent, I. H. Hillier and N. A. Burton, *Dalton Trans.*, 2009, 6201–6207.
54. M. L. Matthews, C. S. Neumann, L. A. Miles, T. L. Grove, S. J. Booker, C. Krebs, C. T. Walsh and J. M. Bollinger, Jr., *Proc. Natl. Acad. Sci. U. S. A.*, 2009, **106**, 17723–17728.
55. T. Borowski, H. Noack, M. Radon, K. Zych and P. E. Siegbahn,*J. Am. Chem. Soc.*, 2010, **132**, 12887–12898.
56. M. G. Quesne and S. P. de Visser,*J. Biol. Inorg. Chem.*, 2012, **17**, 841–852.
57. M. L. Matthews, W. C. Chang, A. P. Layne, L. A. Miles, C. Krebs and J. M. Bollinger*Jr.Nat. Chem. Biol.*, 2014.
58. S. M. Pratter, K. M. Light, E. I. Solomon and G. D. Straganz,*J. Am. Chem. Soc.*, 2014.
59. P. K. Grzyska, T. A. Muller, M. G. Campbell and R. P. Hausinger,*J. Inorg. Biochem.*, 2007, **101**, 797–808.

CHAPTER 19

Isopenicillin N Synthase

PETER J. RUTLEDGE*[a]

[a]School of Chemistry, The University of Sydney, NSW 2006, Australia
*E-mail: peter.rutledge@sydney.edu.au

19.1 Isopenicillin N Synthase and Penicillin Biosynthesis

Isopenicillin N synthase (IPNS) catalyzes the four-electron oxidative double ring closure of the linear tripeptide δ-(L-α-aminoadipoyl)-L-cysteinyl-D-valine (ACV, **1**) to give isopenicillin N (IPN, **2**) (Scheme 19.1), a conversion essential in the biosynthesis of penicillin and cephalosporin antibiotics. From IPN the pathway branches: transacylation affords other penicillins while epimerization at the remote aminoadipoyl stereocentre followed by ring expansion leads to the cephem ring system of the cephalosporins.[1,2]

In converting its linear substrate to the cyclic penam product, IPNS makes both the C–N and C–S bonds of the fused β-lactam/thiazolidine ring system. IPNS does not use 2-oxoglutarate (2OG) or any cosubstrate other than O_2. The full oxidizing potential of molecular oxygen is focused on the substrate, forming the two new C–heteroatom bonds and reducing the oxidant to two equivalents of water. Yet the IPNS sequence is homologous to the 2OG-dependent enzyme family and its structure and mechanism share key features in common with these enzymes.

RSC Metallobiology Series No. 3
2-Oxoglutarate-Dependent Oxygenases
Edited by Robert P. Hausinger and Christopher J. Schofield

Published by the Royal Society of Chemistry, www.rsc.org

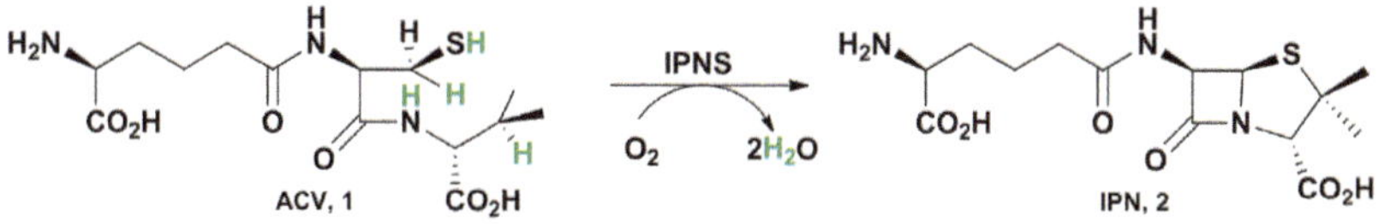

Scheme 19.1 IPNS catalyzes the oxidative double ring closure of linear tripeptide δ-(L-α-aminoadipoyl)-L-cysteinyl-D-valine (ACV, **1**) to isopenicillin N (IPN, **2**). The cosubstrate O_2 is reduced to two equivalents of water.

19.2 Windows on Catalysis

The catalytic cycle of IPNS has been pieced together using a wide range of experiments in a number of different laboratories over many years. Key insights have been gleaned by incubating the solution-phase enzyme with subtly modified substrate analogues and characterizing the modified products.[3,4] Spectroscopic studies to characterize the enzyme active site have exploited the metal centre as a handle for measurements including Mössbauer, electron paramagnetic resonance, extended X-ray absorption fine structure, magnetic circular dichroism and absorption spectroscopy.[5–9] Direct structural insight has been drawn from crystal structures of IPNS with and without bound substrates and substrate analogues,[10–17] and enzymatic turnover experiments in the crystalline protein.[18–21] The structural data have in turn facilitated computational investigations into the enzyme mechanism.[22–27]

Crystal structures for the free enzyme complexed with manganese in place of iron (PDB 1IPS),[10] the anaerobic IPNS•Fe•ACV complex (PDB 1BK0), IPNS•Fe•ACV with nitric oxide bound as a surrogate for molecular oxygen (PDB 1BLZ),[11] and the IPNS•Fe•IPN complex (PDB 1QJE)[12] have all been solved (Figures 19.1 and 19.2). These structures combine to offer a three-dimensional picture of the catalytic cycle, and they have cemented and extended the mechanistic rationale developed from solution phase experiments.

19.3 The Reaction Mechanism

This collective endeavour affords a detailed mechanistic picture of IPNS catalysis (Scheme 19.2), from the substrate-free enzyme **3**, *via* ACV and then O_2 binding to give **5**, β-lactam closure to generate the monocyclic/ferryl (or Fe(IV)-oxo) intermediate **7**, and finally thiazolidine closure to the IPN complex **9**. X-ray crystal structures have been solved for species that correspond to **3**, **4**, **5** and **9** as outlined above, and a monocyclic intermediate closely related to **7** has also been trapped and characterized crystallographically.[10–12]

19.3.1 Substrate Binding

In the resting state of the enzyme, the active site metal ion is six-coordinate: two histidines and one aspartate (the familiar '2-His-1-carboxylate' facial triad),[28] two water molecules and a glutamine are ligated.[10] When ACV binds,

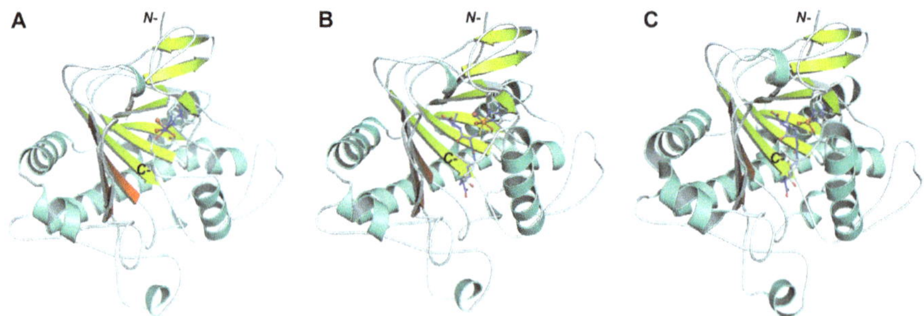

Figure 19.1 X-ray crystal structures of (A) IPNS•Mn (PDB ID: 1IPS);[10] (B) IPNS•Fe•ACV•NO (PDB ID: 1BLZ);[11] and (C) IPNS•Fe•IPN (PDB ID: 1QJE).[12] Mn(II) is in violet, Fe(II) in orange. The conserved β-helix core fold is highlighted in yellow. Adapted from Hamed *et al.*[1]

the glutamine ligand (Gln-330), which is not essential for catalysis, is displaced from the metal.[11] Gln-330 is the penultimate residue in the protein sequence of the *Aspergillus* enzyme, and upon substrate binding the seven *C*-terminal residues take up a conformation that extends the final helix (α-10) to enclose the active site. The cysteinyl sulfur of ACV binds to iron in place of Gln. The *iso*-propyl group of ACV valine displaces one of the water ligands to sit within van der Waals contact of the metal opposite Asp-216, effectively reserving this site for O_2. The substrate is also held by a salt bridge from the aminoadipoyl carboxylate to Arg-87 and hydrogen bonding between the valinyl carboxylate and Tyr-189, Arg-279 and Ser-281. Once ACV has bound, the active site ferrous centre is five-coordinate and, by virtue of thiolate ligation, high-spin.[5]

Coordination of the thiolate to iron reduces the Fe(II)/Fe(III) redox potential and thus primes the enzyme for O_2 binding and reaction. Oxygen binds end-on in the site *trans* to Asp-216, forcing partial rotation of the valinyl *iso*-propyl group. Evidence for this site and mode of O_2 binding comes from a crystal structure of the IPNS•Fe•ACV•NO complex[11] and a combination of spectroscopic and DFT measurements demonstrating that charge donation from the

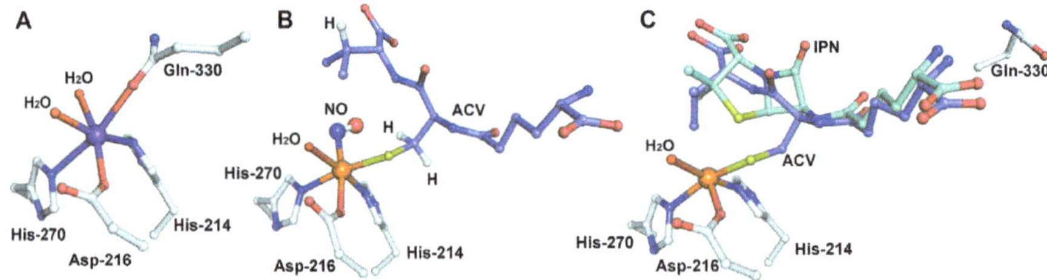

Figure 19.2 Active site region from the X-ray crystal structures of (A) IPNS•Mn (PDB ID: 1IPS),[10] (B) IPNS•Fe•ACV•NO (PDB ID: 1BLZ)[11] and (C) IPNS•Fe•IPN with ACV superimposed (PDB ID: 1QJE).[12] Mn(II) is in violet, Fe(II) in orange. Adapted from Hamed *et al.*[1]

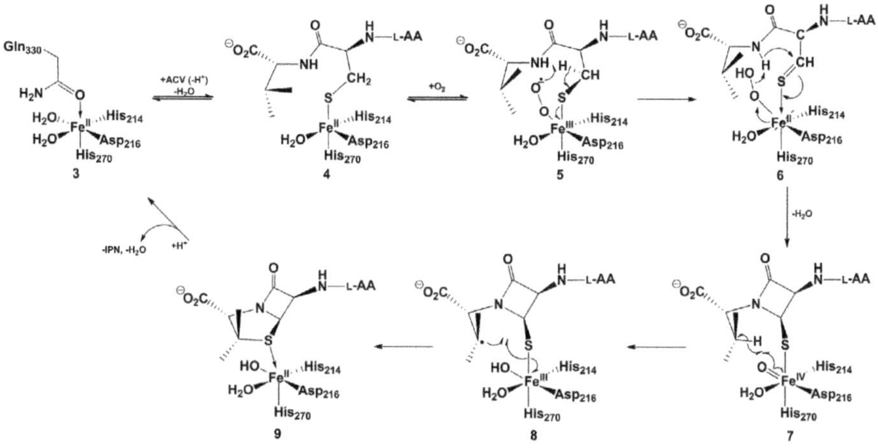

Scheme 19.2 A mechanism for IPNS catalysis, developed on the basis of solution-phase incubation experiments, spectroscopic measurements, X-ray crystallography and computational studies. This scheme shows the 'substrate donor' model for peroxide cleavage (**6** → **7**); the alternative 'ligand donor' model for this step is discussed below and shown in Scheme 19.3. L-AA = δ-(L-α-aminoadipoyl).

ACV thiolate makes formation of an Fe(III)-superoxide species energetically favourable.[5,9] With O_2 binding in this way, IPNS avoids the bridged binding mode invoked for other non-haem iron enzymes. Bridged binding of O_2 generally leads to oxygen insertion (*i.e.* oxygenase activity) in contrast to the oxidase activity that is the primary reaction pathway for IPNS. Once oxygen has bound in the IPNS active site, it selectively removes the pro-*S* hydrogen from the cysteinyl β-carbon of ACV to form the thioaldehyde/Fe–peroxide **6**. This step is irreversible and commits ACV (and IPNS) to the oxidative bicyclization pathway.[29]

19.3.2 β-Lactam Closure

The next step sees deprotonation of the valine N–H and nucleophilic attack of the valinyl nitrogen on the thioaldehyde to close the β-lactam and form monocyclic intermediate **7**. This ring-closing reaction is associated with cleavage of the O–O bond to generate the ferryl species. At the time the IPNS•Fe•ACV•NO structure was first reported, it was proposed that the Fe-bound peroxide abstracts the valine N–H directly to generate the ferryl species and water (as shown in Scheme 19.2).[11] However, more recent computational studies suggest that the barrier to O–O bond cleavage in this manner may be prohibitively high.[22–24] The *in silico* studies favour instead a 'ligand donor' mechanism for peroxide cleavage in which the water ligand at iron provides the proton required for O–O bond fission. This would mean that the ferryl species is generated before β-lactam closure, and that closure of the four-membered ring is coupled instead to a separate proton transfer from the substrate to the water ligand at iron (Scheme 19.3). The ligand donor mechanism is also more in line

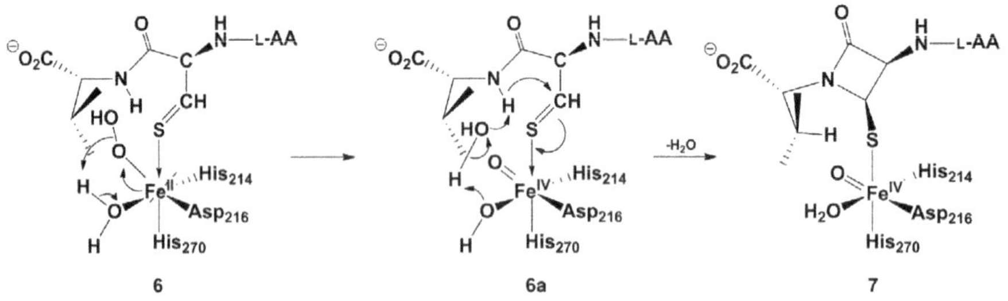

Scheme 19.3 The ligand donor model for hydroperoxide cleavage during formation of the ferryl intermediate 7 is supported by computational studies, and consistent with O–O bond cleavage mechanisms in related non-haem iron enzymes.[22–24] L-AA = δ-(L-α-aminoadipoyl).

with peroxide cleavage pathways seen at other non-haem iron centres.[23] Either way – *via* the ligand donor mechanism for O–O cleavage or the alternative 'substrate donor' mechanism – the same monocyclic ferryl intermediate **7** arises.

Significantly, a monocyclic β-lactam intermediate akin to **7** has been structurally characterized within the IPNS active site by trapping the high-valent ferryl intermediate as a sulfoxide using the substrate analogue δ-(L-α-aminoadipoyl)-L-cysteinyl-D-*S*-methylcysteine (ACmC).[12] Applying hyperbaric oxygenation to initiate the reaction,[30] the IPNS•Fe•ACmC complex **10** was turned over to the monocyclic β-lactam sulfoxide complex **11** *in crystallo* (Scheme 19.4), providing direct evidence for initial β-lactam closure in the IPNS reaction.[12]

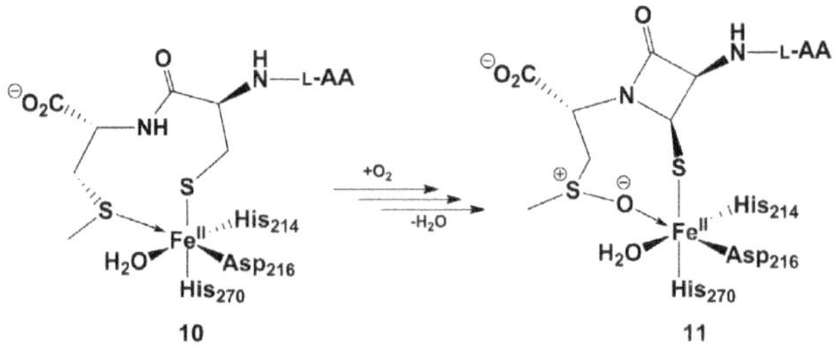

Scheme 19.4 Conversion of the IPNS•Fe•ACmC complex **10** to the β-lactam sulfoxide species **11** provides compelling evidence for initial β-lactam formation and the involvement of monocyclic intermediate 7 in IPNS catalysis.[12] L-AA = δ-(L-α-aminoadipoyl).

19.3.3 Thiazolidine Closure

Closure of the β-lactam is thought to facilitate rotation of the *iso*-propyl group to bring the valinyl βC–H close to the ferryl moiety,[11] which can then drive thiazolidine closure by initiating reaction with this C–H bond. Abstraction of the tertiary C–H from valine generates a reactive substrate radical that closes onto the cysteinyl sulfur with exquisite regio- and stereocontrol. This chemical step generates the five-membered ring and affords the completed IPN product complex **9**. Incubation of ACV selectively ^{13}C-labelled at one of the valinyl methyl positions has confirmed that thiazolidine closure occurs with retention of configuration.[31]

The crystal structure of the IPNS•Fe•IPN complex indicates that the cysteinyl sulfur atom shifts its position during thiazolidine closure,[12] moving towards the valine β-carbon and out of the iron-binding site opposite His-270. The IPNS•Fe•IPN structure also shows that the product carboxylate groups remain bound to the protein. However, bicyclic IPN is contracted relative to linear ACV. It is thought that the shorter product does not accommodate the three-point attachment to iron, Arg-87 and Tyr-189 as well, so is primed to dissociate from the active site and thus complete the catalytic cycle.[12]

19.4 Other Modes of Reactivity

One of the most interesting aspects of IPNS catalysis is the diverse range of substrate variants that this enzyme will convert to oxidized products. IPNS is unusually promiscuous, even within the 2OG oxygenase superfamily, and demonstrates both oxidase and oxygenase modes of reactivity in turning over these many different substrates. Because its native substrate is a peptide, systematic, modular generation of substrate analogues can be achieved relatively easily using the tools of peptide chemistry. Accordingly IPNS has been challenged with a very wide array of substrate analogues, driven by the dual goals of probing reaction mechanism and creating new antibiotics. This chemistry has been reviewed in detail elsewhere,[3,4] and only a selection of the most interesting oxidative chemistry is discussed here.

19.4.1 Modified Alkyl Side Chains Illuminate Second Ring Closure

When modified alkyl side chains are introduced at the third residue of the substrate, variations on the standard oxidase mode of reaction are generally observed, providing support for the proposed radical mechanism of thiazolidine closure. For example, the sterically less demanding analogue δ-(L-α-aminoadipoyl)-L-cysteinyl-D-α-aminobutyrate **12** (ethyl replaces *iso*-propyl of ACV) affords three β-lactam products upon incubation with IPNS: the epimeric penams **13** and **14** and cepham **15** in a 1 : 7 : 3 ratio (Scheme 19.5(A)).[32,33] In contrast, incubation of the marginally bulkier analogue δ-L-α-aminoadipoyl-L-cysteinyl-D-isoleucine **16** (*sec*-butyl replaces *iso*-propyl) yields penam **17** as the major product, along with small amounts of the corresponding cephams **18** and **19** (Scheme 19.5B).[34,35]

The product mixtures observed with these analogues are consistent with a radical route to thiazolidine closure, and suggest that discrimination

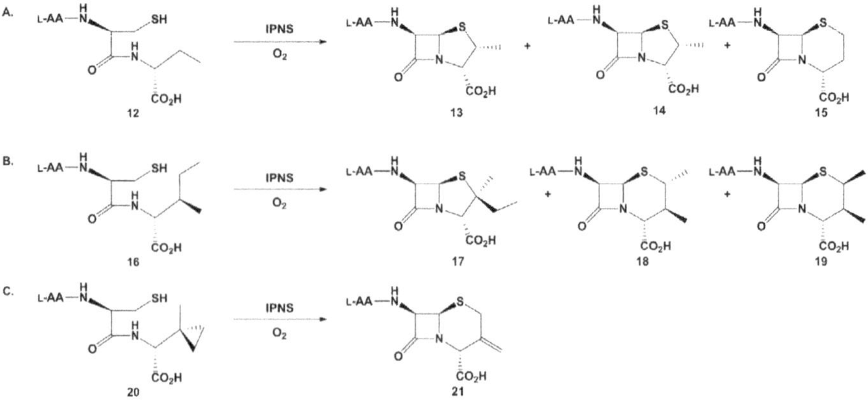

Scheme 19.5 IPNS accepts a wide range of alkyl side chains in place of the valinyl *iso*-propyl group of its natural substrate, including (A) the ethyl group of δ-(L-α-aminoadipoyl)-L-cysteinyl-D-aminobutyrate **12**; (B) the *sec*-butyl groups of δ-(L-α-aminoadipoyl)-L-cysteinyl-D-isoleucine **16** and *allo*-isoleucine; and (C) cyclopropyl groups of radical traps such as δ-(L-α-aminoadipoyl)-L-cysteinyl-D-cyclopropylglycine **20**. L-AA = δ-(L-α-aminoadipoyl).

between penam and cepham products hinges on the selectivity of hydrogen abstraction by the ferryl intermediate. From **12**, penam products are formed when the ferryl reacts with a secondary C–H, while the cepham arises when this reactive species abstracts a primary C–H. With **16** the distinction is tertiary C–H (giving the penam) *versus* secondary C–H (leading to the cepham). In contrast the natural substrate presents tertiary and primary C–Hs, and the greater energy difference between these different C–H bonds ensures the complete selectivity for five-ring formation that is observed with ACV.

Further investigation with specifically deuterated aminobutyrates provides additional support for a radical mechanism for second ring closure,[36] and this proposal is strengthened further by experiments using substrates that incorporate cyclopropyl radical traps such as δ-(L-α-aminoadipoyl)-L-cysteinyl-β-methyl-D-cyclopropylglycine **20**, which is converted by IPNS to the [4,7] fused bicyclic product **21** (Scheme 19.5C).[17,37]

19.4.2 Unsaturated Side Chains Evoke Oxygenase Activity

With unsaturated substrate analogues, IPNS exhibits oxygenase activity, sometimes alongside oxidase activity akin to that seen with the natural substrate and saturated analogues.

When δ-(L-α-aminoadipoyl)-L-cysteinyl-D-vinylglycine **21** is incubated with IPNS, hydroxymethylpenam **22** is the only isolated product (Scheme 19.6A).[18,38] Incubation of δ-(L-α-aminoadipoyl)-L-cysteinyl-D-isodehydrovaline **23** gives the closely related hydroxymethyl compound **24** as well as cepham **25**.[38] Incorporation experiments with $^{18}O_2$ show that the hydroxy group of the mono-oxygenase product **24** derives from molecular oxygen (Scheme 19.6B).[39]

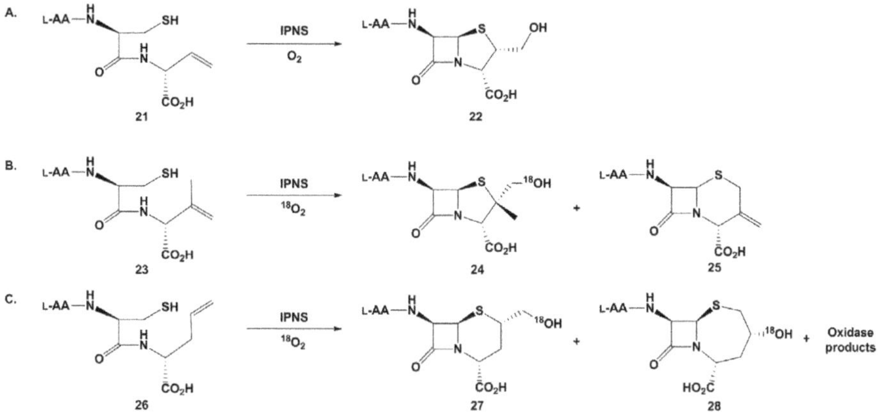

Scheme 19.6 IPNS demonstrates oxygenase activity with unsaturated substrate analogues such as (A) δ-(L-α-aminoadipoyl)-L-cysteinyl-D-vinylglycine **21**; (B) δ-(L-α-aminoadipoyl)-L-cysteinyl-D-dehydrovaline **23**; and (C) δ-(L-α-aminoadipoyl)-L-cysteinyl-D-methyl-D-allylglycine **26**. L-AA = δ-(L-α-aminoadipoyl).

With δ-(L-α-aminoadipoyl)-L-cysteinyl-D-allylglycine **26**, IPNS shows both oxygenase and oxidase activity, leading to hydroxylated cepham (**27**)/homocepham (**28**) products and vinyl penams, respectively (Scheme 19.6C).[40]

19.4.3 Depsipeptide Substrates Redirect the Fe–Peroxide

The mechanism of the β-lactam forming step has been probed by treating IPNS with a variety of analogues, most recently by depriving the Fe–peroxide intermediate **6** (or **6a**) of its natural reaction partner, the valinyl amide N–H that is removed during β-lactam formation. This is achieved by replacing the amide link between these two residues with an ester, *i.e.* creating depsipeptide substrate analogues. δ-(L-α-Aminoadipoyl)-L-cysteine D-α-hydroxyisovaleryl ester (AC*O*V, **28**) differs from ACV only in that an oxygen atom replaces the valinyl NH and was the first depsipeptide analogue to be synthesized, crystallized with IPNS, and turned over within the crystalline protein.[13] AC*O*V is converted to the thiocarboxylic acid **29** in the IPNS active site (Scheme 19.7(A)), *via* a mono-oxygenation reaction pathway in which the hydroperoxide intermediate attacks the thioaldehyde species directly. The transformation **28** → **29** involves removal of both hydrogen atoms from the cysteinyl β-carbon. When either of these hydrogen atoms is replaced with a methyl group – as in the (3*R*)-methyl-L-cysteine depsipeptide analogue **30** – mixed oxidase/oxygenase activity is observed, leading to the hydroxylated ene-thiol product **31** within the crystalline enzyme (Scheme 19.7(B)).[19,21] Combining the depsipeptide strategy with the sulfide → sulfoxide strategy deployed with ACmC (see above) leads to the substrate analogue δ-(L-α-aminoadipoyl)-L-cysteine (1-(*S*)-carboxy-2-thiomethyl)ethyl ester **32** (Scheme 19.7C). This hybrid analogue is converted by IPNS to the *S*-oxidized product **33**, within

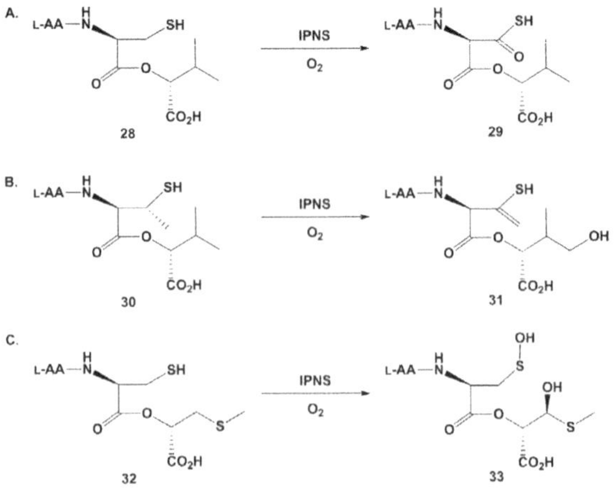

Scheme 19.7 *In crystallo* reaction of IPNS with depsipeptide analogues (A) δ-(L-α-aminoadipoyl)-L-cysteine D-α-hydroxyisovaleryl ester **28**; (B) δ-(L-α-aminoadipoyl)-(3*S*-methyl)-L-cysteine D-α-hydroxyl isovaleryl ester **30**; and (C) δ-(L-α-aminoadipoyl)-L-cysteine (1-(*S*)-carboxy-2-thiomethyl)ethyl ester **32**. L-AA = δ-(L-α-aminoadipoyl).

the IPNS active site.[20] In the mechanism proposed for this transformation, IPNS incorporates both oxygen atoms from O_2 into the substrate at different positions, although the origin of the oxygen atoms in the product has not been confirmed.

19.5 Future Outlook

IPNS is amongst the most studied and best characterized proteins in the non-haem iron oxidase/oxygenase family. It has been challenged with a diverse array of substrate analogues, probed using multifarious spectroscopic techniques, and was the first member of the family to be characterized crystallographically. Yet work continues to confirm the last outstanding details of its reaction mechanism – including the exact timing of bond formation and cleavage events during Fe–peroxide formation and β-lactam closure, and direct observation of a radical intermediate in thiazolidine formation – and towards the challenging goals of mimicking IPNS catalysis in small-molecule models and determining the position of this 'queen of enzymes' in the hierarchy of protein evolution.

References

1. R. B. Hamed, J. R. Gomez-Castellanos, L. Henry, C. Ducho, M. A. McDonough and C. J. Schofield, *Nat. Prod. Rep.*, 2013, **30**, 21.
2. J. E. Baldwin and C. J. Schofield, in *The Chemistry of β-Lactams*, ed. M. I. Page, Blackie, Glasgow, 1992, pp. 1–78.

3. J. E. Baldwin and E. P. Abraham, *Nat. Prod. Rep.*, 1988, **5**, 129.
4. J. E. Baldwin and M. Bradley, *Chem. Rev.*, 1990, **90**, 1079.
5. V. J. Chen, A. M. Orville, M. R. Harpel, C. A. Frolik, K. K. Surerus, E. Munck and J. D. Lipscomb, *J. Biol. Chem.*, 1989, **264**, 21677.
6. A. M. Orville, V. J. Chen, A. Kriauciunas, M. R. Harpel, B. G. Fox, E. Munck and J. D. Lipscomb, *Biochemistry*, 1992, **31**, 4602.
7. C. R. Randall, Y. Zang, A. E. True, L. Que Jr., J. M. Charnock, C. D. Garner, Y. Fujishima, C. J. Schofield and J. E. Baldwin, *Biochemistry*, 1993, **32**, 6664.
8. A. Hadfield and J. A. Hajdu, *J. Appl. Crystallogr.*, 1993, **26**, 839.
9. C. D. Brown, M. L. Neidig, M. B. Neibergall, J. D. Lipscomb and E. I. Solomon, *J. Am. Chem. Soc.*, 2007, **129**, 7427.
10. P. L. Roach, I. J. Clifton, V. Fulop, K. Harlos, G. J. Barton, J. Hajdu, I. Andersson, C. J. Schofield and J. E. Baldwin, *Nature*, 1995, **375**, 700.
11. P. L. Roach, I. J. Clifton, C. M. H. Hensgens, N. Shibata, C. J. Schofield, J. Hadju and J. E. Baldwin, *Nature*, 1997, **387**, 827.
12. N. I. Burzlaff, P. J. Rutledge, I. J. Clifton, C. M. H. Hensgens, M. Pickford, R. M. Adlington, P. L. Roach and J. E. Baldwin, *Nature*, 1999, **401**, 721.
13. J. M. Ogle, I. J. Clifton, P. J. Rutledge, J. M. Elkins, N. I. Burzlaff, R. M. Adlington, P. L. Roach and J. E. Baldwin, *Chem. Biol.*, 2001, **8**, 1231.
14. A. J. Long, I. J. Clifton, P. L. Roach, J. E. Baldwin, P. J. Rutledge and C. J. Schofield, *Biochemistry*, 2005, **44**, 6619.
15. A. J. Long, I. J. Clifton, P. L. Roach, J. E. Baldwin, C. J. Schofield and P. J. Rutledge, *Biochem. J.*, 2003, **372**, 687.
16. A. C. Stewart, I. J. Clifton, R. M. Adlington, J. E. Baldwin and P. J. Rutledge, *ChemBioChem*, 2007, **8**, 2003.
17. A. R. Howard-Jones, J. M. Elkins, I. J. Clifton, P. L. Roach, R. M. Adlington, J. E. Baldwin and P. J. Rutledge, *Biochemistry*, 2007, **46**, 4755.
18. J. M. Elkins, P. J. Rutledge, N. I. Burzlaff, I. J. Clifton, R. M. Adlington, P. L. Roach and J. E. Baldwin, *Org. Biomol. Chem.*, 2003, **1**, 1455.
19. A. Daruzzaman, I. J. Clifton, R. M. Adlington, J. E. Baldwin and P. J. Rutledge, *ChemBioChem*, 2006, **7**, 351.
20. W. Ge, I. J. Clifton, J. E. Stok, R. M. Adlington, J. E. Baldwin and P. J. Rutledge, *J. Am. Chem. Soc.*, 2008, **130**, 10096.
21. W. Ge, I. J. Clifton, A. R. Howard-Jones, J. E. Stok, R. M. Adlington, J. E. Baldwin and P. J. Rutledge, *ChemBioChem*, 2009, **10**, 2025.
22. M. Wirstam and P. E. M. Siegbahn, *J. Am. Chem. Soc.*, 2000, **122**, 8539.
23. M. Lundberg, P. E. M. Siegbahn and K. Morokuma, *Biochemistry*, 2007, **47**, 1031.
24. M. Lundberg, T. Kawatsu, T. Vreven, M. J. Frisch and K. Morokuma, *J. Chem. Theory Comput.*, 2008, **5**, 222.
25. M. Lundberg and K. Morokuma, *J. Phys. Chem. B*, 2007, **111**, 9380.
26. C. D. Brown-Marshall, A. R. Diebold and E. I. Solomon, *Biochemistry*, 2010, **49**, 1176.
27. Y. Zhang and E. Oldfield, *J. Am. Chem. Soc.*, 2004, **126**, 9494.
28. E. L. Hegg and L. Que Jr., *Eur. J. Biochem.*, 1997, **250**, 625.

29. J. E. Baldwin, R. M. Adlington, N. G. Robinson and H.-H. Ting, *J. Chem. Soc., Chem. Commun.*, 1986, 409.
30. P. J. Rutledge, N. I. Burzlaff, J. M. Elkins, M. Pickford, J. E. Baldwin and P. L. Roach, *Analyt. Biochem.*, 2002, **308**, 265.
31. J. E. Baldwin, R. M. Adlington, B. P. Domayne-Hayman, H.-H. Ting and N. J. Turner, *J. Chem. Soc., Chem. Commun.*, 1986, 110.
32. G. A. Bahadur, J. E. Baldwin, J. J. Usher, E. P. Abraham, G. S. Jayatilake and R. L. White, *J. Am. Chem. Soc.*, 1981, **103**, 7650.
33. J. E. Baldwin, E. P. Abraham, R. M. Adlington, B. Chakravarti, A. E. Derome, J. A. Murphy, L. D. Field, N. B. Green, H.-H. Ting and J. J. Usher, *J. Chem. Soc., Chem. Commun.*, 1983, 1317.
34. G. Bahadur, J. E. Baldwin, L. D. Field, E.-M. M. Lehtonen, J. J. Usher, C. A. Vallejo, E. P. Abraham and R. L. White, *J. Chem. Soc., Chem. Commun.*, 1981, 917.
35. J. E. Baldwin, R. M. Adlington, N. J. Turner, B. P. Domayne-Hayman, H.-H. Ting, A. E. Derome and J. A. Murphy, *J. Chem. Soc., Chem. Commun.*, 1984, 1167.
36. J. E. Baldwin, E. P. Abraham, R. M. Adlington, J. A. Murphy, N. B. Green, H.-H. Ting and J. J. Usher, *J. Chem. Soc., Chem. Commun.*, 1983, 1319.
37. J. E. Baldwin, R. M. Adlington, B. P. Domayne-Hayman, G. Knight and H.-H. Ting, *J. Chem. Soc., Chem. Commun.*, 1987, 1661.
38. J. E. Baldwin, R. M. Adlington, A. Basak, S. L. Flitsch, A. K. Forrest and H.-H. Ting, *J. Chem. Soc., Chem. Commun.*, 1986, 273.
39. J. E. Baldwin, R. M. Adlington, S. L. Flitsch, H.-H. Ting and N. J. Turner, *J. Chem. Soc., Chem. Commun.*, 1986, 1305.
40. J. E. Baldwin, R. M. Adlington, A. E. Derome, H.-H. Ting and N. J. Turner, *J. Chem. Soc., Chem. Commun.*, 1984, 1211.

CHAPTER 20

1-Aminocyclopropane-1-Carboxylic Acid Oxidase†

A. JALILA SIMAAN*[a] AND MARIUS RÉGLIER[a]

[a]Aix Marseille Université, CNRS, Centrale Marseille, iSm2 UMR 7313, 13397, Marseille, France
*E-mail: jalila.simaan@univ-amu.fr

20.1 ACC Oxidase and Ethylene Biosynthesis in Plants

Ethylene, the simplest olefin, is an essential hormone involved in many aspects of plant life including root development, germination, senescence, fruit ripening, stress responses and defence mechanisms.[1] The biosynthesis of ethylene was elucidated by the group of S. F. Yang who showed that ethylene is directly produced from the non-canonical cyclic amino acid, 1-aminocyclopropane-1-carboxylic acid (ACC), a metabolite of methionine.[2,3] ACC is produced from *S*-adenosyl-methionine in a reaction catalysed by ACC synthase (ACCS), a pyridoxal phosphate-dependent enzyme (Scheme 20.1). ACCS also produces 5-methylthioadenosine, which is subsequently transformed into methionine in a recycling pathway known as the Yang cycle. This cycle ensures a high rate of ethylene production even when the pool of methionine is low. The final step in ethylene biosynthesis is catalysed by ACC oxidase (ACCO). It consists of the two-electron oxidation of ACC into ethylene in the presence of dioxygen and electrons (Scheme 20.1). In addition to ethylene, the products are cyanide and carbon dioxide.[4] Ascorbate is used as the reductant *in vitro*; given its high

†Dedicated to Prof. David R. Dilley

RSC Metallobiology Series No. 3
2-Oxoglutarate-Dependent Oxygenases
Edited by Robert P. Hausinger and Christopher J. Schofield

Published by the Royal Society of Chemistry, www.rsc.org

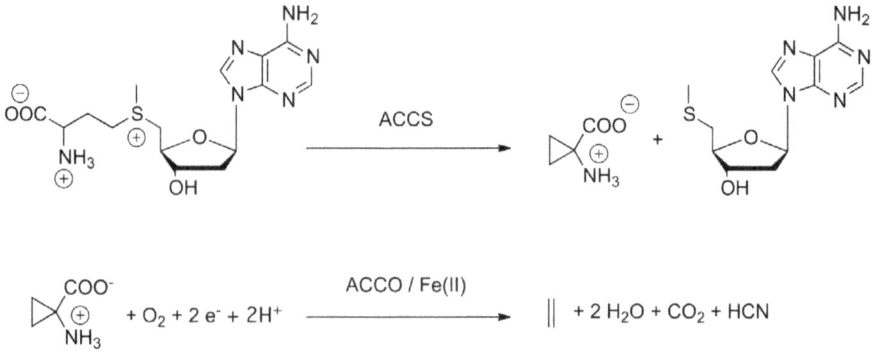

Scheme 20.1 Reaction catalysed by ACCS (top) and ACCO (bottom). The two electrons necessary for the latter reaction can be provided by ascorbate *in vitro*.

concentrations (2–25 mM) in plant cells it appears also as a plausible source of electrons *in vivo* although no direct evidence for this has been reported.[5] Finally, for unclear reasons, ACCO also requires CO_2 (or bicarbonate ions) for activity.[6–8] Although ACCO does not use 2-oxoglutarate (2OG) as a cofactor, this non-haem iron(II)-containing enzyme displays the sequence motifs conserved in most 2OG-dependent oxygenases (2OG oxygenases).

20.2 Structural Aspects

A structure of ACCO from *Petunia hybrida* has been solved (PDB 1WA6).[9] As crystallized, the enzyme is found in a tetrameric form. The structure of each monomer reveals a core composed of the classical distorted double-stranded β-helix (DSBH) fold typical of the other members of the 2OG oxygenases (Figure 20.1 and Chapter 2).[10,11]

The Fe(II) is coordinated to the side chains of His-177, His-234 and Asp-179 in the traditional 2-His, 1-Asp 'facial triad'(Figure 20.2). The conformations of the side chains of the facial triad of iron ligands in ACCO are similar to those found in related enzymes with Asp-179 binding to the metal ion *via* a single oxygen of the carboxylate function. The metal–ligand distances are in the range of 2.2–2.5 Å. A phosphate (or sulfate) ion from the crystallization buffer is also found coordinated to the metal at a distance of 2.1 Å.

It is likely that this structure does not represent an active conformation of the enzyme. Indeed, ACCO is known to be active as a monomer, but it forms a homotetramer in the crystal with the *C*-terminus of each monomer directed away from the active site and interlocking with the *C*-terminus of an adjacent monomer.[9] In related 2OG-dependent enzymes from the deacetoxycephalosporin C synthase subfamily, the *C*-terminal portion of the protein generally forms a 'lid' above the active site; interestingly, conformational changes of this *C*-terminal region during catalysis have been observed for several members of the family.[12] This observation raises the important questions of the

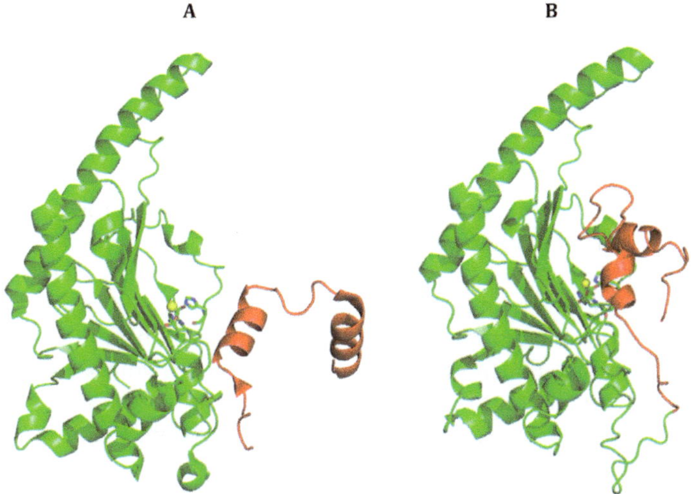

Figure 20.1 (A) View from a crystal structure of ACCO from *P. hybrida* (PDB ID: 1WA6). (B) Modelled structure obtained for the apple ACCO. The metal ion is in yellow and the *C*-terminal part of the protein is in red. Adapted with permission from Zhang *et al.*[9] and Yoo *et al.*[13]

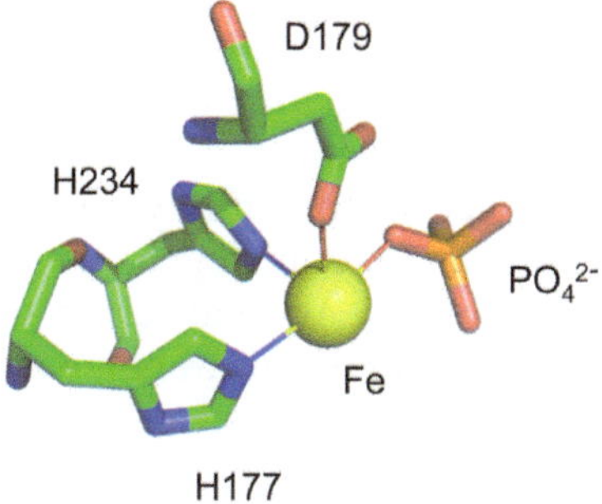

Figure 20.2 Iron(II) active site of ACCO from *P. hybrida*. Adapted with permission from Zhang *et al.*[9]

active conformation of ACCO and the possible flexibility of this *C*-terminal portion during catalysis. Results from mutagenesis studies have indicated that several amino acids of the *C*-terminus of ACCO are strictly necessary for activity.[13] Taking into account the importance of the *C*-terminal region in catalysis, a model of the structure in which this part folds above the active centre has been constructed from the sequence of apple ACCO (Figure 20.1).[13] The reported crystal structure also reveals that Arg-244, which belongs together with Ser-246 to the highly conserved 'RXS' motif involved in 2OG binding in some 2OG oxygenases, adopts a conformation where its side chain is directed away from the metal-binding site.[9] Therefore, in the modelled structure the side chain of Arg-244 was positioned towards the active site.

20.3 Substrate/Cofactor Interactions in the Active Site

To date, no crystallographic data are available depicting the binding of the different substrates/cofactors at the ACCO active site and most of the information has been obtained from mutagenesis, spectroscopic, and modelling studies. Electron nuclear double resonance (ENDOR) spectroscopy of the ternary ACCO:Fe(II):ACC(or alanine):NO adduct have generated data consistent with the proposal that the substrate ACC binds iron(II) in a bidentate fashion *via* the amine and the carboxylate groups.[14,15] The O-atom from the substrate is thought to coordinate *trans* to the aspartate ligand, a binding position also supported by molecular modelling experiments.[16] For comparison, the 2OG cofactor binds the metal ion in a bidentate fashion in 2OG oxygenases with (in most cases) its 2-oxo group *trans* to the carboxylate ligand of the facial triad.[11]

From mutagenesis studies, several conserved charged residues have been proposed to be important for activity and as possible interaction sites for the substrates/cofactors including CO_2: Arg-175, Lys-158, Arg-244, Ser-246, Lys-297 and Arg-300.[17–19] So far, it is not fully clear whether CO_2 or HCO_3^- is the active form of the cofactor, although inhibition studies using borate suggest that bicarbonate or carbonate could be the form at the catalytic centre.[20] Using magnetic circular dichroism spectroscopy, Zhou *et al.* have shown that neither CO_2 nor ascorbate bind the iron(II) ion.[21] It has been proposed that ascorbate binds Arg-244 and Ser-246 from the conserved RXS motif involved in 2OG binding in the related enzymes.[13,22] However, substitutions of these two residues have little effect on the affinity for ascorbate and inhibition studies using α-aminophosphonate analogues suggest a relatively remote binding site for this substrate.[16,17] Alternatively, it has been proposed that Arg-244 and Lys-158 residues bind the bicarbonate cofactor.[20,23] On the basis of kinetic and modelling studies, a scheme of interactions at the active site has been proposed.[16] In this model (Scheme 20.2), ACC binds the metal ion in a bidentate fashion as predicted from ENDOR studies and does not display any additional interaction with active site residues; HCO_3^- is found at H-bond distance from ACC and interacts with Lys-158, Arg-244, Tyr-162 and Ser-246 near the active site, and with Arg-300 in the *C*-terminal region. Although less clear from the calculations, the position of ascorbate is predicted to be located away from ACC and interacting with Arg-175 and Arg-300.

20.4 Mechanistic Aspects

20.4.1 Conversion of ACC into Ethylene: A Stepwise Two-Electron Oxidation

The mechanism of conversion of ACC into ethylene has been the subject of intense studies.[24] In particular, the stereochemistry of the reaction was examined with deuterated amino acids.[25–28] Substrate analogues have also

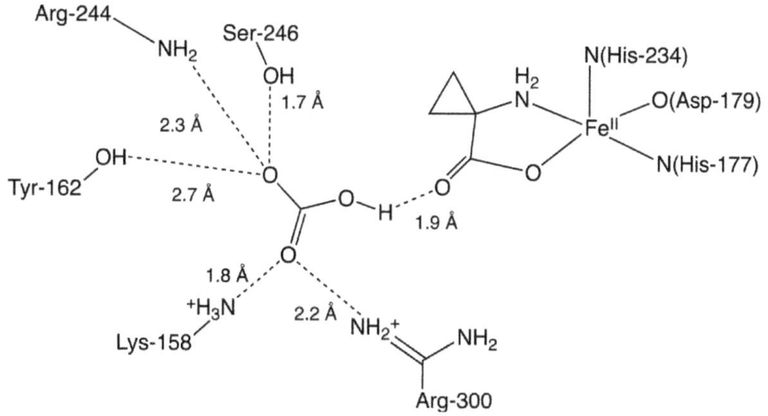

Scheme 20.2 Schematic representation of the possible interactions between ACC, HCO_3^- and ACCO obtained from modelling experiments. Adapted with permission from Brisson *et al.*[16]

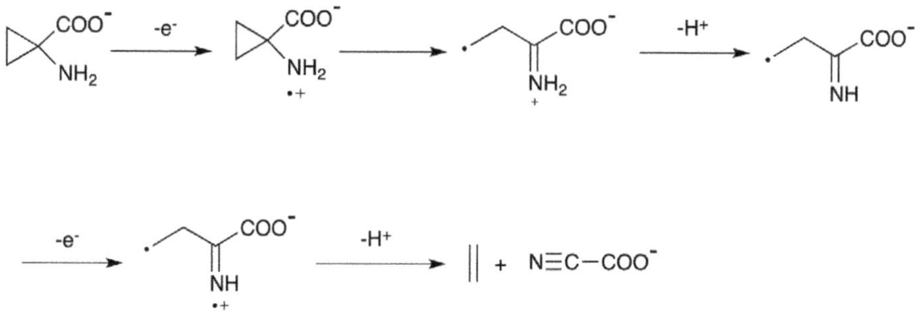

Scheme 20.3 Stepwise proposed radical mechanism for the conversion of ACC into ethylene as catalysed by ACCO.

been used to investigate the ACCO mechanism.[29–31] It is now generally agreed that the oxidation of ACC into ethylene follows a stepwise radical mechanism consisting of two single-electron transfer steps (Scheme 20.3). The first oxidation leads to the aminyl radical cation that undergoes rapid ring opening and subsequent deprotonation.[32,33] A second oxidation leads to the formation of ethylene and cyanoformate. Cyanoformate diffuses out of the active pocket and undergoes decomposition into carbon dioxide and cyanide, preventing the potentially toxic reaction of cyanide with the iron(II) active site to inactivate the enzyme.[34] Cyanide is then metabolized, primarily by β-cyanoalanine synthase to produce the non-proteogenic amino acid β-cyanoalanine, which can be further converted into asparagine.[4,35,36]

20.4.2 Other Substrates

ACCO is able to oxidize various cyclic analogues of ACC into ethylene or the corresponding alkenes as well as various acyclic amino acids into the corresponding carbonyl compound, CO_2 and ammonia.[29,31,37,38] Like in the case of ACC, the oxidation reactions proceed *via* a radical mechanism and the formation of aminyl radicals. Depending on the structure of the amino acid substrate, there are different pathways to form products (Scheme 20.4). The oxidation of cyclic analogues follows the mechanism described in Scheme 20.3, *i.e.* ring opening followed by the second oxidation and then decarboxylation (Path A, Scheme 20.4). Oxidation of the cyclic analogue, 1-aminocyclobutane-1-carboxylic acid (ACBCH) by ACCO only leads to dehydroproline (Δ^1-pyrroline-2-carboxylic acid) following the same mechanism (Path A).[29] In the case of non-cyclic analogues, the aminyl radicals follow a decarboxylation pathway and only the carbonyl products arising from the hydrolysis of the corresponding imines are detected (Path B).[31,37,38]

20.4.3 Possible Mechanism

Given the number of substrates and cofactors required for ACCO activity, it has been difficult to clearly establish an order of binding using steady-state kinetic analysis. An ordered process where ACC binding to ACCO precedes O_2 binding has been proposed.[39] In this sequence, the binding order of ascorbate is unclear and it has been suggested that it could happen first before ACC or third after O_2.[39,40]

In solution, the iron(II) ion is six-coordinate distorted octahedral in the resting state of the enzyme, with the coordination sphere being most probably completed by three water molecules.[21] The substrate ACC binds in a bidentate fashion, most likely followed by O_2 binding, providing an iron(III)–superoxo intermediate (Scheme 20.5). After O_2 binding, several mechanisms have been proposed that all involve an Fe(IV)=O intermediate similar to that

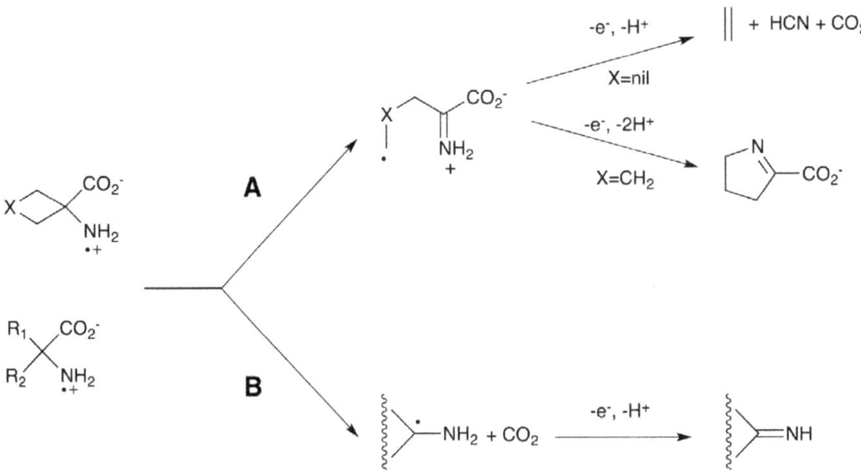

Scheme 20.4 Pathways leading to the different products after oxidation of amino acids to aminyl radicals by ACCO. ACC and ACBCH are oxidised following path A; non-cyclic amino acids are oxidised following path B.

observed in several 2OG oxygenases but differing in the electron transfer sequence and the intermediates responsible for the successive substrate oxidation steps. In one mechanism (not depicted in Scheme 20.5), Rocklin *et al.*[20] used results from single-turnover experiments to propose that one electron reduction forms an Fe(III)–peroxo intermediate responsible for the first ACC oxidation, the concomitantly formed Fe(IV)=O intermediate is responsible for the second oxidation step, and electron transfer occurring after product release completes the catalytic cycle. A crucial role for bicarbonate in assisting a proton transfer step was proposed. Density functional theory (DFT) calculations based on this mechanism allow better understanding of several steps of the mechanism and provide insight into the electronic structure of some possible intermediates.[23] More recently, on the basis of kinetic studies and isotopic effects, Klinman and colleagues have proposed an ascorbate-dependent formation of Fe(IV)=O intermediate, which is responsible for substrate oxidation.[41] Interestingly, the binding of ascorbate to the enzyme was proposed to be random and dependent on its concentration, thus rationalizing the results of previous steady-state kinetic studies.[39,40] This latest proposed O_2 activation mechanism, with two electron-transfer steps from ascorbate at the beginning of the catalytic cycle leading to the Fe(IV)=O intermediate, is presented in Scheme 20.5.

20.4.4 Inactivation Processes

Like other iron-containing oxidases, including several members of the 2OG oxygenases,[42] ACCO is particularly sensitive to metal-catalysed oxidative modifications, which leads to instability of the enzyme.[43,44] The enzyme displays

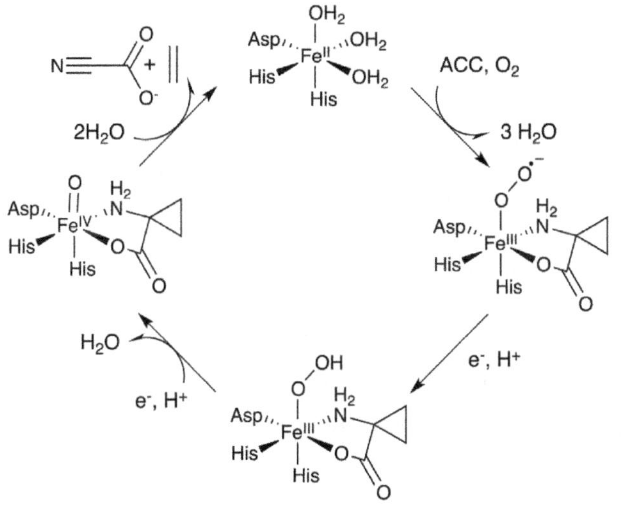

Scheme 20.5 Possible O_2 activation mechanism by ACCO. The electrons are provided by ascorbate.

short lifetimes under catalytic conditions and the loss of activity is followed by partial proteolysis. Inactivation processes were observed in the presence of ferrous ion and ascorbate. This particular combination of cofactors is known to generate reactive oxygen species, which may account for oxidative modifications including proteolysis.[45] This ascorbate-dependent enzyme inactivation is partly prevented by the addition of catalase. In addition to deleterious reactions by the excess metal ions, mutagenesis studies have shown that some inactivation processes are mediated by the iron ion at the active site,[46] and specific inactivation processes related to different fragmentation patterns are observed in the presence of the substrate ACC.[44] Interestingly, these active site mediated inactivation processes are hindered by the presence of bicarbonate. In addition to its ability to promote enzyme activity, this cofactor thus displays a protective effect against auto-oxidation processes.

20.4.5 Bioinspired Model Complexes

The use of model complexes is a powerful strategy to gain better insight into a metalloenzyme's catalytic mechanism. To the best of our knowledge, few structural data are available on iron–ACC complexes,[47] and no complex displaying the bidentate coordination of the amino acid on iron(II) has been reported. A functional model based on the [Fe^{III}(Salen)Cl] (Salen = *N*,*N*′-bis(salicylidene)-ethylenediaminato) is able to efficiently catalyse the oxidation of ACC and various amino acid substrates in the presence of several oxidants (such as H_2O_2).[48,49] However, no data on the interaction between the metal ion and the substrate were obtained. In addition, the oxidation of ACBCH led to a mixture of products indicating that several competing mechanisms occur. In contrast, several Cu(II)–ACC complexes in which ACC is bound in a bidentate mode on the metal ion have been described.[50,51] These complexes display ACCO-like activity in the presence of hydrogen peroxide, with ethylene produced from the bound substrate. Oxidation of amino acid and aminophosphonic acid analogues of ACC were also reported and the oxidation of ACBCH only led to products arising from the ring-opening pathway (path A, Scheme 20.4).[52] The nature of the oxidizing species involved in these copper-based systems is probably different from that involved in the Fe(II)-containing enzyme and the development of iron-based models for acquiring structural, spectroscopic and mechanistic data would thus be of great interest to understand the catalytic mechanism of ACCO.

20.4.6 On the Role of Bicarbonate

It was observed in early studies of ACCO that CO_2 is required for the enzyme activity. As mentioned previously, this cofactor has both an activation role and a protection role against catalytic inactivation processes.[6,44] It has been proposed that bicarbonate may be involved in a proton transfer step during dioxygen activation, giving a plausible explanation for its role in the promotion of the enzymatic activity.[20,23] On the other hand, it has been proposed that this cofactor stabilizes a six-coordinated ACCO:Fe(II):ACC complex that is unable to react with dioxygen until the reductant (ascorbate) is present, thus preventing uncoupled reactions leading to auto-oxidation of the enzyme.[21]

Interestingly, a 3- to 5-fold improved binding affinity of ACC to *holo*-ACCO in the presence of bicarbonate has been observed.[16,21] This stabilization effect is coherent with the interaction between HCO_3^- and ACC evidenced by modelling experiments (Scheme 20.2).[16] Although speculative, the latter model is consistent with most of the observed effects for bicarbonate (stabilization, activation and protection). A simple mechanical effect of the bicarbonate ion on the positioning of ACC could indeed be crucial to avoid auto-oxidation reactions. In addition, modelling studies imply that bicarbonate, being anchored 5–7 Å away from the metal centre, would be perfectly located to assist a proton transfer during the catalytic mechanism (Scheme 20.2).

20.5 ACC Oxidase and the 2OG-Dependent Oxygenases

ACCO is a non-haem iron(II) enzyme displaying the $HX(D/E)X_mHX_nRXS$ sequence motif, associated with Fe(II) and 2OG binding, and many structural features conserved in 2OG-dependent enzymes, the largest subgroup of the cupin superfamily of structurally related proteins.[53] ACCO is different from the vast majority of other members of the family because it does not require 2OG, but rather it is proposed to use ascorbate as a cosubstrate and is stimulated by CO_2 (or bicarbonate). Although they use different substrates and cofactors, mechanistic similarities between ACCO and the related 2OG oxygenases can be highlighted. One is the bidentate binding of ACC to the metal ion that resembles that of 2OG binding in the related oxygenases. Interestingly, one possible binding site for bicarbonate corresponds to that of the terminal (C-5) carboxylate of the 2OG by the RXS motif.[16] A comparison can be drawn between the positions of ACC/bicarbonate on one hand and 2OG on the other, as shown in Figure 20.3, which compares the interactions in the ACCO and anthocyanidin synthase active sites.[54] It has been observed that in the presence of ACC, the bicarbonate cofactor enhances the binding affinity of the metal ion.[16] It is conceivable that 2OG, which develops interactions with residues from the RXS motif in 2OG oxygenases, similarly provides additional chelating ligands to enhance the iron(II) affinity. Finally, there is similarity in the plausible involvement of an Fe(IV)=O intermediate during the dioxygen activation mechanism in ACCO. Whereas the Fe(IV)=O in 2OG oxygenases is formed by oxidation of the bound 2OG substrate, an ascorbate-dependent dioxygen activation pathway has been proposed in ACCO.

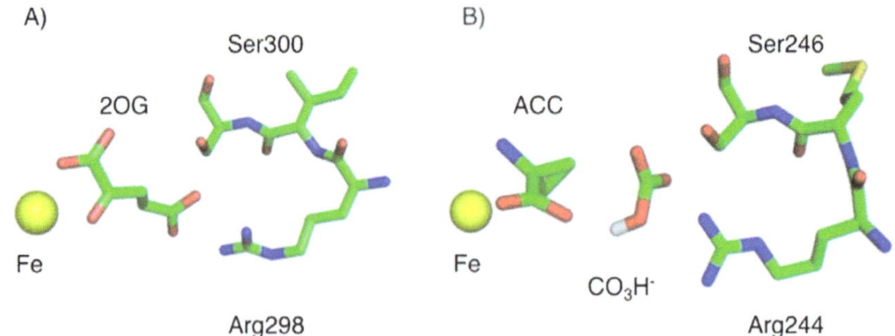

Figure 20.3 Comparison between (A) the interaction of 2OG with anthocyanidin synthase (adapted with permission from Wilmouth *et al.*[54]) and (B) the interactions of ACC and bicarbonate with ACCO predicted from modelling studies.

20.6 Conclusion and Perspectives

Considerable insight has been gained into the ACCO reaction mechanism from X-ray crystallography, steady-state kinetics, site-directed mutagenesis, spectroscopic studies, DFT analyses and modelling studies. However, there are still many unanswered questions regarding this enzyme. One concerns the nature of its active conformation and the possible structural flexibility of the *C*-terminal portion of the protein during catalysis. Moreover, the positions of the substrates/cofactors in the active site remain unclear. In addition, although multiple species (iron-superoxo, iron-peroxo or iron-oxo) have been discussed as possible oxidizing intermediates, spectroscopic evidence is lacking and the catalytic mechanism of ACCO is still debated. Finally, the function is unknown for the intriguing post-translational activities including the phosphokinase and protease activities reported for ACCO.[19] Further analyses may shed light on possible roles for these in the ethylene signal transduction pathway.

Ethylene is one of the most important organic compounds in the chemical industry and it is used as raw material for a large number of commercial synthetic products. Currently, ethylene production in the petrochemical industry is based on the steam cracking of hydrocarbons, a process that requires high pressures and temperatures. There is thus interest in developing technologies for ethylene production from renewable resources including CO_2 and biomass. In addition to plants, a variety of organisms including bacteria and fungi also produce ethylene, probably to take advantage of ethylene-responsive mechanisms in plants and facilitate the successful invasion of plant tissue. Recently an ACCO has been identified in the fungus *Agaricus bisporus*.[55] Remarkably, *Pseudomonas syringae* possesses an ethylene-forming enzyme (EFE) that utilizes 2OG and arginine as substrates.[56] The mechanism of this enzyme remains unclear but EFE appears as a possible biotechnology target because the expression of a single gene is sufficient for ethylene production in the apparent absence of a toxic intermediate.[57] In contrast, while the cellular carbon pool can be diverted into ACC and ethylene by heterologous expression of a fused ACCS-ACCO enzyme in yeast and bacteria,[58] the use of ACCO for biotechnological ethylene production by other organisms may be limited by its low stability (at least for studied variants) and by the production of toxic cyanide during catalysis.

Given the importance of ethylene in plants, control of its level and action is of considerable commercial interest. Accordingly, efforts have been directed towards understanding the signal transduction pathways initiated by the binding of ethylene to its receptors containing a copper cofactor.[59–61] Several strategies to control ethylene production and action have been developed including ethylene removal from the atmosphere by oxidation or adsorption, the use of inhibitors or antagonists interacting at the receptor level, as well as the elaboration of genetically engineered plants with decreased ethylene biosynthesis and sensitivity.[62–64]

References

1. A. B. Bleecker and H. Kende, *Annu. Rev. Cell Dev. Biol.*, 2000, **16**, 1.
2. D. O. Adams and S. F. Yang, *Proc. Natl. Acad. Sci. U. S. A.*, 1979, **76**, 170.
3. S. F. Yang and N. E. Hoffmann, *Annu. Rev. Plant Physiol.*, 1984, **35**, 155.
4. G. D. Peiser, T.-T. Wang, N. E. Hoffman, S.F Yang, H.-W. Liu and C. T. Walsh, *Proc. Natl. Acad. Sci. U. S. A.*, 1984, **81**, 3059.
5. N. Smirnoff, *Curr. Opin. Plant Biol.*, 2000, **3**, 229.
6. J. J. Smith and P. John, *Phytochemistry*, 1993, **32**, 1381.
7. J. G. Dong, J. C. Fernandez-Maculet and S. F. Yang, *Proc. Natl. Acad. Sci. U. S. A.*, 1992, **89**, 9789.
8. D. R. Dilley, J. Kuai, I. D. Wilson, Y. Pekker, Y. Zhu, D. M. Burmeister and R. M. Beaudry, *Acta Hort.*, 1995, **379**, 25.
9. Z. Zhang, J.-S. Ren, I. J. Clifton and C. J. Schofield, *Chem. Biol.*, 2004, **11**, 1383.
10. R. P. Hausinger, *Crit. Rev. Biochem. Mol. Biol.*, 2004, **39**, 21.
11. I. J. Clifton, M. A. McDonough, D. Ehrismann, N. J. Kershaw, N. Granatino and C. J. Schofield, *J. Inorg. Biochem.*, 2006, **100**, 4644.
12. W. Aik, M. A. McDonough, A. Thalhammer, R. Chowdhury and C. J. Schofield, *Curr. Opin. Struct. Biol.*, 2012, **22**, 691.
13. A. Yoo, Y. S. Seo, J.-W. Jung, S.-K. Sung, W. T. Kim, W. Lee and D. R. Yang, *J. Struct. Biol.*, 2006, **156**, 407.
14. A. M. Rocklin, D. L. Tierney, V. Kofman, N. M. W. Brunhuber, B. M. Hoffman, R. E. Cristoffersen, N. O. Reich, J. D. Lipscomb and L. Que Jr., *Proc. Natl. Acad. Sci. U. S. A.*, 1999, **96**, 7905.
15. D. L. Tierney, A. M. Rocklin, J. D. Lipscomb, L. Que Jr. and B. M. Hoffman, *J. Am. Chem. Soc.*, 2005, **127**, 7005.
16. L. Brisson, N. El Bakkali-Taheri, M. Giorgi, A. Fadel, J. Kaizer, M. Réglier, T. Tron, E. H. Ajandouz and A. J. Simaan, *J. Biol. Inorg. Chem.*, 2012, **17**, 939.

17. D. R. Dilley, D. K. Kadyrzhanova and Z. Wang, *Acta Hort.*, 2001, **553**, 143.
18. Y. S. Seo, A. Yoo, J. Jung, S.-K. Sung, D. R. Yang, W. T. Kim and W. Lee, *Biochem. J.*, 2004, **380**, 339.
19. D. R. Dilley, Z. Wang, D. K. Kadirjan-Kalbach, F. Ververidis, R. Beaudry and K. Padmanabhan, *AoB Plants*, 2013, **5**, plt031.
20. A. M. Rocklin, K. Kato, H. W. Liu, L. Que Jr. and J. D. Lipscomb, *J. Biol. Inorg. Chem.*, 2004, **9**, 171.
21. J. Zhou, A. M. Rocklin, J. D. Lipscomb, L. Que Jr. and E. I. Solomon, *J. Am. Chem. Soc.*, 2002, **124**, 4602.
22. T. Iturriagagoitia-Bueno, E. J. Gibson, C. J. Schofield and P. John, *Phytochemistry*, 1996, **43**, 2343.
23. A. Bassan, T. Borowski, C. J. Schofield and P. E. M. Siegbahn, *Chem. Eur. J.*, 2006, **12**, 8835.
24. L. Stella, S. Wouters, F. Baldellon and F. Bull, *Soc. Chim. Fr.*, 1996, **133**, 441.
25. R. M. Adlington, R. T. Aplin, J. E. Baldwin, B. J. Rawlings and D. Osborne, *J. Chem. Soc. Chem. Commun.*, 1982, 1086.
26. R. M. Adlington, J. E. Baldwin and B. J. Rawlings, *J. Chem. Soc. Chem. Commun.*, 1983, 290.
27. M. C. Pirrung, *J. Am. Chem. Soc.*, 1983, **105**, 7207.
28. J. E. Baldwin, D. A. Jackon, R. M. Adlington and B. J. Rawlings, *J. Chem. Soc. Chem. Commun.*, 1985, 206.
29. M. C. Pirrung, J. Cao and J. Chen, *Chem. Biol.*, 1998, **5**, 49.
30. J. E. Bladwin, R. M. Adlington, G. A. Lajoie, C. Lowe, P. D. Baird and K. Prout, *J. Chem. Soc. Chem. Commun.*, 1988, 775.
31. Y.-Y. Charng, S.-J. Chou, W.-T. Jiaang, S.-T. Chen and S. F. Yang, *Arch. Biochem. Biophys.*, 2001, **385**(1), 179.
32. M. C. Pirrung, *Acc. Chem. Res.*, 1999, **32**, 711.
33. M. Ito and H. Tokiwa, *Bull. Chem. Soc. Jpn.*, 2007, **80**(9), 1731.
34. L. J. Murphy, K. N. Robertson, S. G. Harroun, C. L. Brosseau, U. Werner-Zwanziger, J. Moilanen, H. M. Tuononen and J. A. C. Clyburne, *Science*, 2014, **344**, 75.
35. J. M. Miller and E. E. Conn, *Plant Physiol.*, 1980, **65**, 1199.
36. M. C. Pirrung, *Bioorg. Chem.*, 1985, **13**, 219.
37. E. J. Gibson, Z. Zhang, J. E. Baldwin and C. J. Shofield, *Phytochemistry*, 1998, **48**, 619.
38. J. Thrower, L. M. Mirica, K. P. McCusker and J. P. Klinman, *Biochemistry*, 2006, **45**, 13108.
39. J. S. Thrower, R. Blalock and J. P. Klinman, *Biochemistry*, 2001, **40**, 9717.
40. N. M. W. Brunhuber, J. L. Mort, R. E. Christoffersen and N. O. Reich, *Biochemistry*, 2000, **39**, 10730.
41. L. M. Mirica and J. P. Klinman, *Proc. Natl. Acad. Sci. U. S. A.*, 2008, **105**, 1814.
42. M. Mantri, Z. Zhang, M. A. McDonough and C. J. Schofield, *FEBS J.*, 2012, **279**(9), 1563.
43. J. J. Smith, Z. Zhang, C. J. Schofield, P. John and J. E. Baldwin, *J. Exp. Bot.*, 1994, **45**, 274521.

44. J. N. Barlow, Z. Zhang, P. John, J. E. Baldwin and C. J. Schofield, *Biochemistry*, 1997, **36**, 3563.
45. E. R. Stadtman and R. L. Levine, *Amino Acids*, 2003, **25**, 207.
46. Z. Zhang, J. N. Barlow, J. E. Baldwin and C. J. Schofield, *Biochemistry*, 1997, **36**, 15999.
47. W. Ghattas, Z. Serhan, N. El Bakkali-Taheri, M. Réglier, M. Kodera, Y. Hitomi and A. J. Simaan, *Inorg. Chem.*, 2009, **8**, 3910.
48. G. Baráth, J. Kaizer, J. S. Pap, G. Speier, N. El Bakkali-Taheri and A. J. Simaan, *Chem. Commun.*, 2010, **46**, 7391.
49. S. Góger, D. Bogáth, G. Baráth, A. J. Simaan, G. Speier and J. Kaizer, *J. Inorg. Biochem.*, 2013, **123**, 46.
50. W. Ghattas, C. Gaudin, M. Giorgi, A. Rockenbauer, A. J. Simaan and M. Réglier, *Chem. Commun.*, 2006, 1027.
51. W. Ghattas, M. Giorgi, Y. Mekmouche, A. Rockenbauer, T. Tanaka, M. Réglier, Y. Hitomi and A. J. Simaan, *Inorg. Chem.*, 2008, **47**, 4627.
52. J. S. Pap, N. El Bakkali-Tahéri, A. Fadel, S. Góger, D. Bogáth, M. Molnár, M. Giorgi, G. Speier, A. J. Simaan and J. Kaizer, *Eur. J. Inorg. Chem.*, 2014, 2829.
53. J. M. Dunwell, A. Parvis and S. Khuri, *Phytochemistry*, 2004, **65**, 7.
54. R. C. Wilmouth, J. J. Turnbull, R. W. D. Welford, I. J. Clifton, A. G. Prescott and C. J. Schofield, *Structure*, 2002, **10**, 93.
55. D. Meng, L. Shen, R. Yang, X. Zhang and J. Sheng, *Biochim. Biophys. Acta*, 2014, **1840**, 120.
56. H. Fukuda, T. Ogawa, K. Ishihara, T. Fujii, K. Nagahama, T. Omata, Y. Inoue, S. Tanase and Y. Morino, *Biochem. Bioph. Res. Commun.*, 1992, **188**, 826.
57. C. Eckert, W. Xu, W. Xiong, S. Lynch, J. Ungerer, L. Tao, R. Gill, P.-C. Maness and J. Yu, *Biotechnol. Biofuels.*, 2014, **7**(1), 33.
58. N. Li, X. N. Jiang, C. P. Cai and S. F. Yang, *J. Biol. Chem.*, 1996, **271**, 4225738.
59. J. M. Alonso and A. N. Stepanova, *Science*, 2004, **306**, 1513.
60. J. Pirrello, B. C. N. Prasad, W. Zhang, K. Chen, I. Mila, M. Zouine, A. Latché, J. C. Pech, M. Ohme-Takagi, F. Regad and M. Bouzayen, *BMC Plant Biol.*, 2012, **12**, 190.
61. S. N. Shakeel, X. Wang, B. M. Binder and G. E. Schaller, *AoB Plants*, 2013, **5**, plt010, DOI: 10.1093/aobpla/plt010.
62. N. Keller, M.-N. Ducamp, D. Robert and V. Keller, *Chem. Rev.*, 2013, **113**, 5029.
63. E. C. Sisler, *Biotechnol. Adv.*, 2006, **24**, 357.
64. J. C. Czarny, V. P. Grichko and B. R. Glick, *Biotechnol. Adv.*, 2006, **24**, 410.

CHAPTER 21

4-Hydroxyphenylpyruvate Dioxygenase and Hydroxymandelate Synthase: 2-Oxo Acid-Dependent Oxygenases of Importance to Agriculture and Medicine

DHARA D. SHAH[a] AND GRAHAM R. MORAN*[a]

[a]Department of Chemistry and Biochemistry, University of Wisconsin-Milwaukee, 3210 N. Cramer St, Milwaukee, WI 53211-3209, USA
*E-mail: moran@uwm.edu

21.1 Introduction

Among the non-haem Fe(II)-dependent oxygenases, 2-oxoglutarate (2OG) is the most prevalent source of electrons for dioxygen activation.[1] The 2-oxo-acid-dependent oxygenases (2OADOs) utilize a ferrous ion and the oxidative power of dioxygen to collectively perform an extremely broad set of chemistries throughout secondary metabolism and beyond (for a comprehensive list, see Chapter 1 of this volume). The majority of the 2OADOs utilize three substrates: molecular oxygen, 2OG and a substrate that is specific to each enzyme. These enzymes typically catalyse two half-reactions, initial decarboxylation of the 2OG and then oxidation of the specific substrate.

RSC Metallobiology Series No. 3
2-Oxoglutarate-Dependent Oxygenases
Edited by Robert P. Hausinger and Christopher J. Schofield

Published by the Royal Society of Chemistry, www.rsc.org

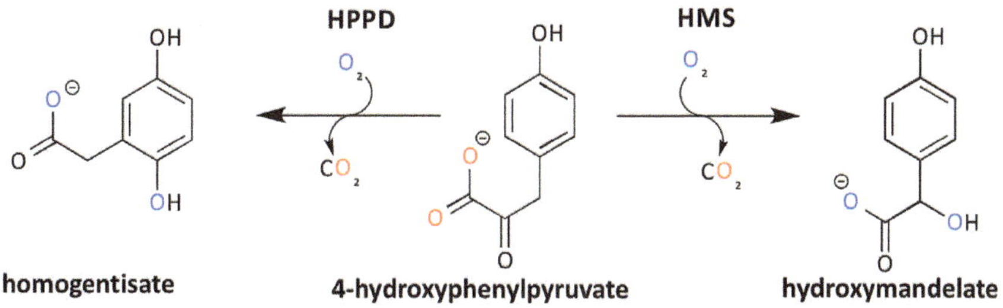

Scheme 21.1 Reactions catalysed by HPPD and HMS.

Grouped within the 2OADO family are two important enzymes that require only two substrates, *i.e.* 4-hydroxyphenylpyruvate dioxygenase (HPPD) and hydroxymandelate synthase (HMS). These sister enzymes accomplish the fundamental catalytic chemistry observed in essentially all 2OADO family members without requiring 2OG. For these enzymes, the requisite 2-oxo acid is supplied from the pyruvate substituent of 4-hydroxyphenylpyruvate (HPP), the organic substrate common to both enzymes. The decarboxylation half-reaction is thought to be largely equivalent in HPPD and HMS, while the oxidation/hydroxylation half-reactions differ in regioselectivity and complexity. HPPD induces oxidative decarboxylation of HPP followed by aromatic hydroxylation and substituent migration (known as the NIH shift), whereas HMS accomplishes decarboxylation and aliphatic hydroxylation at the benzylic carbon (Scheme 21.1).

The ubiquity of the HPPD reaction has meant that it has been enlisted to participate in important added roles in multiple organisms. This has resulted in its inhibition having unique significance in each kingdom of life and has become the basis for the development of herbicides and therapeutics. The objective of this chapter is to offer a concise and current report of the contextual and mechanistic chemistry of both HPPD and HMS.

21.2 Natural Context

HPPD and HMS acquire HPP from the first step of tyrosine catabolism (Scheme 21.2), a pathway that comprises five enzyme activities common to essentially all aerobes and yielding acetoacetate and fumarate for use in energy-yielding metabolism (Scheme 21.2, black structures).

HPPD catalyses the committed second step of the pathway in which HPP is converted to 2,5-dihydroxyphenylacetate (homogentisate, HG).[2] This reaction has generated a great deal of interest due both to the curious chemical transformation catalysed and also as inhibition of this chemistry has unique important outcomes. Tyrosine catabolism metabolites provide biosynthetic stocks for the production of various physiologically important molecules that are unique in each kingdom. Examples are plastoquinone and tocopherols in photosynthetic organisms,[3] melanin in fungi and bacteria,[4] and cofactor

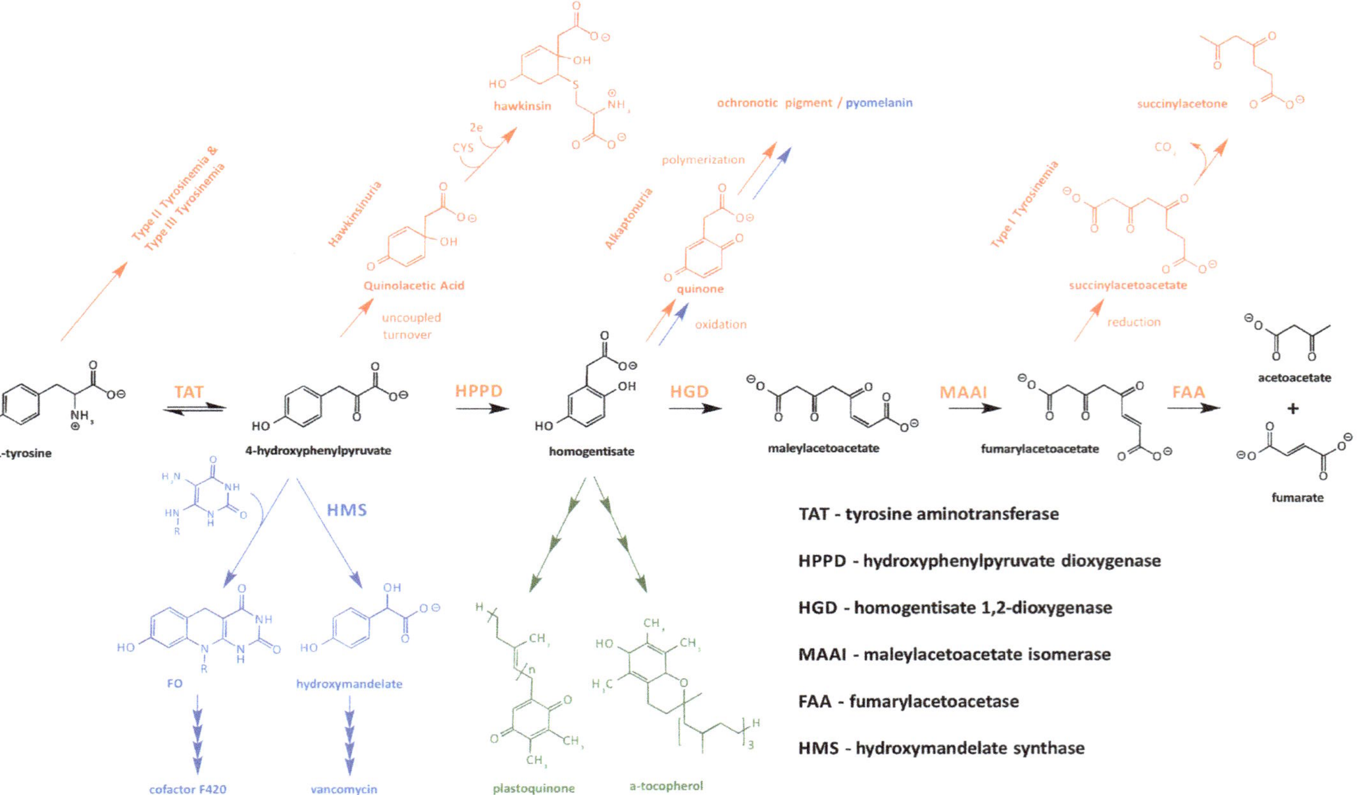

Scheme 21.2 Tyrosine catabolism. Reactions shown in black depict the typical steps of the tyrosine catabolism pathway. Reactions shown in red are mammalian metabolic disorders. Reactions depicted in blue are from bacteria and archaea. Reactions in green occur in photosynthetic organisms.

F420 in mycobacteria and archaea.[5] A relatively new branch from tyrosine catabolism was identified with the discovery of a protein having a similar sequence to that of HPPD, but that acted only on the pyruvate substituent of HPP (Scheme 21.1) to form 4-hydroxymandelate (HMA).[6] It transpired that this enzyme, HMS, catalyses the initial step in the biosynthesis of 4-hydroxyphenylglycine, a building block of various peptide antibiotics.[7]

21.2.1 Role in Photosynthesis

In photosynthetic organisms the HPPD reaction provides HG for the biosynthesis of plastoquinone (a redox cofactor that bridges the photosystems) and tocopherols (antioxidant lipophilic plant hormones).[8] HPPD is therefore required for photosynthesis, and inhibitors of this enzyme have herbicidal activity. Molecules such as leptospermone and usnic acid (Scheme 21.3A) are naturally occurring HPPD inhibitors produced as allelopathic agents.[9–13] Synthetic derivatives based on the leptospermone scaffold have brought new and more effective herbicides. 2-(2-Nitro-4-fluoromethylbenzoyl)-1,3-cyclohexanedione (NTBC) was one of the earliest synthetic HPPD herbicides identified and has remained the paradigm HPPD inhibitor despite never being marketed for agricultural use (see below).[14] Sulcotrione, mesotrione, isoxaflutole and tembotrione are a few examples of highly successful pre- and post-emergent broadleaf herbicides that target HPPD.

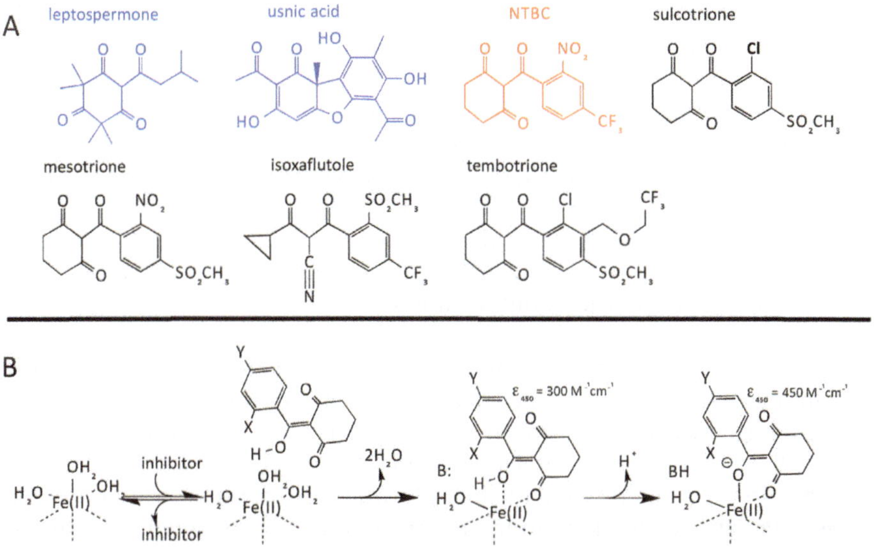

A

leptospermone, usnic acid, NTBC, sulcotrione

mesotrione, isoxaflutole, tembotrione

B

inhibitor; $2H_2O$; H^+; ε_{450} = 300 $M^{-1}cm^{-1}$; ε_{450} = 450 $M^{-1}cm^{-1}$

Scheme 21.3 (A) Examples of HPPD inhibitors. Molecules in blue are naturally occurring. NTBC, in red, is used for therapeutic purposes. Inhibitors shown in black are the active components of commercially available herbicides. (B) A minimal kinetic mechanism for association of inhibitors with HPPD.

21.2.2 Role in Mammals

In mammals tyrosine catabolism is required to diminish blood concentrations of this highly insoluble amino acid. Five inborn errors of tyrosine catabolism result in metabolic disorders (Scheme 21.2, red structures). Type II tyrosinemia is caused by a lack of tyrosine aminotransferase activity.[15] Elevated blood tyrosine levels are observed in patients with this disease resulting in mild intellectual impairment, corneal opacities, and palmoplantar hyperkeratosis.[15] Mutations in the gene encoding HPPD cause type III tyrosinemia and hawkinsinuria. Type III tyrosinemia arises from mutations that render HPPD inactive. Due to the reversibility of the tyrosine aminotransferase reaction, this leads to elevated blood tyrosine. As such, patients with type II and type III tyrosinemias have similar pathologies. Hawkinsinuria results from a specific mutation in the *HPPD* gene yielding an active enzyme that catalyses an abortive reaction.[16,17] This disease is thought to be an autosomal-dominant disorder, because only one mutated allele is required for the symptoms of the disease to manifest. This disease is characterized by quiescence and a general inability to thrive during infancy as a consequence of persistent acidosis, vomiting, diarrhoea, hair growth abnormalities and liver enlargement in some cases.[18] Curiously, the condition seems to self-correct at approximately two years of age for reasons that are unknown.[19,20] The disorder is diagnosed by the presence of hawkinsin in the urine. In addition to hawkinsin, the unusual metabolites quinolacetic acid (QAA), 4-hydroxyphenyllactic acid, 4-hydroxyphenylacetic acid (HPA), 4-hydroxycyclohexylacetic acid and 5-oxoproline are also excreted.[21] The causative mutation for hawkinsinuria was shown to be a missense mutation resulting in an Asn to Ser variant (N241S in human HPPD).[17] This Asn is conserved in HPPD sequences and the mutation at this position was later explored with the structurally equivalent *Rattus norvegicus* N241S HPPD and *Streptomyces avermitilis* N245S HPPD variants. In place of HG, the variant enzyme forms QAA that is prone to react *via* a Michael addition with cellular thiols to form hawkinsin (Scheme 21.2).[22–24] HPPD variants with this mutation are subject to NTBC inhibition in a similar manner to wild-type HPPD, suggesting a role for HPPD inhibitors as therapeutics for hawkinsinuria patients.[22]

Alkaptonuria is the oldest known inherited metabolic disorder. In 1902, Garrod coined the term 'inborn error of metabolism' and speculated that alkaptonuria is caused by inactivity or deficiency of the enzyme involved in the cleavage of the HG aromatic ring.[25,26] In 1958, La Du *et al.* established the deficiency of homogentisate 1,2-dioxygenase (HGD) activity, the third enzyme in the tyrosine catabolism pathway, in the manifestation of the disease.[27] Patients with alkaptonuria excrete dark, melanin-laden urine arising from the oxidation and polymerization of HG. These pigments also accumulate in tissues where they cause ochronosis, a darkening of cartilage and bones, and lead to arthritic joint destruction and deterioration of cardiac valves.[28] HPPD inhibition by NTBC has been tested successfully as a therapy for alkaptonuria patients, causing pronounced abatement of symptoms.[28–30]

The fourth step of the tyrosine catabolism pathway is catalysed by maleylacetoacetate isomerase. Deficiency in the activity of this enzyme is not deleterious as maleylacetoacetate can be converted to fumarylacetoacetate and succinylacetone *via* a glutathione-mediated non-enzymatic bypass,[31] and both of these molecules are reported as substrates for the last enzyme in the pathway, fumarylacetoacetase (Scheme 21.2). The most severe and morbid tyrosinemia is type I. The primary pathological signature is early onset of liver cirrhosis, renal tubular defects and ultimately hepatocellular carcinoma.[32] Type 1 tyrosinemia is caused by the deficiency of fumarylacetoacetase activity. As such, fumarylacetoacetate accumulates and leads to the formation of succinylacetoacetate and succinylacetone, which have been linked with liver and kidney damage possibly due to facile reactions with cellular thiols.[33] Liver transplantation was once the only effective therapy to treat patients with type I tyrosinemia;[34] however, in 1977 inhibition of HPPD activity was suggested as an alternative approach.[33] Patients treated with NTBC for several months show a marked decrease in the plasma succinylacetoacetate and succinylacetone concentrations and improved liver function. These initial trials established the current era of FDA-approved herbicidal treatment of type I tyrosinemia.[35]

In summary, inhibition of HPPD by NTBC has provided established or potential therapies for three of the five inborn errors of tyrosine catabolism. The therapeutic use of NTBC, however, can lead to type II and III tyrosinemia-like symptoms including high blood tyrosine levels, which can result in corneal opacities and suppression of IQ.[36–38]

21.2.3 Role in Microorganisms

In Scheme 21.2, the blue reactions represent tyrosine catabolism branches that are known to occur in bacteria and archaea. These branches function in pathogenesis, redox reactions (coenzymes) and competition (antibiotic production). It was in this latter function that HMS was discovered.[6,7] HMS diverts HPP from the tyrosine catabolism pathway and catalyses the first step in the biosynthesis of 4-hydroxyphenylglycine. This non-proteinogenic amino acid is a substructure for many macrocyclic non-ribosomal peptide antibiotics including vancomycin, complestatin and chloroeremomycin. In another synthetic branch that occurs in mycobacteria and methanogenic archaea, HPP is diverted from tyrosine catabolism and condensed with a pyrimidine dione to form FO, the redox active portion of cofactor F420.[39,40] Numerous pathogenic bacteria (*e.g. Legionella*, *Shewanella* and *Pseudomonas* spp.) are known to produce pyomelanin to aid virulence, and HPPD has been identified as the virulence factor associated with pigment production.[41] It is believed that HG oxidation and polymerization yield pyomelanin and that this over-production is selected for when the bacterial population experiences environmental stresses such as the oxidative onslaught brought on by the immune system.[42] In clinical isolates from cystic fibrosis patients with persistent *P. aeruginosa* infections, bacteria exhibit increased extracellular pyomelanin production resulting from mutation in the *HGD* gene.[43] Increased pyomelanin production is linked to the increased persistence of infecting bacteria.[4,44] *B. cenocepacia* survives in phagocytic vesicles within macrophage cells by the production of pyomelanin that scavenges free radicals. HPPD inhibition through NTBC, however, causes reduced pigment production and increases clearance of pathogens.[45]

Tyrosine catabolism and HPPD have been linked to the establishment of specific fungal infections. For the fungus *Paracoccidioides brasiliensis* the mycelium-to-yeast morphological transition occurring inside the pulmonary epithelium is crucial to establishing the infection in humans. HPPD is overexpressed during this transition, and *in vitro* studies demonstrate that NTBC prevents the change in morphology and halts infection, opening a door to the treatment of dimorphic fungal infections by HPPD inhibition.[46]

21.3 Structure

HPPD is found in almost all aerobic life forms, whereas HMS is found only in specific bacteria. Moreover, organisms that produce HMS also produce HPPD. These factors suggest that the gene encoding HPPD is the progenitor of that for HMS. Secondary structural elements of HPPD and HMS are highly superimposable despite modest (~35%) sequence identity (Figure 21.1A). Both enzymes contain two structural domains, a conserved *C*-terminal domain that includes all residues of the active site plus the ferrous ion cofactor and a less well conserved *N*-terminal domain with apparent

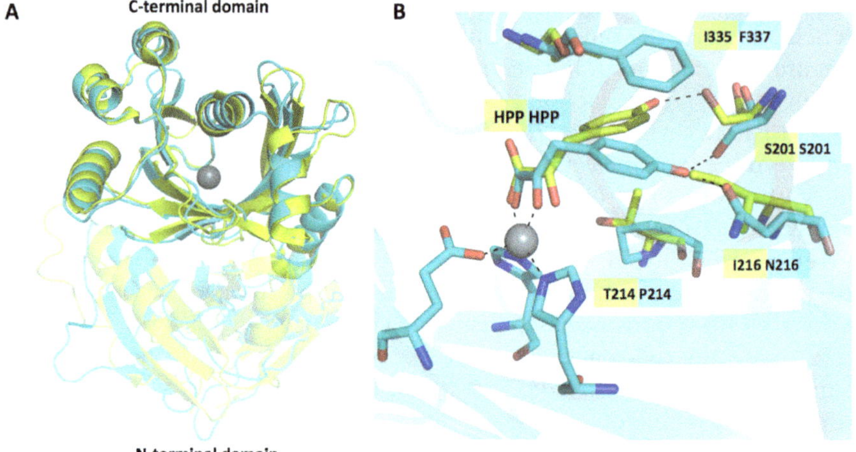

Figure 21.1 (A) Superimposed tertiary structures of *P. fluorescens* HPPD (cyan, PDB ID: 1CJX) and *A. orientalis* HMS (yellow, PDB ID: 2R5V). (B) Unique and mechanistically important residues, including the facial triad, in the active site of HPPD and HMS, with substrate HPP modelled. The positions for the substrate were taken from Brownlee *et al.*[48]

vestigial structural functionality.[47,48] The *N*- and the *C*-terminal domains of each enzyme are also largely superimposable. Each domain forms an incomplete β-barrel comprised of tandem βαβββ units. This overall bi-symmetrical topology suggests two successive gene duplications were required to form the overall structure. This domain topology is that of the vicinal oxygen chelate fold,[49,50] which has no topological resemblance to the structures of other 2OADOs that have modifications to the more common jelly-roll fold.[51] This difference in topology indicates separate evolutionary lineages that yielded independent mechanistic convergence.

It is unsurprising that the majority of sequence conservation in HPPD and HMS occurs in the catalytic *C*-terminal domain and that many of these conserved residues extend inward toward the approximate centre of the β-barrel structure, arrayed toward the metal ion.[47–49] Three of these residues comprise a common structural motif, coined the 2-His-1-carboxylate facial triad, that is present in almost all known mononuclear non-haem Fe(II)-dependent oxygenases.[52,53] This motif, which functions in coordinating Fe(II), has the general spacing of $HX(D/E)X_{50-210}H$ in 2OG-dependent oxygenases, whereas in HPPD and HMS a spacing of $HX_{\sim 80}HX_{\sim 80}E$ is observed, consistent with their very different structures.[54]

21.3.1 The Active Site

As might be expected, HPPD and HMS have numerous active site residues in common; equally expected is that they possess residues that are conserved in only HPPD or HMS sequences.[48] The structural conservation and the similarity of the reactions catalysed has led to attempts to interconvert HPPD and HMS by mutation.[55,56] Figure 21.1B shows a select group of residues (conserved or distinct in these structures) that have been the primary focus of hydroxylation regiospecificity studies. In this figure, HPP is modelled into the substrate-binding pocket to provide reference. This position for the substrate is based on the density observed for HMA in the HMS•Co(II)•HMA structure, as no structure of the HPPD or HMS•Fe(II)•HPP complex has yet been solved.[48] While mutational studies have helped elucidate the chemical mechanisms of HPPD and HMS (see below), few active site residues have been assigned an unambiguous catalytic function.[22,55–58] In HMS, a catalytic tolerance to mutation is observed, *i.e.* variants of HMS with alterations to residues hypothesized to play a key role in catalysis and/or hydroxylation regioselectivity do not exhibit substantially altered catalytic chemistry. The HPPD reaction coordinate, however, proved less durable in that active site variants are observed to frequently uncouple in turnover and accumulate non-native products from abortive pathways. This difference in response to mutation has been suggested to be a consequence of the larger active site cavity in HPPD (~60 Å^3 *vs* 30 Å^3) that confines the phenol ring to a lesser extent than that of HMS.[48]

As detailed in Chapter 3, a select number of 2OADOs have been investigated using magnetic circular dichroism (MCD) spectroscopy to reveal the geometry about the Fe(II). Complexes with liganding 2-oxo acids typically exhibit low-lying metal-to-ligand charge-transfer (MLCT) transitions arising from the coordination of Fe(II) with the carboxylate and keto groups.[59,60] Analysis of the near-IR MCD data for 2OG-dependent oxygenases indicate that resting enzymes show predominantly a six-coordinate (6C) Fe(II) centre (with three endogenous and three water ligands) which remains 6C upon binding 2OG, but switches to five-coordinate (5C) Fe(II) (with three endogenous and two 2OG ligands) upon binding of the specific substrate, thus opening an attack site for dioxygen.[59,61] The resting state HPPD/HMS•Fe(II) and HPPD/HMS•Fe(II)•HPP complexes have similar MLCT transitions centered around ~500 nm, characteristic of bidentate association of the 2-oxo acid moiety to the metal ion.[62–64] Analysis of the resting HPPD/HMS•Fe(II) complex shows a mixture of both 5C (10%) and 6C (90%) ferrous ion geometry. Substrate-bound forms of HPPD and HMS largely retain this 5C : 6C ratio, suggesting that for these enzymes predominant 5C is not required to support their observed rates of reaction with dioxygen.[60,65]

As stated above, no crystal structure of the HPP complex of either HPPD or HMS has been solved. Similar bidentate Fe(II) coordination of HPP occurs in both enzymes, but a curious reversal of transitions was noted in circular dichroism spectra of the HPPD•Fe(II)•HPP and HMS•Fe(II)•HPP complexes.[65] This result suggests unique positions for the HPP phenol in the enzymes and promoted the notion that orientation of the aromatic ring relative to the hydroxylating species is principally responsible for the two observed activities.[65] The structure of the HMS product complex, (HMS•Co(II)•HMA), coupled with the shape similarities of HMA and HPP, aided computer modelling of HPP into the active site of both enzymes (Figure 20.1B).[48] In HMS•Fe(II)•HPP, the 4-hydroxyl group of HPP is a hydrogen bond donor to Ser-201, while in the HPPD•Fe(II)•HPP complex a 4.5 Å displacement of the 4-hydroxyl group allows it to act as a hydrogen bond donor to the amide group of Asn-216 and a hydrogen bond acceptor from Ser-201, simultaneously.[48] This pivot of the aromatic ring presumably dictates proximity of the benzylic carbon with respect to the hydroxylating intermediate after decarboxylation. Density functional theory calculations suggest unique conformations of the phenol are a result of differing conformations of Ser-201 in the HPP/HMS•Fe(IV)=O•HPA complex that dictate the separate reactivities.[66,67]

As described in detail below, a pivot of the *C*-terminal α-helix in response to inhibitor association suggests an active site gating function. Mobile structural elements are commonly observed in structures of 2OADOs.[68–74] It would appear that the 2OADOs employ a strategy of closing the active site with mobile secondary structures to confine the chemistry and exclude the solvent from access to reactive species. Truncation of the last few residues (that are typically disordered in X-ray structures) from the *C*-terminal helix of HPPD slows catalytic activity without affecting HPP binding. This finding strongly suggests that these residues function to define the balance of open and closed forms in solution.[75]

21.4 Inhibition of HPPD and HMS

In terms of cumulative research activity, the catalytic chemistry of HPPD and HMS has been partially eclipsed by the chemistry of HPPD inhibition. While both HPPD and HMS are inhibited by 1,3,3′-triketones, only inhibition of the former has agricultural and medical implications (see below).[76] Despite clear herbicidal potency, one of the earliest inhibitors discovered, NTBC, was not formulated for crop use due to the apparent complications of a long metabolic half-life and limited long-term toxicity data. NTBC, nonetheless, remains the paradigm HPPD inhibitor and has since been shown to have low toxicity and the pharmacological advantage of an extended metabolic half-life.[77] HPPD from *Streptomyces* binds the exocyclic enol of NTBC, with the native ferrous state appearing to bind the inhibitor irreversibly. While both HPPD•Fe(II) and HPPD•Fe(II)•HPP complexes react with molecular oxygen, the HPPD•Fe(II)•NTBC complex does not. Using the kinetics associated with the development of MLCT absorption transitions that arise in the inhibitor complex, a minimal three-step kinetic mechanism for NTBC association was proposed (Scheme 21.3B).[78] In the first step NTBC binds weakly in a non-metal-centered complex. In the second step association of NTBC to the metal ion is observed as an accumulation of MLCT transitions centered about 450 nm. The final and slowest step is proposed to entail an irreversible deprotonation of the exocyclic enol group of the inhibitor to form the enolate complex with more intense MLCT features.[78] MCD spectra of the HPPD•Fe(II)•NTBC complex indicate that, despite its apparent extremely high affinity for HPPD, NTBC association with Fe(II) is weaker than HPP association with the metal ion by ~12 kJ mol.[79] As such, it was concluded that bidentate association of the inhibitor to the active site metal ion does not appreciably contribute to the binding energy for the inhibitor complex.

The first inhibitor-bound HPPD structure reported was for *S. avermitilis* HPPD crystallized in the ferrous oxidation state in complex with NTBC.[72] This complex reveals that the inhibitor makes the previously predicted bidentate interactions to the metal ion *via* its 1-ketone and 3-enol moieties (Figure 21.2). While the overall fold reveals a high degree of similarity to the earlier solved *P. fluorescens* HPPD·Fe(III)·acetate structure,[47] the *C*-terminal α-helix was pivoted about its *N*-terminal end to a position that places it at a ~40° angle compared to the earlier observed position (Figure 21.2). This *C*-terminal α-helix includes Phe-364 that, in combination with Phe-336, makes a staggered π-stacking interaction with the aromatic ring of NTBC (Figure 21.2B). Other mammalian HPPD inhibitor complexes have been solved since that show similar *C*-terminal α-helix displacement and stacking arrangements.[80] Such π-stacking interactions are typically very low-energy associations and unlikely to account for the observed affinity. Because no additional electrostatic or hydrogen bonding interactions are observed in the HPPD•Fe(II)•NTBC complex, the basis for the extremely high affinity is largely unknown; however, it would appear that NTBC has excellent shape complementarity and makes near uniform van der Waals contact with the active site cavity.[81]

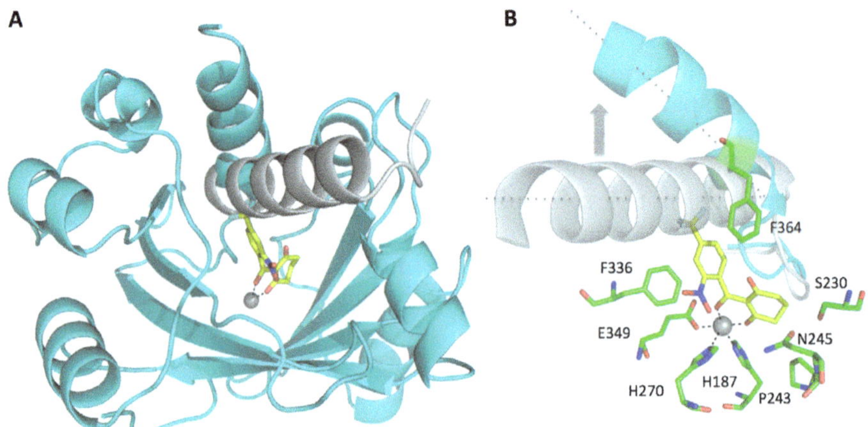

Figure 21.2 Structure of the HPPD•Fe(II)•NTBC complex. (A) The open *C*-terminal domain of HPPD from *S. avermitilis* in complex with NTBC (cyan, PDB ID: 1T47) overlaid with the closed position for the C-terminal α-helix from the *P. fluorescens* HPPD structure solved in the presence of an acetate ligand (grey, PDB ID: 1CJX). (B) The NTBC complex with the active site of HPPD. The residues shown are from the *S. avermitilis* HPPD•Fe(II)•NTBC complex (cyan, PDB ID: 1T47) and are those that contact the metal ion and/or the inhibitor. The displacement of the *C*-terminal α-helix is indicated relative to the position observed in the closed structure from *P. fluorescens* HPPD (grey, PDB ID: 1CJX).

21.5 The Reaction Catalysed

Scheme 21.4 depicts one current mechanistic hypothesis for a single catalytic cycle of HPPD and HMS. In this scheme, reaction steps indicated by black arrows are for the common decarboxylation half-reaction. Those steps indicated with green arrows are for the HPPD hydroxylation half-reaction and those in red are for the HMS hydroxylation half-reaction. In the decarboxylation half-reaction dioxygen reduction occurs after HPP association, and oxidative decarboxylation of the 2-oxo acid moiety yields the reactive Fe(IV)=O (ferryl) intermediate. There is scant experimental evidence for the events of this half-reaction. What is known has been assembled from computational studies and suggests that an Fe(III) superoxo species attacks the 2-oxo carbon atom to initially form a bridged peroxo species that then liberates carbon dioxide and forms a peracid intermediate. Severing of the peroxy bond forms the highly electron-deficient ferryl species coordinated to HPA. From this point the hydroxylation half-reaction ensues. As stated, it is thought that proximity dictates where the Fe(IV)=O species will hydroxylate; in HMS, the benzylic carbon of HPA has greater proximity, while in HPPD it is the aromatic C1 position, each as a direct consequence of the position of the phenol (Figure 21.1B).

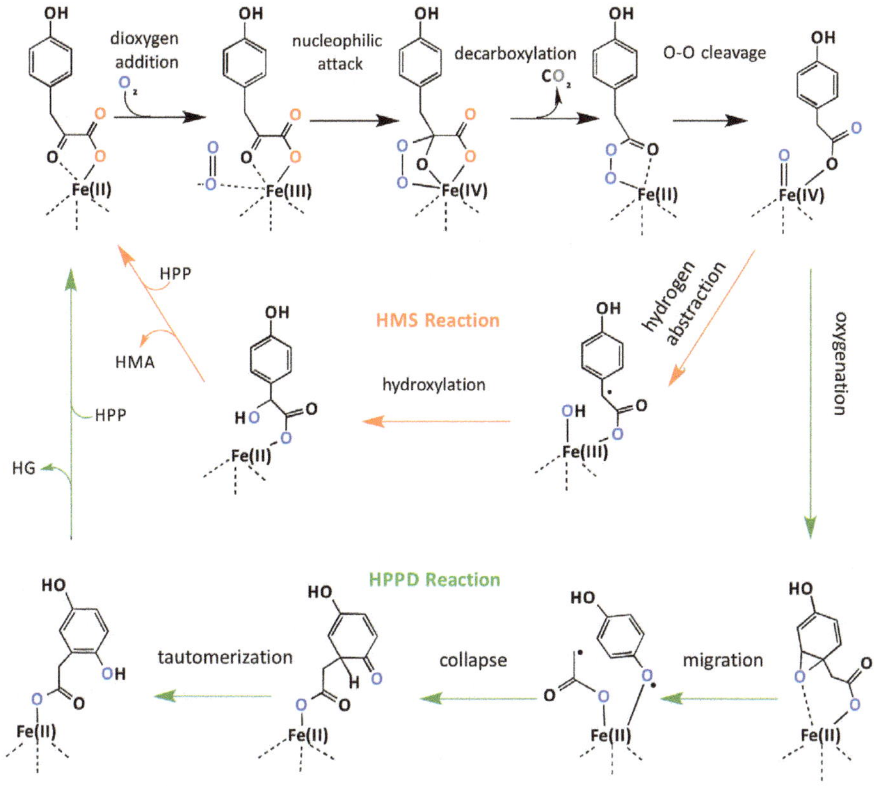

Scheme 21.4 Catalytic cycles of HPPD and HMS.

21.5.1 The Mechanism of Decarboxylation

The decarboxylation half-reaction starts with ordered substrate binding, a commonly observed feature in 2OADOs.[1] As is detailed in previous chapters for the three-substrate 2OADOs, the catalytic cycle starts with the binding of 2OG, followed by the addition of the specific substrate to form a ternary complex that has heightened reactivity toward molecular oxygen[82] as a consequence of the switch from 6C to 5C geometry at the metal ion.[59,61] HPPD and HMS also exhibit ordered substrate addition, with HPP being the first substrate to bind followed by molecular oxygen. Relative to the HPPD/HMS·Fe(II) complexes, the HPPD/HMS•Fe(II)•HPP complexes show reactivities towards molecular oxygen that are several thousand-fold more rapid.[62,63] The basis for these changes in reactivity appears to derive from the conformation of the HPP 2-oxo acid. The absorption spectrum of unliganded HPP shows an $n \rightarrow \pi^*$ transition at ~31 500 cm^{-1} that shifts to ~28 000 cm^{-1} when bound in the active site of HPPD or HMS. The difference in energy of the 2-oxo acid n and π^* orbitals is evidence of the dihedral angle between the carbonyl and carboxyl groups (θ) and this energy difference reaches a minimum when the two groups are coplanar (Figure 21.1B).[59] The planar conformation of the 2-oxo acid maximizes conjugation between the coordinating moieties and the ferrous ion, facilitating metal to dioxygen charge delocalization. The energy of the $n \rightarrow \pi^*$ transition of HPPD/HMS•Fe(II)•phenylpyruvate complexes indicates flattening of the 2-oxo acid only in HMS, consistent with the observation that phenylpyruvate is only a substrate for HMS.[66]

The decarboxylation half-reaction is largely kinetically obscured in single turnover reactions of HPP and HMS. While an early intermediate can be observed to accumulate to some extent, no evidence has yet been offered for the identity of this species.[62–64] Computational data have provided detailed descriptions of the possible mechanisms for the decarboxylation chemistry.[83–85] From such studies it was proposed that the first step after addition of HPP is the stepwise reduction of dioxygen to first form the ferric–superoxo species. Attack on the 2-oxo carbon forms a Fe(IV)–peroxo bridged species that decays with concomitant release of CO_2 to give a ferrous peracid intermediate. This peracid then undergoes heterolytic cleavage of the O–O bond to yield the E•Fe(IV)=O•HPA intermediate. In a variation of this sequence it has also been suggested that the ferric–superoxo species can decay to the peracid *via* a single transition state.[84] The Fe(IV)=O species accumulates in three-substrate 2OADOs as the first intermediate concomitant with CO_2 production.[86–90] The identity of the intermediate was revealed by Mössbauer spectroscopy in conjunction with rapid freeze-quench, facilitated by enhanced accumulation of the intermediate that occurs when a deuterium is substituted at the site of the hydroxylation.[88,91] Fe(IV)=O is a highly electrophilic species that oxidizes a variety of proximal centres and thereby can be used for a range of different chemistries;[1,54] however, the extreme electrophilicity of the ferryl species makes it unruly and prone to hydroxylate nearby active site residues. This phenomenon of self-hydroxylation has been observed in a variety of 2OADO-type enzymes, including HPPD.[92–94]

For HPPD and HMS, multiple intermediates have been observed in rapid mixing, single-turnover reactions. The available data suggest that the ferryl species does not accumulate in these enzymes due to the formation and decay of this intermediate being fast relative to prior and later steps.[63,64,95] Nonetheless, wild-type HPPD and variants of both enzymes make HPA as one of the products in an abortive pathway, consistent with a common decarboxylation half-reaction and indirectly implying the presence of a ferryl species.[22,57,58]

21.5.2 The Mechanisms of Hydroxylation

The fate of the oxygen atom of the ferryl species defines the hydroxylation reactions of HPPD and HMS. Compared to the 2OADOs, the HMS hydroxylation half-reaction is rather conventional. The available data are consistent with the ferryl species abstracting a hydrogen atom from the benzylic carbon of HPA, forming an Fe(III)–OH intermediate. This species is able to quench the benzylic radical formed in the initial abstraction by transfer of •OH in a rebound reaction that restores the Fe(II) state of the metal ion. In contrast to HMS, hydroxylation of the relatively electron-replete aromatic ring of HPA by the Fe(IV)=O species of HPPD induces the quite remarkable displacement of the aceto substituent to the adjacent ring C2 carbon. As such, the overall hydroxylation/rearomatization half-reaction of HPPD is considerably more chemically intricate and numerous hypotheses, based on experiment and/or intuition, have been put forward to account for the chemistry.[2,56–58,84,96–98] Since the first observations of aromatic shift chemistry in the pterin-dependent hydroxylases, there has been a debate over the participation of either a benzylic cation or a 1,2-epoxide as the immediate intermediate arising from oxygenation of an aromatic ring.[99–102] However, this narrow set of possibilities was no doubt confined by precedent for ring oxygen substitutions from organic synthesis studies. Hybrid density functional calculations were used to explore a more extensive set of reaction paths for the attack on the ring C1 carbon by the Fe(IV)=O species, where oxygen addition forms either a radical σ complex, a biradical, an epoxide (arene oxide) or a benzylic cation.[84] In these calculations, the radical σ complex having the Fe(III) state is preferred over the ring cation or the epoxide intermediate. In a separate study, the epoxide was claimed as the first intermediate of the hydroxylation half-reaction based on the accumulation of an oxepinone from turnover of an HPPD active site variant that was thought to be a direct rearrangement of the product epoxide;[56] however, it was later shown that the oxepinone is actually derived from QAA that was rearranged by the inclusion of trifluoroacetic acid during the HPLC identification.[22]

Quite recently methodologies have been applied that successfully probe the hydroxylation half-reactions of both HPPD and HMS. Directed active site variants of either enzyme commonly induce some degree of uncoupling to form both native and non-native products. This outcome has enabled the use of intermediate partitioning to measure kinetic isotope effects (KIEs) on specific steps.[58,103] The products monitored were HG (decarboxylated, doubly oxygenated and rearranged), HMA (decarboxylated and doubly oxygenated), HPA (decarboxylated and singly oxygenated) and QAA (decarboxylated and doubly oxygenated).[22,58] KIEs are derived from the change in ratio of the products in the presence of a strategically placed heavy atom label, the ratio of ratios thus providing the KIE.

Intermediate partitioning KIEs have provided direct insight into both the initial oxygenation by HPPD and HMS and the ensuing NIH shift steps of HPPD. The KIE on the initial oxygenation step of wild-type HPPD when ring deuterons are added to HPP was an inverse value of 0.84, suggesting an sp^2 to sp^3 hybridization change. This hybridization change is predicted only for the formation of a ring epoxide.[58] The ensuing NIH shift must then transiently sever the bond to the aceto substituent in either a heterolytic or homolytic mechanism. The earliest evidence for the mechanism of NIH shift came from the lack of scrambling or inversion at the benzylic carbon when 3′-monodeutero HPP is used as a substrate, and this result was recently confirmed.[57,97] While this lack of scrambling is seemingly consistent with a heterolytic mechanism in which the in-flight benzylic carbon is anionic (sp^3) in character, it was later suggested by calculation to be an inconclusive result. Borowski *et al.* determined by density functional calculations that the barrier to bond rotation to give scrambling is substantially greater than that for recombination and a lack of scrambling cannot by itself be evidence of the mechanism for the shift.[84] 3′,3′-Dideutero HPP yields a KIE of unity for wild-type and most variants of HPPD;[57] however, an N245S variant (corresponding to Asn-216 in Figure 20.1) gives a normal KIE of 1.4 for the formation of HG, suggesting a late transition state moving to sp^2 geometry with this variant. It was therefore concluded that in wild-type HPPD the shift occurs *via* a minimally advanced transition state toward a homolytic mechanism,[57] consistent with the earlier conclusions from computational studies.[84]

Recombination of the aceto group with the aromatic ring forms a dienone that can rearomatize through tautomerization to yield HG. The activation energy barrier calculated for this tautomerization indicates this process could take place either within the enzyme or in solution.[84] Single turnover studies of HPPD have shown that the highly fluorescent HG product forms as the third intermediate ahead of the rate-limiting dissociation of the product, indicating that rearomatization occurs within the active site. Moreover, the rate constant for the formation of this signal exhibits a KIE of 1.7 when ring-perdeutero-HPP is used as a substrate, consistent with a modest primary isotope effect for displacement of a deuteron from the ring 2-carbon during rearomatization.[64] Decay of the third intermediate is sensitive to deuterons derived from the solvent. In HPPD (and HMS) the product release rate constant (k_4) exhibits a solvent kinetic isotope effect of ~2, suggesting a role for an active site base in product release.[63,64]

Hydroxylation in HMS occurs at the benzylic carbon of HPA to give product HMA. Substitution of a deuterium atom(s) at the benzylic position of HPP does not slow any of the steps observed in single-turnover reactions of HMS.[63] This result suggests that the hydroxylation reaction of HMS is fast and unable to be observed directly by conventional kinetic approaches. In HMS, the ferryl species abstracts only the *pro-S* hydrogen atom to form a ferric–hydroxyl species.[63] This Fe(III)–OH then attacks the benzylic radical *via* an oxygen rebound mechanism to form HMA. Like HPPD, HMS variants also show product partitioning, but they form just two products, HPA and HMA. When 3′,3′-dideutero HPP is used as a substrate, the intermediate partitioning KIE is 2.2–2.6. This outcome suggests an H-atom abstraction that is largely devoid of tunnelling.[58] This KIE measured with the dideutero substrate is expected to include a multiplicative component from the non-abstracted deuteron. To dissect each component, the equivalent experiment was carried out with R-3′-monodeutero HPP and the KIE measured. This experiment gave a KIE of 1.06, indicating little geometry change in the transition state during abstraction.[57] The F364I HPPD variant is able to form HMA as a product in 20% of turnovers. The R-3′-monodeutero HPP substrate shows a KIE of 1.92 with this variant, a significantly normal secondary effect, suggesting a hybridization change from sp^3 to sp^2 consistent with a less optimized reaction for the production of HMA in this variant.

21.6 Concluding Remarks

HPPD and HMS are the outliers of the 2OADO family. While they catalyse a mechanistically equivalent set of reactions to those observed in the majority of the enzyme family, they have evolved independently and exhibit unique characteristics that have in some cases elucidated parts of the catalytic cycle that remain obscured in all other family members. The ubiquity of the HPPD reaction has meant that its inhibition has a collectively broad set of consequences. Inhibitors have been found to be highly successful herbicides, targeted therapeutics for in-born errors in tyrosine catabolism, and have shown potential to augment the treatment of microbial infections.

References

1. V. Purpero and G. R. Moran, *J. Biol. Inorg. Chem.*, 2007, **12**, 587–601.
2. S. E. Hager, R. I. Gregerman and W. E. Knox, *J. Biol. Chem.*, 1957, **225**, 935–947.
3. T. W. Goodwin and E. I. Mercer, *Introduction to Plant Biochemistry*, 2nd edn, Pergamon Press, Sydney, 1983.
4. J. D. Nosanchuk and A. Casadevall, *Cell. Microbiol.*, 2003, **5**, 203–223.
5. D. E. Graham, H. Xu and R. H. White, *Arch. Microbiol.*, 2003, **180**, 455–464.
6. B. K. Hubbard, M. G. Thomas and C. T. Walsh, *Chem. Biol.*, 2000, **7**, 931–942.
7. O. W. Choroba, D. H. Williams and J. B. Spencer, *J. Am. Chem. Soc.*, 2000, **122**, 5389–5390.
8. H. Maeda and D. DellaPenna, *Curr. Opin. Plant Biol.*, 2007, **10**, 260–265.
9. R. Hellyer, *Aust. J. Chem.*, 1968, **21**, 2825–2828.
10. F. E. Dayan, S. O. Duke, A. Sauldubois, N. Singh, C. McCurdy and C. Cantrell, *Phytochemistry*, 2007, **68**, 2004–2014.
11. J. G. Romagni, G. Meazza, N. P. Nanayakkara and F. E. Dayan, *FEBS Lett.*, 2000, **480**, 301–305.
12. H. Schindler, *Arzneimittelforschung*, 1957, **7**, 69–72.
13. M. Cocchietto, N. Skert, P. L. Nimis and G. Sava, *Naturwissenschaften*, 2002, **89**, 137–146.
14. R. Beaudegnies, A. J. Edmunds, T. E. Fraser, R. G. Hall, T. R. Hawkes, G. Mitchell, J. Schaetzer, S. Wendeborn and J. Wibley, *Bioorg. Med. Chem.*, 2009, **17**, 4134–4152.
15. R. Huhn, H. Stoermer, B. Klingele, E. Bausch, A. Fois, M. Farnetani, M. Di Rocco, J. Boue, J. M. Kirk, R. Coleman and G. Scherer, *Hum. Genet.*, 1998, **102**, 305–313.
16. K. Tomoeda, H. Awata, T. Matsuura, I. Matsuda, E. Ploechl, T. Milovac, A. Boneh, C. R. Scott, D. M. Danks and F. Endo, *Mol. Genet. Metab.*, 2000, **71**, 506–510.
17. C. B. Item, I. Mihalek, O. Lichtarge, A. Jalan, J. Vodopiutz, A. Muhl and O. A. Bodamer, *Mol. Genet. Metab.*, 2007, **91**, 379–383.

18. F. Endo, *Ryoikibetsu Shokogun Shirizu*, 1998, **18**, 141–143.
19. W. Lehnert, W. Stogmann, U. Engelke, R. A. Wevers and G. B. van den Berg, *Eur. J. Pediatr.*, 1999, **158**, 578–582.
20. B. Wilcken, J. W. Hammond, N. Howard, T. Bohane, C. Hocart and B. Halpern, *N. Engl. J. Med.*, 1981, **305**, 865–868.
21. C. H. Hocart, B. Halpern, L. A. Hick and C. O. Wong, *J. Chromatogr.*, 1983, **275**, 237–243.
22. J. M. Brownlee, B. Heinz, J. Bates and G. R. Moran, *Biochemistry*, 2010, **49**, 7218–7226.
23. A. Niederwieser, S. K. Wadman and D. M. Danks, *Clin. Chim. Acta*, 1978, **90**, 195–200.
24. A. Niederwieser, A. Matasovic, P. Tippett and D. M. Danks, *Clin. Chim. Acta*, 1977, **76**, 345–356.
25. E. A. Garrod, *Lancet*, 1902, **ii**, 1616–1620.
26. A. E. Garrod, *Lancet*, 1908, **ii**, 73–79.
27. V. G. Zannoni and B. N. La Du, *J. Biol. Chem.*, 1959, **234**, 2925–2931.
28. C. Phornphutkul, W. J. Introne, M. B. Perry, I. Bernardini, M. D. Murphey, D. L. Fitzpatrick, P. D. Anderson, M. Huizing, Y. Anikster, L. H. Gerber and W. A. Gahl, *N. Engl. J. Med.*, 2002, **347**, 2111–2121.
29. W. J. Introne, M. B. Perry, J. Troendle, E. Tsilou, M. A. Kayser, P. Suwannarat, K. E. O'Brien, J. Bryant, V. Sachdev, J. C. Reynolds, E. Moylan, I. Bernardini and W. A. Gahl, *Mol. Genet. Metab.*, 2011, **103**, 307–314.
30. P. Suwannarat, K. O'Brien, M. B. Perry, N. Sebring, I. Bernardini, M. I. Kaiser-Kupfer, B. I. Rubin, E. Tsilou, L. H. Gerber and W. A. Gahl, *Metabolism*, 2005, **54**, 719–728.
31. J. M. Fernandez-Canon, M. W. Baetscher, M. Finegold, T. Burlingame, K. M. Gibson and M. Grompe, *Mol. Cell. Biol.*, 2002, **22**, 4943–4951.
32. J. Gentz, R. Jagenburg and R. Zetterström, *J. Pediatr.*, 1965, **66**, 670–696.
33. B. Lindblad, S. Lindstedt and G. Steen, *Proc. Natl. Acad. Sci. U. S. A.*, 1977, **74**, 4641–4645.
34. N. Mohan, P. McKiernan, M. A. Preece, A. Green, J. Buckels, A. D. Mayer and D. A. Kelly, *Eur. J. Pediatr.*, 1999, **158** Suppl 2, S49–S54.
35. P. J. McKiernan, *Drugs*, 2006, **66**, 743–750.
36. S. Ahmad, J. H. Teckman and G. T. Lueder, *Am. J. Ophthalmol.*, 2002, **134**, 266–268.
37. R. P. Wisse, D. Wittebol-Post, G. Visser and A. van der Lelij, *BMJ Case Rep.*, 2012, 2012.
38. E. Thimm, R. Richter-Werkle, G. Kamp, B. Molke, D. Herebian, D. Klee, E. Mayatepek and U. Spiekerkoetter, *J. Inherited Metab. Dis.*, 2012, **35**, 263–268.
39. D. Guerra-Lopez, L. Daniels and M. Rawat, *Microbiology (Reading, England)*, 2007, **153**, 2724–2732.
40. L. D. Eirich, G. D. Vogels and R. S. Wolfe, *J. Bacteriol.*, 1979, **140**, 20–27.
41. E. Wintermeyer, M. Flugel, M. Ott, M. Steinert, U. Rdest, K. H. Mann and J. Hacker, *Infect. Immun.*, 1994, **62**, 1109–1117.

42. S. I. Kotob, S. L. Coon, E. J. Quintero and R. M. Weiner, *Appl. Environ. Microbiol.*, 1995, **61**, 1620–1622.
43. R. K. Ernst, D. A. D'Argenio, J. K. Ichikawa, M. G. Bangera, S. Selgrade, J. L. Burns, P. Hiatt, K. McCoy, M. Brittnacher, A. Kas, D. H. Spencer, M. V. Olson, B. W. Ramsey, S. Lory and S. I. Miller, *Environ. Microbiol.*, 2003, **5**, 1341–1349.
44. A. Rodriguez-Rojas, A. Mena, S. Martin, N. Borrell, A. Oliver and J. Blazquez, *Microbiology (Reading, England)*, 2009, **155**, 1050–1057.
45. K. E. Keith, L. Killip, P. He, G. R. Moran and M. A. Valvano, *J. Bacteriol.*, 2007, **189**, 9057–9065.
46. L. R. Nunes, R. Costa de Oliveira, D. B. Leite, V. S. da Silva, E. dos Reis Marques, M. E. da Silva Ferreira, D. C. Ribeiro, L. A. de Souza Bernardes, M. H. Goldman, R. Puccia, L. R. Travassos, W. L. Batista, M. P. Nobrega, F. G. Nobrega, D. Y. Yang, C. A. de Braganca Pereira and G. H. Goldman, *Eukaryotic Cell*, 2005, **4**, 2115–2128.
47. L. Serre, A. Sailland, D. Sy, P. Boudec, A. Rolland, E. Pebay-Peyroula and C. Cohen-Addad, *Struct. Fold. Des.*, 1999, **7**, 977–988.
48. J. Brownlee, P. He, G. R. Moran and D. H. Harrison, *Biochemistry*, 2008, **47**, 2002–2013.
49. P. He and G. R. Moran, *J. Inorg. Biochem.*, 2011, **105**, 1259–1272.
50. R. N. Armstrong, *Biochemistry*, 2000, **39**, 13625–13632.
51. Z. Zhang, J. Ren, K. Harlos, C. H. McKinnon, I. J. Clifton and C. J. Schofield, *FEBS Lett.*, 2002, **517**, 7–12.
52. E. L. Hegg and L. Que, Jr., *Eur. J. Biochem.*, 1997, **250**, 625–629.
53. K. D. Koehntop, J. P. Emerson and L. Que, Jr., *J. Biol. Inorg. Chem.*, 2005, **10**, 87–93.
54. R. P. Hausinger, *Crit. Rev. Biochem. Mol. Biol.*, 2004, **39**, 21–68.
55. H. M. O'Hare, F. Huang, A. Holding, O. W. Choroba and J. B. Spencer, *FEBS Lett.*, 2006, **580**, 3445–3450.
56. M. Gunsior, J. Ravel, G. L. Challis and C. A. Townsend, *Biochemistry*, 2004, **43**, 663–674.
57. D. D. Shah, J. A. Conrad and G. R. Moran, *Biochemistry*, 2013.
58. D. D. Shah, J. A. Conrad, B. Heinz, J. M. Brownlee and G. R. Moran, *Biochemistry*, 2011, **50**, 7694–7704.
59. E. G. Pavel, J. Zhou, R. W. Busby, M. Gunsior, C. A. Townsend and E. I. Solomon, *J. Am. Chem. Soc.*, 1998, **120**, 743–753.
60. M. L. Neidig, M. Kavana, G. R. Moran and E. I. Solomon, *J. Am. Chem. Soc.*, 2004, **126**, 4486–4487.
61. J. Zhou, W. L. Kelly, B. O. Bachmann, M. Gunsior, C. A. Townsend and E. I. Solomon, *J. Am. Chem. Soc.*, 2001, **123**, 7388–7398.
62. K. Johnson-Winters, V. M. Purpero, M. Kavana, T. Nelson and G. R. Moran, *Biochemistry*, 2003, **42**, 2072–2080.
63. P. He, J. A. Conrad and G. R. Moran, *Biochemistry*, 2010, **49**, 1998–2007.
64. K. Johnson-Winters, V. M. Purpero, M. Kavana and G. R. Moran, *Biochemistry*, 2005, **44**, 7189–7199.

65. M. L. Neidig, A. Decker, O. W. Choroba, F. Huang, M. Kavana, G. R. Moran, J. B. Spencer and E. I. Solomon, *Proc. Natl. Acad. Sci. U. S. A.*, 2006, **103**, 12966–12973.
66. M. L. Neidig, C. D. Brown, M. Kavana, O. W. Choroba, J. B. Spencer, G. R. Moran and E. I. Solomon, *J. Inorg. Biochem.*, 2006, **100**, 2108–2116.
67. A. Wojcik, E. Broclawik, P. E. Siegbahn and T. Borowski, *Biochemistry*, 2012, **51**, 9570–9580.
68. Z. Zhang, J. Ren, D. K. Stammers, J. E. Baldwin, K. Harlos and C. J. Schofield, *Nat. Struct. Biol.*, 2000, **7**, 127–133.
69. I. J. Clifton, L. C. Hsueh, J. E. Baldwin, K. Harlos and C. J. Schofield, *Eur. J. Biochem.*, 2001, **268**, 6625–6636.
70. R. C. Wilmouth, J. J. Turnbull, R. W. Welford, I. J. Clifton, A. G. Prescott and C. J. Schofield, *Structure (Camb)*, 2002, **10**, 93–103.
71. J. R. O'Brien, D. J. Schuller, V. S. Yang, B. D. Dillard and W. N. Lanzilotta, *Biochemistry*, 2003, **42**, 5547–5554.
72. J. Brownlee, K. Johnson-Winters, D. H. T. Harrison and G. R. Moran, *Biochemistry*, 2004, **43**, 6370–6377.
73. I. Muller, A. Kahnert, T. Pape, G. M. Sheldrick, W. Meyer-Klaucke, T. Dierks, M. Kertesz and I. Uson, *Biochemistry*, 2004, **43**, 3075–3088.
74. M. A. McDonough, K. L. Kavanagh, D. Butler, T. Searls, U. Oppermann and C. J. Schofield, *J. Biol. Chem.*, 2005, **280**, 41101–41110.
75. J. F. Lin, Y. L. Sheih, T. C. Chang, N. Y. Chang, C. W. Chang, C. P. Shen and H. J. Lee, *PLoS One*, 2013, **8**, e69733.
76. J. A. Conrad and G. R. Moran, *Inorg. Chim. Acta*, 2008, **361**, 1197–1201.
77. *USA Patent*, US4780127 A, 1988.
78. M. Kavana and G. R. Moran, *Biochemistry*, 2003, **42**, 10238–10245.
79. M. L. Neidig, A. Decker, M. Kavana, G. R. Moran and E. I. Solomon, *Biochem. Biophys. Res. Commun.*, 2005, **338**, 206–214.
80. C. Yang, J. W. Pflugrath, D. L. Camper, M. L. Foster, D. J. Pernich and T. A. Walsh, *Biochemistry*, 2004, **43**, 10414–10423.
81. G. R. Moran, *Arch. Biochem. Biophys.*, 2014, **544C**, 58–68.
82. M. J. Ryle, R. Padmakumar and R. P. Hausinger, *Biochemistry*, 1999, **38**, 15278–15286.
83. A. R. Diebold, C. D. Brown-Marshall, M. L. Neidig, J. M. Brownlee, G. R. Moran and E. I. Solomon, *J. Am. Chem. Soc.*, 2011, **133**, 18148–18160.
84. T. Borowski, A. Bassan and P. E. M. Siegbahn, *Biochemistry*, 2004, **43**, 12331–12342.
85. M. Costas, M. P. Mehn, M. P. Jensen and L. Que, *Chem. Rev.*, 2004, **104**, 939–986.
86. J. C. Price, E. W. Barr, T. E. Glass, C. Krebs and J. M. Bollinger, Jr., *J. Am. Chem. Soc.*, 2003, **125**, 13008–13009.
87. J. C. Price, E. W. Barr, L. M. Hoffart, C. Krebs and J. M. Bollinger, Jr., *Biochemistry*, 2005, **44**, 8138–8147.
88. J. C. Price, E. W. Barr, B. Tirupati, J. M. Bollinger, Jr. and C. Krebs, *Biochemistry*, 2003, **42**, 7497–7508.

89. P. J. Riggs-Gelasco, J. C. Price, R. B. Guyer, J. H. Brehm, E. W. Barr, J. M. Bollinger, Jr. and C. Krebs, *J. Am. Chem. Soc.*, 2004, **126**, 8108–8109.
90. C. Krebs, D. Galonic Fujimori, C. T. Walsh and J. M. Bollinger, Jr., *Acc. Chem. Res.*, 2007, **40**, 484–492.
91. L. M. Hoffart, E. W. Barr, R. B. Guyer, J. M. Bollinger, Jr. and C. Krebs, *Proc. Natl. Acad. Sci. U. S. A.*, 2006, **103**, 14738–14743.
92. F. C. Bradley, S. Lindstedt, J. D. Lipscomb, L. Que, Jr., A. L. Roe and M. Rundgren, *J. Biol. Chem.*, 1986, **261**, 11693–11696.
93. A. Liu, R. Y. Ho, L. Que, M. J. Ryle, B. S. Phinney and R. P. Hausinger, *J. Am. Chem. Soc.*, 2001, **123**, 5126–5127.
94. I. Müller, C. Stuckl, J. Wakeley, M. Kertesz and I. Usón, *J. Biol. Chem.*, 2005, **280**, 5716–5723.
95. V. M. Purpero and G. R. Moran, *Biochemistry*, 2006, **45**, 6044–6055.
96. B. Lindblad, G. Lindstedt and S. Lindstedt, *J. Am. Chem. Soc.*, 1970, **92**, 7446–7449.
97. R. Leinberger, W. E. Hull, H. Simon and J. Retey, *Eur. J. Biochem.*, 1981, **117**, 311–318.
98. R. A. Pascal, Jr., M. A. Oliver and Y. C. Chen, *Biochemistry*, 1985, **24**, 3158–3165.
99. J. W. Daly, D. M. Jerina and B. Witkop, *Experientia*, 1972, **28**, 1129–1264.
100. D. M. Jerina, J. W. Daly and B. Witkop, *J. Am. Chem. Soc.*, 1968, **90**, 6523–6525.
101. G. J. Kasperek, T. C. Bruice, H. Yagi and D. M. Jerina, *J. Chem. Soc., Chem. Commun.*, 1972, 784–785.
102. G. R. Moran, A. Derecskei-Kovacs, P. J. Hillas and P. F. Fitzpatrick, *J. Am. Chem. Soc.*, 2000, **122**, 4535–4541.
103. P. F. Fitzpatrick, in *Isotope Effects in Chemsitry and Biology*, eds. A. Kohen and H.-H. Limbach, CRC Taylor & Francis, Boca Raton, 2006, pp. 861–873.

Subject Index